CHRISTOPHER JON BJERKNES

# THE MANUFACTURE
## AND SALE OF
# SAINT EINSTEIN

## II

# Christopher Jon Bjerknes

## *The Manufacture and Sale of Saint Einstein*

Volume II

Published by
Omnia Veritas Ltd

www.omnia-veritas.com

© Omnia Veritas Ltd - Christopher Jon Bjerknes - 2019.

All Rights Reserved. No part of this publication may be reproduced, distributed, or transmitted in any form or by any means, including photocopying, recording, or other electronic or mechanical methods, without the prior written permission of the publisher, except in the case of brief quotations embodied in critical reviews and certain other noncommercial uses permitted by copyright law.

# TABLE OF CONTENTS:

## 4 EINSTEIN THE RACIST COWARD ............................................................. 7
### 4.4.7 Lenard Sickens of Einstein's Libels ............................................... 7
### 4.4.8 Let the Debate Begin ................................................................ 24
#### 4.4.8.1 Einstein Disappoints—"Albertus Maximus" is a Laughingstock ........... 26
#### 4.4.8.2 Contemporary Accounts of the Bad Nauheim Debate ....................... 42
### 4.5 EINSTEIN THE GENOCIDAL RACIST ..................................................... 85
### 4.6 RACIST JEWISH HYPOCRISY, INTIMIDATION AND CENSORSHIP .................. 104
### 4.7 EINSTEIN'S TRIP TO AMERICA ........................................................ 129
#### 4.7.1 Einstein Faces Criticism in America ........................................... 139
##### 4.7.1.1 Einstein Hides from Reuterdahl's Challenge to Debate ................. 139
##### 4.7.1.2 Cowardly Einstein Caught in a Lie ...................................... 147
##### 4.7.1.3 Reuterdahl Pursues Einstein, Who Continues to Run ................... 150
#### 4.7.2 Einstein All Hype ................................................................ 156
### 4.8 ASSASSINATION PLOTS ................................................................ 163
### 4.9 WOLFF CRYING, DIRTY TRICKS, CENSORSHIP, SMEAR CAMPAIGNS AND ANONYMOUS THREATS IN THE NAME OF EINSTEIN ....................................... 179

## 5 THE PROTOCOLS OF THE LEARNED ELDERS OF ZION ................................. 226
### 5.1 INTRODUCTION ......................................................................... 226
### 5.2 THE PROTOCOLS OF THE LEARNED ELDERS OF ZION ................................ 228
### 5.3 DID ANYONE BELIEVE THAT THE *PROTOCOLS* WERE GENUINE? ................... 273
#### 5.3.1 Human Sacrifice and the Plan to Discredit Gentile Government—Fulfilled ........................................................................................ 289
#### 5.3.2 The World Awakens to the "Jewish Peril" .................................... 302
#### 5.3.3 America Becomes the "New Jerusalem" ...................................... 315
#### 5.3.4 "The Jewish Peril" ............................................................... 319
#### 5.3.5 The Inhumanity of the Bolsheviks ............................................. 358
### 5.4 INTERNATIONAL ZIONIST AND COMMUNIST INTIMIDATION ......................... 365
#### 5.4.1 Suppression of Free Speech .................................................... 365
#### 5.4.2 Jewish Terrorism ................................................................. 375
### 5.5 ATTEMPTS TO PROVE THE *PROTOCOLS* INAUTHENTIC ............................... 385
#### 5.5.1 Why Did Henry Ford Criticize the Jews? ..................................... 388
#### 5.5.2 Controlled Opposition and "The Trust" ...................................... 389
#### 5.5.3 The Sinking of the "Peace Ship" ............................................... 391
#### 5.5.4 Ford Comes Under Attack—The War Against Pacificism .................. 402
#### 5.5.5 Zionists Proscribe Free Speech ................................................ 428
#### 5.5.6 President Woodrow Wilson Becomes a Zionist Dictator .................. 429
### 5.6 WHY DID THE ZIONISTS TROUBLE THE JEWS? ....................................... 445
#### 5.6.1 The Zionist Myth of the Extinction of the "Jewish Race" Through Philo-Semitism and Assimilation ................................................................. 451
#### 5.6.2 The Zionists Set the Stage for the Second World War... and the Third

------------------------------------------------------------------------------------- *477*
5.7 HENRY FORD FOR PRESIDENT --------------------------------------------------------- 484
5.8 THE "JEWISH MISSION" ------------------------------------------------------------- 488
5.9 JEWISH BANKERS DESTROY RUSSIA AND FINANCE ADOLF HITLER ----------------------- 502

# 4 EINSTEIN THE RACIST COWARD

## 4.4.7 Lenard Sickens of Einstein's Libels

Germany had been very good to the Jews. German Jews were the wealthiest people in the world. In the years following the First World War, the Germans resented the fact that the Jews, Einstein being their chief spokesman, had stabbed the Germans in the back during the war, and then twisted the knife at the peace negotiations in France, where a large contingent of Jews decided Germany's fate, and reneged on Woodrow Wilson's Fourteen Points, one of which assured Germany that it would lose no territory. The Germans had thought that Wilson's pledge would be honored after the Germans had surrendered in good faith. Had not the Germans received this promise of the Fourteen Points, they would not have surrendered and were in a position to continue the war. The promise was broken.

In addition, the Allies insisted that Germany pay draconian war reparations that would forever ruin the nation. Leading Jews in Germany sided with the Allies against their native land. It was obvious that leading Jews were profiteering from the war in every way possible, at the expense of the German nation and its People. Jewish leaders instigated crippling strikes in the arms industry, which left German troops without adequate armaments. Jewish revolutionaries took advantage of Germany's weakened state, which Jews had deliberately caused for the purpose, and created a Soviet Republic in Bavaria and overthrew the monarchy. German-Jewish bankers cut off Germany's access to funds. German-Jewish Zionists moved to London and brought America into the war on the side of the British at the very moment Germany was about to win the war.

Those arms which were produced were often substandard and were peddled by Jews to Jews in the German Government, which also left the German troops without adequate arms, while making Jews immensely wealthy. German-Jewish bankers conspired with German arms manufacturers to produce weapons for both sides. The German-Jewish press, which had initially beat the war drums louder than anyone else, teamed up with leading Jews in the German Government at the end of the war and demanded that Germany submit to the demands of the Allies, give up vast territories and make the reparations payments. The German-Jewish press and Jews in the German Government, many of whom were the same persons who had most boisterously called upon the German People to go to war, insisted that the Germans accept responsibility for causing the war, though they had not caused it. Etc. Etc. Etc.

While millions of Germans were starving to death, many Jews in Germany had never known better times. Whenever anyone revealed the truth of what was happening, the Jewish press immediately smeared them by calling them "anti-Semites". The situation was similar to, though even worse than, the situation in America today.

Many German Jews were very wealthy after the war. They had a great deal of power, and many were very arrogant, especially in their dealings with German

Gentiles. A famous German engineer and physicist, who had anticipated many aspects of the theory of relativity, Rudolf Mewes proved that Einstein was a plagiarist. Mewes demonstrated that Albert Einstein had stolen many of his ideas from German scientists.

Albert Einstein made a great show of ridiculing Germans, though he was born in Germany, lived and earned his living in Germany throughout the war, worked for the Prussian Academy of Sciences in Berlin, and published in German journals. Einstein assisted in, and pushed hard for, plans to punish and oppress German scientists after the war—to punish and oppress his German colleagues while he was feted in the British press as the "Swiss Jew". Einstein's ingratitude and treachery were unbearable and he epitomized the Jewish betrayal of Germany in the First World War.

Rudolf Mewes was not afraid to challenge Einstein, or the "Einstein myth" of the "Jewish Newton" which was based on lies, plagiarism, ingratitude, self-glorification and Jewish racism,

"But then, given the above exposé, one must admit that [Max] Born's contention is correct, that the relativistic ideas were not only first conceived and recorded in the German language, but rather also that they demonstrably derived from pure German scientists, namely Christian Doppler, Wilhelm Weber and Rudolf Mewes, though not from the Semitic Professor and Communist Dr. Albert Einstein. The relationship of Mewes to Einstein can accordingly be briefly characterized by the slogans:

'German versus Jew
Increaser of Knowledge versus Fleecer of Knowledge
Rightful Ownership versus Plagiarism
Monarchist versus Communist'"

"Dagegen muß man nach den vorstehenden Darlegungen die Behauptung Borns als richtig zugeben, daß die relativistischen Ideen zuerst nicht nur in deutscher Sprache gedacht und aufgezeichnet worden sind, sondern auch von rein deutschen Forschern, nämlich Christian Doppler, Wilhelm Weber und Rudolf Mewes, nachweislich herrühren, aber nicht von dem semitischen Professor und Kommunisten Dr. Albert Einstein. Das Verhältnis von Mewes zu Einstein läßt sich demgemäß kurz mit den Schlagworten kennzeichnen:

„Deutscher gegen Jude,
Wissensschöpfer gegen Wissensschröpfer,
Eigentum gegen Diebstahl,
Monarchist gegen Kommunist.'"[1]

Germans then knew far more about the genocidal prophecies of Judaism

---

[1]. R. Mewes, "Geschichtliche Entwicklung der Relativitäts- oder Raumzeitlehre", Chapter 4, "Wissenschaftliche Begründung der Raumzeitlehre oder Relativitätstheorie (1884-1894) mit einem geschichtlichen Anhang", *Gesammelte Arbeiten von Rudolf Mewes*, Volume 1, Rudolf Mewes, Berlin, (1920), pp. 48-78, at 78.

than they do today. They could see them deliberately fulfilled before there eyes. They recognized that Bolshevism and the "Great War", the "War to End All Wars", which prepared the way for the "League of Nations", was largely accomplished under the directorship of Jews and deliberately fulfilled Jewish Messianic prophecy. They knew that leading Jews had lured Germany into the war and then destroyed Germany and profited as much as possible from the destruction.

In addition, an unwise and unproductive rift between British science and German science had existed at least since the time of the Leibnitz-Newton priorities dispute over the invention of calculus, and before that there were strong controversies between the Continent and the Island among Giordano Bruno, Henry More, Isaac Newton, Samuel Clark, René Des Cartes, Christiaan Huyghens, and Gottfried Wilhelm Leibnitz. Einstein sided with the British against the Germans during and after the war, despite the fact he was treated like royalty in Berlin.

Jewish news sources promoted the causes of the Social Democrats, Liberal Democrats, Marxists, Bolsheviks, Anarchists or Chernyshevskiist revolution, and they also promoted Albert Einstein, which inspired suspicion of ethnic bias.[2] The segregationist policies of Albert Einstein, Chaim Weizmann—the political Zionists in general—caused many to suspect that the shameless promotion of Albert Einstein involved a Jewish ethnic bias in favor of Einstein.[3] This unfair and unethical Jewish bias preceded and caused the reactions of Ludwig Glaser, Philipp Lenard, Johannes Stark, Willy Wien, Hugo Dingler, Bruno Thüring, and others who sought to defend themselves, their students and their nation.

Einstein was famously quoted in the forward of the first edition to Lucien Fabre's French book, *Une Nouvelle Figure du Monde: les Théories d'Einstein avec une Préface de M. Einstein*, Payot, Paris, (1921), pp. 15-18; not long after the First World War ended,

> "I am a German (Jew) by birth, but I lived in Switzerland from the age of 15 until I was 35, except for brief interruptions. I earned my degree in Zurich; I am a pacifist in favor of an international agreement and have always faithfully conducted myself according to this ideal."

> "*Je suis Allemand (israélite) de naissance, mais j'ai vécu en Suisse de l'âge de 15 à celui de 35 ans, sauf de courtes interruptions. J'ai conquis mes grades à Zurich; je suis pacifiste, partisan d'une entente internationale et*

---

**2**. *See, for example*, On the occasion of Einstein's 50th birthday, "Die Relativitätstheorie und der dialektische Materialismus", *Arbeiterstimme*, (1929), which is quoted by B. Thüring, "Albert Einsteins Umsturzversuch der Physik und seine inneren Möglichkeiten und Ursachen", *Forschungen zur Judenfrage*, Volume 4, (1940), pp. 134-162, at 144-145. Republished as: *Albert Einsteins Umsturzversuch der Physik und seine inneren Möglichkeiten und Ursachen*, Dr. Georg Lüttke Verlag, Berlin, (1941).

**3**. *Cf.* C. Weizmann, *Trial and Error: The Autobiography of Chaim Weizmann*, Harper & Brothers, New York, (1949).

*resté toujours fidèle dans ma ligne de conduite à cet idéal."*

Einstein's political statements were scripted. He repeated his script and asked others to repeat it. Einstein was quoted in *The Literary Digest* of 16 April 1921, pages 33-34,

> "Dr. Einstein asked whether he could not see a copy of my interview with him before it was printed. I told him that I would not write the interview until after my return to America.
> 
> 'In that event,' he said, 'when you write it, be sure not to omit to state that I am a convinced pacifist, that I believe that the world has had enough of war. Some sort of an international agreement must be reached among nations preventing the recurrence of another war, as another war will ruin our civilization completely. Continental civilization, European civilization, has been badly damaged and set back by this war, but the loss is not irreparable. Another war may prove fatal to Europe.'"

Note that Einstein's scripted statements are classic Jewish propaganda and typify the Jewish method of undermining the sovereignty of the Gentile nations. First, the Zionists caused the war. Then they prolonged it by bringing America into it. Then they threatened the war weary nations with a worse war and offered up what they claimed was the only solution: A world led by Israel with a world government in fulfillment of Judaic Messianic prophecies. The conference Einstein hoped for was a conference where the Zionists could push the Palestine Mandate and demand a nation for the Jews. It was a conference that Einstein knew would be dominated by Jews, who would dictate to the ruined nations their future. Einstein was not concerned for humanity. He was an ardent and thoroughly scripted Jewish Zionist propagandist.

The language used in Einstein's statement in French was somewhat open to interpretation. For example, Stjepan Mohorovičić wrote in 1922,

> "Einstein selbst sagt in dem Vorwort des Werkes von L. Fabre (Anmerk. 30) den Franzosen ausdrücklich, daß er nur in Deutschland geboren sei, sonst sei er ein Jude, Pazifist und Mitglied einer internationalen Verbindung.... Es ist nicht schwer zu raten, warum Einstein dies gerade den Franzosen gegenüber gesagt hat (mit eigener Unterschrift), aber lassen wir das, es ist dies nur Geschmacksache...; unsere Arbeit hier ist eine wissenschaftliche. Es ist traurig genug, daß ich gezwungen bin, dies hier zu erwähnen!"[4]

Einstein's use of the word "entente" might also have been interpreted by Germans as a subtle allusion to the Allies. In 1904, England and France entered

---

[4]. S. Mohorovičić, *Die Einsteinsche Relativitätstheorie und ihr mathematischer, physikalischer und philosophischer Charakter*, Walter de Gruyter & Co., Berlin, Leipzig, (1923), pp. 52-53.

into an "Entente Cordiale"—an agreement between the two governments; which, while resolving colonial disputes between England and France, created tensions with Germany. In 1906 the "Entente" evolved into a military alliance, which came to include Russia in 1907. This alliance was opposed to the "Triple Alliance" of Germany, Austria-Hungary and Italy. England, France and Russia, who fought against Germany in World War I, were often referred to as the "Entente" and it might have appeared from Einstein's statement that he had always been a devout enemy of Germany and a partisan for the enemies of Germany, though he had lived in Germany throughout the war. We know that this was in fact the case, whether or not it was what Einstein meant to say in his scripted letter to Fabre. It was almost certainly not what Einstein meant to say in that letter.

Einstein used the word "Entente" to describe the Allies in many of his letters and should have been more careful with "his" words. For example, in a letter to Paul Ehrenfest of 6 December 1918,

> "Ich werde nächster Tage über die Schweiz nach Paris reisen, um die Entente zu bitten, die hiesige ausgeshungerte Bevölkerung vor dem Hungertod zu retten."[5]

Einstein wrote to Emil Zürcher on 15 April 1919,

> "Wenn die Entente gut orientiert[... .]"[6]

Einstein wrote to Hedwig Born on 31 August 1919,

> "Intervention der Entente in Schlessien"[7]

Einstein wrote to the *Neue Freie Presse* on 6 December 1919,

> "[... ]der Centralmächte und denen der Entente[... ]"[8]

Einstein wrote to Hedwig and Max Born on 27 January 1920,

> "Jedenfalls ist die Wirkekraft ihrer Parole gross, denn die Kriegsgeräte der Entente, welche das deutsche Heer aufgerieben haben, schmelzen in

---

[5]. A. Einstein to P. Ehrenfest, (6 December 1918), *The Collected Papers of Albert Einstein*, Volume 8, Part B, Document 664, Princeton University Press, (1998), pp. 960-961.

[6]. A. Einstein to E. Zürcher, (15 April 1919), *The Collected Papers of Albert Einstein*, Volume 9, Document 23, Princeton University Press, (2004), pp. 35-36, at 36.

[7]. A. Einstein to H. Born, (31 August 1919), *The Collected Papers of Albert Einstein*, Volume 9, Document 97, Princeton University Press, (2004), pp. 142-144, at 143.

[8]. A. Einstein to the *Neue Freie Presse*, *The Collected Papers of Albert Einstein*, Volume 9, Document 193, Princeton University Press, (2004), p. 273.

Russland dahin wie der Schnee in der Märzensonne."⁹

Einstein was careless in "his" letter to Fabre, which letter was quoted in Fabre's book.

Einstein did often assert that he was an internationalist and a pacifist, without implying that he had sided with the Allies in the First World War. However, we learn from Einstein's statements to the Frenchman Romain Rolland, as recorded in Rolland's diary after conversations with Einstein in Switzerland on 16 September 1915, that Einstein was indeed loyal to the Entente, not Germany. Rolland wrote,

> "What I hear from [Einstein] is not exactly encouraging, for it shows the impossibility of arriving at a lasting peace with Germany without first totally crushing it. Einstein says the situation looks to him far less favorable than a few months back. The victories over Russia have reawakened German arrogance and appetite. The word 'greedy' seems to Einstein best to characterize Germany. [***] Einstein does not expect any renewal of Germany out of itself; it lacks the energy for it, and the boldness for initiative. He hopes for a victory of the Allies, which would smash the power of Prussia and the dynasty... . Einstein and Zangger dream of a divided Germany—on the one side Southern Germany and Austria, on the other side Prussia. [***] We speak of the deliberate blindness and the lack of psychology in the Germans."¹⁰

Einstein often spoke in genocidal and racist terms against Germany and for the Jews and England. He betrayed Germany before, during and after the war. For example, Einstein wrote to Paul Ehrenfest on 22 March 1919,

> "[The Allied Powers] whose victory during the war I had felt would be by far the lesser evil are now proving to be *only slightly* the lesser evil. [***] I get most joy from the emergence of the Jewish state in Palestine. It does seem to me that our kinfolk really are more sympathetic (at least less brutal) than these horrid Europeans. Perhaps things can only improve if only the Chinese are left, who refer to all Europeans with the collective noun 'bandits.'"¹¹

---

9. Letter from A. Einstein to the Borns of 27 January 1920, *The Collected Papers of Albert Einstein*, Volume 9, Document 284, Princeton University Press, (2004), pp. 386-390, at 387.

10. R. Romain, *La Conscience de l'Europe*, Volume 1, pp. 696ff. English translation from A. Fölsing, *Albert Einstein: A Biography*, Viking, New York, (1997), pp. 365-367. *See also:* Letter from A. Einstein to R. Romain of 15 September 1915, *The Collected Papers of Albert Einstein*, Volume 8, Document 118, Princeton University Press, (1998); **and** Letter from A. Einstein to R. Romain of 22 August 1917, *The Collected Papers of Albert Einstein*, Volume 8, Document 374, Princeton University Press, (1998).

11. Letter from A. Einstein to Paul Ehrenfest of 22 March 1919, English translation by

Einstein almost certainly was not referring to the Allies when referring to an *entente internationale*, but rather to an international agreement. His wording caused further consternation given that there was the soon to appear *Entente Internationale des Partis Radicaux et des Partis Démocratiques similaires*, a group of liberals from many nations who based their movement on the spirit of the *Plan des Libéraux pour recommencer la révolution*, Paris, (1821); probably in the form of the *Carté*. There was also the First International of Marx and Engels, and its offspring: The International Workingmen's Association, the Second International, the Socialist International, the Third International, the Comintern, the Vienna International, the Two-and-a-half International, the Labor and Socialist International, the Fourth International, the Trotsky International, etc. The *Carté* was founded by Communist Henri Barbusse and Einstein's friend and confidant, pacifist Socialist Romain Rolland. In late 1919 and early 1920, Einstein sought to establish a German chapter of the *Clarté* for the purposes of promoting Internationalism.[12] This in itself troubled many Germans, who had come to believe that "Internationalism" was a code word for "Jewish supremacy". Even before the war, the "Proclamation of the Alliance Against the Arrogance of Jewry" of 1912 stated,

> "The Reichstag elections of 1912 have taken place under the sign of Jewry—that is, under the sign of open and clandestine republicanism and internationalism. 'National is irrational'... was and is the slogan that misled millions of Germans, blinded by the fraudulent Jewish catchwords of international culture and international progress. [***] Jewry is international in the sense of Schopenhauer's phrase: 'The fatherland of the Jews is other Jews.'"[13]

Einstein's declarations of his "tribal"—to use his term—loyalty, his public insults against Germans, and his allegedly privileged Zionist nationalism were viewed as legitimate causes for concern—as was the modern terror of the Internationalism of the Bolsheviks, who had made Bavaria a Soviet Republic for a short span of time.

Many Germans were outraged by Einstein's statement as quoted in Fabre's

---

A. Hentschel, *The Collected Papers of Albert Einstein*, Volume 9, Document 10, Princeton University Press, (2004), pp. 9-10, at 10.

**12**. Letter from A. Einstein to R. W. Lawson of 26 December 1919, *The Collected Papers of Albert Einstein*, Volume 9, Document 234, Princeton University Press, (2004). *See also:* A. Einstein, "Welcoming Address to Paul Colin", *The Collected Papers of Albert Einstein*, Volume 7, Document 27, Princeton University Press, (2002). *See also: The Collected Papers of Albert Einstein*, Volume 9, Documents 222, 230, 237, 249, 275, 297 and 331, Princeton University Press, (2004).

**13**. R. S. Levy, *Antisemitism in the Modern World: An Anthology of Texts*, D. C. Heath and Company, Toronto, (1991), pp. 129-130, at 129.

book,[14] which was an obvious attempt by Einstein to distance himself from Germany (Gentiles) and ingratiate himself to the French, no matter how one translated it—and Einstein and his friends instigated a smear campaign against Fabre in order to deflect attention from Einstein's volatile comments.[15] Einstein's friend Solovine smeared Fabre, claiming that he was an anti-Semite—even though Fabre had written a book which was highly flattering to Einstein.

Einstein charged that Fabre cobbled together the forward from Einstein's statements and published this compilation of quotes without Einstein's approval. Einstein protested that Fabre had no right to designate this compilation as if it were a forward Einstein intended to write for Fabre, because he allegedly had not written it in the form in which it appeared and had not approved its publication as a forward to Fabre's book—though he had made the statements—a fact he appeared to publicly deny. Einstein alleged to Solovine that his words were corrupted in translation though the addition of French *gentillesse* by an acquaintance of his, who Einstein implies wrote the letters.

In the second edition of his book, Fabre stated that he had only given a public expression to Einstein's views to a wanting public, with the best of intentions. Fabre stated that Einstein had repudiated Einstein's own statements. Einstein's friends let Einstein know that Fabre had begun to spread the word after Einstein had attacked Fabre, that Henri Poincaré was the true father of the special theory of relativity. Einstein hid from Fabre's accusation that Einstein had plagiarized Poincaré's theory.[16]

The preface to Fabre's first edition states,

"PRÉFACE

*L'ouvrage de M. Fabre est des plus intéressants et fort bien écrit. Ses explications sur l'œuvre de Newton, de Faraday et de Maxwell sont admirablement réussies. L'auteur est un vrai enthousiaste rempli d'un sentiment vibrant pour la beauté scientifique.*

*L'éloge dont il veut bien honorer mes théories est terriblement exagéré. La théorie de la relativité ne peut ni veut donner aucun système du monde,*

---

**14**. *Cf.* S. Mohorovičić, *Die Einsteinsche Relativitätstheorie und ihr mathematischer, physikalischer und philosophischer Charakter*, Walter de Gruyter & Co., Berlin, Leipzig, (1923), p. 53. Einstein stated in the *Jüdische Pressezentral*, Number 111, (21 September 1920), that it irked him to read that he was a German citizen of Jewish faith. He stated that he was not a German citizen, but was a Jew. *Cf.* B. Thüring, *Albert Einsteins Umsturzversuch der Physik und seine inneren Möglichkeiten und Ursachen*, Dr. Georg Lüttke Verlag, Berlin, (1941), pp. 24-25.
**15**. D. K. Buchwald, et al., Editors, *The Collected Papers of Albert Einstein*, Volume 7, Princeton University Press, (2002), pp. 417-419. A. Einstein, "Zuschriften an die Herausgeber: Zur Abwehr", *Die Naturwissenschaften*, Volume 9, (1921), p. 219. *See also:* L. Fabre's response to A. Einstein's objections: L. Fabre, *Une Nouvelle Figure du Monde: Les Théories d'Einstein*, Payot, Paris, (1922), pp. 15-16.
**16**. D. K. Buchwald, et al., Editors, *The Collected Papers of Albert Einstein*, Volume 7, Princeton University Press, (2002), pp. 417-419.

mais seulement une condition restrictive à laquelle les lois de la nature doivent se soumettre, comme par exemple les deux principaux axiomes de la thermodynamique. Celui-là même qui ne reconnaîtrait pas la théorie de la relativité se voit cependant obligé d'admettre une interprétation physique claire des coordonnées de l'espace et du temps. C'est justement à ce point de vue que pèchent les écrits de certains des savants cités par l'auteur.

L'ouvrage de l'un d'entre eux défend une thèse sans espoir qui, traduite en termes géométriques dirait ceci: «*Parmi toutes les directions X possibles dans l'espace, il n'existe qu'une seule direction de coordonnée X absolue*» (*il s'agit en l'espèce d'un temps absolu devant être préposé aux transformations Lorentz*), entreprise sans espoir appuyée sur quelques ambiguïtés involontaires mathématiques.

Un autre de ces savants ne remarque pas — abstraction faite de ce qu'il oublie d'interpréter physiquement l'espace et le temps — que la vitesse de la lumière conformément à l'expérience joue un rôle spécial. Les deux erreurs étroitement liées se cachent sous une enveloppe épaisse de formules mathématiques. Aucun homme raisonnable n'admettra cependant que le son se propage, relativement à l'air en repos, selon les mêmes lois que relativement à l'air en mouvement. L'expérience nous a appris, par contre, que, seule, la vitesse de la lumière est indépendante de l'état de mouvement du système de coordonnées.

On ne peut pas dire non plus que la théorie générale de la relativité ait abandonné, par rapport à la vitesse de la lumière, le principe de la continuité. La vitesse de la lumière, mesurée avec perche et horloge unitaires, dans l'entourage infinitésimal d'un point est toujours, dans la théorie de la relativité aussi, invariablement la même.

Albert EINSTEIN.

Je crois devoir joindre à cette préface quelques extraits d'une lettre de M. Einstein qui me paraissent éclairer la physionomie du savant allemand.

L. F.

*Cher Monsieur, 5-VII-20*

J'ai reçu, par notre ami Oppenheim, au retour d'un long voyage, votre amicale lettre du 19 juin. J'ai étudié votre intéressant travail et j'y ai pris beaucoup de plaisir (*en particulier dans l'exposé du développement historique de la théorie*)... ... ... ... ... ... ... ...... *Parmi les savants français, Langevin a parfaitement pénétré la théorie de la relativité. C'est un esprit merveilleusement clair et un homme sympathique* ... ... ... ... ... ... ... ...... *Je joins à ma lettre le* **curriculum vitae** *que vous souhaitez.* — *Je suis Allemand (israélite) de naissance, mais j'ai vécu en Suisse de l'âge de 15 à celui de 35 ans, sauf de courtes interruptions. J'ai conquis mes grades à Zurich; je suis pacifiste, partisan d'une entente internationale et resté toujours fidèle dans ma ligne de conduite à cet idéal.*

*Agréez, … … … … …..*
A. EINSTEIN.

Voici les renseignements biographiques fournis par M. Einstein:

Albert Einstein est né à Ulm le 14 mars 1879. Il était âgé de six semaines lorsque ses parents émigrèrent vers Munich où il passa son enfance et fréquenta les écoles jusqu'à sa quatorzième année. A quinze aus il se rendit en Suisse, resta un an au collège de Aarau et y obtint son *abiturium*. Il étudia ensuite les mathématiques et la physique à Zurich. En 1902, Einstein fut attaché au bureau des brevets à Berne et prépara simultanément son examen du doctorat auquel il fut admis en 1905. Il fut appelé comme professeur à l'Université de Zurich en 1909, à celle de Prague en 1911 et retourna à Zurich en 1912 comme professeur au Polytechnikum, qu'il quitta en 1914 pour aller occuper un siège à l'académie royale de Prusse à Berlin. Il est également directeur de l'Institut Kaiser-Wilhelm pour la physique."

Einstein wrote in *Die Naturwissenschaften*, Volume 9, Number 13, (1 April 1921), p. 219, giving the false impression that he had not said what he had said,

"Zuschriften an die Herausgeber.
Zur Abwehr.

Herr *Lucien Fabre* hat im Verlage von Payot in Paris ein Buch „Les théories d'Einstein" mit dem Zussatz „Avec une préface de M. Einstein" herausgegeben. Ich erkläre, daß ich keine Vorrede zu dem Buche geschrieben habe und protestiere gegen diesen Mißbrauch meines Namens. Ich bringe den Protest zu Ihrer Kenntnis in der Hoffnung, daß er aus Ihrer Zeitschrift den Weg in die weitere Öffentlichkeit und im besonderen auch in die Zeitschriften des Auslandes finden wird.
Berlin, 16. März 1921. *Albert Einstein.*"

According to Ernst Gehrcke, Einstein's statement was indeed reprinted in the popular press. Fabre responded with a statement published in the *Neue Züricher Zeitung* on 9 May 1921, and in many other papers, and Gehrcke quoted the following from it:

"Diese Vorrede besteht aus drei Dokumenten: sie enthält biographische Daten, wissenschaftliche Ansichten und zuletzt ein internationalistisches Glaubensbekenntnis. Ich halte aufs entschiedenste folgende Behauptungen aufrecht: 1. Verfasser dieser Vorrede ist Herr EINSTEIN. 2. Er selbst hat sie mir zugeschickt und zwar in der Form von Briefen und als Antwort auf briefliche Anfragen meinerseits. 3. Sie war ausschließlich dazu bestimmt, meinen Lesern, d. h. dem französischen Publikum, die moralische und wissenschaftliche Persönlichkeit dieses Gelehrten vorzustellen. Ich bin bereit, obige Behauptungen durch unwiderlegliche Schriftstücke zu bezeugen... "[17]

Fabre had composed the forward from letters he had received from Einstein, and he still held them as proof that Einstein had made the statements he later disowned.

Fabre wrote in the second edition *Une Nouvelle Figure du Monde: Les Théories d'Einstein. Accrue de notes Liminaires, d'un Exposé des Théories de Weyl, et de Trois Notes de M. M. Guillaume, Brillouin et Sagnac sur Leurs Propres Idées*, Payot, Paris, (1922),

"NOTES LIMINAIRES

*La présente édition de cet ouvrage diffère des précédentes.*

*J'ai procédé à une épuration et à une mise à jour.*

\*
\* \*

*J'ai d'abord purgé mon livre des déclarations de M. Einstein qui lui servaient de préface. Une partie de la presse et des amis qui me sont chers, avaient critiqué la forme et le fond de ces déclarations. Je ne les avais moi-même insérées que pour permettre au savant israëlite allemand de dire publiquement du haut de cette tribune ce qu'il voulait donner comme vrai sur ses opinions politiques, sa vie, sa nationalité, ses sentiments, en un mot, sa physionomie non scientifique, laquelle, on le sait de reste, est extrêmement discutée.*

*Bien que j'eusse laissé à M. Einstein la responsabilité de ses déclarations je m'en sentais un peu complice puisque je leur donnais l'hospitalité. Mais je n'en aurais pas purgé ce livre, même si leur teneur m'eût été démontrée mensongère, car elles donnaient sur ce grand savant le témoignage le plus précieux puisqu'il émanait de lui.*

*L'événement le plus imprévu m'a décidé; M. Einstein a, en effet, renié ses déclarations dans la presse allemand. Je me hâte donc de les retrancher de cet ouvrage qui n'aura à connaître que de la figure purement scientifique du grand théoricien; c'est la seule qu'on puisse considérer avec sérénité et même avec quelque sympathie.*

\*
\* \*

*Il va sans dire que j'ai également indiqué sur le mode dubitatif, ou*

---

[17]. E. Gehrcke, *Die Massensuggestion der Relativitätstheorie: Kulturhistorisch-psychologische Dokumente*, Hermann Meusser, Berlin, (1924), p. 67.

même supprimé, les assertions que j'avais, dans le cours de l'ouvrage, avancées sur la foi des paroles d'Einstein, les autographes de celles-ci demeurant entre mes mains pour exercer la sagacité des psychologues futurs.

$$*\phantom{xx}*\phantom{xx}*$$

*Il m'a semblé indispensable d'ajouter à ce travail un bref exposé des théories de Weyl qui complètent très heureusement celles d'Einstein. Leur audace et leur beauté ne peut guère à l'heure actuelle apparaître qu'aux savants. Il est toutefois dès à présent certain que le disciple égale au moins le maître; et peut-être le dépasse-t-il.*

$$*\phantom{xx}*\phantom{xx}*$$

*Les nombreuses lettres qui me sont parvenues m'ont aussi convaincu de l'intérêt que présente pour le public la question du temps relatif. J'ai donné avec assez de détails le point de vue einsteinien pour n'y pas revenir. Mais j'ai pensé que le lecteur entendrait avec plaisir sur le même sujet la voix de M. Guillaume dont j'avais brièvement exposé les théories. Le savant bernois a bien voulu écrire, spécialement pour le présent ouvrage, la note qu'on lira en appendice. On trouvera agrément et profit à la méditer.*

*M. Brillouin a bien voulu également indiquer lui-même son point de vue aux lecteurs du présent ouvrage; on trouvera sa lettre en appendice.*

*Il faut admirer la sûreté, la clarté de cette belle page bien française. Elle met exactement à sa place scientifique la théorie einsteinienne; elle en dégage la convenance et l'utilité en tant qu'hypothèse; très sobrement, elle met en garde contre les commentaires où se peuvent aventurer ceux qui confondent l'hypothèse et le réel; j'y discerne, sans vouloir engager la pensée de son auteur, une méfiance à l'égard des conceptions philosophiques déduites des travaux einsteiniens.*

*Il n'est pas possible de ne pas souscrire à un jugement si parfaitement lucide; sa réserve et sa sagesse ne diminuent en rien l'enthousiasme que les théories d'Einstein et celles de Weyl, peuvent, indépendamment de leur adéquation au réel, inspirer à qui y recherche un excitant intellectuel.*

$$*\phantom{xx}*\phantom{xx}*$$

*Enfin M. Sagnac, dont on a pu écrire, en faisant allusion à la phrase qui termine ce livre, qu'il était peut-être le nouveau Poincaré, le seul capable de nous donner une réponse définitive sur la valeur des théories einsteiniennes, a accepté de confier à ce petit ouvrage le sort d'une note originale dont l'extraordinaire importance n'échappera à personne.*

*Cette note:*
—*d'une part résume l'effet Sagnac sur la rotation dans l'éther (auquel nous avons fait allusion dans notre ouvrage);*
—*d'autre part institue une théorie générale des champs en translation par une extension de la pure mécanique des petits mouvements.*
*Nous sommes extrêmement heureux de pouvoir donner à nos lecteurs la primeur d'un travail qui nous paraît contenir en germe les plus belles découvertes.*"

Many interesting and telling facts emerge from the affair—smear tactic and vilification used to rescue Einstein by means of personal attack meant to divert attention from the real issue, and Einstein's dependence upon collaborators to write his statements, as well as Einstein's image. The preface to Fabre's book was only one of many of Einstein's anti-German, pro-Allies, and, elsewhere, Anglophilic, statements made public.[18]

Suspicion also fell upon Einstein because the "war to end all wars", *i. e.* the end of war—pacifism, socialism, revolution and economic hardship—which were great concerns of the Germans in the post-war period—were forecast in Ivan Stanislavovich Bloch's book, *The Future of War in Its Technical, Economic, and Political Relations; Is War Now Impossible?*, Doubleday & McClure Co., New York, (1899). Bolch was a hero and an inspiration to many Jews and to many Socialists. He was part of the culture that inspired H. G. Wells, Russell, Lorentz and Einstein; and Einstein was seen as a believer in, and vocal advocate of, this Blochian philosophy. The concept of the "war to end all wars" is also a prophetic and Apocalyptic one of Jewish world leadership foretold in the period of peace of the book of *Enoch*, with its "elect" and "Elect One" (*see also: Isaiah* 65; 66) and in the final war in the Old Testament in, among other places, *Isaiah* 2:1-4:

"1 The word that Isaiah the son of Amoz saw concerning Judah and Jerusalem. 2 And it shall come to pass in the last days, *that* the mountain of the LORD's house shall be established in the top of the mountains, and shall be exalted above the hills; and all nations shall flow unto it. 3 And many people shall go and say, Come ye, and let us go up to the mountain of the LORD, to the house of the God of Jacob; and he will teach us of his ways,

---

**18**. "Time, Space, and Gravitation", *The London Times*, (28 November 1919), pp. 13-14. *See also:* "Meine Antwort", *Berliner Tageblatt*, Morgen Ausgabe, (27 August 1920), pp. 1-2. *See also:* "Einstein and Newton", *The London Times*, (14 June 1921), p. 8. *See also:* "Wie ich Zionist wurde", *Jüdische Rundschau*, (21 June 1921), pp. 351-352; English translation by A. Engel, "How I became a Zionist", *The Collected Papers of Albert Einstein*, Volume 7, Document 57, Princeton University Press, (2002), pp. 234-237. Einstein stated in the *Jüdische Pressezentral*, Number 111, (21 September 1920), that it irked him to read that he was a German citizen of Jewish faith. He stated that he was not a German citizen, but was a Jew. *Confer:* B. Thüring, *Albert Einsteins Umsturzversuch der Physik und seine inneren Möglichkeiten und Ursachen*, Dr. Georg Lüttke Verlag, Berlin, (1941), pp. 24-25.

and we will walk in his paths: for out of Zion shall go forth the law, and the word of the LORD from Jerusalem. 4 And he shall judge among the nations, and shall rebuke many people: and they shall beat their swords into plowshares, and their spears into pruninghooks: nation shall not lift up sword against nation, neither shall they learn war any more."

"Pacifists" often promoted the Apocalyptic prophesy of a "war to end all wars", which would establish a "world government" according to prophecy, one run by Jews in Jerusalem. Albert Einstein was one of the many advocates of this plan. "Pacifists" often sought to provoke the most terrible of wars humankind has yet endured on the false premise that it would end war. What these brutal and genocidal wars instead did was weaken the nations making them vulnerable to Jewish revolution, while simultaneously making the Jewish financiers unimaginably wealthy. Thereby, the Jewish financiers could sponsor revolution, dictatorship and genocide, and could buy up the world at reduced rates. The people were intentionally made so weary of war, that they become vulnerable to the sophistical message that the only means to secure peace is to destroy all nations such that there will be no nations left to war with each other. Some Jews press this message in order to bring to fulfillment the Messianic prophecy that the Jews will destroy all nations and religions, and rule the Earth. The false message that the loss of sovereignty leads to peace was a fundamental theme in Communist régimes. The loss of Gentile sovereignty has instead led to the enslavement and extermination of the Gentile peoples, in fulfillment of Judaic Messianic prophecy.

In the era of the German Enlightenment, Moses Mendelssohn asserted that the "Jewish mission" was to convert the world to monotheism and to instill in all peoples the principles of the Jewish moral code, which according to some initially only applied only to Jews, with the ancient Jews viewing Gentiles as subhuman and therefore undeserving of moral treatment. Einstein's friend Georg Friedrich Nicolai (Lewinstein) stated in 1917,

> "Apart from this strange story of Cain, however, murder is forbidden in the Bible, and very sternly forbidden. But—it is only the murder of Jews. As is natural, considering the period from which it dates, the Bible is absolutely national, in character. Only the Jew is really considered as a human being; cattle and strangers might be slain without the slayer himself being slain. In this case there was a ransom. Accordingly, war was of course allowed also, and the Jews were no more illogical than the Moslem who kills the outlander. Of late years the Jews and the Old Testament have often been reproached for their contempt for those who were not Jews; and in practice even Christ acted in precisely the same way."[19]

---

**19**. G. Nicolai, *Die Biologie des Krieges, Betrachtungen eines deutschen Naturforschers*, O. Füssli, Zürich, (1917); English translation: *The Biology of War*, Century Co., New York, (1918), p. 531.

Mendelssohn's message was not very different from that of Jesus Christ, as expressed in the Gospels; or, indeed, that of Islam, "There is no God but God." The political Zionists tended to be secular and racist, and based their beliefs on biological, Darwinistic principles. Albert Einstein saw Judaism as step away from paganistic Polytheism towards utilitarian and scientific morality, with the objectionable premise in the ancient tradition that one is led to morality through fear of the "imaginary" God.[20] However, all of these movements, which meant to lessen the suspicion among Gentiles of Jewish religious aspirations, perpetuated those aspirations which were always more political and racist in nature, than spiritual. Moses of the ten commandments was little different from Moses Mendelssohn.

Einstein followed the line of thought which sponsored European Liberalism, "such as Jacobinism, Fourierism, Owenism, Fabian Socialism, Marxism, and the like",[21] as essentially adopting the moral values of Judaism and replacing the source of these values, "God", with a quasi-Deistic conception of nature. Many critics of the Jews found this irrational, in that the removal of "God" *a priori* removes the fundamental premise of all that can be deduced from this premise, including codes of moral and just conduct, without providing a substitute premise which rationally deduces their conclusions. These critics sought a more synthetic basis for morality than neo-Platonism, and many arrived at pragmatic Darwinism and Metempsychosis, which they argued were logically consistent and empirically justified. In reality, they was less difference between the two points of view than was apparent on the face of the dispute.

Before Bloch were Bertha von Suttner and Alfred Hermann Fried of the *Friedensbewegung* (peace movement) which attracted pacificist physicist and Einstein-supporter Hans Thirring. Suttner published *Die Waffen nieder!*[22] in 1892, which emphasized the harm done to civilians in modern warfare. The American Civil War had demonstrated the destructive force of modern industry applied to warfare. Friedrich Nietzsche, whose work was well known, predicted the massive destruction this would cause in the Twentieth Century.

Unlike Albert Einstein, Philipp Lenard had expressed his loyalty to Germany during and after the First World War. After Einstein smeared him without cause, Nobel Prize laureate Philipp Lenard demanded a very public personal apology from Albert Einstein, which was not forthcoming. Einstein repeatedly made harshly anti-German and warmly Anglophilic statements before and after the Bad Nauheim debate which outraged many Germans.[23] Einstein

---

[20]. A. Einstein, *The World As I See It*, Citadel, New York, (1993), p. 91.
[21]. R. P. Oliver, "Liberalism", *America's Decline: The Education of a Conservative*, Londinium Press, London, (1981).
[22]. English translation, B. v. Suttner, *Ground Arms!"* = *"Die Waffen nieder!" A Romance of European War*, A.C. McClurg & Co., Chicago, (1906). *See also:* B. v. Suttner, *Martha's Kinder: eine Fortsetzung zu "Die Waffen nieder!"*, E. Pierson, Dresden, (1903).
[23]. "Time, Space, and Gravitation", *The London Times*, (28 November 1919), pp. 13-14. *See also:* "Meine Antwort", *Berliner Tageblatt*, Morgen Ausgabe, (27 August 1920), pp. 1-2. *See also:* L. Fabre, *Une Nouvelle Figure du Monde: Les Théories d'Einstein*, Payot

was member of a commission which intended to investigate and publicize alleged German war atrocities, in 1919, for the purposes of a psychological attack on the German psyche attempting to coerce them into accepting Einstein's view that Germany's defeat was a victory for humanity.[24] Einstein also wanted to increase the hardships on the already starving Germans with foreign boycotts on German products soon after the First World War ended.[25] Many hundreds of thousands, if not millions of Germans had starved to death during a naval blockade in the war. Einstein's, and like minded vindictive spirits', love of punishing Germans made the Germans resentful of the Jews who had stabbed them in the back.

Ethnocentric attacks against German science appeared in America[26] in 1918, and in England[27] in 1919. In addition, English and French scientists, in collusion with traitors like Albert Einstein, took punitive actions against German scientists under the auspices of the International Research Council. Among other punitive sanctions, they excluded German and Austrian nationals from international congresses and banned the Nations of the former Central Powers from membership for a period of ten years. Einstein was marketed to the Allies as a Swiss Jew who had opposed Germany from the beginning of the war and Einstein, the "Swiss Jew", was safe from these vicious attacks on the liberty and dignity of German scientists.

Max Born knew that Hendrik Antoon Lorentz was a friend of the Allies after the First World War and Born disliked him.[28] Einstein, who had lived in Germany throughout the war, in spite of the fact that he hated Germany and wanted to see the nation destroyed, wrote to Lorentz on 1 August 1919,

---

& Cie, Paris, (1921), pp. 15-18. *See also:* "Einstein and Newton", *The London Times*, (14 June 1921), p. 8. *See also:* "Wie ich Zionist wurde", *Jüdische Rundschau*, (21 June 1921), pp. 351-352; English translation by A. Engel, "How I became a Zionist", *The Collected Papers of Albert Einstein*, Volume 7, Document 57, Princeton University Press, (2002), pp. 234-237. Einstein stated in the *Jüdische Pressezentral*, Number 111, (21 September 1920), that it irked him to read that he was a German citizen of Jewish faith. He stated that he was not a German citizen, but was a Jew. *Confer:* B. Thüring, *Albert Einsteins Umsturzversuch der Physik und seine inneren Möglichkeiten und Ursachen*, Dr. Georg Lüttke Verlag, Berlin, (1941), pp. 24-25. *See also Einstein's private correspondence, for example: The Collected Papers of Albert Einstein*, Volume 9, Documents 10, 28, 36, 78, 79, 80, 92, 94, 96 and 108, Princeton University Press, (2004).

**24**. Letter from A. Einstein to H. A. Lorentz of 26 April 1919, *The Collected Papers of Albert Einstein*, Volume 9, Document 28, Princeton University Press, (2004). Letter from A. Einstein to W. de Haas of 9 May 1919, *The Collected Papers of Albert Einstein*, Volume 9, Document 36, Princeton University Press, (2004).

**25**. Letter from A. Einstein to H. A. Lorentz of 21 September 1919, *The Collected Papers of Albert Einstein*, Volume 9, Document 108, Princeton University Press, (2004).

**26**. P. G. Nutting, "National Prestige in Scientific Achievement", *Science*, Volume 48, (1918), pp. 605-608.

**27**. "America and German Science", *Nature*, Volume 102, (1919), pp. 446-447.

**28**. Letter from M. Born to A. Einstein of 28 October 1920, M. Born, *The Born-Einstein Letters*, Walker and Company, New York, (1971), pp. 43-45.

"Exclusion of German scholars from social international scholarly exchanges for a number of years might perhaps be a lesson in humility for them, which will not do much harm at all—and, it is to be hoped, might even help."[29]

Many German scientists resented Einstein's treachery. Indeed, under pressure from Lenard for his anti-German activities and as a result of the economic conditions in Germany, Einstein published an appeal to ease the punitive measures taken against German science, which he himself had initially sponsored.[30] However, racist Zionist Albert Einstein saw to it that no German scientist would be present at the Solvoy Conference in April of 1921. His friend Hendrik Antoon Lorentz invited only one German scientist to attend the conference, Albert Einstein. Racist Zionist Albert Einstein then refused the invitation with the excuse that he was heading for America to exploit his ill-founded fame to raise money for his fellow racist Zionists. Einstein wrote to Lorentz,

"As this venture lies close to my heart, and as I, as a Jew, feel a duty to contribute, as far as I am able, to its success, I accepted."[31]

Fellow German Jew Fritz Haber was outraged at Albert Einstein's racist treachery and disloyalty. Einstein confirmed that he was disloyal and a racist, and was obligated,

"[... ] to step in for my persecuted and morally depressed fellow tribesmen, as far as this lies within my power[.]"[32]

In point of fact, Einstein was instead promoting himself and hiding from his critics.

In response to the Berlin Philharmonic lectures, Einstein and his friends arranged for a discussion of the theory of relativity at the Eighty-Sixth Meeting

---

[29]. Letter from A. Einstein to H. A. Lorentz of 1 August 1919, English translation by A. Hentschel, *The Collected Papers of Albert Einstein*, Volume 9, Document 80, Princeton University Press, (2004), pp. 67-68, at 68. See also Document 108, the letter from Einstein to Lorentz of 21 September 1919, at pages 92–93.

[30]. A. Einstein, *Thoughts on Reconciliation*, Deutscher Gesellig-Wissenschaftlicher Verein von New York, New York, (1920), pp. 10-11; facsimile republished in *The Collected Papers of Albert Einstein*, Volume 7, Document 47, Princeton University Press, (2002), pp. 360-364.

[31]. A. Einstein quoted in: H. Gutfreund, "Albert Einstein and the Hebrew University", J. Renn, Editor, *Albert Einstein Chief Engineer of the Universe: One Hundred Authors for Einstein*, Wiley-VCH, Berlin, (2005), pp. 314-318, at 315.

[32]. A. Einstein quoted in: H. Gutfreund, "Albert Einstein and the Hebrew University", J. Renn, Editor, *Albert Einstein Chief Engineer of the Universe: One Hundred Authors for Einstein*, Wiley-VCH, Berlin, (2005), pp. 314-318, at 316.

of German Natural Scientists in Bad Nauheim in late September of 1920. These were annual gatherings which had been interrupted by the war. Einstein threatened that Lenard and all critics of the theory of relativity would be humiliated. Einstein was known for his childish and evasive responses to criticism. He was known for hiding from criticism. Einstein responded,

> "The best proof that I by no means dodge criticism is that I myself arranged that the theory of relativity be discussed at the meeting of the GDNA in Nauheim."[33]

Einstein stated in his challenge that anyone brave enough should speak in Bad Nauheim.

Einstein, himself, was not brave enough. Contrary to his public bravado, Einstein feared the confrontation he had created and wanted others to speak on his behalf. He knew that he could not defend the theory of relativity and that he had no legitimate defense for his plagiarism. Einstein instead wanted to hide from the criticism directed at him.

Albert Einstein wrote to Arnold Sommerfeld on 6 September 1920 that he wanted to hide from the debate,

> "But I do not on any account want to speak myself[.]"[34]

## 4.4.8 Let the Debate Begin

Einstein, against his better judgement, did speak at Nauheim. The event was highly publicized by Einstein and his supporters and thousands showed up to see the debate. The theory of relativity was hyped beyond all reasonable limits and many were certain that the great hero Einstein would crush his opponents, as advertised. The much anticipated debate between Lenard and Einstein over the general theory of relativity began on Thursday, at 12:45 PM. Einstein's advocates, Max Planck who chaired the session, *et al.*, employed armed police to keep anti-relativists and neutral parties out of the audience and attempted to stack the audience with a pro-Einstein claque. This resulted in a tumultuous protest and unbiased audience members stormed the hall and held their ground.

After long and boring lectures by Einstein and his friends which began at 9:00 AM, the bell sounded at 12:45 PM for the time allotted to Einstein-critics to begin. Einstein and Lenard began to debate.

---

[33]. A. Einstein quoted in *Vossische Zeitung*, Morning Edition, Supplement 4, (29 August 1920), p. 1. English translation from, D. K. Buchwald, et. al. Editors, *The Collected Papers of Albert Einstein*, Volume 7, Princeton University Press, (2002), Note 1, p. 357.
[34]. A. Einstein quoted in R. W. Clark, *Einstein: The Life and Times*, The World Publishing Company, (1971), p. 261; referencing A. Einstein to A. Sommerfeld, in A. Hermann, *Briefwechsel. 60 Briefe aus dem goldenen Zeitalter der modernen Physik*, Schwabe & Co., Basel, Stuttgart, (1968), p. 69.

Though accounts of the meeting are incomplete and vary,[35] Lenard clearly made Einstein look very foolish in a very short time. Einstein was flustered and could not give cogent responses, even though Lenard repeated his questions. In a prearranged maneuver, Max Planck called the session, which had begun at 12:45 PM, to an end at about 1:00 PM, after only a few minutes of debate, so as to let Einstein off the hook and prevent a fuller exposure of Einstein's incompetence. Fifteen minutes before the afternoon session began, Einstein ran away. Gehrcke, who had humiliated Einstein at the Berlin Philharmonic, and whom Einstein had openly challenged to speak at Bad Nauheim, repeatedly demanded time to speak, but Max Planck refused to allow Gehrcke a chance to speak, and delayed Gehrcke until the session was closed. Planck also refused to allow Rudolph, another Einstein critic, time to speak.

Pursuant to Planck's corrupt plan, Einstein's critics were only allotted fifteen minutes to speak, including responses from Einstein and his friends, after hours of pro-Relativity lectures. Planck tried to arrange it so that only pro-Einstein mathematical lectures would occur, which would be entirely uninteresting to the public and to the press.

Max Planck fed Friedrich von Müller, the opening speaker to the Bad Nauheim gathering, a prepared speech Planck and Arnold Sommerfeld had written lauding Einstein and unfairly degrading his opponents. Planck arranged it so that armed guards would intimidate anti-Einstein participants and prevent them from attending the meeting hall and attempted to stack the audience and the stage with a pro-Einstein claque. Planck not only limited the time of the anti-Relativists at the Thursday meeting to a few minutes, Planck also greatly restricted their time at the Friday meeting to 12 minutes including discussion—a meeting which Einstein and his cronies did not attend. Einstein hid from his opponents and ran away from the debate, even after Max Planck had arranged it so that Einstein would have every conceivable advantage.

Albert Einstein was ashamed of the fact that he had run away. He wrote to Max Born in October of 1920,

> "I will live through all that is in store for me like an unconcerned spectator and will not allow myself to get excited again, as in Nauheim. It is quite inconceivable to me how I could have lost my sense of humour to such an

---

**35**. A biased and heavily redacted version of the discussion appeared in: *Physikalische Zeitschrift*, Volume 21, (1920), pp. 666-668. That this version is incomplete and biased is proven in: P. Lenard, *Über Relativitätsprinzip, Äther, Gravitation*, Third Edition, S. Hirzel, Leipzig, (1921); **and** "Zur zweiten Auflage. Ein Mahnwort an deutsche Naturforscher.", *Über Äther und Uräther*, Second Edition, S. Hirzel, Leipzig, (1922), pp. 5-10. E. Gehrcke, "Die Relativitätstheorie auf dem Naturforschertage in Nauheim", *Umschau, Wochenschrift über die Fortschritte in Wissenschaften und Technik*, Volume 25, (1921), p. 99; **and** "Zur Relativitätsfrage", *Die Umschau*, Volume 25, (1921), p. 227. *Berliner Tageblatt*, Evening Edition, (24 September 1920), p. 3. *Vossische Zeitung*, Evening Edition, (24 September 1920), p. 1-2.

extent through being in bad company."[36]

### 4.4.8.1 Einstein Disappoints—"Albertus Maximus" is a Laughingstock

Einstein's cowardice and incompetence did not go unnoticed. Johannes Riem ridiculed Albert Einstein,

## "Amerika über Einstein
Von
Professor Dr. Johannes Riem.

Es ist kaum anzunehmen, daß Einstein mit reiner Freude an seine amerikanische Rundreise zurückdenken wird. Ein großer Teil der dortigen Physiker und Astronomen stand von vornherein ablehnend da, vor allem der bekannte Michelson, dessen berühmtes Experiment in seiner falschen Deutung mit den Anlaß für die Relativitätstheorie gegeben hat. Vor mir liegen zwei Zeitungsblätter, „The Minneapolis Sunday Tribune", 1921 May 22, und „The St. Pauly Daily News", 1921 May 8. Beide beschäftigen sich mit der Relativitätstheorie und Einsteins Auftreten drüben. Zunächst die Feststellung, daß Einstein gleichzeitig mit der Abordnung der Zionisten drüben ankam, und daß die Presse davon in ausgedehntem Maße Kenntnis nahm. Doch habe man sehr bald dies als bezahlte Mache erkannt, und die ganze Einsteinsche Reise von Beginn an als einen Bluff erfaßt.

Die Amerikaner wären denn doch zu skeptisch gewesen, ihn ohne weitere Beweise für größer als Kopernikus und Newton zu halten, bloß, weil seine Lehre unverständlicher sei. Denn die Wahrheit sei einfach und verständlich. Man habe die Relativitätstheorie deswegen als einen Schwindel zurückgewiesen, und Reuterdahl vom College St. Paul bezeichnet Einstein als den „Barnum der wissenschaftlichen Welt, der die ganze Welt mit seiner mythischen Theorie zum Narren halte". Derselbe Reuterdahl hat Einstein zu einer Erörterung aufgefordert, auch ihm ist es ergangen, wie voriges Jahr den Gegnern Einsteins in Nauheim, denn Einstein zog sich beizeiten zurück, so daß Reuterdahl die ganze Einsteinfahrt für eine von vornherein abgekartete Geschäftsreise erklärt.

Er führt des längeren aus, daß Leute, wie Mewes, Gehrcke und andere durchaus recht hätten, wenn sie Einstein des Plagiates beschuldigen. Er hat seine Gedanken zum Teil den Arbeiten Zieglers in Bern entnommen, wo ja Einstein früher wohnte, dessen Gedanken aber von der Wissenschaft unterdrückt seien, ferner von Gerber, dessen Arbeiten auch schwer zugänglich waren. Die Zeitungen sind beide über die Gelehrten bei uns gut unterrichtet, die gegen Einstein arbeiten, Lenard, Gehrcke, Fricke.

Der Reklamefeldzug, den die Presse vor einiger Zeit mit und für

---

[36]. M. Born, *The Born-Einstein Letters*, Walker and Company, New York, (1971), p. 41.

Einstein machte, wird den Amerikanern als eine Art Film vorgeführt, der aber für die deutsche Wissenschaft, für ihre Ehre und Förderung wenig nützlich gewesen sei. Es sei sehr zu bedauern, daß die Deutsche Wissenschaft durch einen ihrer Vertreter selbst lächerlich gemacht werde. Lodge, Reuterdahl, Heidenreich und andere haben drüben vorher gewarnt, man solle den Einsteinismus nicht so ohne weiteres annehmen. Natürlich zuerst vergeblich, denn dieser neue Ismus rollte wie eine Flutwelle ungehemmt dahin, aber die Ernüchterung kam bald.

Man geht gegen Einstein vor als den Goliat des Skeptizismus. Vorlesungen dagegen werden veranstaltet. In scharfsinniger Weise wird in einem viel gelesenen Buche „Relativität oder innere Abhängigkeit" die Unhaltbarkeit der Relativitätstheorie nachgewiesen. Der Einwand Einsteins, dies sei nur eine besondere Form des Antisemitismus, wird sehr energisch zurückgewiesen, und mit der Anerkennung Spinozols beantwortet.

Man ist sich darüber klar, daß es sich dabei vor allem darum handelt, mit allen Mitteln die Grundlagen der Theorie zu bekämpfen, da diese fehlerhaft, unvollständig und geeignet ist, das Universum in mechanistische Ideen aufzulösen. Es ist eine widerrechtliche Besitzergreifung durch die Mathematik. Der Astronom Glanville bezeichnet die Relativitätstheorie als eine neue Droge, die als ein neues Allheilmittel angepriesen wird. Dr. Skidmore hat die Sache richtig erfaßt, wenn er sagt, daß die Relativitätstheorie ausgehe von der Nichteuklidischen, sogenannten Metageometrie, sie bestehe aus rein gedanklichen Konstruktionen, die durchaus subjektiv sind und denen in der Natur nichts entspricht. Sehr hübsch ist folgendes Bild: Man nehme der Relativitätstheorie den mathematischen blauen Dunst, in den sie sich hüllt, dann bleibt nur ein lebloses Skelett und dessen Einsteinscher Schädel grinst andauernd seine Zehen an, die auf der Grundlage Galileis stehen. Man stelle sich das einmal vor!"[37]

On 22 April 1922, the *Luzerner Neueste Nachrichten* ridiculed Einstein's flight from the debate (Einstein would often repeat the cliché that great truths are simple, as if he were the first to make use of it),

"'Americans have too much common sense for that. They know that all the great truths are simple and easily understood, and are, therefore, justly suspicious of the unintelligible theory of relativity of Einstein. More than that they have rejected it as a swindle. Just for example Reuterdahl, dean of engineering of the College of St. Thomas, St. Paul, Minnesota, calls Einstein a 'Barnum of the scientific world who is trying to fool the whole world with a mythical theory.' It is further reported that Reuterdahl has challenged Einstein to a debate, into which he is as likely to enter as in the debate announced last year at the meeting for scientific investigation in Bad

---

[37]. J. Riem, "Amerika über Einstein", *Deutsche Zeitung*, Abend Ausgabe, (1 July 1921).

Nauheim, where he preferred to withdraw himself quietly before the announced opponents of his theory could say what they had to say. To these opponents was expressed the regret that Mr. Einstein was unable, because of circumstances, to answer them. This, of course, was another prearranged matter of his general trafficking. It is very likely that he is acting in a similar manner towards Reuterdahl. The more so because the latter has accused him of scientific theft, for Reuterdahl maintains that Einstein has taken the fundamentals of his theory from a work which appeared in 1866 under the pseudonym of 'Kinertia.'"[38]

"Dazu haben die Amerikaner noch zu viel gesunden Menschenverstand. Sie sind sich der großen Tatsachen bewußt, daß alle großen Mehrheiten auch einfach und leicht verständlich sind, und bringen daher der unverständlichen Relativitätslehre Einsteins ein durchaus gerechtfertigtes Mißtrauen entgegen. Ja, mehr als das: sie lehnen sie als Schwindel ab. So nennt Reuterdahl, der Dekan des St. Thomas College in Minneapolis, Einstein „einen Barnum in der wissenschaftlichen Welt", der mit seiner mystischen Theorie alle Welt zum Besten halte. Auch soll Reuterdahl Einstein zu einer Disputation aufgefordert haben, zu welcher sich dieser aber wohl ebenso wenig stellen dürfte, wie zu der an der letztjährigen deutschen Naturforscher-Versammlung in Bad Nauheim angekündigten, wo er es vorzog, sich in aller Stille zu drücken, bevor die zum Worte vorgemerkten Gegner seiner Theorie an die Reihe kamen. Man drückte ihnen dann das Bedauern aus, daß ihnen Herr Einstein nicht habe Rede und Antwort stehen können. Das war natürlich eine abgekartete Sache seines Klüngels. Aehnlich dürfte er sich nun auch gegenüber Reuterdahl verhalten, umso mehr, als ihn dieser des wissenschaftlichen Diebstahls bezichtigt. Reuterdahl behauptet nämlich, Einstein habe die Grundlage seiner Theorie einem Werke entlehnt, welches 1866 unter dem Pseudonym „Inertia" erschien."

J. E. G. Hirzel wrote in the *Luzerner Neueste Nachrichten* of 20 September 1921,

### "Albertus Maximus und die Blamage der Schulweisheit.

Warum Maximus? — In Amerika gefeiert und herausgefordert. — Seine Vorläufer als Duellanten: Reuterdahl in Amerika und Dr. J. H. Ziegler in der Schweiz. — Der Reklameturm von Potsdam.

Am 1. April dieses Jahres wurden in Neuyork die letzten Vorbereitungen zum Empfang des größten Genies getroffen, welches die

---

[38]. From A. Reuterdahl, *The Minneapolis Sunday Tribune*, (22 May 1921). Reuterdahl translates parts of "Professor Einsteins „Triumphzug" durch Amerika", *Luzerner Neueste Nachrichten*, (22 April 1921).

Welt bisher hervorzubringen imstande war. Wenigstens hieß es allgemein, daß alle großen Denker und Entdecker, denen unsere Wissenschaft und Kultur ihr Dasein verdanken, in Zukunft nur noch als bescheidene Vorläufer oder als Herolde jenes größern Genies gelten könnten, so daß fortan Namen wie die eines Heraklit, Giordano Bruno, Kopernikus, Kepler, Newton und wie sei sonst noch heißen mögen die großen Leuchten des Menschengeschlechts, neben dem seinigen ihren Glanz verlören. Dieses alles überstrahlende Gestirn am Himmel der heutigen Wissenschaft heißt Albert Einstein. Ein findiger Berliner Journalist fand jedoch diesen Namen zu bürgerlich und nannte ihn kurz Albertus Maximus. So heißt er jetzt im Hinblick auf jenen berühmten Zeitgenossen des Roger Baco, welcher den Gelehrten seiner Zeit allgemein als doctor mirabilis bekannt war und als der gelehrteste von allen galt, Albertus Magnus: dem großen Lehrer des Kirchenvaters Thomas Aquinas, dem doctor angelicus und eigentlichen Begründer der thomistisch-aristotelischen Philosophie, welche die Wissenschaft das ganze Mittelalter hindurch bis auf die Neuzeit beherrschte. Da diese beiden gewaltigen Männer bekanntlich später von der katholischen Kirche kanonisiert wurden, so erwarteten die Amerikaner den ihnen avisierten ganz Großen mit einer Art heiliger Scheu, auch schon deshalb, weil seine Lehre noch schwerer verständlich sein sollte, als die des heiligen Thomas, welche bereits den gelehrten Theologen schon genug harte Nüsse zu knacken gegeben hatte. Von der Lehre Einsteins hieß es allgemein, sie sei nur für die größten Mathematiker verständlich. Den meisten Amerikern genügte es darum, den Namen dieser Wunderlehre zu kennen, und man war praktisch genug, sich nicht auch noch um ihren Inhalt zu kümmern. Trotzdem war man allgemein von ihr entzückt, und zwar eben deshalb, weil sie so geheimnisvoll war. Nach ihr sollte es überhaupt nichts Absolutes mehr geben, alles sollte nur noch relativ sein. Aber Einstein sagte nicht, warum. Doch nannte er sie die allgemeine Relativitätstheorie. Sie bedeutet die vollste Freiheit im Denken und Handeln, denn sie befreit alle von jeder absoluten Verpflichtung. Der Glaube an das Absolute ist mit ihr erledigt. Er gehörte zu den Grundirrtümern einer veralteten Weisheit, welche einst durch den Teufel in die Welt gekommen sein mußten. Einstein wollte nun gründlich damit aufräumen. Darum die große Spannung. Man hoffte in ihm den kommenden Erlöser aus der Not des Unverstandes, des Zweifels und Irrtums begrüßen zu dürfen, und den Schlichter jeglichen Streites, den Friedensfürsten, welcher im Glorienschein schon vollbrachter und noch zu vollbringender Wundertaten der geplagten Menschheit den geistlichen und weltlichen Frieden bringen und das Reich Gottes auf Erden errichten werde. Einstein aber hatte ganz eigene Absichten. Der Verkünder der Relativitätstheorie wußte, daß alles nur relativ sei, also auch seine Messiasmission, und daß es deshalb am klügsten für ihn sei, dies den Amerikanern nicht zu sagen. Er wollte ihnen im Bluff einmal den Meister zeigen.

Am 1. April ließ er sie hangen und bangen, aber am 2. erschien er, vorläufig aber erst im Hafen von Neuyork. Da die Ankunft programmgemäß auf einen Samstag fiel, so halten Einstein und seine Begleiter dadurch Gelegenheit, ihren frommen Landslauten in New-Jerusalem gleich einen Beweis ihrer orthodoxen Frömmigkeit zu geben. Man wartete deshalb mit der Ausschiffung noch bis zum Sabbath-Ende. Dann erst ließ man sich von einem mit der amerikanischen und jüdischen Flagge versehenen, vom Bürgermeister extra zur Verfügung gestellten majors cutter ans Land setzen. Umgeben von einer zionistischen Delegation, unter Führung des Oberzionisten Weizmann und dessen Adjutanten Ussischkin und Mossinsohn betrat der neue Messias den Boden des gelobten Goldlandes Dollarika. Bei der Fahrt durch die Stadt (so berichtet die jüdische Pressezentrale vom 15. April) harrte ihrer eine unabsehbar Menge — ein Bericht spricht sogar von einer Million — von der sie enthusiastisch akklamiert wurde, so daß der Einzug Einsteins in New-Jerusalem den einfachen von Christus in Alt-Jerusalem vollständig in den Schatten stellte. Offenbar war er viel besser gemanaged. Alles schrie Hosiannah, denn alle Zuschauer waren Juden. Einstein selbst berichtet, er habe in Neuyork zum erstenmal jüdische Volkshaufen gesehen. Aber diese streuten keine Palmblätter, sondern, was den Zionisten viel lieber war, Banknoten und Schecks auf die Bank von England. Denn die jüdische Delegation hatte es nicht auf die Bekehrung der Yankees abgesehen, sondern nur auf die Erleichterung ihrer Börsen. Sie spekulierte nicht auf Seelenfang, sondern auf Gold, und dieses war nach alttestamentlicher Tradition am reichlichsten in Amerika zu finden. Schon Salomo hatte seine Knechte mit denen Hirams nach dem Lande Ophir geschickt, welches nach Mewes mit Peru identisch ist, und sie hatten ihm von dort 450 Zentner Gold zurückgebracht. Jetzt brauchte man es nicht mehr im rohen Zustande. Für die in Jerusalem zu gründende Welt-Universität dienten solide Papiere noch besser, und diese waren in Nordamerika leichter zu beschaffen. Und wirklich brachten die Zionisten hier mit Einstein als „great attraction" in ebenso viel Monaten, als Salomos Knechte Jahre gebraucht hatten, 23 Millionen Dollars zusammen, womit für derartige Expeditionen ein neuer Weltrekord aufgestellt war. Einstein brauchte dabei nicht einmal zu reden. Erstens geriet so sein Geheimnis weniger in Gefahr und zweitens verstärkte sein Schweigen den Nimbus seiner Theorie. Auch wäre ohnedies niemand genial genug gewesen, ihn zu verstehen. Denjenigen, die ihn durchaus hören wollten, spielte er etwas auf seiner Geige vor. Der Präsident und der Vizepräsident der Union bezeugten ihm für seine Leistungen ihre Anerkennung dadurch, daß sie sich mit ihm zusammen photographieren ließen.

Leider wurde Einstein vor seiner Abreise noch ein schlimmer Streich gespielt, ohne den er seinen lukrativen Aufenthalt wahrscheinlich noch erheblich verlängert hätte. Ich erwähnte bereits, daß seine Mission mehr darin bestand, den Amerikanern einen Propheten zu zeigen, als ihnen seine Theorie auseinanderzusetzen. Reden ist Silber, Schweigen ist Gold. Seine Abneigung gegen das Disputieren hatte Einstein schon an der Naturforscher-Versammlung in Bad Nauheim gezeigt. Ueberhaupt läßt sich kein Prophet, der an sich glaubt, aufs Disputieren ein und einer, der es nicht tut, noch viel weniger. Leider hatte nun aber ein amerikanischer Professor hiefür weder das richtige Verständnis, noch das nötige Zartgefühl. Dieser wollte nicht begreifen, daß eine wertvolle Lehre unverständlich sein müsse, sondern meinte, alle großen Wahrheiten müßten notwendig auch einfach und leicht verständlich sein. Aus diesem Grunde forderte er Herrn Einstein auf, diese Meinungsverschiedenheit mit ihm auf dem Wege einer öffentlichen Disputation auszutragen. Eine derartige Zumutung einem öffentlich beglaubigten Genie gegenüber erscheint etwas brutal und erinnert beinahe an den Boxermatsch Dempsen-Carpentier. Da aber dem Friedensfürsten jede Art von Streit ein Greuel ist, so strafte er die taktlose Herausforderung des Professors Arvid Reuterdahl mit stiller Verachtung. Vielleicht fürchtete er auch, er könnte in der Hitze des Zweikampfes seinem Gegner mit seiner übermenschlich-geistigen Kraft schweren Schaden zufügen. Sei dem, wie ihm wolle, jedenfalls verbot ihm seine Menschenliebe den Zweikampf. Aber die Amerikaner verkannten die hohe Moralität Einsteins und glaubten, er fürchte sich vor Reuterdahl und wäre deshalb vor ihm ausgekniffen. Und so fingen sie an, ihn plötzlich und von allen Seiten so grausam zu verhöhnen und lächerlich zu machen, daß sie dabei sogar den guten Ton verletzten und ihre gute Erziehung vergaßen. Das mußte Einstein noch tiefer schmerzen. Denn jetzt kamen sogar die „guten Eindrücke" in Gefahr, welche er von den Amerikanern empfangen hatte. Um diese zu retten, brach er nun schleunigst seine Tournee ab und schiffte sich so rasch als möglich nach England ein, wo er sich dann von Lord Haldane, einem gefühlvollen Stammesgenossen, über die gehabte Enttäuschung trösten ließ.

So endigte das anfängliche Hosiannah auch bei Einsteins Messiade mit einem Kreuziget ihn! Doch ist es heute nicht mehr Brauch, seine Ueberzeugung durch das Martyrium zu bekräftigen. Darum drückte sich der Prophet, bevor seine Sache eine tragische Wendung nahm. Erst, als er sich in Berlin ganz in Sicherheit wußte, stellte er wieder seinen Mann, machte den Amerikanern eine lange Nase und plimperte mit dem Geld in seiner Tasche. Es klang wie fröhliches Kichern. So endigte sein Triumphzug durch Amerika fast genau so, wie es die „Luzerner Neuesten Nachrichten am 22. April vorausgesagt hatten.

Und Reuterdahl? Nun, Reuterdahl konnte sich darüber trösten, daß ihn Einsteins Flucht um den Triumph gebracht hatte, ihm in öffentlicher Disputation die Richtigkeit seiner famosen Relativitätstheorie zu beweisen und ihm dabei die Denkermaske vom Gesicht zu reißen und dem Publikum nur dasjenige eines schlauen wissenschaftlichen Schiebers zu zeigen. Reuterdahl brauchte diesen Triumph nicht. Als Dekan der Ingenieur- und Architektenabteilung des St. Thomas College in St. Paul (Minnesota) genoß er schon Ansehen genug, auch stand sein Ruf als tiefer Denker und bedeutender Mathematiker längst zu fest, als daß er seiner bedurft hätte. Ernsten Forschern liegt nur die Wahrheit am Herzen und sie verachten die Reklame. Die Flucht Einsteins war das schmachvolle Eingeständnis seiner

Niederlage. Nach der hochgeachteten Monatsschrift „The Dearborn Independent" vom 30. Juli sollen bei Einsteins Abfahrt von Neuyork nur noch ein halbes Dutzend Freunde zugegen gewesen sein. Ein stilles Leichenbegängnis! Die Hunderttausende, welche den Ankömmling begrüßt hatten, blieben zu Hause. Viele von ihnen studierten bereits Reuterdahls Werk „Wissenschaftlicher Deismus gegen Materialismus". Die Tendenz dieses Buches ist eine rein absolutistische, radikal antirelativistische, wenn man den Relativismus im Einsteinschen Sinne versteht. Reuterdahl zeigt darin, daß die heutige agnostische Wissenschaft bloß auf vereinbarten Unbestimmtheiten beruhrt, „scientific unknowns", und daß diesem unsichern Zustande nur durch die sichere Bestimmung der notwendig absolut einfachen Grundlage abgeholfen werden könne. Dieses Absolute nennt er, so wie es die Religion tut, Gott. Aber als Mann der Wissenschaft begnügt er sich nicht mit dem unbestimmten Begriff von Gott. Vielmehr faßt er das Prinzip des allmächtig alles Bewirkenden und Durchwirkenden wieder ähnlich auf, wie es früher die beiden gelehrten Jesuiten Athanasius Kircher und Pater Joseph Boskowich getan hatten. Der letztere starb als Professor der Philosophie, Physik, Astronomie und Mathematik im Jahre 1787 in Mailand. Auch war er Verfasser einer Atomistik. Das ewige Grundprinzip von allen Weltlichen bestand nach ihm aus lauter Kraftzentren. Zu eben diesem Schlusse kam auch Reuterdahl. Er vereinigt aber damit ferner auch die beiden Grundbegriffe von Raum und Zeit. Alle zusammen bilden den absoluten Urgrund, auf dem oder woraus sich dann alles Relativ in verständlicher Weise entwickelt. Damit sichert er diesem von Anfang an ein festes System, während in einer bloßen Relativität ohne Voraussetzung eines bestimmten Absoluten selbstverständlich alles systemlos bleibt, so wie es bei Einstein Lehre der Fall ist. Diese ist darum nicht nur unverständlich, sondern sogar höchst gefährlich. Sie ist absolut ordnungswidrig, nihilistisch und negativ. Beidenkapp nannte sie bolschewistisch. Und sie wirkt deshalb nur zersetzend auf Religion und Wissenschaft ein, anstatt stützend und fördernd. Beiden entzieht sie den festen Boden. Bei Reuterdahl ist das Gegenteil davon der Fall. Darum stimmt er aufs Beste mit den Lehren und Bestrebungen J. H. Zieglers überein, dessen Werk er in seiner jüngsten Schrift: „Einstein and The New Science" mit unverhehlter Freude rühmt und als grundlegend für die neue und wahre Wissenschaft anerkennt. Zieglers System fußt bekanntlich ebenfalls auf den drei Begriffen von Urkraft, Urraum und Urzeit, deren Einheit nachzuweisen ihm gelungen ist. Einstein spricht dagegen die Zeit als vierte Dimension des Raumes an! Reuterdahl und Ziegler, der Mathematiker und der Chemiker, ergänzen sich gegenseitig. Einstein dagegen bringt nur mißtönende Anklänge an die Theorie des letztern vor. Immerhin muß man ihm eines lassen. Niemand hat mehr wie er und seine zionistischen und nichtzionistischen Freunde zum Sturze der agnostischen Wissenschaft beigetragen. Denn nichts konnte ihre innere Hohlheit der Menschheit besser zum Bewußtsein bringen, als das marktschreierische Treiben der Einsteinianer. Dieses Treiben lenkte erst die Aufmerksamkeit auf den Schaden und machte sie auf dem ganzen Erdenrund lächerlich und unhaltbar. Das war nun allerdings nicht beabsichtigt, aber es ebnete der neuen, wahren Wissenschaft den Weg. Einstein wurde dadurch nolens volens, zwar nicht zu ihrem Begründer, aber doch wenigstens zu ihrem Herold. Es geht eben oft anders, als man denkt. Das müssen jetzt auch die Koryphäen der alten Wissenschaft erfahren, denn damit, daß sie sich wie ein Mann hinter einen Nachtreter stellen, um mit ihm den ihnen unbequemen Hauptbegründer der neuen Wissenschaft gemeinsam an die Wand zu drücken, gerieten sie nur noch tiefer in den Sumpf einer bodenlosen Relativität, wobei sie ihre Autorität gänzlich einbüßen. Sie suchen sie jetzt vergeblich zu retten; alle Kniffe werden ihnen nichts mehr helfen. In diesen Tagen tauften sie gelegentlich eines Astronomen-Kongresses in Potsdam ein dort errichtetes Observatorium auf den Namen Einsteins und ließen dieses welterschütternde Ereignis sofort durch den Telegraphen urbi et orbi bekannt machen. Der Einsteinturm paradiert daher schon heute in jeder illustrierten Zeitung als aktuellste Sehenswürdigkeit. Er soll dazu dienen, die öffentliche Aufmerksamkeit von den ruhig und still vor sich gehenden Hauptereignissen abzulenken. Ob er aber den Ruhm des großen Mannes verewigen werde, ist daher noch fraglich. Dieser Reklameturm dürfte meines Erachtens in Zukunft eine weiser gewordene Menschheit an die ungeheure Geistesverwirrung unserer agnostischen Zeit erinnern. Der Einsteinturm wäre demnach nur mehr ein Denkmal für ihre letzte Torheit und größte Blamage.

J. E. G. Hirzel."

Artur Fürst and Alexander Moszkowski stated in 1916 that Einstein was the Galileo of the Twentieth Century. They suggested that since the designation *Albertus Magnus* was already taken (by Albert Graf von Bollstädt), the title "*Albertus Maximus*"[39] might be reserved for Einstein:

"So ist auch das jenseitige Ufer der neuen Theorie, der Relativität, nur unter Gefahr zu gewinnen. Aber der Wagemutige, der hinüberkommt, sieht sich in einer unermeßlichen neuen Welt, in der auf Schritt und Tritt ungeahnte Wahrheitswunder erblühen. Und mit Bewunderung gedenkt er der Männer, stie ihm diesen Weg wiesen. Zu ihnen gehören die Physiker und Mathematiker Lorentz und Minkowski, vor allen aber der gewaltige Baumeister des neuen Relativitätsgebäudes, der Galilei des zwanzigsten Jahrhunderts: Albert Einstein.

Vor sieben Jahrhunderten lebte ein Wundermann der Naturlehre, der Graf von Bollstädt, der sich den Namen eines Großen, Albertus Magnus, errang. Die Bezeichnung Albertus Maximus ist noch frei. Es könnte sein, daß dieser Titel für Albert Einstein vorbehalten bleibt und ihm dereinst

---

[39]. Rudolf Peters picked up on the ridiculous title "Albertus Maximus". *See: The Collected papers of Albert Einstein*, Volume 9, Document 388, Princeton University Press, (2004), p. 523, note 2.

verliehen wird."[40]

Fürst and Moszkowski were copying Eugen Karl Dühring's pronouncement that Robert Mayer was the "Galileo of the Nineteenth Century" in Dühring's book *Robert Mayer, der Galilei des neunzehnten Jahrhunderts. Eine Einführung in seine Leistungen und Schicksale*, E. Schmeitzner, Chemnitz, (1880).

The feature article Hirzel referred to was published in the *Luzerner Neueste Nachrichten* on 22 April 1921:

"Feuilleton.

Professor Einstein „Triumphzug"
durch Amerika.

In Nr. 164 vom 9. April brachte die „Vosissche Zeitung" folgende überseeische Depesche: „Prof. Albert Einstein und die gleichzeitig mit ihm eingetroffene zionistische Delegation wurden bei ihrer Ankunft in Neuyork sehr warm begrüßt. Die gesamte Neuyorker Presse widmet dem Ereignis als solchem und der Persönlichkeit Einsteins ausführliche Artikel." Man sieht auf den ersten Blick, daß es sich hiebei wieder um eine bestellte Reklame handelt, wie denn überhaupt das ganze Einsteinsche Unternehmen von Anfang an auf den Bluff berechnet war. Diesmal sollten nun die Amerikaner „dran glauben". Aber die Yankees scheinen weniger naiv zu sein, als die guten Deutschen und Schweizer und sich nicht so leicht zum Glauben an den neuen Propheten kommandieren zu lassen. Sie sind zu skeptisch, um ohne weiteres zu glauben, daß er ein größeres Genie sei, als Kopernikus und Newton, bloß weil er unverständlicher sei als diese. Dazu haben die Amerikaner noch zu viel gesunden Menschenverstand. Sie sind sich der großen Tatsachen bewußt, daß alle großen Mehrheiten auch einfach und leicht verständlich sind, und bringen daher der unverständlichen Relativitätslehre Einsteins ein durchaus gerechtfertigtes Mißtrauen entgegen. Ja, mehr als das: sie lehnen sie als Schwindel ab. So nennt Reuterdahl, der Dekan des St. Thomas College in Minneapolis, Einstein „einen Barnum in der wissenschaftlichen Welt", der mit seiner mystischen Theorie alle Welt zum Besten halte. Auch soll Reuterdahl Einstein zu einer Disputation aufgefordert haben, zu welcher sich dieser aber wohl ebenso wenig stellen dürfte, wie zu der an der letztjährigen deutschen Naturforscher-Versammlung in Bad Nauheim angekündigten, wo er es vorzog, sich in aller Stille zu drücken, bevor die zum Worte vorgemerkten Gegner seiner Theorie an die Reihe kamen. Man drückte ihnen dann das Bedauern aus, daß ihnen Herr Einstein nicht habe Rede und Antwort stehen können. Das war natürlich eine abgekartete Sache seines Klüngels. Aehnlich dürfte er sich nun auch gegenüber Reuterdahl verhalten, umso

---

[40]. A. Fürst and A. Moszkowski, *Das Buch der 1000 Wunder*, A. Langen, München, (1916), pp. 263-264.

mehr, als ihn dieser des wissenschaftlichen Diebstahls bezichtigt. Reuterdahl behauptet nämlich, Einstein habe die Grundlage seiner Theorie einem Werke entlehnt, welches 1866 unter dem Pseudonym „Inertia" erschien. Da indessen dieses Werk in Europa kaum bekannt geworden ist, so dürfte Beschuldigung grundlos sein. Aehnliche Beschuldigungen wurden übrigens auch schon von deutschen Gelehrten, wie dem Ingenieur Rudolf Mewes, Prof. E. Gehrke, Paul Weyland u. a. erhoben. Nach ihnen soll sich Einstein aus einer schwer zugänglichen Veröffentlichung vom Jahre 1898 des verstorbenen Oberlehrers Gerber stillschweigend eine Formel angeeignet haben. Wie es sich damit tatsächlich verhält, wird schwer festzustellen sein. Immerhin gibt schon das eigentümliche Gebaren Einsteins und die ungebührliche und auffällige Reklame seines Klüngels genügend Anlaß, seiner Sache nicht ganz zu trauen. Doch scheinen die meisten auf falscher Fährte zu sein, weil sie die Umstände, welche bei der Entstehung der Einsteinschen Lehre herrschten und darauf Einfluß haben konnten, nicht genügend kennen. Und doch können eigentlich nur diese den äußerst verdächtigen Widerspruch erklären, der uns in Einsteins Lehre von Anfang an entgegentritt und darin besteht, daß sie sich einerseits auf eine zwar durchaus richtige, aber von Einstein gar nicht näher begründete, sondern rein hypothetische Annahme abstellt, nämlich auf die Konstanz der Lichtgeschwindigkeit im Vakuum, währenddem anderseits seine weitern Begründungen dermaßen verworren und widerspruchsvoll sind, daß sie einem ganz andern Geiste entslossen zu sein scheinen. Diese sonderbaren Begründungen und die noch sonderbareren daraus gezogenen Schlüsse wurden von vielen Gelehrten, speziell von Prof. Lenard, einem der frühern Nobelpreisträger für Physik, gerügt. Lenard bemerkte ganz richtig, daß sie dem gesunden Menschenverstand direkt ins Gesicht schlügen. Was dagegen die Annahme von der Konstanz der Lichtgeschwindigkeit betrifft, welche Einstein als feststehendes Bezugsobjekt im uferlosen Ozean seiner Relativitätstheorie annimmt, so scheint es damit eine eigene Bewandtnis zu haben. Sie ist schon deshalb verdächtig, weil die Physiker zu jener Zeit die Existenz eines absolut leeren Raumes bestimmt leugneten und als unmöglich hinstellten, sie aber dann mit der Annahme von Einsteins Hypothese ohne weiteres zugaben und ihm diese zudem als eine hervorragende geniale Tat anrechneten. Tatsächlich scheint sie aber eine Beraubung der nur fünf Jahre früher von J. H. Ziegler aufgestellten universellen Lichtlehre zu sein. Das würde den Verzicht Einsteins auf ihre nähere Begründung zur Genüge erklären. Es gibt aber auch noch andere Gründe, welche mit größter Wahrscheinlichkeit darauf hindeuten, daß die Lehre Zieglers der verborgene Quell der Einsteinschen Entdeckung war, u. a. den, daß sie damals besonders in Bern, wo Einstein domiziliert war, stark diskutiert worden war. Zieglers Lehre gründet sich auf den unwiderleglichen Beweis, das die Gundlage der Welt in dem Urgegensatz von der Masse der unbedingt vollen Urlichtatome, dem Urlicht, und von der Masse des unbedingt leeren Raumes gebildet ist, deren gegenseitiges aktiv-passives Durchdringungsverhältnis Ziegler als Urzeit bezeichnet. Ziegler sprach deshalb von einer Dreieinigkeit von Kraft, Raum und Zeit, einer Dreieinigkeit, welche dann auch Herr Einstein, allerdings in verschleierter Form, brachte. Da die klare und einfache Lehre Zieglers, wonach alle Wirkungen der ewigen Wirklichkeit, d. h. alle Naturerscheinungen, lediglich Mischformen des strahlenden Urlichts und des bewegten Leeren sind, den Vertretern der offiziellen Physik sehr unbequem war, weil sie so ziemlich das Gegenteil von den lehrte, was diese bis anhin gelehrt hatten, so suchten sie dieselbe von Anfang an zu unterdrücken und totzuschweigen, und schufen so einen Zustand, der einem schlauen und geschickten Plagiator die günstigste Gelegenheit zur Aneignung ihrer Hauptlehren darbieten mußte. Ja, ein solcher konnte dabei sogar des Beifalls und der Unterstützung der Physiker sicher sein, besonders für den Fall, daß er sein Plagiat in einer nur ihrer Zunft verständlichen, dem großen Publikum aber unverständlichen Form vortrug. Dazu eignete sich die Mathematik am besten. Wer in ihrer Sprache schreibt, kann nur vom Mathematiker und Physiker verstanden werden, und diese haben dann volle Freiheit, der Laienwelt davon mitzuteilen, was sie für gut halten. Die gewöhnliche, gebildete Welt ist dann ganz von ihnen abhängig. Der Chemiker und Nichtmathematiker Ziegler aber hatte den „Fehler" gemacht, allgemein verständlich zu schreiben und dadurch auch die heutige Physik öffentlich bloßzustellen. Darum erschien Einstein den Physikern wie ein Deus ex machina. Er wurde zum Retter aus der Not. Kein Wunder, daß man ihn denn auch sofort auf den Schild erhob und ihm vor allem Volke als dem längst ersehnten Messias, d. h. dem wahren Lichtbringer, huldigte. Sein Ruhm wurde durch die Zeitungen in alle Weltteile ausposaunt. Das Volk mußte überall an ihm glauben und glaubte auch schließlich an ihn, weil es seine Lehre ja doch nicht selbst auf ihren Wahrheitsgehalt prüfen konnte. Es sah und hörte nur, wie der große Einstein in der Hierarchie der Physiker mit unglaublicher Schnelligkeit von Stufe zu Stufe stieg. Dies wirkte überzeugend, und die große internationale Presse, welche sich fast ganz in den Händen der Volksgenossen Einsteins befindet, bestärkt es fortwährend in dieser Ueberzeugung. Von dem Schweizer Ziegler hörte dagegen niemand etwas. Und so stände denn alles schön und herrlich für die Einsteinianer, hätte die Sache ihres Helden eben nicht auch ihre Achillesferse. Ziegler hatte seine Lehre nicht immer so ausführlich ausgedrückt, daß sie jeder bei oberflächlicher Kenntnisnahme sofort richtig verstehen könnte. Dadurch bot sie Anlaß zu allerlei Mißverständnissen. Und so wird es leicht verständlich, woher die vielen Irrtümer der Relativitätslehre herrühren. Wie sollte sie einheitlich und klar sein können, wenn sie nur einem Mixedpickles aus vielen, mehr oder weniger irrigen Plagiaten gleicht. Daß sie der Zieglerschen Lichtlehre von Jahr zu Jahr ähnlicher wurde, ist auch kein Gegenbeweis dafür, daß man die letztere nicht als den Urquell für die Einsteinsche Weisheit zu betrachten habe, so wenig als der schon seit zwanzig Jahren andauernde Boykott, in den die Einstein-Presse Ziegler getan hat. Davon wissen nun zwar die Herren Amerikaner nichts. Wenn sie Einstein ablehnen, so dürfte es vielmehr nur aus dem Grunde geschehen, daß sie sich darüber ärgern, für dumm genug gehalten zu werden, um die größten wissenschaftlichen Entdeckungen auch für die unverständlichsten zu halten. Die Amerikaner wissen ganz genau, daß das Gegenteil davon der Fall ist. Und schon darum dürfte sich die Geschäftsreise des falschen Propheten im Lande Dollarika wohl kaum zu einem Triumphzuge gestalten. —G—"

Another newspaper article notable for its mention of the Bad Nauheim debate wrote,

## "Wie steht's um Einstein?
### Jüdische Propaganda. — Astronomen in Potsdam. — „Silbersteine" des Einsteinturms. — Die Verschobene Rot-Linie. — Konzessionierter Aether. — Kneip-Knippe in Nauheim und Amerika. — Schlichte Presse.

Wie es vom alten Odysseus heißt, daß er der vielgewandte und erfindungsreiche war, der vieler Menschen Länder und Städte gesehen hatte, und dessen Name bis zum Himmel reichte, so haben wir gegenwärtig in Einstein einen Mann, von dem die ihm nahestehende Presse das gleiche behauptet, — daß er die größten Größen der Wissenschaft, Kopernikus, Kepler Newton bei weitem überträffe, — deren Werke haben bis in die Gegenwart gedauert, das Gedankenwerk Einsteins aber währe in alle Zeiten!

Merkwürdig, daß man das schon voriges Jahr so genau wußte! Jetzt wäre manch' einer froh, es nicht geschrieben zu haben. Vorschußlorbeeren sind immer ein Ding mit zwei verschiedenen Seiten. Denn nachdem die Einsteinpresse das Lob ihres Heros gar zu laut gesungen hatte, so daß die Gegner sich der Sache gründlicher annahmen, da wandte sich das Blatt. Eine lange Reihe von Denkern wurden genannt, bis Descartes zurück, die das, was an der Relativitätstheorie richtig ist, schon lange von Einstein gefunden hatten, daß aber die Theorie in der Form, die ihr Einstein gegeben hat, den allerheftigsten Widerspruch herausfordert.

In Einsteins Gegenwart, und ohne daß dieser oder ein anderer der Seinen etwas dagegen sagen konnte, ist auf der Astronomenversammlung in Potsdam im August dieses Jahres gezeigt worden, daß weder die Beobachtungen der Sterne bei totalen Sonnenfinsternissen, noch die Bewegungen des Planeten Merkur irgendwie eine Beweiskraft für die Relativitätstheorie haben. Die beobachteten Größen finden ihre befriedigende Erklärung auf andere einfache Weise.

Aber hoch ragt jetzt in Potsdam der Einsteinturm, dessen Baugerüst gerade am Tage des Besuches der Astronomenversammlung abgenommen wurde, damit die Fachmänner ihn besuchen konnten. Wie am Vormittag in einem Vortrage gesagt wurde, soll damit eine Messungsreihe gemacht werden, die die Theorie unmittelbar bestätigen würde. Der Turm dient also den Theorien von Einstein, beobachten wird daran Freundlich, erbaut hat den Turm der Architekt Mendelsohn, und das Geld soll, wie erzählt wurde, stammen von der Firma Silberstein. So ist es denn auch ein Bauwerk geworden, was den andern einheitlich gestalteten Bauwerken des astrophysikalischen Observatoriums gegenüber sich verhält, wie der Geist Einsteins zum Geiste von Vogel und Lohse, Müller, Kempf und den andern Astronomen, die die Anstalt berühmt gemacht haben. Es sieht aus wie der Vorderteil eines Kriegsschiffes, von der Seite gesehen. Einer nannte es Bismarckturm, da Freundlich gesagt hatte, seine Formgebung entspräche modernen Anschauungen, ein anderer den Tempel Salomonis, denn wir fanden, daß der unterirdische Raum sieben Vorhöfe hatte!

Aber es ist nur gut, daß die Einrichtung vielseitig gebraucht werden kann, denn es ist unzweifelhaft nachgewiesen, daß der gewünscht Betrag einer Verschiebung der Spektrallinien nach Rot nicht vorhanden. — Sehr peinlich! Denn Einstein sagt, daß mit dieser Verschiebung seine Theorie stehe und falle.

Die ganze Theorie gleicht überhaupt einem Proteus, sie nimmt dauernd neue Formen an: zuerst die spezielle, dann die allgemeine Relativitätstheorie; gegenwärtig hat sie wieder eine neue Gestalt. So ist sie unfaßbar, unverständlich, weil sie nach Gehrcke unverstehbar ist! Eine Massensuggestion!

Bekannt ist die Leugnung des Aethers. Jetzt hat ihn Einstein unter anderer Form wieder in der Theorie drin. Und Lenard sagt, daß bei einer vernünftigen Aethertheorie überhaupt gar kein Raum mehr für die Relativitätstheorie in der Physik bleibe; sie habe gewissermaßen von den Lücken in unserer Erkenntnis gelebt. Daher auch das Verhalten Einsteins den Gegnern gegenüber in der Oeffentlichkeit. Man erinnere sich an Nauheim voriges Jahr, wo er versprochen hatte, in öffentlicher Diskussion Rede und Antwort zu stehen. Als es so weit war, erschien er nicht, und die Geschäftsordnung machte die Gegner mundtot. In Amerika hat er es ebenso gemacht; der als Mathematiker, Physiker und Philosoph bekannte Prof. Reuterdahl von St. Thomas College hat Einstein bei seiner Amerikafahrt aufgefordert, eine Erörterung öffentlich stattfinden zu lassen. Der Erfolg war der gleiche wie in Nauheim, er paßte nicht in das Reiseprogramm. Dadurch ist die amerikanische Presse sehr ernüchtert worden. Als Einstein drüben ankam, waren gegen $150\text{'`}00$ Menschen am Schiff, darunter zahllose Photographen, bei der Abreise ein halbes Dutzend! Es trat eben gar zu kraß hervor, daß die ganze Fahrt eine Verherrlichung das jüdischen Geistes sein sollte. Die Ankunft gleichzeitig mit den Vertretern der Zionisten, der Kreis von jüdischen Lokalkomitees, der den Gefeierten umschloß, die Kritik amerikanischer Zustände durch Einstein nach seiner Rückreise haben bewirkt, daß die dortige Presse mit einer Deutlichkeit sich über den erst Gefeierten ausdrückt, die uns erstaunlich vorkommt. Hält man sich dies vor Augen, dazu die Einblicke in seine Gedankenwelt, wie sie Moszkowski gibt, politisch und wissenschaftlich, dazu die Tatsache, daß er mit der Sowjetregierung Beziehungen hat und gleichzeitig Mitglied der preußischen Akademie der Wissenschaft ist, so sagt man mit dem echten Berliner: Das ist wirklich allerhand! R."

Ernst Gehrcke wrote in 1924,

"Auf dem Deutschen Naturforschertag in Nauheim, wo Tausende aus allen Teilen Deutschlands und viele ausländische Besucher zusammenströmten, wurde von den Anhängern der Relativitätstheorie eine „Diskussion über die Relativitätstheorie" in die Wege geleitet. Am 20. September stellte der Vorsitzende der *Gesellschaft Deutscher Naturforscher und Aerzte* in seiner Einführungsrede diese mit neugieriger Spannung erwartete Relativitätsdiskussion in Aussicht, wobei er gleich seine Meinung dahin äußerte, daß die Physik «die größten Veränderungen ihrer wissenschaftlichen Grundlage» erlitten habe, indem «der Begriff des Äthers im Weltall verschwindet und durch die Relativitätstheorie Einsteins die Begriffe von Raum und Zeit wandelbar wurden.» (Bericht der *Frankfurter Zeitung* vom 20. September 1920). Diese Aussprache begann am 23. September. Sie wurde von EINSTEIN eröffnet, der zu drei vorher gehaltenen Vorträgen anderer Redner (WEYL, GREBE, v. LAUE) Stellung nahm: «EINSTEIN lehnte die WEYLsche Theorie» (eine von der Einsteinschen verschiedene, formale Relativitätstheorie) «ab, wogegen dieser von EINSTEIN den Beweis für seine Theorie aus den Naturgesetzen verlangte» (Bericht des *Berliner Lokal-Anzeigers* vom 24. September 1920). Besonderen Eindruck machte der öffentliche Meinungsaustausch zwischen EINSTEIN und dem berühmten Heidelberger Physiker LENARD. «LENARD ... wandte sich gegen die allgemeine Relativitätstheorie, nach welcher jede Art von Bewegung für uns unerkennbar sein soll, und wir nicht entscheiden können, ob wir uns zum Beispiel in drehender Bewegung befinden oder die gesamte Umwelt sich gegen uns drehe» (aus dem Bericht der *Frankfurter Zeitung* vom 24. September 1920). Eine Einigung zwischen LENARD und EINSTEIN wurde nicht erzielt, und nachdem noch andere Redner für (z. B. Prof. BORN) und wider (Prof. PALAGYI-Budapest) die Relativitätstheorie gesprochen hatten, wurde die weitere Erörterung vertagt, da, wie der Vorsitzende der Sitzung, der berühmte Physiker PLANCK aus Berlin, bemerkte, «die Relativitätstheorie es leider bisher noch nicht fertig gebracht habe, die für die Sitzung verfügbare absolute Zeit von neun bis ein Uhr zu verlängern» *(Kölnische Zeitung* vom 30. September 1920).—Die vertagte Diskussion wurde dann ohne EINSTEIN beendet, der eine Viertelstunde vor Beginn der Nachmittagssitzung abgereist war. Ein mit großen Erwartungen ins Werk gesetztes Ereignis war vorübergegangen, das Pendel der relativistischen Massenbewegung hatte geschwankt und eine Dämpfung erfahren, ohne aber schon zur Ruhe zu kommen."[41]

Philipp Lenard was surprised by Albert Einstein's poor performance. Lenard was hoping for a stimulating debate that might challenge his beliefs. Einstein was instead evasive and ill-prepared, then ran away. When Einstein hid

---

[41]. E. Gehrcke, *Die Massensuggestion der Relativitätstheorie*, Hermann Meusser, Berlin, (1924), pp. 16-17.

from Prof. Arvid Reuterdahl's challenge to debate the following year, many likened it to his flight from Bad Nauheim—this after all the hype assuring the public that Einstein would humiliate the opponents of relativity theory. Lenard wrote after the debate,

> "Auch sonst war ich schließlich erstaunt, wie wenig Herr Einstein auf die Beantwortung meiner Fragen vorbereitet zu sein schien — die doch schon zwei Jahre lang mit seiner Kenntnis gedruckt vorgelegen haben, — während von seiner Seite und auch von einem andern Fachmann Zeitungslesern gegenüber ganz ausdrücklich der Anschein der unbedingten Überlegenheit meinen Gedankengängen gegenüber erweckt worden war. Da ich weder Anhänger noch Gegner irgendeines Prinzips bin, sondern nur Naturforscher sein möchte — wie auf S. 12 schon zu erkennen gegeben, — hätte ich den Nachweis, daß und an welcher Stelle meine Überlegungen nicht genügend gründlich waren, als Gewinn entgegennehmen müssen, wenn er geführt worden wäre (vgl. auch Note k, S. 23), zumal in der rein auf die Sache gerichteten Form, in welcher die Nauheimer Aussprache ablief. Die einzige Aufklärung, welche ich von der Diskussion mitgenommen habe, stammt von seiten des Herrn Mie; sie wird im weiter Folgenden bezeichnet werden."[42]

Einstein lost all credibility at the debate and knew that the scientific community was against him. He undoubtedly wanted only to flee Germany and retreat from the public eye. As happened after Einstein's public humiliation at the Berlin Philharmonic, the Einstein sycophants and the ethnically biased pro-Einstein Jewish press came to his rescue after his public humiliation at Bad Nauheim and carried him through this time of criticism as he traveled the world promoting himself, relativity theory and Zionism, until his second rush of fame, which came with the announcement of the award of his Nobel Prize in late 1922. Many found the award scandalous, given that Einstein was a proven sophist and plagiarist.

Lorentz, Born, von Laue and the others were loyal to Einstein. The acceptance of their fatally flawed theories hinged on the cult of personality which was created for Einstein. If Lorentz exposed Einstein, Lorentz' beliefs and legacy would suffer. The relativists were, and are, so pernicious in their suppression of opposing views, because they were, and are, so insecure and politically motivated. They were, and are, so vicious in their defense of Einstein, because their mythologies are so easily defeated. The theory attacks gullible persons who are willing to accept irrational arguments and who act out of hero worship. Therefore, it is not surprising that these same individuals behave in an unscrupulous and adolescent manner when confronted with the facts.

Knowing they had lost at the debate, Einstein and his friends sought a

---

**42**. P. Lenard, *Über Relativitätsprinzip, Äther, Gravitation*, Third Edition, S. Hirzel, Leipzig, (1921), Note 1, p. 39.

rapprochement with Lenard which would dull the sting of Einstein's humiliation at Nauheim. Tragically, Lenard and Stark, (Nobel Prize laureates each) who were initially very helpful to Einstein in the early years of the special theory of relativity, after witnessing the corruption in the press and in the German Physical Society, after witnessing the Zionist betrayal of Germany, succumbed to the racial mythologies of the National Socialists and became outspoken advocates of Nazism, and in so doing were yet again the victims of Zionist Jews, though they did not realize it. Einstein's actions played no small rôle in elevating Adolf Hitler to power, in that the Nazis exploited Einstein as an example to stereotype millions of innocent people. The Nazis also exploited Einsteinian racist Zionist mythology to promote their own racial myths, which they imposed on the German People at the behest of Jewish Zionists who wanted assimilating Jews segregated from the allegedly inferior "Goyim".[43]

This was, and is, a common practice among Zionists and anti-Semites. They promote one another's common racism. This compounds the problem by creating an incentive for non-racists to forgive the intolerable behavior of characters like Einstein and to refuse to speak out against it for fear of having that behavior generalized in a sense unfavorable to them. An article in the *Patriot* of 18 July 1929, stated,

> "When Ambassador Page was editor of the *Atlantic Monthly* he gave the following advice to a young journalist: *'The most interesting fellow in America is the Jew: but don't write about Jews: without intending it, you may precipitate the calamity America should be most anxious to avoid—I mean Jew-baiting.'* Incidentally we may mention that an English book which happened to contain that quotation was suppressed, soon after birth, by a very obvious withdrawal of the usual advertising nourishment."[44]

The young journalist was Rollin Lynde Hartt.[45] This censorship further results in a group dynamic whereby one member of the group who speaks out against another is chastised for "betraying" the group which will allegedly be

---

**43**. D. Eckart and A. Hitler, *Der Bolschewismus von Moses bis Lenin: Zwiegespräch zwischen Adolf Hitler und mir*, Hoheneichen-Verlag, München, (1924); English translation by W. L. Pierce, "Bolshevism from Moses to Lenin", *National Socialist World*, (1966). URL: <http://www.jrbooksonline.com/DOCs/Eckart.doc> p. 7. J. Klatzkin, *Krisis und Entscheidung im Judentum; der Probleme des modernen Judentums*, Jüdischer Verlag, Berlin, (1921). Heinrich Class under the pseudonym Daniel Frymann, *Wenn ich der Kaiser wär': politische Wahrheiten und Notwendigkeiten*, Dieterich, Leipzig, (1912); English translation, R. S. Levy, "If I were the Kaiser / Daniel Freymann", *Antisemitism in the Modern World: An Anthology of Texts*, Chapter 14, D.C. Heath, Toronto, (1991).
**44**. D. Fahey, *The Mystical Body of Christ in the Modern World*, Browne and Nolan Limited, London, (1935), p. 254.
**45**. R. L. Hartt, "New York and the Real Jew", *Independent* (New York), (25 June 1921). *Cf.* "Jews Are Silent, the National Voice Is Heard", THE *DEARBORN INDEPENDENT*, (30 July 1921).

unfairly stereotyped by the exposure of the behavior of an individual like Albert Einstein. *Numbers* 16:22 states,

> "And they fell upon their faces, and said, O God, the God of the spirits of all flesh, shall one man sin, and wilt thou be wroth with all the congregation?"

Of course, it is human nature to think in symbols and to generalize, especially when viciously and unfairly attacked and threatened, as were the anti-Relativists Lenard and Stark. Beyond this, Jews have long taught that a good Jew never speaks out against another Jew, and a good Jews does not praise a Gentile. *Leviticus* 19:17-18 states,

> "17 Thou shalt not hate thy brother in thine heart: thou shalt in any wise rebuke thy neighbour, and not suffer sin upon him. 18 Thou shalt not avenge, nor bear any grudge against the children of thy people, but thou shalt love thy neighbour as thyself: I *am* the LORD."

### 4.4.8.2 Contemporary Accounts of the Bad Nauheim Debate

As many have recognized,[46] Max Born and others gave a very unrealistic portrayal of the events which took place in Germany in the 1920's and 1930's, vilifying Lenard, Gehrcke and Weyland with falsehoods; which accounts, while dramatic and shocking, simply do not agree with the facts. It is probably best to reproduce contemporary accounts from the period in order to obtain a realistic picture of what occurred at Nauheim.

The *Physikalische Zeitschrift*, Volume 21, (1920), pp. 666-668 gave a partial account of the debate between Lenard and Einstein:

> "Allgemeine Diskussion über Relativitätstheorie.
>
> Lenard: Ich habe mich gefreut, heute in einer Gravitationstheorie vom Äther sprechen gehört zu haben. Ich muß aber sagen, daß, sobald man von der Gravitationstheorie auf andere als massenproportionale Kräfte übergeht, sich der einfache Verstand eines Naturforschers an der Theorie stößt. Ich

---

**46**. *Confer:* A. Unsöld, "Albert Einstein — Ein Jahr danach", *Physikalische Blätter*, Volume 36, (1980), pp.337-339; **and** Volume 37, Number 7, (1981), p. 229. L. R. B. Elton, "Einstein, General Relativity, and the German Press, 1919-1920", *Isis*, Volume 77, Number 1, (March, 1986), pp. 95-103; **and** "Letters: Einstein and Germany", *Physics Today*, Volume 40, Number 7, (July, 1987), pp. 15, 106. W. Krause, "Letters: Einstein and Germany", *Physics Today*, Volume 40, Number 7, (July, 1987), pp. 106, 108. H. Goenner, "The Reaction to Relativity Theory I: The Anti-Einstein Campaign in Germany in 1920", *Science in Context*, Volume 6, (1993), pp. 107-133. M. Janssen *et al*, Editors, "Einstein's Encounters with German Anti-Relativists", *The Collected Papers of Albert Einstein*, Volume 7 (Hardbound), Princeton University Press, (2002), pp. 101-113.

verweise auf das Beispiel vom gebremsten Eisenbahnzug. Damit das Relativitätsprinzip gilt, werden bei Benutzung nicht massenproportionaler Kräfte die Gravitationsfelder hinzugedacht. Ich möchte sagen, daß man sich im physikalischen Denken zweier Bilder bedienen kann, die ich als Bilder erster und zweiter Art bezeichnet habe. In den Bildern erster Art sprach z. B. Herr Weyl, indem er alle Vorgänge durch Gleichungen ausdrückt. Die Bilder zweiter Art deuten die Gleichungen als Vorgänge im Raume. Ich möchte lieber die Bilder zweiter Art bevorzugen, während Herr Einstein bei der ersten Art stehen bleibt. Bei den Bildern zweiter Art ist der Äther unentbehrlich. Er war stets eines der wichtigsten Hilfsmittel beim Fortschritt in der Naturforschung, und seine Abschaffung bedeutet das Abschaffen des Denkens aller Naturforscher mittels des Bildes zweiter Art. Ich möchte zuerst die Frage stellen: Wie kommt es, daß es nach der Relativitätstheorie nicht unterscheidbar sein soll, ob im Falle des gebremsten Eisenbahnzuges der Zug gebremst oder die umgebende Welt gebremst wird?

Einstein: Es ist sicher, daß wir relativ zum Zug Wirkungen beobachten und wenn wir wollen, diese als Trägheitswirkungen deuten können. Die Relativitätstheorie kann sie ebensogut als Wirkungen eines Gravitationsfeldes deuten. Woher kommt nun das Feld? Sie meinen, daß es die Erfindung des Herrn Relativitätstheoretikers ist. Es ist aber keine freie Erfindung, weil es dieselben Differentialgesetze erfüllt wie diejenigen Felder, die wir als Wirkungen von Massen aufzufassen gewohnt sind. Es ist richtig, daß etwas von der Lösung willkürlich bleibt, wenn man einen begrenzten Teil der Welt ins Auge faßt. Das relativ zum gebremsten Zug herrschende Gravitationsfeld entspricht einer Induktionswirkung, die durch die entfernten Massen hervorgerufen wird. Ich möchte also kurz zusammenfassend sagen: Das Feld ist nicht willkürlich erfunden, weil es die allgemeinen Differentialgleichungen erfüllt und weil es zurückgeführt werden kann auf die Wirkung aller fernen Massen.

Lenard: Herrn Einsteins Ausführungen haben mir nichts Neues gesagt; sie sind auch nicht über die Kluft von den Bildern erster Art zu den anschaulichen Bildern zweiter Art hinweggekommen. Ich meine, die hinzugedachten Gravitationsfelder müssen Vorgängen entsprechen und diese Vorgänge haben sich in der Erfahrung nicht gemeldet.

Einstein: Ich möchte sagen, daß das, was der Mensch als anschaulich ansieht, und was nicht, gewechselt hat. Die Ansicht über Anschaulichkeit ist gewissermaßen eine Funktion der Zeit. Ich meine, die Physik ist begrifflich und nicht anschaulich. Als Beispiel über die wechselnde Ansicht über Anschaulichkeit erinnere ich Sie an die Auffassung über die Anschaulichkeit der galileischen Mechanik zu den verschiedenen Zeiten.

Lenard: Ich habe meine Meinung in der Druckschrift „Über Relativitätsprinzip, Äther, Gravitation" zum Ausdruck gebracht, daß der Äther in gewissen Beziehungen versagt hat, weil man ihn noch nicht in der rechten Weise behandelt hat. Das Relativitätsprinzip arbeitet mit einem nichteuklidischen Raum, der von Stelle zu Stelle und zeitlich nacheinander verschiedene Eigenschaften annimmt; dann kann nun eben in dem Raum ein Etwas sein, dessen Zustände diese verschiedenen Eigenschaften bedingen, und dieses Etwas ist eben der Äther. Ich sehe die Nützlichkeit des Relativitätsprinzips ein, solange es nur auf Gravitationskräfte angewandt wird. Für nicht massenproportionale Kräfte halte ich es für ungültig.

Einstein: Es liegt in der Natur der Sache, daß von einer Gültigkeit des Relativitätsprinzips nur dann gesprochen werden kann, wenn es bezüglich aller Naturgesetze gilt.

Lenard: Nur wenn man geeignete Felder hinzudichtet. Ich meine, das Relativitätsprinzip kann auch nur über Gravitation neue Aussagen machen, weil die im Falle der nichtmassenproportionalen Kräfte hinzugenommenen Gravitationsfelder gar keinen neuen Gesichtspunkt hinzufügen, als nur eben den, das Prinzip gültig erscheinen zu lassen. Auch macht die Gleichwertigkeit aller Bezugssysteme dem Prinzip Schwierigkeiten.

Einstein: Es gibt kein durch seine Einfachheit prinzipiell bevorzugtes Koordinatensystem; deshalb gibt es auch keine Methode, um zwischen „wirklichen" und „nichtwirklichen" Gravitationsfeldern zu unterscheiden. Meine zweite Frage lautet: Was sagt das Relativitätsprinzip zu dem unerlaubten Gedankenexperiment, welches darin besteht, daß man z. B. die Erde ruhen und die übrige Welt um die Erdachse sich drehen läßt, wobei Überlichtgeschwindigkeiten aufheben?

Der erste Satz ist keine Behauptung, sondern eine neuartige Definition für den Begriff „Äther".

Ein Gedankenexperiment ist ein prinzipiell, wenn auch nicht faktisch ausführbares Experiment. Es dient dazu, wirkliche Erfahrungen übersichtlich zusammenzufassen, um aus ihnen theoretische Folgerungen zu ziehen. Unerlaubt ist ein Gedankenexperiment nur dann, wenn eine Realisierung prinzipiell unmöglich ist.

Lenard: Ich glaube zusammenzufassen zu können: 1. Daß man doch besser unterläßt, die „Abschaffung des Äthers" zu verkünden. 2. Daß ich die Einschränkung des Relativitätsprinzips zu einem Gravitationsprinzip immer noch für angezeigt halte, und 3., daß die Überlichtgeschwindigkeiten dem Relativitätsprinzip doch eine Schwierigkeit zu bereiten scheinen; denn sie heben bei der Relation jedes beliebigen Körpers auf, sobald man dieselbe nicht diesem, sondern der Gesamtwelt zuschreiben will, was aber das Relativitätsprinzip in seiner einfachsten und bisherigen Form als gleichwertig zuläßt.

Rudolph: Daß sich die allgemeine Relativitätstheorie glänzend bewährt hat, ist kein Beweis gegen den Äther. Die Einsteinsche Theorie ist richtig, nur ihre Ansicht über den Äther ist nicht richtig. Auch wird sie erst annehmbar mit der Weylschen Ergänzung, geht dann aber sogar aus der Ätherhypothese hervor, wenn zwischen den beim Fließen verschobenen Ätherwänden Lücken bleiben, die durch Schleuderkraft infolge Richtungsänderung der Sternfäden leer gehalten werden.

Palagyi: Die Diskussion zwischen Einstein und Lenard hat auf mich einen tiefen Eindruck gemacht. Man begegnet hier wieder den alten historischen Gegensätzen zwischen experimentaler und mathematischer Physik, wie sie schon z. B. zwischen Faraday und Maxwell bestanden. Herr Einstein sagt, daß es kein ausgezeichnetes Koordinatensystem gibt. Es gibt eins. Lassen Sie mich biologisch denken. Dann trägt jeder Mensch sein Koordinatensystem in sich. In der Verfolgung dieses Gedankens ist eine Widerlegung der Relativitätstheorie enthalten.

Einstein weist darauf hin, daß kein Gegensatz zwischen Theorie und Experiment besteht.

Born: Die Relativitätstheorie bevorzugt sogar die Bilder zweiter Art. Ich betrachte als Beispiel die Erde und die Sonne. Wäre die Anziehung nicht, liefe die Erde geradlinig davon usw.

Mie: Daß die Ansicht, der Äther sei der greifbaren Materie wesensgleich, erst durch die Relativitätstheorie als unmöglich erkannt sein solle, habe ich nie verstehen können. Das war doch schon lange vorher durch Lorentz in seinem Buch „Elektrische und optische Erscheinungen in bewegten Körpern" geschehen. Auch Abraham hat in seinem Lehrbuch schon damals, als er der Relativitätstheorie noch ablehnend gegenüberstand, gesagt: „Der Äther ist der leere Raum."

Ich bin der Ansicht, daß man auch bei Annahme der Einsteinschen Gravitationstheorie doch ganz scharf unterscheiden muß zwischen den bloß fingierten Gravitationsfeldern, die man nur durch die Wahl des Koordinatensystems in das Weltbild hineinbringt, und den wirklichen Gravitationsfeldern, die durch den objektiven Tatbestand gegeben sind. Ich habe kürzlich einen Weg gezeigt, wie man zu einem „bevorzugten" Koordinatensystem kommen kann, in welchem von vornherein alle bloß fingierten Felder ausgeschlossen sind.

Einstein: Ich kann nicht einsehen, wieso es ein bevorzugtes Koordinatensystem geben soll. Höchstens könnte man daran denken, solche Koordinatensysteme zu bevorzugen, in bezug auf welche der Minkowskische Ausdruck für $ds^\wedge$ annähernd gilt. Aber abgesehen davon, daß es für große Räume solche Systeme gar nicht geben dürfte, sind diese Koordinatensysteme sicherlich nicht exakt, sondern nur approximater definierbar.

Kraus weist auf eine erkenntnistheoretische Differenz zwischen den Bildern erster und zweiter Art hin, indem er die Bilder erster Art für höherwertig als die Bilder zweiter Art hält.

Lenard: Es ist soeben das Schwerpunktsprinzip hineingebracht worden; ich glaube jedoch, daß das auf prinzipielle Fragen keinen Einfluß haben kann."

The *Berliner Tageblatt* published a report on 24 September 1920, which fills in some of the gaps in the incomplete account presented in the *Physikalische Zeitschrift*,

## "Die Einstein-Debatte

### auf dem Naturforschertag.
Vier physikalisch-mathematische Vorträge. — Ein Rededuell Einstein-Lenard.
(Telegramm unseres Sonderkorrespondenten.)
G. G. Bad Nauheim, 23. September.

Vorläufiger Bericht. Heute vormittag fand vor dichtgefülltem Saale unter dem Vorsitze des Geheimrats Planck und in Gegenwart sämtlicher großen Physiker und auch der Berliner Einstein-Gegner die Einstein-Sitzung der mathematischen und physikalischen Abteilung des Naturforschertages statt. Die Vorträge behandelten zumeist den Gegenstand in streng mathematischer Weise. Es sprachen hintereinander: Weyl (Zürich), Mie (Halle), Laue (Berlin), Grebe (Bonn). Dieser berichtete über Vergleichsmessungen der Sonnenspektren und irdischer Spektre, die sich auf die dritte experimentelle Bestätigung der Relativitätstheorie beziehen. Bei der Diskussion, in welcher u. a. Laue und Mie eingriffen, entspann sich ein lebhaftes Rededuell zwischen Einstein und Lenard. Dieser warf ein, daß die Einsteinsche Theorie der Anschaulichkeit für den gesundes Menschenverstand entbehre. Seine Einzelargumente, die Einstein die willkürliche Annahme irrealer Gravitationsfeldes vorwarfen und Widerspruch der Theorie in sich über die Lichtgeschwindigkeit behaupteten, widerlegte Einstein. Die spannende Diskussion zog sich durch mehrere Stunden hin. (Siehe auch Seite 4.)

[***]

### Ein neuer Beweis für die Einstein-Theorie.
### Das Rededuell Einstein-Lenard.
Die Rotverschiebung im Sonnenspektrum.
(Telegramm unseres Sonderberichterstatters.)
G. S. Bad Nauheim, 23. September.

Wie wir schon gemeldet haben, spielte sich heute unter ungeheurem Interesse die mit Spannung erwartete große Einstein-Debatte des Naturforscherkongresses ab. Der Saal des Badehauses war bis auf die letzte Ecke gefüllt.

Alle unsere großen Physiker, auch die Physikochemiker und eine Menge Interessierter aus anderen Wissensgebieten hatten sich eingefunden. Der scharfe Mathematikerkopf Plancks blickt vom Vorstandstich her. Ihm gegenüber sitzt in der vordersten Reihe der, um dessen Werk es geht, Einstein. Was die Physiker in Erwartung und zur abwehr des kolossalen Ansturms angekündigt hatten, bewahrheitete sich: ,,Die Sitzung wird die Theorie in rein wissenschaftlicher, streng mathematischer Form behandeln.'' Die Einzelheiten der Darlegungen und der vorgebrachten Beweisführung entziehen sich denn auch der summarischen Wiedergabe in eiliger Berichterstattung. Als erster spricht Weyl (Zürich) über seine Theorie von ,,Elektrizität und Gravitation'', dann Professor Mie (Halle) über ,,das elektrische Feld eines um ein Gravitationszentrum rotierenden geladenen Partikelchens'', endlich v. Laue (Berlin) über ,,neue Versuche zur Optik bewegter Körper''. Es hagelt jetzt Differentiale,

Koordinateninvarianz, elementare Wirkungsquanten, Transformationen, Vectorialsysteme usw. Gespannt lauschen die Fachleute, Einstein seelenruhig, Rubens mit seinem bezeichnenden Kopfnicken, Nernst erhobenen Hauptes, Frank interessiert lächelnd, Haber in bequemer Stellung die Decke betrachtend. Dem Laien aber graut es. Einzelne verlassen den Saal, die meisten aber harren in der Schwüle tapfer der Dinge, die da kommen sollen. Und sie werden nicht betrogen.

Professor Grebe aus Bonn ergreift jetzt das Wort. Und was er berichtet, ist des Aufhorchens wert: „Einsteins Theorie hat ihre vorläufige Bestätigung erfahren durch die gelungene Berechnung der Merkurbahn und der Lichtablenkung im Gravitationsfeld der Sonne. Es fehlte noch der Nachweis der von Einstein geforderten Rotverschiebung der Spektrallinien der Sonne. Dazu muß das Absorptionsspektrum der Sonne mit einem irdischen Emissionsspektrum verglichen werden. Mannigfache Einflüsse machen die Messungen schwierig. Wir fanden aber schließlich im Bandenspektrum des Stickstoffes, dem früher so genannten Cyanspektrum, ein gut verwertbares Spektrum. Unser Vergleichsspektrum wurde im Kohlenlichtbogen erzeugt. An jeder einzelnen Linie wurden zwanzig bis vierundzwanzig Messungen gemacht." Es folgt ein Projektionsbild, das in mehreren Linienpaaren die Abweichungen zwischen Sonnen- und irdischen Spektrallinien, zugleich aber auch die Schwierigkeiten der Beobachtung und die vielfachen gegenseitigen Störungen der Linien zeigt. Redner fährt fort: „Der von uns gefundene Unterschied in der Lage der Linien stimmt gut überein mit dem anderer, amerikanischer Beobachtungen. Jedoch war die Verschiebung bei den einzelnen Linien verschieden. Berücksichtigt man aber die gegenseitigen Beeinflussungen, so kommt man zu einem Wert von etwa 0,66, der mit dem Einsteinschen Wert für die Verschiebung von 0,62 bis 0,68 übereinstimmt. Zweifellos müssen auch noch weitere Experimente gemacht werden. Aber wir haben jetzt schon guten Grund zu der Annahme, daß die von der Einsteinschen Theorie verlangte Rotverschiebung wirklich vorhanden ist."

Nun eröffnet Planck die Diskussion. Einstein ist der erste Redner. Unwillkürlich tritt feierliche stille ein. Einstein bespricht die Weylsche Theorie. Weyl, Mie, Laue sprechen weiterhin. Es handelt sich zuerst um die vorhin gehaltenen Vorträge. Dann kommt die Generaldiskussion über die Relativitätstheorie überhaupt. Sie ist ein Zwiegespräch zwischen Geheimrat Lenard (Heidelberg) und Einstein, der sein eigener Anwalt ist. Jetzt kann auch der nicht auf den Höhen der Wissenschaft Thronende wieder leidlich folgen. Es kommt Leben in die Menge. Die zerstreuten Blicke konzentrieren sich jetzt auf die beiden Gegner. Es ist wie ein Turnier. Lenard läßt nicht locker, aber Einstein pariert vorzüglich. Hinter mir steht Weyland, der Berliner Einstein-Töter. Auf dem Boden dieser wissenschaftlichen Versammlung hält er sich im Hintergrunde der Ereignisse und gibt sein Interesse nur durch nervöses Schütteln der Mähne und leise Beifallsrufe bei Lenards worten zu erkennen. Dieser sagt: „Ich bewege mich nicht in Formeln, sondern in den tatsächlichen Vorgängen im Raume. Daß ist die

Kluft zwischen Einstein und mir. Gegen seine spezielle Relativitätstheorie habe ich gar nicht. Aber seine Gravitationslehre? Wenn ein fahrender Zug brennt, so tritt doch die Wirkung tatsächlich nur im Zuge auf, nicht draußen, wo alle Kirchtürme stehen bleiben!"

Einstein: „Die Erscheinungen im Zuge sind die Wirkungen eines Gravitationsfeldes, das induziert ist durch die Gesamtheit der näheren und ferneren Massen."

Lenard: „Ein solches Gravitationsfeld müßte doch auch anderweitig noch Vorgänge hervorrufen, wenn ich mir sein Vorhandensein anschaulich machen will!"

Einstein: „Was der Mensch als anschaulich betrachtet, ist großen Aenderungen unterworfen, ist eine Funktion der Zeit. Ein Zeitgenosse Galileis hätte dessen Mechanik auch für sehr unanschaulich erklärt. Diese „anschaulichen" Vorstellungen haben ihre Lücken, genau wie der viel zitierte „gesunde Menschenverstand". (Heiterkeit.)

Lenard: „Diese Diskussion wird unfruchtbar. Eine andere Frage: Wenn die Erde rotiert, so sagt Einstein, man könne genau so gut sagen, die Erde ruhe, und alle Materie rotiere um sie. Dann kommt man aber für die fernsten Gestirne zu Geschwindigkeiten, die weit über Lichtgeschwindigkeit liegen. Diese soll nach der Theorie aber eine Grenzgeschwindigkeit sein. Das ist ein Widerspruch in sich."

Einstein: Nein, die Lichtgeschwindigkeit ist Grenzgeschwindigkeit nur für die geradlinig gleichförmigen Bewegungen der speziellen Relativität; bei beliebig bewegten Systemen können beliebige Geschwindigkeiten des Lichts auftreten."

Es griffen dann noch verschiedene Herren in die Debatte ein, der Wert und Sinn von Gedankenexperimenten, die „Kluft" zwischen mathematischen und praktischen Physikern, philosophische und erkenntnistheoretische Fragen werden gestreift. Da aber, wie Professor Planck humorvoll bemerkt, die Versammlung nicht beschließen kann, daß die absolute Zeit von 9-1 länger als vier Stunden dauert, so muß man sich schließlich trennen."

*Vossische Zeitung* reported on 24 September 1920,

## "Der Kampf um Einstein.
### Die Auseinandersetzung auf dem Naturforschertag.

Dr. B. Bad Nauheim, 23. September.

Die Einzelheiten der Relativitätstheorie führen in schwierige Gebiete, die nur mit der Kenntnis der höheren Mathematik zu bewältigen sind. Man sollte daher glauben, einer Diskussion über ihre Grundlagen würden andere, als Fachphysiker und Mathematiker, kein besonderes Interesse entgegenbringen. Aber durch die bekannten Vorgänge in Berlin, wo man die Leistungen Einsteins in öffentlichen Versammlungen angreift und sich auch zu persönlichen Beschimpfungen des Gelehrten versteigt, ist die allgemeine Aufmerksamkeit noch mehr, als durch die Erfolge der Theorie bei der jüngsten Sonnenfinsternis, auf sie gelenkt worden.

Kein Wunder, daß auch auf der Naturforscherversammlung die Sitzung der Physikalischen und Mathematischen Abteilung, in der über Dinge, die mit der Relativitätstheorie zusammenhängen, gesprochen werden sollte, das größte Interesse erregte. Um zu verhindern, daß die Physiker und Mathematiker selbst von einem Publikum verdrängt würden, dessen Sensationsluft bei dieser wissenschaftlichen Behandlung sicher nicht befriedigt werden konnte, wurden zunächst nur Mitglieder der Physikalischen und Mathematischen Gesellschaft als Hörer zugelassen und dann erst der Eingang für weitere Besucher geöffnet. Schnell war der große Raum völlig gefüllt, der zusammen mit der Galerie wohl 500 bis 600 Personen faßte.

In nüchtern fachlicher Weise, seine Ausführungen reichlich mit mathematischen Formeln erläuternd, trug nun Weyl-Zürich seine Erweiterung der allgemeinen Relativitätstheorie vor, durch die er neben der Gravitation auch die elektrischen Erscheinungen umfassen will. Es folgte Mie-Greifswald, der das allgemeine Relativitätsprinzip lieber durch ein Prinzip der Relativität der Gravitation ersetzen will. Dann leitet Laue-Berlin rechnerisch aus den Grundlagen der Theorie die bekannte Folgerung ab, daß ein Lichtstrahl in einem Gravitationsfeld sich krümmen müsse, also z. B. beim Vorbeipassieren an der Sonne, und daß die Spektrallinien in einem solchen Gravitationsfeld sich noch dem roten Ende des Spektrums verschieben müßten. Schließlich berichtete Grebe-Bonn über seine gemeinsam mit Herrn Bachem angestellten Versuche, diese Rotverschiebung der Spektrallinien als wirklich zu erweisen.

Nachdem einige Einzelheiten dieser Vorträge noch besprochen waren, wurde die allgemeine Erörterung über die Relativitätstheorie eröffnet. In ihrer Art erinnerte sie an die Wettkämpfe mittelalterlicher Gelehrter, denn in ihrem Hauptteil gestaltete sie sich zu einer Zwiesprache zwischen dem bedeutenden Experimentalphysiker Lenard-Heidelberg und Einstein. Sie konnte, wie vorauszusehen war, zu keinem Ergebnis führen. Lenard stellte zum Schluß fest, daß weder er überzeugt sei, noch wohl auch seinen Gegner überzeugt habe. Es handle sich um den Gegensatz zwischen experimentellen und mathematischen Physikern, der nicht zu überbrücken sei, wenn der mathematische Physiker nicht von den Bildern erster Art, nach Lenards Ausdruck, in denen er zu denken gewohnt sei, zu den Bildern zweiter Art übergehe, den anschaulichen Bildern, in denen der Experimentalphysiker denke.

Von anderen Rednern wurde das Vorhandensein eines solchen Gegensatzes lebhaft bestritten; der mathematische Physiker fasse vielmehr die Erscheinungen, die der Experimentalphysiker erforsche, unter einheitlichen Gesichtspunkten zusammen. Mie hob lebhaft hervor, daß Einstein keineswegs nur als Mathematiker zu betrachten sei, sondern durchaus als Physiker, der seine bedeutende mathematische Geschicklichkeit mit großem physikalischen Blick verbinde.

Einstein selbst bemerkte, die Meinung, was anschaulich oder was nicht anschaulich sei, habe sich im Wechsel der Zeit sehr beträchtlich gewandelt, sie sei im wahrsten Sinne selbst eine Funktion der Zeit. Die Physik sei eben ihrem Wesen nach begreiflich und nicht anschaulich. Den Zeitgenossen Galileis war dessen Mechanik gewiß recht wenig anschaulich, heute aber, und zwar schon lange vor Begründung der Relativitätstheorie betrachtet man die elektrischen Felder als die elementarsten Gebilde, mit denen man arbeitet. Es gibt sogar Elektriker, die sich mechanische Vorgänge erst mit Hilfe der elektrischen Felder anschaulich machen können. Lenard führte das Beispiel des plötzlich gebremsten Eisenbahnzuges an, in dem der darin Sitzende eine gewaltige Erschütterung erleide; es würde jedem gesunden Menschenverstand widersprechen, wenn man annehmen wollte, nicht der Mensch sei in Bewegung gewesen, sondern die gesamte Umwelt.

Einstein warnte vor dem Operieren mit dem „gesunden Menschenverstand", der sehr leicht in die Irre gehe; es komme darauf an, ein für die Rechnung bequemes Koordinatensystem zu wählen, an sich gäbe es in der Welt kein bevorzugtes Koordinatensystem. Das erwiderte er auch auf den Vorhalt, daß bei der Annahme, die Erde ruhe und um sie bewege sich die gesamte Umwelt, man für gar nicht so weit entfernte Massen zu Ueberlichtgeschwindigkeiten kommen müsse. Einstein scheut sich nicht vor diesen Geschwindigkeiten, die keineswegs dem allgemeinen Relativitätsprinzip widersprächen, er sieht in ihnen keinen Grund, ein Koordinatensystem zu verwerfen, wenn nur sonst bei seiner Wahl die Rechnung einfach werde.

In diesem Punkte trat Mie den Einwänden Lenards bei; auch er will die fingierten Gravitationsfelder fortlassen. Sie haben, meint er, keinen Erkenntniswert; ihm kämen diese Dinge als „zu feinspintisiert" vor, er wolle demgegenüber doch lieber an dem gesunden Menschenverstand festhalten. Er glaube auch, daß es tatsächlich ein bevorzugtes Koordinatensystem gäbe. Aber auf die Frage Einsteins, wodurch denn eine solche Bevorzugung eines Koordinatensystmes verständlich gemacht werden sollte, mußte er die Antwort schuldig bleiben.

Am deutlichsten wird für den Leser der Gegensatz der Anschauungen vielleicht, wenn man sich erinnert, daß Lenard immer und immer wieder betont, an dem „Aether" müsse festgehalten werden, der Aether könne gar nicht abgeschafft werden, der „Aether" sei keine Hypothese, sondern Wirklichkeit, denn wenn es keinen „Aether" gäbe, könne man ja die Welt nicht mechanisch begreifen, dann könne man nicht alle physikalischen Erscheinungen auf Bewegungsvorgänge zurückführen. Demgegenüber muß doch betont werden, daß fast alle modernen Physiker die Forderung von der mechanischen Begreifbarkeit der Natur längst aufgegeben haben — es sei nur an den glänzenden Vortrag Plancks auf der Königsberger Naturforscherversammlung vor 10 Jahren erinnert. Es ist eben eine unbegründete Forderung, daß die Natur mechanisch begreifbar sein soll. Der Physiker hat an die Natur keine Forderungen, sondern nur Fragen zu stellen und zu sehen, was die Natur auf diese Fragen antwortet. In Verkennung dieses Verhältnisses hat man lange Jahre von der Natur ihre mechanische Begreifbarkeit gefordert. Die Natur ist aber nicht so liebenswürdig gewesen, diese Forderung zu erfüllen.

Im Verfolg der Erörterungen hob Mie mit Nachdruck hervor, daß die Abschaffung des Aethers ja gar nichts mit der Relativitätstheorie zu tun habe, er sei vielmehr schon in den 80er Jahren des vorigen Jahrhunderts durch die grundlegenden Arbeiten von Lorenth beseitigt worden.

Professor Born-Göttingen meinte, daß gerade die Relativitätstheorie das Bedürfnis nach Anschaulichkeit befriedige. Nach der Newtonschen Auffassung werde die Erde bei den Lauf um die Sonne von der Anziehung der Sonne und der Trägheit in ihrer Bahn gehalten, denke man sich die Sonne weg, so müßte die Erde in grader Linie weitergehen. Warum aber denn in grader Linie und wohin, müss man doch fragen. Hier sage nun die Einsteinsche Theorie, selbst wenn die Sonne weggedacht wird, so bleibt in der Umwelt noch eine große Massenverteilung übrig, und diese wirkt auf der Erde, so daß die Erde in eine gradlinige Bahn gezwungen wird. Im Grunde gebe die Newtonschen Anschauung dem leeren Raum bestimmte Eigenschaften, während die Einsteinsche Theorie nur Wechselwirkungen kennt. Daß die Einsteinsche Theorie darüber hinaus noch zu den Beziehungen der Anziehung zwischen Sonne und Erde komme, und sie erklären könne, obwohl sie gar nicht ihren Voraussetzungen stecke, sei eine glänzende Leistung.

So weit das Wesentliche der Erörterungen.

Ein dem Berichterstatter nahestehendes Lehrbuch aus dem Jahre 1892 beginnt mit den Worten „Die Physik hat die Aufgabe, die Erscheinungen der Natur als Bewegungsvorgänge zu beschreiben". Auf Grund der seitherigen Erfahrungen über Elektrizität hat der Verfasser diese Auffassung preisgegeben. Aus dem Festhalten an ihr kann man die Gegnerschaft gegen Einsteins Theorie verstehen. Aus ihrer Preisgabe leiten sich die Denkrichtungen Einsteins und seiner Anhänger ab.

*

Einsteins Ernennung zum Leydener Professor. Aus dem Haag meldet „Holl. Nieuwsbüro": Die Regierung genehmigte die Ernennung von Professor Dr. Einstein zum „außerordentlichen Professor" der Naturwissenschaften an der Universität in Leyden. (Die Meldung ist in der vorliegenden Form geeignet, Anlaß zu Mißverständnissen zu geben. Prof. Einstein hat sich, wie bereits vor längerer Zeit berichtet, auf Ersuchen der Leydener Universität bereit erklärt, dort in jedem Jahre während einiger Frühjahrswochen Vorlesungen über Relativitätstheorie und andere Kapitel der theoretischen Physik zu halten. Wohl um diese Verpflichtung äußerlich zu kennzeichnen, hat man die Form der Ernennung zum Honorarprofessor gewählt; von einer dauernden Uebersiedelung des berühmten Gelehrten an die holländische Hochschule kann kein Rede sein. D. Red.)"

The *Frankfurter Zeitung* reported,

## "86. Versammlung deutscher Naturforscher

## und Aerzte.

Bad Nauheim, 24. September.

Die Einsteinsche Relativitätstheorie wurde gestern vor dem zuständigen Forum, in den vereinigten mathematischen und physikalischen Abteilungen der deutschen Naturforscher- und Aerzteversammlung behandelt. Da es bekannt war, daß Professor Einstein selbst das Wort zu den Referaten den Professoren Dr. Weyl-Zürich, Laue-Berlin, Mie-Halle und Grebe-Bonn über seine Theorie in der Aussprache nehmen werde, hatte sich eine zahlreiche Zuhörerschaft eingefunden. Der geräumige Saal des Badehauses 8 und die Galerie waren gedrängt voll. Ganz auf dem Standpunkt Einsteins stand das Referat von Mie und auch Grebe-Bonn vertrat die Ansicht, daß sich für die von ihm angestellten Spezialstudien über die Cyanbande des Sonnenspektrums die Eisnteinsche Theorie mit den von ihm gefundenen Werten decken. Professor Weyl-Zürich und Lau-Berlin stimmten zwar nicht vollständig mit Einstein überein, lehnten ihn aber keineswegs prinzipiell ab. Das tat nur Professor Lenard-Heidelberg. Einstein selbst ging auf jeden erhobenen Einwand der Reihe nach ein und tat das in vornehmer, bescheidener, ja fast schüchterner und gerade dadurch überlegener Weise. Zum Schluß trat noch der erst jüngst von Frankfurt nach Göttingen berufene Physiker Professor Dr. Born in entschiedener Weise für Einstein ein, der auf alle Fälle die große Mehrheit der Versammlung auf seiner Seite hatte. Wir geben aus der Aussprache Folgendes wieder:

Weyl-Zürich sprach über eine von ihm vorgenommene Erweiterung der allgemeinen Relativitätstheorie, die auch die elektrischen Erscheinungen mitumfassen und aus allgemeinen Grundlagen erklären will. Dann trug Mie die Durchrechnung eines Spezialproblems vor, demzufolge er lieber von der Relativität der Gravitation als von der allgemeinen Relativität sprechen will. Hierauf leitete Laue-Berlin die Ablenkung eines Lichtstrahls durch ein Gravitationsfeld und die Rot-Verschiebung der Spektrallinien in einem solchen aus der Theorie her, und schließlich berichtete Grebe-Bonn über seine gemeinsam mit Bachem ausgeführten Messungen, die diese von der Theorie geforderte Rot-Verschiebung der Spektrallinien auf der Sonne wirklich zeigen. Die sich anschließende Diskussion mußte streng auf diese Vorträge selbst beschränkt bleiben. Erst nach ihrer Erledigung wurde in eine allgemeine Diskussion über die Relativitätstheorie eingetreten. Sie gestaltete sich sehr lebendig, in der Hauptsache zu einer Diskussion zwischen Einstein und Professor Lenard. Lenard bekannte sich zu einem Anhänger der speziellen Relativitätstheorie, nach welcher eine vollkommen gleichförmige Translationsbewegung durchaus unerkennbar sein muß, dagegen wandte er sich gegen die allgemeine Relativitätstheorie, nach welcher jede Art von Bewegung für uns unerkennbar sein soll und wir nicht entscheiden können, ob wir uns zum Beispiel in drehender Bewegung befinden oder die gesamte Umwelt sich gegen uns drehe, oder ob wir, wenn wir in einem plötzlich gebremsten Eisenbahnzug eine schwere Erschütterung erleiden, diese erleiden zufolge einer Veränderung der Bewegung des Eisenbahnzuges oder nicht vielmehr durch die entsprechend entgegengesetzte Bewegung der Erde. Das letztere widerspricht nach seiner Meinung jedem gesunden Menschenverstand, den der Physiker gerade so gut braucht und anwenden muß wie jeder andere. Auch die Abschaffung des Aethers durch die Relativitätstheorie lehnt Lenard ab, er hält seine Existenz vielmehr für durchaus erwiesen, weil wir ohne ihn die physikalischen Erscheinungen nicht restlos als mechanische Bewegungsvorgänge erklären können — eine Forderung, die notwendig sei, um die Erscheinungen anschaulich begreifen zu können. In Bezug auf diese letzte Bemerkung erwiderte Einstein, was der Mensch als anschaulich oder nicht anschaulich betrachtet, das hat im Laufe der Zeit beträchtlich gewechselt, die Physik ist eben ihrem Wesen nach begrifflich und nicht anschaulich. Den Zeitgenossen Galileis war dessen Mechanik gewiß recht unanschaulich, heute aber, und zwar schon lange vor der Relativitätstheorie betrachtet man die elektrischen Felder als die elementarsten Gebilde, mit denen man arbeitet; dem Elektriker ist das elektrische Feld das anschaulichste, was nicht überholen werden kann, und es gibt Elektriker, die sich mechanische Vorgänge erst mit Hilfe der elektrischen Felder anschaulich machen können. Was den gebremsten Eisenbahnzug betrifft, so handelt es sich eben um die Wechselwirkung zwischen diesem und allen übrigen in der Welt vorhandenen Massen, wobei es ganz gleichgültig ist, welche von beiden gegen die andere bewegt wird. Mit dem gesunden Menschenverstand zu operieren, sei sehr gefährlich. Für die mathematische Behandlung gibt es eben kein an sich bevorzugtes Koordinatensystem und man wird daher jedesmal das für die Durchführung der Rechnung bequemste wählen. Das gleiche gilt von den Rotationsbewegungen. Wenn man bei der Annahme, die Umwelt bewege sich rotierend, und die Erde stehe still, zu Ueberlicht-Geschwindigkeiten komme, so sei das auch kein Widerspruch gegen die allgemeine Relativitätstheorie, die garnicht wie die spezielle eine konstante Lichtgeschwindigkeit fordere. In Bezug auf die Abschaffung des Aethers betonte Professor Mie, daß sie nichts mit der Relativitätstheorie zu tun habe. Schon in den 80er Jahren ist der Aether durch die grundlegenden Arbeiten von Lorentz abgeschafft worden. Im übrigen bekannte sich Mie zwar als begeisterten Anhänger der Relativitätstheorie, trat aber in einem Punkte Herrn Lenard bei, nämlich, daß er glaube es gäbe wirklich ein bevorzugtes Koordinatensystem und man könne fingierte Gravitationsfelder fortlassen. Es scheine ihm nicht als ob ihre Einführung erkenntnistheoretischen Wert habe, es komme ihm vor, als ob man da zu sein spintisiere demgegenüber lobt er sich doch immer unseren gesunden Menschenverstand. Inwiefern es aber ein bevorzugtes Koordinatensystem in der Welt geben soll, konnte er Herrn Einstein nicht sagen. Lenard meinte, die Diskussion habe zu einer Einigung der abweichenden Anschauungen und zu einer gegenseitigen Ueberzeugung ihrer Vertreter nicht führen können, weil der Gegensatz der experimentellen und mathematischen Physiker hier zum Ausdruck komme, eine Meinung, der von anderer Seite lebhaft widersprochen wurde, denn der mathematische Physiker stehe nicht im Gegensatz zum Experimentalphysiker, sondern stelle die von diesem erforschten

Erscheinungen unter einheitlichen Gesichtspunkten dar."

The *Frankfurter Zeitung*, on 21 September 1921, and the *Berliner Tageblatt*, Evening Edition, 20 September 1920, had reported on the Eighty-Sixth Meeting of German Natural Scientists. In the opening address to the meeting of natural scientists, Friedrich von Müller performed a staged and scripted homage to Einstein, and slandered anyone and everyone who disagreed with Einstein. Max Planck and Arnold Sommerfeld provided Müller with the speech. Planck and Sommerfeld also made certain that their personal attacks against Einstein's critics would be accompanied by scripted applause from Einstein's friends.[47] The *Frankfurter Zeitung* stated on 21 September 1920, first morning edition:

## "Versammlung deutscher Naturforscher und Aerzte.
(Privattelegramm der „Frankfurter Zeitung".)

L—z Bad Nauheim, 20. Septbr.
Mit einem phantastischen Schmuck bunter Herbstfarben hat sich das mit Naturreizen so überaus reich versehene Bad Nauheim bekleidet, um die Teilnehmer der 86. Versammlung Deutscher Naturforscher und Aerzte zu begrüßen. Der große Saal des Konzerthauses und seine Galerien sind dicht besetzt mit Männern und Frauen, als bald nach 9 Uhr der Geschäftsführer der 86. Versammlung, Prof. Dr. Grödel (Bad Nauheim) die Erschienenen begrüßt. Dabei gedenkt er nicht nur der Auslandsdeutschen, sondern auch der wenigen Ausländer, die zur Versammlung gekommen sind, und betont, daß die Wissenschaft bei uns keine nationalen Grenzen kenne. Zugleich weist er auf den Unterschied dieser Versammlung gegenüber den früheren hin, der in der veränderten allgemeinen Lage begründet ist. Diese Tagung soll eine Tagung des Ernstes sein. — Als zweiter Redner begrüßte der Präsident des hessischen Bildungsamtes Dr. Strecker die Versammlung. Er bezeichnet die Versammlung als ein Symbol des Aufbaus. Insbesondere sei eine der wichtigsten Aufgaben der deutschen Aerzteschaft, den physischen Wiederaufbau der Bevölkerung zu leiten und zu ermöglichen. Dem Naturforscher und Wissenschaftler im allgemeineren Sinne liegt der geistige Wiederaufbau ob. Die Bedeutung der Natur als Lehrerin bei unserm Nachwuchs zur Geltung zu bringen, sei seine wichtigste Aufgabe. Aus den allgemeinen Betrachtungen heraus fällt das Wort, daß wir nicht nur die Kräfte der Natur beherrschen lernen müssen, sondern auch die im Menschen lebenden Naturkräfte. — Hatte diese politische Anspielung schon den Beifall der Versammlung hervorgerufen, so nimmt die Teilnahme der Zuhörer außerordentlich zu, als nach einigen kurzen Begrüßungsworten des Ministerialrats Balsen als Vertreter des hessischen Finanzministeriums, des Hausherrn der Versammlung als Besitzerin des staatlichen Bades Nauheim, und des Bürgermeisters der Stadt Nauheim Dr. Kaiser der Rektor der

---

**47**. *Cf.* D. K. Buchwald, et al. Editors, *The Collected Papers of Albert Einstein*, Volume 7, Princeton University Press, (2002), p.108.

hessischen Landesuniversität Gießen im Namen der vier benachbarten Hochschulen Marburg, Gießen, Frankfurt und Darmstadt das Wort ergreift. Er nennt als führenden Namen der Hochschulen auf dem Gebiete der Naturwissenschaften Ehrlich für Frankfurt, Behring für Marburg, Liebig für Gießen und Merck für Darmstadt und löst den ersten Beifall aus, als er wünscht, daß nun auch ein leider scheinbar abhanden gekommenes Gefühl sich wieder einstellen möge, das Gefühl des Stolzes, ein Deutscher zu sein. Deutsche Forschung und Wissenschaft kann uns nicht genommen werden; sie müssen zwar darben, aber können nicht untergehen. Helmholtz, Virchow und Haber kann man nicht wegleugnen und annektieren.

Der Vorsitzende der Gesellschaft Deutscher Naturforscher und Aerzte, Prof. Dr. Friedrich v. Müller (München), der nunmehr die eigentlichen Arbeiten der Versammlung einleitet, gedenkt zunächst der zahlreichen Toten, die die Gesellschaft, besonders der Vorstand, in den sechs Fahren, in denen die Versammlungen unterbrochen waren, zu beklagen hat. Er bezeichnet dann den Beschluß, schon in diesem Jahre eine Naturforscherversammlung abzuhalten, als eine mutige Tat, deren Ausführung besonders durch Ernährungs- und Unterkunftsschwierigkeiten in Gefahr geriet. Deshalb mußte Hannover als Versammlungsort aufgegeben werden, und dem hessischen Staat wie der Stadt Nauheim sei besonderer Dank dafür abgestattet, daß sie die Abhaltung der Versammlung durch ihr außerordentliches Entgegenkommen ermöglicht haben. Der Redner streift dann die Aufgaben der Versammlung und deren besondere Bedeutung in den heutigen Tagen. Die Seuchenbekämpfung ist während des Krieges dank unserer medizinischen Wissenschaft und den Männern des Kriegssanitätsdienstes in großem Maße möglich gewesen, so daß wir vor schweren Seuchen bewahrt geblieben sind. Aber drei furchtbare Seuchen gilt es zu bekämpfen: Grippe, Schlafkrankheit und Syphilis. Diesen Krankheiten werden die Arbeiten der Versammlung besonderes Augenmerk widmen. Unter den Naturwissenschaften haben Chemie und Physik in dieser Zeit die größten Veränderungen ihrer wissenschaftlichen Grundlage erlitten: die Chemie dadurch, daß der Grundsatz der Unteilbarkeit der Atome zu Fall gekommen ist, die Physik dadurch, daß der Begriff des Aethers im Weltall verschwindet und durch die Relativitätstheorie Einsteins die Begriffe von Raum und Zeit wandelbar wurden. Damit ist dem Redner Gelegenheit gegeben, in ausdrucksvollen Worten gegen die Berliner Vorgänge zu protestieren. Die außerordentlichen geistigen Taten eines Einstein gehören nicht vor das Forum einer mit Schlagworten und aus politischen Motiven arbeitenden öffentlichen Versammlung, sondern eines Berufskreises von Gelehrten. — Diese offene und deutliche Ehrung Einsteins erweckt lauten Beifall. Müller kommt dann auf die weiteren großen Probleme, deren Behandlung der Versammlung obliegt, zu sprechen: Stickstoff und Eiweiß und die Fragen des Unterrichts. Er betont den Wert der humanistischen Bildung und warnt vor einer Geichmachung des geistigen Besitzes in Anlehnung an die Bestrebungen zur Ausgleichung materiellen Besitzes. Die Beziehungen zum Ausland bezeichnet der Redner als noch gering. Die Zeit für internationale Kongresse ist noch nicht für uns gekommen. Diese sind auch nicht so nötig wie die fremde Literatur. Die Zeitschriften- und Büchernot ist eine große Gefahr für die Wissenschaft. Die Aufrichtung einer absperrenden Mauer gegen unsere geistigen Erzeugnisse erscheint dem Redner weniger gefährlich. Sie spreche eher für eine eistige Armut dessen, der sie aufrichtet. Denn geistig positive Völker vertragen keinen Abschluß, sie brauchen die andern Völker für die Publikation ihrer geistigen Tätigkeit. Von den allgemeinen Betrachtungen gleitet der Redner dann aber ab, als er auf die frühere Gewohnheit, des Landesherren bei solchen Anlässen zu gedenken, hinweist. Diese Gewohnheit habe nun in Fortfall kommen müssen. Aber er halte es für seine Pflicht, der deutschen Fürsten als Förderer der Wissenschaften zu gedenken. Setzt bei diesen Worten schon ein starker Beifall ein, so steigert er sich noch, als der Redner sagt, die Monarchie pflege, die Republik schütze die Wissenschaft, die Revolution zerstöre. Er erinnert dabei an die Hinrichtung Lavoisiers während der französischen Revolution und die sie begleitenden Worte des Richters: *nous n'avons plus besoin de savants*. Aber er hofft, ebenso wie im Frankreich der Revolution ein gewaltiger geistiger Aufschwung folgte, daß auch wir neben dem materiellen einen geistigen Aufschwung erreichen. — Der langdauernde Beifall der Versammlung sprach dafür, daß der Redner mit seiner kleinen Abschweifung auf politisches Gebiet doch sehr den Zuhörern aus dem Herzen gesprochen hat, und das mag bei einer Versammlung von wissenschaftlich gebildeten Zuhörern doch von Bedeutung sein.

Im Anschluß an diese einführenden Worte sprachen Dr. Bosch, der Direktor der Badischen Anilin- und Sodafabriken, Prof. Ehrenberg (Göttingen) und Geheimrat Rubner (Berlin) zu dem Thema des Stickstoffes, worüber weiterer Bericht folgt."

Paul Weyland redressed the dishonest press reports disseminated by Einstein's friends in a statement Weyland published in "Die Naturforschertagung in Nauheim. Erdrosselung der Einsteingegner!", *Deutsche Zeitung*, Number 449, (26 September 1920), Morgen-Ausgabe, 1. Beiblatt, p. 1;[48] reprinted as "Die Naturforschertagung in Nauheim", *Politisch-Anthropologische Monatsschrift für praktische Politik, für politische Bildung und Erziehung auf biologischer Grundlage*, Volume 19, (1920), pp. 365-370:

"Die Naturforschertagung in Nauheim.

Weyland.

Begünstigt von blendend schönem Wetter, gefördert durch den

---

[48]. S. Grundmann, "Das moralische Antlitz der Anti-Einstein-Liga", Wissenschaftliche Zeitschrift der Technischen Universität Dresden, Volume 16, pp. 1623-1626.

Opfersinn von Bevölkerung und Badeverwaltung, tagte in dieser Woche in dem unvergleichlich schönen Bad Nauheim die 86. Versammlung Deutscher Naturforscher und Ärzte. Seit der 85., die in Wien stattfand, wo im Jahre 1913 der greise Kaiser Franz Joseph es sich nicht nehmen ließ, den wissenschaftlichen Gästen seine Hofburg zur Verfügung zu stellen, liegt der Weltkrieg, der hemmend in die Wissenschaft eingriff und nur die Gebiete der Kriegs-Chirurgie und Kriegsmedizin befruchtend beeinflußte. Lediglich die Physik hatte neben der Medizin eine Frage von weitgehender wissenschaftlicher Bedeutung zu erörtern, und dieses war die Relativitäts-Theorie, die seit 1911 und 1915 von Einstein eingeführt wurde. So ist es denn kein Wunder, daß sich mangels jeder anderen wissenschaftlichen Ausbeute dieser fünf Jahre das Hauptinteresse auf die Donnerstag- und Freitags-Sitzung konzentrierte, in welcher Einstein seiner wachsenden Opposition Rede und Antwort zu stehen hatte.

Um es gleich vorweg zu nehmen: er hat nicht sehr glänzend abgeschnitten, wenngleich die unter Einsteinschem Einfluß stehenden Presse-Referate der Deutschen physikalischen Gesellschaft völlig entstellte Berichte in die Welt jagten, die natürlich ein einseitiges Bild der Situation geben. Wir wollen versuchen, so kurz wie möglich die wichtigsten Vorträge herauszugreifen und müssen dabei leider bemerken, daß tatsächlich in diesen fünf Jahren außer der mathematischen Abstraktion der Relativitätstheorie nichts Neues hervorgebracht wurde, es sei denn, daß man als Fortschritt feststellt, daß die physikalische Forschung im Sinne ihrer jetzigen geistigen Leitung völlig zum Sklaven mathematischer Abstraktionen herabgesunken ist und jedes vernunftgemäße Forschen ausschaltet. Einstein hat denn auch eine Art Glaubensbekenntnis abgelegt, indem er die denkwürdigen Worte aussprach: „Gesunden Menschenverstand in die Physik einzuführen, ist gefährlich." Der einzige positive Gewinn dieser Naturforschertagung ist denn auch der, daß die Scheidung der Geister sich vollzogen hat und unter der Leitung Lenards die Vergewaltigung der Physik durch mathematische Dogmen abgelehnt wird, während auf der anderen Seite die Einsteinophilen auf ihrem Standpunkt beharren und hurtig den Parnaß ihres Formelkrames zu erklimmen versuchen ... bis sie von ihren „eisigen Höhen" einmal jäh herabfallen werden.

Schon in der Eröffnungssitzung wies Herr von Müller darauf hin, das diese Versammlung im Zeichen der Relativitätstheorie steht, indem er in einem ihm von dem Einsteinleuten unterschobenen Konzept bemerkte, daß von Einstein eine der größten Geistestaten geschehen ist: er hat ja den Äther abgeschafft. Im übrigen wies Herr von Müller in seiner glänzenden Rede auf die Errungenschaften der Kriegsmedizin und Chirurgie hin, gedachte der Toten der deutschen Naturforscher und leitete in taktvoll feinen Worten die Versammlung ein. Als Vertreter der Regierung Hessens sprach der ehemalige Patriot und jetzige Linksmann Professor Strecker einige Begrüßungsworte, indem er um sich einige Phrasen verbreitete, daß die Naturforscher der Wahrheit dienen sollen und nun auch dafür zu sorgen hätten, daß die Wahrheit auch in uns Deutschen selbst einzudringen hat, daß nicht wieder durch deutsches Verschulden ein solcher Krieg entsteht. Diese versuchte Politisierung wurde merkwürdigerweise schweigend hingenommen und von einem Teil der Versammlung beklatscht. Als aber der Rektor der Gießener Universität Kalbfleisch sich in einer kernigen deutschen Rede an das Auditorium wandte und den famosen Vorredner glatt abfallen ließ, brauste ein nicht endenwollender Beifall durch das Haus. Ein erhebendes Bekenntnis zum Deutschtum lag in dieser Akklamation, und als ferner Herr von Müller in einem weiteren Referat mit Wehmut feststellte, daß man zum ersten Male, so lange die deutschen Naturforscher tagen, nicht mehr des Kaisers gedenken darf und es der Versammlung anheimstellte, in Dankbarkeit der deutschen Fürsten zu gedenken, unter deren Fürsorge die deutsche Wissenschaft blühte und gedieh, zog es wie schmerzlich durch die so zahlreich erschienenen aufrechten deutschen Männer, und mancher gedachte der schönen Zeiten, wo deutsche Wissenschaft an der Spitze aller Wissenschaft stand und die deutschen Institutsleiter nicht von Herrn Haenisch mit Androhung von Disziplinarstrafen belästigt wurden, wenn sie nicht mit ihrem Friedensetat auskamen. Wohl selten hat der Theatersaal einen derartigen Sturm des Beifalls erlebt, wie er durch die Worte von Müllers, der deutschen Fürsten zu gedenken, ausgelöst wurde.

Die allgemeinen Vorträge behandelten die Atom- und Molekulartheorie, welche hauptsächlich von Debye, Frank und Kossel referiert wurden. Das Ernährungsproblem wurde von Bosch, Ehrenberg, von Grube und Paul behandelt.

Neue fundamentale Tatsachen wurden in diesen Vorträgen nicht verkündet. Lediglich des jungen Debyes blendender Vortragskunst gelang es, auch den Wissenden zu fesseln und sein Sammelreferat über Atomstruktur als Plus zu verbuchen. Er gipfelte *summa summarum* in der Andeutung, daß sich die Welt wahrscheinlich aus Vielheiten des Wasserstoffatoms zusammensetzt, wie dies die letzten Rutherfordschen Untersuchungen gezeigt haben, so daß also mit Wahrscheinlichkeit anzunehmen ist, daß die mehr als hundertjährige Proutsche Hypothese wieder zu Ehren gelangt und wahrscheinlich auch Goethes Standpunkt in der Farbenlehre von seinem oppositionellen Standpunkt gegen Newton wieder zur Anerkennung gelangt. Die Vorträge von Frank und Kossel bewegten sich in ähnlichem Rahmen und bestätigten auf anderem Wege die Ausführungen Debyes. In der Medizin war es besonders Sudhoff, dessen greiser Charakterkopf überall in der Versammlung auffiel, der durch eine mit seltener Liebe und Sorgfalt zusammengebrachte Vesal-Ausstellung zu Ehren des 400 jährigen Geburtstages des Begründers der deutschen Anatomie fesselte. Lehmann erfreute sein dankbares Auditorium mit kinematographischen Aufnahmen über die neuesten Ergebnisse in der Forschung der flüssigen Kristalle, und Rinne löste Beifallsstürme seiner Zuhörerschaft aus, die er in seiner liebenswürdigen humoristischen Art mit blendendem Material an sein Thema über Kristallgitter fesselte.

Sehr zu erwähnen ist ferner der von außerordentlicher Fachkenntnis getragene Vortrag von Steuer über die Geologie der Nauheimer Quellen.

Es waren dies ungefähr die Höhepunkte der allgemeinen Vorträge, wenn man von den naturwissenschaftlichen Filmen absehen will, welche die „Ufa" durch Adam vortragen ließ, auf die wir vom pädagogischen Standpunkt aus noch einmal zurückkommen werden. Mittwoch nachmittag begannen die Spezialsitzungen der einzelnen Fakultäten, welche der Öffentlichkeit nichts Bemerkenswertes boten und über die zu referieren zu weit führen würde. Es sei nur bemerkt, daß allein die Physiker z. B. 56 solcher Vorträge zu erledigen hatten, die jedoch samt und sonders nicht über den Rahmen üblicher Laboratoriumstätigkeit hinausgingen und auch ohne Naturforschertag in Zeitschriften ihre Erledigung hätten finden können. So nahte der Donnerstag nachmittag mit seiner Hauptsitzung heran, wo sich zahlreiche Opponenten gegen Einstein gemeldet hatten. Diese Sitzung ist nun wohl eine von den denkwürdigsten, die in der Geschichte der deutschen Naturforschung stattgefunden hat. Obwohl es jedem Tagesteilnehmer freistand, mit seinem Ausweis jeden Vortrag zu besuchen, hatte der Vorstand der Deutschen physikalischen Gesellschaft die Stirn, an der Eingangstür eine scharfe Siebung vorzunehmen, um nur diejenigen hineinzulassen, welche ihm genehm waren. Es erhob sich ein gewaltiger Tumult, das empörte Auditorium schob die wissenschaftliche Polizei beiseite, stürmte den Saal und behauptete sich. Auf diesem Wege gelangten auch andere als Einstein-Freunde hinein. Und nun geschah das Unglaubliche. Statt daß es zu einer wissenschaftlichen Auseinandersetzung kam, wurde von der Vorstandsleitung unter dem Vorsitz von Max Planck dafür gesorgt, daß die Opposition einfach mundtot gemacht wurde. In stundenlangen Reden verbreiteten sich Weyl, Mie, von Laue und Grebe über das Relativitätsprinzip, während den gegnerischen Rednern einschließlich Diskussion 15 Minuten zugebilligt wurden. Um 1 Uhr sollte die Sitzung beendet sein, um $^3/_4$ 1 Uhr war man noch mit der Diskussion der Einstein-Vorträge beschäftigt, und der Apparat der Erdrosselung klappte so vorzüglich, daß tatsächlich die Diskussion ausschließlich von Einstein-Leuten geführt wurde, hauptsächlich von Einstein selbst. Gehrcke-Berlin, der sich mehrfach energisch zum Wort meldete, wurde bis zuletzt gelassen, um ihm dann mitzuteilen, daß die Diskussion geschlossen sei. Rudolph-Koblenz versuchte, wenigstens im Wege einer Geschäftsordnungsbemerkung zu Worte zu kommen: ihm wurde von Planck bedeutet, daß er nicht das Wort habe. Lenard-Heidelberg wurde schon nach drei Sätzen von Planck in die Parade gefahren, so daß Lenard auf das Wort verzichtete. Palagyi-Ofenpest, von dem hauptsächlich neben Mach Einstein seine Weisheit bezog, wurde $^1/_2$ Minute Redezeit bewilligt (in Worten eine halbe Minute), die dann auf 3 Minuten ausgedehnt wurde (!!!) und ähnlich Anmutigkeiten mehr. Der ehrwürdigen und geachteten Persönlichkeit Lenards, über den sich selbst ein Planck nicht hinwegzusetzen vermochte, gelang es schließlich, sich mit aller Energie Gehör zu verschaffen und Einstein zur Rede zu stellen. Er führte kurz aus, daß es nach seiner Auffassung wohl zwei Möglichkeiten physikalischer

Forschung gäbe, nämlich die logisch verständliche und die mathematisch abstrakte. Er richtete an Einstein die klar präzisierte Frage und die dringende Bitte, ihm vernünftig zu erklären, wie es denn komme, daß beim plötzlichen Anrücken des berühmten Eisenbahnzuges nicht der Kirchturm des benachbarten Dorfes umfalle, sondern der Mann im Zuge, welche Voraussetzungen durch die Einsteinsche Theorie gegeben seien. Einstein drückte sich in seinen bekannten gewundenen Erklärungen und billigen Witzeleien um die Beantwortung der Frage herum, was Lenard zu weiterer zweimaliger Anfrage an Einstein veranlaßte, ihm Rede und Antwort zu stehen. Als es ihm nicht gelang, von Einstein eine sachliche Antwort zu erlangen, verzichtete Lenard auf das Wort mit der Feststellung, daß es ihm nicht gelungen sei, eine Übereinstimmung zwischen Einstein und ihm in dem Sinne zu erzielen, daß Einstein eine an ihn klar gerichtete Frage ebenso klar beantworten konnte. Mie trat Lenard zur Seite und erklärte, daß die vernünftige Anschauungsweise nicht ausgeschaltet werden dürfe. Hierauf gefiel sich Einstein in der denkwürdigen Bemerkung, daß es gefährlich sei, mit dem menschlichen Verstand zu operieren, womit er vor aller Welt kundgab, daß er mit der Vernunft nichts mehr zu tun hat. Die im vorhergehenden mitgeteilten Tatsachen finden sich nun nicht in dem offiziellen Pressebericht der Naturforschertagung, der selbstverständlich von den Einsteinleuten herausgegeben wurde. Es verdient hiermit festgenagelt zu werden, in welcher geradezu korrupten Art und Weise die Berichterstattung dieser Leute vonstatten geht und die freie wissenschaftliche Meinung systematisch geknebelt wird. Daß ein Max Planck sich zu derartigen Machenschaften hergab, ist bedauerlich, aber wohl dadurch verständlich, daß er sich, wie die anderen Spitzen der deutschen physikalischen Gesellschaft, mit Einstein wissenschaftlich und noch anders zu eng liiert hat, um anders handeln zu können.

Die zu Wort gemeldeten Gegner Einsteins wurden auf den Freitag versetzt, wo ihnen 12 Minuten Redezeit einschließlich Diskussion bewilligt wurde. Selbstverständlich war es am Freitag nachmittag nicht möglich, fünf Vorträge in einer Stunde à 12 Minuten wissenschaftlich zu erledigen, sie gaben nur Bruchstücke oder wurden schon in der Einleitung vom Vorsitzenden abgesetzt. Wir werden die Berichte jedoch nach dem Manuskript an dieser Stelle später behandeln.

Zu bemerken ist ferner, daß weder Einstein noch seine Freunde diesen Vorträgen beiwohnten.

Zusammenfassend kann man sagen, daß die Art und Weise der freien Forschung, wie sie von der Deutschen physikalischen Gesellschaft verstanden wird, ein in der Geschichte der deutschen Wissenschaft beispielloser Skandal ist und daß es wohl die höchste Zeit wird, daß in dieses Rattennest wissenschaftlicher Korruption einmal frische Luft kommt. Wenn man bedenkt, daß Einstein sogar Weyl ablehnt, weil dessen Mathematik wieder zur einfachen euklidischen Geometrie hinüberführt, so versteht man wohl, daß es sich nicht darum handelt, in der Deutschen physikalischen Gesellschaft der Wissenschaft noch zu dienen, sondern daß es nur gilt, ihrem Papste Einstein die Tiara zu erhalten. Mit einem Gefühl tiefster Beschämung mußte man diese Versammlung verlassen, und auf der Kurpromenade und allen Gängen, wo das Thema besprochen wurde, gab es nur ein Wort der Entrüstung über das unerhörte Gebaren des Vorstandes, besonders seines Vorsitzenden Max Planck. Forscher von Ruf versichern mir, in dieser Gesellschaft kein Wort mehr zu sprechen.

Im übrigen verlief die Tagung in vollster Harmonie, kleine technische Mängel, die ja schließlich überall vorkommen, waren vorhanden. Die Ausstellung war glänzend beschickt, besonders von den optischen Firmen. Hier ragten insbesondere die Stände von Goerz, Leitz und Winkel hervor. Besonders Leitz fesselte durch ein neues dermatologisches Mikroskop, welches durch einfaches Aufsetzen auf den menschlichen Organismus, z. B. durch einfaches Auftragen einer Immersionsflüssigkeit das Leben des Gewebes erkennen ließ und die Blutkörperchen in Vene und Arterie deutlich machte. Höchst beachtenswert war ferner der neue Helldunkelfeldkondensator, welcher der biologisch-bakteriologischen Forschung neue Wege zu weisen berufen ist."

Franz Kleinschrod, who had a theory and an agenda of his own to promote, wrote,

"Die Einsteinsche Relativitätslehre ist bereits zur *cause celèbre* der Wissenschaft geworden. Noch vor wenigen Monaten nur der nächsten Umgebung bekannt, ist heute der Name Einstein im Munde, man darf sagen, wohl der gesamten Wissenschaft. Es dürfte wohl wenig wissenschaftliche Persönlichkeit geben, die in so kurzer Zeit den höchsten Gipfel wissenschaftlicher Popularität ersteigen. Man kann es verstehen, wenn man die Behauptungen und die schrankenlose Begeisterung seiner Anhänger liest: „Damit ist aber die alte Newtonsche Mechanik durch das Relativitätsprinzip über den Haufen geworfen. Das RP greift somit in alle durch Alter geheiligten Denkgewohnheiten ein, es zerstört alle Begriffe, mit denen wir aufgewachsen sind, und es verlangt von uns außerdem eine Fähigkeit zur Abstraktion, gegen die selbst die Anforderungen der vierdimensionalen Mathematik ein Kinderspiel sind. Aber als Gegengabe beschert uns das RP eine Fülle neuer Einsichten; es beschert uns Tag, wo vordem Dämmerung oder Nacht war. Kurz, es ist eine geistige Befreiung, wie die Tat des Kopernikus." (Das Einsteinsche Relativitätsprinzip. A. Pflüger. 2. Aufl. 1920. Cohen-Bonn.) Im ähnlichen Tone ergehen sich alle Anhänger.—

Aber bald erhob sich auch dagegen, wie vorauszusehen war, die Kritik und setzte mächtig ein. Mit großer Spannung erwartete man auf der Naturforscherversammlung in Nauheim die Aussprache der Gegner mit Einstein. Sie verlief, wie auch hier vorauszusehen war, resultatlos. Es stand wohl der größere Teil der Gelehrten auf Seite von Einstein, aber Einstein konnte seine Gegner, besonders seinen Hauptgegner, Lenard (Heidelberg), nicht widerlegen, — aber die Gegner konnten auch Einstein nicht widerlegen. So blieb der Streit unentschieden und wird es auch bleiben, denn beide Parteien schossen mit ihren Angriffen immer dicht an dem Ziel vorbei. Keiner traf den andern richtig. [***] „Ja, selbst die Begriffe von Raum und Zeit, die wir seit Jahrtausenden als feststehend anzusehen gewohnt sind, sind wandelbar geworden durch die Relativitätstheorie." Mit diesen Worten eröffnete Friedr. von Müller die 86. Naturforscherversammlung deutscher Naturforscher und Aerzte zu Nauheim 1920."[49]

Philipp Lenard commented on the Bad Nauheim debate in the third edition of his booklet *Über Relativitätsprinzip, Äther, Gravitation*, S. Hirzel, Leipzig, (1921), pp. 36-44:

"Zusatz,
betreffend die Nauheimer Diskussion über das
Relativitätsprinzip.

Während der Vorbereitung der vorliegenden Neuauflage hat am 23. Sept. d. J. die Diskussion über das Relativitätsprinzip bei der Nauheimer Naturforscherversammlung stattgefunden. Es hat dabei Herr Einstein auf die in dieser Schrift hervorgehobenen Schwierigkeiten einzugehen und die dabei sich ergebenden Fragen zu beantworten versucht, nachdem die Herren Weyl und Mie in ihren Vorträgen über Elektrizität und Gravitation besondere Anregungen gegeben hatten.

Der Eindruck, welchen die Aussprache hinterließ, an welcher außer den genannten Herren auch andere Vertreter der Mathematik und der Physik sich beteiligten, ging nach meinem Urteil im allgemeinen dahin, daß in der Tat an den in dieser Schrift gekennzeichneten Stellen Schwierigkeiten und Fragen vorliegen, deren Erledigung nicht ohne weiteres in befriedigender Weise gelingt und deren Hervorhebung also wohl berechtigt war. Es darf wohl scheinen, daß das Weitereingehen auf dieselben bei Überwindung der vorhandenen Hindernisse eine Weiterführung der Theorie mit Beseitigung ihrer gegenwärtigen Härten ergeben sollte, wie denn auch besonders die von Herrn Mie gelieferten Beiträge nach einer Weiterführung strebten, und zwar nicht ohne teilweises Abgehen von Herrn Einsteins ursprünglichem Wege [*Footnote*: Vgl. in verwandtem Sinne auch E. Wiechert, Astron. Nachr. Bd. 211, Nr. 5054, S. 275, 1920, woselbst auch auf eine bevorstehende weitergehende Veröffentlichung desselben Verfassers über Gravitation in den Annalen der Physik hingewiesen wird. (Erschienen während der Drucklegung des Vorliegenden in Bd. 63, S. 301.)]. Die Hindernisse gegen

---

**49**. F. Kleinschrod, "Das Lebensproblem und das Positivitätsprinzip in Zeit und Raum und das Einsteinsche Relativitätsprinzip in Raum und Zeit", *Frankfurter Zeitgemäße Broschuren*, Volume 40, Number 1-3, Breer & Thiemann, Hamm, Westphalen, (October-December, 1920), pp. 1-2, 63-64.

volles Eingehen auf die von mir hervorgehobenen Schwierigkeiten und Fragen liegen, wie auch bei der Diskussion wieder erkennbar wurde, in der Kluft, welche für gewöhnlich zwischen den Benutzern der beiden auf Seite 25 des Vorliegenden erläuterten Bilderarten besteht.

[*Page 25:* Daß Andere den Äther in ihrem Gesamtbilde und auch bei ihrer Arbeit entbehren können, beweist nichts gegen den Äther, sondern ist vollkommen selbstverständlich, wenn man die Zweifachheit der Bilder bedenkt, die der Menschengeist von der (unbelebten) Natur bisher sich zu machen verstand. Es sei gestattet, diese Zweifachheit hier mit schon einmal gebrauchten Worten zu erläutern [*Footnote:* „Über Äther und Materie", Heidelberg (C. Winter) 1911, S. 5.]: „Nun sind aber diese Bilder des Naturforschers doch von zweierlei Art. Quantitativ sind sie immer; sie können aber — und das ist die erste Art — sich sogar ganz darin erschöpfen, quantitative Beziehungen zwischen beobachtbaren Größen zu sein. In diesem Falle sind sie vollkommen darstellbar in Gestalt mathematischer Formeln, meist Differentialgleichungen. Dies ist der Weg, den Kirchoff und Helmholtz bevorzugt haben, von Kirchoff die mathematische Beschreibung der Natur genannt. Die denknotwendigen Folgen der Bilder, in deren Entwicklung die Benutzung und zugleich die Prüfung der Bilder besteht, sind dann die mathematischen Folgen jener Gleichungen, und auch weiter nichts. Man kann aber weitergehen — und dies ergibt die zweite Art der Bilder —, indem man sich von einer Überzeugung leiten läßt, ohne welche die Naturforschung sicherlich nie Erfolg gehabt hätte. Von der Überzeugung nämlich, daß alle Vorgänge in der Natur — in der unbelebten Natur wenigstens — bloße Bewegungsvorgänge sind, d. i. nur in Ortsveränderungen ein für allemal gegebenen Stoffes bestehen. Dann würde es sich in jedem Falle um Mechanismen handeln, und die Gleichungen, welche wir uns als Bilder erster Art gemacht haben, müssen Gleichungen der Mechanik sein, sie müssen ganz bestimmten Mechanismen entsprechen, und dann können wir auch geradezu diese Mechanismen als die Bilder betrachten, die wir uns von den Naturvorgängen gemacht haben. Wir haben dann mechanische Modelle, dynamische Modelle der Dinge als Bilder derselben in unserem Geiste. Die mechanischen Modelle und die Gleichungen, also die beiden Bildarten, sind, wenn die beide richtige Bilder sind, einander in den Resultaten, welche sie ergeben, vollkommen gleichwertig" [*Footnote:* Man sieht aus dieser Erörterung, daß ich die Bilder zweiter Art als höherstehend betrachte, gegenüber denen erster Art, da sie, wenn vollendet, eine Weiterentwicklung der letzteren sind, obgleich sie in den Anfängen auch umgekehrt oft einleitend diesen letzteren vorausgehen. Allerdings kommt es aus diesem in der Entwickelung liegenden Grunde stellenweise vor, daß bereits gute Bilder erster Art vorhanden sind, wo die Herstellung vollendeter Bilder zweiter Art noch nicht gelungen ist, und dies verleiht den Bildern erster Art an solchen Stellen Überlegenheit.]]

Die Benutzer der Bilder erster Art, zu welchen besonders auch Herr Einstein zählt, scheinen zumeist nicht geneigt, sich nach dem Standpunkt der Bilder zweiter Art zu begeben, um die Schwierigkeiten und Fragen, die von dort aus am deutlichsten zu erkennen sind, überhaupt genügend ins Auge zu fassen. Unzweifelhaft ist es aber, daß eine Theorie, mag sie auf Bilder erster oder zweiter Art gegründet sein, erst dann als einwandfrei gelten kann, wenn sie von beiden Standpunkten aus standhält; denn beide Standpunkte haben sich im Fortschreiten der Naturforschung als voll berechtigt gezeigt, und alle bisherigen gut bewährten Theorien sind von beiden Standpunkten aus widerspruchsfrei erschienen. Wer freilich die „Abschaffung des Äthers" verkündet

[*Footnote:* Die „Abschaffung des Äthers" wurde in Nauheim in großer Eröffnungssitzung wieder als Resultat verkündet (zur früheren Verkündung in Salzburg, von Herrn Einstein selbst, siehe das Zitat in Note 17, S. 27). {*Footnote 17, Pages 27-28:* Als das Überspringen eines Abgrundes konnte wohl seinerzeit die Entdeckung der Lichtquanten erscheinen: Auf der einen Seite waren die Wellen des Lichtes, auf der anderen die neuartigen Lichtquanten, und die Kluft zwischen ihnen wurde leer gelassen, was allerdings dem kühnen Springer selber niemand verdenken wird. Weitergehend war aber, nach der negativen Seite hin, der an diese Entdeckung geknüpfte Ausspruch (Naturforscherversammlung zu Salzburg am 21. September 1909, Verh. d. D. Phys. Ges. S. 482, Physik. Zeitschr. Bd. 10, S. 817, 1909): „Heute aber müssen wir wohl die Ätherhypothese als einen überwundenen Standpunkt ansehen", was zu einer nachträglichen Überbrückung der Kluft, die doch im Interesse der Wissenschaften zu wünschen war, nicht eben ermunterte. Ich habe dennoch eine solche Überbrückung versucht und bin dabei zu dem Resultat gelangt, daß die Lichtquanten dasselbe seien, was man als kohärente Lichtwellenzüge schon lange vorher ins Auge gefaßt hatte, allerdings mit dem wesentlichen neuen Zusatze der Konzentrierung der Energie auf einen Strahl von bestimmter Richtung, welches letztere ich durch die auch sonst naheliegende Annahme nur eines elektrischen Kraftlinienringes (gedacht als diskreter Ätherwirbelring) in jeder durch die Schwingung eines einzelnen Elektrons emittierten Lichtwelle erklärte (S. „Über Äther und Materie", Heidelberg 1911, S. 19 u. f. und die Untersuchung über Phosphoreszenz, Heidelb. Akad. 1913 A 19, S. 34 Fußnote 61. Als kohärente Wellenzüge hat, wie ich nachträglich finde, auch bereits H. A. Lorentz die Lichtquanten erklärt; Physikal. Zeitschr. Bd. 11, S. 353, 1910). Man sieht aus solcher Erklärungsmöglichkeit, was für das Gesamtbild des Naturforschers doch nicht unwichtig ist, daß die Lichtquanten nichts Umstürzendes für die Theorie des Lichtes sind, namentlich auch, daß sie für oder gegen die „Ätherhypothese" überhaupt gar nichts aussagen, sondern daß sie in der Hauptsache eine besondere, bis dahin unbekannt gewesene Eigenschaft der lichtemittierenden Atome betreffen, nämlich die, auf kohärente Wellenzüge von bestimmtem mit der Schwingungsdauer zusammenhängenden Energieinhalt eingerichtet zu sein.

Die Vorstellung, daß das Lichtquant ein kohärenter Wellenzug sei, dessen Länge demnach in jedem Falle durch optische Interferenzversuche feststellbar wäre, hat durch neuartige Versuche von Herrn W. Wien (Annalen d. Phys., Bd. 60, S. 597, 1919) eine augenfällige Bestätigung erfahren, indem die Zeitdauer der Emission des Lichtquants gemessen wurde. Sehr bemerkenswert ist dabei die hier als unmittelbares Beobachtungsergebnis auftretende Erkenntnis, daß die Energie des Lichtquants ungleichmäßig über die Länge des Wellenzugs verteilt ist, indem ein allmähliches Abklingen des emittierenden Atoms stattfindet (nach einer Exponentialfunktion, wie beim akustischen Wellenzuge einer angeschlagenen Glocke), so daß eine bestimmte Länge des Wellenzuges nur dann sich ergibt, wenn man festsetzt, in welchem Stadium des Abklingens man das Ende als erreicht ansehen will. Setzt man beispielswese das Ende bei 1/7 (genauer $1/e^2$) der Anfangsintensität fest, so ergibt sich nach Herrn W. Wiens Messungen die Länge des Lichtquants zu rund 10 m, und zwar gilt diese Länge — was an sich wieder sehr bemerkenswert ist — nach den bisherigen Messungen für Lichtquanten aller Wellenlängen, trotz des verschiedenen Energieinhalts der Lichtquanten verschiedener Wellenlänge. Es käme das darauf hinaus (wenn man bei diesen neuartigen Versuchen schon jetzt verallgemeinern darf), daß die Energie jeder einzelnen Welle irgendeines Lichtquants bei gleichem Abstande vom Anfange des Wellenzuges die gleiche ist. Der verschiedene Energieinhalt verschieden weit vom Anfange abstehender Wellen bestünde dabei in unserer Vorstellung in verschieden großer senkrecht zum Strahl gemessener Breite des elektrischen Kraftlinienringes dieser Wellen.} Man hat nicht dazu gelacht. Ich weiß nicht, ob es anders gewesen wäre, wenn die Abschaffung der Luft verkündet worden wäre.]

und vertritt, der will die Bilder zweiter Art hinwegleugnen (vgl. S. 27); er kann dann allerdings nicht in der Lage sein, auf deren Standpunkt sich zu begeben, und von ihm ist dann die Lösung der Schwierigkeiten und der damit verbundene Fortschritt auch nicht zu erwarten. Es wäre unnütz, hierauf weiter eingehen zu wollen, und es war dankenswert, daß die Aussprache an diesem Punkte in Nauheim von selber abbrach;
[*Footnote:* Die Frage des vierdimensionalen Raumzeitbegriffes war in der Diskussion von vornherein außer Spiel geblieben. Es wäre in Gegenwart so vieler Mathematiker (die oft dem mathematischen Hilfsmittel ebensoviel Bedeutung beilegen, als dem physikalischen Sinn) nicht förderlich gewesen, den mir als Naturforscher (der aber nicht nur die materielle Welt sehen will) allein annehmbar erscheinenden diesbezüglichen Standpunkt (vgl. S. 7 u. Anm. 7, S. 14) zu betonen, da es als Geschmackssache betrachtet werden kann, wieviel Denkfreiheit man zugunsten der „Relativierung der Zeit" opfern will.]

man findet sich hier von der zu Bescheidenheit mahnenden Erkenntnis der ganz außerordentlichen Ansprüche, welche an dieser Stelle der Entwicklung an den Geistesumfang des Naturforschers gestellt werden. Große mathematische Begabung, welche die Bilder erster Art mit Leichtigkeit meistert, scheint nicht oft in demselben Kopfe mit der Leichtigkeit der inneren dynamischen, physikalischen Anschauung verbunden zu sein, welche mehr Vorliebe für die Bilder zweiter Art verleiht, — und umgekehrt [*Footnote:* Man kann hieraus wohl auch ermessen, wie wenig Zweck es hat, wenn volkstümliche Schriften oder Vortragende von einseitigem Standpunkt aus das Relativitätsprinzip vor die Öffentlichkeit bringen, wobei auch der Verdacht kaum abzuweisen ist, daß die Einseitigkeit um des größeren Aufsehens willen, das sie hervorbringt, geliebt wird. Es ist das eine bedauerliche Erscheinung; aber sie besteht, und es wäre ein ungesundes Zeichen, und als solches sicherlich noch viel bedauerlicher, wenn darauf nicht Gegenwirkung einträte. Die „Relativisten" müßten aber eine von ihnen selbst hervorgerufene Gegenwirkung jederzeit ruhig hinzunehmen wissen.].

Im Einzelnen ergab die Aussprache etwa das Folgende:

Es wurden zwei Fragen gesondert diskutiert, deren Zusammenhang aber doch so wesentlich sich zeigte, daß wir sie hier der Kürze halber teilweise zusammenfassen können, nämlich 1. die Frage (vgl. S. 15, 16): Wie ist es im Beispiel des gebremsten Eisenbahnzuges, wo die Folgen der ungleichförmigen Bewegung nur innerhalb des Zuges sich zeigen, möglich, den Sitz der ungleichförmigen Bewegung trotz dieser Einseitigkeit der Erscheinung für unauffindbar erklären zu wollen, wie es die allgemeine Relativitätstheorie tut? Und 2. die Frage des unerlaubten Gedankenexperiments (vgl. Note 10, S. 16, 17): Bedeutet nicht das Auftreten von Überlichtgeschwindigkeiten im Falle einer Drehung der Gesamtwelt, z. B. um die Erde, die von der allgemeinen Relativitätstheorie als eine mit der Drehung irgendeines Körpers, z. B. der Erde, bei ruhender Gesamtwelt gleichwertige Annahme angesehen wird, einen inneren Widerspruch, da doch Überlichtgeschwindigkeiten nach eben derselben Theorie ausgeschlossen seien?

Es wurde von Herr Einsteins Seite selbstverständlich Gewicht auf die Gravitationsfelder gelegt, welche in seiner Theorie jeden Fall ungleichförmiger Bewegung begleiten müssen; aber es blieb doch dabei, daß diese Felder zunächst nur zu dem Zwecke hinzugenommen seien, um das Relativitätsprinzip allgemeingültig erscheinen zu lassen und auf alle Fälle anwenden zu können, woraus aber noch nicht hervorgeht, daß diese Felder weitere Beziehungen zur Wirklichkeit haben, die die Notwendigkeit ihrer Einführung den sie begleitenden Härten gegenüber erweisen (vgl. S. 22). Dabei sollte nicht bezweifelt sein, daß jedes Auftreten einer ungleichförmigen Bewegung mit gewissen Zuständen des Äthers (des „Raumes" liebt die Relativitätstheorie zu sagen, vgl. S. 28) in ihrer Umgebung verbunden sei; aber so lange die Einsteinschen Gravitationsfelder mit ihrem Zubehör den gesunden Verstand nicht befriedigen, wird man zweifeln dürfen, ob sie diese Zustände des Äthers ganz allgemein richtig abbilden. Vergeblich mahnt hierbei Herr Einstein zu

Mißtrauen gegenüber dem gesunden Verstand: Eine Theorie, die nicht in der Lage ist, auf so einfache Fragen, wie die obigen beiden es sind, eine entsprechende einfache, den gewöhnlichen Verstand befriedigende Antwort zu geben, ist nicht einwandfrei. Sie kann Erfolge haben und man kann solche bewundern, sie kann verbesserungsfähig, ja vielleicht schon in Verbesserung begriffen sein, aber sie darf nicht mit den üblichen weit gesteigerten Ansprüchen auftreten, welche wir in der vorliegenden Schrift getadelt haben, und sie darf das am allerwenigsten vor der Allgemeinheit tun, die als nicht sachkundig leicht beliebig irre zu führen ist. Es ist besser, der Allgemeinheit neben den Resultaten auch die Zweifel vorzuführen, um ihr den Ernst der Forschung zu zeigen, — oder aber gar nichts.

Auf die zweite Frage ist übrigens überhaupt keine entscheidende Antwort erfolgt [*Footnote:* Auch sonst war ich schließlich erstaunt, wie wenig Herr Einstein auf die Beantwortung meiner Fragen vorbereitet zu sein schien — die doch schon zwei Jahre lang mit seiner Kenntnis gedruckt vorgelegen haben, — während von seiner Seite und auch von einem andern Fachmann Zeitungslesern gegenüber ganz ausdrücklich der Anschein der unbedingten Überlegenheit meinen Gedankengängen gegenüber erweckt worden war. Da ich weder Anhänger noch Gegner irgendeines Prinzips bin, sondern nur Naturforscher sein möchte — wie auf S. 12 schon zu erkennen gegeben, — hätte ich den Nachweis, daß und an welcher Stelle meine Überlegungen nicht genügend gründlich waren, als Gewinn entgegennehmen müssen, wenn er geführt worden wäre (vgl. auch Note k, S. 23), zumal in der rein auf die Sache gerichteten Form, in welcher die Nauheimer Aussprache ablief. Die einzige Aufklärung, welche ich von der Diskussion mitgenommen habe, stammt von seiten des Herrn Mie; sie wird im weiter Folgenden bezeichnet werden.], und man darf daher wohl sagen, daß die Überlichtgeschwindigkeiten des unerlaubten Gedankenexperiments der allgemeinen Relativitätstheorie in der Tat eine Schwierigkeit bereiten [*Footnote:* Man muß immer bedenken, daß jeder beliebige rotierende Körper auf Erden, mag er auch nur eine Umdrehung in 3000 Jahren ausführen, Überlichtgeschwindigkeit schon der Orionsterne, vielhundertfache Lichtgeschwindigkeit der vielhundertfach ferneren Nebelsysteme ergibt, sobald man die Rotation nicht absolut dem Körper, sondern also der Umwelt zuschreiben will.]. Dies bedeutet aber nicht weniger, als daß diese Theorie in sich selbst — ganz abgesehen von ihrer Übereinstimmung oder Nichtübereinstimmungen mit der Wirklichkeit, — d. i. logisch nicht in Ordnung ist. Der innere Widerspruch, welchen sie enthält, fällt weg, wenn man nach Herrn Mies Vorschlag gewisse, von ihm „vernunftgemäß" genannte Koordinatensysteme für bevorzugt erklärt [*Footnote:* Vgl. G. Mie, Physikal. Zeitschr. 18, S. 551, 574, 596, 1917 und Annalen d. Physik 62, S. 46, 1920.] und die anderen möglichen Koordinatensysteme ausschließt [*Footnote:* Ganz im Sinne der auf S. 15 des Vorliegenden Gesagten; vgl. besonders auch die Note 8a.] Gleichzeitig wäre damit auch die erste Frage erledigt; man braucht nur ein mit dem Eisenbahnzug verbundenes Koordinatensystem als ruhend gedachtes

Bezugssystem auszuschließen und dafür das mit dem Erdboden verbundene Koordinatensystem als vernunftgemäß in Benutzung zu nehmen, um der Schwierigkeit der Frage enthoben zu sein. Aber dieser Ausweg bedeutet nicht eine Rettung, sondern eine Vernichtung des Relativitätsprinzips in seiner allgemeinsten, von Herrn Einstein aufgestellten, einem einfachen und zugleich allumfassenden Naturgesetz entsprechenden und daher das besondere philosophische Interesse in Anspruch nehmenden Form. Denn das Prinzip sagt in dieser Form aus, daß der Ablauf allen Naturgeschehens — die Formulierung der allgemeinen Naturgesetze — unabhängig ist von der Wahl des Bezugssystems [*Footnote:* Dies ist auch wirklich nach dem Ursprung des Prinzips sein einfacher Sinn, wenn überhaupt einer vorhanden ist. Es nützte in philosophischer Beziehung nichts, kompliziertere, verklausulierte Fassungen einzuführen; sind solche notwendig, so hat damit das Prinzip nicht zwar seinen möglichen Wert als Hilfmittel der Naturforschung, aber doch seine Ansprüche auf Wichtigkeit für das allgemeine Denken, für die Naturauffassung im Ganzen verloren.], wodurch es in allen Fällen unmöglich würde, durch irgendwelche Naturbeobachtungen absolut über Vorhandensein von Ruhe oder Bewegung zu entscheiden. Es müßten dann alle Bezugssysteme durchaus gleichwertig sein für die Schlüsse die sie ergeben (weshalb auch Herr Einstein die verschiedenen Koordinatensysteme, auch die, welche zu den offensichtlichsten Schwierigkeiten oder zu inneren Widersprüchen führen, immer wieder als prinzipiell gleichwertig hinstellen will), [*Footnote:* Nur praktische, nicht prinzipielle Gründe sollten nach Herrn Einsteins Äußerung von der Wahl gewisser Koordinatensysteme abhalten. Hierin liegt aber, wenn man sich vergegenwärtigt, daß gewisse, durch das Prinzip selbst gar nicht gekennzeichnete Koordinatensysteme in die Irre führen, eben der (wenn auch versteckte) Hinweis auf die Nichtigkeit der höchsten theoretischen Ansprüche des Prinzips; ganz unbeschadet natürlich seines etwaigen heuristischen und auch entwicklungsfördernden Wertes.] was aber nicht der Fall ist, wie die Beispielsfälle unserer beiden Fragen und in strengerer Form Herrn Mies Untersuchungen zeigen.

Man kann dann also — wie die Sache bis heute steht — das allgemeine Relativitätsprinzip nicht als Naturgesetz in strengem Sinne hinnehmen, und zwar, wie aus den Untersuchungen von Herrn Mie hervorzugehen scheint — und was hier als über den Inhalt der vorstehenden Teile dieser Schrift hinausgehend besonders hervorzuheben ist, — selbst dann nicht, wenn man seine behauptete Allgemeingültigkeit einschränken will auf massenproportionale Kräfte (Gravitationsprinzip, vgl. S. 18);

[*Footnote:* Das allgemeine Relativitätsprinzip ohne Einschränkung scheitert, wenn wirklich ernst genommen, an beiden oben ausgesprochenen Fragen. Das Gravitationsprinzip (die von mir vorgeschlagene Einschränkung des allgemeinen Relativitätsprinzips) ist dagegen allerdings fern von jeder Schwierigkeit der ersten Frage gegenüber (da es sich auf deren Fall gar nicht bezieht), zeigt aber doch der zweiten Frage gegenüber den inneren Widerspruch, der, wie es nun scheint, jeder Anwendung des Relativitätsprinzips auf ungleichförmige Bewegungen gefährlich werden muß, wenn nicht geeignete Kunstgriffe dagegen schützen. Man könnte danach sagen, daß das Gravitationsprinzip zwar in höherem Grade einwandfrei erscheint als das allgemeine Relativitätsprinzip, daß es aber doch ebenfalls nicht völlig und ohne weiteres einwandfrei ist. Immerhin erscheint der Unterschied in den Mängeln der beiden Prinzipien groß genug, um die in der vorliegenden Schrift geschehene Einführung und Hervorhebung des Gravitationsprinzips zu rechtfertigen.]

sondern man kann es — will man Irreführung vermeiden — nur als ein heuristisches Prinzip hinstellen (vgl. Note 11, S. 17), dessen Anwendung von der Hinzunahme nicht in dem Prinzip liegender Festsetzungen oder von besonderem Geschick oder Glück in Nebenannahmen begleitet sein muß, um das Ausmünden in falsche Resultate zu vermeiden, als ein Prinzip also, das unter Umständen richtige, wertvolle, ganz neue Zusammenhänge beobachtbarer Dinge liefern kann, wobei aber doch der wirkliche Beweis für die Richtigkeit der so vorausgesagten Zusammenhänge nur in noch hinzuzunehmender Erfahrung zu suchen wäre, mit der sie besonders verglichen werden müssen, nicht in mathematisch noch so einwandfreier Ableitung aus dem Prinzip.

[*Footnote:* Man bemerkt hier einen Unterschied gegenüber den sonstigen physikalischen Prinzipien, beispielweise dem Energieprinzip. Die aus solchen Prinzipien bei richtiger Beachtung der zugehörigen Begriffe mathematisch fehlerlos gezogenen Schlüsse darf man ohne weiteres für ebenso zutreffend halten wie die Gesamtheit der Erfahrungen, welche dem Prinzip zugrunde liegen und an welchen es bereits bewährt ist. Der Unterschied mag an der Neuheit des Relativitätsprinzips liegen (vgl. S. 14), die noch nicht genügend Klarheit hat aufkommen lassen über Gültigkeitsbereich oder über Zusatzbedingungen, welche bei der Anwendung einzuhalten und also als wesentlich zum Prinzip gehörig zu betrachten sind. Jedenfalls scheint mir bei dieser Sachlage im Falle der Perihelverschiebung des Merkur doch immer noch Gerbers „Ableitung" des richtigen quantitativen Zusammenhanges (sei sie auch nur Scheinableitung gewesen) mit Berücksichtigung der Frühzeitigkeit nennenswert zu bleiben gegenüber der nach dem Gesagten doch auch nur scheinbar aus strenger Anwendung eines Prinzips allein hervorgegangenen Ableitung Einsteins (vgl. S. 10-12 u. 30). Ganz abgesehen ist dabei inbezug auf Gerber davon, daß es mir durchaus unzulässig erscheint, einem längst Verstorbenen, der einen für richtig gehaltenen Zusammenhang (nämlich die Endgleichung für die Perihelverschiebung), also etwas Nützliches gebracht hat (mit dem Ungeschick der Hinzufügung eines anfechtbaren Beweises, aber auch ohne jedes Streben damit hervorzutreten), Pfuscherei oder dergleichen vorzuwerfen, wie es geschehen ist. Ich glaube, daß man den Pythagoräischen Lehrsatz, wenn ihn Pythagoras bloß veröffentlich und nicht bewiesen hätte, doch heute noch nach ihm benennen würde — damaliges genügend schnelles Bekanntwerden des Satzes angenommen, — da er richtig und wertvoll ist.]

Ein möglicherweise praktisch wertvolles Prinzip ist das Relativitätsprinzip also, aber keines, auf das eine neue Weltanschauung sich gründen ließe, oder das berufen sein könnte, bewährte anders geartete Wege der Naturforschung nun auf einmal als abgetan erscheinen zu lassen, wenn es auch selber einen neuen, augenblicklich vielbeschrittenen Weg eröffnet hat.

[*Footnote:* Man kann dann auch wohl sagen, daß es sich beim verallgemeinerten Relativitätsprinzip um ein durch Mathematik in quantitative Bahnen gedämmtes System des Erratens von Naturvorgängen handelt. Solches Erraten unter Aufwand eines ziemlich ausgedehnten mathematischen Apparats spielt auch sonst in der gegenwärtigen Physik eine früher nicht in gleichem Maße dagewesene Rolle, z. B. bei den quantentheoretischen Betrachtungen, und das Verfahren hat sich als sehr förderlich erwiesen, insofern die Kontrolle durch die Beobachtung nicht fehlte. Aber es wäre doch falsch, wenn man — wie einige Mathematiker es tun — nun eine Verwandlung der Physik in einen Nebenzweig der Mathematik als Endziel der Entwicklung vor sich sehen wollte. Die Natur, deren Erforschung Aufgabe der Physik ist, wird mit ihren Wundern, die jederzeit auch tiefsinnigste Forscher überrascht haben, noch nicht so bald zu Ende sein. — Offenbar ist es auch nur Geschmackssache, ob man lieber mit oder ohne mathematische Ableitung sich auf neue, der erfahrungsmäßigen Prüfung wert erscheinende Thesen bringen läßt, wenn die Ableitung nicht exakten Anschluß der Thesen an Erfahrungsresultate und an Annahmen von einfacher physikalischer Bedeutung liefert.]

Der mögliche praktische Wert des Prinzips kann umso höher bemessen werden, als es vielleicht richtige Zusammenhänge hat angeben helfen, die auf die Gravitation sich beziehen, auf eine Kraft, der man seit Newton und Cavendish, also über 100 Jahre lang nicht mehr weiter systematisch hat beikommen können [*Footnote:* Wozu, wenn solche Leistungen in Frage stehen, noch — genau besehen — übertriebene Ansprüche stellen?] Es liegen in dieser Beziehung bekanntlich drei Resultate vor: Die (schon von Gerber angegebene) Perihelverschiebungsgleichung, die Lichtstrahlenkrümmung und die Rotverschiebung der Spektrallinien bei Gravitationszentren, und es handelt sich um deren Prüfung an der Erfahrung, die auch über den mehr oder weniger großen Wert der Theorie entscheiden muß.

Der gegenwärtige Stand dieser Prüfung ist für die beiden erstgenannten Zusammenhänge, Perihelverschiebung und Lichtstrahlenkrümmung, im Vorliegenden bereits besprochen worden (S. 19, 20), und es kann hier der Lage der Sache nach auch nicht so schnell neue Erfahrung hinzukommen. Die Frage des drittgenannten Zusammenhangs, der Rotverschiebung (vgl.

Note 6, S. 19), ist dagegen augenblicklich mehr in Fluß. Es scheint dabei fast, als ob die mit besten Mitteln und von bewährtesten Seiten bisher ausgeführten Beobachtungen zu negativem Resultat sich vereinigten. [*Footnote:* Siehe die reichhaltige Zusammenstellung der in Betracht kommenden Veröffentlichungen in der auf S. 36 zitierten, soeben in den Annalen der Physik erschienenen Arbeit von E. Wiechert.] Jedenfalls erschien es bei der hierauf bezüglichen Diskussion in Nauheim nicht günstig für einwandfreien Überblick, daß nur die Bonner Beobachter (mit positivem Resultat) zu Wort kommen konnten, deren Hilfsmittel, so weit bekannt, weniger vollkommen waren als die der amerikanischen Beobachter, deren Resultat ebenso wie das kürzlich noch hinzugekommene von Julius in Utrecht [*Footnote:* W. H. Julius u. P. H. van Cittert, Kon. Akad. van Wetenschappen te Amsterdam, 29. Mai 1920.] aber negativ war. [*Footnote:* Die in bezug auf die Bonner Beobachtungen noch vorhandenen Zweifel erinnern mich an zwei Fälle, die zeigen, daß im Bonner Physikalischen Institut bei spektralanalystischen Beobachtungen nicht gerade traditionelles Glück vorhanden ist. Man vergleiche die gänzlich unrichtigen Angaben über die räumliche Verteilung der spektralen Lichtemission in den Alkalibogenflammen, die noch heute in nicht genügend kritisch bearbeiteten Werken eine irreführende Rolle spielen (s. dazu Heidelb. Akad. 1914 A 17, Fußnote 94, S. 48, auch Starks Jahrb. 13, S. 234, 1916) und ebenso die Beobachtungen über spektrale Erregungsverteilungen von Phosphoreszenzbanden, die ebenfalls mit der Annahme in die Irre gingen, bereits vorhandene Beobachtungen an Feinheit übertroffen zu haben (siehe dazu Heidelb. Akad. 1913 A 19, Fußnote 1, S. 3.]

Man kann daher bei der Rotverschiebung gegenwärtig noch von keiner experimentellen Bestätigung reden. Die beiden anderen Zusammenhänge sind zwar bestätigt, jedoch — wie auf S. 19, 20 erläutert — so, daß es noch fraglich blieb, ob diese Bestätigung überhaupt auf das Gravitationsprinzip sich beziehen läßt. Weiteres muß erst die Zukunft zeigen. Man wird dann sehen können, wie weit das Gravitationsprinzip — neben dem schon durch einfachste alltäglich Erfahrung widerlegten allgemeinen Relativitätsprinzip — wenigstens heuristischen Wert bewährt."

Hermann Weyl defended Einstein, though Einstein did not agree with Weyl's work.[50] Weyl repeatedly demonstrated dishonesty and his unscientific, unfair and adolescent pro-Einstein bias. In addition to being unfair to Gehrcke, Weyl intentionally underrated David Hilbert's priority for the generally covariant field equations of gravitation of the general theory of relativity. Though Weyl acknowledged Hilbert's work, he failed to emphasize Hilbert's priority as the first to deduce the generally covariant field equations of gravitation of the general theory of relativity. Weyl committed this vile act over

---

**50**. *See, for example: The Collected Papers of Albert Einstein*, Volume 9, Documents 26, 52, 59, 189, 207, 216, Princeton University Press, (2004).

Hilbert's objections, in Weyl's book *Space-Time-Matter*.[51]

Weyl published an article in *Die Umschau*, Volume 24, Number 42, (23 October 1920) pp. 609-610, which was not accessible to your author up to time of this publication. Other references to contemporary accounts which do not appear herein include: "Einladung zur 86. Vers. Dt. Naturforscher.", *Die Naturwissenschaften*, Volume 37, IV; and *Deutsche Allgemeine Zeitung*, 25 September 1920) Morning edition, p. 2.

Ernst Gehrcke redressed Hermann Weyl's (and Kleinschrod's) statement regarding the Bad Nauheim debate,

> "Der in der Umschau vom 23. Oktober 1920, Seite 610, erstattete Bericht von WEYL über die Relativitätssitzung in Nauheim bedarf in mehrfacher Hinsicht der Ergänzung.
>
> Ein nicht ganz unwichtiger Punkt, der auf der Nauheimer Tagung mit bemerkenswerter Deutlichkeit hervortrat, ist dem Berichte von Herrn WEYL nachzutragen: EINSTEIN hat nämlich unzweideutig und klar in der Diskussion seine Mißbilligung der WEYLschen Theorie zum Ausdruck gebracht und die Erklärung abgegeben, daß eine aus rein mathematischen Forderungen der Symmetrie aufgebaute Theorie, wie die von WEYL, abzulehnen sei. Wenn Herr WEYL es unternimmt, seine Gedanken der Öffentlichkeit näher zu führen, so sollte er einen so interessanten Punkt wie den der Stellungnahme EINSTEINs zur WEYLschen Theorie nicht unerwähnt lassen, damit in der Öffentlichkeit von vornherein keine irrige Meinung darüber entstehen kann, wie der Urheber der Relativitätstheorie zur species Relativismus von WEYL steht.
>
> Herr WEYL glaubt in seinem Bericht konstatieren zu dürfen, daß LENARD den Sinn der Relativitätstheorie nicht erfaßt habe. Dies ist nur eine Zurückgabe der von LENARD auf der Nauheimer Tagung gemachten Feststellung, daß die Relativisten kein Verständnis für die Erfordernisse der Wirklichkeitsforschung in der Physik gezeigt hätten, und daß sie keinen Versuch machen, die „Kluft" zu überbrücken. WEYL sollte bedenken, daß auch wenn jemand als Mathematiker virtuose Geschicklichkeit in der Handhabung mathematischer Symbole besitzt, er doch für andere Abstraktionen als Größenbeziehungen der Mathematik einen Mangel an Verständnis bezeigen kann, von dem universeller begabte Naturen frei sind. An Hand der WEYLschen Schriften würde sich leicht eine Liste von erkenntnistheorestischen Schnitzern und begrifflichen Wirrnissen anlegen lassen; es sei in diesem Zusammenhang übrigens auch auf die kürzlich erschienene Schrift von RIPKE-KÜHN: KANT contra EINSTEIN, Verlag von KEYSER-Erfurt, verwiesen.
>
> Der von Herrn WEYL in seinem Bericht näher ausgeführte Punkt in der Diskussion zwischen EINSTEIN und LENARD hinsichtlich dessen

---

**51**. T. Sauer, "The Relativity of Discovery: Hilbert's First Note on the Foundations of Physics", *Archive for History of Exact Sciences*, Volume 53, Number 6, (1999), pp. 529-575, at 568, note 156.

Beispiel des gebremsten Eisenbahnzuges läßt den wesentlichen, von LENARD näher erläuterten Einwand vermissen, daß zur Erzeugung eines Gravitationsfeldes doch nach unseren heutigen physikalischen Kenntnissen Massen da sein sollten, die das Gravitationsfeld hervorbringen. Im Falle des Eisenbahnunglücks, wo nach Angabe des Relativisten nicht der Zug, sondern die ganze Umgebung gebremst worden sein soll, ist keine Massenanordnung und nichts ersichtlich, was das zur Bremsung der Umgebung erforderliche Gravitationsfeld erzeugt haben könnte. Der Relativist wurde denn auch in Nauheim veranlaßt, ausdrücklich Gravitationsfelder ohne erzeugende, gravitierende Massen anzunehmen, wobei er allerdings u. a. offen ließ, woher die Energie dieser Gravitationsfelder genommen wird. Von all dem berichtet uns Herr WEYL nichts.

Endlich hat die Diskussion in Nauheim die Erklärung EINSTEINs gezeitigt, daß nach der allgemeinen Relativitätstheorie der Körper jede beliebige Geschwindigkeit, größer als die Lichtgeschwindigkeit, besitzen dürfen. Auch diese in ihren Folgerungen hier nicht weiter zu behandelnde Angelegenheit erwähnt Herr WEYL nicht. „Ergebnislos" war die Debatte in Nauheim also keineswegs."[52]

Weyl answered *Die Umschau* a.k.a.*Die Umschau; Wochenschrift über die Fortschritte in Wissenschaft und Technik*; a. k. a. *Umschau in Wissenschaft und Technik*, Volume 25, (1921), p. 123.

Ernst Gehrcke wrote,

"Ich möchte hier zum Ausdruck bringen, daß EINSTEIN auf der Nauheimer Naturforscherversammlung die Möglichkeit der Überlichtgeschwindigkeiten vom Standpunkt seines allgemeinen Relativitätsprinzips zugestanden hat. Wenn Herr WEYL dies leugnen zu können glaubt, so ist nur ein neuer Widerspruch zwischen ihm und EINSTEIN — wenigstens zur Zeit der Nauheimer Tagung — festzustellen. Die Erklärung EINSTEINs über die Überlichtgeschwindigkeiten, so unbefriedigend sie sein mag, ist tatsächlich abgegeben worden, und Herr WEYL hätte besser getan, das Beweismaterial zu prüfen, als einen Irrtum LENARDS anzunehmen."[53]

Hermann Weyl wrote in 1921:

## "Die Relativitätstheorie auf der Naturforscherversammlung in Bad Nauheim.

---

[52]. E. Gehrcke, "Die Relativitätstheorie auf dem Naturforschertage in Nauheim", *Die Umschau*, Volume 25, (1921), p. 99.
[53]. E. Gehrcke, "Zur Relativitätsfrage", *Die Umschau*, Volume 25, (1921), p. 227.

Von H. WEYL in Zürich.

Auf Veranlassung der Deutschen Mathematikervereinigung war auf der letztjährigen Naturforscherversammlung in Bad Nauheim die Relativitätstheorie in einer kombinierten Sitzung der mathematischen und physikalischen Sektion zum Mittelpunkt einer Reihe von Vorträgen und einer allgemeinen Diskussion gemacht worden; darüber sei hier — nach reichlich langer Zeit, die aber vielleicht der Klärung und ruhigen Beurteilung der Sachlage zugute kommt — Bericht erstattet.

Den ersten Teil der Sitzung bildeten vier Vorträge aus dem Gebiete der Relativitätstheorie: 1. H. Weyl, Elektrizität und Gravitation; 2. G. Mie, Das elektrische Feld eines um ein Gravitationszentrum rotierenden geladenen Partikelchens; 3. M. v. Laue, Theoretisches über neuere optische Beobachtungen zur Relativitätstheorie; 4. L. Grebe, Über die Gravitationsverschiebung der Fraunhoferschen Linien. Den vier Vorträgen folgte die auf ihren Inhalt sich beziehende „Spezial"-Diskussion. Der letzte und dramatischste Teil, die allgemeine Diskussion über die Relativitätstheorie, gestaltete sich im wesentlichen zu einem Zweikampf zwischen Einstein und Lenard. Mit großem Geschick, Strenge und Unparteilichkeit waltete Planck seines Amtes als Vorsitzender; ihm war es nicht zum wenigsten zu danken, daß dieses „Nauheimer Relativitätsgespräch", in welchem entgegengesetzte erkenntnistheoretische Grundauffassungen der Wissenschaft aufeinanderstießen, einen würdigen Verlauf nahm.

Auf den Inhalt der Vorträge werde hier nur insoweit eingegangen, als er mit den prinzipiellen Fragen der Relativitätstheorie in Zusammenhang steht. Nach der speziellen Relativitätstheorie beruht der *Dopplereffekt* auf den folgenden beiden Tatsachen: 1. Die Frequenzen der von zwei Atomen der gleichen Konstitution, etwa zwei Wasserstoffatomen, ausgesendeten Spektrallinien sind einander gleich, wenn jede von ihnen gemessen wird in der dem Atom eigentümlichen *Eigenzeit*. 2. Die Frequenz einer Lichtwelle ist im ganzen Raum überall die gleiche, wenn sie gemessen wird in der „kosmischen" Zeit $t$ die zusammen mit den drei Raumkoordinaten ein System linearer Koordinaten für die ganze Welt bildet. Wie übertragen sich diese beiden Tatsachen in die allgemeine Relativitätstheorie? Hier wird die Eigenzeit nach Einstein definiert durch die „metrische Fundamentalform" $ds^2 = \sum g_{ik} dx_i dx_k$, eine quadratische Differentialform der vier willkürlichen Weltkoordinaten $x_i$ vom Trägheitsindex 3; und das Analogon zu 1. lautet: für zwei Atome gleicher Konstitution hat das Integral $\int ds$, erstreckt über eine volle Periode, den gleichen Wert. Fragt man indes danach — um der Sache etwas mehr auf den Grund zu gehen —, wodurch das $ds^2$ physikalisch bestimmt ist, wodurch insbesondere der Vergleich der Maßeinheiten des $ds$ an verschiedenen Weltstellen ermöglicht wird, so antwortet Einstein, daß dazu die Atomuhren das Mittel bilden (auch starre Maßstäbe oder, physikalisch etwas strenger gesprochen, die Gitterabstände in einem Kristall können zum gleichen Zwecke dienen): kommt die

Atomuhr im Laufe ihrer Geschichte vom Weltpunkt $O'$ nach dem Weltpunkt $O$ und legt sie beim Passieren von während einer Periode die unendlichkleine Weltstrecke s, beim Passieren von $O'$ während einer Periode die unendlichkleine Weltstrecke s' zurück, so hat *definitionsgemäß* s' die gleiche Länge *ds* wie s. 1. ist danach keine erklärungsbedürftige Tatsache, sondern *ds* ist physikalisch so definiert, daß 1. zutrifft. Dennoch schließt die Möglichkeit dieser Festsetzung über den Transport der Maßeinheit eine physikalische Grundtatsache ein, nämlich die folgende: Haben zwei Atomuhren, die sich an derselben Weltstelle $O$ befinden, dort die gleiche Frequenz und treffen sie, nachdem sie verschiedene Wege in der Welt durchlaufen haben, in einem anderen Weltpunkt $O'$ wieder zusammen, so haben sie auch dort gleiche Frequenz. Meine Theorie von Elektrizität und Gravitation, auf einer Weltgeometrie beruhend, in welcher die Übertragung einer Strecke durch kongruente Verpflanzung längs eines Weges vom Wege abhängig ist, war von den Physikern meist dahin mißverstanden worden, als wolle ich an dieser Tatsache rütteln. Der Hauptzweck meines Vortrages in Nauheim war, dem entgegenzutreten. Ich akzeptiere jene Grundtatsache so gut wie Einstein; wir weichen voneinander ab in ihrer theoretischen Deutung. Nach Einstein ist die metrische Struktur des Äthers von der Art, wie sie Riemann annimmt, die Streckenübertragung vom Wege unabhängig. Die Frequenzen der Atomuhren folgen dieser kongruenten Verpflanzung; die Erhaltung der Frequenz beruht also auf einer von Augenblick zu Augenblick infinitesimal wirksamen *Beharrungstendenz*. Im Gegensatz dazu scheint mir die einzig mögliche physikalische Deutung jener Grundtatsache die zu sein, daß sich die Frequenz durch *Einstellung* auf eine gewisse Feldgröße (von der Dimension einer Länge) bestimmen muß: zufolge ihrer *Konstitution* hat die Atomuhr an einer beliebigen Feldstelle eine Periode, die im Verhältnis zu jener Feldgröße einen bestimmten numerischen Gleichgewichtswert besitzt. [*Footnote:* In einer jüngst erschienenen Note (Berliner Sitzungsberichte 1921, S. 261). akzeptiert Einstein, wenn ich ihn recht verstehe, diesen Standpunkt, nicht aber meine weltgeometrische Deutung der Elektrizität.] In der Tat ergeben die Naturgesetze, daß sich die materiellen Körper so verhalten, und zwar ist die Feldgröße, auf welche sich die Längen einstellen, der aus der skalaren Krümmung des Feldes zu berechnende Krümmungsradius. Die aus dem Verhalten der materiellen Körper in der geläufigen Weise abgelesene Maßgeometrie ist also mit der metrischen Struktur des Äthers nicht identisch, sondern geht aus ihr hervor, indem die kongruente Verpflanzung ersetzt wird durch die Einstellung auf den Krümmungsradius. In der anschließenden Diskussion wurde der beiderseitige Standpunkt klar und knapp zum Ausdruck gebracht, ohne daß einer den andern zu bekehren oder zu widerlegen suchte. [*Footnote:* Eine ausführliche Darstellung meiner Auffassung wurde von mir gerade jetzt veröffentlicht in zwei Arbeiten in den Ann. d. Physik **65** und der Physik. Zeitschrift **22** unter den Titeln: „Feld und Materie", „Über die physikalischen Grundlagen der erweiterten Relativitätstheorie".]

Ich komme zu der oben erwähnten Tatsache 2. und ihrer Übertragung in die allgemeine Relativitätstheorie. Davon handelte der Lauesche Vortrag. Ein *statisches* Gravitationsfeld ist dadurch gekennzeichnet: man kann die vier Weltkoordinaten $x_0=t$, $x_1x_2x_3$ (statische Koordinaten) so wählen, daß sich Zeit ($t$) und Raum ($x_1x_2x_3$) vollständig trennen und die Beschaffenheit des Feldes zeitlich konstant ist; d. h. es wird $ds^2 = f^2 dt^2 - d\sigma^2$, wo $f$ die Lichtgeschwindigkeit, und $d\sigma^2$ die metrische Fundamentalform des Raumes, nur von dem Raumkoordinaten $x_1x_2x_3$ abhängen; $d\sigma^2$ ist positiv-definit. In einem solchen statischen Gravitationsfeld haben die Maxwellschen Gleichungen (komplexe) Lösungen von folgender Art: das elektromagnetische Feld ist gleich einem zeitlich konstanten Felde multipliziert mit dem von der Zeit abhängigen rein periodischen Term $e^{ivt}$; $v$ ist die konstante Frequenz. Sind derartige „einfache Schwingungen", wie wir es annehmen wollen, für den tatsächlichen Vorgang der Lichtausbreitung maßgebend, so heißt das: 2. In einem statischen Gravitationsfeld ist die Frequenz der von einem ruhenden Körper ausgesendeten Lichtwelle überall im Raum die gleiche, gemessen in der kosmischen Zeit $t$ der Zeitkoordinate im System der vier statischen Koordinaten. Aus den beiden Tatsachen 1. und 2. ergibt sich mit Notwendigkeit die von Einstein behauptete *Rotverschiebung der Spektrallinien* in der Nähe großer Massen, die ja nach dem Äquivalenzprinzip mit dem Dopplerschen Prinzip auf engste zusammenhängt; denn im statischen Gravitationsfeld hat $f$ in der Nähe großer Massen einen kleineren Wert als fern von ihnen. — Außerdem leitete Laue in seinem Vortrag nach dem Muster des von Debye für die klassische Elektrodynamik vorgeschlagenen Verfahrens aus den Maxwellschen Gleichungen als erste Näherung für hohe Frequenzen das Grundgesetz der geometrischen Optik her, daß ein Lichtsignal eine geodätische Nullinie beschreibt. Man macht den Ansatz, daß alle Feldkomponenten multiplikativ den Term $e^{ivE}$ enthalten mit einem sehr großen konstanten $v$, und erhält dann für die „Eikonalfunktion" $E$ die partielle Differentialgleichung

$$\sum_{ik} g^{ik} \frac{\partial E}{\partial x_i} \frac{\partial E}{\partial x_k} = 0,$$

deren Charakteristiken die geodätischen Nullinien sind.

An das eben aufgestellte Prinzip 2. sei es gestattet, hier eine kritische Bemerkung anzuknüpfen. Das Prinzip ist eindeutig, wenn durch die Forderung der statischen Koordinaten die Zeit $t$ bis auf eine lineare Transformation in sich, die drei Raumkoordinaten $x_1x_2x_3$ bis auf eine willkürliche Transformation untereinander festgelegt sind. Im allgemeinen ist das der Fall, aber nicht immer. Die gravitationslose Welt der speziellen Relativitätstheorie:

$$ds^2 = dt^2 - \left(dx_1^2 + dx_2^2 + dx_3^2\right)$$

ist ein Beispiel dafür. Doch wird hier unter den linearen Koordinatensystemen eine bestimmte kosmische Zeit $t$ dadurch ausgezeichnet, daß man fordert, der licht-aussendende Körper solle ruhen; und so gestatten in diesem Falle unsere beiden Forderungen 1. und 2. die Lichtwellen zu vergleichen, die von zwei relativ zueinander bewegten Körpern ausgehen (Dopplersches Prinzip). Ein anderes wichtiges Beispiel ist die leere Welt, wie sie sich ergibt, wenn man in den Gravitationsgleichungen das Einsteinsche kosmologische Glied mitberücksichtigt. Nach de Sitter [*Footnote:* On Einsteins theory of gravitation and its astronomical consequences III, Monthly Notices of the R. Astron. Society, Nov. 1917.] ist diese leere Welt ein „Kegelschnitt" $\Omega(x) = a^2$ in einem 5-dimensionalen Euklidischen Raum mit dem Linienelement $ds^2 = -\Omega(dx)$;

$$\Omega(x) = x_1^2 + x_2^2 + x_3^2 + x_4^2 - x_5^2$$

Durch die Substitution

(*) $\qquad x_4 = z \cdot \mathfrak{Cos}\,\dfrac{t}{a}, \quad x_5 = z \cdot \mathfrak{Sin}\,\dfrac{t}{a}$

kommt man hier auf statische Koordinaten $t, x_1 x_2 x_3$ es wird nämlich

$$-ds^2 = \left(dx_1^2 + dx_2^2 + dx_3^2 + dz^2\right) - \frac{z^2}{a^2}dt^2$$

mit

$$z^2 = a^2 - r^2, \quad r^2 = x_1^2 + x_2^2 + x_3^2$$

$f^2 = 1 - \left(\dfrac{r}{a}\right)^2$ nimmt vom Werte 1 im Nullpunkt bis zum Werte 0 auf dem Äquator ab. Ist diese statische Zeit für die Ausbreitung des Lichtes maßgebend, so würden also die Spektrallinien von Sternen um so stärker nach dem Rot verschoben sein, je weiter sie vom Nullpunkt entfernt liegen. De Sitter hat die Möglichkeit erwogen, auf diese Weise die tatsächlich vorhandene systematische starke Rotverschiebung in den Spektren der Spiralnebel kosmologisch zu deuten. Nun ist aber $t$ offenbar keineswegs die einzige „statische Zeit"; zu dem Spiralnebel als Nullpunkt wird ebenso eine solche Zeit gehören wie zu der bisher als Nullpunkt angenommenen Sonne. In der Tat kann man ja vor Ausführung der Substitution (*) die Koordinaten $x_1 \ldots x_5$ einer willkürlichen linearen Transformation unterwerfen, welche $\Omega(x)$ invariant läßt; dann bekommt man ein ganz anderes $t$. Welches soll nun nach dem Prinzip 2. maßgebend sein für die Ausbreitung des Lichtes? Die durch (*) eingeführten statischen Koordinaten stellen nicht den ganzen de Sitterschen Kegelschnitt, sondern nur den Keil $x_4^2 - x_5^2 > 0$ reell dar. Ist die wirkliche Welt der ganze de Sittersche Kegelschnitt, so ist also das Prinzip 2. völlig unberechtigt. Wenn aber die Welt nur aus einem derartigen Keil besteht, wie Einstein es annimmt, ist natürlich dasjenige, bis auf eine lineare Transformation eindeutig bestimmte $t$ zu nehmen, welches diesem Keil entspricht. Steht das im Einklang mit der Wirklichkeit, so ist also auf die Ausbreitung einer Lichtwelle vom Moment ihrer Entstehung an der

Zusammenschluß der Welt im Ganzen von Einfluß, während man doch erwarten sollte, daß die Lichtwelle darauf erst reagieren kann, wenn sie den ganzen Weltraum durchlaufen hat. Mit der in den retardierten Potentialen zum Ausdruck kommenden alten Hertzschen Vorstellung von der Entstehung einer Lichtwelle ist das gewiß unverträglich. So bedarf das Prinzip 2., der Mechanismus der Übertragung der Frequenz in einer Lichtwelle, noch sehr der physikalischen Aufklärung.

Inwieweit die nach Einstein zu erwartende *Rotverschiebung* der Fraunhoferschen Linien im Sonnenspektrum gegenüber den von irdischen Lichtquellen stammenden Linien durch die *Experimente* bestätigt wird, darüber berichtete Grebe. Die Messungen sind angestellt worden von Schwarzschild, dann von Evershed und Royds, später von St. John, schließlich von Bachem und Grebe. Namentlich die mit den schärfsten Hilfsmitteln ausgeführten Beobachtungen von St. John sprachen *gegen* das Vorhandensein des Einsteineffektes. Alle Beobachter stellen aber übereinstimmend fest, daß verschiedene Linien verschiedene Verschiebungen aufweisen. Grebe und Bachem machten nun darauf aufmerksam, daß für die Erklärung dieser Unregelmäßigkeiten vor allem der Umstand in Betracht fällt, daß unmittelbar benachbarte Linien sich gegenseitig in der Lage ihrer Intensitätsmaxima stören. Sie sonderten deshalb auf Grund mikrophotometrischer Aufnahmen aus den von ihnen gemessenen 36 Linien der sogenannten Cyanbande 11 aus, die sie als störungsfrei glaubten in Anspruch nehmen zu dürfen; diese zeigen nun im Mittel eine Rotverschiebung, welche dem Einsteineffekt ungefähr entspricht. Ebenso ergab sich als Mittel der Verschiebungen von 100 *aufeinanderfolgenden* Cyanbandenlinien *ohne jede Auswahl* — wo man erwarten darf, daß die gegenseitigen Störungen sich ausgleichen — nahezu derselbe Wert. Wenn man diese Untersuchungen auch noch kaum als eine definitive experimentelle Bestätigung des Einsteineffektes ansprechen darf, so verstärken sie doch die Wahrscheinlichkeit seines wirklichen Vorhandenseins erheblich. In der seit der Nauheimer Tagung verflossenen Zeit hat sich die Situation in dieser Hinsicht durch neue Beobachtungen noch weiter verbessert.

Um Sinn und Tragweite des Einsteinschen *Äquivalenzprinzips* durch ein vollständig zu übersehendes, nicht triviales Beispiel zu illustrieren, berechnete Mie nach diesem Prinzip das elektrische Feld eines geladenen Teilchens, das um ein elektrisch neutrales Gravitationszentrum unter dem Einfluß der Gravitation eine Kreisbahn beschreibt. Die statischen Koordinaten, in welchen das kugelsymmetrische Gravitationsfeld die von Schwarzschild angegebene Form besitzt, bezeichnet Mie als das vernünftige Koordinatensystem. In einem gewissen „künstlichen" Koordinatensystem, in welchem sowohl das Teilchen ruht wie auch das Gravitationsfeld stationär ist, haben die Maxwellschen Gleichungen eine von der Zeit unabhängige Lösung, welche in der unmittelbaren Nähe des Teilchens mit der elektrostatischen Lösung identisch ist. Transformiert man sie auf das vernünftige Koordinatensystem, so erhält man diejenige Lösung des

Problems, welche nach dem Äquivalenzprinzip dem elektrostatischen Feld eines ruhenden Teilchens gleichwertig ist. Das Feld ist in unendlichgroßer Entfernung nicht von solcher Art, daß eine Ausstrahlung von Energie stattfindet, sondern man erhält es dort, wenn einem nach den Liénard-Wiechertschen Formeln berechneten ausstrahlenden Feld ein einstrahlendes von gleicher Stärke superponiert wird. Zweifellos ist das eine mit den uns bekannten Feldgesetzen verträgliche Lösung; dennoch ist es sicher, daß das wirkliche Verhalten eines elektrisch geladenen Körpers, der um ein Gravitationszentrum rotiert, nicht ihr entspricht, sondern eine elektromagnetische Welle ausstrahlt und dadurch selber in seiner Bewegung modifiziert wird. Die *tatsächlichen* Vorgänge bei Ruhe und Rotation sind also *nicht* einander äquivalent. Mie äußert sich darüber so: Man denke sich ein Einsteinsches Kupee, welches auf einer Kreisbahn um das Gravitationszentrum herumfährt; die Beobachter stellen an einem mitgeführten elektrischen Teilchen Beobachtungen an. Bestehen die Wandungen des Kupees aus Metall, so daß das von dem Teilchen erregte elektrische Feld dort endigt, so gilt das Äquivalenzprinzip; bestehen die Wandungen jedoch aus isolierendem Material, so können die Beobachter im Kupee ihre Bewegung feststellen; die Feldlinien des Teilchens sind sozusagen Fühler, die sie aus dem Kupee heraus ins Unendliche strecken. Damit kann man sich sehr wohl auch vom Einsteinschen Standpunkt aus einverstanden erklären. Solange man mit einem unendlichen Raum operiert, hat man immer den unendlich fernen Saum dieses Raumes zu berücksichtigen, über den gewissermaßen ein das Feld bestimmendes Agens ebenso herüberwirkt wie über die inneren Feldsäume, welche den verschiedenen Materieteilchen entsprechen. Mathematisch äußert sich das darin, daß nur solche Koordinaten zulässig sind, für welche im Unendlichen das $ds^2$ die Gestalt der speziellen Relativitätstheorie hat. In Einsteins geschlossenem Raum aber fällt der unendlich ferne Saum weg, an seine Stelle treten die weit entfernten Massen.

Der Durchrechnung dieses speziellen Problems schickte Mie einige grundsätzliche Bemerkungen voraus, welche zeigen, daß er in einigen Punkten einen andern Standpunkt einnimmt als Einstein. Insbesondere glaubt er an ein ausgezeichnetes „vernunftgemäßes" Koordinatensystem. Nun ist ja zuzugeben, daß sich in speziellen Problemen oft aus der Beschaffenheit des metrischen Feldes heraus ein besonders einfaches und zweckmäßiges Koordinatensystem definieren läßt. So kann man im Schwarzschildschen Fall des statischen kugelsymmetrischen Gravitationsfeldes die Raumkoordinaten $x_1 x_2 x_3$ derart wählen, daß, wenn man mit ihrer Hilfe den wirklichen Raum auf einen Cartesischen abbildet, das lineare Vergrößerungsverhältnis für Linienelemente, welche senkrecht zu den Radien im Bildraum stehen, $= 1$ wird (für radiale Linienelemente wird es dann, wie aus den Gravitationsgleichungen hervorgeht, $= 1/f$, und $f^2$ ist $= 1 - \frac{2a}{r}$; $\alpha$ eine Konstante, $r$ die im Bildraum gemessene Entfernung von Zentrum). Aber gerade in diesem Fall kann man über die radiale Maßskala z. B. doch auch so verfügen, daß die Abbildung auf den

Cartesischen Bildraum konform ist (dann wird das Vergrößerungsverhältnis für alle Linienelemente $\left(1 + \frac{a}{r}\right)^2$ und $f$ ist $\frac{r-a/2}{r+a/2}$ Hier ist gar nicht abzusehen, warum man das eine dieser beiden Koordinatensysteme als „vernunftgemäßer" ansprechen soll denn das andere. Die Frage nach der Existenz eines vernunftgemäßen Koordinatensystems hängt aufs engste mit der andern zusammen, inwiefern es berechtigt ist, zu behaupten: die wahre Geometrie des Raumes sei die *euklidische;* daß materielle Maßstäbe nicht die Relationen erfüllen, welche diese Geometrie für den idealen starren Körper angibt, liege daran, daß die materiellen Körper durch das Gravitationsfeld in bestimmter Weise deformiert werden. Dieser Standpunkt, den z. B. Dingler und Hamel vertreten [*Footnote:* Dingler: Der starre Körper, Physik. Zeitschr. 1920 S. 487; Hamel: Sitzungsber. d. Berl. Mathem. Gesellschaft 1921. S. 65.], ist zunächst natürlich gegenüber der Gravitation physikalisch ebenso berechtigt wie gegenüber der Temperatur (Einstein selbst zieht diese Parallele in seiner populären Schrift über die Relativitätstheorie): kein Mensch behauptet, daß auf einer ungleichförmig erwärmten Platte eine nichteuklidische Geometrie gilt, sondern daß die zur Ausmessung verwendeten Maßstäbe durch die verschiedenen Temperaturen verschiedene Ausdehnungen erfahren. Aber in diesem Fall existiert eine absolut ausgezeichnete Reduktion, die Reduktion auf „gleiche Temperatur", durch welche das Verhalten der Maßstäbe mit der euklidischen Geometrie in Einklang gebracht wird. Im Fall der Gravitation existiert zwar auch eine „Reduktion auf Euklid" (das ist sogar selbstverständlich), aber unter den unendlich vielen möglichen derartigen Korrekturvorschriften, deren jede zu andern Resultaten führt, ist keine physikalisch so ausgezeichnet, daß sie sich zwingend als die „einzig richtige" aufdrängt. Darum ist es hier wertlos, den an den materiellen Körpern abgelesenen Maßzahlen durch Korrektur eine euklidische Geometrie zu supponieren. Vielleicht hat der Philosoph immer noch Recht mit seiner Ansicht, daß man ohne einen idealen euklidischen Anschauungsraum nicht auskomme; ihm entspräche in der mathematischen Darstellung die Notwendigkeit, ein Koordinatensystem zu verwenden. Aber seine Beziehung auf das Ordnungsschema der physikalischen Ereignisse ist wie die Wahl des Koordinatensystems in hohem Maße willkürlich. Die universelle Konstruktion, welche Mie selber für das vernunftgemäße Koordinatensystem andeutet (mit Hilfe einer Einbettung des vierdimensionalen wirklichen Raumes in einen zehndimensionalen euklidischen) ist vieldeutig und ohne inneres Vorzugsrecht. Es ist gar nicht einzusehen, welche Erleichterung dadurch für die Beschreibung der physikalischen Vorgänge geschaffen werden soll; sie läßt sich ja immer mittels invarianter Begriffe vollziehen. — Noch in einem andern Punkte weicht Mie von Einstein ab; er meint, man dürfe nicht von allgemeiner Relativität, sondern nur von einer Relativität der Gravitationswirkungen sprechen, da man nach der Einsteinschen Theorie das Verhalten eines beschleunigt bewegten materiellen Systems aus dem des ruhenden nur dann berechnen kann, wenn die wirkende Kraft die eines Gravitationsfeldes ist. Mir scheint, das ist kein Einwand gegen die Allgemeinheit des

Relativitätsprinzips, sondern eine Bemerkung über seine Tragweite: nur für die im „Führungsfeld" neben der Trägheit mitenthaltenen Kräfte (Zentrifugalkraft, Gravitation), die man an ihrer Massenproportionalität erkennt, ist dieses Prinzip ausreichend, ihre Wirkungsweise a priori aus dem Galileischen Trägheitsprinzip abzuleiten.

Die beiden zuletzt erörterten Punkte kamen auch in der *allgemeinen Diskussion,* die vor allem von Lenard benutzt wurde, zwischen Lenard und Einstein zur Sprache. Es sei um der Übersichtlichkeit willen gestattet, aus diesem Wechselgespräch zunächst noch zwei weitere Streitfragen herauszuschälen, die neben der am Schluß zu besprechenden Hauptdifferenz nur von nebensächlicher Bedeutung sind. Das ist erstens die *Existenz des Äthers.* Lenard meint, Einstein habe, bei Aufstellung der speziellen Relativitätstheorie, allzu voreilig die Abschaffung des Äthers verkündet. In der Tat kann er ja darauf hinweisen, daß Einstein heute wieder in der allgemeinen Relativitätstheorie von einem Äther spricht. [*Footnote:* Siehe namentlich die Leidener Antrittsvorlesung Einsteins über Äther und Relativitätstheorie, Springer 1920.] Man darf sich doch aber durch das gleichlautende Wort nicht über die Verschiedenheit der Sache täuschen lassen! Der alte Äther der Lichttheorie war ein *substantielles* Medium, ein dreidimensionales Kontinuum, von welchem sich jede Stelle *P* in jedem Augenblick *t* in einem bestimmten Raumpunkt *p* (oder an einer bestimmten Weltstelle) befindet; die Wiedererkennbarkeit derselben Ätherstelle zu verschiedenen Zeiten ist dabei das Wesentliche. Durch diesen Äther löst sich die vierdimensionale Welt auf in ein dreifach unendliches Kontinuum von eindimensionalen Weltlinien; infolgedessen gestattet er, *Ruhe* und *Bewegung* absolut voneinander zu unterscheiden. *In diesem Sinne,* etwas anderes hat Einstein nicht behauptet, ist der Äther durch die spezielle Relativitätstheorie abgeschafft; er wurde ersetzt durch die affingeometrische Struktur der Welt, welche nicht den Unterschied zwischen Ruhe und Bewegung festlegt, sondern die *gleichförmige Translation* von allen andern Bewegungen absondert. Der substantielle Äther war von seinen Erfindern als etwas Reales, den ponderablen Körpern Vergleichbares gedacht. In der Lorentzschen Elektrodynamik hatte er sich in eine rein geometrische, d. h. ein für allemal feste, von der Materie nicht beeinflußte Struktur verwandelt. In Einsteins spezieller Relativitätstheorie trat an ihre Stelle eine andere, die affingeometrische Struktur. In der allgemeinen Relativitätstheorie endlich verwandelte sich die letztere, als „affiner Zusammenhang" oder „Führungsfeld", wieder zurück in ein mit der Materie in Wirkungszusammenhang stehendes Zustandsfeld von physikalischer Realität. Und darum hielt es Einstein für angezeigt, das alte Wort Äther für den vollständig gewandelten Begriff wieder einzuführen; ob das zweckmäßig war oder nicht, ist weniger eine physikalische als eine philologische Frage.

Zweitens: die *Überlichtgeschwindigkeit.* Lenard meint, die allgemeine Relativitätstheorie führe die Überlichtgeschwindigkeit wieder ein, da sie als Bezugssystem z. B. die rotierende Erde zuläßt; in hinreichend großen

Entfernungen treten dabei Überlichtgeschwindigkeit auf. Dies ist ein offenbares Mißverständnis. Sind $x_1 x_2 x_3$ die in bezug auf die rotierende Erde gemessenen Raumkoordinaten, $x_0$ die zugehörige „Zeit" (auf ihre präzise Definition kommt es jetzt nicht an), so werden die Koordinatenlinien $x_0$ auf denen bei konstanten $x_1 x_2 x_3$ nur $x_0$ variiert, nicht alle zeitartige Richtung haben, d. h. es wird in diesen Koordinaten nicht überall $g_{00} > 0$ sein. Nun behauptet Einstein allerdings, daß auch solche Koordinatensysteme zulässig sind; auch in solchen Koordinatensystemen gelten seine allgemein invarianten Gravitationsgesetze. Dagegen hält er durchaus daran fest, daß die *Weltlinie eines materiellen Köpers* stets zeitartige Richtung besitzt, daß an einem materiellen Körper (und als „Signalgeschwindigkeit") keine Überlichtgeschwindigkeit auftreten kann. Ein Koordinatensystem von der oben angegebenen Art läßt sich infolgedessen nicht in seiner ganzen Ausdehnung durch einen „Bezugsmollusken" wiedergeben, d. h. man kann sich kein materielles Medium denken, dessen einzelne Elemente die Koordinatenlinien $x_0$ jenes Koordinatensystems als Weltlinien beschreiben.—

Aber es wird Zeit, daß ich auf den entscheidenden Gegensatz zwischen Lenard und Einstein zu sprechen komme. Lenard behauptet, daß die Einsteinsche Theorie mit *fingierten Gravitationsfeldern* operiere, zu denen sich keine erzeugende Materie nachweisen ließe und welche nur dem Relativitätsprinzip zuliebe eingeführt würden. Das anschauliche Lenardsche Beispiel des durch einen entgegenfahrenden Zug plötzlich gebremsten Eisenbahnzuges diene auch hier als Unterlage der Diskussion. Warum, fragt Lenard, geht der Zug in Trümmer und nicht der Kirchtum neben dem Zug, da doch nach Einstein ebensogut von ihm wie von dem Eisenbahnzug gesagt werden kann, daß er gebremst werde? Hierauf scheint mir die Antwort leicht. In der Einsteinschen Theorie gibt es so gut wie nach alter Auffassung das *Führungsfeld,* dem ein Körper nach dem Galileischen Prinzip folgt, solange auf ihn keine Kräfte wirken. Die Katastrophe ereignet sich am Zuge und nicht am Kirchturm, weil der erstere durch die Molekularkräfte des entgegenfahrenden Zuges aus der Bahn des Führungsfeldes herausgeworfen wird, der Kirchturm hingegen nicht. Diese Antwort ist auch vollkommen im Einklang mit dem „gesunden Menschenverstand", der von Herzen damit einverstanden ist, die sich den Kräften entgegenstemmende Beharrungstendenz des Führungsfeldes mit Einstein als eine physikalische Realität anzusehen. Die Frage ist jetzt aber weiter die: ist dieses Führungsfeld eine Einheit oder lassen sich in ihr zwei Bestandteile, die „Trägheit" und die „Gravitation", grundsätzlich voneinander trennen, derart daß die erste von selber ein für allemal vorhanden ist als affinlineare Struktur der vierdimensionalen Welt und nur die zweite durch die Materie erzeugt wird? Hier, für die Gleichberechtigung aller Bewegungszustände, ist die Sachlage eine ganz analoge wie für die Gleichberechtigung aller Richtungen im Raum. Nach Demokrit gibt es an sich ein absolutes Oben-Unten; die wirkliche Fallrichtung eines Körpers setzt sich zusammen aus dieser absoluten Richtung und einer aus physikalischen Ursachen entspringenden Abweichung davon. Demokrit könnte etwa gegen Newton, der die Fallrichtung als Einheit ansieht, genau so argumentieren wie Lenard gegen Einstein: Macht man eine andere als jene wahre Richtung zur Normalrichtung, so muß man außer ihr und der wirklichen Abweichung drittens noch eine überall gleiche und nicht in der Materie verankerte fingierte Abweichung einführen; und das nur, um dem Prinzip von der Gleichberechtigung aller Richtungen im Raume zu genügen. Sobald man die absolute Richtung Oben-Unten zugibt, kann man scheiden zwischen wirklicher und fingierter Abweichung; sobald man ein ausgezeichnetes, „vernunftgemäßes" Koordinatensystem annimmt, muß man (mit Mie und Lenard) scheiden zwischen wirklichen und fingierten Gravitationsfeldern. Auf dem Relativitätsstandpunkt hingegen wird eine solche Scheidung unmöglich. Wenn wir aber mit Newton gegen Demokrit die Unzerlegbarkeit der wirklichen Fallrichtung in ein absolutes Oben-Unten und eine Abweichung davon behaupten, so müssen wir auch nicht nur für die *Abweichung,* sondern für die *Fallrichtung als Ganzes eine physikalische Ursache* angeben; genau so hat Einstein die Verpflichtung, zu zeigen, *wie und nach welchem Gesetz das Führungsfeld als Ganzes durch die Materie erzeugt wird*. Das verlangt Lenard mit vollem Recht von ihm, und das ist der tiefste und eigentlich entscheidende Punkt seiner Einwände. Es muß unverhohlen zugegeben werden, daß hier für die Relativitätstheorie bei ihrer jetzigen Formulierung noch ernstliche Schwierigkeiten vorliegen. Einstein weist zur Beantwortung auf seine *Kosmologie* der räumlich geschlossenen Welt hin; er erwidert Lenard: Das Feld ist nicht willkürlich erfunden, weil es die allgemeinen Differentialgleichungen erfüllt und weil es zurückgeführt werden kann auf die Wirkung aller fernen Massen. Solange man überhaupt an dem Gegensatz von Materie und Feld festhält (und nur dann ist ja die Forderung, daß die Materie das Feld erzeuge, sinnvoll und berechtigt), bedeutet die Einsteinsche Kosmologie dies, daß neben den inneren Säumen des Feldes, über welche die einzelnen Materieteilchen feldbestimmend herüberwirken, nicht noch ein weiterer unendlichferner Saum als ein das Feld im Unendlichen bestimmendes Agens hinzukommt; an seine Stelle ist die Gesamtheit der fernen Massen getreten. Das Mitdrehen der Ebene des Foucaultschen Pendels mit dem Fixsternhimmel macht das ganz sinnfällig. Behoben ist damit die Schwierigkeit aber noch nicht. Erstens ist zu sagen, daß von Einstein nur die Gesetze angegeben werden, welche den inneren differentiellen Zusammenhang des Feldes binden, aber noch keine klare Formulierung der Gesetze vorliegt, nach welchen die Materie das Feld determiniert (das liegt übrigens beim elektromagnetischen Feld nicht wesentlich anders). Zweitens aber und vor allem ist es sogar ganz ausgeschlossen, daß die Materie das Feld eindeutig bestimmen kann, wenn man als Charakteristika der Materie, wie kaum anders möglich, *Masse, Ladung* und *Bewegungszustand* ansieht. Man kann nämlich in der Welt ein solches Koordinatensystem einführen, daß für die dadurch bewirkte Abbildung der Welt auf einen vierdimensionalen Cartesischen Bildraum nicht nur der Weltkanal *eines* Teilchens, sondern *aller* Teilchen simultan vorgegebene Gestalt annimmt, z. B. alle diese Kanäle vertikale Geraden werden. Im Vergleich zu Mach, dessen Bezugskörper stets ein starrer Körper ist, hat sich bei Einstein das Koordinatensystem so „erweicht", daß es sich simultan den Bewegungen aller Teilchen anschmiegen kann, daß man alle Teilchen zugleich auf Ruhe transformieren kann; es hat also hier nicht einmal einen Sinn mehr, vom *relativen* Bewegungszustand verschiedener Körper gegeneinander zu sprechen. Diese Schwierigkeit hat neuerdings Reichenbächer deutlicher hervorgehoben. [*Footnote:* Schwere und Trägheit, Physik. Zeitschr. 22 (1921), S. 234-243.] Das Prinzip, daß die Materie das Feld erzeuge, wird sich danach nur aufrechterhalten lassen, wenn der Begriff der Bewegung ein dynamisches Moment mit in sich aufnimmt; nicht um den Gegensatz *absolut* oder *relativ,* sondern *kinematisch* oder *dynamisch* dreht es sich bei der Analyse des Bewegungsbegriffs. —

In einer zweiten Sitzung am andern Tage demonstrierte F. P. Liesegang (Düsseldorf) einige treffliche Schaubilder zur Darstellung der Zeitraumverhältnisse in der speziellen Relativitätstheorie, und es verlas H. Dingler (München), wie es schien nur zu formalem Protest gegen die Relativitätstheorie, ohne sich um das Publikum zu kümmern, seine kritischen Bemerkungen zu den Grundlagen der Theorie; es ist sonderbar, daß sich bei Dingler mit seinem an Poincaré orientierten konventionalistischen Standpunkt die dogmatische Halsstarrigkeit des geborenen Apriorikers verbindet. Daß der Tragödie am Schluß das Satyrspiel nicht fehle, entwickelte Hr. Rudolph eine phantastische Äthertheorie mit „Lücken" zwischen fließenden Ätherwänden, Sternfäden usw., mit Hilfe deren er aus Nichts die Sonnenmasse auf eine beliebige Anzahl von Dezimalen genau bestimmte ...

Ich habe hier in freier Weise die Fragen kennzeichnen wollen, die in der Nauheimer Diskussion zur Sprache kamen, nicht aber einen objektiven Bericht über den Verlauf der Sitzung erstatten wollen; für eine gekürzte, aber sinngetreue Wiedergabe der Vorträge und der Diskussion sei der Leser auf das Dezemberheft 1920 der Physikalischen Zeitschrift verwiesen.

(Eingegangen am 29. 8. 21.)"[54]

Bruno Thüring wrote,

"Im selben Jahre 1920 fand in Bad Nauheim auf der dortigen Naturforschertagung die berühmt gewordene Diskussion zwischen Philipp Lenard und Albert Einstein statt. In dieser Diskussion, welche in echt jüdischer Weise zu einer Sensation aufgebauscht wurde, verglich Einstein sein Werk mit demjenigen Galileis und tat, als sich Lenard auf den gesunden Menschverstand berief, die Äußerung, daß es gefährlich sei, den gesunden

---

[54]. H. Weyl, "Die Relativitätstheorie auf der Naturforscherversammlung in Bad Nauheim", *Jahresbericht der Deutschen Mathematiker-Vereinigung*, Volume 31, (1922), pp. 51-63.

Menschenverstand in der Physik zur Anwendung zu bringen. Diese seltsame Argumentation ist dann auch in die populärwissenschaftliche Literatur eingegangen.

Im übrigen kam es bei dieser Tagung auch zu tumultuarischen Szenen. Der Vorsitzende Max Planck sah es als seine Hauptaufgabe an, die Einsteinpartei gegen ihre wissenschaftlichen Gegner möglichst gleich durch organisatorische Maßnahmen zu schützen. Er ließ, wie aus Presseveröffentlichungen hervorgeht, an der Eingangstüre eine Siebung vornehmen, um ihm nicht genehme Personen fernzuhalten. Darauf erhob sich zwar ein großer Tumult, und das empörte Auditorium stürmte den Saal. Planck erreichte seinen Zweck schließlich dadurch, daß er die Relativisten in stundenlangen Vorträgen sich verbreiten ließ, während den antirelativistischen Rednern einschließlich Diskussion insgesamt nur 15 Minuten zugebilligt werden sollten. Unter den Rednern dieser Tagung befand sich auch der im Kampf gegen Einstein an vorderster Stelle stehende Hugo Dingler.

Freilich erlag die Opposition gegen den relativistischen Wissenschaftsbetrieb in der Folgezeit der Übermacht der jüdischen Pressepropaganda und der staatlichen Schutzmaßnahmen. Bald wurde Einsteins Lehre als eine „Selbstverständlichkeit" bezeichnet, und die maßgebenden Männer der internationalen Gelehrtenrepublik hielten nach Möglichkeit jeden von einem Lehrstuhl fern, der sich gegen das relativistische Dogma — sei es auch in der wissenschaftlich-sachlichsten Weise — ausgesprochen hatte. So wurden diese Dogmatismen an die junge Physikergeneration so gut wie widerspruchslos weitergegeben."[55]

## 4.5 Einstein the Genocidal Racist

Albert Einstein was himself a racist; and, therefore, a hypocrite when criticizing the racism of others. John Stachel wrote,

> "While he lived in Germany, however, Einstein seems to have accepted the then-prevalent racist mode of thought, often invoking such concepts as 'race' and 'instinct,' and the idea that the Jews form a race."[56]

On 8 July 1901, Einstein wrote to Winteler,

> "There is no exaggeration in what you said about the German professors. I

---

[55]. B. Thüring, "Albert Einsteins Umsturzversuch der Physik und seine inneren Möglichkeiten und Ursachen", *Forschungen zur Judenfrage*, Volume 4, (1940), pp. 134-162, at 159. Republished as: *Albert Einsteins Umsturzversuch der Physik und seine inneren Möglichkeiten und Ursachen*, Dr. Georg Lüttke Verlag, Berlin, (1941), pp. 59-60.
[56]. J. Stachel, "Einstein's Jewish Identity", *Einstein from 'B' to 'Z'*, Birkhäuser, Boston, Basel, Berlin, (2002), pp. 57-83, at 68.

have got to know another sad specimen of this kind — one of the foremost physicists of Germany."⁵⁷

Einstein wrote to Besso sometime after 1 January 1914,

"A free, unprejudiced look is not at all characteristic of the (adult) Germans (blinders!)."⁵⁸

After the war Einstein and some of his friends alluded to much earlier conversations with Einstein, where he had correctly predicted the eventual outcome of the war. In his diaries, Romain Rolland recorded his conversations with Einstein in Switzerland at their meeting of 16 September 1915,

"What I hear from [Einstein] is not exactly encouraging, for it shows the impossibility of arriving at a lasting peace with Germany without first totally crushing it. Einstein says the situation looks to him far less favorable than a few months back. The victories over Russia have reawakened German arrogance and appetite. The word 'greedy' seems to Einstein best to characterize Germany. [***] Einstein does not expect any renewal of Germany out of itself; it lacks the energy for it, and the boldness for initiative. He hopes for a victory of the Allies, which would smash the power of Prussia and the dynasty... . Einstein and Zangger dream of a divided Germany—on the one side Southern Germany and Austria, on the other side Prussia. [***] We speak of the deliberate blindness and the lack of psychology in the Germans."⁵⁹

Jews often sought to Balkanize nations so as to weaken the power of any faction within a nation and to created perpetual agitation between the nations which could be exploited for profit and other Jewish gains. For example, the Rothschilds created the American Civil War and profited from the debts it generated. They hoped to divide America into two nations and to pit these against one another. They were successful. Jews had long been pitting North German Protestants against South German and Austrian Catholics. Jews were the motive force behind the *Kulturkampf*. After creating these divides and

---

57. A. Einstein to J. Winteler, English translation by A. Beck, *The Collected Papers of Albert Einstein*, Volume 1, Document 115, Princeton University Press, (1987), pp. 176-177, at 177.
58. A. Einstein, English translation by A. Beck, *The Collected Papers of Albert Einstein*, Volume 5, Document 499, Princeton University Press, (1995), pp. 373-374, at 374.
59. R. Romain, *La Conscience de l'Europe*, Volume 1, pp. 696ff. English translation from A. Fölsing, *Albert Einstein: A Biography*, Viking, New York, (1997), pp. 365-367. **See also:** Letter from A. Einstein to R. Romain of 15 September 1915, *The Collected Papers of Albert Einstein*, Volume 8, Document 118, Princeton University Press, (1998); **and** Letter from A. Einstein to R. Romain of 22 August 1917, *The Collected Papers of Albert Einstein*, Volume 8, Document 374, Princeton University Press, (1998).

promoting perpetual agitations amongst neighbors, Jewry could then fund one side against the other to destroy it whenever Jewry decided to wreck a given nation.

Einstein's dreams during the First World War remind one of the "Carthaginian Peace" of the Henry Morgenthau, Jr. plan for the destruction of Germany following the Second World War. Morgenthau worked with Lord Cherwell (Frederick Alexander Lindemann), Churchill's friend and advisor, who planned to bomb German civilian populations into submission. Lindemann studied under Einstein's friend, Walther Nernst, who worked with Fritz Haber, a Jewish developer of poisonous gas. James Bacque argues that the Allies, under the direction of General Eisenhower, starved hundreds of thousands, if not millions of German prisoners of war to death. Dwight David Eisenhower was called "the terrible Swedish-Jew" in his yearbook for West Point, *The 1915 Howitzer*, West Point, New York, (1915), p. 80. He was also called "Ike", as in... Eisenhower? The Soviets also abused and murdered countless German POW's after the Second World War.[60]

Einstein often spoke in genocidal and racist terms against Germany, and for the Jews and England, and he betrayed Germany before, during and after the war. Einstein wrote to Paul Ehrenfest on 22 March 1919,

> "[The Allied Powers] whose victory during the war I had felt would be by far the lesser evil are now proving to be *only slightly* the lesser evil. [***] I get most joy from the emergence of the Jewish state in Palestine. It does seem to me that our kinfolk really are more sympathetic (at least less brutal) than these horrid Europeans. Perhaps things can only improve if only the Chinese are left, who refer to all Europeans with the collective noun 'bandits.'"[61]

While responsible people were trying to preserve some sanity in the turbulent period following World War I, Zionists like Albert Einstein sought to validate and encourage the racism of anti-Semites. The Dreyfus Affair taught them that anti-Semitism had a powerful effect to unite Jews around the world. The Zionists were afraid that the "Jewish race" was disappearing through assimilation. They wanted to use anti-Semitism to force the segregation of Jews from Gentiles and to unite Jews, and thereby preserve the "Jewish race". They hoped that if they put a Hitler-type into power—as Zionists had done in the past, they could use him to herd up the Jews and force the Jews into Palestine against their will. This would also help the Zionists to inspire distrust and contempt for Gentile government, while giving the Zionists the moral high-ground in international affairs, despite the fact that the Zionists were secretly behind the

---

**60**. J. Bacque, *Other Losses: An Investigation into the Mass Deaths of German Prisoners at the Hands of the French and Americans after World War II*, Stoddart, Toronto, (1989).
**61**. Letter from A. Einstein to Paul Ehrenfest of 22 March 1919, English translation by A. Hentschel, *The Collected Papers of Albert Einstein*, Volume 9, Document 10, Princeton Univsersity Press, (2004), pp. 9-10, at 10.

atrocities. In 1896, Theodor Herzl wrote his book *The Jewish State*,

> "Great exertions will not be necessary to spur on the movement. Anti-Semites provide the requisite impetus. They need only do what they did before, and then they will create a love of emigration where it did not previously exist, and strengthen it where it existed before. [\*\*\*] I imagine that Governments will, either voluntarily or under pressure from the Anti-Semites, pay certain attention to this scheme; and they may perhaps actually receive it here and there with a sympathy which they will also show to the Society of Jews."[62]

Albert Einstein wrote to Max Born on 9 November 1919. Einstein encouraged anti-Semitism and advocated segregation (one must wonder what rôle Albert's increasing racism played in his divorce from Mileva Marić—a Gentile Serb),

> "Antisemitism must be seen as a real thing, based on true hereditary qualities, even if for us Jews it is often unpleasant. I could well imagine that I myself would choose a Jew as my companion, given the choice. On the other hand I would consider it reasonable for the Jews themselves to collect the money to support Jewish research workers outside the universities and to provide them with teaching opportunities."[63]

In 1933, the Zionists publicly declared their allegiance to the Nazis. They wrote in the *Jüdische Rundshau* on 13 June 1933,

> "Zionism recognizes the existence of the Jewish question and wants to solve it in a generous and constructive manner. For this purpose, it wants to enlist the aid of all peoples; those who are friendly to the Jews as well as those who are hostile to them, since according to its conception, this is not a question of sentimentality, but one dealing with a real problem in whose solution all peoples are interested."[64]

On 21 June 1933, the Zionists issued a declaration of their position with respect to the Nazi régime, in which they expressed a belief in the legitimacy of the Nazis' racist belief system and condemned anti-Fascist forces.[65]

---

[62]. T. Herzl, *A Jewish State: An Attempt at a Modern Solution of the Jewish Question*, The Maccabæan Publishing Co., New York, (1904), pp. 68, 93.
[63]. M. Born, *The Born-Einstein Letters*, Walker and Company, New York, (1971), p. 16.
[64]. English translation in: K. Polkehn, "The Secret Contacts: Zionism and Nazi Germany, 1933-1941", *Journal of Palestine Studies*, Volume 5, Number 3/4, (Spring-Summer, 1976), pp. 54-82, at 59.
[65]. L. S. Dawidowicz, "The Zionist Federation of Germany Addresses the New German State", *A Holocaust Reader*, Behrman House, Inc., West Orange, New Jersey, (1976), pp. 150-155. ***See also:*** H. Tramer, Editor, S. Moses, *In zwei Welten: Siegfried Moses zum*

Michele Besso wrote that it might have been Albert Einstein's racism and bigotry which caused him to separate from his first wife Mileva Marić in 1914. Besso wrote to Einstein on 17 January 1928,

> "[... ]perhaps it is due in part to me, with my defense of Judaism and the Jewish family, that your family life took the turn that it did, and that I had to bring Mileva from Berlin to Zurich[.]"[66]

The hypocrisy of racist Zionists often manifested itself. As another example, consider the fact that racist Zionist Moses Hess was married to a Christian Gentile prostitute named Sybille Pritsch.

Einstein may have been affected by his mother's early racist opposition to his relationship with Marić. Another factor in the Einsteins' divorce was, of course, Albert's incestuous relationship with his cousin Else Einstein, and his desire to bed her daughters, as well as Albert's general promiscuity—some believe he was a whore monger. Albert Einstein opposed his sister Maja's marriage to the Gentile Paul Winteler on racist grounds and thought they should divorce. Albert Einstein wrote to Michele Besso on 12 December 1919 and stated that, "No mixed marriages are any good (Anna says: oh!)"[67] Besso, himself, was married to a Gentile, Anna Besso-Winteler. Denis Brian wrote,

> "When asked what he thought of Jews marrying non-Jews, which, of course, had been the case with him and Mileva, [Albert Einstein] replied with a laugh, 'It's dangerous, but then all marriages are dangerous.'"[68]

On 3 April 1920, Einstein wrote, criticizing assimilationist Jews,

> "And this is precisely what he does *not* want to reveal in his confession. He talks about religious faith instead of tribal affiliation, of 'Mosaic' instead of 'Jewish' because the latter term, which is much more familiar to him, would emphasize affiliation to his tribe."[69]

After declaring that Jewish children segregate due to natural forces and that they

---

*fünfundsiebzigsten Geburtstag*, Verlag Bitaon, Tel-Aviv, (1962), pp. 118.ff; cited in K. Polkehn, "The Secret Contacts: Zionism and Nazi Germany, 1933-1941", *Journal of Palestine Studies*, Volume 5, Number 3/4, (Spring-Summer, 1976), pp. 54-82, at 59.

**66**. English translation quoted from J. Stachel, "Einstein's Jewish Identity", *Einstein from 'B' to 'Z'*, Birkhäuser, Boston, Basel, Berlin, (2002), pp. 57-83, at 78. Stachel cites M. Besso, A. Einstein, *Correspondance, 1903-1955*, Hermann, Paris, (1972), p. 238.

**67**. Letter from A. Einstein to M. Besso of 12 December 1919, English translation by A. Hentschel, *The Collected Papers of Albert Einstein*, Volume 9, Document 207, Princeton University Press, (2004), pp. 178-179, at 179.

**68**. D. Brian, *The Unexpected Einstein: The Real Man Behind the Icon*, Wiley, Hoboken, New Jersey, (2005), p. 42.

**69**. A. Einstein, English translation by A. Engel, *The Collected Papers of Albert Einstein*, Volume 7, Document 34, Princeton University Press, (2002), pp. 153-155, at 153.

are "different from other children",[70] not due to religion or tradition, but due to genetic features and "heritage", Einstein continued his 3 April 1920 statement,

> "With adults it is quite similar as with children. Due to race and temperament as well as traditions (which are only to a small extent of religious origin) they form a community more or less separate from non-Jews. [***] It is this basic community of race and tradition that I have in mind when I speak of 'Jewish nationality.' In my opinion, aversion to Jews is simply based upon the fact that Jews and non-Jews are different. [***] Where feelings are sufficiently vivid there is no shortage of reasons; and the feeling of aversion toward people of a foreign race with whom one has, more or less, to share daily life will emerge by necessity."[71]

Einstein made similar comments in a document dated sometime "after 3 April 1920". Einstein was in agreement with Philipp Lenard that a "Jewish heritage" (read for "heritage", "racial instinct") could be seen in intellectual works published by Jews. Einstein stated,

> "The psychological root of anti-Semitism lies in the fact that the Jews are a group of people unto themselves. Their Jewishness is visible in their physical appearance, and one notices their Jewish heritage in their intellectual works, and one can sense that there are among them deep connections in their disposition and numerous possibilities of communicating that are based on the same way of thinking and of feeling. The Jewish child is already aware of these differences as soon as it starts school. Jewish children feel the resentment that grows out of an instinctive suspicion of their strangeness that naturally is often met with a closing of the ranks. [***] [Jews] are the target of instinctive resentment because they are of a different tribe than the majority of the population."[72]

Albert Einstein often referred to Jews as "tribesmen" and Jewry as the "tribe". Fellow German Jew Fritz Haber was outraged at Albert Einstein's racist treachery and disloyalty. Einstein confirmed that he was disloyal and a racist, and was obligated,

> "[... ] to step in for my persecuted and morally depressed fellow tribesmen, as far as this lies within my power[.]"[73]

---

[70]. A. Einstein, English translation by A. Engel, *The Collected Papers of Albert Einstein*, Volume 7, Document 34, Princeton University Press, (2002), pp. 153-155, at 153.
[71]. A. Einstein, English translation by A. Engel, *The Collected Papers of Albert Einstein*, Volume 7, Document 34, Princeton University Press, (2002), pp. 153-155, at 153-154.
[72]. A. Einstein, English translation by A. Engel, *The Collected Papers of Albert Einstein*, Volume 7, Document 35, Princeton University Press, (2002), pp. 156-157.
[73]. A. Einstein quoted in: H. Gutfreund, "Albert Einstein and the Hebrew University", J. Renn, Editor, *Albert Einstein Chief Engineer of the Universe: One Hundred Authors for*

Einstein bore no such loyalty to Germans, who had feed him and made him famous. In fact, Einstein wanted to exterminate the Germans.

In a draft letter of 3 April 1920, Einstein wrote that children are conscious of "racial characteristics" and that this alleged "racial" gulf between children results in conflicts, which instill a sense of foreigness in the persecuted child. Einstein wrote,

> "Unter den Kindern war besonders in der Volksschule der Antisemitismus lebendig. Er gründete ich auf die den Kindern merkwürdig bewussten Rassenmerkmale und auf Eindrücke im Religionsunterricht. Thätliche Angriffe und Beschimpfungen auf dem Schulwege waren häufig, aber meist nicht gar zu bösartig. Sie genügten immerhin, um ein lebhaftes Gefühl des Fremdseins schon im Kinde zu befestigen."[74]

Einstein's racism was perhaps a defense mechanism to depersonalize the attacks he faced as a child and to counter the hurt with a sense of communal love and communal hatred, which was sponsored by his racist mother. Like Adolf Stoecker before him,[75] Albert Einstein advocated the segregation of Jewish students. Peter A. Bucky quoted Albert Einstein,

> "I think that Jewish students should have their own student societies. [***] One way that it won't be solved is for Jewish people to take on Christian fashions and manners. [***] In this way, it is entirely possible to be a civilized person, a good citizen, and at the same time be a faithful Jew who loves his race and honors his fathers."[76]

Einstein stated,

> "We must be conscious of our alien race and draw the logical conclusions from it. [***] We must have our own students' societies and adopt an attitude of courteous but consistent reserve to the Gentiles. [***] It is possible to be [***] a faithful Jew who loves his race and honours his fathers."[77]

---

*Einstein*, Wiley-VCH, Berlin, (2005), pp. 314-318, at 316.

[74]. Letter from A. Einstein to P. Nathan of 3 April 1920, *The Collected Papers of Albert Einstein*, Volume 9, Document 366, Princeton University Press, (2004), p. 492. Also: *The Collected Papers of Albert Einstein*, Volume 1, Princeton University Press, (1987), p. *lx*, note 44.

[75]. P. W. Massing, *Rehearsal for Destruction: A Study of Political Anti-Semitism in Imperial Germany*, Howard Fertig, New York, (1967), pp. 278-294.

[76]. P. A. Bucky, Einstein, and A. G. Weakland, *The Private Albert Einstein*, Andrews and McMeel, Kansas City, (1992), p. 88.

[77]. A. Einstein, *The World As I See It*, Citadel, New York, (1993), pp. 107-108.

On 5 April 1920, Einstein repeated what he had heard from his political Zionist friends who believed that anti-Semitism was necessary to the preservation of the "Jewish race",

> "Anti-Semitism will be a psychological phenomenon as long as Jews come in contact with non-Jews—what harm can there be in that? Perhaps it is due to anti-Semitism that we survive as a race: at least that is what I believe."[78]

and,

> "I am neither a German citizen, nor is there in me anything that can be described as 'Jewish faith.' But I am happy to belong to the Jewish people, even though I don't regard them as the Chosen People. Why don't we just let the Goy keep his anti-Semitism, while we preserve our love for the likes of us?"[79]

This letter was published in the *Israelitisches Wochenblatt für die Schweiz*, on 24 September 1920, on page 10. It became famous and was widely discussed in newspapers and was used as a political issue. Einstein's racism had already become a weapon for Jewish critics to wield against German Jews who were loyal to the Fatherland. Einstein ridiculed the *Central-Verein deutscher Staatsbürger jüdischen Glaubens*, an organization that combated anti-Semitism and vigorously defended and celebrated Jews, because Einstein sought to promote anti-Semitism and because Einstein believed that being "Jewish" was a racial, not a religious, state. Einstein knew quite well that the letter had been published. The *C. V.* contacted him about it and published a statement regarding it in their periodical *Im deutschen Reich* in March of 1921,

> "So wurde auch in einzelnen Versammlungen der bekannte Brief des Naturforschers Professor Einstein, den dieser an den Central-Verein gerichtet hat, und in welchem er die Bestrebungen des Central-Vereins ablehnt, weil sie zu national-deutsch und zu wenig jüdisch orientiert seien, zum Gegenstand der Erörterungen gemacht. Dieser Brief hat in der öffentlichen Erörterung der jüdischen und judengegnerischen Presse in den letzten Monaten und auch bei den Wahlen eine gewisse Rolle gespielt und Anlaß zu den verschiedenartigsten Betrachtungen je nach der Parteistellung

---

[78]. A. Einstein, English translation by A. Engel, *The Collected Papers of Albert Einstein*, Volume 7, Document 37, Princeton University Press, (2002), p. 159.

[79]. A. Einstein quoted in A. Fölsing, English translation by E. Osers, *Albert Einstein, a Biography*, Viking, New York, (1997), p. 494; which cites speech to the *Central-Verein Deutscher Staatsbürger Jüdischen Glaubens*, in Berlin on 5 April 1920, in D. Reichenstein, *Albert Einstein. Sein Lebensbild und seine Weltanschauung*, Berlin, (1932). This letter from Einstein to the Central Association of German Citizens of the Jewish Faith of 5 April 1920 is reproduced in *The Collected Papers of Albert Einstein*, Volume 9, Document 368, Princeton University Press, (2004).

der Versammlungsredner und der verschiedenen Zeitungen gegeben. So hat sich z. B. die jüdisch-nationale „Wiener Morgenzeitung" veranlaßt gesehen, den Central-Verein in wenig vornehmer Weise anzugreifen und ihn wegen seines nationaldeutschen Standpunktes zu verdächtigen. Diese Angriffe würden durch die Auffassung von Professor Einstein nicht gedeckt worden sein, wenn die „Wiener Morgenzeitung" gewußt hätte, daß Professor Einstein ohne nähere Kenntnis der Bestrebungen und der Arbeit des Central-Vereins seinen Brief geschrieben und keineswegs an eine Veröffentlichung, die nur durch eine Indiskretion erfolgt ist, gedacht hat. Erst nach der Veröffentlichung hat er von der Art und Weise der Tätigkeit des Central-Vereins Kenntnis erhalten und hat, wie mit gutem Grund versichert werden kann, infolge dieser Kenntnis eine wesentlich andere Auffassung vom Werte der Arbeit unseres Central-Vereins gewonnen. Auch dieser Vorfall sollte Anlaß geben, Urteile in der Oeffentlichkeit erst dann zu fällen, wenn die Sachlage einigermaßen geklärt ist."[80]

On 24 May 1931, the *Sunday Express* of London published an interview it claimed it had had with Einstein while he was visiting Oxford. The interview contained inflammatory statements similar to those published in the *Israelitisches Wochenblatt für die Schweiz* on 24 September 1920. These statements were repeated in several German language newspapers across Europe together with scathing editorial indictments of Einstein. Einstein claimed that no interview had taken place and the quotations were taken from a letter he had written eleven years prior. Einstein stated in a letter to Michael Traub of 22 August 1931 that this letter had never been published,[81] though it had been published and Einstein knew quite well that it had been published.

Einstein accused the *Central-Verein deutscher Staatsbürger jüdischen Glaubens e. V.* of instigating the "forgery". The C.V. denied that it was behind the publication in the *Sunday Express* and invited Einstein to respond in their official organ the *Central-Verein Zeitung*. Einstein took the opportunity and stated, "Es wurden mir schon wiederholt Auszüge aus einem Artikel der „Sunday Expreß" zugesandt, aus denen ich ersehe, daß es sich um eine glatte Fälschung handelt. Ich habe in Oxford überhaupt kein einziges Zeitungsinterview gegeben. Der Inhalt ist eine böswillige Entstellung eines vor elf Jahren geschriebenen, nicht für die Oeffentlichkeit bestimmten Briefes."[82] He affirmed in 1931 that he had made the statements in 1920 and did not repudiate them.

In 1932, Einstein stated, referring to the "deplorably high development of nationalism everywhere"—his own rabid Zionism apparently excepted,

---

[80]. "Zeitschau", *Im deutschen Reich*, Volume 27, Number 3, (March, 1921), pp. 90-97, at 92.
[81]. D. K. Buchwald, *et al.*, Editors, *The Collected Papers of Albert Einstein*, Volume 7, Document 37, Princeton University Press, (2002), p. 304, note 8.
[82]. "Professor Einstein erklärt das „Sunday Expreß"-Interview für gefälscht", *Central-Verein Zeitung*, Volume 10, Number 37, (11 September 1931), p. 443.

"The introduction of compulsory service is therefore, to my mind, the prime cause of the moral collapse of the white race, which seriously threatens not merely the survival of our civilization but our very existence. This curse, along with great social blessings, started with the French Revolution, and before long dragged all the other nations in its train."[83]

Einstein had a reputation as a rabid anti-assimilationist—here again Einstein merely parroted the racist anti-assmilationism of his Zionist predecessors, like Solomon Schechter who dreaded assimilation more than pogroms—and Zionists encouraged pogroms in order to discourage assimilation.

Zionists were by no means alone in the anti-assimilationist panic that struck the western world at the end of the Nineteenth Century. In 1906, Chaim Weizmann had persuaded Arthur James Balfour to become a racist Zionist.[84] In 1908, Balfour published a racist and nationalistic lecture on the subject of race degeneration and stagnation called *Decadence*.[85] In America, Theodore Roosevelt had an enduring interest in racial questions and feared "racial suicide" and the decline of a race like the decline of an organism in old age.[86] On 5 March 1908, Roosevelt wrote to Balfour, later signatory of the Balfour Declaration,

"Most emphatically there is such a thing as 'decadence' of a nation, a race, a type; and it is no less true that we cannot give any adequate explanation of the phenomenon. Of course there are many partial explanations, and in some cases, as with the decay of the Mongol or Turkish monarchies, the sum of these partial explanations may represent the whole. But there are other cases, notably, of course, that of Rome in the ancient world, and, as I believe, that of Spain in the modern world, on a much smaller scale, where the sum of all the explanations is that they do not wholly explain. Something seems to have gone out of the people or peoples affected, and what it is no one can say."[87]

*The London Times* wrote on 12 February 1919 on page 9, confirming that Balfour's Declaration was based on precisely the same racist myths of "Blut und Boden" the Nazis would later assert to justify the racism of Nazi Germany,

---

**83**. A. Einstein, translated by A. Harris, "The Disarmament Conference of 1932. I." *The World As I See It*, Citadel, New York, (1993), pp. 59-60.
**84**. "Mr. Balfour on Zionism", *The London Times*, (12 February 1919), p. 9.
**85**. Arthur James Balfour, Earl of Balfour, *Decadence: Henry Sidgwick Memorial Lecture*, Cambridge, University Press, (1908).
**86**. T. G. Dyer, *Theodore Roosevelt and the Idea of Race*, Louisiana State University Press, Baton Rouge, (1992).
**87**. *The Works of Theodore Roosevelt*, Volume 24, Memorial Edition, C. Scribner's Sons, New York, (1923-1926), p. 122. J. B. Bishop, *Theodore Roosevelt and His Time Shown in His Own Letters*, Volume 2, Charles Scribner's Sons, New York, (1920), pp. 104-110, at 105.

## "MR. BALFOUR ON ZIONISM.
### THE CASE FOR A NATIONAL HOME.

Mr. Balfour, in whose hands has been placed the interests of Palestinian Jewry at the Peace Conference, has written a preface to the History of Zionism, shortly to be published from the pen of M. Sokolow, one of the four leaders of the Zionist Executive Committee.

Mr. Balfour says that convinced by conversations with Dr. Weizmann in January, 1906, that if a home was to be found for the Jewish people, homeless now for nearly 1900 years, it was vain to seek it anywhere but in Palestine. Answering the question why local sentiment is to be more considered in the case of the Jew than (say) in that of the Christian or the Buddhist, Mr. Balfour says:—'The answer is, that the cases are not parallel. The position of the Jews is unique. For them race, religion, and country are interrelated, as they are interrelated in the case of no other race, no other religion, and no other country on earth. By a strange and most unhappy fate it is this people of all others which, retaining to the full its racial self-consciousness, has been severed from its home, has wandered into all lands and has nowhere been able to create for itself an organized social commonwealth. Only Zionism—so at least Zionists believe—can provide some mitigation of this great tragedy.

'Doubtless there are difficulties, doubtless there are objections—great difficulties, very real objections... . Yet no one can reasonably doubt that if, as I believe, Zionism can be developed into a working scheme, the benefit it would bring to the Jewish people, especially perhaps to that section of it which most deserves our pity, would be great and lasting.'

The criticism that the Jews use their gifts to exploit for personal ends a civilization which they have not created, in communities they do little to maintain, Mr. Balfour declares to be false. He admits, however, that in large parts of Europe their loyalty to the State in which they dwell is (to put it mildly) feeble compared with their loyalty to their religion and their race. How, indeed, could it be otherwise? he asks. 'In none of the regions of which I speak have they been given the advantages of equal citizenship; in some they have been given no right of citizenship at all.'

'It seems evident that Zionism will mitigate the lot and elevate the status of no negligible fraction of the Jewish race. Those who go to Palestine will not be like those who now migrate to London or New York... . They will go in order to join a civil community which completely harmonizes with their historical and religious sentiments; a community bound to the land it inhabits by something deeper even than custom; a community whose members will suffer from no divided loyalty nor any temptation to hate the laws under which they are forced to live. To them the material gain should be great; but surely the spiritual gain will be greater still.'

Mr. Balfour goes on to consider the position of those, though Jews by descent, and often by religion, who desire wholly to identify themselves with the life of the country wherein they have made their home, many of them distinguished in art, medicine, politics, and law. 'Many of this class,' he says, 'look with a certain measure of suspicion and even dislike upon the Zionist movement. They fear that it will adversely affect their position in the country of their adoption. The great majority of them have no desire to settle in Palestine. Even supposing a Zionist community were established, they would not join it... .

'I cannot share these fears. I do not deny that, in some countries where legal equality is firmly established, Jews may still be regarded with a certain measure of prejudice. But this prejudice, where it exists, is not due to Zionism, nor will Zionism embitter it. The tendency should surely be the other way. Everything which assimilates the national and international status of the Jews to that of other races ought to mitigate what remains of ancient antipathies; and evidently this assimilation would be promoted by giving them that which all other nations possess—a local habitation and a national home."

Others repeated Theodor Herzl's theme, that Jews could not assimilate, because the presence of Jews in a host nation ultimately led to anti-Semitism due to Jewish parasitism—according to Herzl. Hilaire Belloc was a strong advocate of the view that Jews should not integrate. Belloc published a book on the subject entitled *The Jews* in 1922, and expressed similar convictions in *G. K.'s Weekly* in the 1930's. Belloc wrote biographies of men who had fallen under the influence of Zionists, like Oliver Cromwell and Napoleon. Belloc, however, was strongly opposed to Nazism. Douglas Reed took a similar Zionist stance on the alleged unassimilability of Jews in the late 1930's,[88] though he later opposed Zionism.

Racist Zionist Solomon Schecter stated, in harmony with numerous political Zionists, though in opposition to the vast majority of Jews,

> "It is this kind of assimilation [the death of a "race" through integration], with the terrible consequences indicated, that I dread most; even more than pogroms."[89]

On 15 March 1921, Kurt Blumenfeld wrote to Chaim Weizmann,

> "Einstein [***] is interested in our cause most strongly because of his revulsion from assimilatory Jewry."[90]

Einstein stated in 1921,

---

[88]. D. Reed, *Disgrace Abounding*, Jonathan Cape, London, (1939).
[89]. S. Schechter, *Zionism: A Statement*, Federation of American Zionists, New York, (1906); reprinted in the relevant part in A. Hertzberg, *The Zionist Idea*, Harper Torchbooks, New York, (1959), p. 507.
[90]. J. Stachel, *Einstein from 'B' to 'Z'*, Birkhäuser, Boston, (2002), p. 79, note 41.

"To deny the Jew's nationality in the Diaspora is, indeed, deplorable. If one adopts the point of view of confining Jewish ethnical nationalism to Palestine, then one, to all intents and purposes, denies the existence of a Jewish people. In that case one should have the courage to carry through, in the quickest and most complete manner, entire assimilation. We live in a time of intense and perhaps exaggerated nationalism. But my Zionism does not exclude in me cosmopolitan views. I believe in the actuality of Jewish nationality, and I believe that every Jew has duties towards his coreligionists. [***] [T]he principal point is that Zionism must tend to strengthen the dignity and self-respect of the Jews in the Diaspora. I have always been annoyed by the undignified assimilationist cravings and strivings which I have observed in so many of my friends."[91]

In 1921, Einstein declared, referring to Eastern European Jews, "These men and women retain a healthy national feeling; it has not yet been destroyed by the process of atomisation and dispersion."[92]

Einstein wrote in the *Jüdische Rundschau*, on 21 June 1921, on pages 351-352,

"This phenomenon [*i. e.* Anti-Semitism] in Germany is due to several causes. Partly it originates in the fact that the Jews there exercise an influence over the intellectual life of the German people altogether out of proportion to their number. While, in my opinion, the economic position of the German Jews is very much overrated, the influence of Jews on the Press, in literature, and in science in Germany is very marked, as must be apparent to even the most superficial observer. This accounts for the fact that there are many anti-Semites there who are not really anti-Semitic in the sense of being Jew-haters, and who are honest in their arguments. They regard Jews as of a nationality different from the German, and therefore are alarmed at the increasing Jewish influence on their national entity. [***] But in Germany the judgement of my theory depended on the party politics of the Press[.]"[93]

Einstein also stated,

---

**91**. A. Einstein, "Jewish Nationalism and Anti-Semitism", *The Jewish Chronicle*, (17 June 1921), p. 16.
**92**. J. Stachel, "Einstein's Jewish Identity", *Einstein from 'B' to 'Z'*, Birkhäuser, Boston, (2002), p. 65. Stachel cites, *About Zionism: Speeches and Letters*, Macmillan, New York, (1931), pp. 48-49. For Zionist Ha-Am's use of the image of atomisation and dispersion, *see:* A. Hertzberg, *The Zionist Idea*, Harper Torchbooks, New York, (1959), p. 276.
**93**. A. Einstein, "Jewish Nationalism and Anti-Semitism", *The Jewish Chronicle*, (17 June 1921), p. 16.

"The way I see it, the fact of the Jews' racial peculiarity will necessarily influence their social relations with non-Jews. The conclusions which—in my opinion—the Jews should draw is to become more aware of their peculiarity in their social way of life and to recognize their own cultural contributions. First of all, they would have to show a certain noble reservedness and not be so eager to mix socially—of which others want little or nothing. On the other hand, anti-Semitism in Germany also has consequences that, from a Jewish point of view, should be welcomed. I believe German Jewry owes its continued existence to anti-Semitism."[94]

Nazi Zionist Joseph Goebbels, sounding very much like political Zionist Albert Einstein, was quoted in *The New York Times*, on 29 September 1933, on page 10,

"It must be remembered the Jews of Germany were exercising at that time a decisive influence on the whole intellectual life; that they were absolute and unlimited masters of the press, literature, the theatre and the motion pictures, and in large cities such as Berlin, 75 percent of the members of the medical and legal professions were Jews; that they made public opinion, exercised a decisive influence on the Stock Exchange and were the rulers of Parliament and its parties."

On 1 July 1921, Einstein was quoted in the *Jüdische Rundshau* on page 371,

"Let us take brief look at the *development of German Jews* over the last hundred years. With few exceptions, one hundred years ago our forefathers still lived in the Ghetto. They were poor and separated from the Gentiles by a wall of religious tradition, secular lifestyles and statutory confinement and were confined in their spiritual development to their own literature, only relatively weakly influenced by the forceful progress which intellectual life in Europe had undergone in the Renaissance. However, these little noticed, modestly living people had one thing over us: *Every one of them belonged with all his heart to a community*, into which he was incorporated, in which he felt a worthwhile member, in which nothing was asked of him which conflicted with his normal processes of thought. Our forefathers of that era were pretty pathetic both bodily and spiritually, but—in social relations— in an enviable state of mental equilibrium. Then came emancipation. It offered undreamt of opportunities for advancement. The isolated individual quickly found their way into the upper financial and social circles of society. They eagerly absorbed the great achievements of art and science which the Occidentals[95] had created. They contributed to the development with

---

**94**. A. Einstein, A. Engel translator, "How I became a Zionist", *The Collected Papers of Albert Einstein*, Volume 7, Document 57, Princeton University Press, (2002), pp. 234-235, at 235.
**95**. At the time Einstein made his statement, Jews and Gentiles often referred to Jews as

passionate affection, and themselves made contributions of lasting value. They thereby took on the lifestyle of the Gentile world, turning away from their religious and social traditions in growing masses—took on Gentile customs, manners and mentality. It appeared as if they were being completely dissolved into the numerically superior, politically and culturally better organized host peoples, such that no trace of them would be left after a few generations. The complete eradication of the Jewish nationality in Middle and Western Europe appeared to be inevitable. However, it didn't turn out that way. It appears that racially distinct nations have instincts which work against interbreeding. The adaptation of the Jews to the European peoples among whom they have lived in language, customs and indeed even partially in religious practices *was unable to eliminate all feelings of foreignness* which exist between Jews and their European host peoples. In short, this spontaneous feeling of foreignness is ultimately due to a loss of energy.[96] For this reason, *not even well-meant arguments can eradicate it*. Nationalities do not want to be mixed together, rather they want to go their own separate ways. A state of peace can only be achieved by mutual tolerance and respect."

Einstein stated that Jews should not participate in the German Government,

"I regretted the fact that [Rathenau] became a Minister. In view of the attitude which large numbers of the educated classes in Germany assume towards the Jews, I have always thought that their natural conduct in public should be one of proud reserve."[97]

Einstein merely parroted the Zionist Party line. Werner E. Mosse wrote,

"While the leaders of the CV saw it as their special duty to represent the interests of the German Jews in the active political struggle, Zionism stood for... systematic Jewish non-participation in German public life. It rejected as a matter of principle any participation in the struggle led by the CV."[98]

---

"Orientals".
[96]. Einstein repeatedly spoke of the Germans as "greedy" to acquire territory and of the "loss of energy" when different "races" attempted to live together. He have been speaking literally. Georg Friedrich Nicolai wrote of the struggle of life to aquire the energy of the sun and he applied this struggle to humanity. G. Nicolai, *Die Biologie des Krieges, Betrachtungen eines deutschen Naturforschers*, O. Füssli, Zürich, (1917); English translation: *The Biology of War*, Century Co., New York, (1918), pp. 36-39, 44-53.
[97]. R. W. Clarck, *Einstein, the Life and Times*, World Publishing Company, USA, (1971), p. 292. Clarck refers to: *Neue Rundschau*, Volume 33, Part 2, pp. 815-816.
[98]. W. E. Mosse, "Die Niedergang der deutschen Republik und die Juden", *The Crucial Year 1932*, p. 38; English translation in: K. Polkehn, "The Secret Contacts: Zionism and Nazi Germany, 1933-1941", *Journal of Palestine Studies*, Volume 5, Number 3/4, (Spring-Summer, 1976), pp. 54-82, at 56-57.

In 1925, Einstein wrote in the official Zionist organ *Jüdische Rundschau*,

"By study of their past, by a better understanding of the spirit [Geist] that accords with their race, they must learn to know anew the mission that they are capable of fulfilling. [***] What one must be thankful to Zionism for is the fact that it is the only movement that has given many Jews a justified pride, that it has once again given a despairing race the necessary faith, if I may so express myself, given new flesh to an exhausted people."[99]

On 12 October 1929, Albert Einstein wrote to the *Manchester Guardian*,

"In the re-establishment of the Jewish nation in the ancient home of the race, where Jewish spiritual values could again be developed in a Jewish atmosphere, the most enlightened representatives of Jewish individuality see the essential preliminary to the regeneration of the race and the setting free of its spiritual creativeness."[100]

Einstein's public racism eventually waned, but he continued to publicly express his segregationist philosophy in the same terms as anti-Semites, as well as his belief that Jews "thrived on" and owed their "continued existence" to anti-Semitism. Einstein stated in December of 1930 to an American audience,

"There is something indefinable which holds the Jews together. Race does not make much for solidarity. Here in America you have many races, and yet you have the solidarity. Race is not the cause of the Jews' solidarity, nor is their religion. It is something else—which is indefinable."[101]

Einstein's confusing public statement perhaps resulted from his desire to promote multi-culturalism in America, which had the benefit of freeing up Jewish immigration to the United States.[102] Einstein was also likely parroting, or trying to parrot, a fellow anti-assimilationist political Zionist whose pamphlet was well known in America, Solomon Schechter and his *Zionism: A Statement*, Federation of American Zionists, New York, (1906), in which Schechter states,

---

[99]. English translation by John Stachel in J. Stachel, "Einstein's Jewish Identity", *Einstein from 'B' to 'Z'*, Birkhäuser, Boston, (2002), p. 67. Stachel cites, "Botschaft", *Jüdische Rundschau*, Volume 30, (1925), p. 129; French translation, *La Revue Juive*, Volume 1, (1925), pp. 14-16.
[100]. J. Stachel, "Einstein's Jewish Identity", *Einstein from 'B' to 'Z'*, Birkhäuser, Boston, (2002), p. 65. Stachel cites, *About Zionism: Speeches and Letters*, Macmillan, New York, (1931), pp. 78-79.
[101]. A. Einstein quoted in "Einstein on Arrival Braves Limelight for Only 15 Minutes", *The New York Times*, (12 December 1930), pp. 1, 16, at 16.
[102]. E. A. Ross, *The Old World in the New: The Significance of past and Present Immigration to the American People*, Century Company, New York, (1914), p. 144.

among other things, "Zionism is an ideal, and as such is indefinable."[103]
Einstein stated in 1938,

### "JUST WHAT IS A JEW?

The formation of groups has an invigorating effect in all spheres of human striving, perhaps mostly due to the struggle between the convictions and aims represented by the different groups. The Jews, too, form such a group with a definite character of its own, and anti-Semitism is nothing but the antagonistic attitude produced in the non-Jews by the Jewish group. This is a normal social reaction. But for the political abuse resulting from it, it might never have been designated by a special name.

What are the characteristics of the Jewish group? What, in the first place, is a Jew? There are no quick answers to this question. The most obvious answer would be the following: A Jew is a person professing the Jewish faith. The superficial character of this answer is easily recognized by means of a simple parallel. Let us ask the question: What is a snail? An answer similar in kind to the one given above might be: A snail is an animal inhabiting a snail shell. This answer is not altogether incorrect; nor, to be sure, is it exhaustive; for the snail shell happens to be but one of the material products of the snail. Similarly, the Jewish faith is but one of the characteristic products of the Jewish community. It is, furthermore, known that a snail can shed its shell without thereby ceasing to be a snail. The Jew who abandons his faith (in the formal sense of the word) is in a similar position. He remains a Jew.
[***]
### WHERE OPPRESSION IS A STIMULUS
[***]

Perhaps even more than on its own tradition, the Jewish group has thrived on oppression and on the antagonism it has forever met in the world. Here undoubtedly lies one of the main reasons for its continued existence through so many thousands of years."[104]

Albert Einstein was parroting racist political Zionist leader Theodor Herzl, who wrote in his book *The Jewish State*,

"Oppression and persecution cannot exterminate us. No nation on earth has survived such struggles and sufferings as we have gone through. Jew-baiting has merely stripped off our weaklings; the strong among us were invariably

---

**103**. Reprinted in the relevant part in A. Hertzberg, *The Zionist Idea*, Harper Torchbooks, New York, (1959), p. 505.
**104**. A. Einstein, "Why do They Hate the Jews?", *Collier's*, Volume 102, (26 November 1938); reprinted in *Ideas and Opinions*, Crown, New York, (1954), pp. 191-198, at 194, 196. Einstein expressed himself in a similar way to Peter A. Bucky, P. A. Bucky, Einstein, and A. G. Weakland, *The Private Albert Einstein*, Andrews and McMeel, Kansas City, (1992), p. 87.

true to their race when persecution broke out against them. This attitude was most clearly apparent in the period immediately following the emancipation of the Jews. Later on, those who rose to a higher degree of intelligence and to a better worldly position lost their communal feeling to a very great extent. Wherever our political well-being has lasted for any length of time, we have assimilated with our surroundings. I think this is not discreditable. Hence, the statesman who would wish to see a Jewish strain in his nation would have to provide for the duration of our political well-being; and even Bismarck could not do that. [***] The Governments of all countries scourged by Anti-Semitism will serve their own interests in assisting us to obtain the sovereignty we want. [***] Great exertions will not be necessary to spur on the movement. Anti-Semites provide the requisite impetus. They need only do what they did before, and then they will create a love of emigration where it did not previously exist, and strengthen it where it existed before. [***] I imagine that Governments will, either voluntarily or under pressure from the Anti-Semites, pay certain attention to this scheme; and they may perhaps actually receive it here and there with a sympathy which they will also show to the Society of Jews."[105]

In 1938, Einstein stated in his essay "Our Debt to Zionism",

"Rarely since the conquest of Jerusalem by Titus has the Jewish community experienced a period of greater oppression than prevails at the present time. [***] Yet we shall survive this period too, no matter how much sorrow, no matter how heavy a loss in life it may bring. A community like ours, which is a community purely by reason of tradition, can only be strengthened by pressure from without."[106]

Einstein avowed *circa* 3 April 1920, that,

"If what anti-Semites claim were true, then indeed there would be nothing weaker, more wretched, and unfit for life, than the German people".[107]

Einstein often avowed that the anti-Semites' beliefs were true, and, hence, Einstein wished the Germans dead. When discussing the meaning of life, Einstein spoke to Peter A. Bucky about persons and creatures who "[do] not deserve to be in our world" and are "hardly fit for life."[108] Einstein's language is

---

**105**. T. Herzl, *A Jewish State: An Attempt at a Modern Solution of the Jewish Question*, The Maccabæan Publishing Co., New York, (1904), pp. 5-6, 25, 68, 93.
**106**. A. Einstein, "Our Debt to Zionism", *Out of My Later Years*, Carol Publishing Group, New York, (1995), pp. 262-264, at 262.
**107**. A. Einstein, English translation by A. Engel, *The Collected Papers of Albert Einstein*, Volume 7, Document 35, Princeton University Press, (2002), pp. 156-157.
**108**. P. A. Bucky, Einstein, and A. G. Weakland, *The Private Albert Einstein*, Andrews and McMeel, Kansas City, (1992), p. 111.

quite similar to the language of Hitler's "T4" *"Euthanasia-Programme"*.

After siding with Germany's enemies in the First World War—while living in Germany, and after intentionally provoking Germans into increased anti-Semitism, which he thought was good for Jews, and after defaming German Nobel Prize laureates in the international press to the point where they felt obliged to join Hitler's cause, which cause eventually resulted in the genocide of Europe's Jews; Einstein sponsored the production of genocidal weapons to mass murder Germans, whom he had hated all of his life, in the famous letter to President Roosevelt that Einstein signed urging Roosevelt to begin the development of atomic bombs—before the mass murder of Jews had begun.[109]

Einstein callously asserted that the use of atomic bombs on civilian populations was "morally justified". I quote Einstein without delving into the question of who first bombed civilian centers,

> "It should not be forgotten that the atomic bomb was made in this country as a preventive measure; it was to head off its use by the Germans, if they discovered it. The bombing of civilian centers was initiated by the Germans and adopted by the Japanese. To it the Allies responded in kind—as it turned out, with greater effectiveness—and they were morally justified in doing so."[110]

Einstein advocated genocidal collective punishment,

> "The Germans as an entire people are responsible for these mass murders and must be punished as a people if there is justice in the world and if the consciousness of collective responsibility in the nations is not to perish from the earth entirely."[111]

and,

> "It is possible either to destroy the German people or keep them suppressed; it is not possible to educate them to think and act along democratic lines in the foreseeable future."[112]

Albrecht Fölsing has assembled a compilation of post-WW II quotations

---

**109**. A. Unsöld, "Albert Einstein — Ein Jahr danach", *Physikalische Blätter*, Volume 36, (1980), pp.337-339; **and** Volume 37, Number 7, (1981), p. 229.

**110**. A. Einstein, "Atomic War or Peace", *Atlantic Monthly*, (November, 1945, and November 1947); *as reprinted in:* A. Einstein, *Ideas and Opinions*, Crown, New York, (1954), p. 125.

**111**. A. Einstein, "To the Heroes of the Battle of the Warsaw Ghetto", *Bulletin of the Society of Polish Jews*, New York, (1944), reprinted in *Ideas and Opinions*, Crown, New York, (1954), pp. 212-213.

**112**. A. Einstein, quoted in O. Nathan and H. Norton, *Einstein on Peace*, Avenel Books, New York, (1981), p. 331.

from Einstein, which evince Einstein's lifelong habit of stereotyping people based on their ethnicity. Einstein expressed his hatred in the horrific post-Holocaust context—a temptation Max Born had resisted,

> "With the Germans having murdered my Jewish brethren in Europe, I do not wish to have anything more to do with Germans, not even with a relatively harmless Academy. [***] The crimes of the Germans are really the most hideous that the history of the so-called civilized nations has to show. [***] [It was] evident that a proud Jew no longer wishes to be connected with any kind of German official event or institution. [***] After the mass murder committed by the Germans against my Jewish brethren I do not wish any publications of mine to appear in Germany."[113]

Einstein wrote to Born on 15 September 1950 that his views towards Germans predated the Nazi period,

> "I have not changed my attitude to the Germans, which, by the way, dates not just from the Nazi period. All human beings are more or less the same from birth. The Germans, however, have a far more dangerous tradition than any of the other so-called civilized nations. The present behavior of these other nations towards the Germans merely proves to me how little human beings learn even from their most painful experiences."[114]

and on learning that Born would return to Germany, Einstein wrote on 12 October 1953,

> "If anyone can be held responsible for the fact that you are migrating back to the land of the mass-murderers of our kinsmen, it is certainly your adopted fatherland — universally notorious for its parsimony."[115]

## 4.6 Racist Jewish Hypocrisy, Intimidation and Censorship

Sigmund Freud used prominent Gentiles, or "Goyim" as Freud called them, to promote his theories of psychology. He did this to give himself and the theories he plagiarized from Plato and others credibility in the broader "Gentile world". Though Freud thought that Gentiles were inferior to Jews, Freud was after fame.

Freud was another feted Jewish racist, who believed that the Jews were a

---

[113]. A. Einstein quoted in A. Fölsing, *Albert Einstein: A Biography*, Viking, New York, (1997), pp. 727-728.
[114]. M. Born, *The Born-Einstein Letters*, Walker and Company, New York, (1971), p. 189.
[115]. M. Born, *The Born-Einstein Letters*, Walker and Company, New York, (1971), p. 199.

superior race. Kevin MacDonald wrote in his book *The Culture of Critique*,

"Freud's powerful racial sense of ingroup-outgroup barriers between Jews and gentiles may also be seen in the personal dynamics of the psychoanalytic movement. We have seen that Jews were numerically dominant within psychoanalysis, especially in the early stages when all the members were Jews. 'The fact that these were Jews was certainly not accidental. I also think that in a profound though unacknowledged sense Freud wanted it that way' (Yerushalmi 1991, 41). As in other forms of Judaism, there was a sense of being an ingroup within a specifically Jewish milieu. 'Whatever the reasons—historical, sociological—group bonds did provide a warm shelter from the outside world. In social relations with other Jews, informality and familiarity formed a kind of inner security, a 'we-feeling,' illustrated even by the selection of jokes and stories recounted within the group' (Grollman 1965, 41). Also adding to the Jewish milieu of the movement was the fact that Freud was idolized by Jews generally. Freud himself noted in his letters that 'from all sides and places, the Jews have enthusiastically seized me for themselves.' 'He was embarrassed by the way they treated him as if he were 'a God-fearing Chief Rabbi,' or 'a national hero,'' and by the way they viewed his work as 'genuinely Jewish' (in Klein 1981, 85; see also Gay 1988, 599).

As in the case of several Jewish movements and political activities reviewed in Chapters 2 and 3 (see also *SAID*, Ch. 6), Freud took great pains to ensure that a gentile, Jung, would be the head of his psychoanalytic movement—a move that infuriated his Jewish colleagues in Vienna, but one that was clearly intended to deemphasize the very large overrepresentation of Jews in the movement during this period. To persuade his Jewish colleagues of the need for Jung to head the society, he argued, 'Most of you are Jews, and therefore you are incompetent to win friends for the new teaching. Jews must be content with the modest role of preparing the ground. It is absolutely essential that I should form ties in the world of science' (in Gay 1988, 218). As Yerushalmi (1991, 41) notes, 'To put it very crudely, Freud needed a goy, and not just any goy but one of genuine intellectual stature and influence.' Later, when the movement was reconstituted after World War I, another gentile, the sycophantic and submissive Ernest Jones, became president of the International Psychoanalytic Association."[116]

The aggressive rôle that the "Shabbas Goy" Max von Laue played in

---

**116**. K. MacDonald, *The Culture of Critique*, Praeger, Westport, Connecticut, London, (1998), pp. 113-114; **citing:** E. A. Grollman, *Judaism in Sigmund Freud's World*, Bloch, New York, (1965); **and** D. B. Klein, *Jewish Origins of the Psychoanalytic Movement*, Praeger, New York, (1981); **and** P. Gay, *Freud: A Life for Our Time*, W. W. Norton, New York, (1988); **and** Y. H. Yerushalmi, *Freud's Moses: Judaism Terminable and Interminable*, Yale University Press, (1991); **and** K. MacDonald, *Separation and Its Discontents: Toward an Evolutionary Theory of Anti-Semitism*, Praeger, Westport, Connecticut, (1998).

personally attacking Einstein's critics was a part of this pattern.[117] He put a Gentile face on the assault against the rights of Einstein's critics to hold their own opinions and express them in public. Laue championed a smear campaign against Einstein's critics in the full knowledge that Einstein had plagiarized the works of Poincaré and Lorentz, and in full knowledge of the fact that the experimental evidence which had allegedly confirmed the general theory of relativity, did not confirm it, but rather disproved it.

Laue must have known that Einstein was an outspoken Jewish racist, but instead of condemning Einstein for his racism, Laue let himself be used to miscast the scientific and ethical critique of Einstein as if it were an expression of anti-Jewish racism. Einstein played a central rôle in corrupting the universities, the journals and the popular press of his day with Jewish racists and sycophantic Gentiles, who would promote him and the theories he appropriated from others.

Freud did not invent the field of psychology. He was a career plagiarist and he largely deprived the field of its synthetic scientific basis, which appeared in the earlier work of Spencer and James. Freud converted psychology into an introspective metaphysical analysis of his own mental maladies. Freud abused the pseudoscientific doctrines he plagiarized, and the fame he had achieved through the Jewish community, to make political attacks against persons whom he hated, and against Rome—against the Catholic Church. Largely under the directorship of Jews, the field of psychology degenerated into a sadistic house of tortures and mutilation. It was exploited as a means to suppress dissent, especially in Marxist countries, and particularly in the hands of Jews. Psychology, under Freud, also become a means to enrich psychiatrists by providing sick persons with someone with whom they could talk, and giving them the false hope that this panacea of talk would cure them of their physical ailments.

Max Born intimated in his 16 July 1955 lecture in Bern (as had Moszkowski and Freundlich) that the hype promoting Einstein in 1919 was intended, in part, as a *rapprochement* between Great Britain and Germany after the war. Eddington wrote to Einstein on 1 December 1919,

> "It is the best possible thing that could have happened for scientific relations between England and Germany. I do not anticipate rapid progress towards official reunion, but there is a big advance towards a more reasonable frame of mind among scientific men, and that is even more important than the renewal of formal associations. [***] [T]hings have turned out very fortunately in giving this object-lesson of the solidarity of German and British science even in time of war."[118]

---

[117]. See: Letter from M. Planck to W. Wien of 9 July 1922 in J. L. Heilbron, *Max Planck: Ein Leben für die Wissenschaft 1858-1947. Mit einer Auswahl der allgemeinverständlichen Schriften von Max Planck*, S. Hirzel, Stuttgart, (1988), p. 127.

[118]. Letter from A. S. Eddington to A. Einstein of 1 December 1919, *The Collected Papers of Albert Einstein*, Volume 9, Document 186, Princeton University Press, (2004),

Others wrote of their excitement that the eclipse sensation would promote better international relations.[119]

This indicates that the eclipse "observations" signified a political maneuver, not a legitimate experiment. At the time much was made of the fact that Einstein's book had been translated into English and was the first book to be translated from German to English after the war.[120] Einstein's correspondence regarding this translation and his article for the *The London Times* also reveal some of the political motives of *rapprochement* behind the Einstein hype of 1919, and beyond.[121]

In 1955, Born stated that the eclipse expeditions of 1919 created an undescribable stir around the world,

"EINSTEIN became at once the most famous and popular figure, the man who had broken through the wall of hatred and united the scientists to a common effort, the man who had replaced ISAAC NEWTON's system of the world by another and better one. But at the same time an opposition, which had already been apparent while I was in Berlin, grew under the leadership of PHILIPP LENARD and JOHANNES STARK. It was springing from the most absurd mixture of scientific conservatism and prejudice with racial and political emotions, due to EINSTEIN's Jewish descent and pacifistic, antimilitaristic convictions."[122]

Born also stated,

"[... ]EINSTEIN's theory was new and revolutionary, an effort was needed to assimilate it. Not everybody was able or willing to do so. Thus the period after EINSTEIN's discovery was full of controversy, sometime of bitter strife."[123]

Nobel Prize laureates Philipp Lenard (1905 Nobel Prize for Physics) and Johannes Stark (1919 Nobel Prize for Physics) had initially sponsored Einstein and his work, and it was only after Einstein played the race card—publicly and internationally smearing Philipp Lenard without cause, that race became an issue

---

pp. 262-263, at 263.
[119]. *The Collected Papers of Albert Einstein*, Volume 9, Documents 203, 220, 227, 238, 249, 253, Princeton University Press, (2004).
[120]. *See, for example:* "Literarische Mitteilungen", *Jüdische Rundschau*, Volume 25, Number 33, (21 May 1920), p. 254.
[121]. *The Collected Papers of Albert Einstein*, Volume 9, Documents 177, 180, 182, 185, 186 and 194, Princeton University Press, (2004).
[122]. M. Born, "Physics and Relativity", *Physics in my Generation*, second revised edition, Springer, New York, (1969), p. 110-111.
[123]. M. Born, "Physics and Relativity", *Physics in my Generation*, second revised edition, Springer, New York, (1969), p. 100.

in the debate over relativity theory—mostly for Einstein, Max von Laue and Max Born, who had a financial interest in the Einstein myth, and for the press people who smeared Einstein's opponents. They desperately wanted to change the subject from the legitimate claims of Einstein's plagiarism, legitimate arguments against the irrationality of the theory of relativity and the shameless hype and misrepresentation of experimental evidence by Einstein and his friends, to name-calling and racial strife provoked by them.

Lenard and Stark initially opposed Einstein on purely scientific and ethical grounds related to Einstein's sophistry, self-promotion and plagiarism. They later embraced Nazism and its racial mythologies.

Einstein eventually succeeded in bringing racial politics into the debate, though it was initially a larger issue for him than for his opponents. Einstein most often outright refused to discuss his plagiarism or purely scientific, non-political critiques of the theory of relativity; but he did not hesitate to name-call and smear his critics. He could not win in a dispute over the scientific and historical facts, so he provoked a race war over relativity theory in order to avoid legitimate criticism. It was a war everyone would ultimately lose.

Einstein's complaints were hypocritical. He himself sought ethnically segregated educational institutions and an ethnically segregated society and often stated that anti-Semitism was both correct and good for Jews. Einstein had bad experiences early in his youth[124] and always bore a stereotypical prejudice against Gentile Germans, which is consistent with the racism inherent in genocidal Judaism.

Max Born, himself, "played the race card" and misrepresented events at the Bad Nauheim debate. Born stated,

> "[Philipp Lenard] directed sharp, nasty attacks against Einstein, with a blatantly anti-Semitic tendency. Einstein became agitated and answered him sharply, and I believe I remember that I supported him."[125]

Born took pride in his biased and unfair efforts to quash any opposition to Einstein's mythologies. Born stated,

> "There appeared attacks against EINSTEIN by well-known scientists and philosophers in the *Frankfurter Zeitung* which aroused my pugnacity. I answered in a rather sharp article."[126]

Born's contradictory claim that Einstein had concurrently united and

---

**124**. J. Stachel, "Einstein's Jewish Identity", *Einstein from 'B' to 'Z'*, Birkhäuser, Boston, Basel, Berlin, (2002), pp. 57-83, at 59.
**125**. M. Born quoted and translated in: D. A. Buchwald, *et al.* Editors, "Einstein's Encounters with German Anti-Relativists", *The Collected Papers of Albert Einstein*, Volume 7, Princeton University Press, (2002), p. 109, footnote 52.
**126**. M. Born, "Physics and Relativity", *Physics in my Generation*, second revised edition, Springer, New York, (1969), p. 112.

divided scientists indicates Born's blindness to his own hypocrisy and the magnitude of the zealotry he felt for his political cause, which he believed would make him rich. While Born and his ilk boasted of their opposition to anti-Semitism, they themselves were elements in the atmosphere which created Hitler's tragic ascent to power, and for them to pretend to victory among that horror, greatly dishonors the innocent lives lost in the Holocaust. Political Zionists, Einstein among them—Born not, saw anti-Semitism as a good thing and promoted segregation and racial tension. Some even delighted in the fact that forced segregation would bring more Jews into the political Zionist camp.

Albert Einstein was one of the world's leading political Zionists. Political Zionism was a new form of racism that emerged at the end of the Nineteenth Century. It held that Jews were a pure race that could not coexist with non-Jews. Einstein had many powerful friends in the Zionist and Socialist press. Einstein's friends and supporters, in what political Zionist founder Theodor Herzl called the "Jewish papers",[127] libeled those who opposed Einstein or the theory of relativity and deflected attention from Einstein's plagiarism by misrepresenting any criticism of Einstein as if it were anti-Semitism, *per se*.[128]

There was also an anti-Einstein press and an unbiased press which documented Einstein's plagiarism and his scientific and philosophical defeats. Like radicals in general, radical Socialists, Zionists and Communists had well-deserved reputations as defamers, which manifested itself in their vitriolic attacks on Jewish leaders who refused to fund their schemes; or, in the case of Zionism, opposed their racist agenda. Einstein stated, "But in Germany the judgement of my theory depended on the party politics of the Press[.]"[129] German newspapers had well-deserved reputations as being organs for the many political parties which were active in Germany in the Teens of the Twentieth Century. They brought politics into science in a way not previously known.

Einstein took advantage of the political climate after World War I to change the subject from the accusations of plagiarism against him, which were easily

---

[127]. Racist political Zionist Theodor Herzl wrote on 12 June 1895, "Jewish papers! I will induce the publishers of the biggest Jewish papers (*Neue Freie Presse, Berliner Tageblatt, Frankfurter Zeitung,* etc.) to publish editions over there, as the *New York Herald* does in Paris."—T. Herzl, English translation by H. Zohn, R. Patai, Editor, *The Complete Diaries of Theodor Herzl*, Volume 1, Herzl Press, New York, (1960), p. 84.
THE DEARBORN INDEPENDENT, praised the *New York Herald*. "When Editors Were Independent of the Jews", THE DEARBORN INDEPENDENT, (5 February 1921). ***See also:*** T. Herzl, English translation by H. Zohn, R. Patai, Editor, *The Complete Diaries of Theodor Herzl*, Volumes 1 and 2, Herzl Press, New York, (1960), pp. 37, 97, 170, 455, 457, 480. ***See also:*** A. Elon, *Herzl*, Holt, Rinehart and Winston, New York, (1975), pp. 167-168. ***See also:*** *The Collected Papers of Albert Einstein*, Volume 7, Document 35, Princeton University Press, (2002), pp. 296-297, note 8.
[128]. "Prof. Einstein Here, Explains Relativity", *The New York Times*, (3 April 1921), pp. 1, 13, at 1.
[129]. A. Einstein, "Jewish Nationalism and Anti-Semitism", *The Jewish Chronicle*, (17 June 1921), p. 16.

proven, to racial politics, which were explosive at the time. It is tragic that the search for the truth in Physics, and in Ethics related to priorities, became a political issue centered on "the Jewish question", but Einstein succeeded in making it one.

Political Zionists, Einstein and his friends among them, have earned a reputation throughout their history for preventing free and open public dialog about important issues they would rather not see discussed. They have often had open access to the press to publish their smears and the means to largely prevent those who have been wronged from responding. They accomplish these feats by: spuriously presuming to speak for all persons of Jewish descent, organized intimidation, boycott, smear tactic, intensive letter writing campaigns which give an inflated appearance that their views are widely held, threats and acts of violence, etc.

Even the disciples of Christ are said to have feared Jewish tribalism and Jewish religious intolerance, for example *John* 7:1 tells that,

"After these *things* Jesus walked in Galilee: for he would not walk in Jewry, because the Jews sought to kill him."

*John* 7:13 states:

"Howbeit no *man* spake openly of him for fear of the Jews."

*John* 19:38 states:

"And after this Joseph of Arimathaea, being a disciple of Jesus, but secretly for fear of the Jews, besought Pilate that he might take away the body of Jesus: and Pilate gave *him* leave. He came therefore, and took the body of Jesus."

*John* 20:19 states:

"Then the same day at evening, being the first *day* of the week, when the doors were shut where the disciples were assembled for fear of the Jews, came Jesus and stood in the midst, and saith unto them, Peace *be* unto you."

In 1914, Edward Alsworth Ross, a Professor of Sociology at the University of Wisconsin, wrote in his book, *The Old World in the New: The Significance of Past and Present Immigration to the American People*, The Century Co., New York, (1914), pages 143 and 165,

"IN his defense of Flaccus [*Pro Flaccus*, Chapter 28], a Roman governor who had 'squeezed' his Jewish subjects, Cicero lowers his voice when he comes to speak of the Jews, for, as he explains to the judges, there are persons who might excite against him this numerous, clannish and powerful element. With much greater reason might an American lower his voice today in discussing two million Hebrew immigrants united by a strong race consciousness and already ably represented at every level of wealth, power, and influence in the United States. [***] This cruel prejudice—for all lump condemnations are cruel—is no importation, no hang-over from the past. It appears to spring out of contemporary experience and is invading circle after circle of broad-minded. People who give their lives to befriending immigrants shake their heads over the Galician Hebrews. It is astonishing how much of the sympathy that twenty years ago went out to the fugitives from Russian massacres has turned sour. Through fear of retaliation little criticism gets into print; in the open the Philo-semites have it all their way. The situation is: Honey above, gall beneath. If the Czar, by keeping up the pressure which has already rid him of two million undesired subjects, should succeed in driving the bulk of his six million Jews to the United States, we shall see the rise of the Jewish question here, perhaps riots and anti-Jewish legislation. No doubt thirty or forty thousand Hebrews from eastern Europe might be absorbed by this country each year without any marked growth of race prejudice; but when they come in two or three or even four times as fast, the lump outgrows the leaven, and there will be trouble."

Cicero's *Pro Flaccus*, Chapter 28, states,

"XXVIII. The next thing is that charge about the Jewish gold. And this, forsooth, is the reason why this cause is pleaded near the steps of Aurelius. It is on account of this charge, O Lælius, that this place and that mob has been selected by you. You know how numerous that crowd is, how great is its unanimity, and of what weight it is in the popular assemblies. I will speak in a low voice, just so as to let the judges hear me. For men are not wanting who would be glad to excite that people against me and against every eminent man; and I will not assist them and enable them to do so more easily. As gold, under pretence of being given to the Jews, was accustomed every year to be exported out of Italy and all the provinces to Jerusalem, Flaccus issued an edict establishing a law that it should not be lawful for gold to be exported out of Asia. And who is there, O judges, who cannot honestly praise this measure? The senate had often decided, and when I was consul it came to a most solemn resolution that gold ought not to be exported. But to resist this barbarous superstition were an act of dignity, to despise the multitude of Jews, which at times was most unruly in the assemblies in defence of the interests of the republic, was an act of the greatest wisdom. 'But Cnæus Pompeius, after he had taken Jerusalem, though he was a conqueror, touched nothing which was in that temple.' In the first place, he acted wisely, as he did in many other instances, in leaving no room for his detractors to say anything against him, in a city so prone to suspicion and to evil speaking. For I do not suppose that the religion of the Jews, our enemies, was any obstacle to that most illustrious general, but that he was hindered by his own modesty. Where then is the guilt? Since you nowhere impute any theft to us, since you approve of the edict, and confess that it was passed in due form, and do not deny that the gold was openly sought for and produced, the facts of the case themselves show that the business was executed by the instrumentality of men of the highest character. There was a hundredweight of gold, more or less, openly seized at Apamea, and weighed out in the forum at the feet of the prætor, by Sextus Cæsius, a Roman knight, a most excellent and upright man; twenty pounds weight or a little more were seized at Laodicea, by Lucius Peducæus, who is here in court, one of our judges; some was seized also at Adramyttium, by Cnæus Domitius, the lieutenant, and a small quantity at Pergamus. The amount of the gold is known; the gold is in the treasury; no theft is imputed to him; but it is attempted to render him unpopular. The speaker turns away from the judges, and addresses himself to the surrounding multitude. Each city, O Lælius, has its own peculiar religion; we have ours. While Jerusalem was flourishing, and while the Jews were in a peaceful state, still the religious ceremonies and observances of that people were very much at variance with the splendour of this empire, and the dignity of our name, and the institutions of our ancestors. And they are the more odious to us now, because that nation has shown by arms what were its feelings towards our supremacy. How dear it was to the immortal gods is proved by its having been defeated, by its revenues having been farmed out to our contractors, by its being reduced to a state of subjection."[130]

United States Army Captain Montgomery Schuyler reported on 1 March 1919,

"It is probably unwise to say this loudly in the United States but the Bolshevik movement is and has been since its beginning guided and controlled by Russian Jews of the greasiest type[... ]"[131]

Senator Ernest F. Hollings argued before the United States that his position was being mischaracterized, when he put America's interests ahead of the Neo-Conservatives' plan for providing Israel with hegemony in the Mid-East and was called "anti-Semitic". Senator Hollings' comments appear in the *Congressional Record* (Proceedings and Debates of the 108th Congress, Second Session), Volume 150, Number 72, (20 May 2004), pages S5921-S5925; which includes Senator Hollings' article, "Bush's Failed Mideast Policy is Creating More Terrorism", *Charleston Post and Courier*, 6 May 2004, which article has appeared in several websites. The *Congressional Record* is also available online. At pages S5921-S5925, Senator Hollings states, *inter alia*,

"Mr. HOLLINGS. Mr. President, I thank my distinguished colleagues. I

---

**130**. M. T. Cicero, *Pro Flaccus*, Chapter 28; translated by C. D. Yonge, *The Orations of Marcus Tullius Cicero*, Volume 2, George Bell & Sons, London, (1880), pp. 454-455.
**131**. K. A. Strom, Editor, *The Best of Attack! and National Vanguard Tabloid*, National Alliance, Arlington, Virginia, (1984), p. 66.

have, this afternoon, the opportunity to respond to being charged as anti-Semitic when I proclaimed the policy of President Bush in the Mideast as not for Iraq or really for democracy in the sense that he is worried about Saddam and democracy. If he were worried about democracy in the Mideast, as we wanted to spread it as a policy, we would have invaded Lebanon, which is half a democracy and has terrorism and terrorists who have been problems to the interests of Israel and the United States. [***] I want to read an article that appeared in the Post and Courier in Charleston on May 6; thereafter, I think in the State newspaper in Columbia a couple days later; and in the Greenville News—all three major newspapers in South Carolina. You will find that there is no anti-Semitic reference whatsoever in it. [***] But in any event, the better way to do it is go right in and establish our predominance in Iraq and then, as they say, and I have different articles here I could refer to, next is Iran and then Syria. And it is the domino theory, and they genuinely believe it. I differ. I think, frankly, we have caused more terrorism than we have gotten rid of. That is my Israel policy. You can't have an Israel policy other than what AIPAC [American Israel Public Affairs Committee] gives you around here. I have followed them mostly in the main, but I have also resisted signing certain letters from time to time, to give the poor President a chance. I can tell you no President takes office—I don't care whether it is a Republican or a Democrat—that all of a sudden AIPAC will tell him exactly what the policy is, and Senators and members of Congress ought to sign letters. I read those carefully and I have joined in most of them. On some I have held back. I have my own idea and my own policy. I have stated it categorically. [***] Again, let me read: Bush thought tax cuts would hold his crowd together and that spreading democracy in the Mideast to secure Israel would take the Jewish vote from the Democrats. Is there anything wrong with referring to the Jewish vote? Good gosh, every 1 of us of the 100, with pollsters and all, refer to the Jewish vote. That is not anti-Semitic. It is appreciating them. We campaigned for it. I just read about President Bush's appearance before the AIPAC. He confirmed his support of the Jewish vote, referring to adopting Ariel Sharon's policy, and the dickens with the 1967 borders, the heck with negotiating the return of refugees, the heck with the settlements he had objected to originally. They had those borders, Resolution No. 242—no, no, President Bush said: I am going along with Sharon, and he was going to get that and he got the wonderful reception he got with the Jewish vote. There is nothing like politicizing or a conspiracy, as my friend from Virginia, Senator ALLEN, says—that it is an anti-Semitic, political, conspiracy statement. That is not a conspiracy. That is the policy. I didn't like to keep it a secret, maybe; but I can tell you now, I will challenge any 1 of the other 99 Senators to tell us why we are in Iraq, other than what this policy is here. It is an adopted policy, a domino theory of The Project For The New American Century. Everybody knows it because we want to secure our friend, Israel. If we can get in there and take it in 7 days, as Paul Wolfowitz says, then we would get rid of Saddam, and when we got rid of Saddam, now all they can do is fall back and say: Aren't you getting rid of Saddam? Let me get to that point. What happens is, they say he is a monster. We continued to give him aid after he gassed his own people and everything else of that kind. George Herbert Walker Bush said in his book All The Best in 1999, never commit American GIs into an unwinnable urban guerrilla war and lose the support of the Arab world, lose their friendship and support. That is a general rephrasing of it. The point is, my authority is the President's daddy. I want everybody to know that. I don't apologize for this column. I want them to apologize to me for talking about anti-Semitism. They are not getting by with it. I will come down here every day—I have nothing else to do—and we will talk about it and find out what the policy is. [***] We are losing the terrorism war because we thought we could do it militarily under the domino policy of President Bush, going into Iraq. That is my point. That is not anti-Semite or whatever they say in here about people's faith and ethnicity. I never referred to any faith. I should have added those other names from the Project For The New American Century, but I picked out the names I had quotes for. And for space, I left other things out. Mr. President, on May 12 of this year, I had printed in the RECORD the article in its entirety. I diverted from the reading of the article several times, so for the sake of accuracy I wanted the whole article printed. This particular op-ed piece appeared in the Post and Courier. Never would they have thought, having read it, if it was anti-Semitic, that they would have ever put it in there. Nor would the Knight Ridder newspapers in Columbia, SC. Nor would the Metro Media newspapers in Greenville, SC. But the Anti-Defamation League picked it up and now they have given it to my good friend, Senator ALLEN of Virginia. I have his particular admonition how I am anti-Semitic and I cannot let that stay there. [***] Come on. So we have to go out and not speak sense with respect to policy, and when you want to talk about policy, they say it is anti-Semitic. Well, come on the floor, let's debate it. Because my friend from Virginia admonishes me. Referring to me he says, 'I suggest he should learn from history before making accusations.' I didn't make any accusations. I stated facts. That is their policy. That is not my policy."

Former Illinois Congressman Paul Findley experienced first hand the ability and willingness of Zionists in more recent times to defame those who call for open public debate on issues the Zionists would rather suppress, or would have told from their heavily biased perspective and from their perspective only. Findley has written several books exposing the Zionists' ability to unfairly smear him and others, and to force silence through intimidation on any who would otherwise side with Findley in his efforts to involve the American people in an honest and open dialog about the rights of Palestinians.[132] Just as the Zionists

---

**132**. P. Findley, *They Dare to Speak Out: People and Institutions Confront Israel's Lobby*, Lawrence Hill, Westport, Connecticut, (1985); **and** *Deliberate Deceptions: Facing the Facts about the U.S.-Israeli Relationship*, Lawrence Hill Books, Chicago, (1993); **and** *Silent No More: Confronting America's False Images of Islam*, D : Amana

have often sought to suppress public discussion of the Palestinians' rights and an honest discussion of what is in America's best interests, as opposed to the Zionists' perceived self-interests, political Zionists—and indeed like minded Marxist-leaning Socialists—have often obstructed public debate about Einstein's plagiarism from the moment Einstein became their most famous and important spokesman.

Many have been wrongfully and viciously smeared as alleged "anti-Semites" because they refuse to discriminate in their opposition to racism, including but not limited to, their opposition to political Zionist racism. The vast majority of Jews initially opposed political Zionism due to its expressed racism. Their leaders were smeared. After the founding of Israel, debate was largely stifled.

Prof. Tony Martin was attacked when he added the book *The Secret Relationship Between Blacks and Jews*[133] among his offerings in the school bookstore at the university at which he taught. In his book, *The Jewish Onslaught: Despatches from the Wellesley Battlefront*, Majority Press, Dover, Massachusetts, (1993); Prof. Martin details the organized attacks he faced when exposing Jewish involvement in the slave trade and Jewish racism towards blacks. Prof. Martin exposits upon the fact that the Hamitic myth, the "curse of Ham", which condemns Blacks to perpetual slavery and degrades the stereotypical phenotype of a black person or "Canaanite", stems from the story of Noah and his son Ham in the Old Testament (*Genesis* 9:20-27); and from the racist Talmudic interpretations of this story; as well as their misuse to justify the injustice and inhumanity of Black slavery, which was a profitable industry for Jews, especially the trade to Brazil, where the Jews also profited from agriculture—in particular sugar cane.[134]

---

Publications, Beltsville, Maryland, (2001).
**133**. Historical Research Department of the Nation of Islam (Chicago), *The Secret Relationship Between Blacks and Jews*, Chicago, Latimer Associates, (1991). **For counter-argument, see:** H. D. Brackman, *Ministry of Lies: The Truth behind the Nation of Islam's The Secret Relationship between Blacks and Jews*, Four Walls Eight Windows, New York, (1994); **and** "Jews Had Negligible Role in Slave Trade", *The New York Times*, (14 February 1994), p. A16. **Contrast these with Brackman's own statements in his PhD dissertation:** H. D. Brackman, PhD Dissertation, University of Californian, Los Angeles, *The Ebb and Flow of Conflict—History of Black-Jewish Relations Through 1900*, University Microfilms International (Dissertation Services), Ann Arbor, Michigan, (1977); **and see:** T. Martin, *The Jewish Onslaught: Despatches from the Wellesley Battlefront*, Majority Press, Dover, Massachusetts, (1993). *See also:* L. Brenner, Letter to the Editor, *The New York Times*, (28 February 1994), p. A16; **and** "Harold Brackman Believes in Recycling Garbage", *New York Amsterdam News*, (11 March 1995). *See also:* M. A. Hoffman II, *Judaism's Strange Gods*, Independent History and Research, Coeur d'Alene, Idaho, (2000), pp. 66-67.
**134**. H. D. Brackman, PhD Dissertation, University of Californian, Los Angeles, *The Ebb and Flow of Conflict—History of Black-Jewish Relations Through 1900*, University Microfilms International (Dissertation Services), Ann Arbor, Michigan, (1977), pp. 163-164.

*Genesis* 9:20-27:

"20 And Noah began *to be* an husbandman, and he planted a vineyard: 21 And he drank of the wine, and was drunken; and he was uncovered within his tent. 22 And Ham, the father of Canaan, saw the nakedness of his father, and told his two brethren without. 23 And Shem and Japheth took a garment, and laid *it* upon both their shoulders, and went backward, and covered the nakedness of their father; and their faces *were* backward, and they saw not their father's nakedness. 24 And Noah awoke from his wine, and knew what his younger son had done unto him. 25 And he said, Cursed *be* Canaan; a servant of servants shall he be unto his brethren. 26 And he said, Blessed *be* the LORD God of Shem; and Canaan shall be his servant. 27 God shall enlarge Japheth, and he shall dwell in the tents of Shem; and Canaan shall be his servant."

Harold Brackman wrote of the evolution of the Hamitic myth in his PhD dissertation in 1977,

"The opening centuries of the Christian era constituted an interregnum in the native African record of historical achievement separating Cush's era of ancient prominence from the medieval accomplishments of the great Negro states of the Sudan. These same centuries formed the seedbed of rabbinic Judaism. And this fateful coincidence goes far toward explaining why they also formed such fertile soil for the growth of Jewish lore demeaning the Negro. The most famous of these anti-Negro legends cluster about Ham and Noah's cursing of Canaan [***] There is no denying that the Babylonian Talmud was the first source to read a Negrophobic content into the episode by stressing Canaan's fraternal connections with Cush [***] The Talmudic glosses of the episode added the stigma of blackness to the fate of enslavement that Noah predicted for Ham's progeny [***] According to it, Ham is told by his outraged father [Noah] that, because you have abused me in the darkness of the night, your children shall be born black and ugly; because you have twisted your head to cause me embarrassment, they shall have kinky hair and red eyes; because your lips jested at my exposure, theirs shall swell; and because you neglected my nakedness, they shall go naked[.]"[135]

---

**135**. H. D. Brackman, PhD Dissertation, University of Californian, Los Angeles, *The Ebb and Flow of Conflict—History of Black-Jewish Relations Through 1900*, University Microfilms International (Dissertation Services), Ann Arbor, Michigan, (1977), pp. 79-81. *Cf.* T. Martin, *The Jewish Onslaught: Despatches from the Wellesley Battlefront*, Majority Press, Dover, Massachusetts, (1993). L. Brenner, Letter to the Editor, *The New York Times*, (28 February 1994), p. A16; **and** "Harold Brackman Believes in Recycling Garbage", *New York Amsterdam News*, (11 March 1995). M. A. Hoffman II, *Judaism's Strange Gods*, Independent History and Research, Coeur d'Alene, Idaho, (2000), pp. 66-67.

The racist Talmud states in *Sanhedrin* 70*a*,

"'Ubar the Galilean gave the following exposition: The letter *waw* [*and*][4] occurs thirteen times in the passage dealing with wine: And *Noah began to be an husbandman,* and *he planted a vineyard:* And *he drank of the wine and was drunken;* and *he was uncovered within his tent.* And *Ham the father of Canaan, saw the nakedness of his father,* and *told his two brethren without.* And *Shem and Japheth took a garment,* and *laid it upon their shoulders,* and *went backward* and *covered the nakedness of their father, and their faces were backward,* and *they saw not their father's nakedness.* And *Noah awoke from his wine,* and *knew what his younger son had done unto him.*[5] [With respect to the last verse] Rab and Samuel [differ,] one maintaining that he castrated him, whilst the other says that he sexually abused him. He who maintains that he castrated him, [reasons thus;] Since he cursed him by his fourth son,[6] he must have injured him with respect to a fourth son.[7] But he who says that he sexually abused him, draws an analogy between *'and he saw'* written twice. Here it is written, *And Ham the father of Canaan saw the nakedness of his father;* whilst elsewhere it is written, *And when Shechem the son of Hamor saw her* [*he took her and lay with her and defiled here*].[8] Now, on the view that he emasculated him, it is right that he cursed him by his fourth son; but on the view that he abused him, why did he curse his fourth son: he should have cursed him himself?— Both indignities were perpetrated.'"[136]

The racist Talmud states in *Sanhedrin* 108*b*,

"Our Rabbis taught: Three copulated in the ark, and they were all punished—the dog, the raven, and Ham. The dog was doomed to be tied, the raven expectorates [his seed into his mates mouth], and Ham was smitten in his skin. [*Footnote:* I.e., from him descended Cush (the negro) who is black-skinned.]"[137]

The racist *Midrash Rabbah* (Genesis 36:7) states,

"7. AND NOAH AWOKE FROM HIS WINE (IX, 24): he was sobered from his wine.

AND KNEW WHAT HIS YOUNGEST SON HAD DONE UNTO HIM. Here it means, his worthless son, as you read, *Because the brazen altar that was before the Lord was too little to receive the burnt-offering,* etc. (I Kings VIII, 64).[1]

---

**136**. I. Epstein, Editor, Sanhedrin 70a, *The Babylonian Talmud*, Volume 28 (Sanhedrin II), The Soncino Press, (1935), pp. 477-478.
**137**. I. Epstein, Editor, Sanhedrin 108b, *The Babylonian Talmud*, Volume 28 (Sanhedrin II), The Soncino Press, (1935), p. 745.

AND HE SAID: CURSED BE CANAAN (IX, 25): Ham sinned and Canaan is cursed! R. Judah and R. Nehemiah disagreed. R. Judah said: Since it is written, *And God blessed Noah and his sons* (Gen. IX, 1), while there cannot be a curse where a blessing has been given, consequently, HE SAID: CURSED BE CANAAN. R. Nehemiah explained: It was Canaan who saw it [in the first place] and informed them, therefore the curse is attached to him who did wrong.

R. Berekiah said: Noah grieved very much in the Ark that he had no young son to wait on him, and declared, ' When I go out I will beget a young son to do this for me.' But when Ham acted thus to him, he exclaimed, ' You have prevented me from begetting a young son to serve me,² therefore that man [your son] will be a servant to his brethren!' R. Huna said in R. Joseph's name: [Noah declared], 'You have prevented me from begetting a fourth son, therefore I curse your fourth son.'³ R. Huna also said in R. Joseph's name: You have prevented me from doing something in the dark [sc. cohabitation], therefore your seed will be ugly and dark-skinned. R. Hiyya said: Ham and the dog copulated in the Ark, therefore Ham came forth black-skinned while the dog publicly exposes its copulation. R. Levi said: This may be compared to one who minted his own coinage⁴ in the very palace of the king, whereupon the king ordered: I decree that his effigy be defaced and his coinage cancelled. Similarly, Ham and the dog copulated in the Ark and were punished.⁵"[138]

Moses Maimonides, a famous Jewish philosopher and a racist, wrote in the Twelfth Century in his *Guide of the Perplexed*,

"Now I shall interpret to you this parable that I have invented. I say then: Those who are outside the city are all human individuals who have no doctrinal belief, neither one based on speculation nor one that accepts the authority of tradition: such individuals as the furthermost Turks found in the remote North, the Negroes found in the remote South, and those who resemble them from among them that are with us in these climes. The status of those is like that of irrational animals. To my mind they do not have the rank of men, but have among the beings a rank lower than the rank of man but higher than the rank of the apes. For they have the external shape and lineaments of a man and a faculty of discernment that is superior to that of the apes."[139]

The racist cabalistic doctrine of the *Zohar*, I, 73a, associates Blacks with the racist Jewish legend that Eve copulated with the serpent and produced a demonic race that descends from Cain, who slew his brother Abel. Racist Jews claimed

---

**138**. H. Freedman and M. Simon, Editors, *Midrash Rabbah*, Volume 1, The Soncino Press, London, (1939), pp. 292-293.
**139**. M. Maimonides, *The Guide of the Perplexed*, University of Chicago Press, (1963), pp. 618-619.

that the dark skin of Blacks was the "mark of Cain" (*Genesis* 4:10-12, 15), and the "curse of Ham". The *Zohar* states,

> "Of the three sons of Noah that went forth from the ark, Shem, Ham, and Japheth, Shem is symbolic of the right side, Ham of the left side, whilst Japheth represents the 'purple', which is a mixture of the two. AND HAM WAS THE FATHER OF CANAAN. Ham represents the refuse and dross of the gold, the stirring and rousing of the unclean spirit of the ancient serpent. It is for that reason that he is designated the 'father of Canaan', namely, of Canaan who brought curses on the world, of Canaan who was cursed, of Canaan who darkened the faces of mankind. For this reason, too, Ham is given a special mention in the words, 'Ham, the father of Canaan', that is, the notorious world-darkener, whereas we are not told that Shem was the father of such-a-one, or that Japheth was the father of such-a-one. No sooner is Ham mentioned, than he is pointed to as the father of Canaan. Hence when Abraham came on the scene, it is written, 'And Abraham passed through the land' (Gen. xii, 6), for this was before the establishment of the patriarchs and before the seed of Israel existed in the world, so that the land could not yet be designated by this honoured and holy name. Observe that when Israel were virtuous the land was called by their name, the Land of Israel; but when they were not worthy it was called by another name, to wit, the Land of Canaan. Hence it is written: AND HE SAID, CURSED BE CANAAN, A SERVANT OF SERVANTS SHALL HE BE UNTO HIS BRETHREN, for the reason that he brought curses on the world, in the same way as the serpent, against whom was pronounced the doom, 'Cursed art thou among all cattle' (Gen. III, 14)."[140]

The stigmata of the "mark of Cain", which Jewish racists placed on Blacks, had a lasting destructive effect and was used to justify slavery in the Americas and anti-miscegenation laws. A black slave named Phillis Wheatley published a poem in 1773, which evinces the racist accusation that blacks bear the mark of Cain,

### "On being brought from AFRICA to AMERICA.

'TWAS mercy brought me from my *Pagan* land,
Taught my benighted soul to understand
That there's a God, that there's a *Saviour* too:
Once I redemption neither sought nor knew,
Some view our sable race with scornful eye,
'Their colour is a diabolic die.'
Remember, *Christians, Negros*, black as *Cain*,

---

[140]. Translated by H. Sperling and M. Simon, *The Zohar*, Volume 1, The Soncino Press, London, New York, (1933), pp. 246-247.

May be refin'd, and join th' angelic train."[141]

Congressman Paul Findley stated, among his many revealing remarks about Zionist influence,

> "Journalist Harold R. Piety observes that 'the ugly cry of anti-Semitism is the bludgeon used by the Zionists to bully non-Jews into accepting the Zionist view of world events, or to keep silent.' In late 1978 Piety, withholding his identity in order not to irritate his employer, wrote an article on 'Zionism and the American Press' for *Middle East International* in which he decried 'the inaccuracies, distortions and— perhaps worst—inexcusable omission of significant news and background material by the American media in its treatment of the Arab-Israeli conflict.'
>
> Piety traces the deficiency of U.S. media in reporting on the Middle East to largely successful efforts by the pro-Israel lobby to 'overwhelm the American media with a highly professional public relations campaign, to intimidate the media through various means and, finally, to impose censorship when the media are compliant and craven.' He lists threats to editors and advertising departments, orchestrated boycotts, slanders, campaigns of character assassination, and personal vendettas among the weapons employed against balanced journalism."[142]

Former Mossad agent Victor Ostrovsky wrote in his book *The Other Side of Deception: A Rogue Agent Exposes the Mossad's Secret Agenda* (note that a "Sayanim" is a disloyal and deceitful Jew, who is prepared to betray his or her neighbors at any time in order to advance a perceived Israeli interest),

> "The American Jewish community was divided into a three-stage action team. First were the individual *sayanim* (if the situation had been reversed and the United States had convinced Americans working in Israel to work secretly on behalf of the United States, they would be treated as spies by the Israeli government). Then there was the large pro-Israeli lobby. It would mobilize the Jewish community in a forceful effort in whatever direction the Mossad pointed them. And last was B'nai Brith. Members of that organization could be relied on to make friends among non-Jews and tarnish as anti-Semitic whomever they couldn't sway to the Israeli cause. With that sort of one-two-three tactic, there was no way we could strike out."[143]

Prof. Norman G. Finkelstein writes in his book, *Beyond Chutzpah: On the*

---

**141**. P. Wheatly, "On Being Brought from Africa to America", *Poems on Various Subjects, Religious and Moral*, A. Bell, London, (1773), p. 18.
**142**. P. Findley, *They Dare to Speak Out: People and Institutions Confront Israel's Lobby*, Lawrence Hill & Company, Westport, Connecticut, (1985), p. 296.
**143**. V. Ostrovsky, *The Other Side of Deception: A Rogue Agent Exposes the Mossad's Secret Agenda*, Harper Collins, New York, (1994), p. 32.

*Misuse of Anti-Semitism and the Abuse of History*, University of California Press, Berkeley, (2005), pp. 21-22, 32, and 66,

> "THE LATEST PRODUCTION of Israel's apologists is the 'new anti-Semitism.' [***] The main purpose behind these periodic, meticulously orchestrated media extravaganzas is not to fight anti-Semitism but rather to exploit the historical suffering of Jews in order to immunize Israel against criticism. [***] Finally, whereas in the original *New Anti-Semitism* marginal left-wing organizations like the Communist Party and the Socialist Workers Party were cast as the heart of the anti-Semitic darkness, in the current revival Israel's apologists, having lurched to the right end of the political spectrum, cast mainstream organizations like Amnesty International and Human Rights Watch in this role. [***] WHAT'S CURRENTLY CALLED the new anti-Semitism actually incorporates three main components: (1) exaggeration and fabrication, (2) mislabeling legitimate criticism of Israeli policy, and (3) the unjustified yet predictable spillover from criticism of Israel to Jews generally. EXAGGERATION AND FABRICATION The evidence of a new anti-Semitism comes mostly from organizations directly or indirectly linked to Israel or having a material stake in inflating the findings of anti-Semitism."[144]

In 2006, Professors John J. Mearsheimer and Stephen M. Walt wrote in their paper, "The Israel Lobby and U. S. Foreign Policy",

> "No discussion of how the Lobby operates would be complete without examining one of its most powerful weapons: the charge of anti-Semitism. Anyone who criticizes Israeli actions or says that pro-Israel groups have significant influence over U. S. Middle East policy—an influence that AIPAC celebrates—stands a good chance of getting labeled an anti-Semite. In fact, anyone who says that there is an Israel Lobby runs the risk of being charged with anti-Semitism, even though the Israeli media themselves refer to America's 'Jewish Lobby.' In effect, the Lobby boasts of its power and then attacks anyone who calls attention to it. This tactic is very effective, because anti-Semitism is loathsome and no responsible person wants to be accused of it."[145]

Jimmy Carter, Thirty-Ninth President of the United States of America, wrote in his book *Palestine: Peace not Apartheid*, Simon & Schuster, New York, (2006), page 209,

---

**144**. *See also:* N. G. Finkelstein, *The Holocaust Industry: Reflections on the Exploitation of Jewish Suffering*, Second Edition, Verso, London, New York, (2003).
**145**. J. J. Mearsheimer and S. M. Walt, *The Israel Lobby and U. S. Foreign Policy*, Faculty Research Working Papers Series, Harvard University, John F. Kennedy School of Government, (March, 2006), p. 23.

"Two other interrelated factors have contributed to the perpetuation of violence and regional upheaval: the condoning of illegal Israeli actions from a submissive White House and U.S. Congress during recent years, and the deference with which other international leaders permit this unofficial U.S. policy in the Middle East to prevail. There are constant and vehement political and media debates in Israel concerning its policies in the West Bank, but because of powerful political, economic, and religious forces in the United States, Israeli government decisions are rarely questioned or condemned, voices from Jerusalem dominate in our media, and most American citizens are unaware of circumstances in the occupied territories. At the same time, political leaders and news media in Europe are highly critical of Israeli policies, affecting public attitudes. Americans were surprised and angered by an opinion poll, published by the *International Herald Tribune* in October 2003, of 7,500 citizens in fifteen European nations, indicating that Israel was considered to be the top threat to world peace, ahead of North Korea, Iran, or Afghanistan."

Jimmy Carter also stated,

"You and I both know the powerful influence of AIPAC, which is not designed to promote peace. I'm not criticizing them, they have a perfect right to lobby, but their purpose in life is to protect and defend the policies of the Israeli government and to make sure those policies are approved in the United States and in our Congress—and they're very effective at it. I have known a large number of Jewish organizations in this country [that] have expressed their approval for the book and are trying to promote peace. But their voices are divided and they're relatively reluctant to speak out publicly. And any member of Congress who's looking to be re-elected couldn't possibly say that they would take a balanced position between Israel and the Palestinians, or that they would insist on Israel withdrawing to international borders, or that they would dedicate themselves to protect human rights of Palestinians—it's very likely that they would not be re-elected."[146]

In an interview with George Stephanopoulos on ABC television, in February of 2007, President Carter stated,

"It's almost politically suicidal in the United States for a member of the Congress who wants to seek reelection to take any stand that might be interpreted as anti-policy of the conservative Israeli government, which is equated, as I've seen it myself, as anti-Semitism."[147]

---

**146**. J. Carter, as quoted by E. Clift, "Last Word: Jimmy Carter Revisiting 'Apartheid'", *Newsweek International*, (25 December 2006—1 January 2007).
<http://www.msnbc.msn.com/id/16240761/site/newsweek/>
**147**. Jimmy Carter, Thirty-Ninth President of the United States of America, in an

There is nothing new about fabricated accusations of anti-Semitism. The Judeans who fabricated the Old Testament fabricated a history of Egyptian tyranny which never occurred, and which fictions recklessly defamed the Egyptians as anti-Semites. Esau was defamed as an hereditary anti-Semite for daring to be angry at Jacob for stealing the Covenant from him.[148] Jewish historians defamed Caligula for not tolerating Judean intolerance (etc. etc. etc.).

Douglas Reed, who was a British journalist, but was forced out of the profession, because he reported on Zionist brutality, wrote in December of 1950,

"More important still, during all that period and to the present time, it was not possible freely to report or discuss a third vital matter: Zionist Nationalism. In this case the freedom of the press has become a fallacy during the past two decades. Newspaper-writers have become less and less free to express any criticism, or report any fact unfavourable to this new ambition of the Twentieth Century. When I eventually went to America I found that this ban, for such it is in practice, prevailed even more rigidly there than in my own country.

Today an awakening is supposed to have occurred in the matter of Communism. During the most fateful and decisive years of the Second War, when the things were being done which obviously set the stage for a third one, it was in fact almost impossible for any independent writer to publish any reasonable criticism, supported by no matter what evidence, about Soviet Communism and its intentions. Now, when the damage is done, Communism is much attacked, but even so the mass of Communist writers who were planted in the American and British press during those years has by no means been displaced; and the attentive newspaper-reader in either country may see for himself how the most specious Communist sophistries are daily injected into the editorial arguments and the news-columns of newspapers professing the most respectable principles.

In the matter of Zionist Nationalism, which I hold to be allied in its roots to Soviet Communism, the ban is much more severe. In my own adult lifetime as a journalist, now covering thirty years, I have seen this secret ban grow from nothing into something approaching a law of lèse majesté at some absolute court of the dark past. In daily usage, no American or British newspaper, apparently, now dares to print a line of news or comment unfavourable to the Zionist ambition; and under this thrall matters are reported favourably or non-committally, if they are reported at all, which if they occurred elsewhere would be denounced with the most piteous cries of outraged morality. The inference to me is plain: the Zionist Nationalists are powerful enough to govern governments in the great countries of the

---

interview on, "This week with George Stephanopoulos," ABC, as quoted by Yitzhak Benhorin, "Balanced stand on ME is political suicide, says Carter", www.ynetnews.com, http://www.ynetnews.com/articles/0,7340,L-3369679,00.html, (26 February 2007).

**148**. D. Duke, *Jewish Supremacism: My Awakening on the Jewish Question*, Free Speech Press, Covington, Louisiana, (2002), pp. 200-205.

remaining West!

I believe Zionist Nationalism to be a political movement organized in all countries, which aims to bring all Jews under its thrall just as Communism enslaved the Russians and National Socialism the Germans. I hold it to be as dangerous as both of those, and when I recall the results that came of the subtle suppression of information in the cases of Stalinism and Hitlerism, I judge that the consequences of this even more rigorous suppression will not be less grave.

I think it a cardinal error to identify 'Jews' with Zionist Nationalism, 'Russians' with Communism, or 'Germans' with National Socialism. I saw the enslavement of Germans and Russians and know different. I believe that the astonishingly powerful attempt to prevent any discussion of Zionist Nationalism by dismissing it as the expression of an aversion to Jews, as Jews, is merely meant to stop any public discussion of its objects, which seem to me to be as dangerous to Jew as to Gentile. Of the three groups which have appeared, like stormy petrels, to presage the tempests of our century, the Zionist Nationalists appear to me the most powerful. National Socialism, I think, was but a stooge or stalking horse for the pursuit of Communist aims. Communism is genuinely tigerish, and was strong enough to infest governments everywhere and distort the policies which were pursued behind the screen of military operations; but, if forced into a corner by the rising unease of their peoples, Western politicians are prepared in the last resort to turn against it.

But Zionist Nationalism! ... That is a different matter. Today American Presidents and British Prime Ministers, and all their colleagues, watch it as anxiously as Muslim priests watch for the crescent moon on the eve of Ramadan, and bow to it as the faithful prostrating themselves in the mosque at Mecca. The thing was but a word unknown to the masses forty years ago; today Western politicians hardly dare take the seals of office without first, or immediately afterwards, making public obeisance towards this strange new ambition."[149]

Gore Vidal wrote,

"Currently, there is little open debate in the United States on any of these matters. The Soviet Union must be permanently demonized in order to keep the money flowing to the Pentagon for 'defense,' while Arabs are characterized as subhuman terrorists. Israel may not be criticized at all. (Ironically, the press in Israel is far more open and self-critical than ours.) We do have one token Palestinian who is allowed an occasional word in the press, Professor Edward Said, who wrote (*Guardian*, December 21,1986): since the '1982 Israeli invasion of Lebanon ... it was felt by the Zionist lobby that the spectacle of ruthless Israeli power on the TV screen would have to

---

[149]. D. Reed, *Somewhere South of Suez*, Devin-Adir, U. S. A., (1951), pp. 8-10.

be effaced from memory, by the strategy of incriminating the media as anti-Semitic for showing these scenes at all.' A wide range of Americans were then exuberantly defamed, including myself."[150]

Robert I. Friedman wrote in 1987,

"Indeed, Americans have very little idea about how severely troubled Israel is, or how critical many Israelis are of their own government's policies, such as arming the contras, Khomeini's Iran, and South Africa. And some prominent U.S. editors and publishers who have dropped all pretense of objectivity to become public-relations advisors for the Israeli government hope to keep it that way. [***] And many others who have tried to defy this orthodoxy have come under unrelenting attack from the Israel lobby—a coalition of editors and publishers, pro-Israel PACs, and wealthy businessmen—which tries to silence dissidents with accusations of anti-Israel bias or anti-Semitism. [***] Yet these tactics of intimidation in the service of Israel may backfire. 'It is precisely the fact that it is the job of the national press to be fair and objective that gets these superoverheated Jews foaming,' said the *Washington Post's* Stephen Rosenfeld. 'They want 100 percent. They don't want fairness: they want unfairness on their side, and when they don't get it they accuse the press of being unfair. Most journalists get so much uninformed, unfair whining from the organized Jews that Jewish organizations—and ultimately Israel—may lose their credibility.'"[151]

Arvid Reuterdahl wrote to William L. Fisher on 17 October 1931,

"My dear Mr. Fisher,
　　Dr. Erich Ruckhaber recently sent you a letter of Aug. 29, 1931, addressed here to me for consideration.
　　Having lived through the Einstein Battle, I am well aware of all the difficulties which opposition to Einsteinism meets with everywhere, and not the least in the United States. I have had articles refused by Scientific Societies of which I am a member, because they clearly exposed the Einsteinian Sham.
　　It would be a great stroke for truth if we could find the means of getting '100 Autoren Gegen Einstein' published in the English. I managed to get a reference in a St. Paul Paper, and another indirect reference in the Kansas

---

**150**. G. Vidal, *Imperial America*, Nation Books, New York, (2004), pp. 76-77; originally, *The Observer*, London, (15 November 1987), "But written as of March 1987 In *The Nation*."
**151**. R. I. Friedman, "Selling Israel in America: The Hasbara Project Targets the U.S. Media", *Mother Jones*, (February/March, 1987), pp. 1-9; reprinted "Selling Israel to America", *Journal of Palestine Studies*, Volume 16, Number 4, (Summer, 1987), pp. 169-179, at 170, 178.

City Star, on the occasion of a visit to Kansas City. I enclose a copy of the latter. Through friends, elsewhere, I tried to get newspaper notices, but without success.

The forces behind Einstein have excellent control over the press and scientific journals. They control our mathematical and scientific departments (indirectly) in our universities and colleges—a most deplorable condition. I know, by actual experience, whereof I speak.

I fear that no American publishing house will lend its name to '100 Autoren', because of possible boycott and persecution (financial). Hence the publication involves raising the required funds independently and creating a marketing organization. Where the funds can be raised, at the present time of depression, is a stupendous problem. I too know Dr. Dayton C. Miller through correspondence—a splendid gentleman and true scientist. I have had correspondence with Dr. Charles Lane Poor and he knows of my efforts against Einsteinism. But,—are they in a position to back such a venture? My prolonged illness has incapacitated me financially.

I have seen references to the stand taken by Dr. L. J. Moore of Cincinnati, and he is sound on the Einstein fiasco. There are others. There are other U[niversity] scientists—a few besides these three—who are aware of the Einsteinian nonsense, but many are afraid of losing scientific caste, and perhaps their positions.

Since you are personally acquainted with Dr. Dayton C. Miller, it may be possible for you to approach him on the subject in order to learn his reaction. From his answer, conclusions may be drawn which will be of solid and practical value.

If you will kindly take this step, then we can confer again by correspondence. You may, of course, mention my name to Dr. Miller, stating my position in reference to the urgent need of an English translation of '100 Autoren --'.

If a fearless champion can be found who has the financial resources, then '100 Autoren --' can be gotten to the intelligent public and the days of Einsteinism in the U. S. will soon be numbered—such is the power of '100 Autoren' as I appraise it.

Of course, I am ready to serve in such way as Dean in order to bring this most desirable purpose to a realization.

With best wishes, I remain,<br>Most cordially yours,"[152]

Stjepan Mohorovičić wrote,

"Eine vorzügliche und sehr scharfsinnige Kritik veröffentlichte G. v. GLEICH 1930, wo er alle seine diesbezüglichen Arbeiten gesammelt und

---

[152]. Courtesy of the Department of Special Collections, University of St. Thomas, St. Paul, MN.

geordnet hatte, obwohl das 'Relativitätssyndikat' mit allen Mitteln trachtete, das Erscheinen dieses Werkes zu verhindern. Nun es war sehr schwer die Kritik gänzlich zu unterdrücken, da man in der Wahl der Mittel nicht kleinlich war. Alle, für die Relativitätstheorie ungünstigen Arbeiten wurden einfach kurzerhand als unrichtig, fehlerhaft oder falsch bezeichnet oder als unwichtig (heutzutage ein sehr beliebtes Wort!) oder wenigstens als uninteressant verschwiegen. Von den Philosophen erhielten nur die Applaudierenden das Wort, den kritisch Gesinnten warf man ihre mathematischen Unkenntnisse vor; wer sich darüber unterrichten will, sollte die offenen Briefe des bekannten Philosophen O. KRAUS nachlesen,

[*Endnote:* Vgl. Lit. [*Oskar Kraus : Offene Briefe an Albert Einstein u. Max v. Laue über die gedanklichen Grundlagen der speziellen und allgemeinen Relativitätstheorie. Wien u. Leipzig 1925.*] S. 78 u. ff., dann S. 96 u. ff. So sagte beispielsweise O. KRAUS wörtlich S. 94-95: 'Herr EINSTEIN selbst ist philosophisch Laie... Mit der Zuwendung zu Reichenbachs radikalem Konventionalismus hat er, scheint es, nun den Standpunkt erreicht, der seiner Theorie kongenial ist... Der Konventionalismus fälscht den Wahrheitsbegriff pragmatistisch. Diesem Niveau entspricht die Relativitätstheorie vom philosophischen Standpunkt aus.' (O. KRAUS war Professor an der deutschen Universität in Prag zu gleicher Zeit wie auch A. EINSTEIN).]

und doch haben die Philosophen die Grundlage der Rechnung, nicht aber die Rechnung selbst untersucht. Aber die Relativisten haben übersehen, daß die modernen Relativitätstheorien, ähnlich wie die moderne Musik, voll von Dissonanzen sind, (eine solche Musik entzückt den heutigen Snob außerordentlich und er kann nicht begreifen, daß es gebildete Leute gibt, welche die moderne Musik nicht ausstehen können, aber dafür muß man das Ohr und die richtige musikalische Erziehung haben!). O. KRAUS hat besonders den Umstand hervorgehoben (1. c. S. 96.), 'daß jeder Quark, der für die Theorie zu sein scheint, von den Relativisten mit freundlicher Gebärde begrüßt wird... wahrend eine ernste Kritik mißhandelt wird'.

[*Endnote:* Ein erschreckendes Beispiel ist z. B. der beschleunigte Tod des verdienstvollen 80-jährigen Physikers. C. ISENKRAHE, (vgl. 317 [*Oskar Kraus : Offene Briefe an Albert Einstein u. Max v. Laue über die gedanklichen Grundlagen der speziellen und allgemeinen Relativitätstheorie. Wien u. Leipzig 1925.*] S. 96-97); dann wie M. ABRAHAM behandelt wurde; oder, wenn man einen Physiker als den Gegner der modernen Relativitätstheorien bezeichnet, so sind dann alle seine wissenschaftlichen Verdienste umsonst und ein jeder Stümper bildet sich ein, er habe das Recht ihn zu verleumden.— Ein anderes Beispiel ist der weltbekannte und große deutsche Philosoph HUGO DINGLER; in [*Hans Wagner : Hugo Dinglers Beitrag zur Thematik der Letztbegründung. Kantstud. 47, 148-167, 1955-56. Sonderdruck, Köln 1956.*] S. 1. lesen wir folgendes über den von ibm geführten Kampf für die strenge Wissenschaft: '... ein Kampf, der unter schweren äußeren Bedingungen hatte geführt werden müssen — erst unter dem Vorwurf des Antisemitismus, seit er der Einsteinschen Relativitätstheorie entgegengetreten war, nach 1933 unter dem Vorwurf der Semitophilie, welcher ihn alsbald auch seinen Darmstädter Lehrstuhl kostete, 1945 unter dem Vorwurf einer Verbundenheit mit dem Ungeist des Hitlerreichs, der ihn abermals von der Lehrtätigkeit verwies und über ihn die aktuelle Gefahr eines buchstäblichen Hungertodes heraufführte, schließlich nach seiner Rehabilitierung unter der Last eines schweren Augenleidens.' usw. usw. Der Verfasser könnte noch vieles aus eigener Erfahrung beifügen, aber man wird das alles nach seinem Tode erfahren... (vgl. Anm. 90 [Dies alles sage ich aus eigener Erfahrung. Was ich z. B. persönlich in dieser Beziehung erlebt und zu ertragen habe, wird man erst nach meinem Tode erfahren. Dies wird eine wahre Anklage gegen die relativistischen unerhörten Kampfmethoden sein, welche nur mit der mittelalterlichen Inquisition verglichen werden können.])). Siehe auch [*Wilhelm Krampf : Die Philosophie Hugo Dinglers. München 1955.*] u. [A. FRITSCH, G. BARTH, S. MOHOROVIČIĆ: Hugo Dingler Gedenkbuch zum 75. Geburtstag. Wissen im Werden 2, H. 4, 169-183, 1958 (und als selbständige Broschüre München 1959).].

Dies wirkte aber verhängnisvoll und diese modernen Theorien wurden größtenteils ein Tätigkeitsfeld pour ceux qui savent vivre ... oder wie ein lachender Philosoph sagte:

[*Endnote:* * * * Demokritos oder hinterlassene Papiere eines lachenden Philosophen. 4. Aufl. Bd. VII., Stuttgart 1853., S. 322.—Wir müßten ebenfalls mit JULIAN APOSTATA eine Rede gegen die ungebildeten... halten.—Siehe auch [*Clyde R. Miller : Kunstgriffe der Propaganda (Das Institut für Propaganda-Analyse d. Columbia University). Neue Auslese 3, 93-97; 1948 (übersetzt aus d. Jb. 'New Directions', New York).*—*Hier lesen wir folgendes (S. 96): 'Mit falschen Karten spielen ist ein Kunstgriff, bei dem der Propagandist alle Künste der Täuschung und des Truges anwendet, um unsere Unterstützung für sich selbst, seine Gruppe, Nation, Rasse, Politik, Methoden und Ideale zu gewinnen. Er entstellt bewusst die Wahrheit. Er übertreibt oder 'untertreibt', um sich um Diskussionen zu drücken und den Tatsachen aus dem Weg zu gehen. Er 'vernebelt' eine peinlich Angelegenheit, indem er mit grossem Trara eine neue Streitfrage aufs Tapet bringt. Er liefert Halbwahrheiten unter der Maske der Wahrheit (von uns unterstrichen). Durch den Kunstgriff der 'falschen Karten' wird ein mittelmässiger Kandidat als ein Genie hingestellt; ... Zu dieser Art von Falschspielerei gehören Täuschung, Heuchelei und Unverschämtheit'.*]

'... an Höfen ist Höflichkeit der Verstand und die Münze... '."[153]

## 4.7 Einstein's Trip to America

Einstein was discredited in Germany in late 1920. In early 1921, Einstein desperately needed a boost and a break. Zionist Kurt Blumenfeld arranged for Einstein to take a trip to America in order to spread propaganda for political Zionism and to raise money for the cause, on the deceitful premise that the money would go to fund an university in Jerusalem, the "Jewish university"[154] or "Hebrew University". Einstein was deceived. The real goal of the Zionists who took advantage of him was to exploit Einstein's fame for profit.

Elements of the American press again promoted Einstein as the greatest genius of all time. For Jewish racists, this provided helpful racist propaganda claiming that all important contributions to the world of thought were made by Jews. The racist political Zionist United States Supreme Court Justice Louis Dembitz Brandeis wrote in a letter dated 1 March 1921,

> "You have doubtless heard that the Great Einstein is coming to America soon with Dr. Weizmann, our Zionist Chief. Palestine may need something more now than a new conception of the Universe or of several additional dimensions; but it is well to remind the Gentile world, when the wave of anti-Semitism is rising, that in the world of thought the conspicuous contributions are being made by Jews."[155]

Viktor G. Ehrenberg, Hedwig Born's father, wrote to Einstein on 23 November 1919,

> "So it uplifts the heart and strengthens one's faith in the future of mankind when one sees the researchers of all nations prostrating themselves before a man of Jewish blood, who thinks and writes in the German language, in full recognition of his greatness."[156]

---

**153.** S. Mohorovičić, "Raum, Zeit und Welt", in two parts in K. Sapper, Editor, *Kritik und Fortbildung der Relativitätstheorie*, Akademische Druck- u. Verlagsanstalt, Graz, (1958/1962), Part 1 in Volume 1, (1958), pp. 168-281, at 277, 279, notes 317, 352, 364, 365; Part 2 in Volume 2, (1962), pp. 219-352, at 273, 317, 319, 329, notes 90, 108, 109, 110, 637.
**154.** Letter from A. Einstein to H. Bergman of 5 November 1919, English translation by A. Hentschel, *The Collected Papers of Albert Einstein*, Volume 9, Document 155, Princeton University Press, (2004), pp. 132-133, at 132. *See also:* H. N. Bialik, "Bialik on the Hebrew University", in A. Hertzberg, *The Zionist Idea*, Harper Torchbooks, New York, (1959), pp. 281-288, at 284-285.
**155.** L. D. Brandeis, M. I. Urofsky and D. W. Levy, Editors, *Letters of Louis D. Brandeis* Volume 4, State University of New York Press, Albany, New York, (1975), pp. 536-537.
**156.** Letter from V. G. Ehrenberg to A. Einstein of 23 November 1919, English translation by A. Hentschel, *The Collected Papers of Albert Einstein*, Volume 9, Document 173,

Paul Ehrenfest wrote to Einstein that he had heard that the Zionists were using Einstein to promote the myth that he was a "Jewish Newton" and a Zionist. Ehrenfest was tortured by the fact that his character would not allow him to participate in the dishonest promotion of Einstein to the public. He believed it would ultimately be destructive to Jews. Ehrenfest committed suicide in 1933.

In 1905 and 1906, Paul Ehrenfest considered Lorentz' 1904 paper[157] on special relativity and Poincaré's 1905 Rendiconti paper[158] on space-time to be the most significant work (both historically and scientifically) on the subject of the principle of relativity. Ehrenfest and his wife Tatiana attended David Hilbert's 1905 Göttingen seminars on electron theory, which described Lorentz' and Poincaré's work on special relativity. They knew that Einstein did not create the theory of relativity. Paul Ehrenfest wrote to Albert Einstein on 9 December 1919,

> "I hear, for ex., that your accomplishments are being used to make propaganda, with the 'Jewish Newton, who is simultaneously an ardent Zionist' (I personally haven't *read* this yet, but only *heard* it mentioned). [***] But I cannot go along with the propagandistic fuss with its *inevitable* untruths, precisely *because* Judaism is at stake and *because* I feel myself so

---

Princeton University Press, (2004), p. 145.
**157**. H. A. Lorentz, "Electromagnetische Verschijnselen in een Stelsel dat Zich met Willekeurige Snelheid, Kleiner dan die van Het Licht, Beweegt", *Koninklijke Akademie van Wetenschappen te Amsterdam, Wis- en Natuurkundige Afdeeling, Verslagen van de Gewone Vergaderingen*, Volume 12, (23 April 1904), pp. 986-1009; translated into English, "Electromagnetic Phenomena in a System Moving with any Velocity Smaller than that of Light", *Proceedings of the Royal Academy of Sciences at Amsterdam* (*Noninklijke Nederlandse Akademie van Wetenschappen te Amsterdam*), 6, (May 27, 1904), pp. 809-831; reprinted *Collected Papers*, Volume 5, pp. 172-197; a redacted and shortened version appears in *The Principle of Relativity*, Dover, New York, (1952), pp. 11-34; a German translation from the English, "Elektromagnetische Erscheinung in einem System, das sich mit beliebiger, die des Lichtes nicht erreichender Geschwindigkeit bewegt," appears in *Das Relativitätsprinzip: eine Sammlung von Abhandlungen*, B. G. Teubner, Leipzig, (1913), pp. 6-26.
**158**. H. Poincaré, "Sur la Dynamique de l'Électron", *Rendiconti del Circolo matimatico di Palermo*, Volume 21, (1906, submitted July 23rd, 1905), pp. 129-176; reprinted in H. Poincaré, *La Mécanique Nouvelle: Conférence, Mémoire et Note sur la Théorie de la Relativité / Introduction de Édouard Guillaume*, Gauthier-Villars, Paris, (1924), pp. 18-76; reprinted *Œuvres*, Volume IX, pp. 494-550; redacted English translation by H. M. Schwartz with modern notation, "Poincaré's Rendiconti Paper on Relativity", *American Journal of Physics*, Volume 39, (November, 1971), pp. 1287-1294; Volume 40, (June, 1972), pp. 862-872; Volume 40, (September, 1972), pp. 1282-1287; English translation by G. Pontecorvo with extensive commentary by A. A. Logunov with modern notation, *On the Articles by Henri Poincaré ON THE DYNAMICS OF THE ELECTRON*, Publishing Department of the Joint Institute for Nuclear Research, Dubna, (1995), pp. 15-78.

thoroughly a Jew."[159]

Immediately upon his arrival at America's shores, Einstein mischaracterized any and all opposition to him and the theory of relativity as if it were anti-Semitism, *per se*.[160] After Einstein returned to Europe and after these Zionists bilked many generous Americans in the name of ethnic pride and duty, the promised funding of the university did not materialize. The nationalists allegedly could not agree on the final form this ethnically segregated school should take.[161] We learn from American Zionist Louis Dembitz Brandeis' letters that the University was nothing but a "side show",

> "The University, important & dear to us, is merely a side show. It can wait. Nothing must be done in relation to it which would embarrass or confuse the main issue. It should be taken up—if and only if it would be helpful in furthering our fight on the main issue."[162]

And where did the money go, which good-hearted Americans had donated for a university? Again, Brandeis' letters provide us with some likely answers,

> "In telling [Einstein] of the misappropriation of which we learned in London, I mentioned the diversion also of a University Fund & our apprehension as to further diversion."[163]

The editors of Brandeis' letters wrote,

> "It was L[ouis] D[embitz] B[randeis]'s belief that the funds earmarked for the Hebrew University had been used for various projects in the Haifa area, and he wanted deHaas to provide whatever information they had on the matter to Einstein."[164]

Zionist racists set the tone for the racist "Aryan Physics" movement that would soon follow the political Zionists' smear campaigns against Germans,

---

[159]. Letter from P. Ehrenfest to A. Einstein of 9 December 1919, English translation by A. Hentschel, *The Collected Papers of Albert Einstein*, Volume 9, Document 203, Princeton University Press, (2004), pp. 173-175, at 174.
[160]. "Prof. Einstein Here, Explains Relativity", *The New York Times*, (3 April 1921), pp. 1, 13, at 1.
[161]. *Cf.* Schlomo Ginossar, a. k. a. Simon Ginsburg, a. k. a. Salomon Ginzberg, "Early Days", *The Hebrew University of Jerusalem, 1925-1950*, Universitah ha-'uvrit bi-Yerushalayim, Jerusalem, (1950), pp. 71-74.
[162]. L. D. Brandeis, M. I. Urofsky and D. W. Levy, Editors, *Letters of Louis D. Brandeis* Volume 4, State University of New York Press, Albany, New York, (1975), p. 555.
[163]. L. D. Brandeis, M. I. Urofsky and D. W. Levy, Editors, *Letters of Louis D. Brandeis* Volume 4, State University of New York Press, Albany, New York, (1975), p. 556.
[164]. L. D. Brandeis, M. I. Urofsky and D. W. Levy, Editors, *Letters of Louis D. Brandeis* Volume 4, State University of New York Press, Albany, New York, (1975), p. 556.

which followed centuries of active discrimination against Jews which was only then beginning to lessen, and so the cycle of hatred continued. These political Zionists had little respect for the truth or for the innocents they bilked. Einstein's "secretary" on the trip, Salomon Ginzberg, later wrote,

> "It was also hoped that the University, being a non-political institution of great spiritual appeal, would find supporters among the wealthier non-Zionist Jews who might not contribute to Zionist funds proper."[165]

Salomon Ginzberg, a. k. a. Simon Ginsberg, was the son of the famous Zionist Ha'am. Ginzberg apparently thought that Einstein was a somewhat ridiculous person. Ginzberg mocked Einstein's "speech"—a Goebbels-like plea for ethnic unity behind a lone *Führer*.[166] Einstein declared to the Zionists of America,

"You have one leader — Weizmann. Follow him and no other!"[167]

Jewish lore had long inspired a desire among Jews for a charismatic leader, be it another Moses, or the Messiah King. In the 1600 and 1700's many would-be messiahs appeared and some, like Shabbatai Zevi and Jacob Frank, attracted large followings numbering in the millions. Graetz famously called for a charismatic leader to the lead the Jews in the modern world. On the Zionists' quest to find a "great man" to be their "dictator" and on the naturalness of dictatorships to Zionists, *see:* N. Goldman, "Zionismus und nationale Bewegung", *Der Jude*, Volume 5, Number 4, (1920-1921), pp. 237-242, at 240-242; which was part of a series including: "Zionismus und nationale Bewegung", *Der Jude*, Volume 5, Number 1, (1920-1921), pp. 45-47; and "Zionismus und nationale Bewegung", *Der Jude*, Volume 5, Number 7, (1920-1921), pp. 423-425.

---

**165**. *Cf.* Schlomo Ginossar, a. k. a. Simon Ginsburg, a. k. a. Salomon Ginzberg, "Early Days", *The Hebrew University of Jerusalem, 1925-1950*, Universitah ha-'uvrit bi-Yerushalayim, Jerusalem, (1950), pp. 71-74, at 72.
**166**. *See, for example*, J. Goebbels, "Der Führer", *Aufsätze aus der Kampfzeit*, Zentralverlag der NSDAP, Munich, (1935), pp. 214-216; **and** "Goldene Worte für einen Diktator und für solche, die es werden wollen", *Der Angriff*, (1 September 1932); reprinted in: *Wetterleuchten: Aufsätze aus der Kampfzeit*, Zentralverlag der NSDAP., Franz Eher Nachf., München, (1939), pp. 325-327. On the Zionists' quest to find a "great man" to be their "dictator", *see:* N. Goldman, "Zionismus und nationale Bewegung", *Der Jude*, Volume 5, Number 4, (1920-1921), pp. 237-242, at 240-242; which was part of a series including: "Zionismus und nationale Bewegung", *Der Jude*, Volume 5, Number 1, (1920-1921), pp. 45-47; and "Zionismus und nationale Bewegung", *Der Jude*, Volume 5, Number 7, (1920-1921), pp. 423-425.
**167**. *Cf.* Schlomo Ginossar, a. k. a. Simon Ginsburg, a. k. a. Salomon Ginzberg, "Early Days", *The Hebrew University of Jerusalem, 1925-1950*, Universitah ha-'uvrit bi-Yerushalayim, Jerusalem, (1950), pp. 71-74, at 73. ***See also:*** J. Stachel, *Einstein from 'B' to 'Z'*, Birkhäuser, Boston, (2002), p. 79, note 41.

When Albert Einstein traveled to America in April of 1921 to promote his Zionist agenda he had received a triumphant welcome, but soon met with great and growing opposition. Einstein was lampooned and humiliated in certain segments of the international press. Einstein left America in defeat. He expressed his bitterness towards America in an interview for the *Nieuwe Rotterdamsche Courant*. Einstein stated, as reported in *The New York Times* on 8 July 1921 on page 9,

"BERLIN, July 7.—Dr. Albert Einstein, the famous scientist, made an amazing discovery relative to America on his trip which he recently explained to a sympathetic-looking Hollander as follows:

'The excessive enthusiasm for me in America appears to be typically American. And if I grasp it correctly the reason is that the people in America are so colossally bored, very much more than is the case with us. After all, there is so little for them there!' he exclaimed.

Dr. Einstein said this with vibrant sympathy. He continued:

'New York, Boston, Chicago and other cities have their theatres and concerts, but for the rest? There are cities with 1,000,000 inhabitants, despite which what poverty, intellectual poverty! The people are, therefore, glad when something is given them with which they can play and over which they can enthuse. And that they do, then, with monstrous intensity.

'Above all things are the women who, as a literal fact, dominate the entire life in America. The men take an interest in absolutely nothing at all. They work and work, the like of which I have never seen anywhere yet. For the rest they are the toy dogs of the women, who spend the money in a most unmeasurable, illimitable way and wrap themselves in a fog of extravagance. They do everything which is the vogue and now quite by chance they have thrown themselves on the Einstein fashion.

'You ask whether it makes a ludicrous impression on me to observe the excitement of the crowd for my teaching and my theory, of which it, after all, understands nothing? I find it funny and at the same time interesting to observe this game.

'I believe quite positively that it is the mysteriousness of what they cannot conceive which places them under a magic spell. One tells them of something big which will influence all future life, of a theory which only a small group, highly learned, can comprehend. Big names are mentioned of men who have made discoveries, of which the crowd grasps nothing. But it impresses them, takes on color and the magic power of mystery, and thus one becomes enthusiastic and excited.

'My impressions of scientific life in America? Well, I met with great interest several extraordinarily meritorious professors, like Professor Milliken [*sic*]. I unfortunately missed Professor Michelson in Chicago, but to compare the general scientific life in America with Europe is nonsense.'"[168]

This is but a part of a longer polemic interview, in which Einstein also smeared all Germans as corrupt. Einstein repeated some of what Gehrcke had said, though Einstein had called Gehrcke "anti-Semitic" for saying the same thing. The full interview of 29 June 1921 is reproduced in Dutch and English, together with an interpretation initially published in German in the *Berliner Tageblatt* on 7 July 1921, in *The Collected Papers of Albert Einstein*, Volume 7, Appendix D, (2002), pp. 620-627.

Einstein's comments met with much criticism and a damage control apparatus quickly began to repair the harm he had done to his reputation, by denying that he had said what he had said.[169] Some Americans stepped forward to say, "I told you so!" *The Minneapolis Evening Tribune* wrote on 8 July 1921,

**"Einstein Has No Valid
Cause to Congratulate
            Self, Reuterdahl Says**

In Calling Americans 'Lot of
Bored Low Brows,' He

---

**168.** *The New York Times*, (8 July 1921), p. 9.
**169.** N. Robbins, *Baltimore Evening Sun*, (29 April 1921). "Americans Tremendously Bored, Einstein Says, Explaining 'Exaggerated Welcome'", *Minneapolis Morning Tribune*, (8 July 1921). "Einstein Has No Valid Cause to Congratulate Self, Reuterdahl Says", *Minneapolis Evening Tribune*, (8 July 1921), p. 10. "The Amused Mr. Einstein", *Minneapolis Morning Tribune*, (9 July 1921). "Reuterdahl Sees No Cause for Einstein's Slurs on Americans", *The Minneapolis Morning Tribune*, (9July 1921). "Chicago Women Resent Einstein's Opinions", *The New York Times*, (9 July 1921), p. 7. "Probably He Did Say It All", *The New York Times*, (9 July 1921), p. 8. K. W. Payne, "Einstein on Americans, wherein the Eminent Scientist Failed to Understand Us", *The New York Times*, Section 2, (10 July 1921), p. 2. Response, "Einsteins amerikanische Eindrücke. Was er wirklich sah", *Vossische Zeitung*, Morning Edition, Supplement 1, (10 July 1921), Front Page. A transcription is found in *The Collected Papers of Albert Einstein*, Volume 7, Appendix E, Princeton University Press, (2002), pp. 628-630. "A Product of His Education", *The New York Times*, (11 July 1921), p. 10. "Explanation Rather than Denial", *The New York Times*, (12 July 1921), p. 12. "Prohibition Stays, Says Dr. Einstein", *The New York Times*, Section 2, (31 July 1921), p. 4. An anti-Semitic article appeared in *The Dearborn Independent*, "Relatively Unimportant, Extremely Typical", (30 July 1921), p. 14. Einstein had declared America "violently" "anti-German", which statement also brought criticism. *See:* "Dr. Einstein Found America Anti-German. Violently So, He Says, Though He Noted That a Reaction Was Setting In", *The New York Times*, (2 July 1921), p. 3. "A Genius Makes a Mistake", *The New York Times*, (4 July 1921), p. 8. *New York Herald Magazine*, (26 June 1921). J. Riem, "Amerika über Einstein", *Deutsche Zeitung*, (Berlin), (1 July 1921); and "Zu Einsteins Amerikafahrt", *Deutsche Zeitung*, (Berlin), (13 September 1921).

Forgets the Ungullible.

Makes No Mention of Terrific
Lampooning He Received at
Hands of His Critics.

Professor Albert Einstein's lofty conception of the American people as a lot of bored lowbrows who couldn't find intellectual amusement elsewhere and so turned to his theory of relativity without understanding it, drew a sharp rejoinder today from Prof. Arvid Reuterdahl, dean of the department of engineering and architecture at St. Thomas college. The remarks by the scientist whose recent visit to the United States attracted nation-wide attention, were cabled last night from Berlin.

'Doctor Einstein has omitted all reference to the terrific lampooning to which he was subjected by the Eastern newspapers during the last week of his sojourn with us,' Professor Reuterdahl remarked. 'He has no valid reason to congratulate himself while smiling at the unsophistication and gullibility of the American people.

### Einstein Appeared Amused.

'The radio dispatch from Berlin, which appeared in The Minneapolis Morning Tribune today, conveys the impression that Doctor Einstein was greatly amused by his recent reception in the United States,' he continued. 'He attributes the exaggerated enthusiasm shown him to the fact that our people are bored. In that connection he points out that we have theaters to alleviate the weariness of our dull existence but he intimates that we, nevertheless, welcome new thrills. His remarks indicate that he believes that he furnished us with a new 'thrill,' which accounts for the alleged enthusiasm.

'Professor Einstein found this attitude very comical and consequently confirmative of his pre-established conviction that Americans are lacking in intelligence. However, Doctor Einstein did not hesitate to come to our shores in order to lend zest to the financial campaign of the Zionists, who do not underestimate the advertising value of an international celebrity. This remark is not intended to be derogatory to the Zionist movement, which, undoubtedly is a worthy cause. Nevertheless, we cannot avoid feeling like a man who, having been outwitted in a trade, must remain impassive while the victor laughs at him.

### Entire Tale Untold.

'Dr. Einstein, however, has not told the entire tale. He has adroitly omitted all reference to the terrific lampooning to which he was subjected by the eastern newspapers during the last week of his sojourn with us. Never before has a man been subjected to such colossal ridicule. He was even likened to the notorious Dr. Cook and Friedmann.

'Mr. Nelson Robbins, in the Baltimore Evening Sun, April 29, 1921, says: 'But the proletariat having forgotten Friedmann and his unexplainable discoveries, it hasn't forgotten a host of men like him. Remembering them, the proletariat will be ding-busted if it will swear allegiance to any idea that it cannot understand and which is labeled unexplainable by the 'mentally equipped,' who tap the individual inquirer on the head and, with kindly smile, tell him to run along and not bother his little brain about things he cannot understand.'

'Dr. Einstein, therefore, has no valid reason to congratulate himself enthusiastically while smiling outwardly at the unsophistication and gullibility of the American people.'"

Einstein's feigned amusement is belied by his bitterness at being mocked in America. Contrast Einstein's later remarks, after he had left America, with an interview he gave to *The New York Times* while in America, which was published in *The New York Times Book Review and Magazine* on 1 May 1921 on page 50. In this interview Einstein appears as an especially odd and childlike man, who had wondered from his script. On 15 March 1921, Zionist Kurt Blumenfeld had warned Zionist Chaim Weizmann that it would be unwise to let Einstein make speeches during his trip to America, "Einstein is a poor speaker and often says things out of naiveté that are unwelcome to us[.]"[170] The "secretary" who broke into the conversation during the interview was the son of Zionist Ha'am, Salomon Ginzberg. Many of Einstein's comments are reminiscent of the spirit of Zionist Israel Zangwill's play *The Melting-Pot: Drama in Four Acts*, Macmillan, New York, (1909); and Einstein may have been encouraged to promote the melting-pot idea in order to promote the immigration of Eastern European Jews to America. Einstein's interview:

## "Einstein on Irrelevancies
### *By DON ARNALD*

How comfortable you make everything in the hotel! Every door, every window, is perfect; nothing is out of order. It is all so well planned and well organized. I never saw such rooms; such care for details; such hotel lobbies, with so many to serve you. Everything—everything is systematized, down to the bathrooms. You people in America are very practical. I like the way you light up the windows with the signs. I like the cheerful way you arrange the electricity up and down the streets.'

So spoke Professor Albert Einstein, apostle of relativity, in the course of a talk about his experiences in New York.

'What was it that impressed you most when you arrived?' the interviewer asked.

'Ah! I see so many nationalities living together so well. America is a country of many different peoples at peace with one another. Then, too, I like the restaurants with the 'color' of the nations in the air. Each has its own

---

**170**. J. Stachel, *Einstein from 'B' to 'Z'*, Birkhäuser, Boston, (2002), p. 79, note 41.

atmosphere. It is like a zoological garden of nationalities, when you go from one to the other.

'Are you a bit disappointed not to find some beer in our dining rooms?'

'I cannot say alcohol is as bad as people think it is,' replied the professor. 'It may not be so good for men to spend all their wages on drinking. But it is more an economic question than a question of health. Some workmen must have liquor, it seems. We must not take everything away. Prohibition shows the strength of your democratic Government against private interests. In a corrupt State this could not be done.'

'Do you consider it against personal liberty to take liquor away?'

'How could that be in America? You have a republic. You have no dictator who makes slaves of people. Nothing is done by a democratic Government could be done against freedom. I think you will find it best for the economic welfare of the people in the end.'

'How about tobacco?' was the next question. 'Some people want to take that away, too.'

Dr. Einstein drew back in surprise. 'Oh, my, no! I never heard of it. So some one is starting this? Who is doing this?'

'Some temperance organization here in the United States.'

The professor said: 'If I do not wish to smoke, I say it is excellent to take my tobacco away. But I do wish to smoke, so I say I do not like you to do that.'

'But they say it is not healthful.'

'If you take our tobacco and everything else away, what have you left?' cried Professor Einstein. 'It may be healthful to take away tobacco, but it is mighty lonesome.' He thought a moment. 'But this is economic, too,' he said at last. 'The men spend too much money on cigars, and their wives kick; therefore, they take it way. They say it costs too much money to smoke. I do not know! I have never heard of such a thing as taking away a man's smoking! I'll stick to my pipe. I do not care who will not smoke. I will! If you take everything away, life is not worth while!'

'And the blue laws—how about them?'

'Blue laws? Blue laws? I never heard of those blue laws in my life. What are you saying?' The professor fairly blazed with consternation.

'They want to pass laws to close up all places of amusement on Sunday,' the interviewer explained. 'All theatres, music shows, baseball and other places will be shut down, including everything for relaxation, even amusement parks and the movies.'

'For Heaven's sake. More laws? I never heard of such a thing. Here's what I say: Men must have rest, yes? But what is the right rest? You cannot make a law to tell people how to do it. See—some people have rest when they lie down and go to sleep. Others have rest when they are wide awake and are stimulated. They must work or write or go to amusements to find rest. If you pass one law to show all people how to rest, that means you make everybody alike. But everybody is not alike. No, I do not care for these blue laws. They will do no good for the country or the people.

'Many workmen want to go to movies on Sunday because they have no time during the week days, so they find rest there,' he continued. 'And that is very good.'

'What do you think of our movies and the theatres?'

'I've been so busy that I haven't had much time, but I have never in my life seen such theatres—everything for your taste, all sorts of plays, comedy, tragedy, romance, pageants. And the movies? I am enthusiastic about them—I mean for the presentation of living moving things. They will develop more and more. In general, the pictures shown now are not so artistic, but they will get better, very much better, all the time. The art is not high enough now, but soon you will have science through this art, as well as you are now having art through this science. I see how the movies will be used in the future for science in bacteriology and technology. Perhaps not so soon for astronomy, because the motions of the heavenly bodies are too quick for measurement. But the movies must only be fitted well, and they can be used most adequately for instruction in all science! I think, all in all, the movies are only in their infancy. They are very beautiful, but they will get better, until the best plays can be shown. You deserve much credit for doing such fine pictures. I compliment you, and I hope for more artistic plays right along.'

At this point his wife, a charming little gray-haired lady, slipped into the room and sat by her husband's side.

'Maybe I can help you,' she said kindly. 'I speak English, and I can interpret for him.' The interview up to that point had been in German.

'Perhaps you can tell me something about the professor's life,' I asked. Dr. Einstein laughed heartily.

'He does not want my life,' said he. 'That is of no use to him. Why should he care for that. He is asking what I think of New York. I tell him glorious! I tell him I see here the greatest city in the world, like Paris, like London, only better! I tell him here all people of all nationalities are melted together—and are happy. I tell him the stranger comes here and is full of joy because he goes to his people at once and feels at home.'

'But your book on relativity translated into English, maybe he wants that,' queried Mrs. Einstein.

'No, why that?' said the professor. 'He doesn't come here for relativity. He comes here to see me. I want to say something to the people, how I like the restaurants and the theatres and the movies and the hotels, and how I do not like the blue laws—and if they take away my tobacco—I do not know what I'll do, but I'll take America anyway, no matter what they do.'

At this the secretary arrived. He wanted to add a word on the professor's mission in America. He said:

'I suppose you know Professor Einstein is here to help the University of Palestine. Its foundation stone was laid by Dr. Weizmann in 1918, and since then the university site has been expanded. There is also a library with more than 3,000 volumes and rapidly growing. Plans have been worked out both for the complete university of the future and for a comparatively modest beginning. The time has now come for us to make a foundation fund, part of which will go to the university. American people play a great part in world politics, showing that their aspirations are noble, and we have come from sick and suffering Europe with feelings of hope, convinced that our spiritual aims will command the full sympathy of the American Nation.'

Dr. Einstein broke in: 'We will receive their enthusiastic approval, we are sure, but the people know all this. This gentleman asks me other things, and I tell him what I think of New York.'

He slapped me on the back and added: 'You greet for me all the good people of America and you say, 'I feel at home here among people, many different people from all the nations in the world.'"

## 4.7.1 Einstein Faces Criticism in America

Though Einstein had hoped to run away from his critics, he had an international reputation as a coward, a plagiarist and a scientific fraud. Things we not as easy for Einstein in America as he had hoped they would be.

### 4.7.1.1 Einstein Hides from Reuterdahl's Challenge to Debate

On 10 April 1921, *The Minneapolis Sunday Tribune* reported Prof. Arvid Reuterdahl's charges against Einstein,

> *"Einstein Branded Barnum of Science,*
> *Minnesota Man Calls Relativity 'Bunk'*

St. Thomas Dean of Engineering Challenges German to Debate.

Teuton's Pet 'Cult' Born 13 Years Before Him, Says Professor.

Reuterdahl Cites Passages in 1914 Treatise to Back Assertions.

Branding Prof. Albert Einstein as a sophist, a dealer in 'might-have-beens' and the Barnum of the scientific world, Prof. Arvid Reuterdahl, dean of the Engineering school of St. Thomas College, St. Paul, yesterday challenged the German savant to a written debate on his theory of relativity.

Professor Reuterdahl, who has been exploring the worlds conquered by Einstein since 1902, declared that he was willing to meet the much-heralded mathematician at any time in a written debate, and that he was prepared to prove that Einstein's theory is largely 'bunk.' Professor Reuterdahl used the scientific word for it, but that is what he meant.

**'Work Antedated by Another.'**

Coupled with his challenge to a debate, Professor Reuterdahl declared Einstein was not only deceiving scientists with a mythical theory, but that he was either a plagiarist, or his work has been antedated by another without his knowledge.

'Einstein is at liberty to accept either horn of the dilemma,' he said.

That the Einstein theory of relativity in its gravitational aspects was advanced in 1866, 13 years before Einstein was born, by a scientist known under the pen name of 'Kinertia' is the contention of Professor Reuterdahl, in a statement in which he gives the life history of both men, and gives references and dates to prove his charge. While not accepting the theory, he gives 'Kinertia' credit for its origin.

### American Scientists 'Jolted.'

Professor Reuterdahl, however, gives credit to Einstein for one thing, which, he says, more than justifies his claim to prominence. The German savant, he says, has broken down the barriers of set ideas in science, and made it possible for a hearing for new ideas.

'The American scientists,' said Professor Reuterdahl, 'are the most clannish and orthdox in the world. In the Old World the scientific journals publish articles advancing new theories. Here they will not consider anything except that which is based on their own knowledge and belief. If Einstein has done anything, he has jolted American scientists into accepting something new.' Professor Reuterdahl paid tribute to Einstein's genius as a mathematician, declaring him to be one of the greatest in the world.

### Magazine Articles Cited.

Professor Reuterdahl refers to 11 articles which appeared in Harper's Weekly in 1914 giving 'Kinertia' credit for originating the so-called Einstein theory of gravitation.

'If it is true that 'Kinertia' actually considered the Einsteinian problem in these essays,' he says, 'then the question of priority is inevitably raised and the unparalleled originality claimed for Einstein's work becomes a debatable matter.'

Einstein's investigation of his theory is traced by articles which appeared in German publications.

'The year 1905 is considered, by most authorities on Einstein's work,' he says, 'as the birth year of the theory of relativity.

### Theory Announced in 1915.

'Careful search, however, has revealed a paper on this subject which was published in Berlin during the year 1904 in the journal 'Sitzungsberichte.' That portion of Einstein's theory which deals with the phenomenon of gravitation is a later development. Einstein first gave his attention to the problem of gravitation in 1911, when he developed the principle of equivalence of gravitational and accelerative fields.

'Other phases of this subject were dealt with in papers which appeared in the years 1912 and 1913. A further elaboration, the joint work of Einstein and Marcel Grossman, appeared in 1914. The theory in its final and complete form was announced in the year 1915.

### Historical Summary.

'A brief historical summary of the work of 'Kinertia' is now in order. Lord Kelvin first aroused 'Kinertia's' interest in the problem of gravitation. That was in the year 1866, when 'Kinertia' was a student under Lord Kelvin. 'Kinertia' even then did not agree with the Newtonian theory of force as presented by Lord Kelvin. Incidentally, we desire to call the reader's attention to the fact that Albert Einstein was born in 1879 in Ulm, Germany, 13 years later.

'During the period from 1877 to 1881, 'Kinertia' became convinced that acceleration was the basic cause of what we generally speak of as 'weight.'

### 'Kinertia' Ridiculed in U. S.

'The reader undoubtedly is aware of the fact that acceleration plays the fundamental role in Einstein's theory of gravitation. 'Kinertia' corresponded with Kelvin, Tait and Niven of Cambridge with the hope that he would be able to interest these men in his startling theory. This attempt met with little or no sympathy.

'His attempts, dating from the year 1899, to persuade our stubborn American scientists that the Newtonian theory of gravitation must be revised met with nothing but ridicule and indifference. To Harper's Weekly and its managing editor, Mr H. D. Wheeler, belongs the credit of having published 'Kinertia's' series of articles entitled 'Do Bodies Fall?' The first article appeared in the issue of August 29, 1914, Vol. 59.

### Similarity of Views Pointed Out.

The final article is dated November 7, 1914. From the preceding it is evident that 'Kinertia' derived his norm of gravitation before Einstein was born.

Professor Reuterdahl quotes from the writing of Einstein and 'Kinertia' to prove the similarity of their views, and says:

'It is noteworthy that the only real difference between these two citations is that Einstein derives his conclusions from a hypothetical case, whereas 'Kinertia' draws his conclusions from an actual experiment upon himself.'

Further quotations are from Prof. A. S. Eddington's 'Space Time Gravitation,' published by the Cambridge University Press in 1920; from an article by Prof. Edwin B. Wilson of the Massachusetts Institute of Technology, and from 'Kinertia's' articles.

### Striking Similarity.

These quotations, he says. 'show the striking similarity existing between Einstein and 'Kinertia' when they consider the relation between acceleration and gravitation, a similarity which extends not only to intent but affects even the very words.'

The following quotation from Einstein's 'Relatively' illustrates that scientist's theory as to the relation between acceleration and gravitation, according to Professor Reuterdahl:

'We imagine a large portion of empty space, so far removed from stars and other appreciable masses that we have before us aproximately the conditions required by the fundamental law of Galilei.

## Hypothetical Example.

As reference body let us imagine a spacious chest resembling a room with an observer inside who is equipped with apparatus. Gravitation naturally does not exist for this observer. He must fasten himself with strings to the floor, otherwise the slightest impact against the floor will cause him to rise slowly toward the ceiling of the room.

'To the middle of the lid of the chest is fixed externally a hook with rope attached, and now a 'being' (what kind of a 'being' is immaterial to us) begins pulling at this with a constant force. The chest, together with the observer, then begins to move upwards with a uniformly accelerated motion. In course of time their velocity will reach unheard of values, provided that we are viewing all this from another reference body which is not being pulled with a rope.

## Viewpoint of Man in Chest.

'But how does the man in the chest regard the process? The acceleration of the chest will be transmitted to him by the reaction of the floor of the chest. He must therefore take up this pressure by means of his legs if he does not wish to be laid out full length on the floor. He is then standing in the chest in exactly the same way as anyone stands in a room of a house on our earth. If he releases a body which he previously had in his hand, the acceleration of the chest will no longer be transmitted to this body, and for this reason the body will approach the floor of the chest with an accelerated motion.

The observer will further convince himself that the acceleration of the body towards the floor of the chest is always of the same magnitude, whatever kind of body he may happen to use for the experiment.'

## 'Kinertia' Quoted.

'Kinertia's' theory of the relation between acceleration and gravitation is set forth in the following quotation from 'Do Bodies Fall?' and is used by Professor Reuterdahl in building up his argument:

'I set to work to find out by experiment whether bodies actually did fall with the acceleration which the force of attraction was said to produce. Years before that, when in England, where some of our coal mines had vertical shafts about 1,500 feet deep, I had studied the cause of weight by having the hoisting engine drop me down with the full acceleration for about 500 feet. Then, by retardation during the lowest 500 feet, I could experience increase of weight all over me so marked that my legs could hardly support me.

## Weight Not a Force.

'That taught me that acceleration was the proximate cause of weight, but at the time of these experiments I still thought the acceleration of the falling cage was really caused by the earth's attraction.

'Weight is not a kinetic force because it cannot produce acceleration. If a body were accelerated in proportion to its weight, then weight would be a force.'

'Laying aside the right of Einstein to claim originality for his theory,' said Professor Reuterdahl yesterday, 'he is a sophist, and the world will know him as such in due time. He is dealing with mythical beings. They are 'might-have-beens.'

'His fourth dimension is a composite of time and space. That cannot be, because time and space never can be one. Space may be referred to as the distance between two points, A and B. We may travel from A and B, and return to find the same permanent objects in their places. We may require a certain amount of time to make the journey, but when we turn back that time is gone.

'I demand that Einstein show me his proof. I believe in dealing in the physical things of this world. In other words, I am from Missouri. I shall be glad to meet Professor Einstein at any time or place and debate this subject. But I shall demand an actual demonstration of his theory, not a journey into the realm of the mythical. That demonstration he can never give.'"

The story of Reuterdahl's challenge to Einstein was covered by newspapers around the world. *The New York Times* reported on 10 April 1921,

## "CHALLENGES PROF. EINSTEIN

St. Paul Professor Asserts Relativity
Theory Was Advanced in 1866.
*Special to The New York Times.*

MINNEAPOLIS, April 9.—Professor Arvid Deuterdahl, Dean of the College of Engineering of St. Thomas College, St. Paul, yesterday challenged Prof. Albert Einstein to a written debate on his theory of relativity.

That the Einstein theory was advanced in 1866, thirteen years before he was born, by a scientist known under the pen name of 'Kinertia,' is the contention of Professor Reuterdahl, in a statement in which he gives the life history of both men, and gives references and dates to support his contention.

Professor Reuterdahl, however, says the fact that Professor Einstein has broken down the barriers of set ideas in science and made it possible for a hearing for new ideas more than justifies his claim to prominence.

'The American scientists,' said Professor Reuterdahl, 'are the most clannish, I should say the most pig-headed, in the world. In the Old World the scientific journals publish articles advancing new theories. Here they will not accept anything that is not based on their own knowledge and belief. If Einstein has done anything he has jolted American scientists into accepting something new.'

Professor Reuterdahl refers to eleven articles which appeared in Harper's Weekly in 1914, in giving 'Kinertia' credit for originating the Einstein theory.

'Kinertia,' Professor Reuterdahl says, is the nom de plume of a professor believed to be living in California now."

*The Chicago Tribune* (European Edition, Paris) reported on 11 April 1921,

## "AMERICAN CALLS EINSTEIN 'BARNUM'

(Special Cable to The Tribune.)
MINNEAPOLIS, April 10.—Professor Arvid Reuterdahl, dean of the college of engineers at St. Thomas college, has styled Dr. Einstein, discoverer of the theory of relativity, 'the Barnum of the scientific world' and challenges him to a written debate on his theory.

Dr. Reuterdahl asserted that Einstein is not only 'fooling scientists with his mystical theory' but is a plagiarist. He declares the 'Einstein theory' was advanced in 1866 by a scientist under the pen name of 'Inertia.'"

On 11 April 1921, *The Sun* of New York reported,

## "Challenges Einstein, Calls Him Plagiarist

MINNEAPOLIS, April 11. — Not only has Einstein's theory of relativity been challenged but the scientist himself has been charged with being a plagiarist and the 'Barnum of Science' by Prof. Arvid Reuterdahl, dean of the Engineering School of St. Thomas's College, St. Paul. He has issued a challenge to the German scientist to meet him in a written debate.

The gravitational aspects of the Einstein theory were presented in 1866 in *Harper's Weekly* by a writer who called himself 'Kinertia,' Prof. Reuterdahl asserts. But the professor does give Prof. Einstein credit for blazing a new trail in thought for American scientists whom Dr. Reuterdahl declares to be more orthodox than European scientists."

On 11 April 1921, the *New York American* wrote,

## "EINSTEIN CHARGED WITH PLAGIARISM

St. Paul Educator Says Theory of
Relativity Was Advanced in
Harper's Weekly in 1866.

**Special Dispatch to the New York American.**
MINNEAPOLIS, April 10.—That the Albert Einstein theory of relativity in its gravitational aspects was advanced in 1866, thirteen years

before Einstein was born, by a scientist known under pen name of 'Kinertia' was the assertion made to-day by Professor Arvid Reuterdahl, dean of the engineering school of St. Thomas College in St. Paul. He challenged the German savant to defend his theories in a written debate.

Professor Reuterdahl declared Einstein was not only deceiving scientists with a mythical theory, but that he was either a plagiarist or his work had been antedated by another without his knowledge.

He then cited 'Kinertia,' whose theory was expounded in eleven articles running in Harper's Weekly in 1914, according to Professor Reuterdahl. These give 'Kinertia' credit for the so-called Einstein theory of gravitation, which is a later development of the theory of relativity.

The theory of relativity itself, says Einstein's challenger, was made public exactly one year before authorities on Einstein's work credit him with having made the discovery. In 1904, says Professor Reuterdahl, there was a paper on this subject, published in Berlin in the Journal Sitzungsberichte."

On 12 April 1921, the *New York American* reported,

## "EINSTEIN REFUSES TO DEBATE THEORY

Dean Reuterdahl's Challenge to
Discuss Relativity Declined as
Detraction from Mission.

Dr. Albert Einstein was interviewed yesterday in his headquarters at the Hotel Commodore regarding the attack on his theory of relativity made by Dean Arvid Renterdahl, of St. Thomas College, St. Paul, Minn.

Dr. Einstein smilingly listened to newspaper accounts of the Reuterdahl attack. Through his secretary he said:

'I came here with one object—the promotion of the establishment of the Hebrew University in Jerusalem. I will not be led into a discussion of my theory with persons who may not understand. There may be some personal intent in the remarks of this gentleman, whom I have not the honor of knowing.

'The great purpose of my mission to this country must not be overshadowed by my theory. I will be here a short time, and all of that time must be devoted to the great Palestine reconstruction project.

'I have consented to deliver a few lectures, but beyond that I do not wish to encroach upon my limited time. It must be seen plainly that I cannot enter into newspaper discussions with persons who doubt or misunderstand my theories or question my integrity.

'I have not had the opportunity to look into this challenge to debate issued by Dean Reuterdahl. Being without knowledge of the person called

'Kinertia' who is said to have written on the subject, I am not prepared to express any opinion.

It was further said for Dr. Einstein that he had no desire to popularize his theory of relativity; that he had writ-[*Unfortunately your author's photocopy of this article lacks the remainder.*]"

Segments of the press came to Einstein's defense. The *World* of New York wrote on 12 April 1921, quoting Einstein,

## "EINSTEIN AMUSED BY A NEW ATTACK

'Being Called P. T. Barnum of
Scientific World Only What
I Get at Home.'

## DECLINES REUTERDAHL'S CHALLENGE TO A DEBATE.

He, Prof. Weizmann and Others
to Be Guests at Jewish
Mass Meeting To-Night.

Prof. Albert Einstein was not greatly disturbed yesterday when he learned that Prof. Arvid Reuterdahl, dean of the engineering school of St. Thomas College, St. Paul, Minn., had called him the 'P. T. Barnum of the scientific world.' In fact, Prof. Einstein was amused.

'It reminds me of home,' he said, 'In Germany I am quite accustomed to being called names by persons who disagree with me.'

Prof. Einstein said he had never heard of Prof. Reuterdahl and that he was not in the least interested in the latter's challenge to a written debate on the subject of relativity. He intimated that he might read an article written by Prof. Reuterdahl if he happened to come across it, but as for entering a controversy, he couldn't waste the time.

The professor's mail is flooded with letters from persons who have pet theories which they wish to put before him, or who wish to argue on the subject of relativity. Several letters have been received from 'Messiahs' with plans for leading the Jews back to Palestine.

Prof. Chaim Weizmann, President of the World Zionist Organization, Prof. Einstein, M. M. Ussishkin, Chairman of the Zionist Commission to Palestine, Dr. Ben Zion Mosesohn, Principal of the Hebrew High School in Jaffe, and Dr. Schmaya Levine, member of the International Zionist

Committee, will be the principal guests at an all-Jewish mass meeting tonight in the 69th Regiment Armory, 25th Street and Lexington Avenue. This reception is in charge of a committee of 100, representing more than 1,800 local Jewish organizations of every variety and type.

Senator Calder and Dr. Butler, President of Columbia University, will be the principal speakers. In addition there will be addresses by prominent Jewish leaders representing the various elements in Jewry. Morris Rothenberg will welcome the guests in behalf of the American Jewish Congress.

Tickets are free and the seats will be reserved for ticket holders until 8 P. M., and after that all the seats will be thrown open to the public. Reservations have been made for a large delegation of Jewish wounded veterans of the World War. They will be brought from the nearby hospitals under an escort of Jewish legionnaires who fought in Palestine under Gen. Allenby."

## 4.7.1.2 Cowardly Einstein Caught in a Lie

Einstein hypocritically called his critics name-callers, when in fact Einstein had been recklessly defaming his critics for years, and had encouraged others to not respond to criticism of relativity theory other than by way of personal attack. The newspaper tried to deflect attention away from Einstein's evasiveness, but their story also unwittingly revealed that Albert Einstein was dishonest. E. Lee Heidenreich wrote in the *Minneapolis Morning Tribune*, on 16 May 1921,

### "Calls Einstein's Statements Irreconcilable.
*To the Editor of The Tribune:*
The scientific world has lately been much entertained and somewhat mystified by the increasing doubts, which have gradually crept into the press, regarding both the authenticity and the reliability of Professor Einsteins much-vaunted theory of relativity.

Professor Arvid Reuterdahl of St. Thomas college has challenged Professor Einstein to a written debate on the latter's theory, but has so far only been met with more or less evasive statements by Professor Einstein, some of which appear to the writer simply irreconcilable.

Thus, the New York World of April 12, 1921, says: 'Professor Einstein said he never heard of Professor Reuterdahl, and that he was not in the least interested in the latter's challenge to a written debate on the subject of relativity. He intimated that he might read an article written by Professor Reuterdahl, if he happened to come across it, but as for entering a controversy, he could not waste his time.'

The writer spent four months in Norway in 1920, and took occasion to give to 'Aftenposten' in Christiania a brief synopsis of Professor Reuterdahl's theory of interdependence, containing also considerable adverse criticism of both the authenticity and reliability of Professor Einstein's theory of relativity. The latter at that time was in Christiania, where he gave a lecture on his relativity.

'Aftenposten,' Christiania, of June 18, 1920, says: 'But what does Professor Einstein say to this? It would be interesting to know whether he is acquainted with the product of Professor Reuterdahl's pen. 'No,' answers Professor Einstein at our question, 'I do not know the name of Professor Reuterdahl and have never heard mentioned that he is said to have worked on the theory of relativity. I have often corresponded with Professor Mittag-Leffler, but he never mentioned any such work'.'

And later, in the same interview, Professor Einstein continues: 'Ein rechter mensch (a man of justice) would not have made the public announcement which Professor Reuterdahl has made through the American press.'

During the 'frequent correspondence' between Professor Mittag-Leffler and Professor Einstein, the original manuscript by Professor Reuterdahl of his space-time potential remained in the hands of Professor Mittag-Leffler for about four years, sometime between 1914 and 1918, and we have to take Professor Einstein's word for it that no discussion of the space-time potential took place during this 'frequent correspondence'—it would not have mattered much—except for the peculiar fact that Professor Einstein so carefully disclaims any notice of Reuterdahl's existence.

In spite of this, on the 12th day of April, 1921, Professor Einstein, in an interview, stated that 'he had never heard of Professor Reuterdahl.'

One might ask why the professor is afraid of admitting that he has heard of Reuterdahl? Does a ghost of a MS held by Mittag-Leffler lurk around somewhere? Have we here a sword of Damocles?

Professor Einstein denies that he has heard of Reuterdahl on April 12, 1921, in New York World, whereas he did hear of him and discussed his statements in Christiana to Aftenposten June 18, 1920, nearly a year earlier!

Either his memory has slipped away into the four dimensional space-time continuum, or for some reason he misrepresents facts.

As one of the remaining champions of materialistic and atheistic science, why does not the professor bravely come forth to defend the moss-grown theories against the onslaught of Scientific theism, and valiantly charge into the shrinking form of his adversary, right in the arena of the public eye? Does it behoove a world acclaimed scientist, a giant of mathematics, to say: 'My arguments you will not understand, I cast not my pearls before swine.'

It reminds one of the old fairy tale by H. C. Anderson, 'The Emperor's New Clothes,' which were so intricately and fearfully spun that they could not be seen by persons who were not wise, or who could not properly serve his majesty—and thus the visibility of the emperor's new clothes became a criterion of intellect of his subjects—only to have the bubble pricked by an unsophisticated street gamin, who cried out in astonishment: 'But the emperor is stark naked!'—tableau!

If someone has said that only seven, or was it twelve, men in the whole world would understand Einstein's theory of relativity, he should add 'as

Einstein dresses it'—for relativity with common sense and logic instead of a lot of sophistic embellishments is not such a formidable study.

The writer was amazed at the spectacular ascendancy of Professor Einstein in the public view and the acquiescent attitude of a seemingly bewildered lot of scientific institutions—an attitude almost similar to the impulsive reception of Dr. Cook of North Pole fame.

When the reaction comes, when Professor Einstein has left the United States, covered with decorations, the professor probably will realize that it were better had he met the questions squarely in the spirit in which they were made, because they now will stand as though cut in granite: Relativity or Interdependence? And must sooner or later be met without beating the devil around a bush with evasive and irreconcilable statements.—E. Lee Heidenreich, Kansas City, Mo."

As Heidenreich had affirmed, the *Aftenposten* of Oslo, Norway wrote on 18 June 1920,

## "Diskussionen om relativitetstheorien.
En amerikansk professor, som gjør krav paa at være theoriens skaber.
### En udtalelse af professor Einstein.

Vi har liggende foran os et eksemplar af den amerikanske avis »St. Paul Sunday Pioneer Press«, som udkommer i St. Paul, Minnesota. Numeret er dateret 1ste februar 1920 og indeholder bl. a. en længere artikel om relativitetstheorien. Bladet giver en fremstilling af det arbeide, som den amerikanske professor Arvid Reuterdahl har nedlagt til udforskning af den saa meget omtalte relativitetstheori. Det dreier sig om en meget mystisk affære, idet det heder, at professor Reuterdahl saa tidlig som i 1902 har skapt theorien, men paa en lidt usandsynlig maade er hans manuskript kommet paa afveie. Hvordan? Jo, historien lyder som følgende i »St. Pauls Pioneer«:

Professor Einstein offentliggjorde sin teori i »Annalen der Physik« for 1905. Reuterdahl foredrog sin theori den 5te april 1902 i »The American Elektrochemical society« ved dets aabningstnøde i Philadelphia. Udviklingen af theorien beskjæftigede ham helt til 1914, da han var færdig med udarbeidelsen. Hans theori vakte straks stor interesse og i februar 1915 gav han forelæsninger over sin theori ved Kansas State Agricultural College og senere ved Kansas universitet.

Den 19de februar 1915 blev professor Reuterdahls manuskript sendt til Norge, hvor det var meningen, at redaktør Oppedal skulde offentliggøre det i »Verdens Gang«. Redaktør Oppedal refererede professor Reuterdahls arbeide til professor Størmer; men presserende arbeide hindrede en undersøgelse og overveielse. Det blev saa refereret for professor Mittag-Leffler i Stockholm. Her mister man ethvert spor af manuskriptet.

Albert Einstein er nu medlem af en tysk videnskabelig kommission. Hans sidste arbeide hader »Time, Space and Gravitation«. Reuterdahls manuskript bærer titelen »Space, Time Potential, a new concept of Gravitation and Electricity«. Postprotokoller viser, at manuskriptet var et sted i Europa i hænde hos en tysk professor i begyndelsen af 1915.

Professor Reuterdahl har nu under udarbeidelse en ny bog om sin theori og denne bog vil blive hans livsværk.

Saavidt vor amerikanske kilde. Alle de forsøg vi har sat igang for at finde sporet efter det forsvundne manuskript er mislykket og nogen berettiget mening om den mystiske affæres vitterlighed skal vi ikke driste os til at have.

Men hvad siger professor Einstein til dette. Det vilde have sin interesse at vide, om han kjender professor Reuterdahls arbeider. »Nei«, svarer professor Einstein paa vor forespørgsel. »Jeg kjender ikke professor Reuterdahls navn og har aldrig hørt tale om, at han skal have arbeidet paa relativitetstheorien. Jeg har ofte korresponderet med professor Mittag-Leffler, men han omtalte aldrig noget saadant arbeide. Jeg vil ikke bestemt paastaa umuligheden i det, som nævnes i den amerikanske avis, men jeg finder det hele lidet sandsynlig. Hvis professor Reuterdahl virkelig har opdaget relativitetstheorien, vilde vi med stor sandsynlighed have faaet underretning om det. Jeg kjender størstedelen af den literatur om dette emne, men noget arbeide af Reuterdahl har jeg ikke truffet paa. Dette er jo ikke bevis«, slutter professor Einstein, og tilføier: »Ein rechter Mensch vilde ikke have gjort den reklame, som professor Reuterdahl har gjort gjennem den amerikanske avis«.

Det var Einsteins svar, som ikke stiller professor Reuterdahls paastand i noget godt lys. Et moment, som taler for den samme antagelse, ligger deri, at hvis professor Reuterdahl havde ret, vilde et universitet som University of Columbia have tildet ham sin store guldmedalje. Som vi tidligere har meddelt, har Columbiauniversitetet tildelt professor Einstein denne medalje."

### 4.7.1.3 Reuterdahl Pursues Einstein, Who Continues to Run

Heidenreich was right, Einstein's refusal to respond to charges that he was a plagiarist haunted Einstein around the world and throughout his lifetime. The *Minneapolis Evening Tribune* wrote on 15 April 1921,

## "Einstein, Jolted
## Out of Silence,
## Defends Theory

Challenged by St. Thomas Mentor,
Scientist Goes Deeper Into
Relativity Explanation.

Mathematician Ignores Charge
That He Is Not Originator of
Deductions Reached.

Professor Albert Einstein has been jolted out of a silence he has maintained since his arrival in America by the challenge of Professor Arvid Reuterdahl of St. Thomas college, according to dispatches today from New York.

**Plagiarism Charge Ignored.**

The charge that the famous mathematician is a plagiarist or at least not the originator of the theory which upset the scientific world is ignored, on the ground that it is not important. Professor Reuterdahl, however, has succeeded in bringing out a specific statement as to a test of the Einstein theory of relativity, and today the St. Thomas professor declared he was ready to meet the assertions concerning that test, and would make a statement later.

**Einstein's Test Stated.**

Professor Einstein's test, upon which he declares he is willing to rest his whole theory, was stated as follows:

'You know the solar spectrum. Everybody has seen it in the rainbow. You have also seen it when the sunlight passes through a triangular glass prism and falls upon a screen.

'Any light-giving body produces a spectrum, but the spectra from a different bodies are not alike. The spectrum from sodium for instance, shows only two yellow lines. The hydrogen spectrum shows only four colors.

**Band With Seven Colors.**

'The solar spectrum is a colored band, showing seven primary and secondary colors, ranging from red at one side to violet at the other.

'My theory demands that the spectrum of solar light, as compared with similar spectra from all other bodies, must be different in this respect.

'The lines of the solar spectrum must be found displaced—that is out of line—in the direction of red. If my theory of relativity is true, then this must be true. Why? Because of the nearness of the original solar light to the great mass which is the sun. If my theory is true, that mass must affect the spectral lines as I have said.'"

The *Minneapolis Morning Tribune* reported on 16 April 1921,

## "Relativity Hit Counter Blow By Reuterdahl

Twin City Man Says Einstein

Cult Has Not Attained
Dignity of Theory.

Conceding that Prof. Albert Einstein, famous mathematician, whose theory of relativity startled the scientific world, has been supported by the results of one experiment, but contending that his theory still is a mere hypothesis without a foundation in fact, Prof. Arvid Reuterdahl of St. Thomas college yesterday renewed his attack upon the theory.

Replying to Professor Reuterdahl's challenge, Professor Einstein gave out a statement in New York, the first since his arrival in America, in which he declared that he was willing to rest his whole theory upon one experiment.

**'Admission Proves Contention.'**

In turn, Professor Reuterdahl declared that the mathematicians' admission that the theory had not been proved substantiated his contention that relativity had not been established and never would be.

One effect of the challenge by Professor Reuterdahl was that the man whom he had called the Barnum of the scientific world was jolted out of a profound silence. To the charge of plagiarism Professor Einstein gave no heed, but he did rush to the defense of his pet theory.

**Einstein's Test Stated.**

Professor Einstein's test, upon which he declares he is willing to rest his whole theory, was stated as follows:

'You know the solar spectrum. Everybody has seen it when the sunlight passes through a triangular glass prism and falls upon a screen.

'Any light-giving body produces a spectrum, but the spectra from different bodies are not alike. The spectrum from sodium, for instance, shows only two yellow lines. The hydrogen spectrum shows only four colors.

**Band With Seven Colors.**

'The solar spectrum is a colored band, showing seven primary and secondary colors, ranging from red at one side to violet at the other.

'My theory demands that the spectrum of solar light, as compared with similar spectra from all other bodies, must be different in this respect.

'The lines of the solar spectrum must be found displaced—that is out of line—in the direction of red. If my theory of relativity is true, then this must be true. Why? Because of the nearness of the original solar light to the great mass which is the sun. If my theory is true, that mass must affect the spectral lines as I have said.'

Professor Reuterdahl's answer to this statement follows:

'Professor Einstein refuses to enter into a written debate with me concerning the correctness of the basic tenets of the theory of relativity for the reason that he is willing to risk the validity of the entire theory on the result of an experiment. The theory of relativity assumes the displacement of the solar spectral lines toward the red will take place when the original solar light is near to a great mass like the sun. Professor Einstein admits that if this displacement does not take place then the general theory of relativity must be abandoned as untenable.

'Upon the results of this experiment Dr. Einstein rests the validity of his entire theory. Many experiments intended to discover this displacement have already been made. Had these experiments been successful Professor Einstein would not have made the statement which has this very day been transmitted to me by The Minneapolis Tribune.

'Professor Einstein's admission of the absence of this verification transforms the entire situation and leaves the theory as an hypothesis yet to be verified.

'Furthermore, Professor Einstein has admitted that it is extremely difficult to observe the deflection, even if it does exist, because of the fact that the predicted displacement is extremely small.

'Moreover, Professor Einstein has conceded the further fact that it is very difficult to make any calculations whatsoever, because of the indefiniteness of the involved facts.

'Now Professor Einstein himself admits that he rests the validity of his entire intellectual structure upon the future results of this extremely delicate experiment involving conditions difficult of realization.

'Professor Einstein, in his reply to my challenge, makes no mention of the significance of the observations made by the English solar expedition and the observed motion of the planet Mercury.

'Apparently he magnanimously waives the right to contend that the result of his predictions and calculations concerning the bending of light rays and the perihelion-perturbation of Mercury has bearing upon the validity of his theory.

'I gladly grant the importance and bearing of these mathematical deductions of Professor Einstein. The granting of these contentions, however, in no way modifies my conviction that the theory of relativity is grounded upon fallacious assumptions, and therefore cannot survive. The history of science shows that one mathematic-physical theory after another has been abandoned because of inadequacy, unnecessary complexities, and untenability in the light of wider knowledge.

'It is true, of course, that this is the price which must be paid for intellectual advancement.

'Nevertheless it is also true that an hypothesis based upon fallacious assumptions contains the leaven of its own ultimate dissolution, despite the fact that some of the results of its applications to physical phenomena may be approximately correct.

'This I am prepared to prove is the status of Professor Einstein's theory of relativity. I am, indeed, surprised that Professor Einstein, while claiming that he had written his book from scientific motives and not for the sake of notoriety, lightly brushes to one side a challenge to a debate upon the validity of his theory. In no better way can the cause of science be served.

'A theory which so completely upsets all common-sense deductions concerning realities cannot hope forever to go unchallenged. Certainly it is not in keeping with the scientific motives of which Professor Einstein claims to be so ardent an exponent, continuously to reiterate the platitude that those who do not accept his theory are incapable of comprehending its alleged profundities.

'I desire to disabuse Professor Einstein of the correctness of the inference that any ulterior personal motive caused me to issue my challenge to him. The matter of nationality of an earnest investigator or any other ulterior motive never has had and never will have any bearing upon my attitude toward the significance and value of his work.'"

*The Kansas City Post* reported on 17 April 1921,

## "*DUBS EINSTEIN 'BARNUM OF SCIENCE' AND 'KIDDER'*

German Savant Challenges
Theorist to Written Debate
on Relativity.


Charges Feted Jew With
Having Plagiarized Material
From the Past.


A 'Barnum of science.'

Thus is Prof. Albert Einstein, German scientist, who at present is making a triumphal visit to the United States, branded by a former Kansas City public school professor, Dr. Arvid Reuterdahl, dean of the engineering school of St. Thomas colege, St. Paul.

While New York hands the celebrated discoverer of the theory of relativity the key to the city, and while savants, scholars, bankers, butchers, hang on his non-understandable words, Dr. Reuterdahl steps out and boldly calls him names.

A 'sophist,' a dealer in 'might have beens,' says Dr. Reuterdahl of Einstein.

The former Kansas City teacher then challenges the widely heralded mathematician to a written debate.

Dr. Reuterdahl, speaking of course in scientific language, has said in effect that he is prepared to prove the Einstein theory largely 'bunk,' and a borrowing from older scientists. It is easy enough, he insinuates, to set forth a theory of any kind, so long as you make it sufficiently abstruse not to be understood.

Long before Einstein announced his visit to America, Dr. Reuterdahl and he had become involved in an international dispute over his theory. The controversy has attracted wide attention in the old world from Norway to Italy.

Dr. Reuterdahl, who was an instructor at the Polytechnic institute here, left Kansas City in 1915. In the fall of the same year he gave lectures at the Kansas State Agricultural college at Manhattan and at Kansas university on 'Space-Time-Potential,' in which he set forth some of the same views enunciated by Einstein, crediting them to scientists who lived before Einstein was born.

At that time Dr. E. Lee Heidenreich of the Heidenreich Engineering company of Kansas City, a friend of Dr. Reuterdahl, wrote the Carnegie institute of Dr. Reuterdahl's lectures, saying:

'It takes a scientific giant to gainsay a Newton and such a giant we have with us today.'

Coupled with his challenge to a debate, Dr. Reuterdahl now asserts that Einstein is deceiving scientists with a mythical theory and that he is a plagiarist, his works being antedated by another.

Dr. Reuterdahl points out that the Einstein theory of relativity in its gravitational aspects was advanced in 1866 by a scientist who wrote under the pen name of 'Kinertia.' The latter, when a student under Lord Kelvin, is said to have questioned the Newton theory of force.

Dr. Reuterdahl gives Einstein credit for breaking down the barriers of set ideas in science and making it possible for hearing new ideas.

'The American scientists,' says Dr. Reuterdahl, 'are the most clannish and orthodox in the world. They will not consider anything but what is based on their own knowledge and belief.'

Dr. Reuterdahl, while giving Einstein credit for being one of the greatest mathematicians in the world, 'calls' him on many parts of his theory.

'I demand that Einstein show me his proof,' says the American professor. 'I believe in dealing in the physical things in the world. In other words, I am from Missouri. I shall be glad to meet Professor Einstein at any time or place and debate this subject. But I shall demand an actual demonstration of his theory, not a journey into the realm of the mythical. That demonstration he can never give.'"

Ernst Gehrcke noted in his book *Die Massensuggestion der Relativitätstheorie: Kulturhistorisch-psychologische Dokumente*, Hermann Meusser, Berlin, (1924), pp. 29-30; that the *Neue Preußische (Kreuz-) Zeitung* wrote on 11 April 1921, together with many other papers,

"EINSTEIN als Plagiator herausgefordert. Aus Paris, 11. April, wird gedrahtet: Aus Minneapolis erfährt die „*Chicago Tribune*" Prof. ARVID REUTERDAUL, der Präsident der Ingenieure der St. Thomas-Universität, erklärt über die Theorie des Professor EINSTEIN, daß dieser der „BARNUM" der Wissenschaft für die Welt sei. Professor REUTERDAUL fordert EINSTEIN zu einer schriftlichen Debatte über die Relativitätstheorie heraus. REUTERDAUL nennt EINSTEIN nicht nur einen verrückten Wissenschaftler mit mystischer Theorie, sondern auch einen Plagiator und behauptete, daß die EINSTEINsche Theorie bereits 1866 von einem Gelehrten unter dem Namen „INERTIA" entdeckt worden sei."

Gehrcke further notes that the *Vorwärts* wrote on 18 April 1921,

"Ein amerikanischer Professor hat die Theorie des Prof. EINSTEIN für eitel Humbug erklärt und ihn als einen Mann hingestellt, der einfach die wissenschaftliche Welt an der Nase herumführe. EINSTEIN ist der Schöpfer von etwas Neuem, nicht Dagewesenem, der Menge vor der Hand Unbegreiflichem, und daß alle neuen und großen Entdeckungen ihre Gegner haben und in der Geschichte stets hatten, scheint beinahe eine Notwendigkeit zu sein."

According to Gehrcke, the *Dresdner Anzeiger* reported on 18 April 1921,

"Professor EINSTEIN äußerte mit Bezug auf das Urteil des amerikanischen Prof. REUTERDAHL vom Thomas-College über seine Relativitätstheorie, sie sei die Leistung eines „Barnum der Wissenschaft", daß solche Angriffe ihn sehr an seine deutsche Heimat gemahnten ... Prof. EINSTEIN lehnte es formell ab, mit Professor REUTERDAHL sich in eine wissenschaftliche Aussprache einzulassen."

*Die Hamburger Woche* wrote on 9 June 1921,

"Jenseits des großen Teiches hat Albert Einstein, der mit seiner Relativitätstheorie raschen Weltruhm gewann, große Ehrungen erfahren. Beim Besuch der Princeton-Universität wurde er in Anwesenheit vieler Gelehrter anderer amerikanischer Hochschulen zum Ehrendoktor ernannt. Von einer anderen amerikanischen Hochschulseite dagegen ist Einstein ein neuer scharfer Gegner erstanden. Professor Arvid Reuterdahl, der Präsident der Ingenieure der St. Thomas-Universität, erklärte über die Theorie des Professors Einstien, daß dieser der „Barnum der Wissenschaft" für die Welt sei. Professor Reuterdahl fordert Einstein zu einer schriftlichen Debatte über die Relativitätstheorie heraus. Reuterdahl nennt Einstein nicht nur einen „verrückten Wissenschaftler mit hysterischer Theorie", sondern auch einen Plagiator und behauptet, daß die Einsteinsche Theorie bereits 1866 von einem Gelehrten unter dem Namen „Inertia" entdeckt worden sei.
    Man darf gespannt sein, welches objektive Endergebnis sich aus den Kämpfen für und wider Einstein die Wissenschaft schließlich herausdestillieren wird! ... "

## 4.7.2 Einstein All Hype

On 27 April 1921, Gertrude Besse King wrote about the publicity campaign for Einstein in *The Freeman* of New York,

> "ALADDIN EINSTEIN. THE popular interest in America in Professor Einstein's theories has astonished the professor. The public who does not know whether the theory of relativity has accounted for the alteration of mercury or of Mercury, waylays his steps, and delights, with the exception of a mere alderman or two, to do him honour. Gifted newspaper-reporters herald him as the originator of the theory of relativity, which, by the way he is not, and question him as to the ultimate nature of space, though only a mathematical physicist who is also a philosopher could understand the professor's answers.
> 
> This general interest in an extremely difficult science is not quite what it seems. Probably Professor Einstein does not realize how sensationally and cunningly he has been advertised. From the point of view of awakening popular curiosity, his press-notices could hardly have been improved. The newspapers first announced his discovery as revolutionizing science. This sounds well, but its meaning, after all, is rather vague. Then they printed a series of entertaining oddities, supposedly deducible from his hypothesis, although most of them could have been equally well deduced from the conclusions of Lorentz or Poincaré: for example, moving objects are shortened in the direction of their motion. This is a gay novelty until one learns the proportion of the reduction, which is calculated to divest the statement of interest to any but scientists. Further, our newspapers told us that if we were to travel from the earth with the speed of light, and could see the clock we left behind, it would always remain at the same moment, permanently pausing, unable to reach the next tick. But we should be unable to travel at the rate of light for a number of reasons, the most interesting and perhaps the most decisive being that such a speed would cause our mass to be infinite! Finally, our informants assert that no point in space, no moment of time can serve as a permanent base for measurement; we can measure only the relations of space, the relations of time, never absolute space or time; and even to measure space-relations, we have to take into account time! What a fascinating dervish-dance of what we used to regard as immutable fixities! Is it possible that these delicious contradictions are serious and accredited doctrines among those who know? Yet so they appear, for though Professor Einstein is always careful in stating that his hypothesis enjoys as yet only a tentative security, his methods are vouched for by the experts, his procedure is according to Hoyle, and the crowd is at liberty to gorge its appetite for marvels untroubled by the ogres of scientific orthodoxy.
> 
> Aside from the fact that Professor Einstein comes as a distinguished and somewhat mysterious foreigner to partake of our insatiable hospitality, his popular welcome is to be accounted for by the spell of wizardry that the press has cast upon his interpretations. For it is the necromancy of these

strange theories, not their science, that catches the gaping crowd. Reporters are often good, practical psychologists. Instinctively they have divined the public eagerness for miracles, without grasping the factors that feed this taste. They know that most of us are essentially children still clamouring for fairy tales. Man is congenitally restless with the prison-house of this too, too solid world. He is always looking for short-cuts to power. Since he can not find them to his mental satisfaction as once he could through the miracles and divine dispensations of the Church, or through the magic and occultism that were his legitimate resources in the Middle Ages, he now turns to the wonders of science and philosophy. Here, even in theories that he does not understand, he can find release for his cramped position, here he can taste the intoxicating freedom of a boundless universe, and renew his sense of personal potency. [... ]"[171]

Thomas Jefferson Jackson See wrote in *The San Francisco Journal* on 27 May 1923,

"If anyone should ask how Einstein managed to get such vast publicity in the matter of relativity, we may observe that he has the habit of a promoter. Mark Twain humorously wrote to the president of the St. Louis exposition in 1904, that he 'would like to attend the exposition and exhibit himself.' So also does Einstein contrive constantly to be seen among men in conspicuous places. When he came to America, with the Zionist committee, some two years ago, he had to go to the White House at Washington and talk relativity to President Harding. The President, with becoming modesty, said he could not understand the subject.

Things in Europe afterwards became uncomfortable for Einstein, and he sought refuge in an Oriental trip. When in Tokyo he called upon the emperor of Japan, and it was advertised over the world that he was without a dress suit. This report is spectacular and like that of a skillful advertiser.

His return trip is duly chronicled by the press. Thus he finally arrives in Egypt, and on reaching Spain addresses the Academy of Science, at a session held in the presence of the king of Spain. If this is not the trumpeting of an organized press agency, what is it?

Einstein is not liked in Germany. A year or so ago, the students at the University of Berlin hooted him down. It was reported that he was in fear of assassination—but it probably was only a ruse to gain public sympathy."[172]

---

**171**. "Aladdin Einstein", *The Freeman* (New York), Volume 3, Number 59, (27 April 1921), pp. 153-154.
**172**. T. J. J. See, "EINSTEIN A TRICKSTER?", *The San Francisco Journal*, (27 May 1923). Peter A. Bucky recalls that others intimated that Einstein's disheveled appearance was meant to attract publicity. Bucky discounted the notion, as did Einstein. P. A. Bucky, Einstein, and A. G. Weakland, *The Private Albert Einstein*, Andrews and McMeel, Kansas City, (1992), p. 4, 111.

*The Minneapolis Sunday Tribune* published a letter from Arvid Reuterdahl on 22 May 1921, which, while not the best work on the subject, is notable for its ridicule of Einstein for running away from the Bad Nauheim debate, as well as Einstein's refusal to debate Reuterdahl. It quotes a Swiss newspaper's statement that Einstein's flight from the Nauheim debate, "was another prearranged matter of his general trafficking." The alleged corruption is proven by Philipp Frank, who described Max Planck's biased control over the debate and his abuse of his power to censor speakers, intimidate the would-be audience and anti-Einstein speakers with armed guards, and restrict the topics of discussion in a way that would favor Einstein and prevent Einstein's having to face criticisms of the Metaphysics in the theory of relativity.[173] Frank wrote,

"[Max Planck] arranged it so that the greatest part of the available time was filled with papers that were purely mathematical and technical. Not much time remained for Lenard's attack and the debate that would ensue. The entire arrangement was made to prevent any dramatic effects. [***] The armed policemen who had watched the building were withdrawn."[174]

The theory of relativity is largely a metaphysical theory, not a scientific theory. In order to oppose the Metaphysics of relativity theory one must, of course, discuss Metaphysics. Proponents of relativity theory often refuse to discuss Metaphysics claiming that Metaphysics has nothing to do with science, and they thereby insulate their theory from criticism. Einstein did not grasp the distinction between Metaphysics and science. He stated in 1930, "Science itself is metaphysics."[175]

Hugo Dingler, a critic of relativity theory, confirmed that severe time restrictions were placed on the opponents of relativity theory at the Bad Nauheim debate. Others complained that Einstein's followers had stacked the audience with a pro-Einstein claque and tried to prevent the admission of neutral "unauthorized" persons into the forum.[176] Philipp Frank admitted that the corruption backfired—every fairminded person smelled a rat, and knew that Einstein and the relativists were avoiding the facts and dodging the issues. Just when Nobel Prize winner Philipp Lenard, Einstein's primary opponent, had cornered Einstein at the debate, Einstein ran away. Max Planck stopped the discussion for a break, and Einstein never returned. It is difficult to believe that this was not a prearranged maneuver to save face for Einstein.

Reuterdahl's article published in *The Minneapolis Sunday Tribune* on 22

---

[173]. P. Frank, *Einstein: His Life and Times*, Alfred A. Knopf, New York, (1947), pp. 163-166.
[174]. P. Frank, *Einstein: His Life and Times*, Alfred A. Knopf, New York, (1947), p. 163.
[175]. A. Einstein quoted in "Einstein on Arrival Braves Limelight for Only 15 Minutes", *The New York Times*, (12 December 1930), pp. 1, 16, at 16
[176]. *Cf.* H. Goenner, "The Reaction to Relativity Theory. I: The Anti-Einstein Campaign in Germany in 1920", *Science in Context*, Volume 6, Number 1, (1993), pp. 107-133, at 125.

May 1921,:

## "Science's 'Baby Guy' Was Simple Child Till Einstein Adopted It

*Clothed in a Garbled Dress of Mathematical Theories, the Youngster, 'Relativity,' Joined Ranks of Unintelligible Genii—Swiss Paper Backs Reuterdal.*

**By Arvid Reuterdahl.**
Dean Department of Engineering and Architecture
the College of St. Thomas.

In a signed statement published in The Minneapolis Morning Tribune, issue of May 16, Dr. E. Lee Heidenreich, the eminent engineer, mathematician, and philosopher of Kansas City, Mo., points out that Dr. Einstein does not hesitate to make irreconcilable statements in order to avoid facing issues squarely. I now have in my possession a copy of the 'Aftenposten' article which was cited by Dr. Heidenreich in his communication to The Tribune. I also have a copy of the New York World interview with Dr. Einstein. The date of the 'Aftenposten" article is June 18, 1920, and the New York World interview is dated April 12, 1921.

There is only one verdict possible when a comparison is made of these two conflicting statements of Professor Einstein, either his statements are relativistic conveniences or his memory has been weakened by relativistic sophistries. Dr. Einstein, it seems, is permitted to say anything he pleases without being held accountable.

**Access to Ziegler's Work.**

From abroad I have received copies of publications which convey the idea, in no uncertain terms, that while Dr. Einstein was in Switzerland he had access to the work of Dr. J. H. Ziegler and that he used the results of this able investigator's work without giving him any credit whatsoever.

I have now in my possession evidence furnished by 'Kinertia,' which shows conclusively that in the year 1903, copies of certain contributions of 'Kinertia' were in the hands of the imperial Prussian academy of science in Berlin. Did Dr. Einstein avail himself of those easily accessible records? Moreover in September, 1904, a well-known American journal published a statement setting forth 'Kinertia's' theory of gravitation.

**Swiss Paper on Einstein.**

The following quotations from the well known Swiss paper, 'the Lucerne Daily News,' of April 22, 1921, should have been interesting reading to Dr. Einstein under the heading, 'Professor Einstein's Triumphal March Through America,' a translation of the article reads:

'Professor Albert Einstein and the Zionist delegation which arrived simultaneously with him, was accorded a very warm welcome on its arrival in New York. The entire New York press devoted a good deal of space to this happening, as well as to the personality of Einstein. One can clearly see that there is again question here of the previously ordered advertising, just as the whole Einstein undertaking has been from its very beginning a bluff. This time the Americans were supposed to believe, but the good Yankee seemed to be less naive than the good Germans and Swiss, and were not so easily forced into a belief in the new prophet. They are too skeptical to believe without a further proof that he is a greater genius than Copernicus and Newton, simply because he is more unintelligible.

### Too Much Common Sense.

'Americans have too much common sense for that. They know that all the great truths are simple and easily understood, and are, therefore, justly suspicious of the unintelligible theory of relativity of Einstein. More than that they have rejected it as a swindle. Just for example Reuterdahl, dean of engineering of the College of St. Thomas, St. Paul, Minnesota, calls Einstein a 'Barnum of the scientific world who is trying to fool the whole world with a mythical theory.' It is further reported that Reuterdahl has challenged Einstein to a debate, into which he is as likely to enter as in the debate announced last year at the meeting for scientific investigation in Bad Nauheim, where he preferred to withdraw himself quietly before the announced opponents of his theory could say what they had to say. To these opponents was expressed the regret that Mr. Einstein was unable, because of circumstances, to answer them. This, of course, was another prearranged matter of his general trafficking. It is very likely that he is acting in a similar manner towards Reuterdahl. The more so because the latter has accused him of scientific theft, for Reuterdahl maintains that Einstein has taken the fundamentals of his theory from a work which appeared in 1866 under the pseudonym of 'Kinertia.'

### Work Little Known In Europe.

'As this work is scarcely known in Europe, the accusation may possibly be groundless. Similar accusations have been made by German scientists, such as the Engineer Rudolph Mewes, Professors E. Gehrke and Paul Weyland, etc. According to them, Emstein is supposed to have secretly taken a formula from a publication of the deceased Professor Gerber which appeared in 1898, and was very inaccessible, and to have made it his own. The facts in the matter are, of course, difficult to prove, nevertheless, the peculiar conduct of Einstein and his sensational advertising campaign lead one to believe that his whole business is very suspicious. However, most of these opponents seem to be upon a wrong scent, because they do not understand the circumstances which existed at the time of the origination of the Einsteinian teaching, and do not sufficiently understand the influences that may have been at work in regard to his theory. He seems to have started with the correct notion of the constancy of the velocity of light, and of vacuum; which potion, however, he did not test out further, but simply accepted hypothetically; whereas, the other teachings of his theory are so tangled and contradictory that they seem to have come from an entirely different source.

### Deductions Criticized.

'These other peculiar assumed proofs, and the still more peculiar deductions made from them have been criticized by many scientists, notably by Professor Lenard, a former Nobel Prize winner in physics. Lenard calls attention to the fact that these suppositions and deductions are contradictory to common sense. Einstein's acceptance of constancy of the velocity of light, which he makes the one stable concept in the shoreless ocean of his theory of relativity, seems to be a special case. It is already suspicious, because the physicists at that time denied the existence of absolute empty space, and regarded such a thing as impossible, but then conceded it without more adieu when they accepted Einstein's hypotheses, and in addition regarded him as having performed a very acceptable thing. As a matter of fact Einstein's theory of velocity of light seems to be a direct theft of the universal theory of light given out by J. H. Ziegler five years previous. There are reasons that seem to point with great probability to the fact that the teaching of Ziegler was the hidden spring of Einstein's discovery.

### The Unmoved Emptiness.

'Just to mention one of them, the findings of Ziegler were very much discussed in Berne, which was at the time Einstein's domicile. Ziegler speaks of the trinity of energy, space and time, a trinity which Mr. Einstein then brought forth in a modified form. The clear and simple teaching of Ziegler, according to which all natural phenomena are mixed forms of radiating source light (urilcht), and unmoved (unergized) emptiness, were very inconvenient to the exponents of accepted physics, and so they tried from the beginning to suppress it. Thus they created an opportunity for a clever and foxy plagiarist to possess himself of these principle teachings. He would get all the greater hearing and support from those physicists if he would proffer his plagiarism in a manner intelligible to them, but unintelligible to the general public. Mathematics served as an excellent medium. The chemist (not the mathematician) Ziegler, had made the mistake of writing intelligibly and of revealing the mistakes in modern physics, thus Einstein appeared to these physicists as a Deus ex Machina, he was a friend in need. It is no wonder that he was hailed as long-expected Messiah of the world of physics, the true bringer of light.

### Ziegler's Name Forgotten.

'Ziegler's name was forgotten in the great propaganda which the papers carried on for Einstein. Ziegler has not always propounded his teachings so clearly that superficial study would lead to a great understanding of it. Thus, there was occasion of all sorts of misconception. Hence the many mistakes of the theory of relativity. How could this theory of relativity be unified and clear when it was only a mixed pickle affair of erroneous plagiarisms. The fact that Einstein's theory approached the Ziegler light theory more and more every year does not disprove that the Ziegler theory is a source of Einsteinian wisdom, even though the Einsteinian press has carefully boycotted Ziegler for 20 years. The Americans, of course, know nothing of this. It they reject Einstein, it is rather because they are angry to be considered so stupid as to regard the greatest scientific discovery as the most unintelligible. The Americans know well enough that the opposite is the case, and for this reason the business trip of the false prophet in the United States will scarcely constitute a triumphal march.

**From German Journal.**

The following excerpts from the Scientific journal 'Weltwissen,' May, 1921, published in Munich, Germany, is significant:

'From numerous sources we have previously received various printed articles and manuscripts directed against Einstein, among others, one from the 'Regierungsrat,' Dr. H. Fricke, 'The Error In Einstein's Theory of Relativity' and from the Engineer A. Patschke, 'The Overthrow of the Einsteinian Theory of Relativity.' The tremendous advertising campaign, which Einstein has for some time conducted throughout the world has been carried on to such an extent as to throw a sort of protective film over his work. Such procedure does not redound to the honor and furtherance of science, in special letters, at the beginning of the year 1920, we called the attention of the University of Berlin and of the minister of education to this horn-tooting for Einstein. It is a very deplorable fact that German science should be laid open to ridicule by one of Germany's own scientists.'

This statement emanated from Dr. Johannes Zacharias of the editorial department of the journal 'Weltwissen.'"[177]

# 4.8 Assassination Plots

Though Theodor Wolff, editor of the *Berliner Tageblatt*, had stated that there was no anti-Semitic movement in the German government in 1915, Wolff spread the rumor in 1922 (which was denied by the German police) that assassins were out to murder him and Albert Einstein. Wolff's pronouncement followed on the heels of the assassination of Walter Rathenau. Rathenau was a German Jew who found a way around the Treaty of Versailles (which he had supported—profiteering off of the reparations payments made by Germany) by restoring Germany's military in Russia with the Rapallo Treaty. It was alleged that he and his friends could financially profit from this venture and that they sought to sponsor Bolshevism. Bolshevism itself stole the wealth of Russia and channeled it other hands. Rathenau was preparing the way for the Second World War.

Wolff's baseless claims of assassination plots may have been a pretext for Einstein's withdrawal from the meetings of the League of Nations, where he would have had to have met with his critic Henri Bergson, and been publicly challenged to debate his positions. Instead of running this risk, Einstein ran around the world promoting himself and advertising the theory of relativity—and Zionism, at a critical point in the history of the Zionist Movement. In this

---

[177]. A. Reuterdahl, *The Minneapolis Sunday Tribune*, (22 May 1921). Reuterdahl translates parts of "Professor Einsteins „Triumphzug" durch Amerika", *Luzerner Neueste Nachrichten*, (22 April 1921).

same period, Wickham Steed prevented Lord Northcliffe, principal owner of *The London Times* and outspoken critic of Zionism, from voicing his objections to the League of Nations Mandate for Palestine of 24 July 1922 (reproduced in the endnote[178]). Perhaps the Zionists sought sympathy for their cause by spreading

---

**178.**
The Palestine Mandate
The Council of the League of Nations:
July 24, 1922
Whereas the Principal Allied Powers have agreed, for the purpose of giving effect to the provisions of Article 22 of the Covenant of the League of Nations, to entrust to a Mandatory selected by the said Powers the administration of the territory of Palestine, which formerly belonged to the Turkish Empire, within such boundaries as may be fixed by them; and

Whereas the Principal Allied Powers have also agreed that the Mandatory should be responsible for putting into effect the declaration originally made on November 2nd, 1917, by the Government of His Britannic Majesty, and adopted by the said Powers, in favor of the establishment in Palestine of a national home for the Jewish people, it being clearly understood that nothing should be done which might prejudice the civil and religious rights of existing non-Jewish communities in Palestine, or the rights and political status enjoyed by Jews in any other country; and

Whereas recognition has thereby been given to the historical connection of the Jewish people with Palestine and to the grounds for reconstituting their national home in that country; and

Whereas the Principal Allied Powers have selected His Britannic Majesty as the Mandatory for Palestine; and

Whereas the mandate in respect of Palestine has been formulated in the following terms and submitted to the Council of the League for approval; and

Whereas His Britannic Majesty has accepted the mandate in respect of Palestine and undertaken to exercise it on behalf of the League of Nations in conformity with the following provisions; and

Whereas by the afore-mentioned Article 22 (paragraph 8), it is provided that the degree of authority, control or administration to be exercised by the Mandatory, not having been previously agreed upon by the Members of the League, shall be explicitly defined by the Council of the League Of Nations; confirming the said Mandate, defines its terms as follows:

ARTICLE 1. The Mandatory shall have full powers of legislation and of administration, save as they may be limited by the terms of this mandate.

ART. 2. The Mandatory shall be responsible for placing the country under such political, administrative and economic conditions as will secure the establishment of the Jewish national home, as laid down in the preamble, and the development of self-governing institutions, and also for safeguarding the civil and religious rights of all the inhabitants of Palestine, irrespective of race and religion.

ART. 3. The Mandatory shall, so far as circumstances permit, encourage local autonomy.

ART. 4. An appropriate Jewish agency shall be recognised as a public body for the purpose of advising and co-operating with the Administration of Palestine in such economic, social and other matters as may affect the establishment of the Jewish national home and the interests of the Jewish population in Palestine, and, subject always to the control of the Administration to assist and take part in the development of the country.

The Zionist organization, so long as its organization and constitution are in the opinion of the Mandatory appropriate, shall be recognised as such agency. It shall take steps in consultation with His Britannic Majesty's Government to secure the co-operation of all Jews who are willing to assist in the establishment of the Jewish national home.

ART. 5. The Mandatory shall be responsible for seeing that no Palestine territory shall be ceded or leased to, or in any way placed under the control of the Government of any foreign Power.

ART. 6. The Administration of Palestine, while ensuring that the rights and position of other sections of the population are not prejudiced, shall facilitate Jewish immigration under suitable conditions and shall encourage, in co-operation with the Jewish agency referred to in Article 4, close settlement by Jews on the land, including State lands and waste lands not required for public purposes.

ART. 7. The Administration of Palestine shall be responsible for enacting a nationality law. There shall be included in this law provisions framed so as to facilitate the acquisition of Palestinian citizenship by Jews who take up their permanent residence in Palestine.

ART. 8. The privileges and immunities of foreigners, including the benefits of consular jurisdiction and protection as formerly enjoyed by Capitulation or usage in the Ottoman Empire, shall not be applicable in Palestine. Unless the Powers whose nationals enjoyed the afore-mentioned privileges and immunities on August 1st, 1914, shall have previously renounced the right to their re-establishment, or shall have agreed to their non-application for a specified period, these privileges and immunities shall, at the expiration of the mandate, be immediately reestablished in their entirety or with such modifications as may have been agreed upon between the Powers concerned.

ART. 9. The Mandatory shall be responsible for seeing that the judicial system established in Palestine shall assure to foreigners, as well as to natives, a complete guarantee of their rights. Respect for the personal status of the various peoples and communities and for their religious interests shall be fully guaranteed. In particular, the control and administration of Wakfs shall be exercised in accordance with religious law and the dispositions of the founders.

ART. 10. Pending the making of special extradition agreements relating to Palestine, the extradition treaties in force between the Mandatory and other foreign Powers shall apply to Palestine.

ART. 11. The Administration of Palestine shall take all necessary measures to safeguard the interests of the community in connection with the development of the country, and, subject to any international obligations accepted by the Mandatory, shall have full power to provide for public ownership or control of any of the natural resources of the country or of the public works, services and utilities established or to be established therein. It shall introduce a land system appropriate to the needs of the country, having regard, among other things, to the desirability of promoting the close settlement and intensive cultivation of the land. The Administration may arrange with the Jewish agency mentioned in Article 4 to construct or operate, upon fair and equitable terms, any public works, services and utilities, and to develop any of the natural resources of the country, in so far as these matters are not directly undertaken by the Administration. Any such arrangements shall provide that no profits distributed by such agency, directly or indirectly, shall exceed a reasonable rate of interest on the capital, and any further profits shall be utilised by it for the benefit of the country in a manner approved by the Administration.

ART. 12. The Mandatory shall be entrusted with the control of the foreign relations of Palestine and the right to issue exequaturs to consuls appointed by foreign Powers. He shall also be entitled to afford diplomatic and consular protection to citizens of Palestine when outside its territorial limits.

ART. 13. All responsibility in connection with the Holy Places and religious buildings or sites in Palestine, including that of preserving existing rights and of securing free access to the Holy Places, religious buildings and sites and the free exercise of worship, while ensuring the requirements of public order and decorum, is assumed by the Mandatory, who shall be responsible solely to the League of Nations in all matters connected herewith, provided that nothing in this article shall prevent the Mandatory from entering into such arrangements as he may deem reasonable with the Administration for the purpose of carrying the provisions of this article into effect; and provided also that nothing in this mandate shall be construed as conferring upon the Mandatory authority to interfere with the fabric or the management of purely Moslem sacred shrines, the immunities of which are guaranteed.

ART. 14. A special commission shall be appointed by the Mandatory to study, define and determine the rights and claims in connection with the Holy Places and the rights and claims relating to the different religious communities in Palestine. The method of nomination, the composition and the functions of this Commission shall be submitted to the Council of the League for its approval, and the Commission shall not be appointed or enter upon its functions without the approval of the Council.

ART. 15. The Mandatory shall see that complete freedom of conscience and the free exercise of all forms of worship, subject only to the maintenance of public order and morals, are ensured to all. No discrimination of any kind shall be made between the inhabitants of Palestine on the ground of race, religion or language. No person shall be excluded from Palestine on the sole ground of his religious belief. The right of each community to maintain its own schools for the education of its own members in its own language, while conforming to such educational requirements of a general nature as the Administration may impose, shall not be denied or impaired.

ART. 16. The Mandatory shall be responsible for exercising such supervision over religious or eleemosynary bodies of all faiths in Palestine as may be required for the maintenance of public order and good government. Subject to such supervision, no measures shall be taken in Palestine to obstruct or interfere with the enterprise of such bodies or to discriminate against any representative or member of them on the ground of his religion or nationality.

ART. 17. The Administration of Palestine may organise on a voluntary basis the forces necessary for the preservation of peace and order, and also for the defence of the country, subject, however, to the supervision of the Mandatory, but shall not use them for purposes other than those above specified save with the consent of the Mandatory. Except for such purposes, no military, naval or air forces shall be raised or maintained by the Administration of Palestine. Nothing in this article shall preclude the Administration of Palestine from contributing to the cost of the maintenance of the forces of the Mandatory in Palestine. The Mandatory shall be entitled at all times to use the roads, railways and ports of Palestine for the movement of armed forces and the carriage of fuel and supplies.

ART. 18. The Mandatory shall see that there is no discrimination in Palestine against the nationals of any State Member of the League of Nations (including companies incorporated under its laws) as compared with those of the Mandatory or of any foreign State in matters concerning taxation, commerce or navigation, the exercise of industries or professions, or in the treatment of merchant vessels or civil aircraft. Similarly, there shall be no discrimination in Palestine against goods originating in or destined for any of the said States, and there shall be freedom of transit under equitable conditions across the mandated area. Subject as aforesaid and to the other provisions of this mandate, the Administration of Palestine may, on the advice of the Mandatory, impose such taxes and customs duties as it may consider necessary, and take such steps as it may think best to promote the development of the natural resources of the country and to safeguard the interests of the population. It may also, on the advice of the Mandatory, conclude a special customs agreement with any State the territory of which in 1914 was wholly included in Asiatic Turkey or Arabia.

ART. 19. The Mandatory shall adhere on behalf of the Administration of Palestine to any general international conventions already existing, or which may be concluded hereafter with the approval of the League of Nations, respecting the slave traffic, the traffic in arms and ammunition, or the traffic in drugs, or relating to commercial equality, freedom of transit and navigation, aerial navigation and postal, telegraphic and wireless communication or literary, artistic or industrial property.

ART. 20. The Mandatory shall co-operate on behalf of the Administration of Palestine, so far as religious, social and other conditions may permit, in the execution of any common policy adopted by the League of Nations for preventing and combating disease, including diseases of plants and animals.

ART. 21. The Mandatory shall secure the enactment within twelve months from this date, and shall ensure the execution of a Law of Antiquities based on the following rules. This law shall ensure equality of treatment in the matter of excavations and archaeological research to the nationals of all States Members of the League of Nations.

(1) "Antiquity" means any construction or any product of human activity earlier than the year 1700 A. D.

(2) The law for the protection of antiquities shall proceed by encouragement rather than by threat. Any person who, having discovered an antiquity without being furnished with the authorization referred to in paragraph 5, reports the same to an official of the competent Department, shall be rewarded according to the value of the discovery.

(3) No antiquity may be disposed of except to the competent Department, unless this Department renounces the acquisition of any such antiquity. No antiquity may leave the country without an export licence from the said Department.

(4) Any person who maliciously or negligently destroys or damages an antiquity shall be liable to a penalty to be fixed.

(5) No clearing of ground or digging with the object of finding antiquities shall be permitted, under penalty of fine, except to persons authorised by the competent Department.

(6) Equitable terms shall be fixed for expropriation, temporary or permanent, of lands which might be of historical or archaeological interest.

(7) Authorization to excavate shall only be granted to persons who show sufficient guarantees of archaeological experience. The Administration of Palestine shall not, in granting these authorizations, act in such a way as to exclude scholars of any nation without good grounds.

(8) The proceeds of excavations may be divided between the excavator and the competent Department in a proportion fixed by that Department. If division seems impossible for scientific reasons, the excavator shall receive a fair indemnity in lieu of a part of the find.

ART. 22. English, Arabic and Hebrew shall be the official languages of Palestine. Any statement or inscription in Arabic on stamps or money in Palestine shall

rumors that Einstein was in danger from those who had murdered Rathenau. They failed to explain how exposing himself in public and traveling abroad safeguarded Einstein.

Einstein's Internationalism and his anti-Germanism did indeed cause some Germans to wish him dead; and a year earlier, in 1921, Rudolph Leibus put a bounty on Einstein's head and Leibus was prosecuted for it. *The New York Times* carried the story reported by the *Chicago Tribune*,

<blockquote>

**"Urged Murder of Einstein,  
Pays $16 Fine in Berlin Court**

Copyright, 1921, by The Chicago Tribune Co.

BERLIN, April 7.—Charged with attempting to incite the murder of

</blockquote>

---

be repeated in Hebrew and any statement or inscription in Hebrew shall be repeated in Arabic.

ART. 23. The Administration of Palestine shall recognise the holy days of the respective communities in Palestine as legal days of rest for the members of such communities.

ART. 24. The Mandatory shall make to the Council of the League of Nations an annual report to the satisfaction of the Council as to the measures taken during the year to carry out the provisions of the mandate. Copies of all laws and regulations promulgated or issued during the year shall be communicated with the report.

ART. 25. In the territories lying between the Jordan and the eastern boundary of Palestine as ultimately determined, the Mandatory shall be entitled, with the consent of the Council of the League of Nations, to postpone or withhold application of such provisions of this mandate as he may consider inapplicable to the existing local conditions, and to make such provision for the administration of the territories as he may consider suitable to those conditions, provided that no action shall be taken which is inconsistent with the provisions of Articles 15, 16 and 18.

ART. 26. The Mandatory agrees that, if any dispute whatever should arise between the Mandatory and another member of the League of Nations relating to the interpretation or the application of the provisions of the mandate, such dispute, if it cannot be settled by negotiation, shall be submitted to the Permanent Court of International Justice provided for by Article 14 of the Covenant of the League of Nations.

ART. 27. The consent of the Council of the League of Nations is required for any modification of the terms of this mandate.

ART. 28. In the event of the termination of the mandate hereby conferred upon the Mandatory, the Council of the League of Nations shall make such arrangements as may be deemed necessary for safeguarding in perpetuity, under guarantee of the League, the rights secured by Articles 13 and 14, and shall use its influence for securing, under the guarantee of the League, that the Government of Palestine will fully honour the financial obligations legitimately incurred by the Administration of Palestine during the period of the mandate, including the rights of public servants to pensions or gratuities.

The present instrument shall be deposited in original in the archives of the League of Nations and certified copies shall be forwarded by the Secretary-General of the League of Nations to all members of the League.

Done at London the twenty-fourth day of July, one thousand nine hundred and twenty-two.

Professor Albert Einstein, who is now in America on a lecture tour, Rudolph Leibus, an anti-Semitic leader, was assessed a fine of $16 by a Berlin Judge.

Leibus recently offered a reward for the murder of Einstein, Professor Foerster and Maximilian Harden, saying that it was a patriotic duty to shoot these leaders of pacifist sentiment."

Jewish anti-Zionist Walter Rathenau was assassinated on 24 June 1922. Both nationalist Germans and political Zionists hated Rathenau. The political Zionists resented Rathenau for being an advocate for, and prime example of, the possibility of assimilation; and for being a vocal anti-Zionist who believed that assimilation was the best means to end anti-Semitism. Rathenau published an article in Maximilian Harden's newspaper *Die Zukunft* in 1897, in which Rathenau called on Jews to assimilate by adopting the Teutonic values of honesty, manhood and integrity, because they were allegedly not an integral part of German society, but were instead an "alien organism in its body."[179] He famously wrote, *inter alia*,

"What a peculiar sight! Amidst German life, a segregated and heterogeneous tribal race, glitteringly and gaudily garnished, with a hot-blooded and restless temperament. An Asiatic horde on the soil of Brandenburg."

"Seltsame Vision! Inmitten deutschen Lebens ein abgesondert fremdartiger Menschenstamm, glänzend und auffällig staffiert, von heißblütig beweglichem Gebaren. Auf märkischem Sand eine asiatische Horde."[180]

Rathenau also famously stated that there was a committee of 300 persons, known to each other, who effectively ruled the world. Some believed that Rathenau was one of them, and that they were the "Elders of Zion". Rathenau was considered one of the many leading Jews who stabbed Germany in the back in the First World War.

The Zionists had stated that it was *impossible* for Jews to assimilate in a Gentile nation and Rathenau's murder bolstered their contention and lent sympathy to their cause. German nationalists believed that Rathenau, who had numerous connections to big business and was the son of the founder of AEG and became its chairman in 1915, had profiteered from the war in his role as Director of Economic Mobilization in control of military spending in the German War Ministry, and had bought inferior goods from Jewish merchants at inflated prices, then at war's end sold off Germany's machinery of war to his Jewish friends. They quoted statements by Rathenau, in which Rathenau declared that he wanted Germany to lose the war. German nationalists resisted Rathenau, who

---

**179**. H. Kessler, *Walter Rathenau: His Life and Work*, Harcourt, Brace, New York, (1930).

**180**. W. Hartenau (W. Rathenau), "Höre, Israel!", *Die Zukunft*, Volume 18, (6 March 1897), pp. 454-462.

became Minister of Reconstruction in 1921 and Foreign Minister in 1922, because he had sponsored the punitive Versailles Treaty and had demanded that Germany pay the oppressive reparations it imposed. Furthermore, they thought that the Rapallo Treaty was but another opportunity for Jews to profit from war and that it aided the Bolshevists.

Anti-Communist *Freikorps* soldier Ernst von Salomon, who served a five year prison sentence for conspiring to assassinate Rathenau, may have believed that Rathenau was one of the alleged Elders of Zion, who wanted to bring Bolshevism to Germany. Rathenau brought about the Rapallo Treaty with the Bolsheviks, and Rathenau had alleged that 300 men controlled the economic destiny of Europe, which 300 some German nationalists assumed were the alleged Elders of Zion. The murder of Rathenau on 24 June 1922, no matter who had committed it and irregardless of the reasons behind it, served as a convenient propaganda tool for the Zionists' promotion of the adoption of the League of Nations Mandate for Palestine on 24 July 1922.

Racist-segregationist and genocidal-Zionist Albert Einstein stated,

"I regretted the fact that [Rathenau] became a Minister. In view of the attitude which large numbers of the educated classes in Germany assume towards the Jews, I have always thought that their natural conduct in public should be one of proud reserve."[181]

Chaim Weizmann wrote,

"[Rathenau's] attitude was, of course, all too typical of that of many assimilated German Jews; they seemed to have no idea that they were sitting on a volcano; they believed quite sincerely that such difficulties as admittedly existed for German Jews were purely temporary and transitory phenomena, primarily due to the influx of East European Jews, who did not fit into the framework of German life, and thus offered targets for anti-Semitic attacks."[182]

The *Berliner Tageblatt*, Morgen-Ausgabe, reported on 5 August 1922,

### "Einsteins Absage an den Naturforschertag.
Auf der Liste der Mörderorganisation.
(Telegramm unseres Korrespondenten.)

Leipzig, 4. August.
Die „Leipziger Neuesten Nachrichten" bringen in ihrer Sonnabendnummer vom 5. August folgende Aufsehen erregende Meldung aus Naturforscherkreisen: Professor Albert Einstein hatte zugesagt, auf der

---

**181**. R. W. Clarck, *Einstein, the Life and Times*, World Publishing Company, USA, (1971), p. 292. Clarck refers to: *Neue Rundschau*, Volume 33, Part 2, pp. 815-816.
**182**. C. Weizmann, *Trial and Error: The Autobiography of Chaim Weizmann*, Harper & Brothers, New York, (1949), p. 289.

Hundertjahrfeier der Gesellschaft deutscher Naturforscher und Aerzte in Leipzig einen Vortrag über die Relativitätstheorie zu halten. Kurz nach der Ermordung Rathenaus teilte aber Einstein dem Vorsitzenden der Gesellschaft, Geheimrat Planck, mit, daß er seine Beteiligung an der Hundertjahrfeier absagen müsse, weil er für mehrere Monate ins Ausland gehe. Diesen plötzlichen Entschluß faßte Einstein, als er erfuhr, daß auch sein Name auf der Liste der Opfer stehe, die von der Mörderorganisation beseitigt werden sollten, der schon Rathenau zum Opfe gefallen ist. Der Entschluß Einsteins, unter diesen Umständen auf längere Zeit ins Ausland zu gehen, war vollkommen zu begreifen. Inzwischen hat sich durch das tatkräftige Eingreifen der Regierung die Lage im Reich erfreulicherweise bedeutend gebessert. Die Mörderorganisation ist aufgedeckt. Alle Schuldigen und Verdächtigen sind in Gewahrsam gebracht worden, so daß nun hoffentlich dem schädlichen Treiben dieser Kreise ein für allemal ein Ende bereitet worden ist. Der Vorsitzende der Gesellschaft deutscher Naturforscher und Aerzte hat nun den Versuch unternommen, Einstein zur Rückkehr nach Deutschland und zur Teilnahme an der Leipziger Hundertjahrfeier zu bewegen, und er bedauert sehr, daß es seinen Bemühungen bisher noch nicht gelungen ist, Einstein zur Rückkehr zu bewegen. Es scheint, daß ein den Gelehrten umgebender engerer Kreis von Freunden und Bewunderern besorgter ist als Einstein selbst. Denn von dieser Seite wird alles getan, die Rückkehr des Gelehrten nach Deutschland zu verhindern oder doch hinauszuschieben. Hoffentlich aber lassen sich noch diese Schwierigkeiten rechtzeitig überwinden, damit Einstein seinen Vortrag über die Relativitätstheorie in Leipzig doch noch persönlich halten kann.

\*

Wie wir erfahren, trifft es zu, daß Professor Einstein an der Leipziger Hundertjahrfeier der Gesellschaft deutscher Naturforscher und Aerzte nicht teilnehmen wird. Gewiß ist es ein tief bedauerlicher Vorgang, daß einer der ersten Gelehrten unserer Zeit an einer Veranstaltung von dem Range der Leipziger Tagung deshalb nicht teilnehmen kann, weil er befürchten muß, in Deutschland, seiner Heimat, statt der Ehrungen, die ihm in der ganzen Welt entgegengebracht worden sind, der Kugel eines Meuchelmörders ausgesetzt zu sein. Die Meldung, die das Leipziger Blatt aus Naturforscherkreisen veröffentlicht, ist gewiß sehr gut gemeint. Wir vermögen auch nicht zu beurteilen, in welchem Grade das Leben und die Sicherheit des großen Gelehrten gefährdet sind. Aber wenn sich auch durch das tatkräftige Eingreifen der Regierung die Lage gebessert hat, so ist doch die Behauptung, daß alle Schuldigen und Verdächtigen in Gewahrsam gebracht seien, etwas kühn und schwerlich zu verantworten. Der Mordbube, der den Anschlag auf Maximilian Harden ausgeführt hat, ist beispielsweise noch nicht gefaßt und Erzbergers Mörder leben in Freiheit und in Saus und Braus. Es ist auch sehr begreiflich, daß die Freunde des Gelehrten in höherem Maße besorgt sind, als er selbst, und es ist sehr bedauerlich, das Rathenau trotz vielfacher Warnung so wenig besorgt gewesen ist. Vielleicht dient dieser Vorgang, dessen tief beschämender Charakter niemandem entgehen kann, endlich dazu, der moralischen Verwilderung, die aus den genügend gekennzeichneten Gründen in weiten Kreisen des Rechtsradikalismus eingerissen ist, durch die entschiedene Abwehr der anständigen Elemente aus allen Lagern im Interesse des deutschen Namens und der deutschen Ehre Einhalt zu tun."

The *Rheinisch-Westfalische Zeitung* (Essen a. Ruhr) reported on 5 August 1922 that the whole affair was contrived as a means to advertize Einstein, whose stardom was fading,

## "Die flüchtige Relativität"

Eine Teilnahme Einsteins am deutschen Naturforscherkongreß in Leipzig ist, wie das B. T. meldet, nicht zu erwarten. Einstein sollte dort einen Vortrag über seine Relativitätstheorie halten. Nach dem Morde Rathenaus ist er aber ins Ausland gereist, da er, wie er erklärte, auf der schwarzen Liste stände.

*

Die Propagierung der Einsteinschen allgemeinen Relativitätstheorie hat zwar einen für das deutsche Kulturleben gemeingefährlichen Charakter, doch hat Einsteins Person damit nichts zu tun. Seine Flucht und die erdachte schwarze Liste sind eins der vielen jetzt auftauchenden republikanischen Propagandamittel, die man sachlich nicht ernst zu nehmen hat. Einsteins Person ist viel zu unwichtig, als daß jemand um ihretwillen sein Leben aufs Spiel setzen wollte. Daß die von ihm in Szene gesetzte Flucht als Reklame auszulegen ist, die seinen schon merklich verblaßten Stern in neuem Glanze erstrahlen lassen soll, dürfte wohl des Pudels Kern in dieser Affäre bedeuten."

Thomas Jefferson Jackson See wrote in *The San Francisco Journal* on 27 May 1923,

"If anyone should ask how Einstein managed to get such vast publicity in the matter of relativity, we may observe that he has the habit of a promoter. Mark Twain humorously wrote to the president of the St. Louis exposition in 1904, that he 'would like to attend the exposition and exhibit himself.' So also does Einstein contrive constantly to be seen among men in conspicuous places. When he came to America, with the Zionist committee, some two years ago, he had to go to the White House at Washington and talk relativity to President Harding. The President, with becoming modesty, said he could not understand the subject.

Things in Europe afterwards became uncomfortable for Einstein, and he sought refuge in an Oriental trip. When in Tokyo he called upon the emperor of Japan, and it was advertised over the world that he was without a dress suit. This report is spectacular and like that of a skillful advertiser.

His return trip is duly chronicled by the press. Thus he finally arrives in Egypt, and on reaching Spain addresses the Academy of Science, at a session held in the presence of the king of Spain. If this is not the trumpeting of an organized press agency, what is it?

Einstein is not liked in Germany. A year or so ago, the students at the University of Berlin hooted him down. It was reported that he was in fear of assassination—but it probably was only a ruse to gain public sympathy."[183]

*The Associated Press* spread Theodor Wolff's rumors of assassination plots. *The New York Times* wrote on 6 August 1922 in Section 2, on page 1,

## *"Einstein Has Fled Temporarily From Germany Because of Threats That He Will Be Killed*

LEIPSIC, Aug. 5 (Associated Press).—Professor Albert Einstein, originator of the theory of relativity, has fled from Germany temporarily because he was threatened with assassination by the group that caused the murder of Dr. Walter Rathenau, German Foreign Minister, according to a letter from Professor Einstein canceling an engagement to address a meeting here.

Efforts to induce the noted scientist to return, in view of the Government's success in coping with the situation, are said to have so far proved unavailing.

Receipt of the letter was announced by the President of the German Physicists' Association, before which Dr. Einstein was to discuss his relativity theory at the organization's 100th anniversary meeting. It was received soon after Dr. Rathenau's assassination, and stated that Dr. Einstein had learned that he also was listed to be killed and had, therefore, decided to go abroad.

It appears that Dr. Einstein's friends and admirers had been more concerned in keeping the scientist safe in this manner than was he himself, and were doing their utmost to prevent, or at least postpone, his return. Dr. Einstein is not accompanying the expedition to Christmas Island, contrary to previously announced plans.

Considerable comment was caused in Geneva early last week by the absence of Dr. Einstein from the meeting of the members of the Intellectual Committee of the League of Nations to begin the work of organization. He had been designated to represent Germany, but did not appear. It was said he was unable to leave his work at the University of Berlin.

Dispatches from Germany soon after the Rathenau murder quoted police authorities there as accusing the notorious 'Consul' organization of

---

**183**. T. J. J. See, "EINSTEIN A TRICKSTER?", *The San Francisco Journal*, (27 May 1923).

having marked twelve leading politicians, editors and financiers of Jewish extraction for assassination, including Dr, Rathenau, Theodor Wolff, editor of the Berliner Tageblatt, and Max Warburg, the Hamburg banker."

*The New York Times* wrote on 8 August 1922 on page 7,

### "URGE EINSTEIN TO HIDE.
#### Friends Fear Because He Is on Anti-Semite Blacklist.

BERLIN, Aug. 7 (Jewish Telegraph Agency).—Friends of Professor Albert Einstein insist upon his remaining abroad, where he is understood to be hiding from the 'Deutsche Nationale' plotters, by whom he has been blacklisted, together with a number of other leading German Jews.

The fear of Professor Einstein's friends is justified, in the opinion of the Berliner Tageblatt, whose editor, Theodor Wolff, is included in the monarchists' blacklist.

'Professor Einstein's continued concealment is advisable,' the Tageblatt says, 'because the assailants of Maximilian Harden and Mathias Erzberger have not been apprehended. Professor Einstein's enforced absence is a blot on the German name and honor.'"

*The New York Times* published a statement on 9 August 1922 on page 10, that perpetuated the myth that anyone who disagreed with Einstein did so out of envy and resultant malice,

### "His Offense Can Be Imagined.

It takes not a little thought to arrive at even a suspicion why anybody wants to assassinate Dr. EINSTEIN. Whoever has seen his picture knows how unlikely he is to excite angry passions in any minds. He is gentleness personified, and it is incredible that he ever gave anybody any of the ordinary forms of offense.

But wait! Not long ago he announced, or at least allowed somebody else, without denial, to announce, that there were not more than twelve people in the world who could understand his new theory of relativity. That, come to think of it, did waken something of animosity in every mind whose possessor lacked the self-confidence to number himself among the so exceptional dozen. Humiliation is an unpleasing sensation, and few if any turn more readily to dislike of him who causes it, and hatred is not far away.

This may not be the basis of the rumored plot against Dr. EINSTEIN, but it is a working hypothesis that will stand until facts are brought forward to prove it untenable."

The German police refuted Wolff's alarmist claims. The *Casseler Allgemeine Zeitung* reported on 12 August 1922, that the alleged "blacklist" did not exist and that the pro-Einstein press was corrupt:

"Eine nicht vorhandene Mordliste. Nach der Ermordung RATHENAUs lief die Meldung durch die Presse, es sei eine Liste der Mörderorginsation aufgefunden worden, auf der ..... die Namen ..... Prof. EINSTEINs u. a. verzeichnet gewesen sein sollen. Jetzt endlich wird von der zuständigen Berliner Stelle versichert, daß die polizeilichen Erhebungen ... . eine derartige Liste nicht ans Licht gefördert haben. Daß die amtlichen Stellen der Veröffentlichung dieser Gerüchte in der gesamten Presse nicht sofort ein Dementi entgegengesetzt haben, kann selbst in der politischen Verwirrung jener Tage keine zureichende Erklärung finden."[184]

There were many more reasons why some suspected that Einstein's flight from the League of Nations, and the Hundertjahrfeier der Gesellschaft Deutscher Naturforscher und Aerzte in Leipzig, on the pretext of unsubstantiated murder plots against him, was a contrived affair to create a false panic over anti-Semitism and to promote sympathy for Einstein, the theory of relativity and Zionism in anticipation of a grand world tour. German science had turned against Einstein. Philipp Lenard and others promised to again embarrass Einstein at the Leipzig meeting as they had done in Bad Nauheim. The racist coward Albert Einstein wanted to hide from them, as Ernst Gehrcke recorded in his book *Die Massensuggestion der Relativitätstheorie: Kulturhistorisch-psychologische Dokumente*, Hermann Meusser, Berlin, (1924), pp. 62-64. Though Einstein was scheduled to deliver a lecture at the centenary of the Association of German Scientists and Physicians in Leipzig, which was overseen by the corrupt sycophant Max Planck, Einstein again took the coward's way out. Max Planck and Max von Laue again rescued Albert Einstein from certain embarrassment. Laue, who was far more competent, though no less childish, than Einstein, delivered a lecture on the theory of relativity, while Einstein again hid from his critics.

Several top Physicists, Mathematicians and Philosophers joined Nobel Prize laureate Philipp Lenard in protesting Max Planck's attempt to deceive the German Public into believing that the scientific community had accepted the theory of relativity as if it were the climax of modern science. These scholars joined together to protect the lay public from the self-aggrandizement and lies of Max Planck and Albert Einstein. Their published protest revealed that the majority of Physicists, Mathematicians and Philosophers considered the theory of relativity to be an unproven hypothesis and a fundamentally flawed, irrational and untenable fiction,

"Die Leitung der „Gesellschaft Deutscher Naturforscher und Ärzte" hat es für richtig gehalten, unter den wissenschaftlichen Darbietungen der Leipziger Jahrhundertfeier Vorträge über Relativitätstheorie auf die Tagesordnung einer großen, allgemeinen Sitzung aufzunehmen. Es muß und soll dadurch wohl der Eindruck erweckt werden, als stelle die

---

**184**. *Casseler Allgemeine Zeitung*, (12 August 1922), as recorded by Ernst Gehrcke in his book: *Die Massensuggestion der Relativitätstheorie: Kulturhistorisch-psychologische Dokumente*, Hermann Meusser, Berlin, (1924), p. 63.

Relativitätstheorie einen Höhepunkt der modernen wissenschaftlichen Forschung dar.
> Hiergegen legen die unterzeichneten Physiker, Mathematiker und Philosophen entschiedene Verwahrung ein. Sie beklagen aufs tiefste die Irreführung der öffentlichen Meinung, welcher die Relativitätstheorie als Lösung des Welträtsels angepriesen wird, und welche man über die Tatsache im Unklaren hält, daß viele und auch sehr angesehene Gelehrte der drei genannten Forschungsgebiete die Relativitätstheorie nicht nur als eine unbewiesene Hypothese ansehen, sondern sie sogar als eine im Grunde verfehlte und logisch unhaltbare Fiktion ablehnen. Die Unterzeichneten betrachten es als unvereinbar mit dem Ernst und der Würde deutscher Wissenschaft, wenn eine im höchsten Maße anfechtbare Theorie voreilig und marktschreierisch in die Laienwelt getragen wird, und wenn die Gesellschaft Deutscher Naturforscher und Ärzte benutzt wird, um solche Bestrebungen unterstützen."

After his crushing defeat at Bad Nauheim and humiliation at the Berlin Philharmonic, Einstein elected to run away and hide from Lenard and Gehrcke at the Hundertjahrfeier der Gesellschaft Deutscher Naturforscher und Aerzte in Leipzig.

The First World War had emancipated all the Jews of the world. Kerensky and the the Bolsheviks had completely liberated the Jews of Russia. Political Zionism was dying a political death. Would not a world tour expose Einstein to greater danger, not less? Einstein had written to the Generalsekretär des Volkerbundes in Genf in July that he was planning to visit Japan.

The Zionist movement was fractionalizing.[185] Even Louis Brandeis was coming to realize that the Jews did not want to emigrate to the Palestinian desert in large enough numbers to form a majority population and American Zionists were softening. Weizmann and Einstein had a tense relationship. Zionism needed a common enemy, real or manufactured, to hold it together. *The New York Times* reported on 20 July 1922 on page 19,

> "JERUSALEM, June 22 (Correspondence of the Associated Press).— The inhabitants of Palestine, both Moslem and Christian, are immeasurably pleased that the British House of Lords yesterday passed the Islington motion disapproving the Balfour declaration of 1917. The native press is jubilant; pan-Arab demonstrations are being held and the local cable office is swamped with congratulatory messages from Arabs to the House of Lords.
> The Balfour declaration pledged the erection of a Jewish homeland in Palestine. The resolution passed yesterday by a vote of 60 to 29 set forth

---

**185**. H. Morgenthau, "Zionism a Surrender, Not a Solution", *The World's Work*, Volume 42, Number 3, (July, 1921), pp. i-viii. "Mr. Zangwill on Zionism", *The London Times*, (16 October 1923), p. 11. I. Zangwill, "Is Political Zionism Dead? Yes", *The Nation*, Volume 118, Number 3062, (12 March 1924), pp. 276-278.

that 'the mandate for Palestine in its present form is unacceptable to this House, because it directly violates the pledges made by his Majesty's Government to the people of Palestine in the declaration of October, 1915, and again in the declaration of November, 1918 (pledges given to the Arabs), and is as at present framed opposed to the sentiments and wishes of the great majority of the people of Palestine. That, therefore, its acceptance by the Council of the League of Nations should be postponed until such modifications have therein been effected as will comply with pledges given by his Majesty's Government.'

The Arabs regard this incident as a great victory. 'It is the bounden duty,' says an Arab call to a demonstration of celebration, 'of all of us to set forth our gratitude to the House of Lords for having proved to the world that God and justice still live in Great Britain.'

Miraat el Shark, a Jerusalem newspaper, says: 'We will win our fight for freedom; we have God and right on our side.' Beit el Makdes, another local paper, says: 'Our victory in the House of Lords is the beginning of the end of political Zionism.'

The Zionists are correspondingly disappointed at the news. They have not failed to cable strong protests to London. The Chairman of the Zionist organization here said to the Associated Press:

'All our hopes have been shattered on the rocks of political expediency. If the House of Commons follows the lead of the House of Lords, then Jews of the world will have been dealt a more staggering blow than that administered by the Emperor Hadrian 1,800 years ago, when his persecutions brought about the last dispersion of the Jewish race.'"

*The New York Times* reported on 26 August 1922, on page 4,

## "ARABS COMING HERE TO OPPOSE ZIONISM

*Declaring Against Palestine Mandate, They Seek American and British Support.*

Copyright 1922, by The New York Times Company.
Special Cable to THE NEW YORK TIMES

CAIRO, Egypt, Aug. 25.—Following the news last night that the Mesopotamian Ministry had resigned because it was unable to agree with the British regarding the Anglo-Irak treaty comes the news today that the situation in the Irak is restive, due to the efforts of extremists to stir anti-British feeling, while excitement is spreading. The Arab delegation meeting in Congress at Nablus reports that hopes for the success of their Palestine cause against the Jews depend largely on sympathetic action from America and England. Feeling in these two countries is to be aroused for protests against Zionism in Palestine, which will be sent from different Moslem countries if the Arab propagandists succeed in inducing the Moslems to produce protests.

America may be interested to learn that the Nablus Congress has decided to send an Arab mission to the United States to collect subscriptions for the Arab organization to enable it to continue the campaign against a Jewish national home in Palestine on the present conditions.

A message from Mecca, which is confirmed by Pilgrims recently at Mecca, says Moslems from all Arab countries met there recently and agreed to organize a movement throughout the Moslem Arab world for the elimination of all foreign political and commercial influence from Moslem Arab countries in the Mid-East. Details of the preliminary organization are to be submitted to the Congress which reassembles at Mecca on the occasion of next year's pilgrimage. The native press of Egypt does not favor the Mecca Congress policy on the ground that an exclusively Ismalic policy nowadays is doomed to react on Islam and to the advantage of Islam's opponents.

JERUSALEM, Aug. 25 (Jewish Telegraphic Agency).—The Arab Congress, meeting at Nablus, 33 miles north of here, has adopted a resolution, rejecting the League of Nations mandate plan for Palestine, refusing Palestinian nationality and declining participation in the elections to the Legislature Council.

The congress instructed the political committee to prepare a national covenant and send missions to all Arab settlements in order to create a union of eastern nations. It was also decided to establish propaganda headquarters in London.

The congress was attended by over 100 delegates from all parts of the country. The deliberations ran quietly, undisturbed by demonstrations. Most of the speakers in a determined tone advised the policy of non-co-operation with the British Administration in Palestine.

## ZIONISTS URGE UNION.

Karlsbad Congress Seeks to Reconcile
Two American Factions.

KARLSBAD, Aug. 25 (Jewish Telegraph Agency)—Many more delegates to the World Zionist Congress are arriving, the total number now reaching over 150, besides many visitors from Europe and America. Dr. Chaim Weizmann, President of the World Zionist Organization, was to preside at the formal opening today, which follows the meetings of executive committees.

A determined effort is being made to effect a reconciliation between the two Zionist factions in the United States. The delegates chiefly interested in this movement are from Germany, France, Holland and Belgium. It is fostered by the strong sentiment for peace existing among the delegates.

Nahum Sokolow, Chairman of the World Zionist Executive Committee, is said to be advocating an immediate settlement of the differences between the two American groups in order to unite all the

Zionist forces in the task of upbuilding Palestine."

It is clear that the Zionists needed a common enemy to unite them, and the alleged murder threats against Einstein, real, contrived or imagined, played a rôle in the promotion of that goal. The Zionists then worked to create economic conditions which would make Germany ripe for a Zionist dictator named Adolf Hitler. The history of the political Zionists' involvement in German wartime politics is discussed in Isaiah Friedman's *Germany, Turkey, and Zionism, 1897-1918*, Clarendon Press, Oxford, (1977).

## 4.9 Wolff Crying, Dirty Tricks, Censorship, Smear Campaigns and Anonymous Threats in the Name of Einstein

The promoters of Einstein and the theory of relativity have employed many of the same tactics and strategies common to such corrupt Jewish political movements as Zionism and Bolshevism. Charles Lane Poor worked hard to expose Einstein as a fraud.[186] Poor complained of terrible censorship of his efforts to expose Einstein and the experiments taken as evidence in support of the theory of relativity. This was and is a common complaint among those who raise concerns about the shameless promotion of the plagiarist Albert Einstein, and who question the metaphysical fallacies and internal contradictions of the theory of relativity.

In 1930, C. L. Poor wrote,

"Thus the claim of Einstein to have found a new law of gravitation and the many assertions that the theory of relativity has worked in accounting for the motions of Mercury and has been conclusively proved by the eclipse

---

**186**. C. L. Poor, "Planetary Motions and the Einstein Theories", *Scientific American Monthly*, Volume 3, (June, 1921), pp. 484-486; **and** "Alternative to Einstein: How Dr. Poor Would Save Newton's Law and the Classical Time and Space Concept", *Scientific American*, Volume 124, (11 June 1921), p. 468; **and** "Motions of the Planets and the Relativity Theory", *Science*, New Series, Volume 54, (8 July 1921), pp. 30-34; **and** "Test for Eclipse Plates", *Science*, New Series, Volume 57, (25 May 1923), pp. 613-614; **and** C. L. Poor and A. Henderson, "Is Einstein Wrong? A Debate", *Forum*, Volumes 71 & 72, (June/July, 1924), pp. 705-715, 13-21; replies *Forum*, Volume 72, (August 1924), pp. 277-281; **and** C. L. Poor, "Relativity and the Motion of Mercury", *Annals of the New York Academy of Sciences*, Volume 29, (15 July 1925), pp. 285-319; **and** "The Deflection of Light as Observed at Total Solar Eclipses", *Journal of the Optical Society of America*, Volume 20, (1930), pp. 173-211; **and** "What Einstein Really Did", *Scribner's Magazine*, Volume 88, (July-December, 1930), pp. 527-538; discussion follows in *Commonweal*, Volume 13, (24 December 1930, 7 January 1931, 11 February 1931), pp. 203-204, 271-272, 412-413. *See also:* "Alternative to Einstein; How Dr. Poor would Save Newton's Law and the Classical Time and Space Concept", *Scientific American*, Volume 124, (11 June 1921), p. 468.

observations and by the displacement of spectral lines are all merely unproved, and, so far, really unsupported illusions. Einstein and his followers have been dwelling in the 'pleasing land of drowsyshed—'; in the land 'Of dreams that wave before the half shut eye.'"[187]

Though the theory of relativity was hyped in the 1920's as a well-proven and perfectly exact, perfectly logical theory, such claims were just that, just hype. There were few people who were competent to try to defend the theory, and the nonexistence of empirical justification for its fantastical claims led to a great insecurity in the academic community—some members of which had stretched out their necks when the press promoted Einstein as the new and improved "Jewish Newton"—and which was worried that the public might discover that Einstein was a fraud and his theories had no rational justification.

Those brave enough to speak out against the degeneration of science into bizarre mysticism, and the demise of professional integrity in science, faced intimidation, censorship, and the classic pernicious political tactics of crowd manipulation by Einstein's supporters. Einstein and his followers were not above employing dirty tricks to suppress opposition and the public disclosure of the truth.

Hubert Goenner tells the story of how Oskar Kraus was scheduled to deliver a speech in Berlin against the theory of relativity on 2 September 1920. Kraus was not able to give his speech, because he was not allowed to go to Germany. Johannes Riem stated that Kraus had wired him a telegram on 2 September 1920, which informed him that Kraus, "was refused a visa for political reasons."[188] Riem complained that,

"In such a way relativity theory is protected by the immigration service."[189]

Goenner notes that Ernst Gehrcke believed that he was censored at Einstein's request[190] from publishing Einstein's verbal assertion that accelerations are absolute in the theory of relativity. Gehrcke, who was a well published and well respected physicist, attempted to draw attention to Einstein's beliefs in the journal *Die Naturwissenschaften*, a Julius Springer publication

---

[187]. C. L. Poor, "What Einstein Really Did", *Scribner's Magazine*, Volume 88, (July-December 1930), pp. 527-538, at 538.
[188]. H. Goenner, "The Reaction to Relativity Theory. I: The Anti-Einstein Campaign in Germany in 1920", *Science in Context*, Volume 6, Number 1, (1993), pp. 107-133, at 118.
[189]. H. Goenner, "The Reaction to Relativity Theory. I: The Anti-Einstein Campaign in Germany in 1920", *Science in Context*, Volume 6, Number 1, (1993), pp. 107-133, at 118-119.
[190]. E. Gehrcke, *Kritik der Relativitätstheorie*, Hermann Meusser, Berlin, (1924), pp. 34-35. *Cf.* H. Goenner, "The Reaction to Relativity Theory. I: The Anti-Einstein Campaign in Germany in 1920", *Science in Context*, Volume 6, Number 1, (1993), pp. 107-133, at 112.

edited by Einstein's friend and supporter Arnold Berliner,[191] which was quick to provide Einstein with an outlet to attack Lucien Fabre,[192] and which published *ad hominem* attacks against anti-relativists in the form of polemic book "reviews" written by Einstein's friends of anti-relativistic literature.[193] Einstein once commented that Springer had "powerful advertising resources",[194] and indeed the publishing house was large, influential and long-lived. Einstein was very well connected and most of his friends looked to him for letters of recommendation and for his intervention to obtain them positions, grants and increased salaries.[195]

Arvid Reuterdahl wrote of the political atmosphere surrounding the corrupt promotion of Einstein,

## "The Academy of Nations—Its Aims and Hopes
## World-Wide Organization of Learned Men Will Study
## Scientific Questions for the Benefit of All Mankind
### By ARVID REUTERDAHL
Dean, Department of Engineering and Architecture, the
College of St. Thomas. St. Paul, Minn.

WE ARE emerging from a period of material and intellectual chaos. Nations have clashed in war. The intellectual world is still in conflict on the fields of knowledge. Never before has the demarcation between intellectual camps been so clearly defined. The meteoric rise of Einstein marks the beginning of this division in the modern kingdom of intellect. The history of civilization shows us that there is nothing exceptional in this condition of things. There were distinct schools of philosophy in ancient India and Greece. The Middle Ages tell the same story of intellectual diversity. In more recent times we find the schools of Descartes, Spinoza, Locke, Berkeley, Hume, Kant, Hegel, Schopenhauer, Comte, Mill, Spencer, Darwin, Lotze, Nietzsche, Bergson and Haeckel.

Now the intellectual world is divided broadly into the Relativistic and Anti-Relativistic schools. Einstein has served as a chemical reagent which

---

**191**. A. Einstein, "In Honour of Arnold Berliner's Seventieth Birthday", *The World As I See It*, Citadel, New York, (1993), p. 14.
**192**. A. Einstein, "Zuschriften an die Herausgeber. Zur Abwehr", *Die Naturwissenschaften*, Volume 9, Number 13, (1 April 1921), p. 219.
**193**. *Die Naturwissenschaften* exhibited a long history of personal attack by "book review". *See, for example: Die Naturwissenschaften*, Volume 11, Number 2, (12 January 1923), pp. 252-256; **and** *Die Naturwissenschaften*, Volume 19, Number 11, (13 March 1931), pp. 252-256.
**194**. Letter from A. Einstein to W. Dällenbach of 27 September 1919, English translation by A. Hentschel, *The Collected Papers of Albert Einstein*, Volume 9, Document 112, Princeton University Press, (2004), pp. 97-98, at 97.
**195**. See, as but one of countless examples, the letter from W. Dällenbach to A. Einstein of 19 September 1919, *The Collected Papers of Albert Einstein*, Volume 9, Document 107, Princeton University Press, (2004).

has precipitated relativity from the present content of knowledge as a mass insoluble to the average man. Never before has the attention of the entire world been drawn to an intellectual system in so short a time. What are the reasons for this unprecedented occurrence? Does the theory of Einstein contain elements of unique value to the human race? These and many other questions come to us as we ponder over the almost miraculous and sudden advent of Einsteinism. No one will dispute the truth of the statement that, as far as the general public is concerned, the theory of Einstein has little or no value. The intricacies of its mathematics and the subtleties of its sophistries are beyond the average man.

## *How Einsteinism Was 'Put Over'*

WE DO not deny that certain features of Einstein's theory cannot fail to fascinate the general public. The world's greatest masters of the art of appeal have, with infallible accuracy, provided sufficient potions from the 'world-of-make-believe' to excite the imagination and interest of even the most prosaic and matter-of-fact individual. Effective advertising when coupled with equally potent measures of suppression of all that might be inimical to the propaganda, together constitute a moving force capable of converting the world in a very brief time. By these doubtful means Einsteinism has conquered the world.

Were the Theory of Relativity sound, upright men must, nevertheless, protest against such questionable means of forcing its acceptance. Hidden forces, inimical to the frank and open discussion of alleged merits of this theory, have been at work in every civilized land.

I am in possession of letters from eminent European scientists describing the deplorable methods employed to hinder and, if possible, completely prevent an unbiased and free discussion of the problem of relativity. In addition to this evidence my own experience is proof conclusive that the known evil effects are not due to accidental causes, but arise from a well defined and strongly organized plan.

Scientific journals and societies in the United States have been loath to accept articles which even mildly criticized Einstein's theories. The advertisement of a book which contains a criticism of relativity, written by a well-known opponent of Einstein, was refused by a journal known for its vigorous publicity campaign in favor of Einsteinism. Two leading American journals, whose main alleged purpose is the unbiased presentation of both sides of every question, have until recently refrained from publishing any statements inimical and detrimental to the theory of relativity. The change of attitude is undoubtedly due to the potent fact that despite the attempted suppression of free discussion, the entire world is now fearlessly and openly challenging the foundations of Einsteinism. A reaction against relativity, of unprecedented proportions and intensity, has set in and Einstein now finds himself on the defensive.

## *Discrimination Against Scientists*

THE writer's article entitled 'Kinertia Versus Enstein' was rejected by a well-known eastern journal. The editor of this journal, after admitting that I had presented a strong case against Einstein, one that would cause something of a sensation, confided that after many misgivings, he, nevertheless, felt that he must return my article.

To draw certain inevitable inferences concerning the real reason for the rejection of the article was undoubtedly justifiable. It was then that THE DEARBORN INDEPENDENT accepted the article for publication.

Many of our scientific societies have discriminated against comparatively unknown scientists. Their papers have been returned without even a hasty perusal, because the writers were not members of the inner controlling circle. This criticism is, moreover, true also in the case of many scientific journals. In certain instances material has been appropriated from the articles before being returned. No credit has, in these cases, been given to the original contributors. The sacred unwritten law that credit should always be freely given to a contributor for even the smallest addition made to our quota of knowledge has been entirely ignored in many cases. The writer does not desire to convey the impression that these corrupt practices are universal; on the contrary, the splendid standards of purity and integrity of some scientific societies and journals constitute ideals which all should emulate.

There is, at the present time, a distressful lack of co-operation between learned societies. This unsound condition inevitably retards intellectual progress. International intellectual co-operation is, as yet, entirely unknown. Many years are required to transmit, through the laborious machinery of scientific approval, results and discoveries made in one country to another isolated from the former by language and geographical location. No common clearing house exists in which the appraisal and valuation of theories may be expeditiously effected. Organized attempts at unification, co-ordination and standardization of systems of knowledge to expedite educational progress are entirely lacking. The general public must oftentimes wait many years before receiving even a small measure of benefit from valuable discoveries because of the absence of organized means of systematic dissemination of accurate knowledge in a simple and easily understood form.

Many of these unfortunate conditions and deficiencies have been emphasized by the arrival of the theory of relativity. The rapid advent of Einsteinism, however, has taught us the lesson that a theory can be speedily 'promoted' by systematic publicity, fortified by a campaign of suppression of honest criticism. There is a twofold aspect to the lesson taught:

First, a benevolent aspect, consisting in the exemplified truth that knowledge can be rapidly disseminated by systematic co-operation.

Second, a malevolent aspect, involving the imposition of unproved hypotheses on the public by coercive means.

The intellectual world should benefit by both aspects of the lesson taught by the rise of relativity. The intellectual world must organize, sanely and safely, for co-operative derivation and dissemination of knowledge by dignified, simple, and accurate means. The world of intellect must protect itself from the evil effects of coercive effort in the 'promotion' of hypotheses.

The crucial question which now faces us may be briefly stated as follows: Can the errors and deficiencies of the *modus operandi* of the intellectual world, forcibly brought to our attention by the advent of Einsteinism, be eliminated and overcome? Have we the remedy at hand which will make impossible the recurrence of these unfortunate and lamentable conditions?

## *Would Keep World Informed*

THE writer herewith presents for the serious consideration of the thinking world a brief outline of the purposes, scope and organization of The Academy of Nations, with the firm conviction that this instrument, when wielded co-operatively by the intellectual world, will transform the existing intellectual chaos into a cosmos of knowledge, advance the general status of education, protect the public against fallacious theories, disseminate knowledge of value to mankind, and enrich the world by the development of the common good.

Before a synopsis of this significant and important movement is presented, it is eminently fitting that a short statement be made concerning its origin.

Dr. Robert T. Browne, one of America's greatest thinkers, and author of the most profound work ever written on the hyperspace movement (The Mystery of Space) in a letter, May 9, 1921, to the writer, indicated that a renaissance in the field of education was not only necessary but inevitable at the present time. This conviction of Dr. Browne's was particularly gratifying to the writer because he had held the same view since that memorable day in 1919, when it became known here that Einstein's theory *seemed* to be confirmed by the results of the observations of the English Solar Expedition.

After some correspondence I submitted a plan for an international organization which met with the unqualified approval of Dr. Browne. At the request of the writer Dr. Browne proceeded to amplify the original outline of the plan with the result that an epoch-making document has been produced. The following excerpts from the original document will convey a brief idea of the causes, purposes and scope of the plan:

'The intellectual world is passing through a period of reconstruction. The entire body of knowledge is being reconstituted. New and radical developments are becoming manifest in science, philosophy, religion, and art; and these are approaching a synthesis hitherto undreamed of, being brought to this consummation by the advent of a movement of far-reaching significance and importance.

'A powerful creative spirit is at work in the world energizing and illumining the minds of men everywhere. The energies of humankind are seeking new and advanced avenues of expression, demanding freedom, certainty, security and the opportunity for the peaceful pursuit of the highest good.

'In the mind of man a new consciousness is broadening; the foundations of a new race of superior men are being laid; the seeds of a higher and better civilization which may bless the nations of the earth are beginning to germinate. The development and fruition of these mighty factors in the advancement of mankind demand the earnest intellectual co-operation of strong men throughout the world to give direction and tendance to the new impulses, which as yet are without adequate determination and means of expression.

'This new order in the world should not and must not be allowed to lose its regenerating power on account of the lack of intelligent co-operation and conscious direction and guidance. The stream of potent human energies must be harnessed and its power utilized for the enrichment of the common good.'

To meet 'the urgency of the call for the accomplishment of these high purposes' an international organization known as *The Academy of Nations* has been formed.

The principal purposes of this organization are:
1. Unification of national effort in the world of knowledge.
2. Discovery, investigation and dissemination of truth.
3. Classification, standardization. and evaluation of the data of science, philosophy, religion and art.
4. Dialectic treatment of data with the view of arriving at synthetic judgment thereon.
5. Publication of findings under the impress of The Academy of Nations.
6. Announcement at prescribed intervals of the status of knowledge in the four major branches, viz: science, philosophy, religion, art.
Note—This to be equivalent to the charting of the bounds of material knowledge.
7. Recognition and encouragement of individual effort amid contributions to the body of knowledge.

## Will Seek Co-operation

UNDER the plan each national unit will publish a journal at suitable intervals. The most important of these contributions will appear in the journal of the academy, which will be published in the languages of all the nations represented. The Year Book of the Academy of Nations will contain announcement of the advance of knowledge (the knowledge status) for the current year of publication. It will be compiled by an international board composed of members elected by the nation units.

The results of this organized work will be made available to the general public, in simple form, through the medium of the public press and by other suitable means.

The Academy of Nations will function in the unification and co-ordination of systems of knowledge, thus procuring the development of synthesized body of knowledge as against the highly specialized conditions now existing. The methods, aims and programs of education will be standardized. Another important function of the academy will be the promotion of the co-operative commonwealth of man in which the wealth-producing energies, the civilizing energies and the energies inherent in the social heritage of humanity shall be co-ordinated and made to yield the maximum value for the welfare of all mankind. Moreover, the academy will promote the use of scientific knowledge as a guiding principle in every department of human endeavor and it will encourage and develop the application of the principles of scientific human engineering to the problems of humanity and to the shaping of its destiny. There will be instituted a world tribunal for the adjudication of controversies in matters connected with theories, philosophical systems, hypotheses, and so on. The academy will be a powerful instrumentality for effecting international solidarity and for the promotion of good will and accord among the nations of the world. It will function also as a supreme centralized authority for the conferring of honors, merits, prizes, degrees, and so on, for distinguished services and for contributions to the body of knowledge. Heretofore, there has been no world society or authority which could bestow academic honors or recognitions on individuals. Affiliations with governments and other national agencies will be established to advance the cause of knowledge and the execution of its programs.

### *Organization Meeting Is Held*

THE above consists in the main of direct quotations, suitably rearranged, from the original classic document.

In this great academy intellectual freedom will be reborn. There will be no arbitrary exclusion of hypotheses, theories, views and beliefs. The academy will ever function as an open and free forum for the discussion of all the great problems of humanity.

One of the first duties to be assigned to the academy will be the adjudication and appraisal of the precise value and merit of the Theory of Relativity definitely to fix its 'knowledge status.'

The organization meeting of the College of Fellows of the Academy of Nations was held December 28 and 30, 1921, in Brooklyn. National institutes of the Academy of Nations are now being formed in Sweden, Germany. Switzerland, Czecho-Slovakia and Spain. Steps are being taken for the organization of institutes in Norway, Denmark, England, Holland, France and Italy. Within the ensuing year national institutes will be organized in every civilized country of the world.

The field of the academy embraces every general and special class of knowledge and its interests will, therefore, be universal."[196]

In the spring of 1922, Edouard Guillaume gave Einstein fair warning that he would debate him in Paris. Guillaume and others had published their findings that the special theory of relativity derives from a particular light sphere in a preferred frame of reference, and that in translational frames of reference this sphere becomes an ellipsoid.[197] Jánossy and others have since published works

---

**196**. A. Reuterdahl, "The Academy of Nations—Its Aims and Hopes", *The Dearborn Independent*, (7 January 1922), p. 14.
**197**. E. Guillaume's letter, translated by A. Reuterdahl, "Guillaume, Barred in Move To Debate Einstein, Calls Meeting Political Reunion", *Minneapolis Journal*, (14 May 1922), p. 14; reprinted with slight modifications, "The Origin of Einsteinism", *The New York Times*, (12 August 1923), Section 7, p. 8. ***See also:*** "Einstein Faces in Paris Grave Blow at Theory", *The Chicago Tribune*, (31 March 1922). ***See also:*** "Dr. Guillaume's Proofs of Einstein Theory's Fallacy Revealed to the Journal", *Minneapolis Journal*, (9 April 1922). ***See also:*** E. Guillaume, "Un Résultat des Discussions de la Théorie d'Einstein au Collège de France", *Revue Générale des Sciences Pures et Appliquées*, Volume 33, Number 11, (15 June 1922), pp. 322-324. ***See also:*** "Les Bases de la Physique moderne", *Archives des Sciences Physiques et Naturelles*, Series 4, Volume 43, (1917), pp. 5-21, 89-112, 185-198; **and** "Sur le Possibilité d'Exprimer la Théorie de la Relativité en Fonction du Temps Universel", *Archives des Sciences Physiques et Naturelles*, Series 4, Volume 44, (1917), pp. 48-52; **and** "La Théorie de la Relativité en Fonction du Temps Universel", *Archives des Sciences Physiques et Naturelles*, Series 4, Volume 46, (1918), pp. 281-325; **and** "Sur la Théorie de la Relativité", *Archives des Sciences Physiques et Naturelles*, Series 5, Volume 1, (1919), pp. 246-251; **and** "Représentation et Mesure du Temps", *Archives des Sciences Physiques et Naturelles*, Series 5, Volume 2, (1920), pp. 125-146; **and** "La Théorie de la Relativité et sa Signification", *Revue de Métaphysique et de Morale*, Volume 27, (1920), pp. 423-469; **and** "Relativité et Gravitation", *Bulletin de la Société Vaudoise des Sciences Naturelles*, Volume 53, (1920), pp. 311-340; **and** "Les Bases de la Théorie de la Relativité", *Revue Générale des Sciences Pures et Appliquées*, (15 April 1920) pp. 200-210; **and** C. Willigens, "Représentation Géométrique du Temps Universel dans la Théorie de la Relativité Restreinte", *Archives des Sciences Physiques et Naturelles*, Series 5, Volume 2, (1920), p. 289; **and** E. Guillaume, *La Théorie de la Relativité. Résumé des Conférences Faites à l'Université de Lausanne au Semestre d'été 1920*, Rouge & Co., Lausanne, (1921); **and** E. Guillaume and C. Willigens, "Über die Grundlagen der Relativitätstheorie", *Physikalische Zeitschrift*, Volume 22, (1921), pp. 109-114; **and** E. Guillaume, "Graphische Darstellung der Optik bewegter Körper", *Physikalische Zeitschrift*, Volume 22, (1921), pp. 386-388; **and** Guillaume's Appendix II, "Temps Relatif et Temps Universel", in L. Fabre, *Une Nouvelle Figure du Monde: les Théories d'Einstein*, Second Edition, Payot, Paris, (1922); **and** E. Guillaume, "Y a-t-il une Erreur dans le PremierMémoire d'Einstein?", *Revue Générale des Sciences Pures et Appliquées*, Volume 33, (1922), pp. 5-10; **and** "La Question du Temps d'après M. Bergson, à Propos de la Théorie d'Einstein", *Revue Générale des Sciences Pures et Appliquées*, Volume 33, (1922), pp. 573-582; **and** Guillaume's introduction in H. Poincaré, *La Mécanique Nouvelle: Conférence, Mémoire et Note sur la Théorie de la Relativité / Introduction de Édouard Guillaume*, Gauthier-Villars, Paris, (1924), pp. V-XVI; **and** H. Bergson, *Durée et Simultanéité, à Propos de la Théorie d'Einstein*, English translation by L. Jacobson, *Duration and simultaneity, with Reference to Einstein's*

which also favor Lorentz' physical interpretation of light speed anisotropy in "moving" frames of reference, without relying solely upon the paradox of the twins.[198]
The *Chicago Tribune* reported on 31 March 1922,

## *"EINSTEIN FACES IN PARIS GRAVE BLOW AT THEORY*

---

*Theory*, The Library of Liberal Arts, Bobbs-Merrill, Indianapolis, (1965); which contains a bibliography at pages xliii-xlv. *See also:* P. Painlevé, "La Mécanique Classique et la Théorie de la Relativité", *Comptes rendus hebdomadaires des séances de L'Académie des sciences*, Volume 173, (1921), pp. 677-680. *See also:* S. Mohorovičić, "Raum, Zeit und Welt. II Teil", in K. Sapper, Editor, *Kritik und Fortbildung der Relativitätstheorie*, Akademische Druck- u. Verlagsanstalt, Graz, Volume 2, (1962), pp. 219-352, at 273-275. *See also:* K. Hentschel, *Interpretationen und Fehlinterpretationen der speziellen und der allgemeinen Relativitätstheorie durch Zeitgenossen Albert Einsteins*, Birkhäuser, Basel, Boston, Berlin, (1990). *See also:* A. Genovesi, *Il Carteggio tra Albert Einstein ed Edouard Guillaume. "Tempo Universale" e Teoria della Relativtà Ristretta nella Filosofia Francese Contemporanea*, Franco Angeli, Milano, (2000). *See also:* Letter from A. Einstein to E. Guillaume of 24 September 1917, *The Collected Papers of Albert Einstein*, Volume 8, Part A, Document 383, Princeton University Press, (1998). *See also:* Letter from E. Guillaume to A. Einstein of 3 October 1917, *The Collected Papers of Albert Einstein*, Volume 8, Part A, Document 385, Princeton University Press, (1998). *See also:* Letter from A. Einstein to E. Guillaume of 9 October 1917, *The Collected Papers of Albert Einstein*, Volume 8, Part A, Document 387, Princetone University Press, (1998). *See also:* Letter from E. Guillaume to A. Einstein of 17 October 1917, *The Collected Papers of Albert Einstein*, Volume 8, Part A, Document 392, Princeton University Press, (1998). *See also:* Letter from A. Einstein to E. Guillaume of 24 October 1917, *The Collected Papers of Albert Einstein*, Volume 8, Part A, Document 394, Princeton University Press, (1998). *See also:* Letter from E. Guillaume to A. Einstein of 25 January 1920, *The Collected Papers of Albert Einstein*, Volume 9, Document 280, Princeton University Press, (2004). *See also:* Letter from M. Grossmann to A. Einstein of 5 February 1920, *The Collected Papers of Albert Einstein*, Volume 9, Document 300, Princeton University Press, (2004). *See also:* Letter from A. Einstein to E. Guillaume of 9 February 1920, *The Collected Papers of Albert Einstein*, Volume 9, Document 305, Princeton University Press, (2004). *See also:* Letter from E. Guillaume to A. Einstein of 15 February 1920, *The Collected Papers of Albert Einstein*, Volume 9, Document 316, Princeton University Press, (2004). *See also:* Letter from A. Einstein to M. Grossmann of 27 February 1920, *The Collected Papers of Albert Einstein*, Volume 9, Document 330, Princeton University Press, (2004). *See also:* Letter from A. Einstein to P. Oppenheim of 29 April 1920, *The Collected Papers of Albert Einstein*, Volume 9, Document 399, Princeton University Press, (2004).
**198**. L. Jánossy, "Über die physikalische Interpretation der Lorentz-Transformation", *Annalen der Physik*, Series 6, Volume 11, (1953), pp. 293-322; **and** *Theory of Relativity Based on Physical Reality*, Akademiai Kiadó, Budapest, (1971). *See also:* S. J. Prokhovnic, *The Logic of Special Relativity*, Cambridge University Press, (1967). *See also:* K. Sapper, Editor, *Kritik und Fortbildung der Relativitätstheorie*, In Two Volumes, Akademische Druck- u. Verlagsanstalt, Graz, Austria, (1958/1962).

[Chicago Tribune Foreign News Service.]

BERNE, March 30.—Edmond Guillaume says he has discovered a fundamental error in the Einstein theory and is en route to Paris to attend the savant's lecture and to challenge the relativity discoverer.

M. Guillaume hopes for a public debate in which he can use his ellipsoid to demonstrate Prof. Einstein's error.

Former Premier Painleve, a celebrated mathematician, has reached the same conclusions as M. Guillaume, but through a different process. M. Guillaume is a cousin of Charles Albert Guillaume, a recent Nobel Prize winner."

*The Minneapolis Journal* wrote on 9 April 1922,

## "DR. GUILLAUME'S PROOFS OF EINSTEIN THEORY'S FALLACY REVEALED TO THE JOURNAL

Professor Reuterdahl of St. Thomas Makes Public
Correspondence With Swiss Savant Disclosing
Latter's Weapons of Attack on Relativity

### BARES FACTS FOR WHICH SCIENTIFIC WORLD NOW EAGERLY WAITS AT PARIS

Simple Experience of Every Day Railroad Operation
Relied On to Show That Man Who Upset
Accepted Laws of Nature Is All Wrong

With the scientific world awaiting Dr. Edmund Guillaume's appearance in Paris to challenge and attempt to destroy the very foundation of the Einstein theory of relativity, Professor Arvid Reuterdahl, dean of the department of engineering and architecture at the College of St. Thomas, Midway, last night revealed to The Journal the purported proof of the fallacy of 'Einsteinism' which Dr. Guillaume will use in his Paris attack.

Professor Reuterdahl all along has contended the Einstein theory was all wrong and is now preparing a book, 'Fallacies of Einstein.' When Einstein was in America Reuterdahl challenged him to a debate without avail. He has been in correspondence with Dr. Guillaume and has received from the noted Swiss scientist a special contribution for his book containing the very matter which Guillaume will use in his forthcoming Paris attack on relativity. Until Professor Reuterdahl disclosed Dr. Guillaume's proofs to The Journal last night, the St. Thomas dean was the only man in the United States who possessed the explanation that is expected by its advocate to knock the whole Einstein theory of relativity into a cocked hat when Professor Einstein is confronted with it at his forthcoming lecture in Paris.

According to a special cable dispatch published in The Journal March

31, Dr. Guillaume claims that the matter now in possession of Professor Reuterdahl and revealed to the public today, discloses a fundamental error in the Einstein theory. The cable dispatch stated that Dr. Guillaume hoped for a public debate with Einstein in which he would have a chance to hurl his proofs at the author of the relativity theory.

'The final death blow to Einsteinism is about to be delivered by the eminent Swiss physicist and mathematician. Dr. Edouard Guillaume when the scientists convene at Paris,' said Professor Reuterdahl last night. 'Dr. Guillaume in two letters written to me and dated July 25 and Aug. 13, 1921, pointed out a fundamental error in the mathematical speculations of Einstein which explodes the entire theory proving that relativity is the greatest scientific fiasco of all times. Dr. Guillaume shows that Einstein, in his first article entitled, 'Zur Elektrodynamik bewegter Koeper,' which appeared in 1905 in Annalen der Physik, volume 17, commits 'the greatest scientific blunder of modern times.'

### Swiss Savant's Proofs Revealed

'Einsteinism stands or falls upon the socalled postulate of the absolute velocity of light. Dr. Guillaume in a brilliant analysis, shows that this very postulate is destroyed by a fatal error in Einstein's mathematics.'

The following is a translation of Professor Guillaume's final summary communicated to Professor Reuterdahl:

'Einstein considers a luminous signal produced, for instance, on a track by means of an electric pocket lamp. A brief signal gives rise to a wave which moves through space and in all directions with a velocity of 300,000 kilometers per second. This wave forms at each moment a spherical surface, the ray of which increases with this velocity and the center of which is motionless. Let us inquire now how the wave appears to an observer carried along with the train. Let us apply the transformation of Lorentz. What is found? Einstein maintains that the wave appears also as a sphere with its center motionless as regards the train, and whose ray grows likewise with the velocity of 300,000 kilometers a second.

### Simple Test Cited

"Die betrachtete Welle,' says Einstein in conclusion, 'ist auch in bewegten System (Wagon) betrachtet eine Kugelwelle von der Ausbreitungsgeschwindigkeit 300,000 km-sec.' But if we look more closely we detect an error in the famous physicist's calculation: the wave seen from the train is not a sphere, but rather an ellipsoid, and the famous principle of the absolute constancy of light vanishes! At the same time collapse all the paradoxes, and at last we are clear of this inextricable web and beyond the reach of the entangling challenges that Einstein has hurled at our good sense, free from what Americans have so well termed 'Einsteinism.''

'Einstein has been challenged to meet Dr. Guillaume at Paris,' said Professor Reuterdahl last night. 'The evidence presented by Dr. Guillaume is so conclusive that Einstein will hasten the death of the already dying theory of relativity by accepting the challenge. If Einstein uses the same caution that he exhibited when challenged by me he will again carefully avoid the issue by veiling himself in sphynx like silence.'"

On 22 April 1922, Edouard Guillaume complained to Arvid Reuterdahl, in a letter which was reproduced in *The Minneapolis Journal*, which newspaper wrote on 14 May 1922,

## "Guillaume, Barred in Move To Debate Einstein, Calls Meeting Political Reunion

Savant, in letter to Professor Reuterdahl of St. Thomas, Says Ideals of Science Were Treated With Ignominy in Paris

Failing in an attempt to force a public debate which they hoped would disclose fundamental errors in the Einstein theory of relativity, scientists in the antirelativity group will continue their fight on 'Einsteinism,' Professor Arvid Reuterdahl of St. Thomas college said last night.

Dean of the department of engineering and architecture at St. Thomas, a prominent figure in the scientific world because of his research work, Professor Reuterdahl has collaborated with Dr. Edouard Guillaume, Swiss savant, in disputing the theory which has brought fame to Einstein.

When Einstein visited the United States Professor Reuterdahl challenged him to an open debate.

### Guillaume Meets Einstein

In Paris recently Dr. Guillaume faced Dr. Einstein on a platform, before French scientists convened at the College of France. His appearance had been awaited eagerly by scientists throughout the world.

'In a letter which I just have received,' Professor Reuterdahl said, 'Dr. Guillaume gives a vivid picture of the scene which ever will remain a blot on the fair escutcheon of science.

Dr. Guillaume had lectured only a few minutes when he was silenced peremptorily in order to give way to the illustrious man of the hour, Einstein, who dismissed the entire matter with the gesture of a conqueror.'

### Floor Given to Einstein

'I had hoped to be permitted quietly to present the results of my researches,' reads the letter from Dr. Guillaume to Professor Reuterdahl. Unfortunately, I had barely lectured for five minutes when I was interrupted in order to give the floor to Einstein, who was forced to acknowledge the fact that an ellipsoid results from his own mathematics.

(Einstein's theory is that a wave surface of light, traveling outward from any luminous body, such as an electric light, is a spherical surface. Dr. Guillaume and Professor Reuterdahl contend that this surface is ellipsoidal under certain conditions.)

'Einstein dismissed the matter,' the letter continues, 'by saying that he was not interested. At this statement of Einstein's the large audience present applauded vociferously. I then saw that it was absolutely impossible to carry on a scientific discussion under these conditions.

'That, my dear Professor Reuterdahl, is the ignominious treatment which the high ideals of science receive at the present time.

**Called Political Reunions**

'Scientific congresses of this kind are nothing more than political reunions. It is urgent that all honest men unite to fight against these deplorable methods, which can only lead to the death of science. You may say definitely in America that all discussion was prevented and made impossible by the fanatic attitude of the relativists.'

When Professor Reuterdahl revealed April 9, through The Journal, the points to be used by Dr. Guillaume in his Paris debate, he predicted that that attempt to force Einstein into an honest discussion of his own theory would prove a total failure.

Professor Reuterdahl now is preparing a book, 'Fallacies of Einstein,' to which Dr. Guillaume has made a contribution. Dr. Guillaume issued a public statement March 31, which was cabled to The Journal, in which he said a fundamental error had been found in the Einstein theory."

Guillaume's letter, which was also reproduced in *The New York Times*, Arvid Reuterdahl, "The Origin of Einsteinism", (12 August 1923), Section 7, p. 8:

"I had hoped to be permitted quietly to present the results of my researches. Unfortunately, I had barely lectured for five minutes when I was interrupted in order to give the floor to Einstein, who was forced to acknowledge that an ellipsoid results from his own mathematics. Einstein dismissed the matter by saying that he was not interested. At this statement of Einstein's the large audience present applauded vociferously. I then saw that it was absolutely impossible to carry on a scientific discussion under these conditions. That, my dear Professor Reuterdahl, is the ignominious treatment which the high ideals of science receive at the present time. Scientific congresses of this kind are nothing more than political reunions. It is urgent that all honest men unite to fight against these deplorable methods, which can only lead to the death of science. You may say definitely in America that all discussion was prevented and made impossible by the fanatic attitude of the relativists."[199]

---

**199**. E. Guillaume's letter, translated by A. Reuterdahl, "Guillaume, Barred in Move To Debate Einstein, Calls Meeting Political Reunion", *Minneapolis Journal*, (14 May 1922), p. 14; reprinted with slight modifications, "The Origin of Einsteinism", *The New York Times*, (12 August 1923), Section 7, p. 8. ***See also:*** "Einstein Faces in Paris Grave Blow at Theory", *The Chicago Tribune*, (31 March 1922). ***See also:*** "Dr. Guillaume's Proofs of Einstein Theory's Fallacy Revealed to the Journal", *Minneapolis Journal*, (9 April 1922). ***See also:*** E. Guillaume, "Un Résultat des Discussions de la Théorie d'Einstein au Collège de France", *Revue Générale des Sciences Pures et Appliquées*, Volume 33, Number 11, (15 June 1922), pp. 322-324. ***See also:*** "Les Bases de la Physique moderne", *Archives des Sciences Physiques et Naturelles*, Series 4, Volume 43, (1917), pp. 5-21, 89-112, 185-198; **and** "Sur le Possibilité d'Exprimer la Théorie de la Relativité en Fonction du Temps Universel", *Archives des Sciences Physiques et Naturelles*, Series 4, Volume 44, (1917), pp. 48-52; **and** "La Théorie de la Relativité en Fonction du Temps Universel",

*Archives des Sciences Physiques et Naturelles*, Series 4, Volume 46, (1918), pp. 281-325; **and** "Sur la Théorie de la Relativité", *Archives des Sciences Physiques et Naturelles*, Series 5, Volume 1, (1919), pp. 246-251; **and** "Représentation et Mesure du Temps", *Archives des Sciences Physiques et Naturelles*, Series 5, Volume 2, (1920), pp. 125-146; **and** "La Théorie de la Relativité et sa Signification", *Revue de Métaphysique et de Morale*, Volume 27, (1920), pp. 423-469; **and** "Relativité et Gravitation", *Bulletin de la Société Vaudoise des Sciences Naturelles*, Volume 53, (1920), pp. 311-340; **and** "Les Bases de la Théorie de la Relativité", *Revue Générale des Sciences Pures et Appliquées*, (15 April 1920) pp. 200-210; **and** C. Willigens, "Représentation Géométrique du Temps Universel dans la Théorie de la Relativité Restreinte", *Archives des Sciences Physiques et Naturelles*, Series 5, Volume 2, (1920), p. 289; **and** E. Guillaume, *La Théorie de la Relativité. Résumé des Conférences Faites à l'Université de Lausanne au Semestre d'été 1920*, Rouge & Co., Lausanne, (1921); **and** E. Guillaume et C. Willigens, "Über die Grundlagen der Relativitätstheorie", *Physikalische Zeitschrift*, Volume 22, (1921), pp. 109-114; **and** E. Guillaume, "Graphische Darstellung der Optik bewegter Körper", *Physikalische Zeitschrift*, Volume 22, (1921), pp. 386-388; **and** Guillaume's Appendix II, "Temps Relatif et Temps Universel", in L. Fabre, *Une Nouvelle Figure du Monde: les Théories d'Einstein*, Second Edition, Payot, Paris, (1922); **and** E. Guillaume, "Y a-t-il une Erreur dans le PremierMémoire d'Einstein?", *Revue Générale des Sciences Pures et Appliquées*, Volume 33, (1922), pp. 5-10; **and** "La Question du Temps d'après M. Bergson, à Propos de la Théorie d'Einstein", *Revue Générale des Sciences Pures et Appliquées*, Volume 33, (1922), pp. 573-582; **and** Guillaume's introduction in H. Poincaré, *La Mécanique Nouvelle: Conférence, Mémoire et Note sur la Théorie de la Relativité / Introduction de Édouard Guillaume*, Gauthier-Villars, Paris, (1924), pp. V-XVI; **and** H. Bergson, *Durée et Simultanéité, à Propos de la Théorie d'Einstein*, English translation by L. Jacobson, *Duration and simultaneity, with Reference to Einstein's Theory*, The Library of Liberal Arts, Bobbs-Merrill, Indianapolis, (1965); which contains a bibliography at pages xliii-xlv. *See also:* P. Painlevé, "La Mécanique Classique et la Théorie de la Relativité", *Comptes rendus hebdomadaires des séances de L'Académie des sciences*, Volume 173, (1921), pp. 677-680. *See also:* S. Mohorovičić, "Raum, Zeit und Welt. II Teil", in K. Sapper, Editor, *Kritik und Fortbildung der Relativitätstheorie*, Akademische Druck- u. Verlagsanstalt, Graz, Volume 2, (1962), pp. 219-352, at 273-275. *See also:* K. Hentschel, *Interpretationen und Fehlinterpretationen der speziellen und der allgemeinen Relativitätstheorie durch Zeitgenossen Albert Einsteins*, Birkhäuser, Basel, Boston, Berlin, (1990). *See also:* A. Genovesi, *Il Carteggio tra Albert Einstein ed Edouard Guillaume. "Tempo Universale" e Teoria della Relatività Ristretta nella Filosofia Francese Contemporanea*, Franco Angeli, Milano, (2000). *See also:* Letter from A. Einstein to E. Guillaume of 24 September 1917, *The Collected Papers of Albert Einstein*, Volume 8, Part A, Document 383, Princeton University Press, (1998). *See also:* Letter from E. Guillaume to A. Einstein of 3 October 1917, *The Collected Papers of Albert Einstein*, Volume 8, Part A, Document 385, Princeton University Press, (1998). *See also:* Letter from A. Einstein to E. Guillaume of 9 October 1917, *The Collected Papers of Albert Einstein*, Volume 8, Part A, Document 387, Princetone University Press, (1998). *See also:* Letter from E. Guillaume to A. Einstein of 17 October 1917, *The Collected Papers of Albert Einstein*, Volume 8, Part A, Document 392, Princeton University Press, (1998). *See also:* Letter from A. Einstein to E. Guillaume of 24 October 1917, *The Collected Papers of Albert Einstein*, Volume 8, Part A, Document 394, Princeton University Press, (1998). *See also:* Letter from E. Guillaume to A. Einstein of 25 January 1920, *The Collected Papers of Albert Einstein*, Volume 9, Document 280,

William Cardinal O'Connell, who had written a letter condemning anti-Semitism and who had signed John Spargo's protest against anti-Semitism,[200] accused Einstein and his clique of promoting atheism in a lecture the Cardinal had given. Cardinal O'Connell was quoted in the 12 April 1929 issue of the *Boston Evening American*,

> "That there is in certain quarters such a heated defense of an unprovable, certainly unproved hypothesis, only again makes it doubly clear that what I said to the students was true—the claque is applauding noisily so as to drown honest criticism. But that has been from all accounts the Einstein method of answer to all who disagree with him."

Other such staged interruptions as happened to Guillaume took place in defense of the indefensible, in defense of Einstein and his metaphysical nonsense. For example, when Arvid Reuterdahl spoke at the University of Wisconsin, Madison, in March of 1926 about the Einstein swindle, the faculty there allegedly disrupted his lecture.[201] The University's newspaper, *The Daily Cardinal*, reported,

> "Not even a tithe of courtesy is being shown Prof. Reuterdahl [***] At the lecture Wednesday night instructors of the mathematics department

---

Princeton University Press, (2004). *See also:* Letter from M. Grossmann to A. Einstein of 5 February 1920, *The Collected Papers of Albert Einstein*, Volume 9, Document 300, Princeton University Press, (2004). *See also:* Letter from A. Einstein to E. Guillaume of 9 February 1920, *The Collected Papers of Albert Einstein*, Volume 9, Document 305, Princeton University Press, (2004). *See also:* Letter from E. Guillaume to A. Einstein of 15 February 1920, *The Collected Papers of Albert Einstein*, Volume 9, Document 316, Princeton University Press, (2004). *See also:* Letter from A. Einstein to M. Grossmann of 27 February 1920, *The Collected Papers of Albert Einstein*, Volume 9, Document 330, Princeton University Press, (2004). *See also:* Letter from A. Einstein to P. Oppenheim of 29 April 1920, *The Collected Papers of Albert Einstein*, Volume 9, Document 399, Princeton University Press, (2004).
**200**. "Issue a Protest on Anti-Semitism", *The New York Times*, (17 January 1921), p. 10.
**201**. "Reuterdahl Gives Mathematic Lectures", *The Daily Cardinal* (University of Wisconsin, Madison), (11 March 1926). "Prof. Reuterdahl Talks Despite All Faculty Efforts", *The Daily Cardinal* (University of Wisconsin, Madison), (12 March 1926), Front Page. "St. Paulite Piqued by Badger Faculty", *The St. Paul Daily News*, Final Pink, (12 March 1926), Front Page. "Intolerance", *The St. Paul Daily News*, (13 March 1926), Front Page or page 2???. "Everything Fine, Reuterdahl Says of Badger Antics", *St. Paul Dispatch*, (13 March 1926), Front Page. "Reuterdahl Says He Had a 'Bully' Time in Madison", *The Minneapolis Sunday Tribune*, (14 March 1926). "Wisconsin U Mathematics Professors 'Act Like Children' Says Reuterdahl; Had a Fine Time", *The St. Paul Daily News*, (14 March 1926), Front Page. "Reuterdahl Takes Fling at Madison", *St. Paul Dispatch*, (19 March 1926), p. 17. "Intellectual Despotism Is Menace to Honest Research in Science, Dr. Reuterdahl Declares", *St. Paul Daily News*, (19 March 1926), p. 2.

interfered with the lecturer so that he was unable to finish his talk. [\*\*\*] Staff Tries To Stop Talk [\*\*\*] members of the instructional staff of the mathematical department tinkered with the water pressure apparatus which operates the projection screen [\*\*\*] and made it impossible for the lecturer to continue [\*\*\*] the members of the department also blinked the lights in the auditorium while the speaker was lecturing, putting the auditorium in darkness temporarily. This is said to have occurred three times."[202]

Johannes Stark alleged that Ernst Gehrcke was denied a full professorship in Germany, because he had argued against the theory of relativity,

"Gehrcke ist der Kampf gegen die Relativitätstheorie übel bekommen; trotz seiner zahlreichen hervorragenden experimentellen Arbeiten wird er von Fakultäten nicht für ein physikalisches Ordinat vorgeschlagen."[203]

In 1882, Franz Mehring quoted a Jewish author who criticized Jews for, among other things, "the malicious gloating when veritable conspiracies deprived of their livelihoods people who were suspected of anti-Jewish feelings[.]"[204] Einstein and his friends sought to stigmatize *any* criticism of him or of the theory of relativity as if it were "anti-Semitism" *per se*.[205] They thereby threatened anyone who dared speak out with career infringement or the absolute inability to find work. Whether or not significant numbers of people interfered with the careers of persons suspected of anti-Jewish feelings for merely questioning Einstein or discussing the facts, the impression that they would existed and had a chilling effect on Einstein's opposition in the debate over the merits of relativity theory and Einstein's obvious plagiarism. This has been very detrimental to the progress of Physics.

Hugo Dingler's alloted time to speak against the theory of relativity at the Bad Nauheim meeting was severely restricted. Ernst Mach wrote of his admiration for Dingler,

"I myself—seventy-four years old, and struck down by a grave malady—shall not cause any more revolutions. But I hope for important progress from

---

**202**. "PROF. REUTERDAHL TALKS DESPITE ALL FACULTY EFFORTS: Instructors Place Auditorium in Darkness in Attempt to Stop Lecture", *The Daily Cardinal* (University of Wisconsin, Madison), (12 March 1926).
**203**. *Cf.* H. Goenner, "The Reaction to Relativity Theory in Germany, III: 'A Hundred Authors against Einstein", *The Attraction of Gravitation: New Studies in the History of General Relativity*, Birkhäuser, Boston, Basel, Berlin, (1993), p. 250. J. Stark, *Die gegenwärtige Krisis in der Deutschen Physik*, Johann Ambrosius Barth, Leipzig, (1922), p. 16.
**204**. English translation from: P. W. Massing, *Rehearsal for Destruction: A Study of Political Anti-Semitism in Imperial Germany*, Howard Fertig, New York, (1967). p. 315.
**205**. "Prof. Einstein Here, Explains Relativity", *The New York Times*, (3 April 1921), pp. 1, 13, at 1.

a young mathematician, Dr. Hugo Dingler, who, judging from his publications, has proved that he has attained to a free and unprejudiced survey of *both* sides of science."[206]

Gehrcke's accusations that Einstein was a plagiarist were fully justified by the facts, and Dingler correctly pointed out several fatal flaws in the metaphysical formulation of the theory of relativity.[207]

Hubert Goenner wrote,

"[Gehrcke] blame[d] Einstein's reply of 27 August [1920] for arousing political and racial instincts and deflecting public attention from the facts of relativity theory."

Paul Weyland made the same charge, that Einstein's defense of his theory and his claims of originality were so weak that he was forced to run away from Germany, and to change the subject to fabricated accusations of anti-Semitism. Arvid Reuterdahl made a similar claim when the *Scientific American* raised the issue of anti-Semitism in the context of Reuterdahl's questioning of Einstein's priority, while being forced to concede that Reuterdahl was factually correct in his arguments.[208] Reuterdahl responded, stating on 18 June 1921, *inter alia*:

IN AN article published in this journal, April 30, 1921, Professor Arvid Reuterdahl presented definite evidence proving the similarity between the work of the unknown scientist 'Kinertia' and the much-advertised Einsteinian Theory of Relativity. The similarity is so pronounced that any fair-minded person at once must wonder if the alleged contributions of Dr. Einstein rest upon borrowed foundations. It is a fact that 'Kinertia's' work antedates that of Einstein. It is difficult to prove a direct charge of plagiarism. This is particularly true whenever the person involved is surrounded by a veritable host of protectors who refuse to permit an honest investigation.

*Professor Reuterdahl's reply to his critics follows in part:*
In the case of 'Kinertia' Versus Einstein the present writer did not state that Einstein is a plagiarist. To make such a bald statement one must have indisputable proofs. I did state and again repeat the statement: 'If Einstein was aware of 'Kinertia's' discovery then the appellation 'plagiarist,' bestowed upon him by his German professional colleagues, is eminently

---

**206**. E. Mach, *The Science of Mechanics*, Open Court, La Salle, Illinois, (1960), p. xxviii.
**207**. H. Dingler, *Die Grundlagen der Physik; synthetische Prinzipien der mathematischen Naturphilosophie*, Second Edition, Walter de Gruyter & Co., Berlin, (1923); **and** *Physik und Hypothese Versuch einer induktiven Wissenschaftslehre nebst einer kritischen Analyse der Fundamente der Relativitätstheorie*, Walter de Gruyter & Co., Berlin, Leipzig, (1921); **and** "Kritische Bemerkungen zu den Grundlagen der Relativitätstheorie", *Physikalische Zeitschrift*, Volume 21, (1920), pp. 668-669.
**208**. "The Anti-Einstein Campaign", *Scientific American*, (14 May 1921). See also: "Getting Back at Einstein", *The Literary Digest*, (4 June 1921).

fitting. If, on the contrary, Einstein was unaware of this work, then he is, nevertheless, antedated by the work of 'Kinertia'. Einstein is at liberty to choose either horn of the dilemma.'

Referring to an editorial criticism in the *Scientific American* of May 14, Professor Reuterdahl continues: 'The *Scientific American* is particularly disturbed by my article entitled "'Kinertia' Versus Einstein.' On the cover of this issue the following question appeared in bold type 'Is Einstein a Plagiarist?' In reference to this question the *Scientific American* states: 'It will be at once understood that according to Professor Reuterdahl he is.' What I actually stated in my article has been again recorded above in order to refresh the memory of the editorial writer. After this perversion of truth a subtle atmosphere is created in order to link, by contrastive suggestion, both the present writer and THE DEARBORN INDEPENDENT with the ambitions of the former Kaiser of Germany. A diversion is thereby adroitly produced which removes the reader's attention from the actual question in hand, that is, "'Kinertia' Versus Einstein,' to an entirely different issue. Moreover, another irrelevant issue is deftly imposed, that is, anti-Semitism.

The present writer emphatically denies and resents both insinuations created in this questionable manner. I am a loyal citizen of the United States. I was born in Sweden. I came to the United States when I was six and a half years of age. Furthermore, the allegation, also by innuendo, that my attack upon the theories of Einstein are due to anti-Semitic feeling, I brand as a gross misrepresentation.

The *Scientific American* editorial then becomes a plea for Professor Einstein's mathematical product. There seems to be urgent need to show that although Einstein has benefitted by 'ideas which have had a rather nebulous existence before him' nevertheless in the hands of this master craftsman they have been mathematically welded into a 'crowning achievement' which 'has never been approached or approximated in any way.'

Suppose, for the sake of argument, that we grant that this concession in no way affects the real issue which we may state in the form of a question: Has Einstein given proper credit to the creators of the 'nebulous ideas' which he used in constructing this supreme masterpiece of the human intellect? We are not aware that he has ever referred to their humble contributions to his stupendous structure. It seems that he has ruthlessly discarded the scaffolding which he used in building his edifice without paying for its use. Do we find the name of Dr. J. H. Ziegler mentioned in any of his writings? Is there any reference to the contributions of 'Kinertia'? Has he ever answered the charges made by Engineer Rudolph Mewes, Professors E. Gehrke and Paul Weyland that he appropriated a formula which appeared in a work published by the late Professor Gerber in the year, 1898? If perchance Professor Einstein should plead ignorance of these contributions at the time when he developed his mathematical analysis, then we demand that he publicly admit their previous existence and definite worth. It remains to be seen if Dr. Einstein will even condescend to comply with this eminently just demand. We trust that we may be permitted to state that what we have granted in the above, for the sake of argument, we do not admit as an actual fact. The writer is prepared to show that Einsteinism is a pernicious fallacy."[209]

Below is the article in *Scientific American*, which Reuterdahl rejoined. The author of the *Scientific American* article dubbed the practice of standing up for ethical practices and giving due credit to those who deserve it, "picking the bones". The author sought to characterize anyone who would assert their priority for ideas Einstein repeated without an attribution, as if a "vulture". Whereas Reuterdahl focused on the facts, the author of the *Scientific American* article launched a hand-waving personal attack against Reuterdahl, *conceding that he was factually correct*, and mischaracterized the general theory of relativity as an exposition on the mechanism and cause of gravitation, which it is not. The author asserted that, "Nobody would claim that Einstein's entire structure is novel[....]" However, that is exactly what Einstein did do by publishing papers completely devoid of references to the work of his predecessors. Daniel Kennefick wrote in his article, "Einstein Versus the *Physical Review*", *Physics Today*, (September, 2005), pp. 43-48, at 46:

"Although it now bears Einstein and Rosen's names, the solution for cylindrical gravitational waves had been previously published by the Austrian physicist Guido Beck in 1925. But Beck's paper was completely unknown to relativists with the single exception of his student Peter Havas, who entered the field in the late 1950's. In a 1926 paper by the English mathematicians O. R. Baldwin and George B. Jeffery, and in the referee's report on Einstein's paper, there was discussion of the fact that singularities in the metric coefficients are unavoidable when describing plane waves with infinite wavefronts. But although such a wave shows some distortion, in the words of the referee, 'the field itself is flat' at infinity.[9]

Clearly, the referee's familiarity with the literature exceeded Einstein's, but then Einstein was notoriously lax in that regard. The published Einstein-Rosen paper contains no direct reference to any other paper whatsoever and only two other authors are even mentioned by name. In response to Infeld's suggestion that he search the literature for previous work, Einstein laughed and said, 'Oh yes. Do it by all means. Already I have sinned too often in this respect.'"[210]

The *Scientific American* of 14 May 1921 stated:

---

[209]. A. Reuterdahl, quoted in "The Pro-Truth Campaign", *The Dearborn Independent*, (18 June 1921).
[210]. Kennefick cites: Note 9, "G. Beck, *Z. Phys.* **33**, 713 (1925); O. R. Baldwin, G. B. Jeffery, *Proc. Phys. Soc. London, Sect. A.* **111,** 95 (1926)"; **and** Note 5, "L. Infeld, *Quest: An Autobiography*, Chelsea, New York, (1980)", p. 277.

## "The Anti-Einstein Campaign

THE intellectual world moves slowly in the matter of extending recognition to those who have consecrated their lives to the cause of reason. Mendel had been dead many years before the remarkable nature of his work was recognized. When we contrast Mendel's case with that of Einstein we are forced to admit that the German physicist's sensational rise is the most extraordinary in the history of science. Barnum, king of advertisers, could not have staged a more effective or expeditious advertising campaign."

With so much of Professor Reuterdahl's article in the *Dearborn Independent* we suppose anyone will agree. But this article is given its real place by the scare-head of the cover, which asks, in ¾-inch letters, "IS EINSTEIN A PLAGIARIST?" It will be at once understood that according to Professor Reuterdahl he is. We expect this sort of thing from the anti-Semites of Germany, and from those of the former Kaiser's loyal supporters who resent Dr. Einstein's refusal to have anything to do with the celebrated Manifesto of the 93 Immortals. But from a reputable American source—even one celebrated for its anti-Semitism—we should look for something a little different.

It is not easy for a layman to form a just estimate of Einstein's work. And whatever temptation to error is presented to him will be in the direction of underestimation. The phrase "relativity of motion" is not new. The Greeks had it, Newton had it, every popular explanation of Einstein starts by reminding us that this is something we have always known but chosen to ignore. It is easy to overlook that Einstein has taken this familiar notion, applied it with a rigor and a consistency and a generality which it has never before enjoyed, given it a significance and got results out of it which it had never before been dreamed lay in it.

Again with the problem of gravitation. We all know that Newton solved this problem empirically only. We all know that he said nothing about the causes or the mechanism of gravitation—for the excellent reason that he could learn nothing of these. We all know that since his time thousands of scientists have searched for the cause and the mechanism. We do not all know what is equally true, that many of these searchers have been led to propose slight modifications in Newton's mathematical law—modifications which were in agreement with this or that observed fact.

All this makes it very easy to accuse Einstein of plagiarism. Not alone is everyone acquainted with classical relativity apt to judge the contents by the label on the container and assume that Einstein's relativity is the same old stuff, but the claim may with some show of plausibility be made that any investigator of gravitation has anticipated Einstein. This claim gains color in the far-from-rare case that its beneficiary can be shown to have attained results which are included in Einstein's, or to have supplied Einstein with some of his material. Nobody would claim that Einstein's entire structure is novel—the sum total of human knowledge is today too large to make it possible for a contribution like his to be made out of whole cloth.

Everyone who possesses enough mathematics to follow Einstein knows that he has made a very material original contribution—that he has formulated mathematically and as a concrete whole ideas which have had a rather nebulous existence before him, cementing the structure with ideas to which he has himself given birth. His crowning achievement is the precise mathematical formulation; this has never been approached or approximated in any way.

We can paraphrase Professor Reuterdahl with some profit. Never in the history of science has anyone ever made an epoch-marking advance, but what the vultures have flocked about his trail, demanding credit for what he has done and claiming ownership of the work which he has put out. But never before has it been the case that the really big men of science have accepted an advance so promptly and so whole-heartedly, and left this business of picking the bones to the small fry whose names will be forgotten fifty years from now."

In 1846, an author in the *Scientific American* had demonstrated an interest in Zionist affairs,

> "THE ISRAELITES IN GERMANY are in great commotion. At Berlin and Frankfort two-thirds of them have separated from the synagogues, to form new societies, and it is thought that their example will be generally followed. The new school are supported by the government; they celebrate the Sabbath of the Christians, and worship with chaunts, the music of the organ, and sermons. Sir Moses Montefiore, backed by the Rothschilds, is about establishing a Jewish colony in Palestine, and has obtained an ukase from the Emperor Nicholas, authorising the emigration thither of ten thousand Russian Jews."[211]

The maltreatment of anyone who disagreed with Einstein, pointed out his plagiarism or questioned the theory of relativity, reminds one of the fanatical and truly vicious abuse political Zionists inflicted upon anyone who dared disagree with them. Albert T. Clay documented the methods of the political Zionists in Palestine in 1921, in an article, "Political Zionism", *The Atlantic Monthly*, Volume 127, Number 2, (February, 1921), pp. 268-279, at 276-277,

> "The old resident Jews of Palestine certainly have other than religious grounds for their indifference toward the efforts of the Political Zionists. Last winter the Council of Jerusalem Jews appointed a commission of representative men holding leading positions, to visit parents who were sending their children to proscribed schools, in order to secure their withdrawal. Among these schools, which included those conducted by the

---

**211**. *Scientific American*, Volume 1, Number 42, (9 July 1846), p. 3.

convents and churches, some of which have existed in Jerusalem for a long time, are the British High School for Girls, the English College for Boys, and the Jewish School for Girls. In the latter, conducted by Miss Landau, an educated English Jewess, all the teachers are Jewish; most of the teaching is in the English language. This school, which is financed by enlightened Jews of England, was denounced more severely than the others, because, not being in sympathy with the programme of the Political Zionists, Miss Landau refused to teach the Zionist curriculum. She was even informed that her school would be closed.

In a series of articles that appeared in *Doar Hayom*, the Hebrew daily paper, last December, it was stated that the parents who refused to comply with the requests of the Commission [of the Council of Jerusalem Jews] were to be boycotted, cast out from all intercourse with Jews, denied share in Zionist funds, and deprived of all custom for their shops and hotels. 'Anyone who refused, let him know that it is forbidden for him to be called by the name of Jew; and there is to be for him no portion or inheritance with his brethren.' They were given notice that they would 'be fought by all lawful means.' Their names were to be put 'upon a monument of shame, as a reproach forever, and their deeds writte unto the last generation.' 'If they are supported, their support will cease; if they are merchants, the finger of scorn will be pointed at them; if they are rabbis, they will be moved far from their office; they shall be put under the ban and persecuted, and all the people of the world shall know that there is no mercy in justice.'

A month later the results of this 'warfare' were reviewed. We were informed that some Jews had been influenced, 'but others—and the greater number, and those of the Orthodox,—those who fear God—having read the letters [signed by the head of its delegates and the Zionist Commission] became angry at the 'audacity' of the Council of Jerusalem Jews 'which mix themselves up in private affairs,' have torn the letter up, and that finished it.'

Then followed a long diatribe against these parents, boys, and girls, in which it was demanded that the blacklist of traitors to the people be sent to 'those who perform circumcision, who control the cemeteries and hospitals'; that an order go forth so that 'doctors will not visit their sick, that assistance when in need, if they are on the list of the American Relief Fund, will not be given to them.' 'Men will cry to them, 'Out of the way, unclean, unclean.' ... They are in no sense Israelites.'

It is to be regretted that only these few paraphrases and quotations from the series of articles published can be presented here.

The work of the Councils Committee met with not a little success; pupils left schools, and teachers gave up their positions. Two instructors in the English College, whose fathers were rabbis, and a third, whose brother was a teacher in a Zionist school, resigned. Another refused to do so, and declared himself ready, in the interests of the Orthodox Jews, who were suffering under this tyranny, which they deplored, to give the fullest testimony to the authorities concerning this persecution. The administration, under Governor Bols, finally intervened, and at least no further public efforts to carry out their programme were made.

If, in this early stage of the development of Political Zionism, even the Palestinian Religious Jews already find themselves under such a tyranny, what will happen if these men are allowed to have full control of the government? And what kind of treatment can the Christian and th Moslem expect in their efforts to educate their children, if the Political Zionists are allowed to develop their Jewish state to such a point that they can dispense with their mandatory and tell the British to clear out? When such things happen under British administration, what will take place if the Jewish State is ever realized, and such men are in full control?"

Prof. Arvid Reuterdahl was quoted in *The St. Paul Daily News* on 8 May 1921,

### *"Einstein's Theory of Relativity Upset by St. Paul Scientist Whose New Book Charges Gross Errors*

World Has Gone Mad About Mythical Unrealities, Declares Prof. Arvid Reuterdahl, Dean of Engineering and Architecture at St. Thomas College—Offers to Debate Question.

Editor's Note.—The visit to the United States of Prof. Einstein has brought on a countrywide discussion of his theory of relativity. Not many persons know anything about relativity, but nevertheless, they are talking about it and Einstein. In St. Paul there is a man, Prof. Arvid Reuterdahl, dean, department of engineering and architecture, St. Thomas college, who disputes the Einstein theory. He is writing a book now called 'The Fallacies of Einstein.' Prof. Reuterdahl is a distinguished scientist, both in America and abroad. He is the author of various scientific works and a frequent contributor to magazines. At the request of The Daily News he has written the following article dealing with the Einstein theory of relativity.

\* \* \*

BY ARVID REUTERDAHL,
Dean, Department of Engineering and
Architecture,

The College of St. Thomas.

AT THE present time we often hear this question asked:

'What is the theory of relativity?'

Whenever the question is asked Einstein's name is invariably mentioned.

To be exact this question should take the following form:

'What Is Einsteinism?'

A complete answer would require a book of many pages.

However, we may answer the latter question briefly as follows:

Einsteinism is a mind-product produced by combining a few consistent concepts with numerous mythical unrealities into a mental world system with the hope it will correspond with the real physical universe.

### 'SWEPT ENTIRE COUNTRY.'

We may say Einsteinism in the United States began with the publication of a dispatch cabled from Berlin Dec. 2, 1919, to the New York Times.

Like an enormous tidal wave Einsteinism then swept from the Atlantic to the Pacific coast.

Mr. Average Man soon began talking about the theory of relativity. Humorous publications gave versions of Einsteinism which for accuracy in presentation oftentimes surpassed the mathematical outbursts of over-enthusiastic savants.

Nowhere could one hear a dissenting voice.

### EXPOSED LAST YEAR.

The first brief exposition of the fallacies of Einstein, published in the United States, appeared in my work, 'Scientific Theism Versus Materialism: the Space-Time Potential.' This book was published in the fall of 1920 by the Devin-Adair Co., New York. Sir Oliver Lodge a few months previously, however, had issued a warning against the too ready acceptance of Einsteinism.

His warning went unheeded and the great wave of Einsteinism rolled on unchecked. I found myself almost alone in the fight against the greatest and most pernicious scientific fallacy of modern times.

However, I was not entirely alone at this time in my battle against the great sophist of all times.

### AIDED BY HEIDENREICH.

In fact, since the year 1914 my dear friend, Dr. E. Lee Heidenreich, the eminent engineer, mathematician and philosopher, had espoused my cause. With the clear vision of a seer, Dr. Heidenreich realized that the old science must give way before a broader cosmic theory based upon sound philosophic principles grounded in fact.

He courageously and fearlessly championed the cause of my Space-Time Potential. He was instrumental in arranging lectures for me at the Kansas state agricultural college and the University of Kansas.

The commendatory letters concerning these lectures which I received from Dr. A. A. Potter, then dean of the agricultural college, and Dr. H. E. Rice, Kansas state university, have been a source of great encouragement to me during my long and arduous fight for the recognition of a broader and more universally consistent view of the physical universe.

Dr. Heidenreich, being a descendant of the Vikings, gloried in the single combat.

Persistently and fearlessly he has championed my cause both in the United States and in Norway.

When Einsteinism overran the world Dr. Heidenreich refused to accept its fallacious tenets and gave vigorous battle to this new intellectual Frankenstein.

In the early part of the year 1921 an able and fearless writer championed my cause in an article entitled 'Relativity or Interdependence.' This article has since been referred to, time and again, as a classic.

Its author, Rev. Prof. John T. Blankart, in no uncertain terms and with keen acumen points out the inherent inconsistencies in Einsteinism. He brings his masterly article to a close with the following statement:

'Einstein has stated, 'If any deduction from it (the theory of relativity) should prove untenable it must be given up. A modification of it seems impossible without destruction of the whole.'

## MORE AID NECESSARY.

'If this article has indicated to the reader that by that statement Einstein has perhaps signed the death warrant of his theory of relativity, the writer shall feel that part of his purpose has been accomplished.'

This exceptionally meritorious contribution exercised a beneficent influence in limited circles. However, one could hardly expect that a lone volume and a single article, without proper publicity, could stem the onrush of the Einsteinistic heresy.

Now, however, the tide is turning. After I issued my challenge to Einstein to a written debate on the theory of relativity I have received letters from prominent scientists and thinkers who assure me they will do their utmost to help vanish this Goliah of skepticism. Prof. Einstein has insinuated that my attack on his theory of relativity is merely a form of anti-Semitic propaganda.

This insinuation is absolutely without foundation in fact.

## REVERES BARUCH SPINOZA.

If the originator of the theory of relativity had been born in Sweden, my native land, I would have denounced the tenets of his theory with no less vigor. The fact that Dr. Einstein is of Jewish extraction is not the reason for my attack on his theory.

I desire that this be distinctly understood now and for all future time.

My challenge to Prof. Einstein is based upon purely intellectual grounds. I contend his theory is a monstrous and dangerous fallacy which leads to absolute skepticism. I have profound reverence for Baruch Spinoza, the great philosopher. Spinoza was a Jew.

Certain erroneous inferences and unjust insinuations have been made concerning the appearance of my article entitled 'Kinertia Versus Einstein' in the Dearborn Independent.

Before I submitted this article to the Dearborn Independent I sent it to a well-known eastern journal.

## MANUSCRIPT RETURNED.

The editor of this journal finally returned my manuscript with a most courteously worded letter in which he expressed his regret that he could not risk its publication, despite the fact he felt confident I had made out a particularly strong case against Einstein. In fact, he went so far as to state my article would create a sensation if published. Evidently it would have been unwise for this eastern journal to publish my article. The path of truth is beset with many thorns.

It grieves me to be forced into the admission that our scientific journals, while professing to be the free and untrammelled vehicles of truth for its own sake, generally manage by means of plausible excuses to permanently prevent the publication of contributions which do not conform with the intellectual welfare of the clique in control.

The journals which are free from this destructive influence are generally too timid to assert their own independence.

### FREEDOM IN DAILY PRESS.

This latter class is composed of journals which depend upon the European scientists to put the stamp of approval or disapproval upon that which is new or disturbing. It would seem there is much more genuine freedom in the daily press.

The spirit of revolt against this czar of science is growing.

Many independent thinkers have joined the anti-Einsteinism ranks. I believe Einstein himself is now beginning to see the handwriting on the wall.

One may be permitted, not without considerable show of justice, to infer his persistent refusal to enter into any controversial discussion is an indication he tacitly admits the relativity bubble is practically ready to collapse.

The following quotation from a letter which I have recently received from Dr. Robert T. Browne, author of the truly great work, 'The Mystery of Space,' is indeed noteworthy:

'The gods of science have placed their imprimatur upon the theory of relativity and consequently it will be exceedingly difficult to break through the iron ring.

### BROWNE PLEDGES AID.

'Primarily, however, I should think with you, as with me, the consideration of greatest importance is not so much with the incidentals of this movement itself. The theory of relativity is but a phase of that deeper and broader movement of mechanistic conceptualism against which you have argued so incontrovertibly in 'Scientific Theism.' The task, then, is not so much to combat the theory, as I see it, as it is to strike with might and main at the vitals, the fundamental premises of that erroneous, fragmentary and biased view which seeks to interpret the universe in terms of mechanistic concepts.'

Dr. Browne concludes his letter to me with the following assurance:

'Please be assured that should the opportunity come my way I shall be allied with you in the fight against this mathematical usurpation.'

### COMPARED TO DRUG.

Dr. W. E. Glanville, the eminent astronomer of Baltimore, who is a member of British, French and American astronomical societies, states:

'The Einstein theory is like a newly discovered drug which is brought forth and acclaimed as a universal scientific panacea.'

Dr. Sydney T. Skidmore of Philadelphia writes:

'It (Einsteinism) is shapen from non-Euclidean, otherwise called meta-geometry, and this consists entirely of mental constructions that are purely subjective and correspond to nothing in nature.'

'Kinertia' states: 'Science wants more than agnosticism; it wants to know the absolute truth before accepting any such theory; even if D'Alembert's ghost is dressed in Hamiltonian functions.'

### QUOTES SWISS BOOK.

I have just received a complimentary copy of an exceptionally meritorious work written by Dr. Edouard Guillaume of the University of Lausanne, Switzerland. The title of this work is 'La Theorie de la Relativite, Et Sa Signification.'

I quote the following from this work:

'We have gradually come to substitute for Descarte's rigid system of relation, systems of unheard of subtleness, to which Einstein has given the picturesque name of 'mollusk systems.' Our mathematical constructions become, as it were, devilfish which strive, while adapting themselves to fasten upon subtle natural manifestations.'

Note the keen rapier thrusts against Einsteinism by this famous scientific 'maitre d'armes.'

### WORK NEARS COMPLETION.

Dr. Guillaume has not been hoodwinked by the delicate sophism of Einstein.

My work entitled 'The Fallacies of Einstein' is now nearing completion.

In this work I have stripped Einsteinism of its mathematical adornment.

Without this mathematical camouflage Einsteinism is scarcely more than a mere devitalized skeleton whose Einsteinian skull is forever grinning at its Galileian toes."

While it is true that *The Dearborn Independent* published broad criticisms of Jews, Reuterdahl's article was not in any way anti-Semitic and an allegation of ethnic bias is not a racist attack, but is rather a defense against racism. Reuterdahl first sought to publish his article elsewhere and it was refused without stated grounds. Reuterdahl asserted that the circulation of Henry Ford's paper was about 750,000 readers, which offered Reuterdahl the opportunity he had been denied elsewhere to bring his message to a wide audience. Jewish racists ought not to be allowed to censor out all open debate on issues they want suppressed and Reuterdahl had a right and an obligation to express his views wherever he could.

Frederick Drew Bond raised the issue of Reuterdahl's publication of articles in *The Dearborn Independent* in a polemic against Reuterdahl in *The New*

*York Times* in 1923.²¹² Bond's second and then current wife was first cousin of the racist Zionist blackmailer United States Supreme Court Justice Louis D. Brandeis, who was an ardent and politically influential Zionist with close connections to President Wilson and Chaim Weizmann, and who attained his seat in the Supreme Court by blackmailing President Woodrow Wilson. Bond, perhaps speaking from a guilty conscience, denied that his connection to Brandeis had anything to do with his attack on Reuterdahl, in private correspondence with Reuterdahl.²¹³ However, it was Bond who raised the issue of his connection to Brandeis, which was not known to Reuterdahl, and Bond's denial was made as an unsolicited confession. Brandeis had expressed an interest in promoting Einstein. The racist Zionist blackmailer United States Supreme Court Justice Louis Dembitz Brandeis wrote in a letter dated 1 March 1921,

"You have doubtless heard that the Great Einstein is coming to America soon with Dr. Weizmann, our Zionist Chief. Palestine may need something more now than a new conception of the Universe or of several additional dimensions; but it is well to remind the Gentile world, when the wave of anti-Semitism is rising, that in the world of thought the conspicuous contributions are being made by Jews."²¹⁴

The series of letters exchanged in *The New York Times* began with a letter from Dr. Harris A. Houghton, M. D., of No. 97/100 Riverside Drive, New York City, dated 13 April 1923; which accused Einstein of publishing a "Newtonian Duplication".²¹⁵ Houghton was involved with U. S. Army Intelligence and had called the attention of the U. S. Government to the *Protocols of the Learned Elders of Zion* in 1918, informing President Wilson and his cabinet of an alleged plot by Zionists to overthrow the governments of the world and to destroy Christianity.²¹⁶ Brandeis, who controlled Wilson, assured the U. S. Government that the document was a forgery.²¹⁷ Houghton published the "Beckwith" English

---

**212**. A. Reuterdahl, "The Origin of Einsteinism", *The New York Times*, Section 7, (12 August 1923), p. 8. Reply to F. D. Bond's response, "Reuterdahl and the Einstein Theory", *The New York Times*, Section 7, (15 July 1923), p. 8. Response to A. Reuterdahl, "Einstein's Predecessors", *The New York Times*, Section 8, (3 June 1923), p. 8. Which was a reply to F. D. Bond, "Relating to Relativity", *The New York Times*, Section 9, (13 May 1923), p. 8. Which was a response to H. A. Houghton, "A Newtonian Duplication?", *The New York Times*, Section 1, Part 1, (21 April 1923), p. 10.
**213**. F. D. Bond, 24 Manhattan Avenue, New York City, letter to A. Reuterdahl dated 10 July 1923, Department of Special Collections, O'Shaughnessy-Frey Library, University of St. Thomas, Minnesota.
**214**. L. D. Brandeis, M. I. Urofsky and D. W. Levy, Editors, *Letters of Louis D. Brandeis* Volume 4, State University of New York Press, Albany, New York, (1975), pp. 536-537.
**215**. H. A. Houghton, "A Newtonian Duplication?", *The New York Times*, Section 1, Part 1, (21 April 1923), p. 10.
**216**. H. Bernstein, *The Truth about 'The Protocols of Zion'*, Ktav Publishing House, New York, (1971), pp. 43-44.
**217**. Letters from L. Brandeis to J. W. Mack, *et al.* of 26 November 1918, 22 March 1920,

translation of the *Protocols* in 1920.[218] Dr. Houghton also wrote to John Spargo, about Louis Marshall's letter to Max Senior of 26 September 1918, in an effort to convince Spargo that Marshall feared Zionists and believed Zionism was a part of a larger Jewish plot—which accusations Marshall denied.[219] Boris Brasol[220] may have been the one who brought the *Protocols* to U. S. Army Intelligence and convinced them of their authenticity, *viz.* Dr. Harris Houghton and Natalie De Bogory.[221] Houghton wrote to Arvid Reuterdahl on 15 July 1923.[222]

Racist Zionist United States Supreme Court Justice Louis Dembitz Brandeis was a Frankist Jew. Frankist Jews were committed to the destruction of Gentile society. They deliberately wormed their way into positions of power in order to subvert Gentile religions and governments and bring them into war and ruin. Brandeis brought America into the First World War in a *quid pro quo* deal with the British in exchange for the Zionist Balfour Declaration by blackmailing Woodrow Wilson with love letters Wilson had written to Mrs. Peck. Brandeis and his leading Jewish friends instituted Rothschild's banking system in America, which led to the Great Depression. Brandeis was known as the most deceitful lawyer in America. His appointment to the United States Supreme Court was the most scandalous event in the Court's history. Like all Frankist Jews, Brandeis returned Gentile generosity with treachery. Arthur Hertzberg discussed Brandeis' Frankist roots,

> "On the surface Brandeis was a strange kind of leader for the Zionists. Born in Louisville, Kentucky, in 1856 to recent immigrants from Bohemia, who were not much involved in Jewish life, Brandeis had a brilliant career at Harvard Law School, and by the late 1880s had become a successful Boston lawyer. True, many of his initial clients were 'German Jews' to whose social

---

17 November 1920 and 18 November 1920, L. D. Brandeis, M. I. Urofsky and D. W. Levy, Editors, *Letters of Louis D. Brandeis* Volume 4, State University of New York Press, Albany, New York, (1975), pp. 365, 452-453, 506-508.

**218**. *The Protocols of the Wise Men of Zion*, Beckwith, New York, (1920). H. Bernstein, *The Truth about 'The Protocols of Zion'*, Ktav Publishing House, New York, (1971), p. 55.

**219**. L. Marshall to J. Spargo, *Louis Marshall: Champion of Liberty; Selected Papers and Addresses*, Volume 1, The Jewish Publication Society of America, Philadelphia, (1957), pp. 351-353.

**220**. B. Brasol, *The Protocols and World Revolution, Including a Translation and Analysis of the "Protocols of the Meetings of the Zionist Men of Wisdom"*, Small, Maynard & Company, Boston, (1920); **and** *Socialism Vs. Civilization*, C. Scribner's Sons, New York, (1920); **and** *The World at the Cross Roads*, Small, Mayhard & Co., Boston, (1921); **and** *The Balance Sheet of Sovietism*, Duffield, New York, (1922).

**221**. S. G. Marks, "Destroying the Agents of Modernity: Russian Antisemitism", *How Russia Shaped the Modern World*, Chapter 5, Princeton University Press, (2003), pp. 140-175; notes 354-358.

**222**. Houghton's letter is found in the Department of Special Collections, University of St. Thomas, St. Paul, Minnesota.

set he inevitably belonged, but he was even more peripheral to the Jewish community than the most assimilated among them. There was some memory in his family of its origins in Prague in a circle that still harbored loyalty to the memory of Jacob Frank, the false messiah who had appeared in Poland in the latter half of the eighteenth century. Brandeis's mother was very opposed to Jewish particularism. In his earliest Boston years, he was to be found, at least once, on the list of contributors to the First Unitarian Church. On the other hand, he had been deeply influenced in his earliest years by an uncle, Louis Dembitz (whose family name he adopted as his own middle name), a learned, Orthodox Jew."[223]

Here is Reuterdahl's 30 April 1921 article, to which an author responded in the *Scientific American* with an obnoxious *ad hominem* attack,

## "'Kinertia' Versus Einstein
By ARVID REUTERDAHL
Dean, Department of Engineering and Architecture.
The College of St. Thomas, St. Paul. Minnesota
Citations That Raise Delicate Question
on Age of Theory of Relativity

THE intellectual world generally moves slowly in the matter of extending recognition to those who have consecrated their lives to the cause of reason. Mendel had been dead many years before the remarkable nature of his work was recognized. When we contrast Mendel's case with that of Einstein we are forced to admit that the German physicist's sensational rise is the most extraordinary in the history of science. Barnum, the king of advertisers, could not have staged a more effective and expeditious advertising campaign. Within the brief period of a few months, Einstein's name became known in every civilized country in the world. The Theory of Relativity afforded cartoonists material for humorous sketches, and the doctor and his doctrine became subjects for mirth and merriment.

After the first volcanic outburst of scientific approval and humorous recognition, rumblings of discontent were heard from Einstein's native land. A group of German scientists, in no uncertain terms, expressed their doubts concerning the precise value and originality of Einstein's theory. There were even those who boldly charged the author with deliberate plagiarism. In England Sir Oliver Lodge and a few other able men cautioned the world against a too hasty acceptance of the new doctrine of relativity. In the United States, however, Einstein's theory met with immediate and complete success. Even at the present time we rarely hear a dissenting voice. This is particularly strange for the reason that in the year 1914 a well-known American journal published a series of articles by an unknown investigator

---

**223**. A. Hertzberg, *The Jews in America: Four Centuries of an Uneasy Encounter: A History*, Simon and Schuster, New York, (1989), p. 218.

who discussed the very same problem which brought fame to Einstein. We refer to the eleven articles written by the unknown 'Kinertia,' which appeared in *Harper's Weekly* under the caption 'Do Bodies Fall?' If it is true that 'Kinertia' actually considered the Einsteinian problem in these essays, then the question of priority is inevitably raised and the unparalleled originality claimed for Einstein's work becomes a debatable matter. Indeed, the presentation of the very facts which raise these questions is the main purpose of this article. Since the matter of priority is involved, the introduction in this article of a brief chronological survey of the work of both Einstein and 'Kinertia' is of the utmost importance.

The most significant contributions of Albert Einstein have been published in *Annalen Der Physik*. His papers deal with the Special Theory of Relativity, Theory of the Brownian Movements, Inertia of Energy, the Quantum Law of the Emission and Absorption of Light, Theory of the Specific Heat of Solid Bodies, and the General Theory of Relativity. The year 1905 is considered, by most authorities on Einstein's work, as the birth-year of the Theory of Relativity. Careful search, however, has revealed a paper on this subject which was published in Berlin during the year 1904 in the journal *Sitsungsberichte*. That portion of Einstein's theory which deals with the phenomenon of gravitation is a later development. Einstein first gave his attention to the problem of gravitation in 1911, when he developed the Principle of Equivalence of gravitational and accelerative fields. Other phases of this subject were dealt with in papers which appeared in the years 1912 and 1913. A further elaboration, the joint work of Einstein and Marcel Grossman, appeared in 1914. The theory in its final and complete form was announced in the year 1915.

'Kinertia's' contribution deals principally with the problem of gravitation. The question of priority of 'Kinertia' over Einstein consequently involves the phenomenon of gravitation in particular. It must be admitted, however, that 'Kinertia' has also considered Einstein's earlier problem which involved the significance of motion in reference to an observer. Einstein distinguishes this earlier problem from his theory of gravitation by the separate designation, 'Special Theory of Relativity.' A brief historical summary of the work of 'Kinertia' is now in order.

Lord Kelvin first aroused 'Kinertia's' interest in the problem of gravitation. That was in the year 1866 when 'Kinertia' was a student under Lord Kelvin. 'Kinertia' even then did not agree with the Newtonian theory of force as presented by Lord Kelvin. Incidentally, we desire to call the reader's attention to the fact that Albert Einstein was born in 1879 in Ulm, Germany, thirteen years later. It is a curious coincidence that both 'Kinertia' and Einstein were engineers. During the period of time from 1877 to 1881, 'Kinertia' became convinced that *acceleration* was the basic cause of what we generally speak of as 'weight.' The reader is undoubtedly aware of the fact that *acceleration* plays the fundamental role in Einstein's theory of gravitation. 'Kinertia' corresponded with Kelvin, Tait, and Niven, of Cambridge, with the hope that he would be able to interest these men in his startling theory. This attempt met with little or no sympathy. Some years later, through an accident, 'Kinertia' was unfortunately deprived of his hearing. This misfortune forced him to abandon his engineering profession for a rancher's life in the state of California. This new occupation gave 'Kinertia' the requisite leisure to complete his investigations which resulted in confirming his supposition that *acceleration* was the great norm of the phenomenon of gravitation. His attempts, dating from the year 1899, to persuade our stubborn American scientists that the Newtonian theory of gravitation must be revised met with nothing but ridicule or indifference. To *Harper's Weekly* and its managing editor (1914), Mr. H. D. Wheeler, belongs the credit of having published 'Kinertia's' series of articles entitled, 'Do Bodies Fall?' The first article appeared in the issue of August 29, 1914, Vol. 59. The final article is dated November 7, 1914. From the preceding it is evident that 'Kinertia' derived his norm of gravitation before Einstein was born. The question of priority is therefore definitely and irrefutably established in favor of 'Kinertia' in the case of the General Theory of Relativity considered as a discussion of the problem of gravitation and acceleration.

We turn our attention now to the content of these two gravitational theories. We propose, by means of direct quotations from the works of these two men, to set forth their remarkable similarity. In the case of Einstein we shall quote from his recent book, 'Relativity' (Henry Holt and Company, 1920), and in 'Kinertia's' case our quotations will be from the *Harper's Weekly* articles.

The following comparative quotations show the striking similarity existing between Einstein and 'Kinertia' when they consider the relation between acceleration and gravitation, a similarity which extends not only to intent but affects even the very words.

*Einstein.*
'We imagine a large portion of empty space, so far removed from stars and other appreciable masses that we have before us approximately the conditions required by the fundamental law of Galilei.—As reference-body let us imagine a spacious chest resembling a room with an observer inside who is equipped with apparatus. Gravitation naturally does not exist for this observer. He must fasten himself with strings to the floor, otherwise the slightest impact against the floor will cause him to rise slowly toward the ceiling of the room.

'To the middle of the lid of the chest is fixed externally a hook with rope attached, and now a 'being' (what kind of a being is immaterial to us) begins pulling at this with a constant force. The chest together with the observer then begin to move 'upwards' with a uniformly accelerated motion. In course of time their velocity will reach unheard of values—provided that we are viewing all this from another reference-body which is not being pulled with a rope.

'But how does the man in the chest regard the process? The acceleration of the chest will be transmitted to him by the reaction of the floor of the chest. He must therefore take up this pressure by means of his legs if he does not wish to be laid out full length on the floor. He is then standing in the chest in exactly the same way as anyone stands in a room of a house on our earth. If he release a body which he previously had in his hand, the acceleration of the chest will no longer be transmitted to this body, and for this reason the body will approach the floor of the chest with an accelerated motion. The observer will further convince himself *that the acceleration of the body toward the floor of the chest is always of the same magnitude, whatever kind of body he may happen to use for the experiment.'*—(*'Relativity,'* pages 78 and 79.)

*'Kinertia.'*

'I set to work to find out by experiment whether bodies actually did fall with the acceleration which the force of attraction was said to produce. Years before that, when in England, where some of our coal mines had vertical shafts about 1,500 feet deep, I had studied the cause of weight by having the hoisting engine drop me down with the full acceleration for about 500 feet. Then, by retardation during the lowest 500 feet, I could experience increase of weight all over me so marked that my legs could hardly support me. That taught me that acceleration was the proximate cause of weight, but at the time of these experiments I still thought the acceleration of the falling cage was really caused by the earth's attraction.' —('Do Bodies Fall?' *Harper's Weekly*, August 29, 1914, page 210). 'Weight is not a kinetic force because it cannot produce acceleration. *If a body were accelerated in proportion to its weight, then weight would be a force.'*—('Do Bodies Fall ?' Harper's Weekly, October 17, 1914, page 383).

It is noteworthy that the only real difference between these two citations is that Einstein derives his conclusions from an hypothetical case, whereas 'Kinertia' draws his conclusions from an actual experiment upon himself.

The interpreters of Einstein furnish us with further corroborative material which we submit as additional evidence in the case of 'Kinertia' versus Einstein. Professor A. S. Eddington's interpretation of Einstein's theory is authoritative. The following quotations are from his work, 'Space, Time and Gravitation' (Cambridge University Press, 1920). These quotations from Eddington's work also consider the equivalence of acceleration and gravitation.

*Eddington.*

'The nature of gravitation has seemed very mysterious, yet it is a remarkable fact that in a limited region it is possible to create an artificial field of force which imitates a natural gravitational field so exactly that, so far as experiments have yet gone, no one can tell the difference. Those who seek for an explanation of gravitation naturally aim to find a model which will reproduce its effects; *but no one before Einstein seems to have thought of finding the clue in these artificial fields, familiar as they are.*

'When a lift starts to move upward the occupants feel a characteristic sensation, which is actually identical with a sensation of increased weight.— In fact, the upward acceleration of the lift is in its mechanical effects exactly similar to an additional gravitational field superimposed on that normally present.'—('Space, Time and Gravitation,' page 64.)

On the eminent authority of Eddington we may therefore state with absolute certainty that Einstein found his clue to the nature of gravitation in the *artificial field* created by acceleration. Eddington's statement, however, that Einstein was the first scientist to think of this *clue* is evidently erroneous in view of the preceding quotations from the work of 'Kinertia.'

The remarkable similarity in thought of the following quotations pertaining to the relative effects produced by accelerated and uniform motion, is of high evidential interest.

*Eddington.*

'The observer in the accelerated lift travels upward in a straight line, say 1 foot in the first second, 4 feet in two seconds, 9 feet in three seconds, and so on. If we plot these points as $x$ and $t$ on a diagram we obtain a curved track. Presently the speed of the lift becomes uniform and the track in the diagram becomes straight. So long as the track is curved (accelerated motion) a field of force is perceived; it disappears when the track becomes straight (uniform motion) .'— ('Space, Time and Gravitation,' page 66.)

*'Kinertia.'*

'The proof that matter can exist without weight depends on the first law of motion; because if a mass moves uniformly in a straight line in space, it cannot have weight. If weight is caused by the mutual attraction of matter, then a mass subject to attraction must move in a curve. If weight is caused by acceleration then it cannot follow Newton's law and move with uniform velocity in a straight line.'—('Do Bodies Fall ?' *Harper's Weekly*, October 10, 1914, page 350.)

The conclusions of Einstein and 'Kinertia' concerning the very existence of the force of gravitational attraction are identical in content. This is apparent from the following citations from an article by Professor Edwin B. Wilson, (Massachusetts Institute of Technology) and 'Kinertia's' basic articles.

*Wilson.*

'But just suppose that somebody tells us that the force of gravity is physically non-existing quite as much as the centrifugal or Coriolis force, and that the reason we think that gravity is real is essentially the same that leads the untutored mind to believe there is a physical force acting to move objects to one side when a train goes around a curve—namely, an unhappily ignorant view of Nature. This is what Einstein asserts.' —('Space, Time and

Gravitation,' the *Scientific Monthly*, March, 1920, page 226.)

*'Kinertia.'*
'But now, since it can be proved that there is no such force in the universe as attraction and that the supposed fall of bodies toward the earth *by that force* is only an illusion of the senses, there will be new ground upon which theologians can meet the Laplace attractionists, and Haeckel and his materialists.'—('Do Bodies Fall?' *Harper's Weekly*, September 19, 1914, page 285.)

The preceding citations are sufficient to establish conclusively the fact that, in underlying essence, 'Kinertia's' theory of gravitation is identical with Einstein's. Both men find the crux of the problem in acceleration, and the development of both theories is based upon the very same experiment.

It will be particularly interesting to compare the conclusions of the two men concerning the nature of the path of the earth's motion in space.

*Eddington.*
'Consider, for example, two events in space-time, namely, the position of the earth at the present moment, and its position a hundred years ago. Call these events $P^2$ and $P^1$ In the interim the earth (being undisturbed by impacts) has moved so as to take the longest possible track from $P^1$ to $P^2$—or, if we prefer, so as to take the longest possible proper-time over the journey. In the weird geometry of the part of space-time through which it passes (a geometry which is no doubt associated in some way with our perception of the existence of a massive body, the sun) this longest track is a *spiral*—a circle in space drawn out into a spiral by continuous displacement in time. Any other course would have had a shorter interval-length.'—('Space, Time and Gravitation,' page 72.)

*Wilson.*
'Draw from the sun perpendicular to the plane of the earth's orbit a line which shall represent the time-axis and disregard the third spatial dimension. Now for each kilometer that the earth moves around in its orbit, it must be considered to move in time by 10,000 kilometers. The path of the earth in space and time on this diagram is therefore a *helix* with an extremely steep pitch winding once a year about the cylinder standing in the earth's orbit but advancing ten thousand billion kilometers while 'circulating' one billion kilometers.'— ('Space, Time and Gravitation.' The *Scientific Monthly*, March, 1920, page 227.)

*'Kinertia.'*
'The possible motion of the sun in space, as adrift with the planets, was anticipated by Newton; but the laws of motion prevented him from reaching the true *corkscrew* path of the planets in space as they revolve round the sun.'—('Do Bodies Fall?' *Harper's Weekly*, September 19, 1914, page

285.)

In this connection we submit as corroborative evidence of the highest import, the illustration of this *corkscrew* path of the earth and moon which was used to elucidate 'Kinertia's' article in *Harper's Weekly*, September 19, 1914, page 285.

This illustration, taken in conjunction with 'Kinertia's' statement, quoted above, proves conclusively that the unknown 'Kinertia' derived the same type of path for the earth's motion in space that Einstein claims as his original contribution.

We introduce the following final quotation in order definitely to fix the date of 'Kinertia's' contribution:

*'Kinertia.'*

'This statement is concerning a discovery in natural science and the ordinary phenomena of daily life, which I discovered about fifteen years ago while engaged in carrying on some experiments to verify what I had previously suspected to be the true physical cause of *Elasticity, Gravity, Weight and Energy.'*—('Do Bodies Fall?' *Harper's Weekly*, August 29, 1914, page 210.)

Since this article bears the date 1914, it is clear that the year 1899, fifteen years earlier, is the date which can safely be regarded as the birth-year of 'Kinertia's' theory of gravitation. We have seen that Einstein's first work on gravitation was done in the year 1911; consequently 'Kinertia' antedates Einstein by twelve years.

We rest the case of 'Kinertia' Versus Einstein on the evidence submitted in this article. If Einstein was aware of 'Kinertia's' discovery then the appellation 'plagiarist,' bestowed upon him by his German professional colleagues, is eminently fitting. If, on the contrary, Einstein was unaware of this work, then he is, nevertheless, antedated by the work of 'Kinertia.' Einstein is at liberty to choose either horn of the dilemma."[224]

On 12 February 1920, Einstein gave a speech at the University of Berlin. He allowed non-students to attend, in direct violation of the University's rules. A similar situation had occurred a year earlier at the University of Zürich, where persons not entitled to attend Einstein's lectures did attend, and those who had purchased tickets, but whose seats were taken by those without tickets, requested a refund.[225] During his lecture in Berlin, Einstein called the student council the "dregs of humanity". Einstein was met again and again with applause and left to

---

[224]. A. Reuterdahl, "'Kinertia' Versus Einstein", *The Dearborn Independent*, 30 April 1921, pp. 2 and 14.
[225]. Letter from T. Vetter to A. Einstein of 28 January 1919, *The Collected Papers of Albert Einstein*, Volume 9, Document 4, Princeton University Press, (2004).

general applause.[226] The only disturbance of any kind was the reaction of the crowd of Eastern European Jews when Einstein spoke of cancelling future lectures should non-students not be permitted to attend, and returning their fees. Eastern European Jews created a series of disturbances,[227] because they wanted to attended the lectures, which the rules would not allow them to attend. Eastern European Jews were noted for producing Zionists, prostitutes, Frankist revolutionaries and for their pronounced tribalism[228]—their appearance and actions identified them, as the *Deutsche Zeitung* noted,

> "[The audience had] a predominantly Asiatic imprint. One saw distinguished matrons, young ladies of questionable quality, schoolboys with the sacred colors of Zion on the blazonry of the Jewish wandering club[.]"[229]

According to Einstein, and the newspaper *Berliner Tageblatt* (14 February 1920), and a petition signed by almost 300 students, nothing anti-Semitic was said or done at the meeting.[230] A young Jewish student, Hans Toby Cohn, wrote to Einstein to apologize for his and his fellow Jews actions, because they were too young to decipher yet whether to be,

> "a Communist or a Monarchist, whether an atheist or a nationalistic Jew."[231]

The uproar did not involve any anti-Semitic statements, but according to Cohn did include such statements as, "'Socialist' and 'money refund' or 'Are we still students?!'"[232] which were made by young Jews. Despite these facts,

---

**226**. D. K. Buchwald, *et al.*, Editors, *The Collected Papers of Albert Einstein*, Volume 9, Document 312, Princeton University Press, (2004), pp. 426-427, at 427, note 3.
**227**. *Deutsche Zeitung*, (17 February 1920), p. 5. *Deutsche Zeitung*, (19 February 1920).
**228**. E. A. Ross, "The East European Hebrews", *The Old World in the New: The Significance of Past and Present Immigration to the American People*, Chapter 7, The Century Co., New York, (1914), pp. 143-167. *See also:* B. J. Hendrick, "The Jews in America: I How They Came to This Country", *The World's Work*, Volume 44, Number 2, (December, 1922), pp. 144-161; **and** "The Jews in America: II Do the Jews Dominate American Finance?", *The World's Work*, Volume 44, Number 3, (January, 1923), pp. 266-286; **and** "The Jews in America: III The Menace of the Polish Jew", *The World's Work*, Volume 44, Number 4, (February, 1923), pp. 366-377; **and** "Radicalism among the Polish Jews", *The World's Work*, Volume 44, Number 6, (April, 1923), pp. 591-601.
**229**. D. K. Buchwald, *et al.*, Editors, *The Collected Papers of Albert Einstein*, Volume 9, Document 311, Princeton University Press, (2004), pp. 425-426, at 426, note 2.
**230**. *Cf.* M. Janssen, *et al*, Editors, *The Collected Papers of Albert Einstein*, Volume 7, Princeton University Press, (2002), pp. 284-288.
**231**. Letter from H. T. Cohn to A. Einstein of 12 February 1920, English translation by A. Hentschel, *The Collected Papers of Albert Einstein*, Volume 9, Document 309, Princeton University Press, (2004), pp. 258-259, at 258.
**232**. Letter from H. T. Cohn to A. Einstein of 12 February 1920, English translation by A. Hentschel, *The Collected Papers of Albert Einstein*, Volume 9, Document 309,

numerous sources have misrepresented the events which took place and misrepresented the disorderly outbursts of Eastern European Jews, as if anti-Semitic attacks by German Gentiles. As with the Berlin Philharmonic affair, it was Einstein and his friends who made an issue of anti-Semitism, where it was not a legitimate issue. It was yet another example of their Jewish racism and Jewish tribalism. Recall that Einstein called the Student Council, the "refuse of humankind".[233]

The newspaper *Vorwärts* published an article on 13 February 1920 and wrote of alleged "excesses of an anti-Semitic student mob" "Exzessen eines antisemitischen Studentenpöbels".[234] The newspaper *8-Uhr Abentblatt* wrote on 13 February 1920,

### "Tumultszenen bei einer Einstein-Vorlesung.
Professor Einstein verzichtet auf weitere Vorlesungen an der Universität. — Rückzahlung der Kollegien an die Studenten.

Bei der gestrigen Vorlesung des Universitätsprofessors Einstein über seine Relativitätstheorie and der Berliner Universität kam es zu unliebsamen Szenen, die eine Unterbrechung der Vorlesung bewirkten und Professor Einstein zwangen, die Studenten aufzufordern, sich die eingezahlten *Kollegiengelder zurückzahlen* zu lassen. Nach einer uns übermittelten Darstellung dieses Zwischenfalles wollte der Studentenausschuß es nicht zulassen, daß die Vorlesungen des Professors *Einstein* außer den imatrikulierten [sic] Studenten auch von *Richtstudenten* besucht werden. Als nun Professor Einstein die gestrige Vorlesung dazu benutzte, um an die Studentenschtft [sic] die Bitte zu richten, ihren Standpunkt zu verlassen, wurde dieses Ersuchen mit einem Tumult beantwortet, bei dem auch *Aeußerungen antisemitischen Charakters* fielen. Professor Einstein sah sich infolge dieses unqualifizierbaren Verhaltens der Studentenschaft gezwungen, die Vorlesung abzubrechen und an seine studentische Zuhörerschaft die Aufforderung zu richten, sich die *Kollegiengelder zrückzahlen* [sic] zu lassen.

### Eine Erklärung Professor Einsteins.

Auf unsere Anfrage teilte uns Herr Professor Einstein über den gestrigen Vorfall folgendes mit:
„Meine populär gehaltenen Vorträge über die Relativitätstheorie besuchten nicht nur Studenten, sondern auch viele andere Leute, die dazu

---

Princeton University Press, (2004), pp. 258-259, at 258.
**233**. Letter from Eduard Meyer to A. Einstein of 13 February 1920, A. Hentschel, translator, *The Collected Papers of Albert Einstein*, Volume 9, Document 312, Princeton University Press, (2004), p. 260.
**234**. C. Kirsten and Hans-Jürgen Treder, Editors, *Albert Einstein in Berlin, 1913-1933 : Teil I, Darstellung und Dokumente*, Akademie-Verlag, Berlin, (1979), p. 202.

eigentlich nicht berechtigt sind. Der Studentenausschuß erklärte deshalb, dies nicht länger zulassen zu wollen. Ich machte darauf aufmerksam, daß der große Saal für alle Platz habe, die zuhören wollen und daß es dadurch zu keinen Unzulänglichkeiten kommen müsse. Der Studentenausschuß hat sich damit jedoch nicht zufrieden gegeben, sondern sich in dieser Frage an den *Rektor* gewandt. Der Rektor schrieb mir einen *Brief*, in dem er darauf hinwies, daß nach der bestehenden Vorschrift jene Leute nicht die Berechtigung haben, den Saal zu betreten. Dies ist *formellrichtig*. Ich habe mich jedoch auf den Standpunkt gestellt, daß es mir widerstrebe, ohne inneren Grund es Leuten unmöglich zu machen, weiter zu hören, und ich habe deswegen gestern, statt zu lesen, eine Besprechung mit meiner Zuhörerschaft veranstaltet, die jedoch zu einem bestimmten Ergebnis nicht führte. Ich habe mich daher veranlaßt gesehen, auf meine weiteren Vorlesungen zu verzichten und der Studentenschaft erklärt, sie könne ihre eingezahlten Kollegiengelder sich zurückzahlen lassen. Ich habe aber nicht die Absicht, meine Vorlesungen überhaupt zu unterlassen, ich werde sie vielmehr in anderer Form wieder aufnehmen. In welchem Saal ist aber noch unbestimmt. Sollte es noch einmal zu solchen Szenen wie gestern kommen, dann höre ich überhaupt auf. Von einem *Skandal*, der sich gestern abgespielt haben soll, kann nicht die Rede sein, immerhin bewiesen manche Aeußerungen, die fielen, eine gewisse animose Gesinnung mir gegenüber. *Antisemitische Äußerungen* als solche fielen nicht, doch konnte ihr *Unterton* so gedeutet werden."

Eduard Meyer, Rector of the University of Berlin, was astonished by these reports of anti-Semitism, which he knew were utterly false. On 13 February 1920, Meyer wrote to the Ministry of Culture, stating, *inter alia*,

"Vorausschicken muß ich, daß ich zu meinem größten Erstaunen durch Herrn Seeberg erfuhr, daß behauptet wird, dabei habe der Antisemitismus eine Rolle gespielt und sei von Judentum u. ä. dei Rede gewesen. Demgegenüber muß ich erklären, daß das völlig unbegründet ist und ich gar nicht begreife, wie solche Behauptungen haben entstehen können. Das Gespräch, das ich gestern mit Herrn Kollegen Einstein über die Sache hatte, ist in der friedlichsten Weise ganz glatt verlaufen, und ebenso erklärt mir der offizielle Vertreter des studentischen Ausschusses, den ich darum befragt habe, daß in den Diskussionen in der gestrigen Vorlesung, an denen er selbst Anteil genommen hat, mit keinem Wort von Antisemitismus, Judentum usw. die Rede gewesen ist."[235]

In 1962, Peter Michelmore conveyed an even more alarming, though also purely fictional, account of the events at the University of Berlin, than had the

---

**235**. C. Kirsten and Hans-Jürgen Treder, Editors, *Albert Einstein in Berlin, 1913-1933 : Teil I, Darstellung und Dokumente*, Akademie-Verlag, Berlin, (1979), p. 201.

Jewish newspapers,

> "A group of black-shirted students broke up one of Einstein's lectures at the University of Berlin. A blond youth screamed above the din, 'I'm going to cut the throat of that dirty Jew.'"[236]

This alarmist script, this Jewish canard, appeared many times and was attributed to many different events. Ernst Gehrcke recorded that the newspaper *Freiheit* changed its story repeatedly after the events at the Berlin Philharmonic of 24 August 1920:

> "[... ]So sprach die *Freiheit*, das Parteiorgan EINSTEINS, am 26. August noch von «wissenschaftlichen Einwänden», am 27. August von der «auf ihre Urheber zurückfallenden, schimpflichen Art, in der der Kampf gegen Professor EINSTEIN und seine Relativitätstheorie geführt wird», am 31. August setzte sich das Blatt über gesellschaftliche und parlamentarische Formen der Berichterstattung hinweg, indem es «einen studentischen Rowdy» sagen läßt, er wolle dem «Saujud EINSTEIN an die Gurgel», und am 4. September: «Die ernsthafte exakte Wissenschaft ist also ein Geschäft, das mit Schiebergewinnen abschließt»."

*Die Umschau*, Volume 24, (1920), page 554, alleged that someone said,

> "man sollte diesem Juden an die Gurgel fahren."[237]

*Vossische Zeitung* reported on 29 August 1920, Morning Edition, Supplement 4, front page, that someone loudly stated,

> "Diesem Saujuden müßte man eigentlich an die Gurgel springen."[238]

Yet another account, again by interested pro-Einstein parties, in 1927, places the alleged incident at an unnamed "public meeting in the spring of 1919."[239]

Johannes Riem, who was not bashful, wrote on 1 July 1921, in reference to Reuterdahl,

> "Man geht gegen Einstein vor als den Goliat des Skeptizismus. Vorlesungen dagegen werden veranstaltet. In scharfsinniger Weise wird in einem viel gelesenen Buche „Relativität oder innere Abhängigkeit" die Unhaltbarkeit

---

**236.** P. Michelmore, *Einstein: Profile of the Man*, Dodd, Mead, New York, (1962), p. 88.
**237.** M. Janssen, *et al.*, Editors, *The Collected Papers of Albert Einstein*, Volume 7, Princeton University Press, (2002), p. 348, note 3.
**238.** M. Janssen, *et al.*, Editors, *The Collected Papers of Albert Einstein*, Volume 7, Princeton University Press, (2002), p. 348, note 3.
**239.** H. Goenner, "The Reaction to Relativity Theory. I: The Anti-Einstein Campaign in Germany in 1920", *Science in Context*, Volume 6, Number 1, (1993), pp. 107-133, at 112.

der Relativitätstheorie nachgewiesen. Der Einwand Einsteins, dies sei nur eine besondere Form des Antisemitismus, wird sehr energisch zurückgewiesen, und mit der Anerkennung Spinozols beantwortet."[240]

Physicist Stjepan Mohorovičić declared that he was intimidated out of opposing Einstein's myths and plagiarism, through fear of being labeled an anti-Semite and by anonymous threats. Johannes Jürgenson writes,

"Ein weiterer Punkt war, daß es Einstein, der selbst Jude war, geschickt verstand, seinen Gegnern Antisemitismus zu unterstellen:
'Die erste Opposition der wissenschaftlichen Welt gegen die neuen Relativitätstheorien hat man einfach gebrochen, indem man sie als eine Folge des Antisemitismus dem breiten Publikum vorgestellt hat' sagte Mohorovicic 1962. Auch er hatte in jener Zeit in Zagreb seine Kritik zurückgestellt, um nicht als Antisemit zu gelten."[241]

Mohorovičić wrote in 1962 in the second volume of *Kritik der Relativitätstheorie*,

"The initial opposition in the scientific world against the new theory of relativity was easily crushed by convincing the general public that it was a product of anti-Semitism, although no one could reliably make such an accusation against M. ABRAHAM, O. KRAUS, O. D. CHWOLSON, etc.! But it disgusts me to speak further of such things; those wanting to learn more about it can glean the facts from many sources, for example [269-270] through [316-317] and others."

"Die erste Opposition in der wissenschaftlichen Welt gegen die neuen Relativitätstheorien hat man einfach gebrochen, indem man sie als eine Folge des Antisemitismus dem breiten Publikum vorgestellt hat, obwohl man dies sicher nicht einem M. ABRAHAM, O. KRAUS, O. D. CHWOLSON, etc. vorwerfen konnte! (usw.). Aber es ekelt mir, über solche Verhältnisse weiter zu sprechen; wer sich darüber unterrichten will, müßte vieles nachlese, wie z. B. [269-270] bis [316-317] und manches andere."[242]

---

[240]. J. Riem, "Amerika über Einstein", *Deutsche Zeitung*, (1 July 1921).
[241]. J. Jürgenson, "Es lebe die Theorie - oder das Recht auf freie Phantasie", *Die lukrativen Lügen der Wissenschaft*, Ewert, (1998), ISBN: 389478699X.

URL:<http://www.unglaublichkeiten.info/unglaublichkeiten/htmlphp/erfindungeneslebedietheorie.html>

This is likely a reference to: S. Mohorovičić, "Raum, Zeit und Welt. II Teil", in K. Sapper, Editor, *Kritik und Fortbildung der Relativitätstheorie*, Akademische Druck- u. Verlagsanstalt, Graz, Volume 2, (1962), pp. 219-352.
[242]. S. Mohorovičić, "Raum, Zeit und Welt", in two parts in K. Sapper, Editor, *Kritik*

Mohorovičić also stated that the "Relativity Syndicate" vehemently obstructed the publication of works which criticized the theory of relativity (your author has personally witnessed such corrupt practices):

> "Eine vorzügliche und sehr scharfsinnige Kritik veröffentlichte G. v. GLEICH 1930, wo er alle seine diesbezüglichen Arbeiten gesammelt und geordnet hatte, obwohl das 'Relativitätssyndikat' mit allen Mitteln trachtete, das Erscheinen dieses Werkes zu verhindern. Nun es war sehr schwer die Kritik gänzlich zu unterdrücken, da man in der Wahl der Mittel nicht kleinlich war. Alle, für die Relativitätstheorie ungünstigen Arbeiten wurden einfach kurzerhand als unrichtig, fehlerhaft oder falsch bezeichnet oder als unwichtig (heutzutage ein sehr beliebtes Wort!) oder wenigstens als uninteressant verschwiegen. Von den Philosophen erhielten nur die Applaudierenden das Wort, den kritisch Gesinnten warf man ihre mathematischen Unkenntnisse vor; wer sich darüber unterrichten will, sollte die offenen Briefe des bekannten Philosophen O. KRAUS nachlesen [108]), und doch haben die Philosophen die Grundlage der Rechnung, nicht aber die Rechnung selbst untersucht. Aber die Relativisten haben übersehen, daß die modernen Relativitätstheorien, ähnlich wie die moderne Musik, voll von Dissonanzen sind, (eine solche Musik entzückt den heutigen Snob außerordentlich und er kann nicht begreifen, daß es gebildete Leute gibt, welche die moderne Musik nicht ausstehen können, aber dafür muß man das Ohr und die richtige musikalische Erziehung haben!). O. KRAUS hat besonders den Umstand hervorgehoben (l. c. S. 96.), 'daß jeder Quark, der für die Theorie zu sein scheint, von den Relativisten mit freundlicher Gebärde begrüßt wird... während eine ernste Kritik mißhandelt wird' [109]). Dies wirkte aber verhängnisvoll und diese modernen Theorien wurden größtenteils ein Tätigkeitsfeld pour ceux qui savent vivre... oder wie ein lachender Philosoph sagte [110]): '...an Höfen ist Höflichkeit der Verstand und die Münze...'."[243]

Mohorovičić stated in 1922 that he had received anonymous threats for opposing relativity theory,

> "Viele wurden von der Behauptung geblendet, daß diese Theorie sich mit der Erfahrung in Übereinstimmung befinde (vgl. II, 4), was von den Anhängern der Einsteinschen Theorie sehr geschickt zu

---

*und Fortbildung der Relativitätstheorie*, Akademische Druck- u. Verlagsanstalt, Graz, (1958/1962), Part 1 in Volume 1, (1958), pp. 168-281; Part 2 in Volume 2, (1962), pp. 219-352, at 317, note 89.
**243**. S. Mohorovičić, "Raum, Zeit und Welt", in two parts in K. Sapper, Editor, *Kritik und Fortbildung der Relativitätstheorie*, Akademische Druck- u. Verlagsanstalt, Graz, (1958/1962), Part 1 in Volume 1, (1958), pp. 168-281; Part 2 in Volume 2, (1962), pp. 219-352, at 273.

Propagandazwecken ausgenutzt wurde. Das letzte (nämlich diese gewissenlose Reklame) ist gerade auch die dunkelste Seite des erwähnten Kampfes, welcher nie in einer so scharfen Form ausgebrochen wäre, wenn nicht diese unglückliche und unerhörte Propaganda gewesen wäre, welche in der Geschichte fast aller Wissenschaften beispiellos ist [*Footnote deleted*]. Alles dies wird noch durch die Tatsache verschärft, daß Einstein und die Mehrzahl seiner ersten Anhänger Juden sind — (ich hätte keinen Grund, die Rasse Einsteins zu erwähnen, wenn nicht Einstein *selbst* so häufig betont hätte, daß er ein Jude sei) [*Footnote:* Einstein selbst sagt in dem Vorwort des Werkes von L. Fabre (Anmerk. 30) den Franzosen ausdrücklich, daß er nur in Deutschland geboren sei, sonst sei er ein Jude, Pazifist und Mitglied einer internationalen Verbindung.... Es ist nicht schwer zu raten, warum Einstein dies gerade den Franzosen gegenüber gesagt hat (mit eigener Unterschrift), aber lassen wir das, es ist dies nur Geschmacksache...; unsere Arbeit hier ist eine wissenschaftliche. Es ist traurig genug, daß ich gezwungen bin, dies hier zu erwähnen!] —, und da die letzteren fast die ganze Weltpresse in den Händen haben, so bereiteten sie für Einstein eine kolossale Reklame und haben fast jede Arbeit, welche gegen diese Theorie gerichtet wurde, zu unterdrücken gesucht. Zu diesem religiös-sozialen Moment kommt noch ein politisches Moment hinzu, worüber ich hier nicht zu reden wünsche. *Ich bin nur überzeugt, daß wir, die wir uns ziemlich welt von diesem Kampfe befinden, viel ruhiger und objektiver über diese neue Richtung urteilen können, und daß wir nicht sofort blind und kritiklos jede neue Richtung, welche zu uns aus dem Ausland gelangt, anzunehmen brauchen.* [*Footnote:* Leider sind diese »Methoden« des Streits auch zu uns gekommen. Mitglieder einer philosophischen Fakultät, die in ihrem fanatischem Abscheu gegen jede sachliche, kritische Stellungnahme zur Relativitätstheorie offenbar ganz vergessen hatten, daß die Wissenschaft eine *über* den Parteien stehende Sache ist, haben sich nicht gescheut, persönliche Gehässigkeit gegen mich als Kritiker der Relativitätstheorie an den Tag zu legen, wie ich mehrfach erfahren mußte. Einige Herren Relativisten haben mir *anonyme Drohbriefe* zugestellt und sich anderer, sonst in wissenschaftlichen Kreisen sehr ungewöhnlicher Mittel bedient. Es ist die höchste Zeit, mit solchen Methoden endlich aufzuhören!]"[244]

Einstein, too, was attacked by lunatics—who made death threats and plots against him, but these were political attacks which were not directly related to the theory of relativity. In the spring of 1921, Rudolph Leibus offered a reward to anyone who murdered Einstein, Harden or Foerster. Theodor Wolff, editor of the *Berliner Tageblatt*, spread the false rumor that Einstein and he were targets of assassins after the murder of Walter Rathenau in 1922. This may have been a

---

**244**. S. Mohorovičić, *Die Einsteinsche Relativitätstheorie und ihr mathematischer, physikalischer und philosophischer Charakter*, Walter de Gruyter & Co., Berlin, Leipzig, (1923), pp. 52-53.

pretext to give Einstein an excuse to back away from his commitment with the League of Nations and the police denied Wolff's charges. *The New York Times* reported on the front page on 19 February 1923 that Prof. Herzen of Lausanne University told a meeting of the Brussels Engineering Association in a discussion on the theory of relativity that Einstein was on a death list. *The New York Times* reported on 1 February 1925 on page 13 that Marie Evgenievna Dickson was arrested after she showed up at the Einstein's home and frightened Mrs. Einstein. Dickson had been expelled from France for planning to murder the Soviet Ambassador Leonid Krassin. Years later, after the World Committee for Help for Victims of German Fascism, for which Einstein was a figurehead, published *The Brown Book of the Hitler Terror*,[245] the rumor spread that the Nazis had put a bounty on Einstein's head.[246]

*Ad hominem* attack and smear campaigns were Einstein's preferred method of response to challenges to Einstein's priority and challenges to relativity theory, as even Einstein's advocates were forced to concede in 1931. Von Brunn, a defender of Einstein, wrote,

> "Even individual fanatic scientific advocates of the Einsteinian theory seem to have finally abandoned their tactic of cutting off any discussion about it with the threat that every criticism, even the most moderate and scrupulous ones, must be discredited as an obvious effluence of stupidity and malice. But even if these monstrous products of the 'Einstein frenzy' [*Einstein-Taumel*] now belong to history and are thus eliminated from consideration, thoroughly respectable reasons for a certain discomfort with relativity theory still do remain[.]"[247]

This was published in a pro-Einstein "review" of *Hundert Autoren gegen Einstein*, which anti-Einstein book stated,

> "It is the aim of this publication to confront the terror of the Einsteinians with an overview of the quality and quantity of the opponents [of the theory of relativity] and opposing arguments."[248]

Sadly, the *ad hominem* attacks against anyone who criticized Einstein or

---

**245**. "Einstein Denies Part in Book on Hitlerism", *The New York Times*, (4 September 1933), p. 2.
**246**. "Price Declared Put on Einstein's Head", *The New York Times*, (7 September 1933), p. 8. "Wants Only Peace", *The New York Times*, (11 September 1933), p. 9.
**247**. A. v. Brunn, quoted in: K. Hentschel, Ed., A. Hentschel, Ed. Ass. and Trans., *Physics and National Socialism: An Anthology of Primary Sources*, Birkhäuser, Basel, Boston, Berlin, (1996), p. 11.
**248**. From the preface of *Hundert Autoren gegen Einstein* translated by: H. Goenner, "The Reaction to Relativity Theory in Germany, III: 'A Hundred Authors against Einstein'", J. Earman, M. Janssen, J. D. Norton, Eds., *The Attraction of Gravitation: New Studies in the History of General Relativity*, Birkhäuser, Boston, Basel, Berlin, (1993), p. 251.

relativity theory were not relegated to history, despite Brunn's claims; and, ironically, one need only read his "review" of *Hundert Autoren gegen Einstein* to see that the so-called "review" was itself an *ad hominem* attack against the authors. *One Hundred Authors Against Einstein* was a response to personal attacks from Einstein and his followers, and largely contained philosophical objections to relativity theory, some better than others.

Charles Lane Poor complained of severe censorship.

Einstein liked to smear his critics. Henri Bergson published a book, which was, according to Abraham Pais, not included in his collected works, and which was a negative critique of relativity theory titled *Duration and Simultaneity*. Pais wrote,

> "In his presentation speech on December 10, 1922, Arrhenius said, 'Most discussion [of Einstein's oeuvre] centers on his theory of relativity. This pertains to epistemology and has therefore been the subject of lively debate in philosophical circles. It will be no secret that the famous philosopher Bergson in Paris has challenged this theory, while other philosophers have acclaimed it wholeheartedly'.
>
> Bergson's collected works appeared in 1970 [B3]. The editors did not include his book *Durée et Simultanéité: A Propos de la Théorie d'Einstein*. Einstein came to know, like, and respect Bergson. Of Bergson's philosophy he used to say, 'Gott verzeih ihm,' God forgive him."[249]

In the 1965 English translation of Bergson's book, *Duration and Simultaneity,* physicist Herbert Dingle wrote an introductory piece detailing the suppression of criticisms of relativity theory. Dingle warned of the dangers of the anti-rational state of awareness induced by Logical Positivism in its pseudo-relativistic adherents, with its celebration of the denial of physical reality, its solipsism, hypocrisy, numerology, and semantics; with the positivists' acceptance of metaphysical fallacy as if fact.

Dingle asked us all to consider the fact that we place our lives in the hands of a class of scientists who see as their goal the denial of the physical world, as for them it is an illusion supplanted by numbers, and who corruptly pursue the unchecked promotion of their myths. Herbert Dingle, whose words were often suppressed, stated, *inter alia*,

> "The facts must be faced. To a degree never previously attained, the material future of the world is in the hands of a small body of men, on whose not merely superficially apparent but absolute, intuitive (in Bergson's sense of the word) integrity the fate of all depends, and that quality is lacking. Where there was once intellectual honesty they have now merely the idea that they possess it, the most insidious and the most dangerous of all usurpers; the substitution is shown by the fruits, which are displayed in unmistakable

---

[249]. A. Pais, *Subtle is the Lord*, Oxford University Press, New York, (1982), p. 510.

clarity in the facts described here. After years of effort I am forced to conclude that attempts with the scientific world to awaken it from its dogmatic slumber are in vain. I can only hope that some reader of these pages, whose sense of reality exceeds that of the mathematicians and physicists and who can command sufficient influence, might be able from the outside to enforce attention to the danger before it is too late."[250]

Under the headline "When a scientist challenges dogma, he's the one who gets mauled", Scott LaFee wrote in the *The San Diego Union-Tribune* of 2 November 1994,

"But unfortunate things can still happen when a novel contention challenges the perceived or popular 'truth.' Instead of receiving an honest but critical evaluation, the new idea can be ridiculed or, worse, ignored, its creator punished professionally and personally.
'I wouldn't do it again,' says Wallace Kantor, a retired local physicist who questioned Einstein's Special Theory of Relativity in several scientific papers and a book. 'Reaction to my work ranged from intense rage to contemptuous pity. It was career-damaging. It wasn't worth it.'"[251]

---

[250]. H. Dingle, in his introduction to H. Bergson's, *Duration and Simultaneity*, Bobbs-Merrill Company, Inc., Indianapolis, New York, Kansas City, (1965), p. xlii. ***See also***: H. Dingle, *Science at the Crossroads*, Martin, Brian and O'Keefe Ltd., London, (1972).
[251]. "The Outsiders: When a scientist challenges dogma, he's the one who gets mauled", *The San Diego Union-Tribune*, Lifestyle Section, (2 November 1994), p. E-1. URL: <http://www.aidsinfobbs.org/articles/rethink/rethink1/434>

# 5 THE PROTOCOLS OF THE LEARNED ELDERS OF ZION

*At the turn of century, Sergei Nilus, a Russian Orthodox theologian of good reputation, published a purported transcript of a Judeo-Masonic conspiracy to take over the world. It received little attention at the time, but when it was republished in 1917, and when numerous translations appeared after the First World War, Europeans and Americans realized that the events of the Russian Revolution and the World War fulfilled many of the plans laid out in the "Protocols of the Learned Elders of Zion", which was first published many years before these events began. This cast suspicion over the Jews of the world, who had long been the primary proponents of revolution and the leading warmongers around the globe.*

> "*The Times* has not as yet noticed this singular little book. Its diffusion is, however, increasing, and its reading is likely to perturb the thinking public. Never before have a race and a creed been accused of a more sinister conspiracy. We in this country, who live in good fellowship with numerous representatives of Jewry, may well ask that some authoritative criticism should deal with it, and either destroy the ugly 'Semitic' bogy or assign their proper place to the insidious allegations of this kind of literature."—THE LONDON TIMES, 8 MAY 1920, PAGE 15

> "For it is the day of the LORD's vengeance, and the year of recompences for the controversy of Zion."—ISAIAH 34:8

> "A more bloodthirsty and vindictive race has never seen the light of day. They regard themselves as the Chosen of the Lord and believe they are destined to annihilate and torture all Gentiles. The first and foremost task they expect their Messiah to accomplish is that he shall murder and slay all human beings with his sword. From the very earliest days they have undertaken all in their power to practically demonstrate this to the Christians and have continued to do so whenever they could."—MARTIN LUTHER[252]

## 5.1 Introduction

We know that the Rothschilds intended for one of theirs to become the King of the Jews. According to Jewish mythology, this King would be the Messiah of the Jews and would own all the wealth of the world and rule over the entire world from Jerusalem. In order for this plan to work without divine intervention, it would require an organized plan.

Jews had been ardent students of politics and political psychology from their beginnings, and their religion is more political, than it is spiritual. The Rothschilds' plan for Messianic rule of the world must have included the incorporation of the ideas of political writers, statesmen, and political sycophants

---

[252]. Bishop M. Sasse, Eisenach, *Martin Luther and the Jews*, Second Printing, Sons of Liberty, Hollywood, California, (1967), p. 5.

like Machievelli and Maurice Jolly. It would not be surprising to find such ideas discussed by the Rothschilds and their Zionist agents.

The Czar of Russia created a secret police force, in large part to counteract the Jewish revolutionaries, who sought to unseat him and destroy Russian society and mass murder the Russian people. This police force employed Jewish spies to watch over the meetings of Jewish leaders and listen in on the lectures Jewish subversives often gave in synagogues and on street corners. The Czar's police probably had a very good notion of what it was that the Rothschilds and their agents had planned for the world. It is possible that a copy of this plan fell into the hands of patriotic Russians. If the Czar's police forged *The Protocols of the Learned Elders of Zion*, as many who dispute the authenticity of the *Protocols* claim, it would still not be likely that they entirely fabricated them through plagiarism. Given that the *Protocols* so closely anticipate the methods of the Jewish Bolsheviks and the Zionist Nazis, it appears that whoever wrote the *Protocols* had a very good knowledge of what the Rothschilds and their Zionist agents had planned for the world.

Christians tend to overlook the fact that the so-called "Jewish conspiracy" to take over the world did not appear for the first time in the allegedly forged *Protocols*, but is Judaism itself. Critics of the Jews did not fabricate the Jewish plans to take over the world, rob it of its wealth, destroy all other religions, rule the world in an autocracy headed by a Jewish King descended from David, and then exterminate the "unrighteous Jews" and the Goyim; which plans are plainly stated in the Old Testament,[253] the *Zohar*, the Talmud, and numerous other Jewish religious writings. Jews created these ancient plans and iterated them in the Hebrew Bible, the Talmud and in their Cabalistic writings. Christians see the Old Testament as the work of God, and whether the individual Christian believes these supposedly divine prophecies have already been fulfilled, or were transferred to Christians to be ultimately fulfilled as in the *Revelation*, or are yet to be fulfilled for the allegedly divine race of Jews—God's chosen people, the Christian has often been duped into becoming an agent of the Jewish plan to destroy humanity, a plan better known as "Judaism".

The *Protocols* were effective in revealing this plan, not because they differ substantially from Judaism—they do not, as is revealed by Michael Higger's book *The Jewish Utopia*[254] and by the Old Testament itself—rather, the *Protocols* effectively alerted Christians, because, like the Talmud and *Zohar*, they appeared after Christianity appeared and ridiculed the Christians, just as the Old Testament ridicules and advocates the genocide of the non-Jew. Judaism has remained consistent in its plans. Christians have accepted its myths, because they

---

**253**. *Exodus* 34:11-17. *Psalm* 2; 72. *Isaiah* 1:9; 2:1-4; 6:9-13; 9:6-7; 10:20-22; 11:4, 9-12; 17:6; 37:31-33; 41:9; 42; 43; 44; 61:6. *Jeremiah* 3:17; 33:15-16. *Ezekiel* 20:38; 25:14. *Daniel* 12:1, 10. *Amos* 9:8-10. *Obadiah* 1:18. *Micah* 4:2-3; 5:8. *Zechariah* 8:20-23; 14:9. *Romans* 9:27-28; 11:1-5.

**254**. M. Higger, *The Jewish Utopia*, Lord Baltimore Press, Baltimore, (1932). Higger's book is analyzed in: R. H. Williams, *The Ultimate World Order—As Pictured in "The Jewish Utopia"*, CPA Book Publisher, Boring, Oregon, (1957?).

believe them to have been made Christian. The Christians' blindness to the Judaic plan for their demise is best unmasked by works the Christians do not view as divinely inspired—even if those works simply repeat the Judaic plan for world domination laid out in the Old Testament.

## 5.2 The Protocols of the Learned Elders of Zion

The following is an English translation of the *Protocols*, which translation was first published in 1920, in the book, *The Protocols and World Revolution Including a Translation and Analysis of the "Protocols of the Meetings of the Zionist Men of Wisdom"*, Small, Maynard & Co., Boston, (1920), pp. 11-73:

"Protocols of the Meetings of the<br>Zionist Men of Wisdom

### PROTOCOL NO. I

LET us put aside phraseology and discuss the inner meaning of every thought; by comparisons and deductions let us illuminate the situation. In this way I will describe our system, both from our own point of view and from that of the GOYS. [*Footnote:* The GOYS—the Gentiles.]

It must be remembered that people with base instincts are more numerous than those with noble ones; therefore, the best results in governing are achieved through violence and intimidation and not through academic discussion. Every man seeks power; every one would like to become a dictator if he possibly could; and rare indeed are those who would not sacrifice the common good in order to attain personal advantage.

What has restrained the wild beasts we call men?

What has influenced them heretofore?

In the early stages of social life they submitted to brute and blind force; afterwards—to the Law, which is the same force but disguised. I deduce from this that according to the laws of nature, right lies in might.

Political freedom is not a fact but an idea. One must know how to employ this idea when it becomes necessary to attract popular forces to one's party by mental allurement if it plans to crush the party in power. The task is made easier if the opponent himself has contradicted the idea of freedom, the so-called liberalism, and for the sake of the idea yields his power. It is precisely here that the triumph of our theory becomes apparent: the relinquished reins of power are, according to the laws of nature, immediately seized by a new hand because the blind force of the people cannot remain without a leader even for one day, and the new power merely replaces the old, weakened by liberalism.

In our day the *power of gold* has replaced liberal rulers. There was a time when faith ruled. The idea of freedom cannot be realized because no one knows how to make reasonable use of it. Give the people self-government for a short time and it will become corrupted. From that very moment strife begins and soon develops into social struggles, as a result of which states are set aflame and their authority is reduced to ashes.

Whether the state is exhausted by internal convulsions, or whether civil wars deliver it into the hands of external enemies, in either case it can be regarded as hopelessly lost: it is in our power. The despotism of capital, which is entirely in our hands, holds out to it a straw which the state must grasp, although against its will, or otherwise fall into the abyss.

To him who, because of his liberal inclinations, would contend that arguments of this kind are immoral, I would propound the question: If a state has two enemies, and if against the external enemy it is permitted and it is not considered immoral to use all methods of warfare, and as a protective measure not to acquaint the enemy with the plans of attack, such as night attacks or attacks with superior forces, then why should the same methods be regarded as immoral when applied to a worse foe, a transgressor against social order and prosperity?

How can a sound and logical mind hope successfully to guide the masses by means of reasonable persuasion or by arguments if there is a possibility of contradiction, even though unreasonable, but which may appear more attractive to the superficially thinking masses? Guided entirely by shallow passions, superstitions, customs, traditions, and sentimental theories, the people in and of the mob become embroiled in party dissensions which prevent all possibility of an agreement, even though it be on a basis of perfectly sound reasoning. Every decision of the mob depends upon the accidental or prearranged majority, which, owing to its ignorance of political secrets, pronounces absurd decisions, thus introducing the seeds of anarchy into the government.

Politics have nothing ill common with morals. The ruler guided by morality is not a skilled politician, and consequently he is not firm on his throne. He who desires to rule must resort to cunning and hypocrisy. The great popular qualities—honesty and frankness—become vices in politics, as they dethrone more surely and more certainly than the most powerful enemy. These qualities must be the attributes of GOY countries ; but we by no means should be guided by them.

Our right lies in might. The word 'right' is an abstract idea, unsusceptible of proof. This word means nothing more than : Give me what I desire so that I may have evidence that I am stronger than you.

Where does right begin? Where does it end?

In a state with a poorly organized government and where the laws are insignificant, and the ruler has lost his dignity as the result of the accumulation of liberal rights, I find a new right, namely, the right of might to destroy all existing order and institutions, to lay hands on the law, to alter all institutions, and to become the ruler of those who have voluntarily, liberally renounced for our benefit the rights to their own power.

With the present instability of all authority our power will be more unassailable than any other, because it will be invisible until it is so well rooted that no cunning can undermine it.

From temporary evil to which we are now obliged to have recourse will emerge the good of an unshakable government, which will reinstate the orderly functioning of the mechanism of popular existence now interrupted by liberalism. The end justifies the means. In laying our plans we must turn our attention not so much to the good and moral as to the necessary and useful. Before us lies a plan in which a strategic line is shown, from which we must not deviate on pain of risking the collapse of many centuries of work.

In working out an expedient plan of action it is necessary to take into consideration the meanness, vacillation, changeability of the mob, its inability to appreciate and respect the conditions of its own existence and of its own well-being. It is necessary to realize that the power of the masses is blind, unreasoning, and void of discrimination, prone to listen to right and left. The blind man cannot guide the blind without bringing them to the abyss; consequently, members of the crowd, upstarts from the people, even were they men of genius but incompetent in politics, cannot step forward as leaders of the mob without ruining the entire nation.

Only the person prepared from childhood to autocracy can understand the words which are formed by political letters.

The people left to themselves, that is to upstarts from among them, are ruined by party dissensions created by greed for power and honors, and by the disorders resulting therefrom. Is it possible for the masses of the people to direct the affairs of the state without rivalries, and without interjecting personal interests? Are they capable of protecting themselves against external enemies?—This is impossible, since a plan divided into as many parts as there are minds in a mob loses its unity, and consequently, becomes incomprehensible and unworkable.

Only an autocrat can outline great and clear plans which allocate in an orderly manner all the parts of the mechanism of the government machinery. From this it is concluded that the government which is the most efficient for the benefit of a country must be concentrated in the hands of one responsible person. Civilization cannot exist without absolute despotism, for government is carried on not by the masses, but by their leader, whoever he may be. A barbarous crowd shows its barbarism on every occasion. The moment the mob grasps liberty in its hands it is speedily changed to anarchy, which is in itself the height of barbarism.

Look at those beasts, steeped in alcohol, stupefied by wine, the unlimited use of which is granted by liberty. Surely you cannot allow our own people to come to this. The people of the GOYS are stupefied by spirituous liquors; their youth is driven insane through excessive study of the classics, and vice to which they have been instigated by our agents—tutors, valets, governesses—in rich houses, by clerks, and so forth, and by our women in the pleasure places of the GOYS. Among the latter I include the so-called 'society women,' their volunteer followers in vice and luxury.

Our motto is Power and Hypocrisy. Only power can conquer in politics, especially if it is concealed in talents which are necessary to statesmen. Violence must be the principle; hypocrisy and cunning the rule of those governments which do not wish to lay down their crowns at the feet of the agents of some new power. This evil is the sole means of attaining the goal of good. For this reason we must not hesitate at bribery, fraud, and treason when these can help us to reach our end. In politics it is necessary to seize the property of others without hesitation if in so doing we attain submission and power.

Our government, following the line of peaceful conquest, has the right to substitute for the horrors of war less noticeable and more efficient executions, these being necessary to keep up terror, which induces blind submission. A just but inexorable strictness is the greatest factor of governmental power. We must follow a program of violence and hypocrisy, not only for the sake of profit, but also as a duty and for the sake of victory.

A doctrine based on calculation is as potent as the means employed by it. That is why not only by these very means, but by the severity of our doctrines, we shall triumph and shall enslave all governments under our super-government.

Even in olden times we shouted among the people the words ' Liberty, Equality, and Fraternity.' These words have been repeated so many times since by unconscious parrots, which, flocking from all sides to the bait, have ruined the prosperity of the world and true individual freedom, formerly so well protected from the pressure of the mob. The would-be clever and intelligent GOYS did not discern the symbolism of the uttered words; did not notice the contradiction in the meaning and the connection between them; did not notice that there is no equality in nature; that there can be no liberty, since nature herself has established inequality of mind, character, and ability, as well as subjection to her laws. They did not reason that the power of the mob is blind; that the upstarts selected for government are just as blind in politics as is the mob itself, whereas the initiated man, even though a fool, is capable of ruling, while the uninitiated, although a genius, will understand nothing of politics. All this has been overlooked by the GOYS.

Meanwhile dynastic government has been based upon this, that the father passed to his son the knowledge of the course of political evolution, so that nobody except the members of the dynasty could possess this knowledge, and no one could disclose the secrets to the governed people. In the course of time the meaning of the dynastic transmission of the true is understanding of politics has been lost, thus contributing to the success of our cause.

In all parts of the world the words ' Liberty, Equality, and Fraternity' have brought whole legions into our ranks through our blind agents, carrying our banners with delight. Meanwhile these words were worms which ruined the prosperity of the GOYS, everywhere destroying peace, quiet, and solidarity, undermining all the foundations of their states. You will see subsequently that this aided our triumph, *for it also gave us, among other things, the opportunity to grasp the trump card, the abolition of privileges; in other words, the very essence of the aristocracy of the GOYS, which was the only protection of peoples and countries against us.*

On the ruins of natural and hereditary aristocracy we built an aristocracy of our intellectual class—the money aristocracy. We have established this new aristocracy on the qualification of wealth, which is dependent upon us, and also upon science, which is promoted by our wise men.

Our triumph was also made easier because, through our connections with people who were indispensable to us, we always played upon the most sensitive chords of the human mind, namely, greed, and the insatiable selfish desires of man. Each of these human weaknesses taken separately is capable of killing initiative and of placing the will of the people at the disposal of the buyer of their activities.

Abstract liberty offered the opportunity for convincing the masses that government is nothing but the manager representing the owner of the country, namely, the people, and that this manager can be discarded like a pair of worn-out gloves.

The fact that the representatives of the nation can be deposed, delivers them into our power and practically places their appointment in our hands.

## PROTOCOL NO. II

IT is necessary for us that wars, whenever possible, should bring no territorial advantages: this will shift war to an economic basis and force nations to realize the strength of our predominance; such a situation will put both sides at the mercy of our million-eyed international agency, which will be unhampered by any frontiers. Then our international rights will do away with national rights, in a limited sense, and will rule the peoples in the same way as the civil power of each state regulates the relation of its subjects among themselves.

The administrators chosen by us from among the people in accordance with their capacity for servility will not be experienced in the art of government, and consequently they will easily become pawns in our game, in the hands of our scientists and wise counselors, specialists trained from early childhood for governing the world. As you are aware, these specialists have obtained the knowledge necessary for government from our political plans, from the study of history, and from the observation of every passing event. The GOYS are not guided by the practice of impartial historical observation, but by theoretical routine without any critical regard for its results. Therefore, we need give them no consideration. Until the time comes let them amuse themselves, or live in the hope of new amusements or in the memories of those past. Let that play the most important part for them which we have induced them to regard as the laws of science (theory). For this purpose, by means of our press, we increase their blind faith in these laws. Intelligent GOYS will boast of their knowledge, and verifying it logically they will put into practice all scientific information compiled by our agents for the purpose of educating their minds in the direction which we require.

Do not think that our assertions are without foundation: note the successes of Darwinism, Marxism, and Nietzscheism, engineered by us. The demoralizing effects of these doctrines upon the minds of the GOYS should be already obvious to us.

It is essential that we take into consideration the modern ideas, temperaments, and tendencies of peoples in order that no mistakes in politics and in guiding administrative affairs may be made. The triumph of our system, parts of whose mechanism must be adapted in accordance with the temperament of the peoples with whom we come in contact, cannot be realized unless its practical application is based upon a résumé of the past as related to the present.

*There is one great force in the hands of modern states which arouses thought movements among the people. That is the press.* The rôle of the press is to indicate necessary demands, to register complaints of the people, and to express and foment dissatisfaction. The triumph of free babbling is incarnated in the press; but governments were unable to profit by this power *and it has fallen into our hands*. Through it we have attained influence, while remaining in the background. Thanks to the press, we have gathered gold in our hands, although we had to take it front rivers of blood and tears.

But it cost us the sacrifice of many of our own people. Every sacrifice on our part is worth a thousand GOYS before God.

## PROTOCOL NO. III

TO-DAY I can tell you that our goal is close at hand. Only a small distance remains, and the cycle of the *Symbolic Serpent*—the symbol of our people—will be complete. When this circle is completed, then all the European states will be enclosed in it as in strong claws.

The modern constitutional scales will soon tip over, for we have set them inaccurately, thus insuring an unsteady balance for the purpose of wearing out their holder. The GOYS thought it had been sufficiently strongly made and hoped that the scales would regain their equilibrium, but the holder—the ruler—is screened from the people by his representatives, who fritter away their time, carried away by their uncontrolled and irresponsible authority. Their power, moreover, has been built up on terrorism spread through the palaces. Unable to reach the hearts of their people, the rulers cannot unite with them to gain strength against the usurpers of power. The visible power of royalty and the blind power of the masses, *separated by us*, have both lost significance, for separated, they are as helpless as the blind man without a stick.

To induce the lovers of authority to abuse their power, we have placed all the forces in opposition to each other, having developed their liberal tendencies towards independence. We have excited different forms of initiative in that direction; we have armed all the parties; we have made authority the target of all ambitions. We have opened the arenas in different states, where revolts are now occurring, and disorders and bankruptcy will shortly appear everywhere.

Unrestrained babblers have converted parliamentary sessions and administrative meetings into oratorical contests. Daring journalists, impudent pamphleteers, make daily attacks on the administrative personnel. The abuse of power is definitely preparing the downfall of all institutions and everything will be overturned by the blows of the infuriated mobs.

The people are shackled by poverty to heavy labor more surely than they were by slavery and serfdom. They could liberate themselves from those in one way or another, whereas they cannot free themselves from misery. We have included in constitutions rights which for the people are fictitious and are not actual rights. All the so-called 'rights of the people' can exist only in the abstract and can never be realized in practice. What difference does it make to the toiling proletarian, bent double by heavy toil, oppressed by his fate, that the babblers receive the right to talk, journalists the right to mix nonsense with reason in their writings, if the proletariat has no other gain from the constitution than the miserable crumbs which we throw from our table in return for his vote to elect our agents. Republican rights are bitter irony to the poor man, for the necessity of almost daily labor prevents him from using them, and at the same time deprives him of his guarantee of a permanent and certain livelihood by making him dependent upon strikes, organized either by his masters or by his comrades.

Under our guidance the people have exterminated aristocracy, which was their natural protector and guardian, for its own interests are inseparably connected with the well-being of the people. Now, however, with the destruction of this aristocracy the masses have fallen under the power of the profiteers and cunning upstarts, who have settled on the workers as a merciless burden.

We will present ourselves in the guise of saviors of the workers from this oppression when we suggest that they enter our army of Socialists, Anarchists, Communists, to whom we always extend our help, under the guise of the rule of brotherhood demanded by the human solidarity of our *social masonry*. The aristocracy which benefitted by the labor of the people by right was interested that the workers should be well fed, healthy, and strong.

We, on the contrary, are concerned in the opposite—in the degeneration of the GOYS. Our power lies in the chronic malnutrition and in the weakness of the worker, because through this he falls under our power and is unable to find either strength or energy to combat it.

Hunger gives to capital greater power over the worker than the legal authority of the sovereign ever gave to the aristocracy. Through misery and the resulting jealous hatred we manipulate the mob and crush those who stand in our way.

*When the time comes for our universal ruler to be crowned, the same hands will sweep away everything which may be an obstacle in our way.*

The GOYS are no longer accustomed to think without our scientific advice. Consequently, they do not see the imperative need of upholding that which we will sustain by all means when our kingdom is established, namely, the teaching in the schools of *the only true science, the first of all sciences—the science of the construction of human life, of social existence, which requires the division of labor and, consequently, the separation of people into classes and castes.* It is necessary that all should know that *equality cannot exist, owing to the different nature of various kinds of work;* that there cannot be the same responsibility before the law in the case of an individual who by his actions compromises an entire caste and another who does not affect anything but his own honor.

The correct science of the social structure, to the secrets of which we do not admit the GOYS, would demonstrate to all that occupation and labor must be differentiated so as not to cause human suffering by the discrepancy between education and work. The study of this science will lead the masses to a voluntary submission to the authorities and to the governmental system organized by them. Whereas, under the present state of science, and due to the direction of our guidance therein, the people, in their ignorance, blindly believing the printed word, and owing to the misconceptions which have been fostered by us, feel a hatred towards all classes whom they consider superior to themselves, since they do not understand the importance of each caste.

This hatred will be still more accentuated by the *economic crisis*, which will stop financial transactions and all industrial life. Having organized a general economic crisis by all possible underhand means, and with the help of gold which is all in our hands, we will throw great crowds of workmen into the street, simultaneously, in all countries of Europe. These crowds will gladly shed the blood of those of whom they, in the simplicity of their ignorance, have been jealous since childhood and whose property they will then be able to loot.

*They will not harm our people because we will know of the time of the attack and we will take measures to protect them.*

We have persuaded others that progress will lead the GOYS into a realm of reason. Our despotism will be of such a nature that it will be in a position to pacify all revolts by wise restrictions and to eliminate liberalism from all institutions.

When the people saw that they obtained concessions and license in the name of liberty, they imagined that they were the masters, and rushed into power; but like every blind person, they encountered innumerable obstacles; *they rushed to seek a leader, with no thought of returning to the old one,* and laid power at our feet. Remember the French Revolution, which we have called 'great'; the secrets of its preparation are well known to us, for it was the work of our hands.

Since then we have carried the masses from one disappointment to another, so that they will renounce even us in favor of *a despot sovereign of Zionist blood, whom we are preparing for the world.*

At present, as an international force, we are invulnerable, because if we are attacked by one state we are supported by other states. The unlimited baseness of the GOY peoples, who grovel before force, who are pitiless towards weakness, who are merciless to misdemeanors and lenient to crimes, who are unwilling to tolerate the contradictions of a free social structure; patient unto martyrdom in bearing with the violence of daring despotism—this is what helps our independence. They tolerate and permit such abuses from their modern premiers—dictators—for the least of which they would behead twenty kings.

How can such a phenomenon be explained, such an illogical conception on the part of the mass of the people towards events of seemingly the same nature? This phenomenon can be explained by the fact that these dictators through their agents whisper to their people that by these abuses they injure the states for a supreme purpose, namely, for the attainment of the happiness of the people, their universal fraternity, solidarity, and equality. Of course, they are not told that this unification will be achieved only under our rule. Thus, the people condemn the just and acquit the unjust, more and more convinced that they can do what they please. Owing to this, the people destroy all stability and create disorder on every occasion.

The word 'Liberty' brings all society into conflict with all authority, be it that of God or Nature. This is why, at the moment of our enthronement, we shall strike this word from the dictionary as being the symbol of brute power, which turns the masses into bloodthirsty beasts. It is true, however, that these beasts go to sleep as soon as they have drunk blood, and then it is easy to shackle them; but if the blood is not given to them they will not sleep and will struggle.

## PROTOCOL NO. IV

EVERY republic passes through several states. The first stage is like the early period of insane ravings of a blind man throwing himself right and left. The second is the demagogy which breeds anarchy, which inevitably leads to despotism, not of a legal and open character and, consequently, responsible, but an unseen and unknown despotism, no less effective because exercised br some secret organization, acting even less ceremoniously because it is hidden under the cover and behind the backs of different agents. The change of these agents will even help the secret organizations, as it will thus be able to rid itself of the necessity of spending money to reward employees of long terms of service.

Who and what can overthrow an unseen power? For such is the character of our power. *External Masonry [Footnote:* The reference is probably to those Masonic Lodges in Continental Europe which, contrary to the fundamental principles of Anglo-Saxon Lodges, have been converted into *quasi* political and anti-Christian organizations. See Encyclopedia Britannica, Eleventh Edition, Article 'Freemasonry,' Vol. XI, p. 84.] *acts as a screen for it and its aims, but the plan of action of this power, and its very headquarters, will always remain unknown to the people.*

Liberty could also be harmless and remain on the state program without detriment to the well-being of the people if it were to retain the ideas of the belief in God and human fraternity, free from the conception of equality for such a conception is in contradiction to the laws of nature which establish subordination. With such a faith the people would be governed by the guardians of the parish and would thrive quietly and obediently under the guidance of their spiritual leader, accepting God's dispensation on earth. It is for this reason that we must undermine faith, tearing from the minds of the GOYS the very principal of God and Soul, and substituting mathematical formulas and material needs.

In order that the minds of the GOYS may have no time to think and notice things, it is necessary to divert them in the direction of industry and commerce. Thus all nations will seek their own profit, and while engaged in the struggle they will not notice their common enemy. But in order that liberty should finally undermine and ruin the GOY'S society, it is necessary to put industry on a basis of speculation. The result of this will be that everything, absorbed by industry from the land, will not remain in the hands of the GOYS, but will be directed towards speculation; that is, it will come into our coffers.

The intense struggle for supremacy, the shocks to economic life, will create, moreover have already created, disappointed, cold, and heartless societies. These societies will have complete disgust for high politics and religion. Their only guide will be calculation, *i.e.*, gold, for which they will have a real cult because of the material delights which it can supply. It will be at that stage that the lower classes of the GOYS, not for the sake of doing good, nor even for the sake of wealth, but solely because of their hatred towards the privileged, will follow us against our competitors for power, the intelligent GOYS.

## PROTOCOL NO. V

WHAT form of government can be given to societies in which bribery has penetrated everywhere, where riches are obtained only by clever tricks and semi-fraudulent means, where corruption reigns, where morality is sustained by punitive measures and strict laws and not by voluntary acceptance of moral principles, where cosmopolitan convictions have eliminated patriotic feelings and religion? What form of government can be given to such societies other than a despotism such as I shall describe?

We will create a strong centralized government, so as to gather the social forces into our power. We will mechanically regulate all the functions of political life of our subjects by new laws. These laws will gradually eliminate all the concessions and liberties permitted by the GOYS. Our kingdom will be crowned by such a majestic despotism that it will be able, at all times and in all places, to crush both antagonistic and discontented GOYS.

We may be told that the despotism outlined by me is inconsistent with modern progress, but I will prove to you that the contrary is the case.

At the time when people considered rulers as an incarnation of the will of God, they subjected themselves without murmur to the autocracy of the sovereigns; but as soon as we inspired them with the thought of their personal rights, they began to regard the rulers as ordinary mortals. The holy anointment fell from the heads of sovereigns in the opinion of the people; and when we deprived them of their belief in God, then authority was thrown into the street, where it became public property and was seized by us. Moreover, the art of governing the masses and individuals by means of cunningly constructed theories and phraseology, by rulers of social life, and other devices not understood by the GOYS, belongs, among other faculties, to our administrative mind, which is educated in analysis and observation, and is also based upon skillful reasoning in which we have no competitors, just as we have none in the preparation of plans for political action and solidarity. Only the Jesuits could be compared to us in this; but we were able to discredit them in the mind of the senseless mob as a visible organization, whereas we, with our secret organization, remained in the dark. After all, is it not the same to the world who will be its master—whether it be the head of Catholicism or our despot of Zionist blood? To us, however, the Chosen People, it is by no means a matter of indifference.

Temporarily, a world coalition of the GOYS would be able to hold us in check, but we are insured against this by roots of dissension so deep among them that they cannot now be extracted. We have set at variance the personal and national interests of the GOYS: we have incited religious and race hatred, nurtured by us in their hearts for twenty centuries. Owing to all this, no state will obtain the help it asks for from any side because each of them will think that a coalition against us will be disadvantageous to it. We are too powerful—*we must be taken into consideration. No country can reach even an insignificant private understanding without our being secret parties to it.*

*Per me reges regnant*—'Through me the sovereigns reign.' The prophets have told us that we were chosen by God himself to reign over the world. God endowed us with genius to enable us to cope with the problem. Were there a genius in the opposing camp, he would struggle against us, but a newcomer is not equal to an old inhabitant. The struggle between us would be of such a merciless nature as the world has never seen before; moreover their genius would be too late.

All the wheels of government mechanism move by the action of the motor which is in our hands, and *that motor is gold*. The science of political economy, invented by our wise men, has long ago demonstrated the royal prestige of capital.

To attain freedom of action, capital must obtain freedom to monopolize industry and trade; this is already being done by an unseen hand in all parts of the world. Such liberty will give political power to traders, and will aid in subjugating the people. At present it is more important to disarm peoples than to lead them to war; it is more important to utilize flaming passions for our purposes than to extinguish them; more important to grasp and interpret the thoughts of others in our own way than to discard them.

*The most important problem of our government is to weaken the popular mind by criticism; to disaccustom it to thought, which creates opposition; to deflect the power of thought into mere empty eloquence.*

At all times both peoples and individuals have mistaken words for deeds, as they are satisfied with the visible, rarely noticing whether the promise is performed in the fields of social life.

Therefore, we will organize ostensible institutions which will prove eloquently their good work in the direction of 'progress.'

We will appropriate to ourselves the liberal aspect of all parties, of all shades of opinion, and we will provide our *orators with the same aspect, and they will talk so much that they will exhaust the people by their speeches and cause them to turn away from orators in disgust.*

*To control public opinion it is necessary to perplex it by the expression of numerous contradictory opinions until the* GOYS *get lost in the labyrinth, and come to understand that it is best to have no opinion on political questions.*

Such questions are not intended to be understood by the people, since only he who rules knows them. This is the first secret.

The second secret necessary for the success of governing consists in so multiplying popular failings, habits, passions, and conventional laws that no one will be able to disentangle himself in the chaos, and consequently, people will cease to understand each other. This measure would help us to sow dissension within all parties, to disintegrate all those collective forces which still do not wish to subjugate themselves to us; to discourage all individual initiative which might in any degree hamper our work.

*There is nothing more dangerous than individual initiative;* if it has a touch of genius it can accomplish more than a million people among whom we have sown dissensions. We must direct the education of the GOY societies so that their arms will drop hopelessly when they face every task where initiative is required. The intensity of action resulting from individual freedom of action dissipates its force when it encounters another person's freedom. This results in heavy blows at morale, disappointments and failures.

*We will so tire the* GOYS *by all this that we will force them to offer us an international power, which by its position will enable us conveniently to absorb, without destroying, all governmental forces of the world and thus to form a supergovernment.* In lieu of modern rulers, we will place a monster which will be called the Super-Governmental Administration. Its hands will be stretched out like pincers in every direction so that this colossal organization cannot fail to conquer all the peoples.

## PROTOCOL NO. VI

WE will soon begin to establish great monopolies—reservoirs of huge wealth, upon which even the large fortunes of the GOYS will depend to such an extent that they will be drowned, together with the governmental credits, on the day following the political catastrophe.

You economists, here present, will please carefully weigh the significance of this scheme! ...

We must develop, by all means, the importance of our supergovernment by representing it as the protector and reward-giver of all those who willingly submit to us.

*The aristocracy of the* GOYS *as a political force is dead. We do not need to take it into consideration; but as landowners they are harmful to its because they can be independent in their resources of life. For this reason we must deprive them of their land at any cost.*

To attain this object, the best method is to increase land taxes—the indebtedness of the land. These measures will keep land ownership in subjection.

The aristocracy of the GOYS, which as a matter of heredity is unable to be satisfied with small things, will soon be ruined.

At the same time it is necessary to patronize trade and industry vigorously, and more important, to encourage speculation, whose function is to act as a counterbalance to industry. Without speculation, industry will increase private capital and tend to the amelioration of land ownership by freeing it from indebtedness created by the loans granted by agricultural banks. It is necessary that industry should suck out of the land both labor and capital and through speculation deliver into our hands all the money of the world, thus throwing all the GOYS into the ranks of the proletarians. Then the GOYS will bow before us in order to obtain the mere right of existence.

To destroy GOY industry we will create among the GOYS as an aid to speculation the strong demand for boundless luxury which we have already developed.

*Let its raise wages, which, however, will be of no benefit to the workers, for we will simultaneously cause the rise in prices of objects of first necessity under the pretext that this is due to the decadence of agriculture, and of the cattle industry.*

*We will also artfully and deeply undermine the sources of production by teaching the workmen anarchy and the use of alcohol, at the same time taking measures to expel all the intelligent GOYS from the land.*

*That the true situation should not be noticed by the GOYS until the proper time, we will mask it by a pretended desire to help the working classes and great economic principles, an active propaganda of which principles is being carried on through the dissemination of our economic theories.*

## PROTOCOL NO. VII

THE intensification of armament and the increase of the police force are essential to the realization of the abovementioned plans. It is necessary that there should be besides ourselves in all countries only the mass of the proletariat, a few millionaires devoted to us, policemen, and soldiers.

We must create unrest, dissensions, and hatred throughout Europe and through European affiliations, also on other continents. In this there is a twofold advantage: First, we will hold all countries under our influence, since they will realize that we have the power to create disorders or to restore order whenever we wish. All countries have come to regard us as a necessary burden. Second, we will entangle by intrigues all the threads stretched by us into all the governmental bodies by means of politics, economic treaties, or financial obligations. To attain these ends we will worm our way into parleys and negotiations, armed with cunning, but in so-called 'official language' we will assume the opposite tactics of seeming honest and reasonable. In this way the peoples and the governments of the GOYS, taught by us to regard only the surface of that which we show them, will look upon us as benefactors and saviors of mankind.

*We must be able to overcome all opposition by provoking* a war by the neighbors of that country which dares to oppose us. Should, however, those neighbors, in their turn, decide to unite against us we must respond by a world war.

Chief success in politics lies in the secrecy of its undertakings. There must be inconsistency between the words and actions of diplomats.

We must influence the GOY governments to action beneficial to our broadly conceived plan, now approaching its triumphant goal, creating the impression that such action is demanded by public opinion which in reality is secretly organized by us with the help of the so-called 'great power,' namely, the press; the latter, however, with few exceptions that need not be considered, is already entirely in our hands.

In short, to sum up our system of shackling the GOY governments of Europe, we will show our power to one of them by assassination and terrorism, and should there be a possibility of all of them rising against us, we will answer them with American, Chinese, or Japanese guns.

### PROTOCOL NO. VIII

WE must provide ourselves with the same arms our enemies can employ against us. We must seek the most subtle expressions and evasions of the legal dictionary to justify those cases in which we will be forced to announce decisions which may seem unnecessarily bold and unjust, for it is important that these decisions should be expressed in terms so forcible that they will appear as the highest moral rules of a legal character.

Our government must be surrounded by all the forces of civilization, in the midst of which it will have to function. It will surround itself with publicists, experienced lawyers, administrators, diplomats, and, finally, people educated along special lines in our special advanced schools.

These people will know all the secrets of social existence; they will know all languages composed of political letters and words; they will be familiar with the reverse side of human nature, with all its sensitive chords, upon which they must know how to play. These chords are the structure of the intellects of the GOYS, their tendencies, their failings, their vices, and their virtues, the peculiarities of classes and castes. It is evident that the highly talented members of our government, to which I refer, will be recruited not from the ranks of the GOYS, accustomed to performing their administrative duties without questioning their aim, and without thinking why they are necessary. The GOY administrators sign papers without reading them and work for profit or for pride.

We will surround our government by a whole world of economists. It is for this reason that economics is the chief science taught to the Jews. We will be surrounded by a crowd of bankers, traders, capitalists, *and most important of all, by millionaires, because in essence everything will be decided by a question of figures.*

Meanwhile, as it is not yet safe to give the responsible government posts to our brother Jews, we will give them to people whose record and whose character are such that there is an abyss between them and the people; also to people for whom, in case of disobedience to our orders, there will remain nothing but condemnation or exile—thus forcing them to protect our interests to their last breath.

## PROTOCOL NO. IX

IN applying our principles, turn your attention to the character of the people in whose countries you will be resident and among whom you will act, for a general similar application of them before the reëducation of a people according to our plan cannot be successful. But by advancing carefully in their application you will see that before ten years have passed the most obstinate character will have changed, and we can then count another people among those who already have submitted to us.

When we are enthroned we will substitute for the liberal words of our Masonic catchword, ' Liberty, Equality, and Fraternity,' another group of words expressing simply ideas, namely, 'the right of Liberty, the duty of Equality, the ideal of Fraternity.' Thus we will speak and... we shall have the goat by the horns... . *De facto,* we have already destroyed all governments except our own, although *de jure* there are still many left. At present, if any of the governments raises a protest against us, it is done only as a matter of form, and at our desire, and by our order, because *their anti-Semitism is necessary to enable us to control our smaller brothers.* I will not further explain this, as it has already been the object of numerous discussions.

In reality there are no obstacles before us. Our supergovernment exists under such extra-legal conditions that it is common to designate it by an energetic and strong word—a Dictatorship.

I can honestly state that at the present time we are lawmakers; we are the judges and inflict punishment; we execute and pardon; we, as the chief of all our armies, ride the leader's horse. We rule by indomitable will because we hold in our hands the fragments of a once strong party now subject to us. We possess boundless ambition, burning greed for merciless revenge, and bitter hatred.

*From us emanates an all-embracing terror. People of all opinions and of all doctrines are in our service; people who desire to restore monarchies, demagogues, socialists, communists, and other utopians.* We have had to put all of them to work; every one of them is undermining the last remnant of authority, is trying to overthrow all existing order. All the governments have been tortured by this procedure; they beg for peace, and for the sake of peace are prepared to make any sacrifice, but we will not give them peace until they recognize our international super-government openly and with submission.

The masses have begun to demand the solution of the social problem by means of an international agreement. *The division into parties has delivered all of them to us, because in order to conduct a party struggle money is required, and we have it all.*

We might fear the union of the intelligent power of the GOYS' rulers with the blind power of the masses, but we have taken all measures against such a possibility. Between the two powers we have raised a wall in the form of mutual terror; thus the blind power of the people continues to be our support, and we alone will act as its leader and, naturally, we will direct it towards our goal.

To prevent the hand of the blind from freeing itself from our guidance, we must from time to time keep in close touch with the masses, if not through personal contact then through our most devoted brethren. When we become a recognized power we will personally address the masses in open places, and we will expound political problems in the desired direction.

How verify what is taught in village schools? But whatever the representative of the government or the ruler himself states will be immediately known to the entire nation, for it will rapidly spread by the voice of the people.

In order not prematurely to destroy GOY institutions, we have touched them with our efficient hands and grasped the ends of the springs of their mechanism. Formerly these springs were in rigid but just order; we have changed it to liberal, disorderly, and arbitrary lawlessness.

We have affected legal procedure, electoral law, the press, personal freedom, and, most important, education, the cornerstone of free existence.

*We have misled, corrupted, fooled, and demoralized the youth of the GOYS by education along principles and theories known by us to be false but which we ourselves have inspired.*

Without changing substantially the existing law we have created stupendous results by distorting the laws through contradictory interpretations. These results first manifested themselves by the fact that interpretation has concealed the law itself, and thereafter has completely hidden it from the eyes of the governments by the impossibility of understanding such complicated jurisprudence.

Hence the theory of the court of conscience. [*Footnote:* This probably means the practice which arose of not adhering to the letter of the law but of judging by conscience. In European countries jurors are not compelled to render their verdict pursuant to the technical provisions of law.]

You may say that there will be an armed rising against us if our plans are discovered prematurely; but in anticipation of this we have such a terrorizing manoeuver in the West that even the bravest soul will shudder.

Underground passages will be established by that time in all capitals, from where they can be exploded, together with all their institutions and national documents.

## Protocol No. X

TODAY I will begin by reiterating what has already been stated. *I beg you to remember that the government and the masses are satisfied with visible results in politics.* How can they examine the inner meaning of things when their representatives consider that pleasure is above everything? It is important to know one detail in our policy. It will help us in discussing division of authority, freedom of speech, of the press, of religion (faith), the right of assembly, equality before the law, inviolability of property and of the home, indirect taxes and the retrospective force of law. All such questions should never be directly and openly discussed before the masses. When it becomes necessary for us to discuss them, they should not be elaborated but merely mentioned, without going into details, pointing out that modern legal principles are being accepted by us. The significance of this reticence lies in the fact that a principle which has not been openly declared gives us freedom of action to exclude unnoticed one point or another, whereas if elaborated the principle becomes as good as established.

The people feel an especial love and admiration towards the political genius, and they always react to their acts of violence as follows:

'Yes, of course it is villainy, but how clever!—It is a trick but cleverly done! So majestically! so impudently! ... '

We count upon attracting all nations to the construction of the foundations of the new edifice which has been planned by us. It is for this reason that it is necessary for us first of all to acquire that spirit of daring, enterprise, and force which, through our agents, will enable us to overcome all obstacles in our path.

*When we accomplish our coup d'état, we will say to the peoples: 'Everything went badly; all of you have suffered. We will abolish the cause of your sufferings, that is to say, nationalities, frontiers, and national currencies. Of course you are free to condemn us, but would your judgment be just if you were to pronounce it before giving a trial to what we will give you?' Thereafter they will exalt us with a sentiment of unanimous delight and hope. The voting system which we have used as a tool for our enthronement, and to which we have accustomed even the most humble members of humanity by organizing meetings and prearranged agreements, will have performed its last service and will make its last appearance in the expression of unanimous desire to become more closely acquainted with us before having pronounced a judgment.*

To attain this we must force all to vote, without class discrimination, to establish the autocracy of the majority, which cannot be obtained from the intellectual classes alone. Through this method of accustoming every one to the idea of self determination, we will shatter the GOY family and its educational importance. We will not allow the formation of individual minds, because the mob, under our guidance, will prevent them from distinguishing themselves or even expressing themselves. The mob has become accustomed to listen only to us who pay it for obedience and attention. We will thus create such a blind power that it will be unable to move without the guidance of our agents, sent by us to replace their leaders.

The masses will submit to this régime because they will know that their earnings, perquisites, and other benefits depend upon these leaders.

The plan of government must emanate already formed from one head, as it would be impossible to put it together if disintegration by many minds into small pieces is allowed. That is why we only are allowed to know the plan of action; but we must not discuss it in order not to affect its ingenuity, the correlation between its component parts, the practical force of the secret meaning of its every clause. Were such a plan to be submitted to and altered by frequent voting, it would reflect the stamp of the misconceptions of every one who has not penetrated its depth and the correlation of its aims. For this reason our plans must be strongly and clearly conceived. Consequently, the inspired work of our leader must not be thrown to the mercy of the mob or even of a limited group.

These plans will not immediately upset contemporary institutions. They will only alter their organization, and consequently the entire combination of their development, which will thus be directed according to the plans laid down by us.

More or less the same institutions exist in different countries under different names, such as representative bodies, ministries, senate, state council, legislative and executive bodies. It is not necessary for me to explain to you the connecting mechanism of these different institutions, as it is well known to you. I only call to your attention that every one of the aforesaid institutions fulfills some important governmental function, and, moreover, I beg you to notice that the word 'important' refers not to the institution but to the function. Consequently, it is not the institutions that are important but their functions. Such institutions have divided among themselves all the functions of government, namely, administrative, legislative, and executive powers; therefore, their functions in the state organism have become similar to those in a human body. If one part of the governmental machine is injured, the state itself falls ill, in the same way as the human body, and then it dies.

When we injected the poison of liberalism into the state organism, its entire political complexion changed; the states became infected with a mortal disease, namely, the decomposition of the blood. It is only necessary to await the end of their agony.

Constitutional governments were born of liberalism, which replaced the autocracy that was the salvation of the GOYS, for the constitution, as you well know, is nothing more than a school for dispute, discussion, disagreement, fruitless party agitation, dissension, party tendencies—in other words, a school for everything which weakens the efficiency of government. The platform no less than the press condemned the authorities to inaction and impotency and thereby rendered them useless and superfluous, for which reason they were overthrown in many countries. The rise of the republican era then became possible, and then we substituted for the ruler a caricature of government—a president chosen from the mob, from among our creatures, our slaves. This was the kind of mine we laid under the GOYS, or, more correctly, under the GOY nations.

In the near future we will make the president a responsible officer, whereupon we will no longer stand on ceremony in carrying out the things for which our dummy will be responsible. What difference does it make to us that the ranks of those aiming at authority will thin out, that confusion will result from inability to find presidents, confusion which will definitely disorganize the country?

To accomplish our plan, we will engineer the election of presidents whose past record contains some hidden scandal, some 'Panama'—then they will be faithful executors of our orders from fear of exposure, and from the natural desire of every man who has reached authority to retain the privileges, advantages, and dignity connected with the position of president. The Chamber of Deputies will elect, protect, and screen presidents, but we will deprive it of the right of initiating laws or of amending them, for this right will be granted by us to the responsible president, a puppet in our hands. Of course then the power of the president will become the target of numerous attacks, but we will give him the means of self-protection by giving him the right of directly applying to the people, for their decision, over the heads of their representatives. In other words, he will turn to the same blind slave—to the majority of the mob. Moreover, we will empower the president to proclaim martial law. We will justify this prerogative under the pretext that the president, as chief of the national army, must control it in order to protect the new republican constitution, which he, as a responsible representative of this constitution, is bound to defend.

It is obvious that under such conditions the keys to the shrine will be in our hands, and nobody except ourselves will be able to guide the legislative power.

We will also take away from the Chamber, with the introduction of the new republican constitution, the right of interpellation in regard to governmental measures, under the pretext that political secrets must be preserved. With the aid of this new constitution we will reduce the number of representatives to the minimum, thus also reducing to the same extent political passions and passion for politics. If, in spite of this, those remaining are recalcitrant, we will abolish them completely by appealing to the majority of the people.

The appointment of the president and vice presidents of the Chamber and Senate will be the prerogative of the president. Instead of continuous parliamentary sessions, we will shorten them to a few months. Moreover, the president, as chief executive, will have the right to convene or dissolve parliament, and in the case of dissolution, defer the appointment of a new parliament. But to prevent the president from being held responsible before our plans are matured for the results of all these essentially illegal actions inaugurated by us, we will give the ministers and other high administrative officials surrounding the president the idea of circumventing his orders by issuing instructions of their own. Consequently, they will be made responsible instead of him. We recommend that the execution of this plan be given especially to the Senate, State Council, or Council of Ministers, and not to individuals. Under our guidance the president will interpret in ambiguous ways such existing laws as it is possible so to interpret. Moreover, he will annul them when the need is pointed out to him by us: he will also have the right to propose temporary laws and even modifications in the constitutional work of government, alleging as the motive for so doing the exigencies of the welfare of the country.

By such measures we will be able to destroy gradually, step by step, everything that, upon entering into our rights, we were obliged to introduce into government constitutions as a transition to the imperceptible abolition of all constitutions, when the time comes to convert all government into *our autocracy*.

The recognition of our autocrat may come even before the abolition of the constitution; the moment for this recognition will come when the people, tormented by dissension and the incompetency of their rulers, incited by us, will exclaim: Depose them, and give us one universal sovereign who will unite us and abolish the causes of dissension—national frontiers, religion, state indebtedness—and who will give us the peace and quiet which we cannot find with our rulers and representatives.

But you know well that to render such a universal expression of desire possible, it is necessary continuously to disturb the relationship between the people and the government in all countries, and so to exhaust everybody by the dissension, hostility, struggle, hatred, and even martyrdom, hunger, inoculation of diseases, and misery, as to make the GOYS see no other solution than an appeal to our money and complete rule.

Should we give the people a rest, however, the longed for moment will probably never arrive.

## PROTOCOL NO. XI

THE Council of State will tend to accentuate the power of the ruler; in the capacity of an ostensible legislative body, it will act as a committee for the drawing up of laws and statutes on behalf of the ruler.

The following is the program of the new constitution which we are preparing. We will make laws and control the courts in the following manner:

1. By suggestions to the legislative body.

2. By means of orders issued by the president as general statutes, decrees of the Senate, and decisions of the Council of State, as regulations passed by the ministries.

3. And when the opportune moment arrives—in the form of a *coup d'état*.

Having thus roughly outlined the *modus agendi,* we will now take up in detail those measures by which we will complete the development of the governmental mechanism in the above direction. By these measures, I mean the freedom of the press, the right of assembly, religious freedom, electoral rights, and many other things which must disappear from the human repertoire, or must be fundamentally altered on the day following the declaration of the new constitution. It is only at this moment that it will become possible for us to announce all our decrees, for at any time in the future every perceptible change would be dangerous, and this for the following reasons: If these changes should be introduced and rigidly enforced, it might cause despair by creating the fear of further changes in a similar direction; if, however, they are made with a tendency to subsequent leniency, then it might be said that we have recognized our mistakes, which would undermine the faith in the infallibility of the new authority; it might also be said that we were frightened, and that we were forced to make concessions for which nobody would be thankful since they would be considered as legitimately due.

Any of these impressions would be detrimental to the prestige of the new constitution. It is necessary for us that, from the first moment of its proclamation, when the people are still dumbfounded by the accomplished revolution and are in a state of terror and surprise, they should realize we are so strong, so invulnerable, and so mighty that we shall in no case pay attention to them, and not only will we ignore their opinions and desires, but be ready to and capable of suppressing at any moment or place any sign of opposition with indisputable authority. We shall want the people to realize, that we have taken at once everything we wanted, and that we shall under no circumstances share our power with them. Then they will close their eyes to everything out of fear and will await further developments.

The GOYS are like a flock of sheep—we are wolves.

Do you know what happens to sheep when wolves get into the fold?

They will also close their eyes to everything because we will promise to return to them all their liberties after the enemies of peace have been subjugated and all the parties pacified.

Is it necessary to say how long they would have to wait for the return of their liberties?

Why have we conceived and inspired this policy for the GOYS without giving them an opportunity to examine its inner meaning if not for the purpose of attaining by a circuitous method what is unattainable for our scattered race by a direct road?

This constituted a base for our organization of s*ecret masonry which is not known to and whose aims are not even suspected by these cattle, the GOYS. They have been decoyed by us into our numerous ostensible organizations, which appear to be Masonic lodges, so as to divert the attention of their coreligionists.*

God has given us, his chosen people, the power to scatter, and what to all appears to be our weakness, has proved to be our strength, and has now brought its to the threshold of universal rule.

Little remains to be built on these foundations.

## PROTOCOL NO. XII

THE word 'Liberty' can be differently interpreted. We will define it as follows:

Liberty is the right to do that which is permitted by law. Such a definition of this word will eventually serve us, because liberty will be in our power; and also because the laws will either destroy or construct only what we desire in accordance with the above mentioned program.

We will deal with the press in the following manner: What is the present rôle of the press? It serves to arouse furious passions or egotistic party dissensions which may be necessary for our purpose. It is empty, unjust, inaccurate, and most people do not understand what end it serves. We will shackle it and keep a tight rein on it. We will also do the same with other printed matter, for what use would it be for us to rid ourselves of attacks on the part of the periodical press if we remain open to criticism through pamphlets and books? We will convert the products of publicity, now so expensive, owing to the need of censorship, into a source of income for our state. We will impose a special stamp tax. When a newspaper printing shop is started, bonds will have to be deposited, which will guarantee our government from all attacks on the part of the press. In case of an attack, we will mercilessly impose fines. Such measures as stamps, bonds, and fines, the payment of which is guaranteed by the bonds, will bring a huge income to the government. It is true that party papers might not fear the loss of money, so we will suppress these after the second attack on us. No one shall touch the prestige of our political infallibility and remain unpunished. The pretext for stopping a publication will be that the publication in question excites public opinion without cause or reason. *I ask you to bear in mind that among those who attack us there will be also organs established by us, but they will attack exclusively those points which we plan to change.*

*Not one notice will be made public without our control.* This is already being done by us, since the news from all parts of the world is received through several agencies in which it is centralized.

These agencies will then be completely in our power and they will publish only such news as we will permit.

If we have already managed to subjugate the minds of the GOYS to such an extent that almost all of them see world events through colored glasses which we put over their eyes; if, even at present, there is not one state which bars our access to state secrets, so termed by the stupid GOYS, then what will it be when we, in the person of our universal sovereign, are the recognized rulers of the world?

Let us return to the future of the press. Anybody who wishes to become an editor, a librarian, or a printer, will be obliged to obtain a diploma, which in case of disobedience will be immediately revoked.

With such measures, *thought will become an educational instrument in the hands of our governmentt, which will not allow the people to be led astray into realms of fancy and dreams about beneficent progress.* Who of us does not know that these fantastic blessings are the direct road to baseless hopes which lead to anarchistic relations between the people and the government? Progress, or better still the idea of progress, has led to the creation of different modes of emancipation without setting any limit to it.

All so-called liberals are essentially anarchists in thought if not in action. Each one of them pursues the phantom of liberty, becoming self-willed, that is to say, falling into a state of anarchy by protesting for the mere sake of protesting.

We will now again refer to the question of the press. We will place stamp taxes secured by bonds on each page of all printed matter, while on books containing less than four hundred and eighty pages we will place a double tax. We will classify them as pamphlets, so as to lessen the number of magazines, which represent the worst printed poison—and on the other hand, to force writers to prepare such long works that they will be little read, especially as they will be expensive. Our own publications, guiding public opinion in the direction we desire, will be cheap and rapidly bought. The tax will discourage the writing of mere leisure literature, whereas punishment will make the writers dependent upon us. Even if there were writers who would like to attack us, they would find no publishers for their works. Before printing any work, the editor or printer will have to apply to the authorities for permission. We will then know beforehand of the attacks that are being prepared against us, and we will destroy them by coming out with advance statements on the subject.

Literature and journalism are the two most important educational forces; for this reason our government will become the owner of most of the periodicals. This will neutralize the injurious influence of the private press and have great influence on the people. If we permit ten periodicals, we ourselves will print thirty, and so forth. This, however, must not be suspected by the public. All the periodicals published by us will seem to be of contradictory views and opinions, inviting trust in us, thus attracting to us unsuspecting enemies, and in this way they will be caught in our trap and made harmless.

The predominant place will be held by periodicals of an official character. They will always stand guard over our interests and consequently their influence will be comparatively limited.

In the second category we will place semi-official organs, whose aim will be to attract the indifferent and little interested.

The third category will be our ostensible opposition, which at least in one of its publications will represent the opposition to us. Our real enemies will mistake this seeming opposition as belonging to their own group and will thus show us their cards.

All our newspapers will represent different tendencies, namely, aristocratic, republican, revolutionary, even anarchistic, so long of course as the constitution lasts. Like the Indian God VISHNU, these periodicals will have one hundred arms, each of which will reach the pulse of every group of public opinion. When the pulse beats faster, these arms will guide opinion toward our aims, since the excited person loses the power of reasoning and is easily led. Those fools who believe that they repeat the opinions expressed by the newspapers of their party will be repeating our opinions or those which we desire them to have. Imagining that they are following the press of their party, they will follow the flag which we will fly for them.

In order that our newspaper militia may carry out our program, we must organize the press with great care. Under the title of the Central Department of the press, we will organize literary meetings at which our agents unnoticed will give the passwords and countersigns. Discussing and contradicting our policies, although always superficially, without touching their essence, our press will conduct an empty fire against official newspapers so as to give us only an opportunity to express ourselves in greater detail than we were able to in our preliminary declarations. This, of course, will be done when it is useful to us.

*These attacks against us will also seem to convince the people that complete liberty of the press still exists, and it will give our agents the opportunity to declare that the papers opposing us are mere wind-bags,* since they are unable to find any real ground to refute our orders.

Such measures, which will escape the notice of public attention, will be the most successful means of guiding the public mind and of inspiring confidence in our government. Thanks to them, we will as the need arises excite or pacify the public mind on political questions. We will be able to persuade or confuse them, sometimes printing the truth, sometimes lies, referring to facts or contradicting them according to the way they are received by the public, always carefully sounding the ground before stepping on it. *We will surely conquer our enemies, because they will not have the press at their disposal in which to express themselves in full.* Moreover, with the above mentioned plans against the press, we will not even need to refute them seriously.

The trial balloons thrown out by us in the third category of our press, we will deny energetically, in case of need, in our semi-official organs.

In French journalism there already exists the Masonic solidarity of a password; all organs of the press are bound by professional secrecy; like the ancient augurs, not one member will disclose his secret if he is not ordered to do so. Not one journalist will dare to disclose this secret, for not one of them is admitted to literary headquarters unless he has a disgraceful action in his past record. The fact would immediately be made public. While these disgraceful actions are known only to a few, the prestige of the journalist attracts opinion throughout the country—he is admired.

Our plans must extend chiefly to the provincial districts. There we must excite hopes and ambitions opposed to those of the capitals, by means of which we may always attack them, presenting such ambitions to the capitals as the inspired views and aims of provincial districts. It is obvious that their source will be ours. It is necessary for us that while we are not yet in full power, the capital should be under the influence of provincial public opinion; that is under the influence of the majority prearranged by our agents. It is necessary for us that at the critical psychological moment the capitals should not discuss an accomplished fact, for the mere reason that it had been accepted by the provincial majority.

*When we reach the phase of the new régime, which is transitory to our accession to power, we must not allow the press to expose social corruption. It must be thought that the new régime has satisfied everybody to such an extent that even criminality has stopped.* Cases of criminal activity must only be known to their victims or their accidental witnesses, and to these alone.

### PROTOCOL NO. XIII

THE need of daily bread forces the GOYS to silence and compels them to remain our obedient servants. The agents taken from among them for our press will discuss the facts they are ordered to publish, when it is inconvenient for us to publish statements openly in official documents. While discussion and dispute are taking place, we will simply pass the measures we desire and present them to the public as an accomplished fact. Nobody will dare to demand the rejection of measures thus passed, and the more so as they will be interpreted as an improvement. At this point the press will divert the thoughts of the people to new problems (we having accustomed the people always to seek new emotions). Those brainless creators of destiny, who heretofore have been unable to understand and do not now understand that they are ignorant of matters which they undertake to discuss, will also hasten to discuss these new problems. Political questions are meant to be understood only by those who have created them and have been directing them for many centuries.

From all this you will realize that by aiming to control the opinion of the mob we will only facilitate the functioning of our mechanism, and you will also notice that we seek approbation, not for actions but for words uttered by us on various occasions. We always declare that we are guided in all our policies by the hope and certainty of serving the general good.

To divert the over-restless people from discussing political problems, we now make it appear that we provide them with new problems, namely, those pertaining to industry. Let them become excited over this subject as much as they like. The masses will consent to remain inactive, to rest from so-called political activity (to which we ourselves accustomed them for the purpose of helping us in our struggle against the GOY government), only on condition of a new occupation in which we can show them supposedly the same political background.

To prevent them from reaching any independent decisions, *we will divert their minds by amusements, games, pastimes, passions, and cultural centers for the people.* We will soon begin to offer prize contests, through the press, in the field of art, and sports of all kinds. Such attractions will definitely deflect the mind from problems over which we would otherwise have to fight with the people. By losing more and more the custom of independent thought, they will begin to talk in unison with us, because we alone will provide new lines of thought through persons with whom of course we will presumably have no connection.

The rôle of liberal Utopians will be definitely terminated when our government is recognized. Until that time, they will do us good service. For this reason we will still direct thought towards different fantastic theories which will appear to be progressive. For it was by the word 'progress' that we have successfully turned the brains of the stupid GOYS. There are no brains among the GOYS to realize that this word is but a cover for digression from the truth, unless it is applied to material inventions, *since there is but one truth and there is no room for progress*. Progress, being a false conception, serves to conceal the truth so that nobody may know it except ourselves, God's elect, who are its guardians.

When our kingdom is established, our orators will discuss the great problems which have stirred humanity for the purpose of bringing it finally under our blessed rule.

Who will then suspect that all those problems were instigated by us, according to a political plan which has not been disclosed by any one during so many centuries.

## PROTOCOL NO. XIV

WHEN we become rulers we will not tolerate the existence of any other religion except our own, which proclaims one God, with whom our fate is bound up because we are the Chosen People, and our fate has determined the fate of the world. For this reason we must destroy all other religions. If the result of this produces modern atheists, as a transitory step, this will not interfere with our plans but will act as an example to those generations which will listen to our teaching of the religion of Moses, which, owing to its solid and thoughtful system, will eventually lead to the domination of all nations by us. We will also lay stress on the mystical truth of Masonic teaching which, we will assert, is the foundation of its whole educative power.

On every possible occasion we will then publish articles in which we will compare our beneficial rule with that of the past. The benefits of peace, although attained through centuries of unrest, will serve to demonstrate the beneficial character of our rule. The mistakes made by the GOYS during their administration will be pictured by us in the most vivid colors. We will cause such disgust towards the administration of the GOYS that the masses will prefer the peace of serfdom to the rights of the much lauded liberty which has so cruelly tortured them and drained from them the very source of human existence, and by which they were exploited by a mass of adventurers, ignorant of what they were doing. *The useless changes of government, to which we ourselves prompted the GOYS, when we were undermining their governmental apparatus, will become such a nuisance to the people by that time, that they will prefer to endure anything from us rather than risk a repetition of former unrest and hardships.* We will, moreover, lay particular stress on the historical mistakes made by the GOY governments, which caused humanity to suffer for many centuries for lack of understanding of all matters pertaining to its true welfare, and because of their search for fantastic schemes of social welfare. The GOYS did not notice that such schemes instead of improving mutual relationship, which is the basis of human existence, have only made it worse.

The whole force of our principles and measures will lie in the fact that they are put forward and interpreted by us as being in sharp contrast to the decayed social order of former times.

Our philosophers will discuss all the shortcomings of the GOY religion, but nobody will ever discuss our religion in the light of its true aspect, and nobody will ever thoroughly understand it, except our own people, who will never dare to disclose its secrets.

*In countries so-called advanced we have created insane, dirty, and disgusting literature.* For a short time after our entrance into power we will encourage its publication in order that the contrast between it and the speeches and programs which will be heard front our heights should be more pointedly marked. Our wise men, trained as guides to the GOYS, will prepare speeches, plans, memoranda, and articles, by which we will influence the minds and direct them towards the conceptions and the knowledge which we wish them to have.

### PROTOCOL NO. XV

WHEN we finally become rulers by means of revolutions, which will be arranged so that they shall take place simultaneously in all countries and immediately after all existing governments shall have been officially pronounced as incapable (which may not happen soon, perhaps not before a whole century), we will see to it that no plots are hatched against us. To effect this, we will kill heartlessly all who take up arms against the establishment of our rule.

The establishment of any new secret society will be met by the death penalty, and those societies which now exist and are known to us and either work or have worked for us, will be disbanded and their members exiled to continents far removed from Europe.

We will deal in the same manner with those Masons among the GOYS who know too much. The Masons whom we may pardon for any reason will be kept under continual fear of exile. We will pass a law whereby all members of secret organizations will be exiled from Europe, that being the center of our government. The decisions of our government will be final and there will be no right of appeal.

In the GOY society, where we have planted such deep roots of dissension and protest, order can only be restored by merciless measures which will serve as evidence that our power cannot be infringed. There is no necessity for regard towards the victims sacrificed for the future good. To attain good, even though by the sacrifice of life, is the duty of every government which realizes that its existence depends not upon privileges alone, but upon the exercise of its duties as well.

The most important means for erecting a stable government is to strengthen the prestige of authority. This is only obtained by its majestic and unshakable power, which will convey the impression that it is inviolable because of its mystical nature, namely, because chosen by God. *Such until recently has been the Russian Autocracy—our only dangerous enemy throughout the world, with, the exception of the Pope.* Remember Italy drowning in blood; she did not touch a hair on the head of Sulla who had shed that blood. Sulla had become powerful in the eyes of the people, although they were tortured by him; his manly return to Italy placed him beyond persecution. The people do not touch those who hypnotize them by bravery and steadfastness of spirit.

Meanwhile, until our rule is established, we, on the contrary, will organize and multiply free masonic lodges in all the countries of the world. We will attract to them all those who are and who may become public-spirited, because in these lodges will be the chief source of information and from them will emanate our influence.

All these lodges will be centralized under one management, known only to us and unknown to all others; these lodges will be administered by our wise men. The lodges will have their own representative in this management in order to screen the above mentioned Masonic government; he will give the password and elaborate the program. We will tie the knot of all revolutionary liberal elements in these lodges. Their membership will consist of all strata of society. The most secret political plans will be known to us and will fall under our leadership on the very day of their origination. *Among the members of these lodges will be almost all the agents of the international and national police,* whose work is indispensable for us, inasmuch as the police not only are able to take independent measures against the rebellious, but may also serve to mask our actions, provoke discontent, and so forth.

Most people who become members of secret societies are adventurers, career makers, and irresponsible persons in general, with whom we will have no difficulty in dealing and who will help us to set in motion the mechanism of the machine planned by us. If this world becomes perturbed, it will only prove that it was necessary for us to disorganize it so as to destroy its too great solidarity. *If a plot is laid, it must be headed by one of our most trustworthy servants.* It is only natural that we want nobody but ourselves to guide the work of the Masons,

[*Footnote:* It is important to point out that *some of the Jews themselves* in their writings have claimed that Masonry is largely controlled by Jewish influence. In this connection the statement of Dr. Isaac M. Wise may be recalled: 'Masonry is a Jewish institution whose history, decrees, charges, passwords and explanations are Jewish, from the beginning to the end, with the exception of only one by-decree and a few words in the obligation.' (Dr. Isaac M. Wise, *The Israelite*, August 3rd and 17th, 1855; quoted by Samuel Oppenheim in his pamphlet 'Jews and Masonry in the United States before 1810,' American Jewish Historical Society, New York, 1910 No. 19, pp. 1, 2.)]

for we know where we are trending, we know the final aim of every action. The GOYS, however, understand nothing, not even the immediate results. They are usually concerned about the momentary satisfaction of their ambitions in achieving their intentions. They do not notice, however, that the intention itself was not initiated by them, but that it was we who gave them the idea.

The GOYS become members of the lodges out of pure curiosity, or hoping to receive their share in the public funds. There are others who come for the purpose of seizing the opportunity of putting before the public their impossible and baseless hopes. They long for the emotion of success and for the applause which we grant them lavishly. We create their success in order to utilize the self-deception that is born with it and by which people, without noticing, begin to follow our suggestions without suspecting them, and being fully convinced that their infallibility originates its own ideas and, therefore, does not need those of others. You have no idea how easy it is to bring even the most intelligent GOYS to a state of unconscious credulity, and, on the other hand, how easy it is to discourage them by the smallest failure, or merely by ceasing to applaud them, thus bringing them into servitude for the sake of achieving new success. *To the same extent as our people ignore success for the sake of carrying out their plans, so are the GOYS ready to sacrifice all their plans for the sake of success.* Their psychology makes the problem of direction easier for us. Those tigers in appearance have the souls of sheep and nonsense filters through their heads. As a hobby we have given them the dream of submerging human individualism through the symbolic idea of *collectivism*.

They have not yet discovered and will not discover that this hobby is a clear infringement on the principal law of nature, which, from the beginning of the world, created a being unlike all others, precisely for the sake of expressing his individuality.

If we were able to lead them to such insane and blind beliefs, does it not obviously prove the low level of development of the GOY mind as compared to our mind? It is precisely the thing which guarantees our success.

How far sighted were our wise men of old when they said that to attain a serious object one must not stop at the means, nor should one count the victims sacrificed to the cause. We have not counted the victims from among the GOYS, those seeds of cattle. Although we have sacrificed many of our own peoples, we have already given them in return a formerly undreamed-of position on earth. The comparatively few victims from among our own people have saved our race from destruction.

Death is the unavoidable end of all. It would be better to accelerate this end for those who interfere with our cause than for our people or for us, ourselves, the creators of this cause to die. *We kill Masons in such a way that none but the brothers suspect, not even the victims; they all die when it is necessary, apparently from a natural death.* Knowing this, even the brethren, in their turn, dare not protest. It is through such measures that we have uprooted the heart of protest against our orders from among the Masons. Preaching liberalism to the GOYS, at the same time we hold our people and our agents under iron discipline.

Through our influence the enforcement of the GOY laws has been reduced to a minimum. The prestige of the law has been undermined by the liberal interpretations introduced by us. The courts decide as we dictate the most important principles, both political and moral, viewing the cases in the light presented by us for the GOY administration. This we accomplished naturally through agents, with whom we have ostensibly no connection, namely, through the press or otherwise. Even senators and high officials blindly follow our advice. The purely animal mind of the GOYS is incapable of analysis and observation, and even less so of foreseeing to what results the development of the principle involved in a case may lead.

It is through this difference in the process of reasoning between us and the GOYS that it becomes possible clearly to demonstrate the stamp of God's elect as compared to the instinctive and bestial mentality of the GOYS. They see, but they cannot foresee, and they cannot invent anything except material things. It is clear, therefore, that nature herself intended us to rule and guide the world.

When the time comes for our open rule, then will be the time to show its benefits, and we will change all the laws. Our laws will be short, clear, irrevocable, and requiring no interpretation, so that everybody will be able to know them thoroughly. The chief point emphasized in them will be a highly developed obedience to authority, which will eliminate all abuses, for all without exception will be responsible before the supreme power vested in the highest authority.

Abuse of power by minor officials will then disappear, because it will be punished so mercilessly that they will lose the desire to experiment with their power. We will closely watch every action of the administration, upon which depends the action of the government machinery, for corruption there creates corruption everywhere; not a single violation of law or act of corruption will remain unpunished. Acts of concealment and willful neglect on the part of governmental officials will disappear after they have seen the first example of severe punishment. The prestige of power necessitates that appropriate, that is to say severe, punishments should be inflicted even for the smallest violations of the sanctity of the supreme authority, committed for the sake of personal gain. The guilty, if punished severely, will be like a soldier who falls on the battlefield of administration for the sake of Authority, Principle, and Law; these principles do not allow any digression from their social function for a personal motive, even on the part of those who rule. For instance: *Our judges ,will know that by attempting to show stupid mercy, they over step the law of justice, which was created solely for exemplary punishment of crimes and not for the manifestation of moral qualities on the part of the judge.* Such qualities are commendable in private, but not in public life, which constitutes the educational forum of human life.

The personnel of our judges will not remain in office after the age of fifty-five. First, because old people adhere more persistently to prejudiced opinions and are less capable of submitting to new commands; and secondly, because that enables us to achieve a certain flexibility of change in the personnel, which will bend more easily under our pressure. He who wishes to retain his position will have to obey blindly.

In general, our judges will be selected only from among those who will clearly understand that they must punish people and enforce the laws, and not indulge in dreams of liberalism at the expense of the educational plan of the government, as is now imagined by the GOYS. The method of changing the personnel will also serve to undermine the collective solidarity of the governmental officials and will attach them to the cause of the government, which decides their fate. The younger generation of judges will be so educated as to prevent any criminal activity which might interfere with the inter-relationship which we have established for our subjects.

At present the GOY judges, lacking a clear conception of the nature of their duties, make exceptions to all kinds of crimes. This occurs because the present rulers, when appointing judges, do not take the trouble to encourage the sense of duty and conscientiousness in the work to be performed by them. As the animal sends out its young in search of prey, so the GOYS are giving their subjects responsible offices without taking the time to explain their functions. Owing to this, their rule is undermined by their own efforts and through the actions of their own administration. Let us use the result of such actions as one more example of the advantage of our own rule.

We will eliminate liberalism from all the important strategic positions in our administration upon which depend the training of our subjects for our social order. These positions will be given only to those who have been trained by us for governmental work.

In answer to a possible remark, that the putting of old officials on the retired list may prove expensive for the treasury, I can state first, that, prior to their dismissal, some private work will be found for them to replace what they are losing, and secondly, I may also remark, that all the world's money will be concentrated in our hands; consequently, our government need not fear expense.

Our autocracy will be consistent in every respect, and consequently every manifestation of our great power will be respected and unconditionally obeyed. We will ignore grumbling and discontent, and all active manifestations of either will be suppressed by punishment, which will serve as an example to the rest of the people.

We will abolish the right of appellate courts to annul judicial decisions, which will become the exclusive prerogative of the sovereign, for we cannot permit the people to think that an incorrect decision may possibly be rendered by the judges appointed by us. Should, however, such an error happen, we ourselves will annul the decision; but the punishment which we will impose upon the judge for misconception of his duties and of his responsibility will be so severe that it will eliminate the very possibility of a recurrence. I repeat that we will watch every step taken by our administration in order to enable us to satisfy the people, for they have a right to demand a good appointee from a good administration.

In the person of our sovereign, our government will bear the appearance of a patriarchal or fatherly tutelage. The people, our subjects, will see in him a father who takes care of every need, every action, and who is concerned with every relationship, both among the subjects themselves and between them and the sovereign.

Thus, they will become imbued with the idea that it is impossible for them to do without this guardian and guide if they wish to live in a world of peace and quiet. *They will recognize the autocracy of our sovereign, whom they will respect and almost deify,* especially when they realize that our agents do not usurp his power, but merely execute his orders blindly. They will be glad that everything is regulated in their lives, as is done by wise parents who wish to educate their children to a sense of duty and obedience. With regard to the secrets of our political plans, both the masses and their administration are like little children.

As you can see for yourselves, I base our despotism upon right and duty; the right of forcing the performance of duty is the direct function of government, acting as the father to its subjects. It is the right of the strong to utilize his power in order to lead humanity towards a social order established by the law of nature, namely, obedience. Everything in the world is subject, if not to some other persons, then to circumstances, or to its own nature; but in any case, to something stronger than itself. Consequently, let us be the strongest for the common good.

We must sacrifice without hesitation those individuals who violate the existing order, for in exemplary punishment of evil there lies a great educational problem.

When the King of Israel [the Jewish Messiah] places the crown offered to him by Europe on his sacred head, he will become the Patriarch of the World. The necessary sacrifices made by him will never equal the number of victims sacrificed to the mania of greatness during the centuries of rivalry between the GOY governments.

Our sovereign will be in constant communication with the people, delivering from tribunes addresses which will be spread to all parts of the world.

### PROTOCOL NO. XVI

FOR the purpose of destroying all collective forces except our own, we will nullify the universities, the first stage of collectivism, by reconstructing them along new lines. *Their directors and professors will be trained for their work through detailed secret programs of action, from which they will not be able to deviate in the least with impunity. They will be appointed with special care and will be so placed as to be completely dependent upon the government.*

We will exclude from the curriculum civic law, as well as all that touches upon political questions. These subjects will be taught only to a few dozen selected for their striking ability from among the initiated. *The universities must not allow the callow youths to graduate who concoct plans of constitutions as they do comedies or tragedies, or who meddle with political matters which even their fathers do not understand.*

Poorly directed study of political questions by a great number of people creates Utopians and poor citizens, as you can judge by the universal education as conducted by the GOYS along those lines. It was necessary for us to infiltrate into their educational system such principles as have successfully broken down their social order. When we are in power, we will eliminate all disturbing subjects from educational systems and will make young people obedient children of their superiors, loving the sovereign as their assurance of hope, peace, and quiet.

For the study of the classics and ancient history, which contain more bad than good examples, we will substitute a program dealing with the future. We will obliterate from the memory of the people all those facts pertaining to former centuries which are not to our advantage, leaving only those which emphasize the mistakes of the GOY governments. The study of practical life, of obligatory social order, of the interrelationship of human beings, the avoidance of evil, egotistical examples that plant the seed of evil, and other questions of a pedagogical nature, will head the educational program. This program will differ for each caste, never allowing education to be of a uniform character. Such a system is of special importance.

Each caste must be educated with strict limitations, according to its particular occupation and the nature of the work. Accidental genius has always been able and always will be able to rise to a higher caste; but, for the sake of this rare exception, to open the door to the inefficient, and to admit them to higher castes or ranks, enabling them to occupy positions of others born and trained to fill them—is absolute insanity. You, yourself, know what happened to the GOYS when they yielded to this nonsense.

In order to implant the sovereign firmly in the minds and hearts of his subjects, it is necessary to acquaint the people, during his term of office, both in schools and in public places, with the importance of his activity and the benevolence of his enterprises.

We will abolish all unlicensed teaching. Students will have the right to gather, with their relatives, in their colleges as if in clubs. During these gatherings, on holidays, the teachers will read supposedly unbiased lectures on problems of human relationship, on the law of imitation, on the cruelty of unrestricted competition, and finally, on new philosophical theories which have not yet been disclosed to the world.

We will promote these theories into dogmatic beliefs, using them as stepping-stones to our faith. After having presented our program of action for the present and for the future, I will read to you the principles of these theories.

In short, knowing from the experience of many centuries that men live and are guided by ideas, that these ideas are imbued only by means of education given to persons of all ages, of course by different methods but meeting with equal success, we will absorb and appropriate to our own advantage the last traces of independent thought, which for a long time have been directed to the goal and to the ideas necessary to us. The system of enslaving thought is already in action through so-called visual education.

This system tends to turn the GOYS into thoughtless, obedient animals, expecting to see in order to understand. In France one of our best agents, Bourgeois, has already announced a new program of visual education.

## PROTOCOL NO. XVII

THE lawyer's profession makes people grow cold, cruel, stubborn and unprincipled, and compels them to take an abstract or purely legal viewpoint in all matters. They have learned to consider solely the personal gain derived from every case they handle and not the possibility of the social benefit of its results. They rarely refuse to take a case and always strive for acquittal at all cost, clinging to minor technical points of a legal nature. In this way they demoralize the courts. Therefore we will limit this profession, converting it into an executive public office. Lawyers will be deprived of the right of contact with their clients on the same basis as are the judges. They will receive their cases only from the court, preparing them on the strength of written reports and documents and defending their clients after they have been examined in court on the basis of the facts obtained during the trial. They will receive a salary, regardless of whether the defense has been successful or not. They will act as simple exponents of the case on behalf of the defense in counterbalance to the public prosecutor, who will act as exponent on behalf of the prosecution. This will shorten legal procedure and establish an honest and impartial defense, conducted not for the sake of personal gain, but based on the personal conviction of the lawyer. This will also eliminate the existing bribery among fellow lawyers and prevent their allowing the side to win which pays.

We have already taken care to discredit the clergy of the GOYS and thus to undermine their function, which at the present time could have been very much in our way. Their influence over the people diminishes daily.

To-day freedom of religion has been proclaimed everywhere; consequently, it is only a *question of a few years before the complete collapse of Christendom*. It will be still easier to deal with other religions, but it is too early to discuss this problem. We will confine clericalism and clericals within such a narrow field that their influence will have an effect opposite to what it used to have.

When the moment comes to annihilate the Vatican completely, an invisible hand, pointing towards this court, will guide the masses in their assault. When, however, the masses attack, we will come forward as defenders to prevent too much bloodshed. By this method we will penetrate its very heart and will not leave it until we have undermined its power.

The King of Israel [the Jewish Messiah] will become the real Pope of the Universe, the Patriarch of the International Church.

But until we have accomplished the re-education of the youth to new transitional religions and finally to our own, *we will not openly attack the existing churches, but will fight them by means of criticism, thus creating dissension.*

In general, our press will denounce governmental activities and religion, and will expose the inefficiency of the GOYS in the most unscrupulous terms, so as to humiliate them to such an extent as only our ingenious race is capable of doing. Our rule will simulate the God Vishnu, who resembles us physically; each of our hundred hands will hold one of the springs of the social machine. We will see everything without the aid of the official police; in its present organization, however, which we have worked out for the GOYS, the police prevent the government from seeing anything. According to our program, one-third of our subjects will watch the others from a pure sense of duty, as volunteers for the government. Then it will not be considered disgraceful to be a spy and an informer; on the contrary, it will be regarded as praiseworthy. Unfounded reports, however, will be severely punished to prevent abuse of this privilege.

Our agents will be recruited both from among the highest and the lowest ranks of society; they will be selected from among the pleasure-loving governmental officials, editors, printers, booksellers, salesmen, workmen, drivers, butlers, etc. This police force will have no official rights or credentials, which give opportunity for the abuse of power, and consequently it will be powerless; it will merely act as observer and will make reports. The verification of such reports and the issue of warrants for arrests will rest with a responsible group of police controllers. The actual arrests, however, will be made by a gendarme corps or the municipal police. In case of failure to report any political matter which has been observed or rumored, the person who should have reported it may be brought to trial for concealment of crime, if it is proven that he is guilty.

*In the same way that our brethren are now under obligation to report on their own initiative on all apostates,* or on any person marked as being opposed to the Kehillah, so in our Universal Kingdom it will be obligatory for all subjects to serve the state in that direction.

Such an organization will eliminate all abuse of power and various kinds of coercion and corruption, in fact, the very things which have been introduced into the customs of the GOYS by our councils and by the theories of the rights of supermen. But how otherwise could we foment the increasing causes for disorder in the midst of their administration? What other means could we use? Among these means, one of the most important is the employment of such agents for the preservation of order as are in a position to manifest their own evil inclinations in the course of their destructive work, namely, their self-will, abuse of authority, and, most important of all, bribery.

## PROTOCOL NO. XVIII

WHEN the time comes for us to strengthen the measures of police protection (the most terrible poison for the prestige of authority), we will artificially organize disorder or simulate the expression of discontent with the aid of experienced orators. These orators will be joined by sympathizers. This will give us the pretext for searches and special restrictions which will be put in force by our servants among the GOY police.

As most conspirators work as amateurs for the sake of chattering we will not disturb them until we see that they are about to take action; but we will introduce in their midst secret service agents. It must be remembered that the prestige of authority diminishes if conspiracies against it are often discovered, for that leads to the presumption of the weakness of the authority, or, what is worse, to the admission of its own mistakes. You are aware that we have destroyed the prestige of the ruling GOYS by frequent attempts made on their lives through our agents, who were but blind sheep of our flock, easily moved, by a few liberal phrases, to crimes, so long as they were of a political nature. *We have forced the rulers to admit their own weakness by adopting open measures of police protection, and thereby we have ruined the prestige of their authority.*

Our sovereign [the Jewish Messiah] will be protected only by the most invisible guard, because we will never allow any one to think that conspiracy might exist against him which he is unable to combat and from which he has to hide himself. If we were to allow this thought to prevail, as it prevails among the GOYS, we would thereby sign the death warrant, if not of the sovereign himself, then of his dynasty in the near future.

Observing strict decorum, our sovereign will use his power only for the benefit of the people, but never for his own good or for that of his dynasty. By strictly adhering to this decorum, his authority will be respected and protected by his subjects; moreover, he will be worshiped, because it will be known that upon his authority depends the well-being of every citizen of the kingdom, and the stability of the social order itself.

*To guard the sovereign openly is equivalent to an admission of the weakness of his governmental organization.*

Our sovereign, when amidst his people, will always appear to be surrounded by a crowd of curious men and women, who will stand beside him as though accidently and will hold back the other people as though through respect for order. This example will implant an idea of self-restraint in others. If there be a person in the crowd trying to present a petition, and working his way through the ranks, the person nearest to him must take the petition and present it to the sovereign in sight of the petitioner himself, so that all may know that the petition presented has reached its destination and consequently that there exists a control of affairs on the part of the sovereign himself. The prestige of authority demands that the people a should be able to say, ' If only the king could know it,' or, 'The king will know about this.'

With the establishment of an official police guard the mystical prestige of authority vanishes at once; with a certain amount of audacity, every one considers himself superior to authority; the assassin realizes his strength and only has to watch his opportunity to make an attempt against an official. We preached differently for the GOYS, but we can see the results to which open methods of protection have led them.

We will arrest criminals upon the first more or less well founded suspicion. Because of the fear of a possible mistake political criminals should not be given the opportunity to escape; indeed towards political crime we will show no mercy. If, in exceptional cases, it may seem possible to allow the investigation of motives which have led to ordinary criminal offences, there is no excuse for those who attempt to deal with matters which no one can understand except the government. Moreover, not even all governments are capable of understanding the right policy.

## PROTOCOL NO. XIX

THOUGH we will not allow individuals to become involved in politics, we will, on the other hand, encourage the submission for the approval of the government of all petitions and reports containing suggestions and plans for bettering the condition of the people. This will bring to our knowledge the shortcomings or merely the fantastic aspirations of our subjects. These suggestions we will answer either by favorable action or by refusals proving the lack of intelligence and the errors of those who have submitted such suggestions.

Sedition is nothing but the barking of a lap dog at an elephant. From the point of view of a government which is well organized, not from the police standpoint but with regard to its social basis, the lap dog barks at the elephant because he does not realize his strength. It is only necessary for the elephant to show his strength once and the dog barks no more; he begins to wag his tail the moment he sees the elephant.

In order to eliminate the prestige of martyrdom from political crime, we will seat the political criminal on the same bench with thieves, murderers, and other disgusting and dirty criminals. Then public opinion will regard that class of criminals as quite as disgraceful as any other, and will brand them with equal contempt.

We have endeavored to prevent, and I hope have succeeded in preventing, the GOYS from using such methods of dealing with seditious activities. In order to attain this end, we have made use of the press and public speeches; indirectly, through cleverly compiled historical textbooks, we have given publicity to martyrdom as though revolutionists had undergone it for the sake of human welfare. Such an advertisement has increased the contingent of liberals and forced thousands of GOYS into the herds of our cattle.

## PROTOCOL NO. XX

TO-DAY we shall deal with the financial program, the discussion of which I have postponed until the end of my report because it is the most difficult, conclusive, and decisive point in our plans. In approaching it, I will remind you that I have already intimated that the result of our actions is measured in figures.

When we become rulers, our autocratic government, for the sake of self-defense, will avoid burdening the people with heavy taxes, and it will not forget the rôle it has to play, namely, that of Father and Protector. But as government organization is costly, it is necessary to raise the means for its maintenance. Consequently, we must carefully work out the plan of a fair distribution of taxation.

In our government the sovereign will have the legal fiction of owning everything in his kingdom (which is easily put into practice), and can resort to legal confiscation of all money in order to regulate its circulation throughout the country. Consequently, the best method of taxation is the levying of a progressive tax on property. Taxes will thus be paid without difficulty or ruin in respective proportion to the amount of property owned. The rich must realize that it is their duty to give a part of their surplus wealth for the benefit of the country as a whole, because the government guarantees inviolability of the remaining part of their property and the right of honest gain. I say *honest* because the control of property will prevent legal theft.

This social reform must come front above, for the time is ripe and it is becoming necessary as a guarantee of peace.

The tax on the poor is the seed of revolution, and it acts detrimentally to the government, which loses the great in its pursuit of the little. Moreover, the taxation of capital will lessen the increase of wealth in private hands, in which at present we have concentrated it as a counterweight to the governmental power of the GOYS, namely, to the state treasury.

Progressive taxation, assessed according to the amount of capital, will produce a much greater revenue than the present system of taxing every one at an equal rate, which is useful to us now only as a means of exciting revolt and discontent among the GOYS. The power of our sovereign will rest mainly in equilibrium and in guarantees of peace. For these, the capitalists must cede a part of their income so as to protect the action of the government machine. Public needs must be met by those who can best afford to do so and by those from whom there is something to take.

Such a measure will eliminate the hatred of the poor towards the rich, as they will be regarded as the financial supporters of the state and the upholders of peace and prosperity. The poor will also see that the rich are providing the necessary means to insure this end.

To prevent intelligent taxpayers from being too discontented with the new system of taxation, they will be furnished with detailed reports of the disbursement of public funds, exclusive of such as are appropriated for the needs of the throne and administrative institutions.

The sovereign will not own property, since everything in the state will seem to belong to him and these two conceptions would contradict each other. Private means would eliminate his right to own everything.

The relatives of the sovereign, aside from his descendants who will also be supported by the state, must join the ranks of government officials, or otherwise work for the right of holding property. The privilege of being of royal blood must not entitle them to rob the state treasury.

Sales, profits, or inheritances will be taxed by a progressive stamp tax. The transfer of property, whether in cash or otherwise, without the required stamp, will place the payment of the tax on the original owner, dating from the time of the transfer until the time of the reported failure to record the transaction. Transfer vouchers must be shown weekly at the local branch of the state treasury, together with a statement of the names, surnames, and the permanent addresses both of the original and of the new owner. The recording of the names of those participating in a transaction will be necessary in all transactions involving more than a certain amount for ordinary expenditure. The sale of prime necessities will be taxed only by a stamp tax, which will represent a certain small per cent of the cost of the particular article.

Just calculate how many times the amount received from such taxes will exceed the income of the GOY governments.

The state bank must keep a definite reserve fund, and all sums in excess must be put back into circulation. The cost of public works will be met out of this surplus fund. The initiative of such works emanating from the government will also tie the working class to the interests of the government and the rulers. Some of this money will be allotted to prizes for inventions and for the purposes of production.

Even small sums in excess of a certain definite and broadly calculated fund, should not be allowed to be kept in the state treasury, because money is intended to circulate, and every impediment to circulation is detrimental to the governmental mechanism, which the money lubricates; the congestion of lubricating substances can stop the proper functioning of the mechanism.

The substitution of bonds for a part of the currency has created just such an impediment. The result of this has already become sufficiently evident.

We will also establish an auditing office, so as to enable the sovereign to find at all times a full account of state revenues and expenses, except for the current month not yet made up, and that of the previous month not yet presented.

The only person who will not be interested in robbing the state treasury will be the sovereign, its owner. This is the reason why his control will prevent the possibility of loss or misappropriation.

Receptions for the purpose of etiquette, which waste the valuable time of the sovereign, will be abolished, because the ruler needs time for control and thought. Then his power will not be frittered away on the people surrounding the throne for the sake of appearance and brilliance, and who have only their own and not the public interest in mind.

The economic crises were created by us for the GOYS only by the withdrawal of money from circulation. Huge amounts of capital were kept idle and were taken away from the nations, which were thus compelled to apply to us for loans. Payment of interest on these loans burdened the state finances and made the states subservient to capital. The concentration of industry having taken production out of the hands of the artisan and put it into the hands of capitalists, sucked all the power out of the people and also out of the state.

The present issue of money generally does not coincide with the need per capita, and consequently it cannot satisfy all the needs of the working classes. The issue of currency must correspond with the increase in population, and children must be reckoned as consumers from the day of their birth. The revision of the issue of currency is an essential problem for the whole world.

You know that gold currency was detrimental to the governments that accepted it, for it could not satisfy the requirements for money, since we took as much gold as possible out of circulation.

We must issue a currency based on the value of the working power, whether it be of paper or wood. We will issue money in proportion to the normal demands of every subject, adding a certain amount at every birth and decreasing it with every death.

Every department (the French administrative divisions), [*Footnote:* The words in parentheses would seem to be a comment of Nilus's.] every district, will be in charge of its own accounts.

To avoid any delay in paying government expenses, the terms of such payments will be decreed by order of the sovereign; this will eliminate any favoritism of the ministry (of finance) [*Footnote:* The words in parentheses are inserted by the editors.] over any other department to the detriment of the others.

The budget of revenues and the budget of expenditure will be placed side by side, in order that they may always be compared with each other.

We will present plans for the reform of the GOY financial institutions and of their principles, as planned by us, in such a manner that nobody will be frightened. We will demonstrate the need of reform by the disorderly twaddle produced by the financial disorganization of the GOYS. We will show that the first reason for this confusion lies in the drafting of rough estimates for the budget, which increases from year to year. This annual budget is with great difficulty made to last during the first half of the year; then a revised budget is demanded and the funds thus allotted are spent in the next three months, after which a supplementary budget is called for and all this is wound up by a liquidation budget. As the budget of the following year is based on the total expenditure of the preceding year, the divergence from the normal reaches fifty per cent annually, so that the annual budget trebles every ten years. Owing to such a procedure, resulting from the carelessness of the GOY governments, their treasuries became empty. The period of loans followed and used up the remainder and brought all the GOY states to bankruptcy.

You can well understand that such a management of financial affairs as we induced the GOYS to pursue cannot be adopted by us.

Every loan proves the impotency of the government and its failure to understand its own rights. Loans, like the sword of Damocles, hang above the heads of the rulers, who instead of placing temporary taxes on their subjects, stretch forth their hands and beg the charity of our bankers. Foreign loans are leeches, which can never be removed from the governmental body until they either fall off themselves or the government itself manages to get rid of them. But the GOY governments instead of throwing them off increase their number, so that these governments must inevitably perish through self-inflicted loss of blood.

Indeed, what is a loan, especially a foreign loan, if not a leech? A loan is the issuance of government obligations which involve the liability to pay interest in proportion to the sum borrowed. If the loan pays five per cent, then in twenty years the government has unnecessarily paid in interest an amount equal to the principal sum borrowed. In forty years it has paid twice; in sixty years it has trebled the sum, while the loan still remains an unpaid debt.

From this calculation it is evident that under the system of universal taxation the government takes the last penny from the poor taxpayers in the form of taxes in order to pay interest to foreign capitalists, from whom the money was borrowed, instead of collecting these same pennies for its needs free from all interest.

So long as the loans were domestic, the GOYS only shifted the money from the pockets of the poor into those of the rich; but when we bribed the proper persons to make the loans foreign, then national riches poured into our hands and all the GOYS began to pay us the tribute of subjects.

The carelessness of the reigning GOYS in statemanship, the corruption of their ministers, the ignorance of other officials of financial problems, has forced their countries into debt to our banks to such an extent that they can never pay off their debts. It should be realized, however, that we have gone to great pains in order to bring about such a state of affairs.

Impediments to the circulation of money will not be allowed by us, and therefore there will be no government bonds, except one per cent bonds, so that the payment of interest should not deliver the power of the state to the sucking of leeches. The right of issuing bonds will be exclusively granted to industrial corporations, which will easily pay the interest out of their profits. The government, however, does not derive profit on borrowed money as these corporations do, since the state borrows money for expenditure and not for production.

Industrial bonds will also be bought by the government, which instead of being, as at present, the payer of tribute on loans, will become a sound creditor. Such a measure will prevent stagnation in the circulation of money, as well as indolence and laziness, which were useful to us so long as the GOYS remained independent, but are not wanted by us in our government.

How apparent is the shortsightedness of the purely bestial brains of the GOYS! It manifested itself when they borrowed money for at interest. It did not occur to the GOYS that, at any rate, this money, with the additional interest on it, would have to be taken from the resources of the country and paid to us. Would it not have been more simple to take the needed money from their own people?

This proves the genius of our distinguished mind, for we were able to present the question of loans to them in such a light that they saw in loans an advantage for themselves.

Our estimates, which we will produce when the time comes, will be based on the experience of centuries, on all those experiments which were conducted by us at the expense of the GOY governments; our estimates will prove to be clear and definite, and will obviously demonstrate the advantage of our new system. They will end all those abuses which made it possible for us to master the GOYS, but which cannot be permitted in our reign.

We will so organize the accounting system that neither the sovereign himself nor the most humble clerk will be able to deflect the smallest sum from its destination or direct it into a different channel from that indicated in our original financial plan.

It is impossible to govern without a definite plan. Traveling along a definite road with an indefinite supply of provisions destroys heroes and knights.

The GOY rulers, to whom we once gave advice to neglect governmental duties for grandiose receptions, etiquette, and pleasures, only concealed our rule. The accounts of the powerful favorites who replaced the sovereign were drawn up by our agents, and they always satisfied the shallow minds by promises that in the future there would be savings and improvements. Savings from what? From new taxes? This might have been asked but was not asked by those who read our reports and plans. You know to what their carelessness has led them, what financial disorganization they have reached in spite of the wonderful diligence of their people.

### PROTOCOL NO. XXI

I WILL, add one more detail regarding domestic loans in addition to the report which I made at the last meeting. I will not speak any more of foreign loans, for they filled our coffers with the national money of the GOYS. There will be no foreigners in our government, nobody outside.

We profited by the corruption of the administrators and by the negligence of the rulers in receiving sums that were doubled, trebled, and even more, loaning the GOY governments money which in reality was not needed by the states at all. Who could do the same with regard to us? Therefore, I will only set forth details in regard to domestic loans.

In announcing such a loan, the governments open a subscription to their bonds. To make them accessible to all, they vary the denomination from one hundred to thousands, and the first subscribers are allowed to buy below face value. The following day the price is artificially raised on the pretext that everybody hurried to buy the bonds. In a few more days there is a pretense that the treasury is filled and that it is not known what to do with the money, which has been oversubscribed. (What was the use of taking it?) The subscription is evidently considerably in excess of the amount asked for. Therein lies the effect, for it is thus demonstrated that the public has confidence in the government obligations.

But after the comedy has been played the fact of the debt appears, and it is usually a heavy one. In order to pay the interest, new loans have to be issued, which do not liquidate but increase the original debt. Then when the borrowing capacity of the government has been exhausted, it becomes necessary to meet the interest on the loan—not the loan itself—by new taxes. These taxes are nothing but a debit used to cover a debit.

Then comes the period of conversions, but these only decrease the payment of interest while they do not annul the debts. Moreover, they cannot be made without the consent of the bondholders. When a conversion is advertised, an offer is made to return the money to those who are not willing to convert their bonds. If everybody were to demand his money, the government would be caught in its own net and would be unable to return all the money. Fortunately, the GOY subjects, ignorant of financial affairs, always preferred to suffer a fall in the value of their securities and a reduction of interest to the risk of new investments; thus, they have given these governments more than one opportunity of throwing off a deficit of several millions. At present, with the existence of foreign loans, the GOYS cannot play such tricks, for they know that we would demand all the money back.

Thus, an avowed bankruptcy will be the best proof of the lack of common interest between the people and their government.

I direct your express attention to the above circumstance, as also to the following: At present all domestic loans are consolidated into so-called floating debts; in other words, into those whose terms of payment are more or less close at hand. Such debts consist of money placed in savings banks. Being at the disposal of the government, for a considerable length of time, these funds vanish in the payment of interest on foreign loans, and they are replaced by an equal amount of government securities. *The latter cover all the deficits in the government treasuries of the Goys.*

When we mount the throne of the universe, such financial expedients, being detrimental to our interests, will vanish. We will also destroy all stock exchanges, for we will not allow the prestige of our authority to be shaken by the shifting of the prices of our securities. We will fix the full price of their value legally without any possibility of its fluctuation. (A rise leads to a fall, and this was precisely what we did to the GOY stocks and bonds at the beginning.)

We will replace the stock exchanges by great government credit institutions, whose functions will be to tax commercial values according to governmental plans. These institutions will be in a position to throw daily on the market 500,000,000 shares of industrial stocks, or to buy up a like amount. Thus all industrial enterprises will become dependent upon us. You can well imagine what power that will give us.

## PROTOCOL NO. XXII

IN all that I have hitherto reported to you I have carefully tried to show you a true picture of the mystery of present events, as also of those of the past, which all flow into the stream of great events, the results of which will be seen in the near future. I have exposed our secret plans which govern our relations with the GOYS, as well as our financial policy. There remains but little to add.

We hold in our hands the greatest modern power—gold. In the course of two days we can get it from our treasuries in any desired quantity.

Is there any more need for us to prove that our rule is decreed by God? Do we not prove by such wealth that all the evil which we were forced to do during so many centuries has served in the end to true happiness—to the restoration of order? Although by means of violence, order will nevertheless be established. We will be able to prove that we are benefactors, who have brought true welfare and individual freedom to the tortured world, insuring at the same time the possibility of enjoying peace, quiet, and dignity of relationships, upon the sole condition, of course, that obedience to the laws established by us is practiced. We will also make it clear that freedom does not mean license and in doing whatever people please, no more than dignity and power imply the right to propound destructive doctrines, like freedom of conscience, equality, and similar things. Individual freedom by no means imports the right of disturbing oneself and others, disgracing oneself by making ridiculous speeches in disorderly gatherings, and implies that true liberty means individual inviolability through an honest and strict obedience to social laws; that moreover, human dignity implies the conception of one's rights as well as the idea of legal inhibitions which prohibit fantastic dreams about the *Ego*.

Our power will be glorious because it will be mighty; it will rule and guide, and not helplessly crawl after leaders and orators, shouting insane words which they call great principles, and which in reality are simply Utopian. Our power will lead to order, which, in turn, brings happiness to the people. The prestige of this power will excite mystical adoration, and the peoples will bow before it. True power does not yield to any right, even be it that of God. None will dare approach it in order to deprive it even of an atom of its might.

## PROTOCOL NO. XXIII

TO teach the people obedience they must be taught modesty, and to accomplish this the production of luxuries must be limited. We will thus improve customs, demoralized by rivalry, resulting from luxury.

We will restore handicraft, which will undermine the private capital of manufacturers. This is necessary, because big manufacturers often influence, although not always consciously, the thoughts of the people against the government.

A people, practicing handicraft, does not know what unemployment means, and this makes them cling to existing conditions and consequently to the power of authority. Unemployment is most dangerous for a government. It will have finished its work for us as soon as authority falls into our hands.

Drunkenness will also be forbidden by law and will be punishable as a crime against human decency, for man becomes bestial under the influence of alcohol.

Once more I state, that people obey blindly only the hand that is strong and entirely independent of them, in which they see a sword of defense and a stronghold against the blows of social misfortune. Why should the sovereign have an angel's heart. They want to see in him the personification of might and power.

The sovereign who will replace the present existing governments, dragging along their existence in the midst of a society demoralized by us, which denies even the power of God and from whose midst rises on all sides the flames of anarchy, must primarily undertake to extinguish this all-consuming fire. Therefore, he must destroy such a society, if necessary drown it in its own blood, in order to resurrect it as a well-organized army, which consciously struggles against the infection of any anarchy affecting the state organism.

He, God's elect, is chosen from above for the purpose of crushing the insane forces that are moved by instinct and not by intellect, by bestiality and not by humanitarianism. These forces are now triumphant, and assume the form of robberies and all kinds of violence exercised in the name of liberty and of right. They have destroyed all social order, so as to establish the throne of the King of Israel; but their rôle will be ended with his coming into power. Then it will be necessary to sweep them from his path, on which not a twig or an impediment shall remain.

Then we will say to the peoples: Pray to God and bow before him who bears the mark of predestination, to whom God Himself showed His Star, so that none but He Himself should free you from all sinful forces and from evil.

## PROTOCOL NO. XXIV

NOW I shall refer to the manner in which we will strengthen the dynastic roots of King David so as to cause this dynasty to endure until the last day [the Jewish Messiahs]. This method will consist chiefly of the same principles which enabled our Wise Men to conserve their power to cope with universal problems and to guide the education of the thoughts of humanity at large.

A few members of the seed of David will train the sovereigns and their successors, who will be selected not by right of inheritance, but according to their personal ability. To them the deep political mysteries and the plan of our rule will be confided, but in such a wise manner that nobody will know these secrets. The aim of this method is to prove to all that power will not be given to the uninitiated in the mysteries of political art.

Only such people will be taught how to apply the above mentioned plans in practice, by comparing them with the experiences of many centuries, and only they will be initiated in the conclusions drawn from all the observations of political, economic, and social movements and sciences; in short, only they will know the true spirit of the laws, irrevocably established by nature for the purpose of regulating human relationship.

Direct descendants of the sovereign will often be prevented from inheriting the throne if, during the period of their study, they show signs of frivolity, lenience, or other tendencies detrimental to authority, which would make them incapable of government and dangerous to the prestige of the Crown.

Only those of an undoubtedly able and firm, even cruel character, will receive the reins of government from our Wise Men.

In case of illness, loss of will-power, or any other form of inefficiency, the sovereigns will be compelled to hand over the reins of government to new and able hands.

The sovereign's immediate plan of action and its application in the future will be unknown even to the so-called closest advisers.

Only the sovereign and his three sponsors will know the future.

In the person of the sovereign, with his immovable will over himself and humanity, all will recognize Fate itself with her mysterious paths. Nobody will know the aims of the sovereign when he issues his orders, and thus nobody will dare oppose him.

Naturally the mental capacity of the sovereign must be equal to the plan of rule herein contained. For this reason he will not mount the throne before a test of his mind is made by the above mentioned Wise Men.

To make people know and love their sovereign, it is necessary that he should address the people in public places, thus establishing harmony between the two forces, now separated from each other by mutual terror. This terror was necessary for us until the time came to make both forces fall under our influence.

The King of Israel [the Jewish Messiah] must not be influenced by his passions, especially by sensuality. No particular element of his nature must have the upper hand and rule over his mind. Sensuality, more than anything else, upsets mental ability and clearness of vision by deflecting thought to the worst and most bestial side of human nature.

The Pillar of the Universe in the person of the World Ruler, sprung from the sacred seed of David, must sacrifice all personal desires for the benefit of his people.

Our sovereign must be irreproachable."

## 5.3 Did Anyone Believe that the *Protocols* were Genuine?

Jews and crypto-Jews instigated and financed the Japanese war against Russia, while concurrently cutting off Russia's access to funds. Jews and crypto-Jews financed and led revolutions against the Czar. Jews and crypto-Jews organized and led massive strikes, which further crippled the Russian economy. Jews and crypto-Jews fought against the Czar's effort to integrate racist Jews into Russian society. When all the havoc Jews and crypto-Jews deliberately caused began to hurt the Russians and the Russian Jews, Jews and crypto-Jews used their media control to blame the Czar for the very things he was desperately trying to prevent, the very things these Jews had deliberately caused. The Jews who were deliberately harming the Russian People turned the Russian People against the Czar who was trying to save them.

Richard B. Spence wrote of the crypto-Jewish spy, financier, warmonger and war profiteer Sidney Reilly, born Salomon Rosenblum, whose adventures fulfilled the plans spelled out in the *Protocols* (it is interesting to note that the author appears to believe that the poor Jewish spies who were out to destroy Russia and to profit from the destruction were inconveniently forced to hide the fact that they were Jews, because the Czar, in his poor paranoia, believed that there were Jewish spies aiming to destroy Russia and profit from its destruction—in reality the practice of crypto-Judaism is already found in the Old Testament story of Hadassah, a. k. a. Esther, *see: Esther* 2:7; and the Jews had long since been accused of war profiteering and revolutionary activity, and the fact that they were doing it again in Russia proved the Czar correct, not incorrect, as is obvious—in addition, the fact that the revolutionaries and fomenters of war were Jewish freemasons lends credence to the genuineness of the *Protocols*, it does not tend to disprove their authenticity),

"It was during 1905, in London or Petersburg, that Reilly first made the acquaintance of (later Sir) George Owens Thurston.[41] The latter was a naval engineer and chief of construction for Vickers [the armaments manufacturer?]. Among his clients worldwide were the Japanese and Russian navies. However, perhaps the most significant thing about him for our purposes is that he was now and for many years to come a close personal friend and advisor to Basil Zaharoff. Thurston certainly forms an important link in the chain linking Reilly and the Greek. Doubtless Thurston, and probably Sir Basil, encouraged Sidney to return to Russia at least partly on their behalf.

Manasevich and Reilly arrived in St. Petersburg around October, just as the revolutionary wave crested and Nicholas' days on the throne seemed numbered. In September, the disastrous Japanese war was brought to end by a treaty negotiated in Portsmouth, New Hampshire. Representing Russia was Sergei Witte who returned the man of the hour. In September a general strike shut down the Imperial capital and other cities. Under pressure from Witte and members of his own family, Nicholas caved in and issued the October Manifesto that promised a constitution and elected parliament, or Duma. Liberals rallied to support the Tsar, while the radical Soviets were crushed. By year's end, Nicholas was again in control.

In the aftermath of war and revolution, Russia stabilized and for the better part of a decade experienced an unprecedented burst of rearmament and economic expansion. It was a wonderful place to play the System. However, there were hazards as well, notably a sharp rise in violent anti-Semitism. The Tsarist regime fanned the flames by condemning the revolutionary disturbances as an insidious Jewish conspiracy. *The Protocols of Zion*, already noted, was an integral part of this counter-propaganda campaign. Bloody pogroms sprang up across the Empire. In 1906, one struck Bialystok, very near Reilly's boyhood home and still the abode of many of his kin. Under the circumstances, it was more important than ever to conceal or compensate for his Jewish antecedents. Thus, in Petersburg he styled himself an English expatriate 'who had become for all intents and purposes Russian.'[42] As such, he set out to assemble and exploited an ever-widening network of contacts in Russia's commercial, political and underground spheres. Before long the name and influence of the mysterious Briton would even penetrate the precincts of the Imperial Court.

In 1906, the directory *Ves' Peterburg* ('All St. Petersburg'), listed a new name among its array of businessmen, professionals and public servants—*Sidnei Georg'evich Raille* doing business as a *komisioner* (commission agent) at #1/2 Kazanskaia Ploshchad (Square).[43] On hand to assist his climb up the social and Secret World ladders were a bevy of old friends and fellow intriguers. In the immediate aftermath of the war, Zaharoff arrived in St. Petersburg to cash in on Russia's rearmament bonanza. Friend Ginsburg was on the scene as well. Having brushed off accusations of treason in Port Arthur, he was ensconced as a 'first guild' tradesman with interests in banking and insurance, both spheres of acute interest to Reilly.[44] Zaharoff and Ginsburg each had links to the Brothers Zhivotovskii, Abram (recently encountered in Port Arthur) and David, ambitious *affairistes* with an eye on high finance and Russia's burgeoning armaments industry.[45] The Zhivotovskiis had their roots in the Grodno-Bialystok region which means they may have known something of Reilly's true origins. However, Abram Zhivotovskii's most interestingly connection was his supposed kinship with one Lev Davidovich Bronshtein, better known as the above-mentioned revolutionary firebrand, Leon Trotsky. Sources cannot agree on just what relationship joined the two, Abram being described variously as Trotsky's brother-in-law, cousin and uncle, but it seems most likely that they were related by marriage.[46]

Besides business, another thing that Reilly, Ginsburg, Abram Zhivotovskii, and Zaharoff (reputedly even Trotsky) had in common was freemasonry. We noted this earlier as a frequent common denominator in Sidney's London associations.[47] In Petersburg it was almost universal among his contacts and cronies. To simplify matters, when first noted, an (M) after the name will indicate known masonic affiliation. The real question, of course, is what difference does that make? In the semi-liberalized atmosphere after 1905, Russian freemasonry emerged from the shadows. By 1914, some forty lodges flourished, including ones in the Duma and the military. While the total number of masons was probably less than 2,000 out of a total population of some 150,000,000, the brethren counted among their number a sizable share of the Empire's, commercial, political and intellectual elite. In the Romanov family itself, no less than five Grand Dukes were reputed brethren of one variety or another.[48] In Moscow, Reilly affiliated with the *Vozrozhdenie* ('Renaissance') lodge whose members included Aleksandr Guchkov, now leader of the center-right Octobrist Party and one of the brightest stars in the Russian political firmament. In Petersburg, Sidney linked himself to the prestigious *Astrea* lodge.

While masonic ideology was not monolithic and factionalism abounded, it would be fair to say that the overwhelming current was liberal and anti-autocratic. On the other hand, frankly revolutionary sentiments could be found as well; both Lenin and Trotsky were alleged to be brethren.[49] There was no 'masonic conspiracy' in Russia, which is not to say that there were no conspiracies among masons. The main lodges were caught up in 'purely political' agendas.⁵⁰ In 1912, for instance, representatives of many lodges constituted the so-called Supreme Council of the Peoples of Russia.⁵¹ Later rumors held that the body spawned a 'shadow government' that plotted to undermine and replace the regime of Nicholas II. What is certain is that among its adherents were many of the men who five years later would constitute the post-Tsarist Provisional Government, among them Guchkov and a young socialist attorney, Aleksandr Kerenskii.⁵²"²⁵⁵

Einstein's "secretary" during his trip to America in the spring of 1921 was Simon Ginsburg (a. k. a. Salomon Ginzberg, a. k. a. Schlomo Ginossar); who was the son of Zionist Usher Ginsburg (a. k. a. Asher Ginberg, a. k. a. Ahad Ha'am), who published under the *nom de plume* "Achad Ha-am". Ginsburg, the Elder, was the secretary for the Odessa Committee for Palestine. Some alleged that he was the voice behind *The Protocols of the Learned Elders of Zion*.²⁵⁶

It is interesting that Ha-am's son spoke for Einstein on Einstein's self-described "propaganda" tour for extreme racist Jewish nationalism in America—a man who, in Einstein's words,

"translated for me only what was essential."²⁵⁷

In February of 1923, when Einstein visited Palestine to generate publicity for himself and for his Zionist colleagues, the Zionist Executive appointed Simon Ginsberg to be "Einstein's official escort" and Ginsberg again told Einstein what to say.

Stranger still, many of Einstein's thoughts sound hauntingly similar to passages in *The Protocols of the Learned Elders of Zion*²⁵⁸ (widely available on

---

**255**. R. B. Spence, *Trust No One: The Secret World of Sidney Reilly*, Feral House, Los Angeles, (2002), pp. 65-67.

**256**. L. Fry, *L'Auteur des Protocols Achad ha-Am et le Sionisme*, Editions de la Vieille-France, Paris, (1921); **Russian** *Akhad-Khem (Asher Geintsberg); Tainyi vozhd' iudeiskii*, Berlin, (1922); **German**, Th v. Winberg, translator, *Achad Cham (Ascher Hinzberg)*, München, (1923). *See also:* N. D. Zhevakhov, *Il retroscena dei Protocolli di Sion: la vita e le opere del loro editore, Sergio Nilus e del loro autore Ascer Ghinsberg*, Unione editoriale d'Italia, Roma, (1939). *See also:* C. Weizmann, *Trial and Error: The Autobiography of Chaim Weizmann*, Harper & Brothers, New York, (1949), pp. 107-108, 266.

**257**. D. K. Buchwald, *et al.*, Editors, *The Collected Papers of Albert Einstein*, Volume 7, Appendix D, Princeton University Press, (2002), p. 623.

**258**. **English:** S. Nilus, *The Protocols and World Revolution Including a Translation and Analysis of the "Protocols of the Meetings of the Zionist Men of Wisdom"*, Small, Maynard & Co., Boston, (1920); **and** *Præmonitus præmunitus. The protocols of the Wise Men of Zion*, Beckwith Co., New York, (1920); **and** *The Jewish Peril: Protocols of the Learned Elders of Zion*, Eyre & Spottiswoode, London, (1920); **and** *The Jewish Peril: Protocols of the Learned Elders of Zion*, The Britons, 62 Oxford Street, London, (1920). **Russian:** G. Butmi, *Vragi roda cheloviecheskago, Izd. Soiuza russkago naroda*, S.-

the internet in many languages), which book portends to be the transcript of a plot by unnamed Jewish leaders, who allegedly controlled the Freemasons, to create a world government by means of the revolutionary activities encouraged by Adam Weishaupt's *Illuminati* and by the Communists, and later the Zionist Nazis.

Much has been written arguing that the *Protocols* are spurious.[259] The

---

Peterburg, Second Edition, ispr. izd., (1906). **German:** Gottfried zur Beek, under the *nom de plume* Ludwig Müller von Hausen, *Die Geheimnisse der Weisen von Zion*, Verlag Auf Vorposten, Charlottenburg, (1919). **French:** *La Conspiration Juive Contre les Peuples; "Protocols", Procès-verbaux de Reunions Secrètes des Sages d'Israël*, La Vieille-France, Paris, (1920). **Italian:** *I "Protocolli" dei "Savi Anziani" di Sion : Versione Italiana con Appendice e Introduzione*, Third Edition, G. Preziosi, (1938). *L'Internazionale Ebraica: I "Protocolli" dei "Savi Anziani" di Sion*, La Vita italiana, rassegna mensile di politica, Roma, (1938). **Swedish:** *Förlåten Faller—: Det Tillkommande Världssjälvhärskardömet Enligt "Sions vises Hemliga Protokoll"*, Enskilt Förlag, Helsingfors, (1919). **Danish:** S. Nilus translated by L. Carlsen, *Jødefaren. De Verdensberygtede Jødiske Protokoller*, Eget Forlag, Kjøbenhavn, (1920). **Hungarian:** *Sion Bölcseinek Jegyzokönyvei; a Bolsevikiek Bibliája*, Az Egyesült keresztény nemzeti liga, Budapest, (1922). **Latvian:** G. A. Kalnins, *Zidu loma cilveces vesture Veca deriba, Talmuds, Brivmurnieki jeb masoni, Cianas gudro protokoli, Ritualas slepkavibas u. c.*, Tautas vairoga, Riga, (1934). **Dutch:** S. Nilus, translated by J. Nijsse, *De Protokollen van de Wijzen van Sion*, Stichting de Misthoorn, Amsterdam, (1941). **Spanish:** *Protocolos de los Jefes de Israel: Un Plan Secreto de los Judios?*, M. Aguilar, Madrid, (1932).

[259]. "'Die Geheimnisse der Weisen von Zion'. Der Schwindelbericht eines russischen Spitzels, *Im deutschen Reich*, Volume 16, Number 5, (May, 1920), pp. 146-153. *See also:* L. Wolf, *The Jewish Bogey and the Forged Protocols of the Learned Elders of Zion*, London, The Press Committee of the Jewish Board of Deputies, (1920); **and** *The Myth of the Jewish Menace in World Affairs; or, The Truth about the Forged Protocols of the Elders of Zion*, The Macmillan company, New York, (1921). *See also:* *An Exposure of the Hoax Which is Being Foisted upon the American Public by Henry Ford in His Weekly Newspaper Entitled "The Dearborn Independent": and in the Pamphlet which He is Distributing Entitled "The World's Foremost Problem"*, Anti-Defamation League, Chicago, (1920). *See also:* W. Hard, *The Great Jewish Conspiracy*, The American Jewish Book Co., New York, (1920). *See also:* W. H. Taft, *Anti-Semitism in the United States*, Anti-defamation League, Chicago, (1920). *See also:* J. Spargo, *The Jew and American Ideals*, Harper & Brothers, New York, London, (1921). *See also:* H. Bernstein, *The History of a Lie, "The Protocols of the Wise Men of Zion"; a Study*, Ogilvie Pub. Co. New York, (1921); **and** *The Truth About "The Protocols of Zion"; a Complete Exposure*, Covici, Friede, New York, (1935). *See also:* E. B. Samuel, *An Examination of the Book Called "The Jewish Peril," or "The Protocols of the Elders of Zion." A Paper Read Before the Prophecy Investigation Society, April, 1921*, C.J. Thynne, London, (1921). *See also:* J. H. Hertz, *The New Anti-semitism: The Official Protests of the British and American Jewish Communities*, Press Committee of the Jewish Board of Deputies, London, (1921). *See also:* C. A. Windle, *The Tyranny of Intolerance*, Iconoclast Pub. Co., Chicago, (1921). *See also:* I. Zangwill, *The Forged Protocols: Complicity of Russian Police*, London, (1921). *See also:* N. Hapgood, "Henry Ford's Jew-Mania", *Hearst's International*, (November, 1922), pp. 106-108. *See also:* S. W. McCall, *Patriotism of the American Jew*, L. Middleditch Company, New York, (1924), pp. 28-41. *See also:* B. W. Segel, *Die Protokolle der Weisen von Zion kritisch beleuchtet; eine Erledigung*, Philo

Verlag, Berlin, (1924); **and** *Welt-Krieg, Welt-Revolution, Welt-Verschwörung, Welt-Oberregierung*, Philo Verlag, Berlin, (1926); **and** *The Protocols of the Elders of Zion, the Greatest Lie in History, by Benjamin W. Segel; authorized translation from the German by Sascha Czazckes-Charles, with Ten Letters of Endorsement from Eminent German Non-Jewish Scholars*, Bloch publishing co., New York, (1934); **and** *La Mentira mas Grande de la Historia: Los Protocolos de los Sabios de Sion*, D.A.I.A., Buenos Aires, (1936). ***See also:*** *Die angeblichen "Protokolle der Weisen von Zion" als weltpolitisches Agitationsmittel*, (1930). ***See also:*** J. B. Rusch, *Protokolle der Weisen von Zion, die grösste Fälschung des Jahrhunderts!*, Ragaz, (1933). ***See also:*** *The Peril of Racial Prejudice*, Jewish Information Bureau, New York, (1933). ***See also:*** *Confrontation der "Geheimnisse der Weisen von Zion". ("Die Zionistischen Protokolle") mit ihrer Quelle "Dialogue aux enfers entre Machiavel et Montesquieu" der Nachweis der Fälschung. ; Dialogue aux enfers entre Machiavel et Montesquien*, Rechtsschutzabteilung des Schweizerischen Israelitischen Gemeindebundes, Basel, (1933). ***See also:*** E. Newman, *The Jewish Peril and the Hidden Hand; The Bogey of Anti-Semitism's International Conspiracy*, Minneapolis, (1934), **and** *The Conflict or the Falsehood of the Ages, Which?*, Minneapolis, (1934); **and** *The Fundamentalists' Resuscitation of the Antisemitic Protocol Forgery*, Minneapolis, (1934). ***See also:*** F. Langer, *Die Protokolle der Weisen von Zion: Rassenhass und Rassenhetze: ein Vortrag*, Saturn-Verlag, Wien, (1934). ***See also:*** S. Federbusch, *Sions vises hemliga protokoll i sanningens ljus ...*, Tilgmanns tryckeri, Helsingfors, (1934). ***See also:*** *Di "Protokoln fun Zikney-Tsien": Di greste Provokatsye in der Yidisher Geshikhte*, Groshn-Bibliotek, Varshe, (1934). ***See also:*** M. Lazarus and A. Levy, *The Challenge*, The Mercantile Press, Port Elizabeth, (1935). ***See also:*** I. Goldberg, *The So-called "Protocols of the Elders of Zion": A Definitive Exposure of One of the Most Malicious Lies in History*, Haldeman-Julius Co., Girard, Kansas, (1936). ***See also:*** A. Rubinstein, *Adolf Hitler: Schüler der "Weisen von Zion"*, Verlagsanstalt "Graphia", (1936). ***See also:*** Pablo Montesinos y Espartero, duque de la Victoria, *Israel manda; Profecias cumplidas*, Publicaciones del Comite Argentino de Defensa Antisemitica, Buenos Aires, (1936). ***See also:*** *Berner Bilderbuch vom Zionisten-Prozess um die "Protokolle der Weisen von Zion."*, U. Bodung-Verlag, Erfurt, (1936). ***See also:*** C. A. Loosli, *Die schlimmen Juden!*, Pestalozzi-Fellenberg-Haus, Bern, (1927). ***See also:*** I. Heilbut, *Les Vrais Sages de Sion*, Denoël, Paris, (1937). ***See also:*** I. Heilbut and G. Keller, *Die öffentlichen Verleumder die "Protokolle der Weisen von Zion" und ihre Anwendung in der heutigen Weltpolitik*, Europa Verlag, Zürich, (1937). ***See also:*** R. M. Blank and P. N. Miliukov, *Adolf Hitler: Ses Aspirations, sa Politique, sa Propagande et les "Protocoles des Sages de Sion"*, L. Beresniak, Paris, (1938). ***See also:*** E. Raas and G. Brunschvig, *Vernichtung einer Fälschung: der Prozess um die erfundenen "Weisen von Zion"*, Verlag "Die Gestaltung", Zürich, (1938). ***See also:*** J. Gwyer, *Portraits of Mean Men; a Short History of the Protocols of the Elders of Zion*, Cobden-Sanderson, London, (1938). ***See also:*** A. Confino, *Les protocoles des Sages de Sion: Conférence*, Editions du comité juif algérien d'études sociales, Alger, (1938). ***See also:*** G. Feige, *Debunking the Protocols, a Catholic Scholar Replies (An Exchange of Letters by Two Catholic Priests in Which the Protocols of Zion Are Shown to Be a Classic Forgery)*, Holemans, New York, (1939). ***See also:*** H. Rollin, "L'Apocalypse de Notre Temps; les Dessous de la Propagande Allemande d'Après des Documents Inédits", *Problèmes et Documents*, Eighth Edition, Gallimard, Paris, (1939). ***See also:*** J. S. Curtiss, *An Appraisal of the Protocols of Zion*, Columbia University Press, New York, (1942). ***See also:*** *Protocols of the Elders of Zion: A Fabricated "Historic" Document*, United States of America Congressional Senate Committee on the Judiciary, Washington, U.S.

similarity between Einstein's comments and the *Protocols* is perhaps due to the racist Zionist *Zeitgeist* and the consistent use of the clichés of early political Zionism, the libertarian *Illuminati*-style views of some political radicals of the period and the influence of Karl Marx and Friedrich Engels' writings on both the authors of the *Protocols* and on Einstein, or perhaps one should say, on Einstein's script-writers. Einstein may also have been influenced by H. G. Wells, who predicted back in 1913 that a benevolent world government would follow nuclear holocaust in the 1940's,[260] or Einstein may have conversed with others about the similar pursuits of some Wellsian Socialists.

The common link in the family tree of all of these factions and various movements for world government is ancient Jewish prophesy, a. k. a. Judaism. The ancient Jews advocated terrorism, subversion and genocide to bring about world rule by a Jewish King, or "Messiah"; or, as the Frankists and their predecessors would have it, a series of incarnations of the "Messiah" in an unbroken string of Jewish kings, who would destroy the Gentiles through attrition.

---

Government Printing Office, (1964). *See also:* N. R. C. Cohn, *Warrant for Genocide: The Myth of the Jewish World-Conspiracy and the Protocols of the Elders of Zion*, Harper & Row New York, (1967); **and** *Die Protokolle der Weisen von Zion*, Baden-Baden, (1998), pp. 267-289; **and** "Der Mythos der 'Protokolle der Weisen von Zion'", *Verschwörungstheorien: Anthropologische Konstanten—historische Varianten*, Fibre, Osnabrück, (2001). *See also:* R. Singerman, "American Career of the *Protocols of the Elders of Zion*", *American Jewish History*, Volume 71, (September, 1981), pp. 48-78. *See also:* M. Hagemeister, "Wer war Sergej Nilus?", *Ostkirchliche Studien*, Volume 40, Number 1, (1991), pp. 49-63; **and** "Die 'Protokolle der Weisen von Zion'", *Russland und Europe. Historische und kulturelle Aspekte eines Jahrhundertproblems*, Leipzig, (1995), pp. 195-206; **and** "Sergej Nilus und die 'Protokolle der Weisen von Zion'", *Jahrbuch für Antisemitismusforschung*, Volume 5, (1996), pp. 127-147. *See also:* U. Lüthi, *Der Mythos von der Weltverschwörung: Die Hetze der Schweizer Frontisten gegen Juden und Freimaurer, am Beispiel des Berner Prozesses um die "Protokolle der Weisen von Zion"*, Helbing & Lichtenhahn, Basel, (1992). *See also:* G. Larsson, *Fact or Fraud?: The Protocols of the Elders of Zion*, AMI-Jerusalem Center for Biblical Studies and Research, Jerusalem, San Diego, (1994). *See also:* R. S. Levy and B. W. Segel, *A Lie and a Libel: The History of the Protocols of the Elders of Zion*, University of Nebraska Press, Lincoln, (1995). *See also:* C. De Michelis, "Les Protocoles des sages de sion: Philologie et histoire", *Cahiers du Monde russe*, Volume 38, Number 3, (1997), pp. 263-306. *See also:* S. E. Bronner, *A Rumor about the Jews: Reflections on Antisemitism and the Protocols of the Learned Elders of Zion*, St. Martin's Press, New York, (2000). *See also:* S. G. Marks, "Destroying the Agents of Modernity: Russian Antisemitism", *How Russia Shaped the Modern World*, Chapter 5, Princeton University Press, (2003), pp. 140-175; notes 354-358. *See also:* S. L. Jacobs and M. Weitzman, *Dismantling the Big Lie: The Protocols of the Elders of Zion*, Simon Wiesenthal Center in Association with KTAV Pub. House, Jersey City, (2003). *See also:* C. G. De Michelis and R. Newhouse, *The Non-Existent Manuscript: A Study of the Protocols of the Sages of Zion*, University of Nebraska Press, Lincoln, (2004).

**260**. H. G. Wells, *The World Set Free: A Story of Mankind*, Macmillan, London, (1914); also publish in Leipzig, Germany by B. Tauchnitz.

Many were struck by the similarity of the plans laid out in the *Protocols* to the later events occurring in the Bolshevist movements, particularly those led by Lev Davidovich Bronstein, a. k. a. "Leon Trotsky", and Aaron Cohen, a. k. a. "Béla Kuhn"—around whom the murderous Jews of Hungary rallied.[261] The Bolshevists, often led by Jews, committed genocide, destroyed Gentile cultures, subverted Gentile governments, destroyed religions, and took horrible vengeance against nations which lagged behind in the movement to emancipate Jews, all of which was prophesied in the Old Testament and reiterated by Jewish authors throughout history, and reiterated in the *Protocols* of 1905.

The Bolshevist movement was immense in the early Twentieth Century. It worked to undermine all societies and was especially active in Europe. Bolshevism had a disproportionately Jewish leadership, and manifested itself most prominently and successfully in nations with large Jewish populations. Jewish influence was especially pernicious, given that it carried out Jewish vengeance[262] and Jewish aggression—carried out the events called for in Jewish Messianic mythology. The fact that Jewish radicals were deliberately fulfilling horrific Jewish Messianic prophecies caused consternation among several governments around the world and provoked a worldwide panic that racist, tribal Jews, including Albert Einstein, were attempting to take over the world and mass murder, or destroy the lives of, non-Jews and assimilatory Jewry in what they viewed as an historic phase of Judaism.

The United States Government investigated the question of whether or not "Russian Jew" and "Bolshevist" were synonymous terms.[263] Did those who were alarmed by the *Protocols*, which foretold the carnage of the First World War, the deaths of tens of millions of Gentiles and the carnage of Bolshevism which threatened to take over the world—the mass murder of hundreds of millions of innocent civilians—the deliberate mass murder of the best of society and of the best of the human gene pool—the utter destruction of Western culture—did those who called attention to the parallels of the events foretold in the *Protocols* published in 1905, and actual unprecedented events which had since occurred from 1914 to 1920, have a right to raise their concerns?

The editors and translators of various editions of the *Protocols* expressed these concerns and published evidence in support of these facts. For example, the Small, Maynard & Company translation of 1920, published in Boston, relied upon an article published in *La Vieille-France*, Number 160, (February, 1920), pp. 10-13, to stress the common belief that,

"The article asserts that Bolshevism is nothing but a phase of Judaism, and also states that the Jewish Bolshevist leaders in Russia were subsidized by

---

[261]. P. S. Mowrer, "The Assimilation of Israel", *The Atlantic Monthly*, Volume 128, Number 1, (July, 1921), pp. 101-110, at 108-109.
[262]. "A Jewish Revolution", *The Maccabean*, New York, (November, 1905), p. 250. W. E. Curtis, "The Revolution in Russia", *National Geographic Magazine*, (May, 1907), p. 313.
[263]. "President Gives Hope to Zionists", *The New York Times*, (3 March 1919), pp. 1, 3.

Jewish banking houses in the United States and Germany."[264]

The book, which also contains the above translation of the *Protocols*, devotes more than half of its pages to proving this thesis, by quoting witnesses and statistics; as well as, in the authors' minds, implausible, disingenuous and easily refuted denials by leading Jews. The editors even quote eminent Jews like Lionel de Rothschild, who took, or pretended to take, his fellow Jews to task for bringing Bolshevism to England.[265] Several references to the predominance of Jews among the Bolsheviks are cited in this translation and exposition, *The Protocols and World Revolution Including a Translation and Analysis of the "Protocols of the Meetings of the Zionist Men of Wisdom"*, Small, Maynard & Co., Boston, (1920); with specific emphasis on testimony from the Overman Committee, as recorded in: *Bolshevik Propaganda. Hearings Before a Subcommittee of the Committee on the Judiciary, United States Senate, Sixty-Fifth Congress, Third Session and Thereafter, Pursuant to S. Res. 439 and 469. February 11, 1919, to March 10, 1919.*, United States Government Printing Office, Washington, D. C., (1919), pp. 47, 69, 111, 114, 116, 132, 135, 142, 269, 270, 310, 321, 424.

Whether or not one believed in the authenticity of the *Protocols*, there was no doubting the world-wide threat posed by Jewish Bolsheviks. On 19 June 1920, *The Chicago Tribune* published an article by John Clayton on the front page, which alleged that an international Jewish organization sought Jewish supremacy, largely through the destruction of the British Empire,

## "TROTZKY LEADS RADICAL CREW TO WORLD RULE

Bolshevism Only a
Tool for His Scheme
BY JOHN CLAYTON.
(Chicago Tribune Foreign News Service.)
(By Special Cable.)
(Copyright: 1920: By the Tribune Company.)

PARIS, June 18.—For the last two years army intelligence officers, members of the various secret service organizations of the entente, have been bringing in reports of a world revolutionary movement other than

---

**264**. S. Nilus, *The Protocols and World Revolution Including a Translation and Analysis of the "Protocols of the Meetings of the Zionist Men of Wisdom"*, Small, Maynard & Co., Boston, (1920), p. 4.
**265**. S. Nilus, *The Protocols and World Revolution Including a Translation and Analysis of the "Protocols of the Meetings of the Zionist Men of Wisdom"*, Small, Maynard & Co., Boston, (1920), p. 134; *which cites:* L. de Rothschild, Swaythling, P. Magnus, M. Samuel, H. S. Samuel, L. L. Cohen, I. Gollancz, J. Monash, C. G. Montefoire and I. Spielmann, "To the Editor of the Morning Post", *London Morning Post*, (23 April 1919).

Bolshevism. At first these reports confused the two, but latterly the lines they have taken have begun to be more and more clear.

Bolshevism aims for the overthrow of existing society and the establishment of an international brotherhood of men who work with their hands as rulers of the world. The second movement aims for the establishment of a new racial domination of the world. So far as the British, French and our own department's inquiry have been able to trace, the moving spirits in the second scheme are Jewish radicals.

### Use Local Hatreds.

Within the ranks of communism is a group of this party, but it does not stop there. To its leaders, communism is only an incident. They are ready to use the Islamic revolt, hatred by the central empires for England, Japan's designs on India, and commercial rivalry between America and Japan.

As any movement of world revolution must be, this is primarily anti-Anglo-Saxon. It sees its greatest task in the destruction of the British empire and the growing commercial power of America. The brains of this organization are in Berlin.

### Trotzky at Head.

The directing spirit which issues the orders to all minor chiefs and finds money for the work of preparing the revolt is in the German capital. Its executive head is none other than Trotzky, for it is on the far frontiers of India, Afghanistan, and Persia that the first test of strength will come. The organization expert of the present Russian state is recognized, even among the members of his own political party, as a man of boundless ambition, and his dream of an empire of the east is like that of Napoleon.

The organization of the world Jewish-radical movement has been perfected in almost every land. In the states of England, France, Germany, Poland, Russia, and the east it has its groups. It is behind the Islamic revolt with all the propaganda skill and financial aid at its command because it hopes to control the shaping of the new eastern empire to its own ends. Sympathy with the eastern nationals probably is one of the chief causes for the victory of the pro-nationals in the bolshevik party, which threw communism solidly behind the nationalist aspirations of England's colonies.

### Out to Grab Trade Routes.

The aims of the Jewish-radical party have nothing of altruism behind them beyond liberation of their own race. Except for this their aims are purely commercial. They want actual control of the rich trade routes and production centers of the east, those foundations of the British empire which always have been the cornerstone of its national supremacy.

They are striking for the same ends as Germany when she entered the war of 1914 to establish Mittel Europa and so give the Germans control of the Bagdad railway. They believe Europe is tired of conflict and that England is too weak to put down a concerted rebellion in part of her eastern possessions. Therein lies the hope of success. They are staking brains and money against an empire.

Westward the course of empire makes its way, but even it swings backward to the old battleground where for countless ages peoples have fought. Nations have risen and crumbled around control of eastern commerce."[266]

The Jewish press tried to make it appear that it was illogical to charge German-Jewish bankers with sponsoring Bolshevism. The following article appeared in *The Jewish Chronicle* on 11 April 1919 on page 8,

**"Jews and Bolshevism.**
WE observe that writers in the Press describe ninety-five per cent. of the new Bolshevist Government in Hungary as Jews. Whether these reports are correct we do not know. The prominence of certain individual Jews in the Russian movement having been established—though it would seem from a letter which appeared in the *Times* the other day and is quoted by a contributor elsewhere in this issue, the Jewish *personnel* has been much exaggerated—long historical tradition inevitably inclines the uncritical to treat all other Bolshevist administrations as Jewish, and to assume that every sympathiser with LENIN must be a Jew with a disguised name. Despite the identification with it of individual Jews we believe that, in essence, Bolshevism is repugnant to average Jewish sentiments as it exists. For good or ill, the Jew is for the most part a 'law and order' man. He hates violence, political equally with civil. He gravitates, in the mass, to Conservative doctrine, as we have seen, in striking fashion, in the political history of British Jewry since the days of emancipation. He has respect for property and an ambition to share the good tidings of the world. So much is this the case, indeed, that the undiscriminating have coined the foolish phrase, 'as rich as a Jew,' and malicious writers have for generations confounded Judaism with Capitalism. Trotzky and his companions, therefore—though no one in reason could deny their right to be Bolsheviks because they are Jews or Jews because they are Bolsheviks—are in no sense whatever representative of Jewish feelings or tendencies. Indeed, if popular notions as to Jewish wealth are only half true, then there is no body of men more concerned in the extirpation of Communist ideas than the Jewish people. The world cannot have it both ways. It cannot at one and the same time hold the Jew up to execration as the symbol of Capitalism and of expropriating Socialism. None the less, the Jewish disciples of Bolshevism are, as has been said, in one sense, essentially Jewish. They are Jewish in their search after an ideal. We may quarrel with that ideal—though we see that, stripped of its barbarism and cruelty, as in Hungary, the Allies do not hesitate to hold converse with it and negotiate with it, while, as we were reminded last week, a great London daily newspaper recently declared Bolshevism in essence to be idealism unmatched since the teachings of JESUS were promulgated. Even though we quarrel with Bolshevism, it cannot be doubted that, to many

---

**266**. *See also:* "World Mischief", *The Chicago Tribune*, (21 June 1920), p. 8.

believers in the theory, it is an ideal, and that, as the writer referred to observed, is the point of attraction for the Jews who are attracted by its doctrines. A people has been exiled from its own soil for centuries and persecuted by the exponents of Nationalism, in every land. Is it really a matter of surprise if, robbed of the national ideal, and schooled to regard it as their worst enemy, some Jews turn away from the jargon of frontiers and armies, and go in quest of some economic ideal? We stress these comments because it is time that the general Press tried to probe deeper into the heart of things, and because we believe they do the interests of this or any other country little good by taking superficial—which too often are harmful—views of current phenomena. The moral will not, we hope, be lost on thinking men—or on thinking Jews."

The following article appeared in *The Jewish Chronicle* on 11 April 1919 on page 13 (note that the statement in the *Jewish World* to which the *Morning Post* responded was also published in *The Jewish Chronicle* on 28 March 1919 on page 11—*see also*: *The Jewish Chronicle*, 2 May 1919 on pages 18 and 19, 9 May 1919 on page 18, 25 July 1919 on page 9,

## "The 'Morning Post' and the 'Jewish World.'

Tuesday's *Morning Post* contained an article entitled 'Bolstering the Bolshevik,' in the course of which that paper said:
We notice that the *Daily Herald* and the *Daily News* are persistently telling the people of this country that we are fighting Bolshevism in obedience to the pressure of the capitalists. Now that is a lie. We are fighting Bolshevism in opposition to a very strong group of German-Jewish and Russian-Jewish capitalists, who are secretly working for the Bolshevik cause. We have mentioned several times the disagreeable fact that the Russian Bolsheviks were Russian Jews. Those Jews are at the present moment in control of the Russian Government, and they have powerful friends in all the Allied countries who are helping them. We have appealed to the British Jews, but appealed so far in vain, to dissociate themselves formally from a cause which is doing the Jewish people terrible harm in all parts of the world. In reply the Jewish Press shower upon us not only abuse but threats. Thus, for example, the *Jewish World* threatens us with the fate of Mordecai: '... we wish it no harm, but we would beg it to recollect,' so it says, '*while yet it has its feet upon the earth* the fate of its anti-Jewish forbear in that narrative, in the hope that it may amend its ways betimes.'
We are aware of the significance of that threat. We fully understand what it means, and the secret Allies upon whom the *Jewish World* reckons when it makes it. We saw them at work in Glasgow and in Belfast. We see them at work now in Budapest, where, it is reported, out of thirty members of the Bolshevik Soviet, twenty-six are Jews. We understand the threat; but we do not propose to be deterred in our duty to the British public by the terrorist methods of the Bolsheviks. And we suggest to the British Jewish

community—most of whom, we believe, are by no means in sympathy with this crusade—that they are being served very badly by their newspapers, which openly threaten Bolshevik methods and scoff at advice which is tendered in a friendly spirit. In secret, we feel certain, the majority of British Jews distrust and dislike the fanatics who are now leading Jewry astray in the cause of a spurious Jewish Imperialism. But they are afraid to dissociate themselves publicly from the dervishes of Judaism. In the meantime these powerful influences are at work in every country, and chiefly in Paris, where they are working powerfully against the cause of Poland. An unseen hand is at this present time stifling the infant Poland in its cradle, and this is being done in the interests of German-Jewish Capitalism. It is a conspiracy which is assisted by so-called Liberal newspapers like the *Daily News* and so-called Labour newspapers like the *Daily Herald*; but it is a conspiracy nevertheless which is directed against the cause of liberty in Poland and in the interests of alien Capitalism.

Wednesday's *Jewish World* trenchantly answered the *Morning Post*, and, it goes without saying, made no little play of its muddling up Mordecai with Haman. It pointed out how the allegations contained in the *Morning Post*, concerning Jews and Bolshevism, were little more than 'a whirling screed of bemused contradictions,' in which Jews are at one and the same time pilloried as Bolsheviks and Capitalists."

If the same Jewish banker can trap some rabbits with a snare in the forest and trap other rabbits with a spring trap in the grass, then the same Jewish banker can sponsor and profit from both Capitalism and Bolshevism at the same time. Jewish leaders have always profited from war and without opposing sides there is no war so it is in their interests to create and sponsor opposing political forces. Indeed the sophistry promoted in the Jewish press that leading Capitalist Jews could not possibly sponsor and profit from Bolshevism is easily refuted by the fact that one of the premier Jewish Capitalists in the world financed the Bolshevik Revolution in Russia, financed Trotsky and Lenin, and closed off the Czar's access to international money markets. That banker was Jacob Schiff, a German-Jewish Capitalist whose family had long had intimate ties to the Rothschild family. What would prevent a German-Jewish banker from paying crypto-Jews to overthrow the Czar so that the German-Jewish Capitalists like the Warburgs and their cohorts could steal the wealth of the Russian nation and commit genocide against the Russian People, whom they expressly despised? Apparently nothing, since that is exactly what German-Jewish Capitalists did do, and Jacob Schiff openly bragged about it.

On 5 June 1916, on page 6, in an article entitled, "Jacob Schiff Quits Jewish Movements", the *New York Times* quoted Schiff's own statement that he had blocked Russia's access to the money markets during several of Russia's wars, wars which the Jews had caused,

"I, who have for twenty-five years singlehanded struggled against the invasion of the Russian Government into American money markets, and to this day stave them off."

There was nothing new about the Jewish campaign to destroy Russia with war and by blocking Russia's access to loan capital. Major Osman Bey wrote in the 1870's that,

"Yet the Jews hate the Russians because so far they have not succeeded in obtaining such a powerful influence in their country, as it has been shown that they possess in England; and therefore the following cablegram, received in the United States, March 29, 1878, from abroad, will explain itself. Here it is:

'Russia's pecuniary troubles increase every day. It is said, that all the Hebrew bankers in London and on the continent have agreed not only to refuse to lend Russia any money, but to prevent the success of any loan she may put on the market. *England, on the contrary, can command practically endless supplies of the sinews of war!* '"[267]

Jewish leaders were very familiar with the Greek and Hegelian notions of the cycles of government and of human history. They sought to control every phase of these cycles and struggles, and there is no contradiction in that fact. They profited from pitting Capitalist nations, which were ultimately under their control, against Bolshevist nations, which were ultimately under their control. The synthesis of these dialectical struggles was gold in their pockets. If it benefitted the Jewish bankers to have a Capitalist revolution, then they had one. If it profited them to instigate a Bolshevist revolution, then they did so. The Jewish ideal is to take over the wealth and the governments of the world. It is not surprising that Jewish bankers have used various means to accomplish that end. It was not unlike Jews to pretend to be of one faith, while espousing another. Nor was it unlike Jews to throw stumbling stones onto the paths of others, or to promise Utopian dreams to Gentiles to manipulate their actions and as a trap to deliberately lead them into disaster.

No one accused the Jewish bankers of personally and sincerely holding opposing views at the same time. The accusation was quite the contrary, that the basic duplicity of Jewish bankers led them to entice others into self-destruction through deliberate lies and unfair and deceitful practices.

In the minds of the authors of numerous translations of the *Protocols*, the resolution of the seeming paradox of the Jew as capitalist and the Jew as Bolshevik, was easily found in the *Protocols*, where politics is said to be amoral and insincere, where actions are paramount, and where liberal political movements are merely a means to weaken Gentile governments, so that Jewish wealth can prevail and fulfill Jewish prophecy. Denis Fahey was one of many who argued that Jewish financiers were behind Marx, Trotsky, Lenin, Stalin, etc. and sought to use Communism as a means to gain absolute Jewish control over

---

[267]. Major Osman Bey, *The Conquest of the World by the Jews: An Historical and Ethnical Essay*, St. Louis Book & News Company, St. Louis, (1878), p. 55.

the world.[268] Liberalism secured Jewish rights, and, thereby, Jewish access to the press and to government. Liberalism destroyed monarchies, which had served as natural barriers to Jewish political domination, and which Jewish prophecy demanded must be abolished.

Altruism was not the motivating force behind organized "Jewish policy", behind Jewish Liberalism, rather it was perceived self-interest. The "Jewish idealism" of Bolshevism was a Trojan Horse, which lured Gentile nations into falling into the trap Bolshevism in name of "liberty, equality and fraternity", which Bolshevism immediately stripped the Gentiles of all their rights and put cruel and murderous Jews into power. When Jewish leaders had sufficiently crippled a society to the point where its members clamored for a dictator to restore order and peace, the principles of Liberalism were not only abandoned by Jewish leadership, they were ridiculed. Jewish Liberalism was not a Jewish ideal, nor an end, but rather a means to obtain absolute Jewish domination. It was the typical Jewish bait of a promised Utopia that once swallowed poisoned its prey. Though the Jewish Bolshevists held out candy in one hand, they clutched a knife behind their backs the entire time they were petitioning for power.

Jewish Capitalism worked in collusion with Jewish Liberalism toward the same end. The concentrated wealth of the Jewish financiers enabled them to create wars, control the press and politicians, and finance revolutions. It also gave them control over international finance so that they could foment wars and then ensure a given nation would collapse in economic, as well as military, ruin. Jewish revolutionaries would instigate strikes, which would further bankrupt the nation. Jewish revolutionaries would then draw the attention of the public to its misery, misery they had caused but which they would blame on the government. Jewish Liberalism and Jewish Capitalism worked together to create international Jewish domination.

In the *Protocols*, Capitalism and Communism, and the strife between them, all serve the end of racist, tribal Jewish wealth accumulation and the acquisition of power—the fulfillment of Jewish prophesy through the weakening of Gentile power, especially Gentile monarchies. There is no more a contradiction in self-interests, to the exclusion of lofty logical consistency, in one tribe concurrently advocating both Communism and Capitalism; than there is in one imperialistic nation concurrently advocating both absolute national sovereignty and colonialism—as so often happens. As the *Protocols* indicate, sophistry and hypocrisy do indeed prevail in politics, where the true motives of the leaders are often not reflected in expressed party ideologies.

The accusation that racist, tribal Jews advocated both Communism and Capitalism was not an accusation that they were sincere in both of these mutually exclusive ideals, but that they were insincere and exploitive of others sincerity and naïveté, and sought to profit from conflict. There is no denying that

---

**268**. D. Fahey, *The Mystical Body of Christ in the Modern World*, Browne and Nolan Limited, London, (1935), pp. 86-93, 99-100.

Communist nations have been robbed of their wealth, deliberately and as a matter of circumstances, and that conflicts between Communist nations and Capitalistic nations have profited international financiers, as can any war, and further that where Capitalism has failed to corrupt a monarchy (or rather failed to spice it with the preferred flavor of corruption), Communism can overthrow it—and Communism did infect Eastern Europe following World War II—and many believed that Jews provoked wars so as to weaken societies and leave them vulnerable to Communist takeover, and/or Capitalistic buyout. All the nations of Europe were under constant attack from Bolsheviks during and after the First World War. For those who saw in this attack a tribal mission by racist Jews, which revolutionary mission is a pervasive theme in Judaism, the *Protocols* served as,

> "Proof that Communism is a Jewish world plot to enslave the Gentiles by creating wars and revolutions, and to seize power during the resulting chaos and to rule with their claimed superior intelligence as the chosen people."[269]

"Part Two" of the 1920, Small, Maynard & Company translation of the *Protocols* starts off with the statement,

## "Part Two
EVIDENCE AS TO ORIGIN AND AUTHENTICITY
I. PARALLELISM BETWEEN THE ACTUAL
POLICIES OF THE BOLSHEVIKI AND
THE PROTOCOLS

THE most striking fact in connection with the Protocols is the close resemblance which their ruthless program bears in many respects to the policies actually put into effect by the Bolsheviki in Russia. Indeed, without this fact before us, the necessity for a serious consideration of the Protocols would be much less apparent. If the evidence shows that the Bolshevist movement is a movement conducted under Jewish leadership and principally controlled by Jews, and, furthermore, that it closely corresponds with the political program outlined in the Protocols, then, indeed, we have facts of grave significance supporting the authenticity of the Protocols."[270]

"Mentor" wrote in *The Jewish Chronicle* on 4 April 1919 on page 7,

> "THERE is much in the fact of Bolshevism itself, in the fact that so many Jews are Bolshevists, in the fact that the ideals of Bolshevism at many points are consonant with the finest ideals of Judaism, some of which went to form

---

**269**. Cover of: S. Nilus, *Protocols of the Meetings of the Learned Elders of Zion*, English translation by V. E. Mardsen, Thunderbolt, Savannah, Georgia, (1960).
**270**. S. Nilus, *The Protocols and World Revolution Including a Translation and Analysis of the "Protocols of the Meetings of the Zionist Men of Wisdom"*, Small, Maynard & Co., Boston, (1920), p. 74.

the basis of the best teachings of the founder of Christianity—these are things which the thoughtful Jew will examine carefully. It is the thoughtless one who looks upon Bolshevism only in the ugly repulsive aspects which all social revolutions assume and which make it so hateful to the freedom-loving Jew—when allowed to be free. It is the thoughtless one that thus partially examines the greatest problem the modern world has been set, and as his contribution to the solution dismisses it with some exclamation made in obedient deference to his own social position, and to what for the moment happens to be conventionally popular."

## 5.3.1 Human Sacrifice and the Plan to Discredit Gentile Government—Fulfilled

Racist Zionist Theodor Herzl secretly wrote in his diary of a conversation he had had with racist Zionist Max Nordau,

"Never before had I been in such perfect tune with Nordau. [***] This has nothing to do with religion. He even said that there was no such thing as a Jewish dogma. But we are of one race. [***] 'The Jews,' he says, 'will be compelled by anti-Semitism to destroy among all peoples the idea of a fatherland.' Or, I secretly thought to myself, to create a fatherland of their own."[271]

After the Nazis had segregated, humiliated and slaughtered millions of Jews at the behest of the Jewish financiers, and had ruined Germany and the image of Gentile government, racist Zionist Albert Einstein wrote, among other things, in 1945,

"[The Jews'] status as a uniform political group is proved to be a fact by the behavior of their enemies. Hence in striving toward a stabilization of the international situation they should be considered as though they were a nation in the customary sense of the word. [***] In parts of Europe Jewish life will probably be impossible for years to come. In decades of hard work and voluntary financial aid the Jews have restored the soil of Palestine to fertility. All these sacrifices were made because of trust in the officially sanctioned promise given by the governments in question after the last war, namely that the Jewish people were to be given a secure home in their ancient Palestinian country. To put it mildly, the fulfillment of this promise has been but hesitant and partial. Now that the Jews—especially the Jews in Palestine—have in this war too rendered a valuable contribution, the promise must be forcibly called to mind. The demand must be put forward that Palestine, within the limits of its economic capacity, be thrown open to Jewish immigration. If supranational institutions are to win that confidence

---

[271]. T. Herzl, English translation by H. Zohn, R. Patai, Editor, *The Complete Diaries of Theodor Herzl*, Volume 1, Herzl Press, New York, (1960), p. 196.

that must form the most important buttress for their endurance, then it must be shown above all that those who, trusting to these institutions, have made the heaviest sacrifices are not defrauded."[272]

Lenni Brenner wrote in his exposé *Zionism in the Age of the Dictators*, "The Wartime Failure to Rescue", Chapter 24, Lawrence Hill Books, Chicago, (1983), pp. 235-238 [Brenner cites in his notes: "22. Michael Dov-Ber Weissmandel, *Min HaMaitzer* (unpublished English translation). 23. Ibid. 24. Ibid. (Hebrew edn), p. 92. 25. Ibid., p. 93."],

**"'For only with Blood Shall We Get the land'**

The Nazis began taking the Jews of Slovakia captive in March 1942. Rabbi Michael Dov-Ber Weissmandel, an Agudist, thought to employ the traditional weapon against anti-Semitism: bribes. He contacted Dieter Wisliceny, Eichmann's representative, and told him that he was in touch with the leaders of world Jewry. Would Wisliceny take their money for the lives of Slovakian Jewry? Wisliceny agreed for 50,000 in dollars so long as it came from outside the country. The money was paid, but it was actually raised locally, and the surviving 30,000 Jews were spared until 1944 when they were captured in the aftermath of the furious but unsuccessful Slovak partisan revolt.

Weissmandel, who was a philosophy student at Oxford University, had Volunteered on 1 September 1939 to return to Slovakia as the agent of the world Aguda. He became one of the outstanding Jewish figures during the Holocaust, for it was he who was the first to demand that the Allies bomb Auschwitz. Eventually he was captured, but he managed to saw his way out of a moving train with an emery wire; he jumped, broke his leg, survived and continued his work of rescuing Jews. Weissmandel's powerful post-war book, *Min HaMaitzer* (From the Depths), written in Talmudic Hebrew, has unfortunately not been translated into English as yet. It is one of the most powerful indictments of Zionism and the Jewish establishment. It helps put Gruenbaum's unwillingness to send money into occupied Europe into its proper perspective. Weissmandel realised: 'the money is needed here – by us and not by them. For with money here, new ideas can be formulated.'[22] Weissmandel was thinking beyond just bribery. He realised immediately that with money it was possible to mobilise the Slovak partisans. However, the key question for him was whether any of the senior ranks in the SS or the Nazi regime could be bribed. Only if they were willing to deal with either Western Jewry or the Allies, could bribery have any serious impact. He saw the balance of the war shifting, with some Nazis still thinking they could win and hoping to use the Jews to put pressure on the Allies, but others

---

**272**. A. Einstein, "Unpublished Preface to a Blackbook", *Out of My Later Years*, Philosophical Library, New York, (1950), pp. 258-259, at 259.

beginning to fear future Allied retribution. His concern was simply that the Nazis should start to appreciate that live Jews were more useful than dead ones. His thinking is not to be confused with that of the Judenrat collaborators. He was not trying to save some Jews. He thought strictly in terms of negotiations on a Europe-wide basis for all the Jews. He warned Hungarian Jewry in its turn: do not let them ghettoise you! Rebel, hide, make them drag the survivors there in chains! You go peacefully into a ghetto and you will go to Auschwitz! Weissmandel was careful never to allow himself to be manoeuvred by the Germans into demanding concessions from the Allies. Money from world Jewry was the only bait he dangled before them.

In November 1942, Wisliceny was approached again. How much money would be needed for all the European Jews to be saved? He went to Berlin, and in early 1943 word came down to Bratislava. For $2 million they could have all the Jews in Western Europe and the Balkans. Weissmandel sent a courier to Switzerland to try to get the money from the Jewish charities. Saly Mayer, a Zionist industrialist and the Joint Distribution Committee representative in Zurich, refused to give the Bratislavan 'working group' any money, even as an initial payment to test the proposition, because the 'Joint' would not break the American laws which prohibited sending money into enemy countries. Instead Mayer sent Weissmandel a calculated insult: 'the letters that you have gathered from the Slovakian refugees in Poland are exaggerated tales for this is the way of the '*Ost-Juden*' who are always demanding money'.[23]

The courier who brought Mayer's reply had another letter with him from Nathan Schwalb, the HeChalutz representative in Switzerland Weissmandel described the document:

> There was another letter in the envelope, written in a strange foreign language and at first I could not decipher at all which language it was until I realised that this was Hebrew written in Roman letters, and written to Schwalb's friends in Pressburg [Bratislava] ... It is still before my eyes, as if I had reviewed it a hundred and one times. This was the content of the letter:
>
> 'Since we have the opportunity of this courier, we are writing to the group that they must constantly have before them that in the end the Allies will win. After their victory they will divide the world again between the nations, as they did at the end of the first world war. Then they unveiled the plan for the first step and now, at the war's end, we must do everything so that Eretz Yisroel will become the state of Israel, and important steps have already been taken in this direction. About the cries coming from your country, we should know that all the Allied nations are spilling much of their blood, and if we do not sacrifice any blood, by what right shall we merit coming before the bargaining table when they divide nations and lands at the war's end? Therefore it is silly, even impudent, on our part to ask these nations who are spilling

their blood to permit their money into enemy countries in order to protect our blood—for only with blood shall we get the land. But in respect to you, my friends, *atem taylu*, and for this purpose I am sending you money illegally with this messenger.'[24]

Rabbi Weissmandel pondered over the startling letter:

> After I had accustomed myself to this strange writing, I trembled, understanding the meaning of the first words which were 'only with blood shall we attain land'. But days and weeks went by, and I did not know the meaning of the last two words. Until I saw from something that happened that the words '*atem taylu*' were from '*tiyul*' [to walk] which was their special term for 'rescue'. In other words: you, my fellow members, my 19 or 20 close friends, get out of Slovakia and save your lives and with the blood of the remainder—the blood of all the men, women, old and young and the sucklings—the land will belong to us. Therefore, in order to save their lives it is a crime to allow money into enemy territory—but to save you beloved friends, here is money obtained illegally.
>
> It is understood that I do not have these letters, for they remained there and were destroyed with everything else that was lost.[25]

Weissmandel assures us that Gisi Fleischman and the other dedicated Zionist rescue workers inside the working group were appalled by Schwalb's letter, but it expressed the morbid thoughts of the worst elements of the WZO leadership. Zionism had come full turn: instead of Zionism being the hope of the Jews, their blood was to be the political salvation of Zionism."

Racist Zionist leader Rabbi Stephen S. Wise boldly stated soon after the First World War and the Bolshevik Revolution in Russia, as quoted in an article, "President Gives Hope to Zionists", *The New York Times*, (3 March 1919), pp. 1, 3.

"The rebuilding of Zion will be the reparation of all Christendom for the wrongs done to Jews."

As Rabbi Wise' must have known, the Old Testament and modern Zionists asserted that Gentiles, "Esau", would fund, labor, and provide the military needed to create, build and maintain Israel. The Zionists believed that it was the prophetic duty of the Gentile to God and to Jacob to slave and die building and fighting for Israel, the "chosen people". It was the assimilatory Jew's prophetic duty to die together with the Gentile.

The following article appeared in *The Jewish Chronicle* on 22 September 1922 on page 31, which states that there would be no peace without a solution of the Jewish question, and that the Palestine Mandate was "reparation to the

Jew for two thousand years of martyrdom",

## "5682.
## THE YEAR'S RETROSPECT.

THE year just closing will be for ever memorable in Jewish annals as the year which saw the confirmation of the Mandate, with its formal and solemn establishment of the Jewish claim to Palestine as the National Home of the race. That one great central, irrevocable fact, however it be construed or whittled down by individual statesmen, stamps 5682 as *annus mirabilis* in Jewish history. It calls a halt to two thousand years of aimless drifting, and sets a definite direction in which the Jew may march with confidence. It comes at a moment of immense opportuneness to lift, if ever so little, an almost intolerable burden of suffering, confusion, and despair. It represents a movement which, whatever deductions may legitimately be made from its value upon this or that ground, is, at all events in essence, constructive. It embodies the recognition by the nation that it has a second problem of 'reparations' to solve—reparation to the Jew for two thousand years of martyrdom; and that the solution of the Jewish question is indispensable to world peace. Whether the Jewish Palestine, as the politicians are at the moment fashioning it, be a great bright light, illuminating the darkness of the Diaspora, or a will-o'-the-wisp full with fatality for the hopes of our people, the world-approved Mandate we cannot away with. Hold destiny what it may, the future of the Jewish People after the Mandate's confirmation can never be like the past. It is that which makes the year now ending a year of years in our people's chequered career, and its story a tale to linger over in the depressing procession of tragedies called Jewish history."

What absolute power did Zionist Jews have to ensure perpetual war if the Gentiles refused to let themselves be coerced into stealing the land of Palestine from its indigenous population and giving it to Zionist Jews who had no right to it? What debt did the English have to pay as "reparation to the Jew for two thousand years of martyrdom"?

Joseph Finn wrote in a Letter to the Editor of *The Jewish Chronicle* published on 22 September 1922 on page 14,

> "We will reach our [Hebrew deleted.] when all wars—military and commercial—shall cease, and in consequence thereof the nations become truly civilised and refined, when they begin to feel sorrow because of the wrongs they have done to us throughout the centuries. Then will *our* day come, when the nations will be eager to compensate us for the wrongs we are suffering and have suffered. Blessed be those who live to see that day!"

Finn speaks of the revenge of the Jews upon the Gentiles for the "Controversy of Zion"—of the prophesied age when the Jews will enslave and then destroy the Gentiles, after the Jewish Messiah passes judgment on non-Jews and assimilated Jews (*Isaiah* 11). The Jewish book of *Zechariah* 8:23 promises the Jews that ten Gentiles will gladly slave for every Jew,

> "Thus saith the LORD of hosts; In those days *it shall come to pass*, that ten men shall take hold out of all languages of the nations, even shall take hold of the skirt of him that is a Jew, saying, We will go with you: for we have heard *that* God *is* with you."

The Talmud at *Shabbath* 32*b* increases the number of Gentile slaves per Jew to 2,800,

> "Resh Lakish said: He who is observant of fringes will be privileged to be served by two thousand eight hundred slaves, for it is said, *Thus saith the Lord of hosts: In those days it shall come to pass, that ten men shall take hold, out of all the languages of the nations, shall even take hold of the skirt of him that is a Jew, saying, We will go with you,* etc."[273]

The Jewish book of *Genesis* 25:23; 27:38-41 promises the Gentiles to the Jews as their slaves and slave soldiers, and gives the Jews an incentive to exterminate the Gentiles because the Gentiles dare to be angry at the Jews for deceiving them and using them as slaves,

> "25:23 And the LORD said unto her, Two nations *are* in thy womb, and two manner of people shall be separated from thy bowels; and *the one* people shall be stronger than *the other* people; and the elder shall serve the younger. [\*\*\*] 27:38 And Esau said unto his father, Hast thou but one blessing, my father? bless me, *even* me also, O my father. And Esau lifted up his voice, and wept. 27:39 And Isaac his father answered and said unto him, Behold, thy dwelling shall be the fatness of the earth, and of the dew of heaven from above; 27:40 And by thy sword shalt thou live, and shalt serve thy brother; and it shall come to pass when thou shalt have the dominion, that thou shalt break his yoke from off thy neck. 27:41 And Esau hated Jacob because of the blessing wherewith his father blessed him: and Esau said in his heart, The days of mourning for my father are at hand; then will I slay my brother Jacob."

Rabbi Wise's statement in the immediate post-WW I era, recalls the Jewish prophecy that Gentiles would be massacred as reparation for the wrongs done to the Jews and that the rebuilding of Zion heralded the event. *Isaiah* 34 states:

> "1 Come near, ye nations, to hear; and hearken, ye people: let the earth hear, and all that is therein; the world, and all things that come forth of it. 2 For the indignation of the LORD is upon all nations, and his fury upon all their

---

**273**. I. Epstein, Shabbath 32*b*, *The Babylonian Talmud*, Volume 7, The Soncino Press, London, (1938), p. 149.

armies: he hath utterly destroyed them, he hath delivered them to the slaughter. 3 Their slain also shall be cast out, and their stink shall come up out of their carcases, and the mountains shall be melted with their blood. 4 And all the host of heaven shall be dissolved, and the heavens shall be rolled together as a scroll: and all their host shall fall down, as the leaf falleth off from the vine, and as a falling fig from the fig tree. 5 For my sword shall be bathed in heaven: behold, it shall come down upon Idumea, and upon the people of my curse, to judgment. 6 The sword of the LORD is filled with blood, it is made fat with fatness, and with the blood of lambs and goats, with the fat of the kidneys of rams: for the LORD hath a sacrifice in Bozrah, and a great slaughter in the land of Idumea. 7 And the unicorns shall come down with them, and the bullocks with the bulls; and their land shall be soaked with blood, and their dust made fat with fatness. 8 For it is the day of the LORD's vengeance, and the year of recompences for the controversy of Zion. 9 And the streams thereof shall be turned into pitch, and the dust thereof into brimstone, and the land thereof shall become burning pitch. 10 It shall not be quenched night nor day; the smoke thereof shall go up for ever: from generation to generation it shall lie waste; none shall pass through it for ever and ever. 11 But the cormorant and the bittern shall possess it; the owl also and the raven shall dwell in it: and he shall stretch out upon it the line of confusion, and the stones of emptiness. 12 They shall call the nobles thereof to the kingdom, but none shall be there, and all her princes shall be nothing. 13 And thorns shall come up in her palaces, nettles and brambles in the fortresses thereof: and it shall be an habitation of dragons, and a court for owls. 14 The wild beasts of the desert shall also meet with the wild beasts of the island, and the satyr shall cry to his fellow; the screech owl also shall rest there, and find for herself a place of rest. 15 There shall the great owl make her nest, and lay, and hatch, and gather under her shadow: there shall the vultures also be gathered, every one with her mate. 16 Seek ye out of the book of the LORD, and read: no one of these shall fail, none shall want her mate: for my mouth it hath commanded, and his spirit it hath gathered them. 17 And he hath cast the lot for them, and his hand hath divided it unto them by line: they shall possess it for ever, from generation to generation shall they dwell therein."

Martin Luther, who had intimate contacts with the Jews of his day, wrote,

"A more bloodthirsty and vindictive race has never seen the light of day. They regard themselves as the Chosen of the Lord and believe they are destined to annihilate and torture all Gentiles. The first and foremost task they expect their Messiah to accomplish is that he shall murder and slay all human beings with his sword. From the very earliest days they have undertaken all in their power to practically demonstrate this to the Christians and have continued to do so whenever they could."[274]

The Bolsheviks' genocide of the people of Russia, of Hungary, and the millions lost in the "Great War", made many people suspicious of the Zionists and the Bolshevists and their desire for reparations for thousand of years of suffering in the form of the fulfillment of genocidal Judaic prophesies—especially since the League of Nations was formed to create a world government by a movement disproportionately populated with, and represented by, Jews. This League sought to establish a few of the policies spelled out in the *Protocols of the Learned Elders of Zion*, such as the proscription that war could not change national borders—that a nation could not acquire new territory by means of warfare and aggression, which would make war a fountain of wealth for Jews without any chance of the formation of an empire which could challenge their dominance.

Ironically, the Security Council of the United Nations later issued Resolution 242 condemning the State of Israel for violating this principle. Israel refuses to comply with United Nations Resolution 242, reiterated in United Nations Resolutions 267, 338, 446, 452, 465, 468, 469, 471, 476, 478, 484, 605, 607, 608, 636, 641, 672, 673, 681, 694, 726, 799, 1073, 1322; and repeatedly ignored by Israel. Israel has been condemned by United Nations Resolutions countless times and has refused to comply with countless other United Nations Resolutions, including 106, 111, 127, 162, 171, 228, 233, 234, 237, 248, 250, 251, 252, 256, 259, 262, 265, 270, 271, 279, 280, 285, 298, 313, 316, 317, 332, 337, 347, 425, 427, 444, 450, 467, 487, 497, 498, 501, 508, 509, 512, 513, 515, 516, 517, 518, 520, 521, 573, 587, 592, 611, 904, and 3379.

Many have argued that this principle, that territory cannot be acquired by war, is imposed on non-Communist Gentile countries, so that war can become a perpetual means for Jews to reap profits from conflict, and in order to prevent the formation of empires not under direct Jewish control. The "Jewish State", on the other hand, does not yet occupy "Greater Israel", the territory from the Nile to the Euphrates. Many Jews have designs on that territory, and go so far as to claim that sorrowful events which befall Israelis today are God's punishment for the Israeli withdrawal from the Gaza strip. They cite Jewish religious writings, which they believe command Jews to never surrender any "Jewish" soil.

Setting aside Jewish religious myths, which prophesy Jewish world dominance and the genocide of the Gentiles—assuming for the sake of argument that Stephen Wise in no wise referred to such things as Jewish prophesy, which Jews had clung to for centuries in hopes of vengeance against the Gentiles—and so stated in their writings—there is no basis for Wise to assert that the reconstruction of Zion represented the sacrifice of anything by Christendom, nor reparations for anything, let alone for historic offenses committed against Jews by Christians—unless one sees, together with the Zionists, the reconstruction of

---

**274**. Bishop M. Sasse, Eisenach, *Martin Luther and the Jews*, Second Printing, Sons of Liberty, Hollywood, California, (1967), p. 5.

Zion as the product of the First World War and as the only means to save Western Civilization from Bolshevism—the only means to save Western Civilization from Jews.

The theft of Palestine was instead an unprovoked crime against the Moslems who lived there. It was the appropriation of territory from the Turkish Empire by warfare and bloodshed. The Romans who destroyed Jerusalem and the Temple, and then caused a very significant phase of the Diaspora, were not Christians. What gain was there to anyone in stealing that land from Turkey and giving it to a diverse group of people who did not want to populate it, unless the Zionists' real plan was to usher in the Messianic Age? For Christian Zionists the end times meant the demise of the Jews, the return of Christ and the ascendence of the Christians. For Jewish Zionists, the end times meant their dominance over the entire world promised to them by themselves by their prophets—profits—reparations?

In addition to the plans set forth in Biblical prophesy, racist Zionist Theodor Herzl believed that the Christians ought to pay the Jews to create Jewish colonies in Palestine and that the Christians ought to fight for the Jewish Zionists, lest they face the wrath of Jewish revolutionaries. Herzl proposed these things in 1896 in his book *The Jewish State*. He reasoned that since the Christians would profit from the expulsion of the Jews, and since the Christians had the military means to take Palestine and defend it, the Christians ought to be the ones to do the dirty work for the Jews.

The same cynical *quid pro quo* Zionist argument Rabbi Wise had made after the First World War—Jewish suffering and the loss of Jewish life in exchange for Palestinian land and a Christian clear conscience—reappeared after the Second World War, and was made by, among others, Albert Einstein.[275] In his book, *The First Holocaust: Jewish Fund Raising Campaigns with Holocaust Claims During and After World War One*, Holocaust Handbook Series, Volume 6, Theses & Dissertations Press, Castle Hill Publishers, Chicago, (October, 2003),[276] Don Heddesheimer proved through citation to primary sources, that Jewish relief efforts during and after the First World War taught Jewish leaders that they could raise enormous sums of money by pitching the idea that six million Jews were in danger of perishing in a "holocaust" in Eastern Europe. After the Holocaust of the Second World War, Zionist leaders sought to finance the founding of the State of Israel with reparation monies taken from Germany. What gave them the right to steal Palestinian land, and why did they want it, if not to fulfill Messianic prophecy?

The sacrifice of Jewish life for blood-monies and land was an old idea. In 1924, racist Zionist Israel Zangwill ironically stated that it would be a wonderful thing if the legions of lost Jewish lives could turn a profit with which to fund the founding of the "Jewish State". Zangwill said,

---

[275]. A. Einstein, "Unpublished Preface to a Blackbook", *Out of My Later Years*, Philosophical Library, New York, (1950), pp. 258-259, at 259.
[276]. URL:<http://vho.org/dl/ENG/tfh.pdf>

"Mussolini demanded of Greece fifty million lire as compensation for a few murdered Italians. If we had the power to impose blood-money for *our* murdered, the financing of Palestine would become child's play."[277]

Two decades later, on 20 September 1945, immediately after the Holocaust of the Second World War; Chaim Weizmann demanded reparations from Germany, which were eventually paid to finance Israel.[278] Weizmann had read Zangwill's article of 1924 and had responded to it in the same issue of *The Nation* in which it had appeared.[279] One has a right to ask if the Zionists had planned the attacks on Jews in part as a means to fund their project, or merely cynically demanded the "blood-money" after they put the Zionist Nazis into power to persecute innocent Jews and force them towards Zionism against their will.

In 1945, after the Nazi atrocities, Albert Einstein callously reminded the world of the Balfour Declaration and the Palestine Mandate in order to exploit the tragedy of the Holocaust, which the Zionist Nazis had perpetrated, as an opportunity to steal the Palestinians' land. Einstein exploited the Holocaust—the suffering of millions of Jews—to justify the fulfilment of his racist pre-Nazi political Zionist agenda. Einstein asserted that the Holocaust proved that the world thought of the Jews as a nation, thereby mocking the dead assimilationist Jews Einstein hated—those who had been mudered by the Zionists' Nazis.

As the *Protocols*, and Max Nordau, forecast, the Zionists caused unimaginable suffering in order to discredit Gentile governments, when in fact all the while it was the Zionists themselves who created the turmoil and took the innocent lives, amny of them innocent Jewish lives. After the Second World War, Germany and much of Europe lay in ruins, and the Zionists obtained their goals of a racist apartheid "Jewish State", a "United Nations" and the discrediting of the idea of a "fatherland" for any human being other than a Jew.

The Zionists promoted the myth that the Germans were the genetic enemies of the Jews, and that the Jews were the innocent victims of Gentile aggression, when it was the Zionists who had deliberately caused the massive suffering of their assimilating Jewish brethren—not that the European Gentiles should be forgiven for their willingness to follow the Zionists' leaders into the abyss. The Zionists created the Nazis. The Zionists put the Nazis in power. The Zionists carried out the war and the Holocaust. Then the Zionists destroyed Germany and plunged Eastern Europe into Jewish Bolshevik tyranny.

Genocidal human sacrifice had long been a Judaic tradition, and in more recent times, Friedrich Engels made it clear that the Communists were comfortable with human sacrifices amounting to ten million lives lost in order

---

[277]. I. Zangwill, "Is Political Zionism Dead? Yes", *The Nation*, Volume 118, Number 3062, (12 March 1924), pp. 276-278, at 276.
[278]. "Reparations, German", *Encyclopaedia Judaica*, Volume 14, Encyclopaedia Judaica, Jerusalem, The Macmillan Company, New York, (1971), cols. 72-73.
[279]. C. Weizmann, "Zionism—Alive and Triumphant ", *The Nation*, Volume 118, Number 3062, (12 March 1924), pp. 279-280.

to prepare the way for revolution and Communist world dominance. In 1887, Frederick Engels knew that the First World War was coming and that it would destroy the empires of Europe and leave them ripe for revolution,

> "No other war is now possible for Prussia-Germany than a world war, and indeed a world war of hitherto unimagined sweep and violence. Eight to ten million soldiers will mutually kill each other off, and in the process devour Europe barer than any swarm of locusts ever did. The desolation of the Thirty Years' War compressed into three or four years and spread over the entire continent: famine, plague, general savagery, taking possession both of the armies and of the masses of the people, as a result of universal want; hopeless demoralization of our complex institutions of trade, industry and credit, ending in universal bankruptcy; collapse of the old states and their traditional statecraft, so that crowns will roll over the pavements by the dozens and no one be found to pick them up; absolute impossibility of foreseeing where this will end, or who will emerge victor from the general struggle. Only *one* result is absolutely sure: general exhaustion and the creation of the conditions for the final victory of the working class."[280]

To this day, some argue that the Holocaust, not the Covenant with Abraham, gives Israel a "birthright", though they fail to explain why the Holocaust, which was created and perpetrated by Zionists in Europe, gave the Jews a right to steal the land of the Palestinians and send the world into perpetual turmoil.

Gideon Levy published an article on *www.haaretz.com*, on 26 February 2006, entitled "Denial Is Not a Reason for Arrest", which stated,

> "Israel's right to exist, as a birthright of the Holocaust, is stronger than all its deniers, including the president of Iran."[281]

In 1945, Einstein wrote, among other things,

> "[The Jews'] status as a uniform political group is proved to be a fact by the behavior of their enemies. Hence in striving toward a stabilization of the international situation they should be considered as though they were a nation in the customary sense of the word. [\*\*\*] In parts of Europe Jewish life will probably be impossible for years to come. In decades of hard work and voluntary financial aid the Jews have restored the soil of Palestine to fertility. All these sacrifices were made because of trust in the officially sanctioned promise given by the governments in question after the last war,

---

**280**. B. D. Wolfe, *Marxism: One Hundred Years in the Life of a Doctrine*, Dial Press, New York, (1965), p. 67. Wolfe cites: "From Engels's introduction to the reissue of a pamphlet by Sigismund Borkheim. Borkheim's pamphlet, *Zur Errinnerung fuer die deutschen Mordspatrioten 1806-07* [\*\*\*] The introduction is reproduced in *Werke*, Vol. XXI, pp. 350-351."
**281**. <http://www.haaretz.com/hasen/spages/687099.html>

namely that the Jewish people were to be given a secure home in their ancient Palestinian country. To put it mildly, the fulfillment of this promise has been but hesitant and partial. Now that the Jews—especially the Jews in Palestine—have in this war too rendered a valuable contribution, the promise must be forcibly called to mind. The demand must be put forward that Palestine, within the limits of its economic capacity, be thrown open to Jewish immigration. If supranational institutions are to win that confidence that must form the most important buttress for their endurance, then it must be shown above all that those who, trusting to these institutions, have made the heaviest sacrifices are not defrauded."[282]

After the war, Zionist racists like Albert Einstein callously demanded Palestine on a *quid pro quo* basis for the human sacrifice of millions of Jews, which the Zionists had wrought.[283] But where was the logic in this? If the Europeans had murdered six million Jews, as the Zionists claimed, why should the Palestinians pay with their lives and their property for the crimes of the Zionist Nazis? In typical fashion, the Zionists exhibited their infamous dishonesty and argued both sides of the same issue as opposing and mutually exclusive arguments suited their needs.

David Ben-Gurion wrote in his *Memoirs* of 1970,

> "I have called the Arab attitude towards Israel irrational. Nevertheless, the Arab world has levelled several concrete accusations against us and it might be well to answer these here.
> They have said, for instance, that the Moslem portion of the globe is paying for Nazism in Europe, that without the holocaust we would never have come here as a mass and never have founded a State. And, complain the Arab propagandists, it isn't fair that this part of the world should pay for the persecutions carried out in Europe.
> I have already gone exhaustively into the reasons for our being here, reasons that I as a pioneer of 1906 can affirm have nothing to do with the Nazis! I think that Hitler did much to retard, not advance, our nationhood. In the middle thirties, it looked as though we were soon to achieve a Jewish State. But with war in Europe looming ever closer, thanks to the Nazis, Britain cracked down on Jewish nationalist aspirations with the famous White Paper of 1939. Ripe as we were for nationhood at that time, we had the greatest difficulty in helping even a fraction of European Jewry escape the gas chambers. Certainly Israel's population contains no massive element of direct victims of Nazism or their descendants. We just were unable to save the majority of these people. And those who did escape from Germany and the other countries didn't always come here as we weren't equipped to

---

**282**. A. Einstein, "Unpublished Preface to a Blackbook", *Out of My Later Years*, Philosophical Library, New York, (1950), pp. 258-259, at 259.
**283**. A. Einstein, "Unpublished Preface to a Blackbook", *Out of My Later Years*, Philosophical Library, New York, (1950), pp. 258-259, at 259.

get them in their hundreds of thousands past the British embargo on immigration or offer them a true nation once they got here.

I would agree, however, that the advent of Nazism and its consequences in Europe did have one direct effect on Israel. It indicated to us all, to every Jew, the potential danger of being without a homeland. Nazism proved that Jews could live for five hundred years in peace with their neighbours, that they could all but assimilate in national society save for a few traditions and separate religious practices. They could believe themselves integral citizens of states professing freedom of belief and granting full rights to all inhabitants. Such was the situation prevailing in Germany, France, Italy, Holland, Denmark, Norway. Yet one raving maniac could blame the world's troubles on a group constituting less than six per cent of Europe's population and the holocaust was at hand!

So, many a Jew realized that to be fully Jewish and fully a human being, and fully safe as both, one had to have a country of one's own where it was possible to live and work for something belonging to a personal cultural heritage. In this sense, Nazism did bring many Jews to Israel, from everywhere on earth. Not as victims of persecution but as believers in the positive good of a Jewish national home.

I have said that personally I was never a victim of anti-Jewish persecution. I have, however, seen and marked the 'outsider' status of the Jews in even the most enlightened countries, as opposed to their full participation in our society here."[284]

The formation of the "Jewish State" was not enough for the Zionists. They continue to exploit and dishonor the dead, whose deaths they caused, by using the Holocaust as a means to intimidate others into surrendering their rights to free speech, even to free thought, and to capture funds. On the post-Holocaust, "Holocaust industry", which has seen Jews exploiting the death and suffering of millions of other Jews to stifle debate and generate personal profits, *see:* Norman G. Finkelstein's books, *The Holocaust Industry: Reflection on the Exploitation of Jewish Suffering*, Verso, London, New York, (2000); and *Beyond Chutzpah: On the Misuse of Anti-semitism and the Abuse of History*, University of California Press, Berkeley, (2005).

Racist Jews continue to segregate themselves. In Israel, racist Jews are constructing an enormous wall to seal in the boundaries of their self-imposed "World Ghetto",[285] just as they did in the Holocaust. The Jewish book of *Numbers* 33:50-56 states,

> "50 And the LORD spake unto Moses in the plains of Moab by Jordan *near* Jericho, saying, 51 Speak unto the children of Israel, and say unto them,

---

**284**. D. Ben-Gurion, *Memoirs*, The World Publishing Company, New York, Cleveland, (1970), pp. 163-164.
**285**. T. Herzl, English translation by H. Zohn, R. Patai, Editor, *The Complete Diaries of Theodor Herzl*, Volume 1, Herzl Press, New York, (1960), p. 172.

When ye are passed over Jordan into the land of Canaan; 52 Then ye shall drive out all the inhabitants of the land from before you, and destroy all their pictures, and destroy all their molten images, and quite pluck down all their high places: 53 And ye shall dispossess *the inhabitants of* the land, and dwell therein: for I have given you the land to possess it. 54 And ye shall divide the land by lot for an inheritance among your families: *and* to the more ye shall give the more inheritance, and to the fewer ye shall give the less inheritance: every man's *inheritance* shall be in the place where his lot falleth; according to the tribes of your fathers ye shall inherit. 55 But if ye will not drive out the inhabitants of the land from before you; then it shall come to pass, *that those* which ye let remain of them *shall be* pricks in your eyes, and thorns in your sides, and shall vex you in the land wherein ye dwell. 56 Moreover it shall come to pass, *that* I shall do unto you, as I thought to do unto them."

Whenever the door to integration and assimilation opens to the Jews, it is racist Jews who rush in to slam it shut. It will be Jews who will covertly promote a rise in anti-Semitism in America. It will be Jews who will covertly promote a rise in anti-Semitism in Russia. It will be Jews who will impose a police state on the world, as they did in Bolshevik Russia and Nazi Germany. Judaism endures. It is the bane of mankind.

## 5.3.2 The World Awakens to the "Jewish Peril"

The title "The Protocols of the Learned Elders of Zion" probably stems from the official published reports of the various Zionist Congresses: *Stenographisches Protokoll der Verhandlungen des* [fill in the number of the congress] *Zionisten-Congresses gehalten zu* [fill in the place] *vom* [fill in the dates]. These official published reports are known to be incomplete and redacted, but do not resemble *The Protocols of the Learned Elders of Zion* in many important respects.

The *Protocols of the Learned Elders of Zion* were published in Russian at least as early as 1901, by Sergei Nilus.[286] They later appeared in English, German, French, Italian, and Japanese translations and led to a rapid rise in international anti-Semitism in the immediate post-World War I period. Many people feared that an international Jewish organization initiated World War I in order to force the nations of Europe to procure Palestine for the Zionists and to create weakness among European states, which would enable revolutionaries to overthrow those states, eliminate all monarchies, destroy Christianity and fully emancipate the Jews, and also to exact vengeance for the pale of settlement in Russia, the Pogroms, the Ghettoes and other offenses committed against Jews by Gentile Europeans. Typical statements of this belief are found in the writings of Henry Ford's *The International Jew: The World's Foremost Problem* of 1920,

---

**286**. Сергей Нилусъ, Великое въ маломъ и Антихристъ, какъ близкая политическая возможность, (1901/1905).

Adolf Hitler's *The International Jew and the International Stock Exchange—Guilty of the World War* of 1923, and Roman Dmowski's *The Jews and the War* of 1924.[287]

The conservative press made a concerted effort to inform the public that the Bolshevik revolution in Russia was part of an overall Jewish conspiracy to take over all of the governments of the world, in order to enslave humanity; and in retaliation against Christians and Gentiles for the Diaspora, the mediaeval ghettoes, the pogroms and the Pale of Settlement. The role the German government came to play in fomenting dissent in Russia during World War I, so as to diminish Russia's capacity to fight against Germany, was not generally emphasized. The involvement of the German Government came at the instigation of Jewish financiers. Both Kaiser Wilhelm II and General Ludendorff stated that they had been dupes of the Jews.

Herman Bernstein—one of many who argued that the *Protocols* are fabrications—witnessed the rise in awareness of "the Jewish Peril" following the Russian Revolution, and the Bolshevik takeover of the revolution, as early as November of 1917. Henry Ford named Herman Bernstein as one of the two Jews on the Peace Ship who explained "the Jewish Peril" to him in 1915. The other Jew was Rosika Schwimmer. Bernstein capsulized the allegations against Jews, which Ford attributed to Bernstein,

> "That leading members of the Jewish faith precipitated the World War. 2. That in the middle of the war they switched their support to the Allies, selling out to the highest bidder, and that their price was the aid of the allied nations in restoring Palestine to the Jewish people as a national home. 3. That they murdered or caused the murder of the Russian Czar and his family. 4. That most of the dangerous and destructive theories of government abroad in the world are of Jewish origin. 5. That they have debased the professions, prostituted the arts and degraded sports and corrupted commerce. 6. That they control and dominate the press, finance, resources, institutions and politics of the United States, and prostitute the same to unlawful and iniquitous purposes and to their own aggrandizement and to the great injury

---

[287]. **English:** *The International Jew: The World's Foremost Problem*, In Four Volumes, (1920-1922); which reproduces articles which first appeared in THE DEARBORN INDEPENDENT. **German:** *Der internationale Jude*, Hammer-Verlag, (1922). **Russian:** Mezhdunarodnoe evreistvo: perevod s angliiskago, (1925). **Italian:** *L'Internazionale Ebraica. Protocolli dei "Savi Anziani" di Sion*, La Vita Italiana, Rassegna Mensile di Politica, Roma, (1921). **Spanish:** B. Wenzel, *El Judío Internacional: Un Problema del Mundo*, Hammer-Verlag, Leipzig, (1930). **Portuguese:** S. E. Castan and H. Ford, O Judeu Internacional, Revisão, Porto Alegre, RS, Brasil, (1989). *See also:* A. Hitler, "The International Jew and the International Stock Exchange—Guilty of the World War", in R. S. Levy, Editor, *Antisemitism in the Modern World: An Anthology of Texts*, D. C. Heath and Company, Lexington, Massachusetts, Toronto, (1991), pp. 213-221. *See also:* R. Dmowski, "The Jews and the War", in R. S. Levy, Editor, J. Kulczycki, translator, *Antisemitism in the Modern World: An Anthology of Texts*, D. C. Heath and Company, Lexington, Massachusetts, Toronto, (1991), pp. 182-189.

of the civilized world. 7. That their alleged wealth and power as a race constitutes a threat to mankind."[288]

Herman Bernstein, who denied having told Ford these things, was with Ford on the famous "Peace Ship" expedition, but withdrew from the mission. Bernstein was born on the border of Germany and Russia in Neustadt-Schwerwindt in 1876 and his family emigrated to the United States in 1893. He married Sophie Friedman in 1901. He was an "insider" among the Jewish elite, who sponsored Woodrow Wilson's presidential campaign and Zionism. Ironically, in 1906, he translated Leo Tolstoy's anti-Zionist appeal "ZIONISM: An Argument against the Ambition for Separate National Existence. A Plea for Devotion to the Idea of Common Humanity" for *The New York Times*, which was published on 9 December 1906, on page SM2.

The explosive rise in awareness of "the Jewish Peril" in the West, which attended the disclosure of Bolshevist atrocities, alarmed Western anti-Zionist Jews. The rise in the assimilation of Jews in Russia following Kerensky's "emancipation proclamation" and Lenin's proscriptions against anti-Semitism, alarmed Zionist Jews who wanted the Jews to be segregated.[289]

This created a dynamic situation for Jewish leadership. Zionists preferred that the Russian Jews suffer from anti-Semitism, which the Zionists hoped would force Russian Jews to emigrate to Palestine and do the dirty work for the wealthier Western Jews, who would then move into palatial estates built by Russian Jewish slave labor. The Zionist knew that wealthy Western Jews were worried about a backlash against them for the atrocities committed by Jewish Bolsheviks. On the other hand, Western Jews were worried about a severe backlash, a "Holocaust", against Russian Jews should the Bolshevik régime fail and the Russians be restored to power. This was the very thing the Zionists

---

**288**. "$200,000 Libel Suit Filed Against Ford", *The New York Times*, (19 August 1923), p. 2.
**289**. V. I. Lenin, "Anti-Jewish Pogroms", *Collected Works*, Volume 29, English translation of the Fourth Russian Edition, Progress Publishers, Moscow, (1972), pp. 252-253. *See also:* D. Fahey, *The Mystical Body of Christ in the Modern World*, Browne and Nolan Limited, London, (1935), p. 251. *See also:* G. B. Shaw, *The Jewish Guardian*, (1931). *See also: Congress Bulletin*, American Jewish Congress, New York, (5 January 1940). *See also: The Jewish Voice*, (January, 1942). *See also:* G. Aronson, *Soviet Russia and the Jews*, American Jewish League against Communism, New York, (1949). *See also:* J. Stalin, "Anti-Semitism: Reply to an Inquiry of the Jewish News Agency in the United States" (12 January 1931), *Works*, Volume 13, Foreign Languages Publishing House, Moscow, (1955), p. 30. *See also:* S. S. Montefiore, *Stalin: The Court of the Red Star*, Vintage, New York, (2003), pp. 305-306. *See also:* "Anti-Semitism", *Great Soviet Encyclopedia: A Translation of the Third Edition*, Volume 2, (1973), pp. 175-177, at 176. *See also:* "Jews", *Great Soviet Encyclopedia: A Translation of the Third Edition*, Volume 9, Macmillan, New York, (1975), pp. 292-293, at 293. *See also:* N. S. Alent'eva, Editor, *Tseli i metody voinstvuiushchego sionizma*, Izd-vo polit. lit-ry, Moskva, (1971). Н. С. Алентьева, Редактор, Цели и методы воинствующего сионизма, Издательство Политической Литературы, Москва, (1971).

wanted and would achieve through the Bolshevik Zionist Nazi régime.

Jewish leaders settled on a plan. They would covertly keep the Bolsheviks in place, while publicly denouncing them in the West. At the same time, they would try to segregate Russian Jews by forming a "Jewish State" in territory under Bolshevik control. If that failed because the Jews did not want to segregate, they would cause a rise in anti-Semitism in order to prevent assimilation. In the West, they would threaten Christians with a choice between Zionism and Bolshevism, while concurrently and irrationally denying that Jews were behind Bolshevism. They accomplished this end by having Jews in high places denounce Bolshevism in England and in America, while low level Zionists and high level Gentile and crypto-Jewish Zionist "anti-Semites" informed the public that the Jews were indeed behind Bolshevism.

On the Continent, they would install a Zionist Bolshevik dictator. Since the Jewish Bolsheviks were unsuccessful in Germany and other Western nations, and further since the Jews of Europe did not want to go to Palestine even after Jewish leaders had destroyed the Turkish Empire, Jewish leaders planned to install crypto-Jewish Bolshevik dictators on the Continent on an anti-Semitic platform, which became easier for them after the Jewish Bolsheviks and Jewish bankers had created anti-Jewish sentiments.

Things really began to heat up in 1917, after the Zionists had arranged for America to enter the war on the side of the British. The Zionists decided to bury Germany and Russia. They had to assure the British and the Americans that this fate did not await them, though it ultimately does.

As Jewish leaders have done so often in the past—in the case of Rome with Caligula and then Nero—in the English Revolution with Cromwell—in the French Revolution with Robespierre—in the Young Turk Revolution; Jewish leaders deliberately threw the Russian Nation into chaos by means of a Jewish led and financed revolution after Jews had deliberately made conditions unbearable in the nation; then, Jews and their agents loudly cried out that the only way to restore order was to install a dictator, one who would covertly do the bidding of Jewish leadership. The entire process made it appear that the Jews were moral and good to the Russian working class, and that it was the Russian Gentiles who bankrupted the nation and led the people into ruin. In fact, the opposite was the case. Jews deliberately made conditions unbearable in the nation. Jews carried out the revolution. Jews installed the dictator. Jews oppressed the masses and conducted genocide—in each instance—as they would later do in the Nazi Revolution with Hitler.

*The New York Times* wrote on 9 November 1917, on the front page and continuing onto page 2, in an article entitled, "Hope Strong Man Will Rule Russia",

> "Herman Bernstein, who was in Petrograd during the Maximalist riots of last July, said that he was confident that Trotzsky was only the agent of Lenine, who from his hiding had been directing this revolt, as he had done the rising of that period.
> 
> 'It can't win,' he said, 'for Lenine and Trotzsky are both extremely

unpopular. They had a better chance last July, when, if they had only had well-laid plans, they would have been able to dominate Petrograd. As it was, they failed at the time, and the popular execration directed against Lenine after the bloodshed of July was such as to convince me that he will never be able to dominate the Russian people.

'But undoubtedly Kerensky cannot continue in his present position. He has tried to be gentle with the Bolsheviki, in the confidence that they would appreciate his position and treat him as he treated them. Now there must be leaders who will know how to handle them. It has been well established that Lenine is in the German pay, and there is no doubt that the present rising is supported by German funds.

'The ideal of Trotzsky and Lenine is what Trotzsky calls 'the permanent revolution,' a revolution continuing until the maximum Socialist program is in force throughout the world. I don't think there is much likelihood that this program will win, but there is certain to be considerable disorder if the reports so far are correct. One thing I am afraid of is that there will be more pogroms. Trotzsky is a Jew, and unfortunately there are a number of Jewish leaders among the most radical faction. Of course, it is very far from being a wholly Jewish affair. Lenine himself, whose real name is Ulyanoff, comes of an old and noble Russian family, and there are plenty of other Russian leaders. But the prominence of a few Jews is, I am afraid, likely to be avenged on the entire race.

'One thing worthy of note is that the Bolsheviki have learned a point from the procedure of the original revolutionists. You will remember that the revolutions of March seized the telegraph and cable offices, so that after a few days of no news from Petrograd there came out of a clear sky the story of the completed revolution and the full list of Ministers of the Provisional Government.

'This had a great effect in bringing into line the provincial cities and the country districts which might have hesitated if there had come full accounts of the indecisive fighting of the first two or three days. Lenine overlooked this point in his July revolt, but Trotzsky's promptitude in seizing the means of communication at present indicates a desire to try to swing the provinces to the support of a fait accompli in the same manner.'"

Note the subtle messages Bernstein was conveying to his readers—the trap he was setting for the Russian People. The terrible Germans were ultimately responsible for the Bolsheviks, though Bernstein knew that Jewish bankers were the true culprits. The noble Jew Kerensky was too good to lead. The terrible Bosheviki left the world no choice but to install a dictator in Russia who could deal with them with a strong hand. But who would that dictator be, after the Gentiles had swallowed the tyrannical bait? Bernstein does not say, though he is suspiciously sympathetic to the Bolshevik leaders he pretends to denounce. History shows that those dictators were none other than the Bolshevik leaders Lenin and Trotsky—and they most certainly did know how to *reign* in the Bolsheviks.

Jewish leaders would use similar treacherous tactics with Hitler, a Zionist Bolshevik, whom Jewish leaders put in power on an anti-Bolshevik, anti-Semitic platform. Jewish leaders destroy Christian churches in a similar way, by putting crypto-Jews and Jewish agents in key positions in those churches to subvert them, often with an anti-Jewish Zionist agenda.

Leading Jews were worried that their Bolshevist scheme might backfire, and that the Russians would retaliate against the Jews for destroying Russia, stealing the Russians' wealth and mass murdering the Russian people. Leading Jews also feared that Western Gentiles would awaken to the "Jewish Peril" and would organize to take back the monies Jewish bankers had been stealing from Gentiles for centuries. *The New York Times* reported on 19 November 1917 on page 2,

### "JEWS AGAINST BOLSHEVIKI.
Maximalists Represent 'Dark Forces'
of Russia, Bernstein Says.

Denouncing as false reports in the European and American newspapers that Jews were leading and supporting the Bolshevik movement in Russia, Herman Bernstein, in an address before the Institutional Synagogue, at the Mount Morris Theatre, in East 116th Street, declared yesterday that the attempt to associate the Jews with the Bolshevik was merely another expression of anti-Semitic propaganda. Far from being the friends and leaders of the Bolsheviki, he said, the Jews of Russia were their avowed enemies, because the Maximalists included in their ranks representatives of the same 'dark forces' that had always advocated the suppression of Jewish freedom.

Mr. Bernstein, who spent three months in Petrograd after the revolution and had seen the Maximalists at work, said their aim was to bring about utter destruction not only of the freedom of the Jews, but also the freedom of all Russia. The fact that there were seven or maybe ten Jews, including Trotzky, among the leaders of the party was not to be taken as an indication, according to Mr. Bernstein, that the Jews of Russia were supporting their efforts.

'In the first place,' declared Mr. Bernstein, 'these men are not Jews in the real sense of the word. They are not in the least sympathetic to Jewish culture or Jewish ideals. Most of them have been converted to other faiths, and the word Jew has no particular significance to them. The great body of Jews in Russia look upon these men, who were once of their faith, as enemies to the race. The Jews of Russia are no more proud of the Bolsheviki of Jewish descent, than the gentiles of Russia are proud of the Bolsheviki of the Christian faith.'"

Though many Jews who were Bolsheviks made an outward show of opposing Bolshevism, many Jews who were not Bolsheviks also felt obliged to do what they could to keep the murderous Bolsheviks in power for fear of retaliation against the Jews of Russia for the Bolsheviks' atrocities. Of course, Jewish leadership put the Bolsheviks in power in Russia and wanted them to stay in power and the Bolsheviks committed their atrocities against Christians because the Jewish bankers told them to commit them. It was widely known that Bolshevism was a Jewish movement led by Jews and financed by Jews. Chaim Weizmann reported to the Fifth Meeting of the Zionist Advisory Committee, in London, on 10 May 1919,

> "Bolshevism covers a multitude of sins, especially in Poland, and we pay the cost. As a result of the official statement issued by the Bolsheviks in Petrograd to join them, 2½ per cent of the Jewish population have joined, 90 per cent have refused. It is quite true that 60 per cent of the Bolshevik officials are Jews. It is simply that they have got to find means of living, and they are the only people who can read and write."[290]

The attempted Russian revolution of 1905 was also widely known to have been the work of Jews, and many Jews took great pride in that fact. *The Maccabean* of London wrote in a November, 1905, article, "A Jewish Revolution", on page 250,

"The revolution in Russia is a Jewish revolution, a crisis in Jewish history. It is a Jewish revolution because Russia is the home of about half the Jews of the world, and an overturning of its despotic government must have a very important influence on the destinies of the millions living there and on the many thousands who have recently emigrated to other countries. But the revolution in Russia is a Jewish revolution also because Jews are the most active revolutionists in the Tsar's empire."[291]

William Eleroy Curtis delivered an address to the National Geographic Society on 14 December 1906, and stated, *inter alia*,

## "THE VENGEANCE OF THE JEWS

Perhaps these reforms are the cause of the present tranquility, because the revolutionary leaders nearly all belong to the Jewish race and the most effective revolutionary agency is the Jewish Bund, which has its headquarters at Bialystok, where the massacre occurred last June. The government has suffered more from that race than from all of its other subjects combined. Whenever a desperate deed is committed it is always done by a Jew, and there is scarcely one loyal member of that race in the entire Empire. The great strike which paralyzed the Empire and compelled the Czar to grant a constitution and a parliament was ordered and managed by a Jew named Krustaleff, president of the workingmen's council, a young man only thirty years old. He was sent to the penitentiary for life, and had not been behind the bars more than three weeks when he organized and

---

[290]. C. Weizmann, *The Letters and Papers of Chaim Weizmann*, Volume 1, Series B, August 1898-July 1931, Transaction Books, Rutgers University, (1983), pp. 241-242.
[291]. L. Fry, *Waters Flowing Eastward: The War Against the Kingship of Christ*, TBR Books, Washington, D. C., (2000), p. 40.

conducted a successful strike of the prison employees.

Maxim, who organized and conducted the revolution in the Baltic provinces, is a Jew of marvelous ability. Last fall he came over here lecturing and collecting money to carry on the revolutionary campaign, but for some reason has vanished and nobody seems to know what has become of him.

Gerschunin, the most resourceful leader of the terrorists, who was condemned to life imprisonment in the silver mines on the Mongolian frontier, has recently escaped in a water cask, and is supposed to be in San Francisco. He is a Polish Jew only twenty-seven years old. I might enumerate a hundred other revolutionary leaders and every one of them would be a Jew. Wherever you read of an assassination or of the explosion of a bomb you will notice in the newspaper dispatches that the man was a Jew. The most sensational and dramatic episode that has occurred since the mutinies was on October 27, when, in the very center of Saint Petersburg, at the entrance of Kazan Cathedral, four Jews held up a treasury wagon and captured $270,000. They passed the package to a woman, who instantly vanished, and no trace of her has ever been found; but they were all arrested and were promptly punished. On the 8th of November a few Jewish revolutionaries entered a treasury car near Ragow, in Poland, got $850,000 and disappeared.

Every deed of that kind is done by Jews, and the massacres that have shocked the universe, and occurred so frequently that the name 'pogrom' was invented to describe them, were organized and managed by the exasperated police authorities in retaliation for crimes committed by the Jewish revolutionists."[292]

The Bolsheviks mass murdered millions of Christian Slavs and terrorized the world. On the Jewish role in Bolshevism and in the persecution of the Russian masses, *see:* I. Shafarevich, И. ШАФАРЕВИЧ, *Трехтысячелетняя загадка.*Алгоритм, Москва, (2005) [*Three Thousand Year Old Riddle*, Algorithm, Moscow, (2005).]; *and* Alexander Solzhenitsyn, А. СОЛЖЕНИЦЫН, *Двести лет вместе*, Русский путь, Москва, (2001) [*Two Hundred Years Together*, Russian Way, Moscow, (2001).].

Jews around the world desperately lied and attempted to downplay the fundamental rôle Jews played in the genocide of millions of Slavic Christians. At the same time as they were denying that Jews were behind the Bolshevist movement, leading Jews did what they could to perpetuate Bolshevism until such time as they could shape the Slavic mind and make the Slavs impotent and subservient to Jewish interests. The outspoken racist Zionist Israel Zangwill provides us with a fitting example. He protested loudly in 1919 that he was against Bolshevism, but that the Allies should not confront the threat of

---

**292**. W. E. Curtis, "The Revolution in Russia", *The National Geographic Magazine*, Volume 18, Number 5, (May, 1907), pp. 302-316, at 313-314.

Bolshevism because it was inevitable that there would be a world government—this while proudly avowing his rabid Zionist nationalism. The racist Zionists felt justified in demanding that the Gentile nations surrender their sovereignty to a genocidal Jewish movement, while concurrently demanding that Palestine be made a "Jewish State", because the racist Zionists were following the racist supremacist precepts of Judaism, which demands the "restoration of the Jews to Israel" and the concurrent ruin of all other Peoples.

Jews had been calling on Western nations to intercede on their behalf in Russia for centuries. They held massive fund raisers for Russian Jews, but leading Jews discouraged the Western nations from interceding on behalf of Russian Christians after the Russian Revolution, which was funded and led by Jews—Christians who were being slaughtered in the millions at the behest of leading Jews. Zangwill tipped his hand when he proclaimed that the "ideal political aim" was to "make the world safe for minorities" and not "majorities". He likely had in mind the destruction of Gentile nations and creation of a "Jewish State" for the Jewish minorities. On 28 March 1919 on page 11, *The Jewish Chronicle* republished an exchange of letters which first appeared in the *Morning Post*,

## "Bolshevism and the Jews.

MR. ISRAEL ZANGWILL AND THE 'MORNING POST.'

The *Morning Post* of Tuesday printed the following letter from Mr. Israel Zangwill:—

In a leader of the 20<sup>th</sup> instant, you called in the *Times* as 'a witness who will not be suspected of partiality' to testify to 'the sentiments and... the demonstrations countenanced by Mr. Zangwill' at the Albert Hall. Suffer me to be amused your idea of the *Times*, for it so happens that this degenerate organ, once the forum of Britain, not merely forbore to publish my true sentiments, but brazenly refused to allow me to correct its suppression of the true and its suggestion of the false.

The fact is, that I was not a silent 'assistant' on the platform. I made the longest speech of the evening, but strictly in reference to the advertised object of the meeting, viz., protestation against intervention in Russia—a policy now apparently the Governmental one—and I began by repudiating Bolshevism and disavowing the irrelevant utterances that had preceded mine. Not to make the world safe for majorities, but to make the world safe for minorities, seems to me the ideal political aim. It is true that I appeared in 'compromising' company, but I would rather be compromised in a good cause than reported *verbatim* by the *Times* in a bad one. And I know no better cause than to save our soldiers and our country from a continuance of the superhumanly prolonged fighting of which Bolshevism, like the influenza plague, is the natural sequel.

That Jews should be immune from either was hardly to be expected. But that even in Russia they are not all on one side is tragically shown by the fact that the girl who wounded Lenin was of the race of Trotsky. And, oddly enough, as I was writing to you, I received a visit from an influential Russian Jew, newly escaped from Petrograd, who is planning an anti-Bolshevist crusade, and who with tears in his eyes and voice, declared he would sacrifice his last rouble, nay, life itself, to save Russia for real democracy. The thought of the thousands dying from hunger—while professional Bolshevists banquetted royally—made him unable, he declared, to swallow his own food. According to him, there is abundant food in Russia, though disorganisation or tyranny prevents its distribution.

But since Bolshevism and the influenza mock at frontiers, it is clear that the world is increasingly becoming one place, and therefore I fail to perceive why you read a lurid Semitic significance into my view that State Sovereignty is a conception 'absurd and antiquated.' That view is surely implicit in the League Of Nations; it was indeed already implicit in Christianity, so that your phrase, 'the nationalism of the Christian nations,' seems as paradoxical to me as it doubtless would appear to Lord Hugh Cecil, if nationalism is to imply an autocratic sovereignty transcending international obligations of Reason and Justice. But whether my view be right or wrong, do, please, allow me elbow-room and breathing-space as an individual writer, without affixing the responsibility for my heresies to my race or community. Are all Christian authors in agreement with one another or with the mass of their fellow-citizens?

Thank you for your sympathetic perception of the dignity of Jewish nationalism, I am, yours, &c.,

ISRAEL ZANGWILL.

Far End, East Preston, Sussex,
March 24th.

The *Morning Post* on Wednesday, in a leading article headed 'Mr. Zangwill Explains,' says: It is a little unfortunate, when he [Mr. Zangwill] saw the sort of company into which be had fallen, and saw also the symbols of Revolution flaunted under his nose, that he did not mark his disapprobation by getting up and leaving the hall. That is how a law-abiding and loyal Englishman might be expected to act in the circumstances. When a public character—as his modesty cannot prevent us regarding Mr. Zangwill—takes his place on the platform of a meeting, he suggests by his presence a certain patronage or approval of its aims. And why, by the way, did this meeting, distinctively Jewish, according to the *Times*, and undeniably Bolshevik, at one and the same time, celebrate the obsequies of Bolsheviks in Germany and protest against Allied intervention in Russia? Was it really, as Mr. Zangwill would have us believe, 'to save our soldiers and our country,' or was it not to save the Bolsheviks? People who hang out red flags draped in black for Rosa Luxemburg and Liebknecht are not likely to be thinking of 'our soldiers and our country.'"

*The London Times* article to which the *Morning Post* referred appeared on 10 February 1919 on page 10; and note that Bertrand Russell, who advocated genocidal world population reduction, was in attendance; and note further that Sinn Fein was a Bolshevist institution which employed Jewish terrorist methods to create perpetual strife between British and Irish, Catholic and Protestant,

## "SOCIALISTS AT THE ALBERT HALL.

A Socialist demonstration was held at the Royal Albert Hall on Saturday night to protest against intervention in Russia and to demand the withdrawal of the Allied troops from that country. Mr. F. C. Fairchild presided, and among those on the platform were Mr. Israel Zangwill, Mrs. Despard, and Miss Sylvia Pankhurst. Messages expressing sympathy with the object of the meeting were read from, among others, the Hon. Bertrand Russell, Mr. Arthur Ponsonby, Mr. E. D. Morel, Mr. Austin Harrison, and Mr. Bernard Shaw.

It was stated on the programme that the cost of the meeting was at least £400. A collection was made to meet this, but the young aliens of Jewish extraction who formed a large part of the audience and corps of stewards did not appear to contribute very liberally, and it is doubtful if anything approaching the sum stated was raised. But it is understood that substantial donations had been received previously by the organizers. The hall was not full, although on Friday it was announced that every seat had been allotted. Accommodation had been provided for the Press, and two of the speakers denounced and warned the 'scribes of the capitalist newspapers' and, incidentally, the 'camouflaged shop stewards of Scotland-yard.' A red flag draped in black commemorated Rosa Luxemburg and Liebknecht. There were also a few Sinn Fein flags on the platform.

Mr. NEIL MACLEAN, M.P., who suggested that the workers should also demand 'Hands off Glasgow,' moved a resolution in accordance with the object of the meeting, and calling on the working class of Great Britain 'to enforce this demand by the unreserved use of their political and industrial power.'

Mr. JOHN MACLEAN, the Bolshevist 'Consul' in Glasgow, demanded the immediate release of the Sinn Feiners, and conscientious objectors and all other political prisoners of 'that brazen-faced scoundrel Woodrow Wilson.'

Mr. W. F. WATSON, the chairman of the London Workers' Committee, deplored the attitude of the great majority of London workmen who were not inclined to come out on strike or remain out very long. As matters stood they must wait for the miners to move and take every possible advantage of every industrial grievance to make industry impossible."

As late as 1924, racist political Zionist Israel Zangwill wrote that the Jews feared the downfall of Bolshevism and therefore had an overwhelming incentive to perpetuate Bolshevism and destroy all Gentiles in its grasp lest they someday retaliate against Jews for the wrongs done by Jews to them,

"National politics is the realm of might, and if, as Dr. Hertz warns us, the menace of massacre still lies over the whole Russian Jewry should the Soviet Government be overthrown, we must face the sad fact that Jewish might does not exist."[293]

America is today being manipulated in the same manner. Jewish media terrifies the American People with a Moslem bogey that does not exist. Many Jews are attempting to create war between Christians and Moslems by asserting that Moslems are attacking Christians, and that elite Christians are pitting Moslems and Jews against each other. These Jews cleverly pit Moslems and Christians against each other by falsely claiming that Moslems are attempting pit Christians against Jews, and that Christian leaders are attempting to pit Moslems and Jews against each other. These Jews deceptively blame others for the strife these same Jews deliberately cause the world.

Jews, Jewish agents and Jewish dupes carry out staged "terrorist attacks" and the American People join the Jewish media's chorus clamoring for war and dictatorship. Most American Jews want nothing of this, but are deliberately being led up to a backlash against them which will force them to Palestine. Jewish war profiteers concentrate the wealth of the world in their hands through war and irrational tax policies. The American economy is being subverted and the world is being led towards a nuclear World War III and a world-wide depression, which will result in a ruined environment, world government and a world-wide police state—Jewish goals from at least the Fifth Century before Christ. The Zionist Jews believe that by taking these steps they are fulfilling Judaic Messianic prophecies and that they will soon enjoy a world without Gentiles in a paradise God will give them on the "New Earth". They are not concerned about the destruction of the environment or the immorality of the genocide of Gentiles, because they believe God will create a new Earth and wants the Gentiles dead, as the Jewish prophets declared. *Isaiah* 65 states (*see also: Enoch*), and note that the "elect", the "remnant" of the "chosen", are the Jews and only the Jews,

"1 I am sought of *them that* asked not *for me;* I am found of *them that* sought me not: I said, Behold me, behold me, unto a nation *that* was not called by my name. 2 I have spread out my hands all the day unto a rebellious people, which walketh *in* a way *that was* not good, after their own thoughts; 3 A people that provoketh me to anger continually to my face; that sacrificeth in gardens, and burneth incense upon altars of brick; 4 Which remain among the graves, and lodge in the monuments, which eat swine's flesh, and broth of abominable *things is in* their vessels; 5 Which say, Stand by thyself, come not near to me; for I am holier than thou. These *are* a smoke in my nose, a fire that burneth all the day. 6 Behold, *it is* written before me: I will not keep

---

[293]. I. Zangwill, "Is Political Zionism Dead?", *The Nation*, Volume 118, Number 3062, (12 March 1924), pp. 276-278, at 276.

silence, but will recompense, even recompense into their bosom, 7 Your iniquities, and the iniquities of your fathers together, saith the LORD, which have burned incense upon the mountains, and blasphemed me upon the hills: therefore will I measure their former work into their bosom. 8 Thus saith the LORD, As the new wine is found in the cluster, and *one* saith, Destroy it not; for a blessing *is* in it: so will I do for my servants' sakes, that *I* may not destroy them all. 9 And I will bring forth a seed out of Jacob, and out of Judah an inheritor of my mountains: and mine elect shall inherit it, and my servants shall dwell there. 10 And Sharon shall be a fold of flocks, and the valley of Achor a place for the herds to lie down in, for my people that have sought me. 11 But ye *are* they that forsake the LORD, that forget my holy mountain, that prepare a table for *that* troop, and that furnish the drink offering unto *that* number. 12 Therefore will I number you to the sword, and ye shall all bow down to the slaughter: because when I called, ye did not answer; when I spake, ye did not hear; but did evil before mine eyes, and did choose *that* wherein I delighted not. 13 Therefore thus saith the Lord GOD, Behold, my servants shall eat, but ye shall be hungry: behold, my servants shall drink, but ye shall be thirsty: behold, my servants shall rejoice, but ye shall be ashamed: 14 Behold, my servants shall sing for joy of heart, but ye shall cry for sorrow of heart, and shall howl for vexation of spirit. 15 And ye shall leave your name for a curse unto my chosen: for the Lord GOD shall slay thee, and call his servants by another name: 16 That he who blesseth himself in the earth shall bless himself in the God of truth; and he that sweareth in the earth shall swear by the God of truth; because the former troubles are forgotten, and because they are hid from mine eyes. 17 For, behold, I create new heavens and a new earth: and the former shall not be remembered, nor come into mind. 18 But be ye glad and rejoice for ever *in that* which I create: for, behold, I create Jerusalem a rejoicing, and her people a joy. 19 And I will rejoice in Jerusalem, and joy in my people: and the voice of weeping shall be no more heard in her, nor the voice of crying. 20 There shall be no more thence an infant of days, nor an old man that hath not filled his days: for the child shall die an hundred years old; but the sinner *being* an hundred years old shall be accursed. 21 And they shall build houses, and inhabit *them;* and they shall plant vineyards, and eat the fruit of them. 22 They shall not build, and another inhabit; they shall not plant, and another eat: for as the days of a tree *are* the days of my people, and mine elect shall long enjoy the work of their hands. 23 They shall not labour in vain, nor bring forth for trouble; for they *are* the seed of the blessed of the LORD, and their offspring with them. 24 And it shall come to pass, that before they call, I will answer; and while they are yet speaking, I will hear. 25 The wolf and the lamb shall feed together, and the lion shall eat straw like the bullock: and dust *shall be* the serpent's meat. They shall not hurt nor destroy in all my holy mountain, saith the LORD."

*Isaiah* 66:22-24 states,

"22 For as the new heavens and the new earth, which I *will* make, *shall* remain before me, saith the LORD, so shall your seed and your name remain. 23 And it shall come to pass, *that* from one new moon to another, and from one sabbath to another, shall all flesh come to worship before me, saith the LORD. 24 And they shall go forth, and look upon the carcases of the men that have transgressed against me: for their worm shall not die, neither shall their fire be quenched; and they shall be an abhorring unto all flesh."

## 5.3.3 America Becomes the "New Jerusalem"

Jewish revolutionaries destroyed Russian society in collaboration with Jewish financiers, by conducting disastrous strikes and denying the Russian economy access to investment capital, while plunging Russia into war. As Russian society collapsed, the Jews blamed the Czar for the problems the Jewish revolutionaries and financiers had caused. Some Jews may even have asked previous Czars to create the Pale of Settlement and to appear anti-Semitic, in order to prevent assimilation, and they may have manipulated the Czars' actions through carefully placed *agents provocateur* like Rasputin. During Napoleon's reign, some Jews betrayed Napoleon's philo-Semitism and encouraged all Jews to side against Napoleon and with an anti-Semitic Czar, because they feared that Napoleon's emancipation of the Jews was leading to assimilation.[294] A Jewish leader of the time, Shneur Zalman, who hated Gentiles, reasoned that,

> "If Bonaparte wins, the wealth of the Jews will increase and their positions will be raised. But their hearts will be estranged from their Father in Heaven. However, if Czar Alexander wins, then although the poverty of the Jews will increase and their position will be lower, their hearts will cleave to and be bonded with their Father in Heaven."[295]

Those Jewish leaders who promoted anti-Semitism were interested in preserving their own power over other Jews, as well as in preventing assimilation. Jewish leaders depend upon wealth concentration and anti-Semitism to maintain their power—just as they are today war profiteering with a false Moslem bogey in America. In 1881, the Nihilists murdered Czar Alexander II. Konstantine Petrovitch Pobiedonostsev (*also:* Constantin Pobedonoszteff), a man of Jewish appearance who won the favor of Alexander III, retaliated with pogroms against the Jews; which, while certainly bad, were

---

**294**. A. M. Dershowitz, *The Vanishing American Jew: In Search of Jewish Identity for the Next Century*, Little, Brown and Company, Boston, New York, Toronto, London, (1997), pp. 2-3.
**295**. *From:* A. Nadler, "Last Exit to Brooklyn: The Lubavitcher's Powerful and Preposterous Messianism", *The New Republic*, (4 May 1992), pp. 27-35, at 34. Nadler appears to quote from: N. Loewenthal, *Communicating the Infinite: The Emergence of the Habad School*, University of Chicago Press, (1990).

exaggerated in the international press. The alleged Czarist persecution of the Jews, which did not occur, was used as a reason to sponsor the emigration of Jews to the West, which emigration had a negative impact on the Russian economy. The Jewish population in the United States steadily rose from about 200,000 in 1880, to several million by 1920. In the period of 1881-1917, the Jews of Russia had their agents, probably including Pobedonostzeff, stage anti-Semitic pogroms where crypto-Jews attacked comparatively small numbers of Jews in order to give the Jews an incentive to migrate to America, the "New Jerusalem", while simultaneously opening up the Pale of Settlement on the West, such that the Jews were encouraged to move to America and to form an American Jewish homeland—or to prepare for one in Palestine.

It is clear that the staged attacks and the "May Laws" against Russian Jews hurt the Russian People and benefitted the Jews, especially the Zionists like Baron Hirsch, who needed bodies to fill his proposed "Jewish State". This fits a broader pattern of Jewish behavior of deliberately instigating anti-Semitism in order to fulfill the plans of Jewish leadership. Dr. Maurice Fishberg wrote enthusiastically about the Russian Jew in "The Russian Jew in America", *The American Monthly Review of Reviews*, Volume 26, Number 3, (September, 1902), pp. 315-318. However, this journal was created by William T. Stead to promote the views of Cecil Rhodes, who was himself a Rothschild agent.[296] Though the article bears the typically anti-Russian pro-Jewish bias of such publications, it is nevertheless useful for the facts it contains. Fishberg wrote, *inter alia*, at pages 315-316,

"THE history of the Jews in America begins with the discovery of the continent by Columbus. It has been established beyond question that at least five Jews were with him on his first voyage. Among the first settlers in South America and Mexico, at the end of the fifteenth century, were many Jews, mostly refugees from Spain and Portugal. Some of these again emigrated to the colonies in North America. Many other Jews came directly from Holland, Spain, and Portugal. There are records showing that there were German and Portuguese Jews in New Amsterdam as early as 1650. At the time of the Revolution the number of Jews in the colonies was comparatively small; in 1818, Mordecai M. Noah estimated their number at 3,000, and Isaac C. Harby put it at 6,000 in 1826. The American Almanac of 1840 speaks of 15,000. The number of Jews in the United States did not materially increase up to 1880, when a committee appointed by the Board of Delegates of the American Israelites estimated them at 230,257. The Russian Jewish immigration began at that time, and in 1888 Isaac Markens estimated the American Jewry at 400,000, nearly double that of eight years before. The American Jewish Year Book for 1901-02 shows that in 1900 there were 1,058,133 Jews in America. The largest number, 400,000, is credited to New York; Pennsylvania, with 95,000 Illinois, with 75,000; Idaho

---

**296**. G. E. Griffin, *The Creature from Jekyll Island: A Second Look at the Federal Reserve*, Fourth Edition, American Media, Westlake Village, California, (2002), p. 208.

and Nevada appear as having the least,—300 Jews each. This estimate is far too low. According to a statistical investigation by Mr. Joseph Jacobs, based on the number of dead interred in Jewish cemeteries, it has lately been calculated that there are at the present time 584,788 Jews in Greater New York, which is 184,788 more than that of the American Jewish Year Book. The same is probably true of Pennsylvania, Illinois, etc. I think that 1,500,000 is nearer the truth. This means that there are more Jews in the United States than in any other country, excepting Russia and Austria-Hungary. Greater New York, with its 584,788 Jews, has more than Prussia (379,716), France (80,000), and Italy (50,000) combined. When the first Russian-American Congregation was organized in New York on June 4, 1852, it had less than two dozen members. But since 1882 the number of Russian Jews has been rapidly increasing, and at present their number in Greater New York is estimated at 367,690.

After Alexander II. was assassinated on March, 14, 1881, repeated anti-Jewish riots broke out in various parts of Russia. Thousands of Jewish homes were destroyed, and many Jews who were rich, or at least in easy circumstances, suddenly found themselves reduced to poverty. The police and the military authorities did not, in the majority of these riots, make any serious attempts to help the Jews, and in many instances it is known they even assisted in the pillaging of Jewish property. The cause of these riots is known to have been purely political. The constant discontent of the Russian peasants, due to incessant oppression by the Russian authorities and unbearable taxation, endangered the stability of the new government under Alexander III. The government and the inspired press used the Jew as a means of distracting the minds of the common people from their discontent and revolutionary tendency. They pointed out that many of the younger Jews participated in the revolutionary movement of the Nihilists, and that the Jews were consequently responsible for the death of the 'Czar-Emancipator.'

The distressing condition of the Jews became absolutely intolerable on May 15, 1885, when the so-called *'May Laws'* were enacted in Russia. These consist essentially of the establishment of the 'Pale of Settlement' of fifteen governments (districts) in Poland, Ukraine, Lithunia,—'All stolen by Russia from other people' (Harold Frederic),—in which the Jews may live, and prohibiting them from living in the interior of Russia. In the 'Pale' the Jews may live only in towns and cities, and not in the villages. All the leases and mortgages held by the Jews on landed estates were canceled by this act. These laws, in addition to older laws exacting from Jews special taxation on property, rents, legacies, breweries, vinegar factories, printing presses, etc., made it practically impossible for the bulk of the Jews to sustain themselves. Even meat killed 'kosher' is taxed in Russia, so that a Jew has to pay for a pound of meat nearly double the price for that which is not 'kosher.' Jewish children are admitted to the high schools and universities to the extent of only 5 per cent. of the population; and, as there are cities in the 'Pale' in which the population consists of more than 50 per cent. of Jews, the benches of the high schools are vacant, while hundreds of the Jewish youth are vainly applying for admission. The result of these restrictions can be easily imagined. The first relief came by emigration. Baron de Hirsch rendered some assistance. He aided many to emigrate to Argentine and to Canada. But the United States, with its great opportunities, attracted most of them, and up to date over 600,000 Russian-Jewish immigrants have settled here. Freedom from oppression was the chief attraction to this country. Then the great opportunities offered in the United States to the Jews, —whose enterprising spirit, tenacity of purpose, and inexhaustible energy are well known,—were other attractions. Here he may engage in any business, trade, follow any vocation, and as long as he does not violate the laws of the country he is not interfered with. The schools and universities are open to him,—a fact which attracted many. I personally know a goodly number who have emigrated to the United States for the last reason alone. All these, and many other minor causes, have been operative in the Jewish immigration to America, and it is predicted that if conditions in Russia keep up in the manner they have for the last twenty years, at least one-half of the Jews in Russia will emigrate to the United States within the next quarter of a century.

## OCCUPATION OF THE JEWS IN RUSSIA.

It has been stated by people who have never been in Russia that the Jews never engage in any occupation requiring manual labor; that they are nearly all merchants, small traders, agents, and solicitors. How false this is can be seen from the statistics gathered by Mr. Joseph Jacobs, showing that 12 per cent. of the entire population of the 'Pale' are artisans (Jewish Encyclopedia, Vol. II., pp. 115-116), which is a higher proportion than in the general communities of either France or Prussia. They work as tailors, shoemakers, furriers, bookbinders, house painters, opticians, diamond setters, glovers, tanners, watchmakers, etc. In fact, I have observed that in many cities in the Pale no work can be done on Saturdays because the Jewish artisans observe the Sabbath; and it is agreed by all who are acquainted with the conditions, that should the Jews leave in a body it would cause an industrial and commercial disaster in Russia from which it would take years to recover. In the 'Pale,' particularly, there would be no skilled artisans to replace them. It is also agreed by all that as skilled artisans they are of the best. In fact, the Russians give them preference on account of their skill, steadiness, and sobriety, the two latter qualities being uncommon among the Russian workmen to the same extent. Besides all these, the Jews are represented in the learned professions to a greater extent than the Russians. There is a considerable number engaged in the practice of medicine, law, architecture, engineering, journalism, and the like. A great number have also achieved international fame as musicians, painters, sculptors, writers, poets, and scientists."

Herbert N. Casson published a warmly philo-Semitic article in 1906, in which he stated,

"Zionists may dream of the return to Palestine, but the destiny of their race is turning in another direction. America is rapidly becoming the Promised Land of the Jews and New York their New Jerusalem. [\*\*\*] Every anti-Semite eruption in Europe has sent thousands of refugees to Castle Garden, until to-day every fourth person in Manhattan and every sixth in Greater New York is a Jew. [\*\*\*] The Jews make good raw material for citizenship, because they are the only immigrants who come to us without a country, without a flag. They have no fatherland to split their allegiance. America is their home, and their only home."[297]

An article had appeared long before, in *The Religious Intelligencer*, Volume 9, Number 26, (27 November 1824), page 411, which stated,

## "PROPOSED RESTORATION OF THE JEWS.

The Gazette of Spires, assures its readers, that the house of Rothschilds [an immensely rich Jewish banking house in London] has recently received proposals from the Turkish government, for a loan to a considerable amount, and an offer of the entire of Palestine as a security for the payment. In consequence, adds the paper, a confidential agent has been despatched by that house to Constantinople, to examine into the validity of the pledge offered by the Turkish Cabinet.

The N. Y. Advocate says, that the Jews will be restored to their former country, and possess it in full sovereignty cannot be doubted.

Our country must be an asylum to the ancient people of God. Here they must reside; here, in calm retirement, study laws, governments, sciences, become familiarly known to their brethren of other religious denominations; cultivate the useful arts; acquire a knowledge of legislation, and become liberal and free. So, that appreciating the blessings of just and salutary laws, they be prepared to possess permanently their ancient land, and govern righteously."

## 5.3.4 "The Jewish Peril"

On 10 May 1920, *The London Times* published a letter to the editor on page 8,

### "'THE JEWISH PERIL.'
[\*\*\*]
TO THE EDITOR OF THE TIMES.

Sir,—In the article in to-days issue of *The Times* the writer says that the Russian Government contains a large percentage of Jews. As I have had an opportunity of perusing a list of the names and nationalities of the principal

---

[297]. H. N. Casson, "The Jew in America", *Munsey's Magazine*, Volume 34, Number 4, (January, 1906), pp. 381-395, at 382, 393, 394.

State functionaries of Russia compiled from Soviet sources, your readers may like to know the exact figures. Out of a total of 556 there are 458 Jews and 17 Russians, the remainder being made up of Letts, Germans, Armenians, and a few other of the non-Russians included within the late Empire.

As Jewry must be represented in 'tous les partis et toutes les patries [all the parties and all the fatherlands],' as the French say, it is interesting to inquire how the 'opposition' to the Bolshevists is made up. The Menshevists and other parties of the opposition comprise six Russians and 55 Jews.
Yours, &c.,
May 8,   J. H. CLARKE."

THE DEARBORN INDEPENDENT, a widely read newspaper published in Detroit, Michigan, which was owned by Henry Ford the automobile manufacturer, published a series of articles beginning in May of 1920 and continuing over the course of many years, which attempted to prove the authenticity, if not of the *Protocols* themselves, then of the alleged plot by some Jewish leaders to rule the world. Many of these articles were reproduced in book form as *The International Jew: The World's Foremost Problem*,[298] which was published in many languages (it is widely available on the internet). When Einstein visited America in 1921 with Chaim Weizmann, they participated with the Jews of Hartford, Connecticut in a parade of over 400 cars. They boycotted Ford automobiles, which had the counterproductive effect of advertising the brand.[299]

In 1839 and 1840, *The London Times*[300] had reported on efforts by the British Government and the Anglican Church to secure Palestine for the Jews. The plans and religious competition between Protestants, Roman Catholics, Russian Orthodox Catholics and Islam spelled out in these reports foretold much of what later occurred in the First World War, and what is occurring today. These reports also demonstrate the foundations of the fanatical Protestant Christian

---

**298**. **English:** *The International Jew: The World's Foremost Problem*, In Four Volumes, (1920-1922); which reproduces articles which first appeared in THE DEARBORN INDEPENDENT. **German:** *Der internationale Jude*, Hammer-Verlag, (1922). **Russian:** Mezhdunarodnoe evreistvo: perevod s angliiskago, (1925). **Italian:** *L'Internazionale Ebraica. Protocolli dei "Savi Anziani" di Sion*, La Vita Italiana, Rassegna Mensile di Politica, Roma, (1921). **Spanish:** B. Wenzel, *El Judío Internacional: Un Problema del Mundo*, Hammer-Verlag, Leipzig, (1930). **Portuguese:** S. E. Castan and H. Ford, O Judeu Internacional, Revisão, Porto Alegre, RS, Brasil, (1989).

**299**. *Schlesische Zeitung* (Breslau), (16 June 1921), quoted in E. Gehrcke, *Die Massensuggestion der Relativitätstheorie: Kulturhistorisch-psychologische Dokumente*, Hermann Meusser, (1924), p. 35.

**300**. "The State and Prospect of the Jews", *The London Times*, (24 January 1839), p. 3. "Restoration of the Jews", *The London Times*, (9 March 1840), p. 3. "Syria.—Restoration of the Jews", *The London Times*, (17 August 1840), p. 3. "Restoration of the Jews: Memorandum/Correspondence", *The London Times*, (26 August 1840), pp. 5-6. "Letters to the Editor", *The London Times*, (26 August 1840), p. 6.

Fundamentalist support for Israel presently found in America and England.

Though the Zionists believed that anti-Semitism played into their hands, they knew that anti-Zionism did not. The *Times* published numerous anti-Semitic statements, but few anti-Zionist statements, in the critical years following the First World War. The *London Times* published parts of the *Protocols* on 8 May 1920, on page 15, together with a call for an investigation:

### "'THE JEWISH PERIL.'
[*Footnote to the title:* THE JEWISH PERIL.
Protocols of the Learned Elders of Zion.
Eyre and Spottiswoode, London, 1920.]

### A DISTURBING PAMPHLET

CALL FOR INQUIRY.

(FROM A CORRESPONDENT.)

*The Times* has not as yet noticed this singular little book. Its diffusion is, however, increasing, and its reading is likely to perturb the thinking public. Never before have a race and a creed been accused of a more sinister conspiracy. We in this country, who live in good fellowship with numerous representatives of Jewry, may well ask that some authoritative criticism should deal with it, and either destroy the ugly 'Semitic' bogy or assign their proper place to the insidious allegations of this kind of literature.

In spite of the urgency of impartial and exhaustive criticism, the pamphlet has been allowed, so far, to pass almost unchallenged. The Jewish Press announced, it is true, that the anti-Semitism of the 'Jewish Peril' was going to be exposed. But save for an unsatisfactory article in the March 5 issue of the *Jewish Guardian* and for an almost equally unsatisfactory contribution to the *Nation* of March 27, this exposure is yet to come. The article of the *Jewish Guardian* is unsatisfactory, because it deals mainly with the personality of the author of the book in which the pamphlet is embodied, with Russian reactionary propaganda, and the Russian secret police. It does not touch the substance of the 'Protocols of the Learned Elders of Zion.' The purely Russian side of the book and its fervid 'Orthodoxy' is not its most interesting feature. Its author—Professor S. Nilus—who was a minor official in the Department of Foreign Religions at Moscow, had, in all likelihood, opportunities of access to many archives and unpublished documents. On the other hand, the world-wide issue raised by the 'Protocols' which he incorporated in his book and are now translated into English as 'The Jewish Peril,' cannot fail not only to interest, but to preoccupy. What are the theses of the 'Protocols' with which, in the absence of public criticism, British readers have to grapple alone and unaided? They are, roughly:—

(1) There is, and has been for centuries, a secret international political organization of the Jews.

(2) The spirit of this organization appears to be an undying traditional hatred of the Christian world, and a titanic ambition for world domination.

(3) The goal relentlessly pursued through centuries is the destruction of the Christian national States, and the substitution for them of an international Jewish dominion.

(4) The method adopted for first weakening and then destroying existing bodies politic is the infusion of disintegrating political ideas of carefully measured progressive disruptive force, from liberalism to radicalism, and socialism to communism, culminating in anarchy as a *reductio ad absurdum* of egalitarian principles. Meanwhile Jewry remains immune from these corrosive doctrines. 'We preach Liberalism to the Gentiles, but on the other hand we keep our own nation in entire subjection' (page 55). Out of the welter of world anarchy, in response to the desperate clamour of distraught humanity, the stern, logical, wise, pitiless rule of 'the King of the Seed of David' is to arise.

(5) Political dogmas evolved by Christian Europe, democratic statesmanship and politics, are all equally contemptible to the Elders of Zion. To them, statesmanship is an exalted secret art, acquired only by traditional training, and imparted to a select few in the secrecy of some occult sanctuary, 'Political problems are not meant to be understood by ordinary people; they can only be comprehended, as I have said before, by rulers who have been directing affairs for many centuries.'

(6) To this conception of statesmanship, the masses are contemptible cattle, and the political leaders of the Gentiles, 'upstarts from its midst as rulers, are likewise blind in politics.' They are puppets, pulled by the hidden hand of the 'Elders,' puppets mostly corrupt, always inefficient, easily coaxed, or bullied, or blackmailed into submission, unconsciously furthering the advent of Jewish dominion.

(7) The Press, the theatre, stock exchange speculations, science, law itself, are, in the hands that hold all the gold, so many means of procuring a deliberate confusion and bewilderment of public opinion, demoralization of the young, and encouragement of the vices of the adult, eventually substituting, in the minds of the Gentiles, for the idealistic aspiration of Christian culture the 'cash basis' and a neutrality of materialistic scepticism, or cynical lust for pleasure.

Such are the main theses of the 'Protocols.' They are not altogether new, and can be found scattered throughout anti-Semitic literature. The condensed form in which they are now presented lends them a new and weird force.

Incidentally, some of the features of the would-be Jewish programme bear an uncanny resemblance to situations and events now developing under our eyes. Professor Nilus's book was, undoubtedly, published in Russia in 1905. The copy of the original at the British Museum bears the stamp of August 10, 1906. This being so, some of the passages assume the aspect of fulfilled prophecies, unless one is inclined to attribute the prescience of the 'Elders of Zion' to the fact that they really are the hidden instigators of these events. When one reads (page 8) that 'it is indispensible for our plans that wars should not produce any territorial alterations,' one is most forcibly reminded of the cry, 'Peace without annexations' raised by all the radical parties of the world, and especially in revolutionary Russia. And, again:—

We will create a universal economic crisis, by all possible underhanded means and with the help of gold, which is all in our hands. Simultaneously we will throw on to the streets huge crowds of workmen throughout Europe. We will increase the wages, which will not help the workmen as, at the same time, we will raise the price of prime necessities... . it is essential for us at all costs to deprive the aristocracy of their lands. To attain this purpose the best method is to force up rates and taxes. These methods will keep the landed interests at their lowest possible ebb.

Nor can one fail to recognize Soviet Russia in the following:—

' ... in governing the world the best results are obtained by means of violence and intimidation.' 'In politics, we must know how to confiscate property without any hesitation, if by so doing we can obtain subjection and power. Our State, following the way of peaceful conquest, has the right of substituting for the terrors of war, executions less apparent and more expedient, which are necessary to uphold terror, producing blind submission.' 'By new laws we will regulate the political life of our subjects as though they were so many parts of a machine. Such laws will gradually restrict all freedom and liberties allowed by the Gentiles.' 'It is essential for us to arrange that, besides ourselves, there should be in all countries nothing but a huge proletariat, so many soldiers and police loyal to our cause' ; 'in order to demonstrate our enslavement of the Gentile Governments of Europe, we will show our power to one of them by means of crime and violence, that is to say a reign of terror' ; 'our programme will induce a third part of the populace to watch the remainder from a pure sense of duty or from the principle of voluntary service.'

Bearing in mind when this was published, we see, 15 years later, a government established in Russia of which a high percentage of the leaders are Jews, whose *modus operandi* follows the principles quoted, and whose mainstay is a Communist Party, which answers to the last quotation. We see this, and it seems uncanny. The trouble is that all this fosters indiscriminate anti-Semitism. That the latter is rampant in Eastern Europe is a fact. That its propaganda in France, England, and America is growing is a fact also. Do we want, and can we afford to add exacerbated race-hatred to all our political, social, and economic troubles? If not, the question of the 'Jewish Peril' should be taken up and dealt with. It is far too interesting, the hypothesis it presents is far too ingenious, attractive, and sensational not to attract the attention of our none too happy and none too contented public. The average man thinks that there is something very fundamentally wrong with the world he lives in. He will eagerly grasp at a plausible 'working hypothesis.'

What are these 'Protocols'? Are they authentic? If so, what malevolent assembly concocted these plans, and gloated over their exposition? Are they a forgery? If so, whence comes the uncanny note of prophecy, prophecy in parts fulfilled, in parts far gone in the way of fulfillment? Have we been struggling these tragic years to blow up and extirpate the secret organization of German world dominion only to find beneath it another more dangerous because more secret? Have we, by straining every fibre of our national body, escaped a 'Pax Germanica' only to fall into a 'Pax Judæica'? The 'Elders of Zion,' as represented in their 'Protocols' are by no means kinder taskmasters than William II, and his henchmen would have been.

All these questions, which are likely to obtrude themselves on the reader of the 'Jewish Peril' cannot be dismissed by a shrug of the shoulders unless one wants to strengthen the hand of the typical anti-Semite and call forth his favorite accusation of the 'conspiracy of silence.' An impartial investigation of these would be documents and of their history is most desirable. That history is by no means clear from the English translation. They would appear, from internal evidence, to have been written by Jews for Jews, or to be cast in the form of lectures, and notes for lectures, by Jews to Jews. If so, in what circumstances were they produced and to cope with what inter-Jewish emergency? Or are we to dismiss the whole matter without inquiry and to let the influence of such a book as this work unchecked?"

Perhaps not coincidently, this article was followed in the same column of the paper by the next article, "Zionist Aspirations. Dr. Weizmann on Future of Palestine."

*The London Times,* and its principal owner, Lord Northcliffe, had been criticized in a letter from "Mentor", which was published in *The Jewish Chronicle,* on 12 December 1919, on pages 9 and 10:

## "AN OPEN LETTER TO LORD NORTHCLIFFE.

By Mentor.

MY LORD,

It is many years since I had the pleasure of your lordship's personal acquaintance. I recollect that it was in days which, although big with your future destiny, must seem to you now like tiny specks of sand from the high eminence from which you now can view them. They were days of your early life in a north-western suburb, when you inhabited a trim-built villa, the rent of which could not have been as much as £40 a year. It was in a road which, if I mistake not, gave the name to one of the numberless industries that your genius has founded. The denomination of the Pandora Publishing Company was evidence of a strong vein in your character, just as was your giving to a printing enterprise of yours the name of the Viscountess, your lady. These apparent trifles are remarkable indications of a splendid quality in you. You have never been unmindful of your own. You have always been loyal and dutiful beyond measure to the members of your family. There never was a better son than you have been to your mother, nor such a brother as you. It is a pride with you that the old friends of your early youth are your friends to-day, if you come into contact with them. Wealth, power, position—all these—have not shaken this splendid trait in you. I am credibly informed that the man who, throughout your career, has had professional charge—and has it still—of your most intimate affairs is a Jew who was one of your schoolboy chums, in the days of long ago to which I have referred. All this disposes me to feel sure that you will not raise the remotest cavil at, but will welcome, my venturing to address you as one of your long-ago friends. Our paths in life have diverged, but I have constantly and closely watched your career, always with the wonderment and sometimes—let me confess it—with the trepidation with which one, standing upon solid earth, notes the way of the aeroplane in the sky, and which, if he had been living to-day, Agur ben Jakeh would have added as the fifth thing that was 'too wonderful' for him.

**A Great Wrong.**

That you will not resent this entirely friendly letter which I am venturing to address to you, I, therefore, take for granted. I believe that as you read it, you will be disposed, as was Ahasuerus when Esther approached his throne, to hold out to me your sceptre of greeting—if not of approval. For, in fact, I am in a humble way trying to fill the part that Esther played so gloriously, with such magnificent heroism, and with the bravery of which only a woman could be capable. I come to you, my Lord, because my heart is heavy and my spirit burdened for the sake of my people. I come to you, because it is in *your* power to stop a great wrong that is being done to Jews, because you possess the means, by mere work of mouth and by your mere decree, to put an end to what I conceive to be a malicious and wicked plot designed for the undoing of Israel. In your name and within your journalistic realm, the forces of your newspaper empire are being employed in a device, which it is not much exaggeration to say could be well described in the Bible terms—for our being 'sold, I and my people, to be destroyed, to be slain and to perish.' That you—at least consciously—have had a hand in this miserable business, I will not believe, and who the Haman is, who, for the purpose, is prostituting the means you have accorded him, I do not stop to enquire. That you know anything of the real meaning of the anti-Jewish campaign of which the *Times* has recently become the medium, is utterly inconceivable to anybody who knows even the little I do of you, your characteristics, and your ambitions.

**An Ancient 'Stunt.'**

Because the *Times* has lent itself during the last week or two, to about as mean and miserable an anti-Jewish campaign as could well be thought of; and you are not the man to do, or to countenance the doing of, anything that is paltry. The campaign, indeed, is the sort that has been indulged in for a long time by rival papers of yours, such as the *Morning Post*, the *Evening Standard*, and other smaller fry up and down the country; and you are not the man to follow journalistic 'stunts.' You are the man who leads them—with originality, courage, bravery, and acumen. To think that you, who devised the brilliant *coup* of a pound-a-week-for-life prize; who contrived the mighty problem of the missing word; upon whose brain there first flashed the idea of a daily picture paper; you, who first realised the 'snap' of saving the people a halfpenny on their morning journal; you, whose wonderful inventiveness conceived the idea of making all England eat Standard bread and plant sweet peas—that you should deign to copy a miserable, thousand times tried and thousand times failed, 'stunt' of an anti-Jewish campaign is well-nigh impossible. You are above all things and in all things up to date, and an anti-Jewish campaign is as old as the hills. Such a campaign waged round the Pyramids when they were four thousand years younger; the mighty King of Persia was worried with one, as my reference to Queen Esther will remind you, twenty centuries ago. An anti-Jewish campaign can be carried on by such empty-headed numskulls as a Beamish or a Fraser, the defendants in the Mond case. But that you should consciously have allowed your marvellous career, your heavenward flight, of abnormal success to nose-dive to such an ancient, discredited sort of newspaper feature—that you would have copied Germans who shone in nothing so much as in their anti-Jewish attacks (and even black can be made to shine)—is to me unbelievable.

**The Jew-Bolshevist Illusion.**

Let me explain to you what the *Times* has been doing. Righteously wrath with the Bolsheviks in Russia and all their works; indignant at the outrages which they are said to have committed; rightly disgusted with the oppression, the looting, the murder—and worse—which has been attributed to them; correctly (to my way of thinking, at least) estimating the hollowness and impracticability of Communism as a form of government, and seeing in Russian Bolshevism (again I am in agreement, and have insisted upon it throughout) not democracy, but the cruellest, the most relentless, the most unfair of autocratic tyrannies; your chief paper has devoted itself to bringing before the English public, what it conceives to be the true nature of the Soviet Government. But by some malign influence, this quite comprehensible and perfectly commendable policy has been diverted into being made a means for whipping the Jews. It may be that this diversion has occurred solely through ineptitude, misunderstanding or even ignorance. In raking over the records of Bolshevism, Jews have been found prominent in the Bolshevist ranks. Several Bolsheviks who were not Jews in any sense of the word, but who bore German-sounding names which were commonly used among Russian Jews, were thought to be Jews, and altogether a grossly exaggerated idea of the part played by Jews in the Bolshevist movement resulted. This is a quite general experience. It takes the presences of only a few Jews among non-Jewish surroundings to cause one to over-estimate in perfect good faith the number of Jews who are actually present. Go into a railway carriage in which there are, say, ten passengers. Let four of those be Jews—persons who by feature and manner are evidently Semitic and not Anglo-Saxon—and you, or anyone else remarking upon the incident, would feel—and if narrating it would say, that you found the carriage was 'full' of Jews. Analogously, if from the window of the *Times* office you were watching the traffic in Queen Victoria Street, and you saw, say half-a-dozen negroes among the passers-by, you would declare that London was 'full' of blacks. And so you would declare it 'full' of Japs, if you saw a dozen natives of the Land of the Rising Sun. There is nothing to wonder at, then, that anyone looking through the records of Bolshevism in Russia, and finding a number of Jews among the Commissaries, or what not, should rush to the conclusion that the whole of Bolshevism was being carried on by children of Israel.

**A Decadent Occupation.**

There are, to be sure, reasons why the number of Jews identified with the Bolshevist administrative offices are proportionately larger than the Jewish population warrants. One of the reasons is that the Jews of Russia have taken care to keep their children educated and have nurtured their intelligence, while the masses of the non-Jewish population have continued sunk in mental darkness, in the ignorance that was directly fostered by Tsarism in the interests of the Tsarist Church. You will surely not have failed to notice how Bolshevism in Russia has by all accounts ushered in an era of educational revival among the masses as part of its efforts for fighting what remains of the spirit of the old *régime*. But allowing for all this, there must have been an influence of sheer Anti-Semitism which could have induced the turning by the *Times* of the instruction—from its point of view—of the English people about Bolshevism into an attack upon the whole of the Jewish people. That a certain number of Jews are Bolshevists is any proof that I am a relentless Shylock, is about as reasonable as to say that because some Irishmen are Sinn Feiners, you are a rebel. And, my lord, you have not reached such a height of your romantic career—the admiration of your friends, as it is the envy of your enemies—in order to reduce the greatest newspaper the world has ever seen to an unreasonable campaign fit for the mentality, perhaps, of some of your competitors or certainly of the obsessed poor-minded creatures whose decadence has reduced them to indulging in the piteous occupation of Jew-baiting. A Northcliffe—a Harmsworth—was obviously devised for something less pusillanimous, something less silly, something more original, something less banal.

**The 'Booby Trap.'**

Then, my lord, just hear what the *Times* has been urging. It has been suggesting that when Bolshevism in Russia fails, the forces that are arrayed against it are going to massacre the Jews, because of the part they have taken in supporting the Bolshevist Movement. There is something, it seems to me, of the spirit of 'don't nail his ear to the pump!' about the grim anticipation here set forth. But let that go. On the pretence of its being anxious to save the poor Jews from massacre, the *Times* has been asking the Jews of this country to walk into its parlour and to give themselves away by, as Jews, forswearing Bolshevism and all its works and denouncing fellow-Jews for having supported both. Having done that, what is going to happen? Does the *Times* think that the hooligans in Russia are going to stay their hands because the Jews here have denounced Bolshevism? Does it suppose that some Russian bandit who would otherwise loot a Jew's property or murder him, would suddenly fling away all the instruments of violence that he was employing, and clasp the Jew to him in tender solicitude upon calling to mind the fact that some of his victims' brethren in Western Europe had declared that they were not Bolsheviks and they did not like Bolshevism? One of the writers in your organ said that Jews were stupid; and, certainly, if they were altogether a wise people they assuredly would not, in the first quarter of the twentieth century, be in the position of being pilloried by your paper. Nor would they have suffered themselves to be, as they have been, the Azazel goat, upon the head of whom the sins of every world-movement have been cast for close upon two thousand years. But so stupid as to think that the acknowledgment which the *Times* wishes to wring out of our people is demanded in the interests of our Russian coreligionists, or that it would subserve these in the least, it is no vain conceit on my part as a Jew, to tell you we are not. For us to proclaim to the world that Bolshevism and Judaism are so intimately associated that it is necessary for Jews to dissociate themselves in the public mind from the Russian Movement, and that the renunciation was going to prevent an otherwise certain holocaust [Note the use of the term "holocaust"—CJB]—no! Lord Northcliffe!—*so* stupid even the Jews whom the writer referred to in your paper so insolently contemns, assuredly are not.

[As quoted above, *The New York Times* published articles about and quoting Herman Bernstein, a man of Jewish descent, on 9 November 1917, and on 19 November 1917, in which Bernstein said what "MENTOR" claimed no Jew would ever say. The predicted Holocaust did occur and was heinously "justified" for the reasons claimed. Bernstein's efforts failed, as did Mentor's refusal to act. One should also note the irony of the author's identifying herself? with Esther, who brought on a genocide much like the vindictive mass murder of the Russian people by revolutionary agents of Jewish financiers. Ironically, Mentor speaks of Jews in general in tribalistic terms, though criticizing others for doing the same.—CJB]

Anyone with half an eye, anyone although bereft of half his senses, any dull fool, could see the trap that the *Times* writer was setting, in this proposal, for us Jews. It was, indeed, a booby trap; so obvious that it could scarcely be missed even by the mentally blind. It was a device without the least cleverness, the least subtlety, the least cunning—employing the words in the most complimentary sense—and no one could have regarded it as the product of a master mind, or have looked for its source of inspiration to a genius such as yours. This again, I say, is fair evidence that your influence and your power you have delegated to hands that have proved unworthy, and I hope you will thank me for calling your attention to the manner in which they have been employed.

**'Epatism.'**

At the moment of writing, it doubtless appears to some that the campaign has been called off and the 'stunt' stopped. 'Verax,' has not 'veraxed' for some days. 'Janus' and 'Philo Judæus' *et hoc genus* have remained silent for over a week, while the contribution of 'Ivan Ivanovitch' read to many like a desperate, final gasp. Frankly, I regard the state of the matter at the moment in a somewhat different light. It occurs to me that the letter of 'Verax' like the one signed 'X,' which purported to be one sent by a British officer serving in Russia to his wife in England (the letter which, by the by, set the ball rolling), formed an essay in what the *Times* itself has termed 'Epatism.' Your paper has explained the word by reference to the phrase of Flaubert's circle—*épater le bourgeois*, to 'startle John Citizen.' It is the art of preparing the public mind by giving it a shock—'shock tactics,' as the German phrase had it in the war. An Epatist, as the *Times* went on to show, 'seeking to achieve something new,' 'takes refuge in distortion and the misuse of colour.' Exact contour and faithful reproduction are outside his scheme, and he deliberately flaunts his carelessness of qualities hitherto accepted as necessary. Epatism, in short, the *Times* says, is 'an affront with a purpose.' This, it occurs to me, gives us the key to the recent attack upon Jews in your paper. As in art, so in literature, as in literature, so in journalism; and the anxiety of those responsible for the anti-Jewish campaign in the *Times* was not, it is surely obvious, for exactness of statement, faithfulness of argument, or correctness as to alleged facts. These did not in the least count, in face of the determination to 'achieve something new.' 'Refuge in distortion and misuse of colour' were merely the manner of the Epatist. And for what purpose was this exercise in Epatism indulged in? There can remain no doubt with anyone who reads the letters which in big type are now (as I write) appearing. By the by, the type in which these contributions are printed is a remarkable contrast to the type in which the letters defending Jews that have been admitted to the *Times* have invariably, with one exception, that of the Chief Rabbi's, been printed—another evidence that your scrupulous fairness to opponents was not in this play, and that the fine traditions of the *Times* had been set aside.

**A Ridiculous Notion.**

That just by the way: What is the burden of these latest contributions to which I refer? It is that Bolshevism is a movement which designs to uproot and throttle Christianity as the world has it. I do not stop to argue whether Bolshevism can, in fact, be reasonably supposed to have that as its objective, or still less whether it has the remotest chances of effecting any such moral revolution among mankind, or whether, again, the same could not have been said of the Russian religious school of thought led by Count Tolstoi, himself surely a Christian from the religious point of view *sans peur et sans reproche*. But I do call your attention to the way in which the *Times*, by means of epatism—of distortion and misuse of colour, of startling John Citizen—has first tried to shock its readers into believing that Bolshevism and Judaism are one, and then followed that up with an impeachment of Bolshevism as a force designed to undermine Christianity. The object manifestly is to 'achieve something new' in the way of a silly bogey—to frighten the readers of the *Times* into an attitude of bitter, relentless, unyielding enmity to the Bolshevists by insidiously impressing upon the readers of the great paper which you own, that Jews have to-day designs against the Christian Church. The object has been to make the people who read the *Times* think that Jews desire Christianity to perish, and that they are banded together in the Russian movement we know as Bolshevism, so that they may wipe away Christianity from off the face of the earth. It would follow that in order to defend Christianity it is necessary to crush Bolshevism. Now, if your people said that in so many words, the statement would have been greeted by a Homeric burst of laughter wherever the words were read or repeated. That is why the spurious nonsense was applied by means of 'Epatism' and insidious suggestion. I say the statement plainly made would have been met with laughter—and not least by Jews, who know so well how religious carelessness and *laisser faire* are eating into the vitals of our people. To such an extent is this so, that it is with anxiety that Jews, who care for Judaism, contemplate the religious future of their faith, and against the enormous forces of indifference are bringing to bear their mightiest efforts in every land. And the *Times* wants us to believe that side by side with this religious indifference there exists the sort of religious zeal that would seek to uproot Christianity, so that Judaism might dominate! How densely ignorant of Jews must be those who imagine this vanity! Why, I do not know of a single Jew to-day, here or abroad, from the far west to the far east, whatever may be the form of Judaism which he favours, whatever may be the politics he supports, whatever may be the shade of Judaism to which he is allied, who would lift his little finger to do damage to the religious faith that is dominant throughout the Western world. There are some Jews who dislike Christians—and will you say without good reason? But there are no Jews who hate Christianity, or indeed care about it at all to the extent of indulging in a campaign against it.

**Judaism and Christianity.**

All Jews, it is true, look forward to the moral prevalence of Jewish doctrine and Jewish teaching. If they did not, their Judaism would necessarily, even in their own estimation, be a poor sort of thing. If they did not think of Judaism as a faith which in God's good time, and by force of moral suasion, will become that of all the world—if they did not conceive the synagogue as a House of Prayer for all nations—we Jews would indeed be a segregated, aloof, religiously and nationally selfish, and hence debased and degraded, people. Judaism is and has always been a faith appealing to all Humanity, and Christianity, so far as it was a triumph over heathenism, was a victory for Jewish doctrine and the Jewish faith. How, then, can anyone (especially one like 'Verax' who pretends to some knowledge of Judaism and sufficient Jewish culture, not know how to transliterate correctly *Beth Hamidrash*) suggest anything so monstrously absurd as that Judaism would, in any sense whatever, fulfil its mission by destroying Christianity at this stage of the world's civilization? And how ridiculous, from a practical point of view! We Jews are a handful of people scattered up and down the Earth, a people than whom there is none more materially forlorn than is, taken as a whole, our poor folk. Of the fifteen or sixteen million of Jews existing to-day, it has been calculated that less than ten thousand can be considered rich in such a sense as, say you my Lord, would deem anybody wealthy, while more than 70 per cent are poor, inasmuch as they are without any capital. Who will believe that such a people in such a position would contemplate the smashing and killing of a religious institution which has been one of the strongest social, moral, political, and religious pillars of the world for generations? The man who could believe it is a fit object less for laughter, when we come to think about it, than for tears of sympathy. Even if Jews could compass the destruction of Christianity in the way these silly people credit them with conniving, what sort of Jew pray would do it? The religious Jew? He certainly would never seek to hurt and destroy an institution, which rightly viewed—however much the Jew sees of fundamental error in, and however false the doctrine, as he perceives it, of Christianity—is the greatest world triumph of the Jew. Is it then the irreligious Jew? Surely he would not trouble himself to pull down Christianity to which he, in so many cases, has a proneness to assimilate for the sake of uprearing in its place Judaism of which he is sometimes so careless, sometimes renegade, and in regard to which such a Jew is always so negligent, that he will not lift a little finger to aid and support it even in his own person? And let me remind you *en passant* that the prominent Bolsheviks that are Jews are not exactly Orthodox adherents to Judaism. Really, this bogey of Christianity in danger—and in danger from Jews!—is the silliest 'fimmel' that ever crept into the brain of a man whose sanity was whole and unimpaired. Frankly, my lord, this cry of alarm would cause me some trepidation only if for a second I could believe it was genuine. For if Christians really imagined that Christianity was in such case that Jews to-day could destroy it, however much they tried, there would be revealed in Christianity a consciousness of inherent weakness deplorable beyond words.

**Duty.**

Now, my lord, I have put our case, and I doubt not what you will do with the facts thus presented to you. In the light of them you will do your duty as a worthy son of the most chivalrous and human-spirited people on earth. You will do your duty as citizen of an Empire which was founded upon Justice and upon Right. You will do your duty as one of the choicest ornaments of a profession which, in its highest and best conception, knows no fear and no favour, but is ever fast allied to public truth and public righteousness. You will, too, I feel sure, do your duty to the finest traditions of the great journal, the securing of the ownership of which was the most brilliant *coup* of your brilliant career. Your duty, my Lord, in all these aspects happily coincides and dovetails with exactly the purpose I have in writing this letter to you. Your duty is to stop at all costs, and at once, and forbid any future recurrence of the campaign of vilification and abuse, the insidious, malicious, underhand war, which someone, misusing the power of your Press, has been carrying on against my people.

                        Believe me to remain,
                        Your obedient Servant,
                        MENTOR.
TO THE RIGHT HON.
THE VISCOUNT NORTHCLIFFE, ETC. ETC."

This sophistical appeal was a reaction to a series of letters which had appeared in *The London Times* following World War I,[301] many of which set forth the allegedly self-fulfilling prophecy that all Jews ought to condemn Bolshevism, because if they failed to condemn it, when Bolshevism fell a holocaust would ensue and the Jews of Eastern Europe would be annihilated—in retaliation for the vindictive Jewish destruction of Russia and the Jewish genocide of Russian Gentiles. The appeal is further evidence that some leading Jews felt a need to perpetuate the genocidal Bolshevist regime in Russia in order to shield Jews from retaliation, which genocidal regime Jewish financiers had put into power and which was disproportionately staffed by Jews, while assimilating Jews sought desperately to distance themselves from Bolshevism, Zionism and Judaism. While these letters in the *Times* may appear meanspirited, they are historically important because they evince the linkage of Bolshevism to Western Jews in general, and the planned and feared reaction that Jews would be attacked in a murderous rampage in order to protect Western Civilization from Bolshevism. This tragic attitude did indeed lead to the Holocaust. However, it was Zionist Jews who intentionally brought it about.

A similar debate took place earlier when the Jewish Young Turks mass murdered Christians. Well intentioned persons in the West pleaded with Western

---

**301**. X, "Flight from Bolshevism", *The London Times*, (14 October 1919), p. 14; **and** "The Horrors of Bolshevism", *The London Times*, (14 November 1919), pp. 13-14. *See also:* I. Cohen, "Jews and Bolshevism", *The London Times*, (21 November 1919), p. 8; **and** "Jews and Bolshevism", *The London Times*, (25 November 1919), p. 8; **and** "Jews and Bolshevism: The Mosaic Law in Politics: Racial Temperament", *The London Times*, (27 November 1919), p. 15; **and** "Jews and Bolshevism: A Further Rejoinder", *The London Times*, (1 December 1919), p. 10. *See also:* Philojudaeus, "Jews and Bolshevism: The Group Round Lenin", *The London Times*, (22 November 1919), p. 8. *See also:* Janus, "Jews and Bolshevism: Revolutionary Elements", *The London Times*, (26 November 1919), p. 8. *See also:* Judaeus, *The London Times*, (26 November 1919), p. 8; **and** "Jews and Bolshevism: A Reply to 'Verax.'", *The London Times*, (28 November 1919), p. 8. Verax, "Jews and Bolshevism: The Mosaic Law in Politics: Racial Temperament", *The London Times*, (27 November 1919), p. 15; **and** "Bolshevism and the Jews: A Larger Issue: The Danger in Russia", *The London Times*, (2 December 1919), p. 10. *See also:* J. H. Hertz, Chief Rabbi, "Jews and Bolshevism: The Chief Rabbi's Reply", *The London Times*, (29 November 1919), p. 8. *See also:* Pro-Denikin, "A Witness from Russia", *The London Times*, (29 November 1919), p. 8. *See also:* An English-Born Jew, *The London Times*, (1 December 1919), p. 10. *See also:* Ivan Ivanovich, "The Jews and Bolshevism", *The London Times*, (6 December 1919), p. 10. **"Epatism" defined in *The London Times*:** "Epatism", *The London Times*, (10 December 1919), p. 15.

Jews to repudiate the actions of the Jews and crypto-Jews who were behind the Young Turks.[302]

The "Holocaust" was planned as a threat to anti-Zionist Jews. The fulfillment of this threat was carried out by vengeful Zionists. Don Heddesheimer, in his book *The First Holocaust: Jewish Fund Raising Campaigns with Holocaust Claims During and After World War One*, Holocaust Handbook Series, Volume 6, Theses & Dissertations Press, Castle Hill Publishers, Chicago, (October, 2003),[303] has proven that several newspapers published articles in the late Teens and early 1920's, which promoted fund raising campaigns for Jewish relief in Eastern Europe. These often exploited the alarmist slogan that six million Jews were on the verge of perishing in a "holocaust". Immense sums of money were raised in these campaigns and Heddesheimer sees in them a pattern of deception and exploitation. This was further evidence of how effective fear was in mobilizing and segregating the Jewish community—in perpetuating their self-image of victimhood and separation.

The evidence supports Mentor's assertions that the vast majority of Western Jews were not out to destroy Christianity, but instead sought to integrate into society. This fact is perhaps rendered most obvious by the many public expressions of disenchantment of the Zionists, who could not persuade a majority of Jews to join them in a march to Palestine, and by the high rates of "intermarriage" of Jews to non-Jews. However, Mentor's motives and sincerity can be questioned based upon an article "Our 'Abandoned' Children" published in *The Jewish Chronicle* on 24 November 1911 on pages 20 and 31. "Mentor" was later identified as the interviewer in that article in a response published by Isaac Goldston, "A Danger that Portends a Doom", in *The Jewish Chronicle* of

---

302. "Jews and the Situation in Albanian", *The London Times*, (11 July 1911), p. 5. *See also:* M. Gaster, "Jews and the Situation in Albanian", *The London Times*, (12 July 1911), p. 5. *See also:* M. A. Syriotis, "The Jews and the Young Turks", *The London Times*, (19 July 1911), p. 5. *See also:* "Jews and the Situation in Albanian", *The London Times*, (27 July 1911), p. 5. *See also:* M. Gaster, "Jews and the Situation in Albanian", *The London Times*, (1 August 1911), p. 11. *See also:* G. F. Abbott, "To the Editor of the Times", *The London Times*, (1 August 1911), p. 11. *See also:* H. C. Woods, "The Adana Massacres", *The London Times*, (3 August 1911), p. 4. *See also:* The Israelite Community of Salonika, "Jews in Turkey", *The London Times*, (4 August 1911), p. 11. *See also:* "Jews and the Situation in Albanian", *The London Times*, (8 August 1911), p. 5. *See also:* "Jews and the Situation in Albanian", *The London Times*, (9 August 1911), p. 3. *See also:* M. Gaster, "To the Editor of the Times", *The London Times*, (9 August 1911), p. 3. *See also:* "The Jews and the Young Turks", *The London Times*, (9 August 1911), p. 9. *See also:* H. C. Woods, "The Jews and the Situation in Albanian", *The London Times*, (11 August 1911), p. 3. *See also:* S. Schiff, "To the Editor of the Times", *The London Times*, (11 August 1911), p. 3. *See also:* A Citizen of London, "To the Editor of the Times", *The London Times*, (11 August 1911), p. 3. *See also:* "The Jews and Albania", *The London Times*, (19 August 1911), p. 3. *See also:* M. A. Syriotis, "The Jews and Young Turks", *The London Times*, (25 August 1911), p. 3.
303. URL:<http://vho.org/dl/ENG/tfh.pdf>

1 December 1911 on pages 18 and 27.

Though Mentor questions "Verax's" sincerity, "Verax" was the pseudonym of a writer for the *Centralverein deutscher Staatsbürger jüdischen Glaubens*, a Jewish organization which combated anti-Semitism and racist political Zionism; and if these "Veraxes" are one, then "Verax" was likely sincere. *See:* Verax, "Jüdische Rundschau", *Im Deutschen Reich* [official organ of the Centralverein deutscher Staatsbürger jüdischen Glaubens], Volume 16, Number 5, (May, 1920), pp. 163-171; and Verax, "Jüdische Rundschau", *Im Deutschen Reich*, Volume 16, Number 6, (June, 1920), pp. 196-205. *See also: Jüdische Rundschau*, Volume 25, Number 38, (11 June 1920), p. 296.

Numerous translations of *The Protocols of the Learned Elders of Zion* presented arguments and evidence that Bolshevism was a Jewish movement, celebrated by some Jews as such, and constituted the fulfilment of a long planned phase of genocidal Judaism, which prophesied the destruction of Gentile governments, religion, and, eventually, peoples. Despite protests to the contrary, there were leading Jews who sought the downfall of Christianity and Judaism teaches that all religions other than Judaism must be destroyed, and that all the governments must be destroyed and replaced by one world government ruled by the Jewish Messiah from Jerusalem.

Jewish plays and writings provide ample evidence of widespread Jewish hostility towards Christians, most especially towards Russian Christians, and the Jews were no less poor when the Jewish Frankists sought to undermine Christianity, than when the Bolsheviks sought to undermine Christianity. After all, it was the immense wealth (obtained through corrupt means) of Jewish financiers, which brought Russia to ruins, and it was the concentration of this wealth which enabled leading Jews to destroy peoples and governments, despite Mentor's suggestion that the concentration of wealth rendered such things impossible. It was the very poverty of average Jews in the East, and their minority status, which drove them to be anti-Christian, and this in no wise prevented them from seeking to undermine Christianity, but instead provided two motivating factors. The poverty of average Eastern European Jews, should they as a group desire the downfall of Christianity, made Bolshevism a necessity for their cause, because it was only by tearing down Christian society that they could terrorize Christians and suppress religion among Gentiles, as their religion taught them to do. Mentor's sophistry is most apparent in her(?) transparent efforts to flatter Northcliffe—though by insulting his intelligence and impugning his character should he find cause for alarm in facts which alarmed many a reasonable person. Try as she might to beguile and deceive Northcliffe, Mentor was no Esther. It should be noted that if the Jews had not concentrated their collective wealth in the hands of the Rothschilds and their agents, the Jews would not have had anywhere near the power they did have. This is to say that if the Rothschilds had shared their concentrated wealth with all the Jews, then there would not have been the pool of monies the Rothschilds used to undermine the governments of the world.

The Rothschilds has religious precedent for their wealth concentration. The Jewish book of *Proverbs* 1:13-14, states,

"13 We shall find all precious substance, we shall fill our houses *with* spoil:
14 Cast in thy lot among us; let us all have one purse:"

The Government of the United States received urgent warnings that the Bolshevists, who were without a doubt mass murderers, were largely led and funded by Jews, and that they openly sought to destroy Christian Civilization in the manner of genocidal Messianic Judaism. This increasingly widespread awareness naturally led to generally "anti-Jewish feelings" through an unfair and unrealistic—though natural—generalization of the actions of leading Jews to all Jews.

The "Report of the Netherland Minister relating to conditions in Petrograd", *Papers Relating to the Foreign Relations of the United States, 1918, Russia*, Volume 1, File Number 861.00/3029, United States State Department Publication Number 222, 65th Congress, 3d Session, House Document Number 1868, United States Government Printing Office, Washington, D. C., (1931), pp. 675-679, at 678-679; states,

"The foregoing report will indicate the extremely critical nature of the present situation. The danger is now so great that I feel it my duty to call the attention of the British and all other Governments to the fact that if an end is not put to Bolshevism in Russia at once the civilisation of the whole world will be threatened. This is not an exaggeration but a sober matter of fact; and the most unusual action of German and Austrian Consuls General before referred to, in joining in protest of neutral legations appears to indicate that the danger is also being realised in German and Austrian quarters. I consider that the immediate suppression of Bolshevism is the greatest issue now before the world, not even excluding the war which is still raging, and unless as above stated Bolshevism is nipped in the bud immediately it is bound to spread in one form or another over Europe and the whole world as it is organised and worked by Jews who have no nationality, and whose one object is to destroy for their own ends the existing order of things. The only manner in which this danger could be averted would be collective action on the part of all powers."

State Department Document Number 861.00/1757, 2 May 1918, states,
"Jews predominate in local Soviet Government, anti-Jewish feeling growing among population which tends to regard oncoming Germans as deliverers."[304]

State Department Document Number 861.00/2205, 5 July 1918, states,

---

[304]. K. A. Strom, Editor, *The Best of Attack! and National Vanguard Tabloid*, National Alliance, Arlington, Virginia, (1984), p. 65.

"Fifty per cent of Soviet Government in each town consists of Jews of worst type, many of whom are anarchists."[305]

United States Army Captain Montgomery Schuyler reported on 1 March 1919,

"It is probably unwise to say this loudly in the United States but the Bolshevik movement is and has been since its beginning guided and controlled by Russian Jews of the greasiest type[... ]"[306]

United States Army Captain Montgomery Schuyler reported on 9 June 1919,

"These hopes were frustrated by the gradual gains in power of the more irresponsible and socialistic elements of the population guided by the Jews and other anti-Russian races. A table made in April 1918 by Robert Wilton, the correspondent of the London Times in Russia, shows that at that time there were 384 'commissars' including 2 negroes, 13 Russians, 15 Chinamen, 22 Armenians and more than 300 Jews. Of the latter number 264 had come to Russia from the United States since the downfall of the Imperial Government."[307]

*The Jewish Chronicle* published the following article on 11 April 1919 on page 10,

### "Percentage of Jewish Bolsheviki in Petrograd.

COPENHAGEN [F. O. C.]

On the trustworthy authority of the well-known Zionist leader, M. Idelson (of Petrograd), I am in a position to state that only two and a-half per cent. of the Jews in Petrograd have declared themselves in sympathy with Bolshevism. Although sixty per cent. of the Bolshevik leaders are Jews, and although a declaration against Bolshevism involves serious sacrifices, the Jews of Petrograd have fearlessly stated their attitude towards the movement. We are, therefore, confronted with the anomaly of the Jews furnishing for the Bolsheviki the majority of their leaders, although a smaller percentage of Jews than of any other nationality approve of Bolshevism."

A. Borisow wrote in an article "'Nep' and the Jews" in *The Jewish Chronicle* on 22 September 1922 on page 16,

"Still I repeat that the 'Nep' in Russia is a persecutor of the Jews.

---

**305**. K. A. Strom, Editor, *The Best of Attack! and National Vanguard Tabloid*, National Alliance, Arlington, Virginia, (1984), p. 65.
**306**. K. A. Strom, Editor, *The Best of Attack! and National Vanguard Tabloid*, National Alliance, Arlington, Virginia, (1984), p. 66.
**307**. K. A. Strom, Editor, *The Best of Attack! and National Vanguard Tabloid*, National Alliance, Arlington, Virginia, (1984), p. 66.

During the whole of the last two years the Jews have not suffered economically so much as they have during the few months since the introduction of the 'Nep.' It is not for nothing that the Jews translate the initials of the 'Nep' as the 'Nestchastnaja' ('luckless') Economic Policy.

What is it that the 'Nep' has brought us?

To begin with, it has reduced the number of officials. Many of the Soviet institutions have been closed down. In most of the others, 50 to 60 per cent. of the staff has been dismissed. Viewed on its merits, this is most welcome. It will mean a decrease in the heavy taxation which went to keep all these officials. But for the Jewish population it is a terrible blow. It is no secret that the Soviet institutions, especially in the cities, were staffed almost entirely by Jews. About three-quarters of the total number of officials were Jews. Tens of thousands of Jewish intellectuals and semi-intellectuals, lawyers, journalists and doctors, managed to earn a crust of bread in the service of the Soviet institutions. They formed the majority of the lettered population. Now they are dismissed, driven out into the streets, condemned to unemployment and to starvation. That is the first blessing which the 'Nep' has brought to the Jews."

Jews tried to justify the fact that Jews ruled the Bolshevik *régimes* by claiming that the Gentiles were too stupid to rule themselves. This was odd, given that the Jewish Bolsheviks promoted Jewish intellectuals, while concurrently mass murdering Gentile intellectuals in the hundreds of thousands, if not millions. Why were not all intellectuals murdered or promoted in proportionate numbers, if there was no ethnic bias, no Jewish genocidal racism involved in the process? "Mentor" wrote in an article entitled "Peace, War—and Bolshevism" in *The Jewish Chronicle* on 4 April 1919 on page 7,

"IT is not difficult to see why a people which has managed to subsist through Tsardom, because of the religious ideals and ideas which it nourished throughout all its classes, and not least among its peasantry, has been attacked by the ideals of Bolshevism, and why, released from Tsardom, it has, pendulum-like, swung into the arms of Lenin, looking to the ideals of his creed, and not to its wickedness or its excesses. The same reason obtains for the number of Jews who are to be found in the Bolshevist ranks. The Jew is an idealist. He will give much for an ideal. He thirst for idealism as a goal of life. This may seem strange to those who associate the Jew with materialism. But the capacity of the Jew for idealism is such that he notoriously idealises even the material. The fact that there are so many of our people who have associated themselves with the ideals of Bolshevism, even although as Jews its excesses must be repugnant to them, has to be placed in conjunction with another fact. These men will be found for the most part unassociated with or dissociated from the Synagogue. In the ordinary way of speaking they are not observing Jews. Is it not patent that the Synagogue, having failed to attract them by its idealism, and no other ideal, not even a material ideal, having been provided for them—for they are not men of wealth and substance, such as are usually to be found among the *bourgeoisie*— they have ranged themselves on the side of Bolshevism, because here was no

Jewish ideal to which these Jews could devote their sentiments and their energies? I cannot understand how people who for generations have, unprotesting, allowed the Jew, particularly in Eastern Europe, in Russia, to suffer pogroms, to be massacred and ill-treated, and tortured and murdered, and for two thousand years have kept our people outside the ambit of the most potent source of idealism that can appeal to men—that associated with National being—now have the hypocrisy, the soulless impertinence, to complain that so many of our people are Bolshevists! That Jews have been chosen to the extent they have to take a leading part in the movement in Russia and in Hungary, is merely because they are heavily endowed with intellectualism and capacity, as compared with the rest of the population. But the world must not surprised that the Jew, who is an idealist or nothing, has turned to the idealism of Bolshevism, which a British writer has declared to be comparable to the idealism preached by the founder of Christianity. It were surprising, really, were it otherwise. You cannot keep a people out of their rightful place amid the nations of the world, and then complain because they take the leading part which their abilities entitle them to in the nations among whom you have scattered them. The fact that a timorous millionaire afraid, and doubtless with good cause, of Bolshevism, which he probably has never taken the trouble, or perhaps has not the capacity to appreciate in full measure, places a ban of religious excommunication upon those Jews who are Bolshevists, is a thing for the gods to laugh at!

THERE is much in the fact of Bolshevism itself, in the fact that so many Jews are Bolshevists, in the fact that the ideals of Bolshevism at many points are consonant with the finest ideals of Judaism, some of which went to form the basis of the best teachings of the founder of Christianity—these are things which the thoughtful Jew will examine carefully. It is the thoughtless one who looks upon Bolshevism only in the ugly repulsive aspects which all social revolutions assume and which make it so hateful to the freedom-loving Jew—when allowed to be free. It is the thoughtless one that thus partially examines the greatest problem the modern world has been set, and as his contribution to the solution dismisses it with some exclamation made in obedient deference to his own social position, and to what for the moment happens to be conventionally popular."

Chaim Weizmann reported to the Fifth Meeting of the Zionist Advisory Committee, in London, on 10 May 1919,

> "Bolshevism covers a multitude of sins, especially in Poland, and we pay the cost. As a result of the official statement issued by the Bolsheviks in Petrograd to join them, 2½ per cent of the Jewish population have joined, 90 per cent have refused. It is quite true that 60 per cent of the Bolshevik officials are Jews. It is simply that they have got to find means of living, and they are the only people who can read and write."[308]

---

**308**. C. Weizmann, *The Letters and Papers of Chaim Weizmann*, Volume 1, Series B, August 1898-July 1931, Transaction Books, Rutgers University, (1983), pp. 241-242.

The book of *Obadiah* verse 8 teaches the Jews to destroy the intellectual class of non-Jews and deprive the Gentiles of knowledge,

"Shall I not in that day, saith the LORD, even destroy the wise *men* out of Edom, and understanding out of the mount of Esau?"

The Bolsheviks mass murdered the educated among the Gentiles, but education was what was claimed to have saved those Jews who replaced them, as if that explained away the fact that Jews predominated the Bolshevik Government. What was it that caused the Jewish Bolsheviks to mass murder highly educated and intellectual Gentiles, while education and intellectualism were the reasons given for the promotion of the Jewish minority and the predominance of the Jews in leadership rôles, if Jews weren't in charge of the Bolsheviks from the outset? A 20 February 1930 article in the *Patriot* stated,

"No one who has paid the slightest attention to the course of Russian events since the Bolshevik accession to power in November, 1917, can have failed to know that, when all the important members of the Russian aristocracy, the learned profession, the Army and Navy, had been executed, or imprisoned, or driven abroad, Red Jews were in possession of the great majority of responsible positions in and under the Soviet. So clear was this that, in the past, Jewish apologists, here and in America, have explained the fact by the true statement that only among the Jews could be found any longer the brains and business experience for filling important posts. Yet in the face of this situation there have been dozens of books published in English, and innumerable articles throughout the Press, and any number of lectures delivered, all with the astounding omission of any mention of Jewish handiwork in Russian Bolshevism. There have been public references to the sufferings of some orthodox non-Communist Jews at the hands of the Soviet."[309]

As late as 1924, racist political Zionist Israel Zangwill wrote that many Jews felt a need to keep the murderous Jewish Bolsheviks in power, those Bolsheviks who came to power through the might of Jewish financiers,[310]

"National politics is the realm of might, and if, as Dr. Hertz warns us, the menace of massacre still lies over the whole Russian Jewry should the Soviet Government be overthrown, we must face the sad fact that Jewish might does not exist."[311]

---

**309**. D. Fahey, *The Mystical Body of Christ in the Modern World*, Browne and Nolan Limited, London, (1935), pp. 256-257; *see also:* pp. 93, 101.
**310**. D. Fahey, *The Mystical Body of Christ in the Modern World*, Browne and Nolan Limited, London, (1935), pp. 86-93, 99-100.
**311**. I. Zangwill, "Is Political Zionism Dead?", *The Nation*, Volume 118, Number 3062,

Robert Wilton published *Russia's Agony*, Longmans, Green & Co.; New York, London, E. Arnold, (1918); and *The Last Days of the Romanovs, from 15th March, 1917: Part I, the Narrative; Part II, the Depositions of Eye-Witnesses*, Thornton Butterworth, London, (1920); in French, with an ethnic analysis of leading figures, *Les Derniers Jours des Romanof. Le Complot Germano-Bolchéviste Raconté par les Documents*, G. Crès & Cie, Paris, (1920); in Russian, *Posliednie dni Romanovykh*, Grad Kitezh, Berlin, (1923); in Polish, *Ostatnie dni Romanowów*, Warszaw. Denis Fahey published a list of Bolshevik crypto-Jews, together with their true names, and revealed an abundance of evidence which proved that Bolshevism was principally led and financed by Jews, which is not the same thing as saying that most Jews were Bolsheviks—they were not.[312]

Many of the common myths unfairly asserted against Jews in general appeared in this era. Brazen Jewish racism typical of the political Zionists also manifested itself. Racist Zionist Jews aggressively responded to other Jews who asserted that Jewishness was a religion, not a race. "An English-Born Jew" wrote in *The London Times*, on 1 December 1919, on page 10:

"TO THE EDITOR OF THE TIMES.

Sir.—Your correspondent 'Judæus' would seem to belong to the class of Jew satirized very recently by a Jewish writer as always anxious to cast overboard any fellow-Jews who are pointed to as inconvenient Jonahs. To-day he is bent upon dissociating himself as an English Jew from his Russian brethren because the latter are involved in Bolshevism. Yesterday he was anxious to dissociate himself from his German brethren because they were involved in Prussian militarism. He is desirous of disclaiming a Trotsky as a fellow-Jew, while doubtless willing to bask in the reflected glory of an Einstein.

But I am more concerned with his curious excursus into the ethnology of the Jew. He would have us believe that the Jew is contradistinguished from his fellow-beings only by religion, and that for the rest he is Russian in Russia, a German in Germany, and an Englishman in England—that race has no bearing upon the Jew as a product, and that we are wholly the result of the environment in which we may happen to be placed. It would be interesting, indeed, if 'Judæus' would tell us how soon he thinks a Skye terrier domiciled in England would become a bulldog, or how long it would take for a race of bulldogs bred in the Celestial Empire to produce Pekinese pups.

Obediently yours,
AN ENGLISH-BORN JEW."

---

(12 March 1924), pp. 276-278, at 276.
**312**. D. Fahey, *The Mystical Body of Christ in the Modern World*, Browne and Nolan Limited, London, (1935).

*The Jewish Chronicle* published a couple of letters to the editor in response to this exchange, on 8 December 1911, on page 38,

**"THE JEWISH POSITION:**
What Mr. Chesterton said.

TO THE EDITOR OF THE "JEWISH CHRONICLE."

SIR,—I hope that others besides myself will write to you to state that Mr. Kisch entirely misrepresents what Mr. Chesterton said on the 26th November at the West End Jewish Literary Society. Mr. Kisch apparently calls Mr. Chesterton reactionary because Mr. Chesterton believes in nationality, but if this is reactionary surely the Jews are the most reactionary people in the world as they have most deliberately insisted on retaining part of their nationality. Mr. Chesterton never said a word about 'attraction towards Zion' ever being a possible danger to the British Empire: he saw a source of demoralisation in those rich cynical Jews who have no enthusiasm for any ideal. He also doubted whether it is possible to have two nationalities which are equal in their claims on an individual and, it anyone will think the matter out, I think they will find that in any testing crisis they could not be. Mr. Kisch may be in favour of a policy of drifting purposelessness and inconsistency: those who are not will welcome all critics who help to clear away the endless humbug of Jews who believe in their mission and are actually missionaries of nothing and do not know what their message is, and who believe in their nationality and do not want self-government.

Yours obediently,
Westbourne Terrace, Hyde Park. ARTHUR D. LEWIS.

**Can Jews be Patriots?**

TO THE EDITOR OF THE 'JEWISH CHRONICLE.'

SIR,—It is not in the least surprising that Mr. Chesterton's lecture to the West End Jewish Literary Society should have proved so unpalatable to the members of that body in general and to your correspondent, Mr. Kisch, in particular.

There are quite a number of ladies and gentlemen with a weathercock cast of mind—the sort of person who though he has never read a single one of M. Bergson's books, can never say anything just now without mentioning his name—who, at prize distributions of Sabbath classes, boys' and girls' clubs, and other functions of the kind, makes it a constant burden of all his speeches, that Jews besides being good Jews should always be good Englishmen. This is the message that the West is repeatedly flashing to the East. When, therefore, a gentleman of Mr. Chesterton's logical cast of mind comes along and very flatly tells them that good Jews cannot be patriotic Englishmen, it is not unnatural that the ladies and gentlemen in question should kick. The patriotism of the Jew is simply a cloak he assumes to please the Englishman and so when Mr. Chesterton is shrewd enough to detect the Jew beneath the Englishman's clothing, the masqueraders become exceedingly angry. They had hoped to placate the Englishman by saying that they loved him and agreed with him. Judge then of their dismay when he turns round and says: I can only accept your love when you hate me and differ from me. The Jew is suspect and he knows it; and in the hope that the suspicion will be drowned in the noise, he becomes most vulgarly loud in his profession of patriotism. This atmosphere of suspicion in which the Jew lives from the moment of his birth, makes him so horribly fidgety, that when he meets a Gentile, the fact that he is a Jew is either the very first or the very last thing he wants to tell him. The Jew never takes the fact that he is one as a matter of course, which shows that he is never sure of himself, since it is only the things we are sure of and easy about that we take as matters of course.

Mr. Kisch seems to think that because some thirty years ago, two eminent men had a quarrel about the question whether good Jews could be patriotic Englishmen that, therefore, the matter has been disposed of at once and for all. To the Jews of this generation, the question is more acute and insistent than ever. We Jews of the younger generation are simply being coerced and intimidated, not through the compulsion of physical force but through the more subtle and insidious compulsion of a tyrannous public opinion, into a profession of patriotism, which, in the nature of things, must always be viewed with distrust and suspicion. I think it can be laid down as a general law, that the more Jews become Englishmen the less they become Jews. That does not imply any moral censure; it is simply a statement of fact, and Jews who pretend that they can at once be patriotic Englishmen and good Jews are simply living lies.

Yours obediently,<br>B. FELZ."

Dietrich Eckart wrote, quoting Adolf Hitler, who capitalized on Jewish racism in order justify anti-Jewish racism, which served to justify more Jewish racism, which served to justify more anti-Jewish racism, and so on (both Dietrich Eckart and Adolf Hitler were working for the Jewish Zionists),

"One doesn't need spectacles to see that. 'I am a British subject but, first and foremost, a Jew,' screamed a Hebrew years ago in a large English-Jewish newspaper. [*Notation:* M.J. Wodeslowsky, *Jewish World*, January 1, 1909.] And another: 'Whoever has to choose between his duties as an Englishman and as a Jew must choose the latter.' [*Notation:* Joseph Cohen, *Jewish World*, November 4, 1913.] And a third: 'Jews who want to be both patriotic Englishmen and good Jews are simply living lies.' [*Notation: Jewish Chronicle*, December 10, 1911.] That they could venture things of that sort so openly indicates how overrun with Jews England already was then."³¹³

The letters by "Verax" and Israel Cohen address most of the issues raised by "Mentor" in her(?) open letter to Lord Northcliffe. Verax and Israel Cohen wrote in *The London Times* on 27 November 1919 on page 15,

## "JEWS AND BOLSHEVISM.

### THE MOSAIC LAW IN POLITICS.

### RACIAL TEMPERAMENT.
### TO THE EDITOR OF THE TIMES.

Sir,—As an old student of Jewish history, Jewish literature, and of the Jewish people themselves, I have read with much interest and sad amusement the correspondence in your columns on the Jews and Bolshevism. The preponderence of Jews, renegade and other, in the development and direction of Bolshevism is too well known to need special demonstration. The letters of Mr. Israel Cohen have, however, a merit in this respect that is conspicuously absent from the letter of 'Judæus.' Mr. Cohen writes of the Jews as a 'race,' whereas 'Judæus' would have us, at this time of day, believe that the Jews are merely a religious 'denomination.' This is the kind of casuistry that so often deprives Jewish apologetics of value. The Jews are, first of all, a race, with a religion suited to their race-temperament. Temperament and religion have acted and reacted upon each other for thousands of years until they have produced a type distinguishable at a glance from any other race-type in the world. Persecution, religious, economic, and political, has had comparatively little to do with the matter. Otherwise, there would surely not exist caricatures more than 2,000 years old of the specifically Jewish types which 'Judæus' and his like would probably have us accept as a consequence of Christian intolerance.

But this, after all, is not the main point. I, for one, cannot find it in me to denounce Trotsky and his associates for the havoc they have wrought in Russia. Knowing something of the Jewish character, its persistence, its intensity, and its inexorable vindictiveness, I can understand that Trotsky and his fellow 'gun men' from New York should delight in trampling upon the Russia that oppressed their race and in destroying every vestige of the system that held millions of Jews in shameful bondage. I can understand, too, how Jews the world over, orthodox and renegade, glory in their heads at the vengeance thus wreaked by men of their own race upon Tsarism and all its works. For the inwardness of Jewry is not solely religion. It is, above

---

**313**. D. Eckart and A. Hitler, *Der Bolschewismus von Moses bis Lenin: Zwiegespräch zwischen Adolf Hitler und mir*, Hoheneichen-Verlag, München, (1924); English translation by W. L. Pierce, "Bolshevism from Moses to Lenin", *National Socialist World*, (1966). URL: <http://www.jrbooksonline.com/DOCs/Eckart.doc> p. 8.

all, pride of race, belief in its superiority, faith in its ultimate triumph, the persuasion that Jewish brains are superior to Gentile brains—the attitude of mind, in short, that corresponds to the inbred conviction that the Jews are the Chosen People destined, one day, to be the rulers and law-givers of mankind.

Whether this conviction was engendered in them by religious doctrine, or whether the doctrine was fashioned to suit the conviction, I cannot say. Nor is it possible to determine whether the Law of Moses, with its eye for an eye and a tooth for a tooth, has given to the Jewish character its hard and tenacious revengefulness, or whether the Law of Moses itself is an expression of that peculiar race-character. Be this as it may, the Jews as a race are as proud of the Law of Moses and as persuaded of its superiority to the Law of Christ, with its doctrine of forgiveness, as they are of the superiority of their blood over that of non-Jewish peoples. Those who may wish to ponder these matters might do worse than betake themselves to the Court Theatre and see the great Jewish actor, Moscovitch, play Shylock. They may then begin to understand many things, and, among others, one thing that students of Jewry too often overlook—the apparently untamable passionateness and the apparently incurable short-sightedness of Jewish minds.

No one who knows the Jews—not a few more or less pleasant, attractive, or brilliant individuals, but Jews in the mass—can doubt that the picture Shakespeare drew of the Jewish temperament in Shylock is true to life. Nor is it doubtful that the most illuminating trait in Shylock's character is not his revengefulness and cruelty, but his stupidity. He pursues his vengeance without ever dreaming that reaction against his conduct may recoil disastrously upon himself and undo him utterly, whereas a little forgiveness, a little comprehension even of the cash value of the 'quality of mercy' would have given him assured prosperity. It is in this respect that Shylock is most typical of the spirit of Jewry—that is to say, of its inability to forgive, or, in other words, its fidelity to the spirit of the Law of Moses as distinguished from the Law of Christ. For the Jews to be revenged on Russia must be sweet indeed, and they may well have felt that no price was too high for the satisfaction of their explicable rancour. Have they not worked and plotted against Russia for generations? Were not the Marxist doctrines, that are the roots of Bolshevism, the fruit of a Jewish brain? Was not the whole revolutionary organization in Russia largely Jewish? Undoubtedly many Jews in Russia who had escaped the rigours of the old *régime*, or had even grown prosperous under it, have opposed Bolshevism and suffered the penalty. Undoubtedly Jews were influential in the Cadet Party and in the Menshevist section of the Russian Socialist Party. Undoubtedly the Zionist organizations in Russia have suffered under Bolshevism because they are an expression of Jewish national feeling and as such are obnoxious to Bolshevism. But the fact remains that the warp and woof of the Bolshevist organization has been Jewish, and that throughout Russia and, indeed, throughout Central Europe, including Hungary and what remains of Austria, Bolshevism and Jewry are regarded as practically synonymous.

Herein lies grave peril for the masses of the Jewish people in Russia. Many Jews now perceive this peril and are endeavouring, on the one hand, now to prove that the connexion between Jewry and Bolshevism is slight, and, on the other, to promote a policy in Allied countries favourable to some agreement with Bolshevism so that the danger of a general massacre of Jews after the overthrow or the collapse of Bolshevism may be averted. These tactics are transparent, short-sighted, and, indeed, stupid. The only sound policy for the Jews would have been, and would still be, for their representative leaders to dissociate themselves whole-heartedly and publicly from Bolshevism and all its works, and to use all their influence, public and private, in favour of its overthrow by the constitutional and democratic forces of Russia, with the support and under the control of the Allies. I can see no other way of escape from the appalling peril that hangs over Jewry in Eastern Europe. Otherwise the Jews may find, when it is too late, that the excess of their vengeance upon Russia has recoiled upon them in terrible fashion and that, to them who have hated much, little, too little, will be forgiven.

I am, Sir, your obedient servant,
VERAX.

TO THE EDITOR OF THE TIMES.

Sir,—In your issue of to-day your correspondent 'Janus' gives a list of 28 'conspicuous Bolshevists' who, he states, 'are either full-blooded Jews or of Jewish extraction.' It is only fair to your readers that they should be informed that as many as 10 names in this list are those either of non-Jews or of anti-Bolshevists or of dead Bolshevists:—

(1-3) Lunacharsky, Chernov, and Bogdanov are pure Russian Bolshevists.

(4) Zagorsky is neither a Jew nor a Bolshevist, but a Russian Radical.

(5-6) Kamkov and Bunakov are Social Revolutionaries—*i.e.*, anti-Bolshevists. Kamkov (-Katz), after his participation in the assassination of Count Mirbach, had to flee from Bolshevist Russia to Archangel.

(7-8) Dan and Martov are the Jewish leaders of the Menshevists—*i.e.*, the most determined opponents of Lenin and his group. They were referred to as anti-Bolshevists in your columns only a few days ago.

(9-10) Uritzky and Volodarsky have both been murdered, the former by the Jew Kannesgiesser.

I have no doubt that 'Janus' has sent you his list in good faith, but the fact that it has to be discounted to such a great extent is typical of the general misrepresentations of the Jewish share in Bolshevism.

Yours faithfully,
ISRAEL COHEN.
77, Great Russell-street, W.C., Nov. 26."

Israel Cohen wrote in *The London Times* on 1 December 1919 on page 10,

## "JEWS AND BOLSHEVISM.

A FURTHER REJOINDER.
TO THE EDITOR OF THE TIMES.

Sir,—I am loth to trespass further upon your space, but the grave indictment of the Jewish people contained in the letter of 'Verax,' who forms with your correspondents 'Philojudæus' and 'Janus' the third element in an accusing Trinity, impels me to invoke the courtesy of your hospitality once again. 'Verax' describes himself as 'an old student of Jewish history, Jewish literature, and of the Jewish people, themselves,' but the whole spirit and contents of his letter betray how superficial and unprofitable, or perhaps, how ancient his studies have been. His presentation of the Jewish character is a gross travesty, and his interpretation of the Jewish part in the Bolshevist movement is fanciful and unfounded. He has shifted the base of attack from the domain of facts and figures, where he finds the position of his fellow-accusers untenable, to the domain of racial psychology; but his arguments, however plausible, will be found upon examination to possess not the flimsiest shred of substance.

Burke once declared that you cannot indict a nation, but 'Verax' thinks he knows better. He maintains that Judaism is founded upon the principle of revenge, and he declares that 'Jews the world over, orthodox and renegade, glory in their hearts at the vengeance thus wreaked by men of their own race upon Tsarism and all its works.' His premise is false, and his conclusion is a calumny. He cites the principle of an eye for an eye and a tooth for a tooth, as though that were ever intended literally. Has 'Verax,' in his studies of Jewish history and literature, ever come across a single case where this was literally applied or even advocated? Does he not know that this principle has always been interpreted by all Talmudical and Rabbinical authorities without exception (*vide* talmud Baba Kama, pp. 83*b* and 84*a*), as meaning simply the rendering of just monetary compensation, an interpretation which is in complete harmony with the canons of modern jurisprudence? Or does 'Verax' also take quite literally the saying in the Sermon on the Mount, 'And if any man shall sue thee at the law, and take away thy coat, let him have thy cloak also?' In support of his thesis he invokes the shade of Shakespeare and points at the pitiable figure of Shylock, but Shakespeare, living in the days of Queen Elizabeth, could not have known any typical Jews, as the residence of Jews in England was then forbidden: and, as 'Verax' can learn from the commentators, Shakespeare simply imputed to a Jew the heartless bargain attributed in the original story to a non-Jew. If anything proves the un-Jewishness of Shylock it is his acceptance of Christianity to save his life. Surely, 'Verax' must know from his study of Jewish history that Jews without number have sacrificed their lives rather than accept the waters of baptism. His antithesis between a Jewish law of revenge and a Christian law of forgiveness is utterly fallacious. The Bible and the Talmud utter repeated warnings against hatred and revenge, and insist upon forgiveness as one of the cardinal bases of human conduct. The law of Moses distinctly states:—'Thou shalt not avenge, nor bear any grudge against the children of thy people, but thou shalt love they neighbour as thyself,' (Levit. xix., 18). And in Talmudic literature 'Verax' can find such noble sayings as:—'Be of the persecuted and not of the persecutors,' and 'Who is strong? He who turns an enemy into a friend.'

Now how does your correspondent's misreading of Jewish psychology apply to Bolshevism? Even if revenge were inculcated by the Law of Moses, we would expect it to be exercised by those to whom the Law of Moses is dear, by the pious or orthodox. But the orthodox Jews, to a man, have eschewed the pernicious doctrine; they have only suffered by it. The Jews who are Bolshevists are opposed to orthodoxy; they are opposed to the Jewish religion in any form; indeed, they are contemptuously hostile to all religion. They will have nothing to do with Judaism as religion, race, or nation. Nor can the Bolshevist *régime* be adduced as proof that the Jews wished to see the downfall of Tsarism and all its works, for that end was already achieved by Kerensky's revolution. When Trotsky first began to play a leading part in Bolshevism, a deputation of the Council of the Petrograd Jewish Community pleaded with him to break off his connexion with the movement, on the ground that it would lead to the shedding of innocent Jewish blood: but he refused, replying that he was not a Jew himself, and did not recognize Jews as such. Attempts have been made by the relatives of other Jewish Bolshevists to wean them from their heresy, but without avail. 'Verax' seems to suggest that Bolshevism is a product of the Jewish mind, heedless of the fact that it was hatched in the brain of Lenin, the pure Russian, who, during the revolution of 1905, returned from Switzerland to his native country as an apostle of Jewish pogroms, by which he thought, through the massacre of the Jewish *bourgeoisie*, he could hasten his Communist paradise! And the thesis of your correspondent involves the further absurdity of supposing that the Jews in Russia would deliberately destroy the foundations of their own material existence; for the Jews in Bolshevist Russia are for the most part merchants, manufacturers, and members of the liberal professions—the very classes against which Lenin and his associates have dealt their direst blows.

'Verax' concludes by declaring that many Jews are now trying 'to promote a policy in Allied countries favourable to some agreement with Bolshevism.' What are his proofs, what are his data? Why does he not at least give one specific instance? Your correspondent appeals to the representative leaders of Jewry to use all their influence in favour of the overthrow of Bolshevism. I have no right to speak in the names of these leaders, but I cannot help recalling that when they appealed a few years ago for intervention in Russia, not for the overthrow of Tsardom, but for the suppression of pogroms, they were told that intervention was impossible. The question, I venture to think, is not one for Jewish leaders, who might afterwards be accused by some other anonymous correspondent of usurping political power—even 'Verax,' in an earlier passage, taunts the Jews with the conviction that they are destined to be the rulers of mankind—but for the Allied and Associated Governments. If these Governments, with all the resources of their collective statesmanship and immeasurable munitions, fail to solve the problem, and there should indeed be a fear of the further massacres which 'Verax' foreshadows, then I hope the Army of Liberation, when it redeems the Bolshevist-ridden country, will act not in the vindictive spirit which he predicts but in that of true Christian charity. And if the millions of Jews whose lives are now menaced have no claim to protection on the mere ground of humanity, may not the memory of the myriads of their fellow-Jews who fought and fell in the War of Liberation, and in the hope of a better era for their persecuted people, serve as a mute yet potent plea on their behalf?

Yours faithfully,<br>
ISRAEL COHEN.

November 27."

In his desire to discredit "Verax", Cohen badly miscalculated the nature and source of the threat. The Nazis were not Christian and painted themselves as victims of the "War of Liberation". Cohen also misrepresented the Judaic proscriptions against attacking one's neighbors, which were meant only for fellow Jews, not Gentiles.[314] The Talmud states in *Sanhedrin* 59*a* (*see also:* folio 57*a*),

"A goyim who studies the Torah must be killed."

and,

"The Law Moses gave unto us as an heritage; it is an heritage for us, not for them."[315]

The Talmud states in *Baba Mezia*, 108*b*,

"'If he sells to a heathen' — because a heathen is certainly not subject to [the exhortation], *'And thou shalt do that which is right and good in the sight of the Lord.'*"[316]

---

314. G. Nicolai, *Die Biologie des Krieges, Betrachtungen eines deutschen Naturforschers*, O. Füssli, Zürich, (1917); English translation: *The Biology of War*, Century Co., New York, (1918). B. Lazare, *Antisemitism: Its History and Causes*, (1894); *L'Antisémitisme, son Histoire et ses Causes*, L. Chailley, Paris, (1894).
315. R. H. Williams, *The Ultimate World Order—As Pictured in "The Jewish Utopia"*, CPA Book Publisher, Boring, Oregon, (1957?), p. 7.
316. I. Epstein, Editor, H. Freedman, Translator and Annotator, *The Babylonian Talmud*, Seder Nezikin, Tractate Baba Mezia, folio 108*b*, The Soncino Press, London, (1935), pp. 619-622, at 619.

The Talmud states in *Baba Mezia*, 114*b*,

"For it has been taught: R. Simeon b. Yohai said: The graves of Gentiles do not defile, for it is written, *And ye my flock, the flock of my pastures, are men;*⁶ only ye are designated *'men'*."³¹⁷

The Talmud states in *Yebamoth*, 60*b*-61*b*,

"It was taught: And so did R. Simeon b. Yohai state [61*a*] that the graves of idolaters do not impart levitical uncleanness by an *ohel*,¹⁵ for it is said, *And ye My sheep the sheep of My pasture, are men;*¹ you are called *men*² but the idolaters are not called *men*.²"³¹⁸

The Talmud states in Tractate *Yebamoth*, folio 98*a*,

"Raba stated: With reference to the Rabbinical statement that [legally] an Egyptian has no father,¹⁰ it must not be imagined that this is due to [the Egyptians'] excessive indulgence in carnal gratification, owing to which it is not known [who the father was], but that if this were known¹ it is to be taken into consideration;² but [the fact is] that even if this is known it is not taken into consideration. For, surely, in respect of twin brothers, who originated in one drop that divided itself into two, it was nevertheless stated in the final clause,³ that they 'neither participate in *halizah* nor perform levirate marriage'.⁴ Thus it may be inferred that the All Merciful declared their children to be legally fatherless,⁵ for [so indeed it is also] written, *Whose flesh is as the flesh of asses, and whose issue is like the issue of horses.*⁶"³¹⁹

The Talmud states in Tractate *Kiddushin*, folio 68*a*,

"AND WHATEVER [WOMAN] WHO CANNOT CONTRACT KIDDUSHIN etc. How do we know [it of] a Canaanitish bondmaid?¹³ — Said R. Huna, Scripture saith, *Abide ye here with* ['im] *the ass*¹⁴ — it is a people ['*am*] like unto an ass.¹⁵ We have thus found that *kiddushin* with her is invalid: [68*b*] how do we know that the issue takes her status? — Because

---

**317**. I. Epstein, Editor, H. Freedman, Translator and Annotator, *The Babylonian Talmud*, Seder Nezikin, Tractate Baba Mezia, folio 114*b*, The Soncino Press, London, (1935), pp. 651-653, at 651.
**318**. I. Epstein, Editor, I. W. Slotki, Translator and Annotator, *The Babylonian Talmud*, Seder Nashim, Tractate Yebamoth, Volume 1, folios 60*b*-61*a*, The Soncino Press, London, (1936), pp. 401-408, at 40-405.
**319**. I. Epstein, Editor, I. W. Slotki, Translator and Annotator, *The Babylonian Talmud*, Seder Nashim, Tractate Yebamoth, Volume 2, folio 98*a*, The Soncino Press, London, (1936), pp. 670-673, at 670-671.

Scripture saith, *the wife and her children shall be her master's.*¹ How do we know [it of a freeborn] Gentile woman? — Scripture saith, *neither shalt thou make marriages with them.* ² How do we know that her issue bears her status? — R. Johanan said on the authority of R. Simeon b. Yohai, Because Scripture saith, *For he will turn away thy son from following me:*³ thy son by⁴ an Israelite woman is called thy son, but thy son by a heathen is not called thy son.⁵"³²⁰

The Talmud states in Tractate *Keritoth*, folio 6*b*,

"OR USES OF ANOINTING. Our Rabbis have taught: He who pours the oil of anointing over cattle or vessels is not guilty; if over heathens or the dead, he is not guilty. The law relating to cattle and vessels is right, for it is written: *Upon the flesh of man* [adam] *shall it not be poured;*⁶ and cattle and vessels are not man. Also with regard to the dead, [it is plausible] that he is exempt, since after death one is called a corpse and not man. But why is one exempt in the case of heathens; are they not in the category of *adam*? — No, it is written: *And ye my sheep, the sheep of my pasture, are* adam [*man*]:⁷ Te are called *adam* but heathens are not called *adam*."³²¹

The Talmud states in Tractate *Berakoth*, folios 58*a-b*,

"R. Hamnuna further said: If one sees a crowd of Israelites, he should say: Blessed is He who discerneth secrets.⁵ If he sees a crowd of heathens, he should say: *Your mother shall be ashamed*, etc.⁶ [\*\*\*] Our Rabbis taught: On seeing the Sages of Israel one should say: Blessed be He who hath imparted of His wisdom to them that fear Him. On seeing the Sages of other nations, one says, Blessed be He who hath imparted of His wisdom to His creatures. On seeing kings of Israel, one says: Blessed be He who hath imparted of His glory to them that fear Him. On seeing non-Jewish kings, one says: Blessed be He who hath imparted of His glory to His creatures. R. Johanan said: A man should always exert himself and run to meet an Israelitish king; and not only a king of Israel but also a king of any other nation, so that if he is deemed worthy,¹ he will be able to distinguish between the kings of Israel and the kings of other nations. [\*\*\*] R. Shila administered lashes to a man who had intercourse with an Egyptian⁵ [footnote 5 in the Soncino Edition states: "Var. lec. Gentile." *Varia lectio* is Latin for: "a variant reading", indicating that Jews do not only consider Egyptians, but all non-Jews to be subhuman animals.—CJB] woman. The

---

**320**. I. Epstein, Editor, H. Freedman Translator and Annotator, *The Babylonian Talmud*, Seder Dashim, Tractate Kiddushin, folio 68*a*, The Soncino Press, London, (1936), pp. 342-345, at 344-345.
**321**. I. Epstein, Editor, I. Porusch, Translator and Annotator, *The Babylonian Talmud*, Seder Kodashim, Tractate Keritoth, folio 6*b*, The Soncino Press, London, (1948), pp. 43-47, at 45.

man went and informed against him to the Government, saying: There is a man among the Jews who passes judgment without the permission of the Government. An official was sent to [summon] him. When he came he was asked: Why did you flog that man? He replied: Because he had intercourse with a she-ass. They said to him: Have you witnesses? He replied: I have. Elijah thereupon came in the form of a man and gave evidence. They said to him: If that is the case he ought to be put to death! He replied: Since we have been exiled from our land, we have no authority to put to death; do you do with him what you please. While they were considering his case, R. Shila exclaimed, *Thine, Oh Lord, is the greatness and the power.*[1] What are you saying? they asked him. He replied: What I am saying is this: Blessed is the All-Merciful who has made the earthly royalty on the model of the heavenly, and has invested you with dominion, and made you lovers of justice. They said to him: Are you so solicitous for the honour of the Government? They handed him a staff[2] and said to him: You may act as judge. When he went out that man said to him: Does the All-Merciful perform miracles for liars? He replied: Wretch! Are they not called asses? For it is written: *Whose flesh is as the flesh of asses.*[3] He noticed that the man was about to inform them that he had called them asses. He said: This man is a persecutor, and the Torah has said: If a man comes to kill you, rise early and kill him first.[4] So he struck him with the staff and killed him. [\*\*\*] Our Rabbis taught: On seeing the houses of Israel, when inhabited one says: Blessed be He who sets the boundary of the widow;[4] when uninhabited, Blessed be the judge of truth. On seeing the houses of heathens, when inhabited, *one says: The Lord will pluck up the house of the proud;*[5] when uninhabited he says: *O Lord, thou God, to whom vengeance belongeth, thou God, to whom vengeance belongeth, shine forth.*[3] [\*\*\*] Our Rabbis taught: On seeing Israelitish graves, one should say: Blessed is He who fashioned you in judgments who fed you in judgment and maintained you in judgment, and in judgment gathered you in, and who will one day raise you up again in judgment. Mar, the son of Rabina, concluded thus in the name of R. Nahman: And who knows the number of all of you; and He will one day revive you and establish you. Blessed is He who revives the dead.[4] On seeing the graves of heathens one says: *Your mother shall be sore ashamed,* etc. [\*\*\*] R. Joshua b. Levi said: On seeing pock-marked persons one says: Blessed be He who makes strange creatures. An objection was raised: If one sees a negro, a very red or very white person, a hunchback, a dwarf or a dropsical person, he says: Blessed be He who makes strange creatures."[322]

Jakob Ecker's *Der „Judenspiegel' im Lichte der Wahrheit: Eine wissenschaftliche Untersuchung*, Bonifacius-Druckerei, Paderborn, (1884), p. 120; states:

---

**322**. I. Epstein, Editor, M. Simon, Translator and Annotator, *The Babylonian Talmud*, Seder Zeraim, Tractate Berakoth, folios 58*a-b*, The Soncino Press, London, (1948), pp. 359-366.

"*Tosaphoth* zu Talmud *Kethuboth 3*, b:
„Sein (des AKUM) Same wird angesehen wie V i e h s a m e."

*Tosaphoth* zu Talmud *Sanhedrin 74*, b:
„*Concubitus AKUM* ist wie *concubitus bestiae.*"

See also: *Ezekiel* 23:20; 34:31.

Johann Andreas Eisenmenger wrote in his *The Traditions of the Jews, Contained in the Talmud and other Mystical Writings*, Volume 1, J. Robinson, London, (1748), pp. 254-255,

"In the Great *Jalkut Rubeni*, in the *Parasha Bereshith*, [*Footnote:* Fol. 10. Col. 1.] we have the following Passage, 'The Skin and the Flesh is the Coat of a Man. The Spirit within is the Man. But the Idolaters (*meaning all the Nations but the Jewish*) are not call'd Men, because their Souls have their Origin from the Unclean Spirit. But the Souls of the *Israelites* are derived from the Holy Spirit.' And a little farther on in the same Treatise, it is said, [*Footnote:* Fol. 10. Col. 2.] 'An *Israelite* is called a Man, because his Soul cometh from the Supreme Man. But an Idolater, whose Soul cometh from the Unclean Spirit, is call'd a Swine. If so, then is an Idolater the Body and Soul of a Swine.' In another Part of the said Treatise, entitled *Shaar olam hattobu* [*Footnote:* Fol. 23. Col. 4.], there is a Passage running thus: 'The Wicked are stiled the Dead in their Life-Time, because they have not a Holy Soul from the Foundation, which is called *Him that liveth for ever*. But they have the Soul from *Kelifa* (i. e. the Shell) by which is meant the Devil) who is call'd Death, and the Shadow of Death: And through the Sparklings of the same they live.'"

The danger of the Jewish-Bolshevik universal generalization, which was immediately apparent to Herman Bernstein's handlers, was very real, and was later exploited by Zionist Jews in order to place their agents in power on an anti-Semitic and anti-Bolshevist platform. "Verax" wrote in *The London Times* on 2 December 1919 on page 10,

## "BOLSHEVISM AND THE JEWS.

### A LARGER ISSUE.

### THE DANGER IN RUSSIA.

TO THE EDITOR OF THE TIMES.

Sir,—I am obliged to the Chief Rabbi for his helpful reply to my letter.

He protests 'with all possible vehemence,' or, as I might have said, 'with untamable passionateness,' against what he calls my 'attack upon the religious doctrines of Judaism and its alleged effects upon 'his' 'people.' He avers that 'the beginning and the end of all Jewish teaching is loving-kindness to all, even to our enemies.' He alleges that even were he to reprint in your columns 'a whole anthology of Bible and Rabbinical texts' in support of his claim, I should, 'at best merely proceed to seek new pretexts to maintain 'my' 'prejudices.' May I assure him that I have no prejudices, but some decades of experience. He adds that the 'breadth of humanity and passion for righteousness' which his anthology would reveal are 'nowhere to be surpassed (even in the Gospels, which, by the way, are also the work of Jews, written by Jews for Jews).'

It is perhaps as well that the Chief Rabbi should refrain from producing his 'Bible and Rabbinical texts,' lest your readers be moved to ask what reason there is to think that, since the Gospels, 'the work of Jews, written by Jews for Jews,' have profited Jewry so little, the Rabbinical and other texts, equally written by Jews for Jews, have been of greater avail. Incidentally, the Chief Rabbi's mention of the Gospels as 'the work of Jews' tends to substantiate both my reference to Jewish pride in the work of Jews, irrespective of their religious faith, and the argument, which 'Judæus' has sought to invalidate, that orthodoxy in Judaism is by no means essential to a Jewish status.

But these matters touch only the fringe of the grave question debated in your columns; and in any case *The Times* is not a Betha Midrash for the solving of pious conundrums or answering the riddle: 'When is a Jew not a Jew!' Nor can the testimony of your hospitable pages be invoked solely to prove that 'during these last five years Jewish citizens of every Allied country have been loyal and true and patriotic to the ideals of freedom and have fought in gladness the battle of righteousness.' To the patriotic conduct of most British and Allied Jews I, who know something of the inner history of the Jewish movement during these same five years, am glad to testify; but your columns have also recorded other things, such as the doings and the downfall of the *Bonnet Rouge* gang in France (Vigo-Almeyreda, Landau, Goldsky, and others), whose work for the Allies was of a quite peculiar sort. This merely as a reminder to the Chief Rabbi that, as I pointed out in my former letter, Jewish minds are prone to short-sightedness.

Mr. Israel Cohen's latest contribution need not detain me, save in one respect. His assertion that 'if anything proves the un-Jewishness of Shylock it is his acceptance of Christianity to save his life' makes me wonder whether he has ever read the lamentable story of the Marranos in the 14th century or that of Sabbatai Zebi, or Zevi, in the 17th. His followers, the Dönmehs, or crypto-Jews, of Salonika are with us to this day.

But, Sir, these matters are really of secondary importance. The real issue which it was the purpose of my letter to raise is: How is the Jewish people in Russia and other parts of Eastern Europe to escape from the wrath that is sure to come when Bolshevism collapses or is overthrown, unless steps be taken now to avert it? Frankly, I am anxious to see these masses of poor Jews saved from massacre. I am convinced, and have reason for my conviction, that they may pay dearly for the indisputable fact that, in wide regions of Central and Eastern Europe, Bolshevism and Jewry are regarded as practically synonymous. I do not say, and have not said, that they are synonymous, but I repeat that they are regarded as being practically synonymous, and that, when the process begins of seeking scapegoats for the unspeakable havoc that Bolshevism has wrought, the masses of poor Jews are likely to pay for the sins of Trotsky and his associates. With the fate of the rich Jews I am not so much concerned, for they usually manage to look after themselves. Therefore I repeat that the only sound policy for the Jews outside Russia, and as far as possible in Russia, would be to dissociate themselves, whole-heartedly and publicly, from Bolshevism and all its works, and to use all their influence, public and private, in favour of its overthrow by the constitutional and democratic forces of Russia with the support and under the control of the Allies.

If this be anti-Semitism, I am an anti-Semite—in company with many prophets of Israel who were sawn asunder, stoned, and crucified for daring to tell Jewry the truth: and I again sign myself. Yours obediently,

VERAX."

## 5.3.5 The Inhumanity of the Bolsheviks

As with "the Terror" of the French Revolution, the Bolshevik revolutionaries committed numerous atrocities against the monarchy and the Russian people. Many believed this genocide was revenge for the Pogroms and for the Pale of Settlement in Russia. In part it was, but in the greater part it was the fulfillment of Judaic Messianic prophecy and a means to keep Gentile empires from posing a threat to Jewish supremacy.

It is interesting to note that the Jews took revenge on the English who had expelled them, with Cromwell under the directorship of the Cabalist Jew Manasseh Ben Israel and others. The Jews also took revenge on the Germans, with Martin Luther's purges under the directorship of Cabalist Jews, and with the slaughter of innocent Germans under Bismarck and continuing through Hitler's régime. The Jews took revenge on the Romans and Christians by burning Rome and blaming the fire on the Christians, under the directorship of Nero's crypto-Jewish wife Poppæa,[323] and Pope St. Clement (88-97AD) recognized that the Jews were responsible for Nero's persecution of the Christians. The Jews

---

[323]. Josephus, "Antiquities of the Jews", Book XX, Chapter 8, *The Works of Flavius Josephus: Comprising the Antiquities of the Jews; a History of the Jewish Wars; and Life of Flavius Josephus, Written by Himself*, S. S. Scranton Co., Hartford, Connecticutt, (1916), pp. 609-613, at 612-613. *See also:* Tacitus, *Annal*, Book XV, in: "Dissertation III", *The Works of Flavius Josephus: Comprising the Antiquities of the Jews; a History of the Jewish Wars; and Life of Flavius Josephus, Written by Himself*, S. S. Scranton Co., Hartford, Connecticutt, (1916), p. 960. *See also:* E. Gibbon, "The Conduct of the Roman Government towards the Christians, from the Reign of Nero to that of Constantine", *The History of the Decline and Fall of the Roman Empire*, Chapter 16, Volume 3, Fred De Fau and Company, New York, (1776).

took revenge on the Spanish who expelled them, with the crypto-Jewish instigators of the Spanish Civil War, and then installed the crypto-Jewish tyrant Francisco Franco. The Jews took revenge on the Turks and Armenians with the revolutionary Young Turks, who were crypto-Jews known as *Dönmeh* Turks.[324] Racist Jews are today taking action against the United States for daring to be a mighty nation, after the creation of the State of Israel; because Jewish mythology demands that the Jews must rule the world from Jerusalem. After the United States' subservient rôle as the sword of this power is completed, it will be destroyed as an empire and the American People will face a genocide and tyranny.

Einstein, himself, wrote to Emil Zürcher on 15 April 1919 that he knew for certain that Bolshevik leaders were stealing the wealth of the Russian Nation and were "systematically" mass murdering everyone who did "not belong to the lowest class."[325] In addition to diminishing their ability to fight for their own interests, this also weakened the genetic stock of the Russian people,[326] and left them unable to conduct a counter-revolution—with the hope of ultimately leaving them unable to fight a counter-revolution against Zionist world domination at any point in the future.[327] The Talmud at *Sanhedrin* 37a teaches the Jews the importance of the fact that taking the life of an individual can also signify the genocide of countless unborn descendants of that individual. The Jews in control of the Bolshevik mass murderers sought to exterminate the better part of the Russian People and leave an inferior and easily managed "race" forever, or at least until they were completely wiped out.

Lenin fulfilled his own murderous ambitions and answered the call for merciless violence of Marxists like Georges Sorel, who published *Réflexions sur la Violence* in 1908.[328] *Circa* 17 October 1919, Heinrich Zangger wrote to Albert Einstein that the Bolsheviks were intentionally destroying food and murdering "all who know anything".[329] He wrote of their hatred, brutality and senseless destruction in their quest for power and of the danger it posed and widespread misery it caused. Trotsky made a point of declaring that the Bolshevik revolution

---

**324.** I. Zangwill, *The Problem of the Jewish Race*, Judaen Publishing Company, New York, (1914), pp. 9, 11; which was first published as an article, "The Jewish Race", *The Independent*, Volume 71, Number 3271, (10 August 1911), pp. 288-295, at 290-291. J. Prinz, *The Secret Jews*, Random House, New York, (1973), pp. 111-112.
**325.** Letter from A. Einstein to E. Zürcher of 15 April 1919, English translation by A. Hentschel, *The Collected Papers of Albert Einstein*, Volume 9, Document 23, Princeton University Press, (2004), p. 19.
**326.** R. H. Williams, *The Ultimate World Order—As Pictured in "The Jewish Utopia"*, CPA Book Publisher, Boring, Oregon, (1957?), p. 6.
**327.** D. Fahey, *The Mystical Body of Christ in the Modern World*, Browne and Nolan Limited, London, (1935), pp. 93, 101, 256-257.
**328.** G. Sorel, *Réflexions sur la Violence*, Librarie de "Pages libres", Paris, (1908). English translation: *Reflections on Violence*, B.W. Huebsch, New York, (1912).
**329.** Letter from H. Zangger to A. Einstein *circa* 17 October 1919, English translation by A. Hentschel, *The Collected Papers of Albert Einstein*, Volume 9, Document 143, Princeton University Press, (2004), pp. 120-121, at 120.

was a world-wide revolution that would eventually touch every human being. All of this serves no other purpose than to deliberately fulfill Jewish Messianic prophecy.

On 16 March 1922 on page 12 *The London Times* published the following Letter to the Editor:

## "BOLSHEVIST EXECUTIONS.

### TO THE EDITOR OF THE TIMES.

Sir,—The *Gaulois* published on December 23 last the following statistics showing the executions which have taken place in Russia during the past four years. The figures, based on the official documents of the Soviet, are as follows:—

The following persons have been executed since October, 1917:—28 Bishops, 1,215 priests, 6,775 schoolmasters and professors, 8,800 physicians, 54,650 officers, 260,000 soldiers, 10,500 officers of the constabulary and police, 48,500 soldiers of the same forces, 12,950 land owners, 355,250 so-called 'intellectual' citizens, 193,350 workmen, 815,100 peasants—total, 1,766,118.

Mr. Lloyd George wishes to arrange a meeting in Genoa with the perpetrators of these terrible crimes, to discuss the means of 'reconstructing' Russia. He might call together on the same occasion several cannibals and discuss with them the possibilities of 'reconstructing' Africa by means of devouring the African people.

Yours faithfully,
H. A. VAN DE LINDE.
4, Fenchurch-avenue, E. C.3, March 15."

Lord Sydenham of Combe informed the House of Lords in 1923 that the Bolshevist murders and the intentional starvation of populations under Bolshevist control resulted in approximately 30 million deaths since the Bolshevists seized power.

*The London Times* published the following report on 14 November 1919 on page 14, which was later released as a pamphlet by *The Times* (note that the accusation that the Bolsheviks tortured people with the "human glove" was reiterated by Dietrich Eckart and Alfred Rosenberg[330] in anti-Semitic Zionist propaganda),

## "THE HORRORS OF BOLSHEVISM.

### SUMMONS TO A CRUSADE.

---

[330]. D. Eckart, "Wovor uns Kapp behueten wollte", *Auf gut Deutsch*, D. Eckardt, München, (16 April 1920). A. Rosenberg, *Pest in Russland!: der Bolschewismus, seine Häupter, Handlanger und Opfer*, E. Boepple, München, (1922), p. 78.

REMARKABLE LETTER BY AN OFFICER.

*We print below a very remarkable letter sent by a British Officer in South Russia to his wife. The letter is notable not only for its revelations of Bolshevist atrocities, but as a human document. The man who has seen what Bolshevism really means cannot rest without enlisting his wife and all his family into a crusade against it and a campaign for the enlightenment of the British public.*

*The letter is published exactly as sent, except that names and dates have been altered, so that the writer and his wife will not be embarrassed. We make no apology in present circumstances for publishing certain passages of a nature generally considered 'unprintable.'*

DEAREST,

This should be your birthday and wedding day letter. I'll send the postal order for your hat and silk stockings and gloves along with this. M., dear, how I shall think of you on this 26th and 28th—or is it 31st by now? I wonder whether you will feel me near you—I shall dedicate these two days to my Molly.

Just fancy, Molly, they've made me a Staff officer! (acting). I shall break out in red tabs all over—that is, if I can get any. Would you draw on Cox and stagger round to the Army and Navy, and buy me a red hat band and one pair staff officer's gorget patches (red)? S-Staff officer's G-horget patches—and two little buttons? They'll take two months to reach me, Molly, but then we'll astonish the natives.

And—I'm going to another army—an army of umpty-thousand Cossacks, all irregular cavalry, splendid wild men, easily the most interesting, in fact rather exciting, crowd, and any amount of scope. And any amount of work to do. They make wild cavalry raids of hundreds of miles.

Do you remember my saying I wonder whether I'd have the chance of getting 'longside some Cossacks? And now I'm going to the one Cossack army of the four.

So I'll write you once more before I go, and I do hope I'll get another mail before I start, for it's a month from here to them, and communication by courier only.

Now, dearest, to the serious part of my letter.

I want you to do war work. <u>WAR</u> WORK. I want you to spend one hour, or, if you cannot, only half an hour, daily, in doing the Bolshevist *harm*. With your typewriter. In thought, word, and deed. I want you to put heart and soul into *helping General Denikin and his cause*. For if ever there was a crusade it is this. I shall put my heart and soul into helping to organize and supply in my area, into creating good feeling and moral values, into actual fighting, and into collecting and forwarding to you such information and photos as I hope will set England blazing with indignation and disgust. Both in the rough and in the letters to Cousin Masterton. And much that is unprintable, but MUST BE KNOWN.

It all goes home officially and gets held up—somewhere.

And I hope and pray that I shall rouse you, and all our friends, to such a white heat of enthusiasm for this crusade and holy hatred for the Bolshevist that you will do everything in your power to enlighten people at home.

### GERMANS' SUBTLE METHODS.

To start with, I want to give *you* a few points on the situation:—

1. The Boche is still fighting us, through the Bolshevist, but in a subtle way, and by underground means which it is hard to counter.

The Germans, in the beginning of the war, hoped to be at France in three months. Detached forces were to drive the contemptible (or contempti*bly*, what does it matter?) little Army into the sea. They then intended to turn on Russia, to defeat her, reconstitute her as a vassal State, firmly allied and bound over to Germany, to organize and utilize her vast resources of men and material as a means of ruling the world.

They did not succeed in breaking the French or us in a short time. They thereupon used every means of peaceful penetration in Russia and had prepared to paralyse Russia's efforts as an effective member of the Alliance. They worked through spies, agents making propaganda, the many German bankers, &c., who had always been German agents, and some unfortunately corruptible Russians. That devil Rasputin was in their pay, but arrangements for his death, merely as getting too big for his boots, were being made by them when he was killed fortuitously, but too late for Russia.

At the same time they made every effort, unfortunately with the greatest success, of discrediting the Tsar and Imperial family in Allied countries.

When it was seen that Russia could not be got out of the war under the *ancien régime*, they helped to bring about the revolution.

When it appeared that Kerensky, a fool, but not *altogether* a knave, and his Government intended to continue the war, they redoubled their efforts to undermine the Army and Navy. I have described some of the means they used often to you.

They succeeded.

They 'sent Lenin to Russia' (*vide* Ludendorff), organized Bolshevism, gained a footing in the Ukraine, commenced exploiting the resources of Russia, and were contemplating the raising of Russian troops for use on the Western front.

### DENIKIN FIGHTING FOR A UNITED RUSSIA.

Since the Armistice they have not lost hope or interest in Russia. They continue to organize Bolshevism and Bolshevist propaganda in Allied countries. They hate Denikin and oppose him, because Denikin is fighting for a *united* Russia, free from German influence and exploitation.

Bolshevist Russia is a channel of communication to the Committee of Union and Progress, to Egypt, India, and Afghanistan.

2. Unless beaten by us, the Bolshies will beat us. It's a side issue for the present, but the danger of their rousing and letting loose the Chinese is not

so very remote.

3. They have declared war on Christianity. The Bible to them is a 'counter-revolutionary' book, and to be stamped out.

They are aiming at raising all non-Christian races against the Christian countries.

The Bolshevists form about 5 per cent. of the population of Russia— *Jews* (80 to 90 per cent. of the commissaries are Jews), Chinese, Letts, Germans, and certain of the 'skilled labour' artisans. The conscripted peasantry, originally captured by the catchwords mentioned in the pamphlets, now often goaded beyond endurance, is rising against them over wide districts. Still conscripted and put up to fight, under severe penalties, they form most of the 'cannon fodder' used by the Bolshies. They desert, often *en masse*, and many a peasant who marched for the Bolsheviks last week is fighting for Denikin in the Volunteer Army to-day.

Ref. Jews.—In towns captured by Bolshevists the only unviolated sacred buildings are the synagogues, while churches are used for anything, from movie-shows to 'slaughter-houses.' The Poles, Galacians, and Petlura have committed 'pogroms' (massacres of Jews). *Not* the Russian Volunteer Armies under Denikin. Denikin has, in fact, been so strict in protecting the Jews that he has been accused by his sympathizers of favouring them.

If, however, a Commissary, steeped in murder, with torture and rape, with mutilation, happens to be a Jew, as most of them are, should he receive exceptional treatment?

The very enemies of General Denikin who have committed pogroms accuse him of all men, and his Volunteer Armies of massacring Jews. It is one more expedient to turn the sympathies of Western countries against Denikin, not very successful, on the whole, and a side issue. I don't know why I wasted so much time on this minor point of the Jews. Possibly because they are one of the largest non-Russian contingents among the Bolshies, and the most influential. The Chinese and Letts act more as executioners and torturers.

## UNPRINTABLE PHOTOGRAPHS.

4. The Bolshevists are devils... . I hope to send you copies of 64 official photos taken by British officers at Odessa when the town was retaken from the Bolshevists. (The French and Greek divisions had cleared out; the Bolshies had taken the town and were finally driven out by Denikin's 'Iron Brigade.' The successful assault was made by a detachment of 413 of the Volunteer Army.)

As no paper will print them I suggest that you should have copies done. If we're too hard up you could pay for them by sending me no parcels, or selling my Caucasian dagger, or Persian book, or something. And I suggest that you should then do with them as you think fit, to make them most widely known.

Their horror may make people realize. They must realize. By God, they shall realize!

They show men who've been crucified with the torture of the 'human glove.' The victim gets crucified, nails through his elbows. The hands are treated with a solution which shrivels the skin. The skin is cut out with a razor, round the wrist, and peeled off, till it hangs by the finger nails, the 'human glove.'

I'm not sparing you. I hope you'll show and send them to everybody we know. People at home, apathetic fools they are, do not deserve to be spared. They must be woken up. John and Katie ought to see them.

Most of the photos are of women. Women with their breasts cut off to the bone. Women with their bodies cut open. One woman with her stomach cut open and unborn twins half dragged out.

It is not surprising that such people can't stand up to Denikin's men in anything like even numbers or equipment.

General Denikin started the war with 403 officers and 200 roubles (£4 11s, 6d.).

With 4,000 he liberated a large area. With 8,000 he walked through over 80,000 Bolshevists.

The worst of it is, that though his armies are numerous now, their equipment and supplies of all kinds are still insufficient. That's where we try to help.

And that his enemies are active in making political trouble for him everywhere. And everybody can do a bit to counteract this, surely, every little bit helps.

### OUTRAGES ON WOMEN.

Two little bits, ref. Bolshevist atrocities, you might type in as many copies as you can. If you and several others left them in different tea-shops every afternoon, it might touch quite a lot of people. I shall send you chapter and verse if I can. If I haven't sent chapter and verse in a month, do your best without. Papers are no good, because papers would put it more delicately.

'We have here at H.Q. passes issued to Bolshevists by commissaries on occupying Ekaterinodar. These passes authorize their holders to arrest any girl they fancy for the use of the soldiery. Sixty-two girls of all classes were arrested like this and thrown to the Bolshevist troops. Those who struggled were killed quite early on. The rest, when used and finished, were mutilated and thrown, dead and dying, into the two small rivers flowing through Ekaterinodar.

'In all towns occupied by Bolshevists and reoccupied by us 'slaughter-houses' are found choked with corpses. Hundreds of 'suspects,' men, women, and children, were herded in these—doors and windows manned and the struggling mass fired into until most of them were dead or dying. The doors were then locked and they were left. The stench in these places, I am told, is hair-raising. These 'slaughter-houses' are veritable plague spots and have caused widespread epidemics.'

I want you to proselytize Robinson and galvanize the Colonel and everybody else you can get hold of. I'd like James to see this and No. 47 and Dorothy. Above all the Mater. For I feel sure, that whatever happens, she and you will be glad that I've come out.

I shall not be able to send you, the Mater, Dorothy, or anyone else any more detailed news. I want to start the letters to the Colonel. If I make the first (to Taranto) cheery and amusing, the second (Constantinople and Black Sea) interesting, I can then start propaganda. So please get your news out of them. And share with the Mater and Dorothy and anybody else who cares.

This has been a full letter for your birthday, dearest, and just when your two dear letters had helped me to find a lighter tone. But these things do move me so.

I've been inoculated and have such a headache. I've got to stop.

Ever yours,  X."

## 5.4 International Zionist and Communist Intimidation

In the early 1920's, Lord Northcliffe, principal owner of *The Times*, doubted the justice of denying the land of Palestine to its majority populations and giving it instead to the political Zionists. Northcliffe was not alone, Zionist Martin Buber capsulized Mahatma Gandhi's statement, "that Palestine belongs to the Arabs and that it is therefore 'wrong and inhumane to impose the Jews on the Arabs.'"[331]

Douglas Reed, who worked for *The London Times*, alleged in his book *The Controversy of Zion*[332] that Lord Northcliffe, principal owner of the *Times* and an anti-Zionist, believed that he was being poisoned. An editor at *The Times*, Wickham Steed, wished to suppress Northcliffe's anti-Zionist views. Northcliffe sought to fire Steed, and Steed hired Northcliffe's own lawyer to defend him—Steed. Northcliffe wanted to take over as editor of *The Times*, and would have spoken out against the Palestine Mandate in the League of Nations. Some Jewish newspapers railed against Northcliffe.[333] An unnamed doctor, at Steed's instigation, declared Northcliffe insane and Northcliffe died soon thereafter, on 14 August 1922. Reed presents the history of events that led to Northcliffe's demise. Lord Northcliffe's reports on Palestine were suppressed in his own newspaper, while the League of Nations ratified the Zionist mandate.

### 5.4.1 Suppression of Free Speech

Spoken statements and written works which criticize Zionist dogmas, as did Reed's, are increasingly being proscribed around the world under pressure from

---

**331**. M. Buber quoted in A. Hertzberg, *The Zionist Idea*, Harper Torchbooks, New York, (1959), p. 463.
**332**. D. Reed, "The End of Lord Northcliffe", *The Controversy of Zion*, Chapter 34, Bloomfield, Sudbury, (1978).
**333**. "Jüdischnationale Paroxismen", *Die Wahrheit* (Wien/Vienna), Volume 38, Number 6, (9 March 1922), pp. 3-4.

Jewish groups, who would prohibit open debate and proscribe free speech—exactly as did the Bolsheviks and the Nazis. They insist that the public obey legislated opinions and be legally barred from doubting state-mandated views, which recalls Hitler's policy of *Gleichschaltung* and Lenin's "democratic centralism". At the time of this writing, several authors are being held in prisons around the world for simply daring to voice opinions these Jewish groups want suppressed—apparently opinions these groups have a hard time refuting. This is not a new phenomenon.

In an article entitled "The Jews" in a paper published by Peter Schmidt of 80 Maiden Lane, New York, *The German Correspondent. By Hermann*, Volume 1, Number 2, (29 February 1820), pp. 9-12, at 12, it states,

> "At Frankfort on the Maine, a work on *Judaism* was published, containing some severe remarks on the Jews. It was suppressed by the police."

In 1850 and 1869, composer Richard Wagner publish an essay which criticized the Jewish influence on the arts.[334] Jews organized to ruin his career, and Wagner was smeared around the world. Under the heading "Foreign Gossip", *The Chicago Tribune* reported on 25 April 1869 on page 5,

> "Richard Wagner's pamphlet against the Jews, who he says are utterly unable to achieve distinction in any branch of art, has created a great commotion in the literary and artistic circles of Germany and France. Some critics even go so far as to assert that the composer of Tannhauser is half insane."

Like Richard Wagner, Eugen Karl Dühring was attacked by an organized Jewish campaign to ruin his career. In 1882, Franz Mehring quoted a Jewish author who criticized other Jews for, among other things,

> "the malicious gloating when veritable conspiracies deprived of their livelihoods people who were suspected of anti-Jewish feelings[.]"[335]

---

**334**. R. Wagner under the *nom de plume* K. Freigedank, "Das Judenthum in der Musik", *Neue Zeitschrift für Musik*, (3 and 6 September 1850); Reprinted with revisions and an appendix, R. Wagner, *Das Judenthum in der Musik*, J. J. Weber, Leipzig, (1869); the original unrevised 1850 article was reprinted in *Gesammelte Schriften und Dichtungen*, Volume 5; English translation of the 1869 version by E. Evans, *Judaism in Music (Das Judenthum in der Musik) Being the Original Essay Together with the Later Supplement*, W. Reeves, London, (1910); Also, "Judaism in Music", *Richard Wagner's Prose Works*, Volume 3, Broude Brothers, New York, (1966), pp. 75-122.; which is a reprint of the original "Judaism in Music", *Richard Wagner's Prose Works*, Volume 3, Kegan Paul, Trench, and Trübner, London, (1894).

**335**. English translation from: P. W. Massing, *Rehearsal for Destruction: A Study of Political Anti-Semitism in Imperial Germany*, Howard Fertig, New York, (1967). p. 315.

Eugen Karl Dühring wrote in the 1880's:

> "In a review which was underhanded and misleading to the public of a scholarly work (incidently suffering from a Kantianising philosophasterish weakness) on *Judaism* (by L. Holst, Mainz, 1821),[336] [Börne] made to the author of the same an explanation which is significant even today for the conduct of the Jews. He brought to his attention that he, Börne, hoped to experience still the time when every such inflammatory writing against the Jews would bring its author either into the prison or the lunatic asylum; Börne died, now, in 1837. [\*\*\*] Even in my personal affairs, that is, however, on the occasion of the battle which was associated with my removal from Berlin University, I could perceive tangibly how many Jewish doctors, who were also litterateurs at the same time, had engaged the unions of professors against me and sought to degrade me before the public with falsehoods and criticisms as well as especially with the imputation of megalomania and persecution mania. Individuals in these camps were so maliciously involved that they were publicly dismissed, even if they were protected by the Jewish papers themselves in which they had written by the nonacceptance of every settlement. In another work *Robert Mayer, der Galilei des 19. Jahrhunderts*, I have more closely elucidated these and other little pieces with the naming of names and provided many facts also on individual newspapers of the most marked Jewishness."[337]

Communist Zionist Nachman Syrkin jokingly wrote in 1898, referring to the generally base nature of anti-Semitic leadership,

> "At least one part of Ludwig Börne's famous saying, that the anti-Semites of the future will be candidates either for the workhouse or for the insane asylum, has been realized."[338]

In 1933, Norman Bentwich wrote in an article entitled, "Is Judaism Doomed in Soviet Russia", *B'nai B'rith Magazine*, (March, 1933),

> "The teaching of the Hebrew Prophets, 'to set free the oppressed and to break every yoke,' was the underlying motive of the Bolshevik revolution. It is certain that the principal prophet of the proletarian movement was the

---

**336**. L. Holst, *Das Judentum in allen dessen Teilen. Aus einem staatswissenschaftlichen Standpunkt betrachtet*, Mainz, (1821).
**337**. E. K. Dühring, *Die Judenfrage als Racen-, Sitten- und Culturfrage: mit einer weltgeschichtlichen Antwort*, H. Reuther, Karlsruhe, (1881); English translation by A. Jacob, *Eugen Dühring on the Jews*, Nineteen Eighty Four Press, Brighton, England, (1997), p. 107, 135. Dühring refers to Ludolf Holst, *Das Judentum in allen dessen Teilen. Aus einem staatswissenschaftlichen Standpunkt betrachtet*, Mainz, (1821).
**338**. N. Syrkin, under the nom de plume "Ben Elieser", *Die Judenfrage und der socialistische Judenstaat*, Steiger, Bern, (1898); English translation in A. Hertzberg, *The Zionist Idea*, Harper Torchbooks, New York, (1959), pp. 333-350, at 340.

German Jew, Karl Marx, whose picture hangs in every public institution and whose book, *Kapital*, is the gospel of the Communist creed; that another German Jew, Ferdinand Lassalle, whose heroic statue adorns the Nevski Prospect of Leningrad, was one of the inspirers of the early revolutionary parties; that Jews have, from the beginning to the present day, played a part in the creation and the maintenance of the revolution; and that for no community has the revolution brought about a greater change of status than for the Jews. Under the Czars their life was outwardly a long humiliation; but it had its compensations in the inner strength of the community and in the national ideal of which the flame burnt eternally. To-day, they have been given complete civic and social equality with the rest of the population; and, indeed, Lenin's saying is constantly quoted, that those peoples which were previously oppressed should be specially favored. [***] The essential feature about their community which strikes the visitor is that the Jews, and particularly the younger generation, feel at home, and part and parcel of the new order. They are proud of their share in the councils of the revolution: of Trotsky, who organized the Red Army (though among non-Jews he is in disgrace and his name is not mentioned), and of the Jews who hold high positions in the Foreign Office and other Ministries, in the Army and the Navy, in the economic councils and academies.

When we landed in Leningrad, our interpreters and guides from the State Tourist Organization were usually Jews and Jewesses. It is the function of the Jew to be the interpreter of Soviet Russia to the world and of the world to Soviet Russia; for he forms the principal element in the proletarian society which has close touch with the Western European culture and languages.... The suppression of the Ghetto and of the Orthodox Church has brought this outward freedom; and the Government punishes severely any outward manifestation of anti-Semitism. [***] In the towns such as Kiev, Odessa, Berdichev, where the Jews are a quarter or more of the whole population, there are Yiddish law courts and Yiddish codes of law, and Yiddish is an official language. But the Rabbinical law which used to regulate Jewish family affairs may not be applied, and the Beth-Din may not function. The academy of higher learning in such centres, which has taken the place of the former university, includes a section for Jewish learning and research."[339]

On 1 March 1946, the *American Hebrew* quoted a sermon by Rabbi Leon Spitz at a Purim festival,

"Let Esau whine and wail and protest to the civilized world, and let Jacob raise his hand to fight the good fight. The anti-Semite... understands but one language, and he must be dealt with on his own level. The Purim Jews stood up for their lives. American Jews, too, must come to grips with our

---

[339]. D. Fahey, *The Mystical Body of Christ in the Modern World*, Browne and Nolan Limited, London, (1935), pp. 250-251.

contemporary anti-Semites. We must fill our jails with anti-Semitic gangsters. We must fill our insane asylums with anti-Semitic lunatics. We must combat every alien Jew-hater. We must harass and prosecute our Jew-baiters to the extreme limits of the laws. We must humble and shame our anti-Semitic hoodlums to such an extent that none will wish or dare to become (their) fellow-travelers."[340]

Karl Ludwig Börne was a crypto-Jewish revolutionary, who attempted to subvert German society. He was born Loeb Baruch on 6 May 1786 in Frankfort. Baruch both changed his name and feigned Christian conversion on 5 June 1818.[341]

Börne's vision of legislation proscribing speech which is offensive to Jews has since become a reality. After the Russian Revolution, it became illegal to criticize Jews, Jewish racism, or to point out the fact that Jewish bankers had brought about the Revolution, or to identify crypto-Jews.[342] Sigmund Freud sought to stigmatize the criticism of Jewish racism as if it were a mental disorder, and thereby set the stage for the notorious political oppression of the Soviet psychoprisons. In America we have "Hate Crimes" laws and the *Global Anti-Semitism Review Act of 2004*. In Europe there are far more stringent laws proscribing certain speech, which include prison time and fines as sanctions against speaking freely; such as Britain's *Race Relations Act* of 1976 Section 5A, as amended in 2000 and 2003; France's Gayssot law; and Germany's *Volksverhetzung* § 130 of the *Strafgesetzbuch*. Austria has proscribed free speech under the pretext of proscribing "Nazi revivalism" with its *Verbotsgesetz*.

---

**340**. As quoted in: R. H. Williams, *The Ultimate World Order—As Pictured in "The Jewish Utopia"*, CPA Book Publisher, Boring, Oregon, (1957?), p. 14.
**341**. "BÖRNE, KARL LUDWIG", *The Jewish Encyclopedia*, Volume 3, Funk and Wagnells Company, New York, (1902), pp. 323-325. "BOERNE, LUDWIG", *Encyclopedia Judaica*, Encyclopedia Judaic, Jerusalem, (1971), cols. 1166-1168.
**342**. V. I. Lenin, "Anti-Jewish Pogroms", *Collected Works*, Volume 29, English translation of the Fourth Russian Edition, Progress Publishers, Moscow, (1972), pp. 252-253. *See also:* D. Fahey, *The Mystical Body of Christ in the Modern World*, Browne and Nolan Limited, London, (1935), p. 251. *See also:* G. B. Shaw, *The Jewish Guardian*, (1931). *See also: Congress Bulletin*, American Jewish Congress, New York, (5 January 1940). *See also: The Jewish Voice*, (January, 1942). *See also:* G. Aronson, *Soviet Russia and the Jews*, American Jewish League against Communism, New York, (1949). *See also:* J. Stalin, "Anti-Semitism: Reply to an Inquiry of the Jewish News Agency in the United States" (12 January 1931), *Works*, Volume 13, Foreign Languages Publishing House, Moscow, (1955), p. 30. *See also:* S. S. Montefiore, *Stalin: The Court of the Red Star*, Vintage, New York, (2003), pp. 305-306. *See also:* "Anti-Semitism", *Great Soviet Encyclopedia: A Translation of the Third Edition*, Volume 2, (1973), pp. 175-177, at 176. *See also:* "Jews", *Great Soviet Encyclopedia: A Translation of the Third Edition*, Volume 9, Macmillan, New York, (1975), pp. 292-293, at 293. *See also:* N. S. Alent'eva, Editor, *Tseli i metody voinstvuiushchego sionizma*, Izd-vo polit. lit-ry, Moskva, (1971). Н. С. Алентьева, Редактор, Цели и методы воинствующего сионизма, Издательство Политической Литературы, Москва, (1971).

Canada, too, has at times sought to proscribe certain forms of political and historical speech and to impose criminal penalties against those who speak freely, if offensively, under the *Spreading False News* statute. Malta proscribes certain classes of speech under Article 82A of the criminal code. Israel also penalizes proscribed speech. Internationally famous historian David Irving languishes in prison in Austria for expressing opinions Jewish organizations want suppressed and proscribed by law. Irving is but one of many who have been imprisoned for speaking about ideas that Jewish organizations do not want expressed. The truth is no defense in these prosecutions, nor are the defendants or their legal counsel permitted the normal due process of law. Instead, thought criminals who offend Jewish organizations are railroaded into prison through procedures which are blatant human rights violations, and the international press, governments and human rights organizations remain silent, while Jewish organizations cheer on the illegal prosecutions and call for broader powers to suppress speech. Whenever those who are persecuted by Jewish organizations dare to point out the fact that Jewish organizations are attacking them and their fundamental human rights in an organized and coordinated effort, those same Jewish organizations who pride themselves on their Jewish heritage call those they persecute "anti-Semitic" for pointing out that self-styled "Jewish" organizations attack them and seek the suppression of their human rights to free speech, freedom of association, due process of law, and liberty itself.

These laws exhibit the power of "Jewish" organizations. Jewish Messianic prophecy calls for the mass murder of those who are not "righteous".[343] Their plan is to first murder off those who do not submit to their mythology, which states that Jews are the God-given masters of the world and that Gentiles must serve the Jews as their slaves and submit to laws which emanate from Jerusalem (*Exodus* 34:11-17. *Psalm* 2; 72. *Isaiah* 2:1-4; 9:6-7; 11:4, 9-10; 42:1; 61:6. *Jeremiah* 3:17. *Joel* 3:16-17. *Micah* 4:2-3. *Zechariah* 8:20-23; 14:9). Ultimately, though, only the Jews will be considered "righteous",[344] and only they will survive.[345] Laws which are enacted at the insistence of Jews, and which make it illegal to question Jewish dogma, are laws which are deliberately "fulfilling" these Jewish Messianic prophecies (*Psalm* 72. *Isaiah* 42; 49; 50; 52; 53; 54; 60; 61, etc. *Daniel* 12. *Malachi* 4).

There is an old political tactic, employed long ago against Caligula and

---

[343]. *Exodus* 34:11-17. *Psalm* 2; 72. *Isaiah* 1:9; 2:1-4; 6:9-13; 9:6-7; 10:20-22; 11:4, 9-12; 17:6; 37:31-33; 41:9; 42; 43; 44; 61:6. *Jeremiah* 3:17; 33:15-16. *Ezekiel* 20:38; 25:14. *Daniel* 12:1, 10. *Amos* 9:8-10. *Obadiah* 1:18. *Micah* 4:2-3; 5:8. *Zechariah* 8:20-23; 14:9. *Romans* 9:27-28; 11:1-5.

[344]. *Exodus* 34:11-17. *Psalm* 2; 72. *Isaiah* 1:9; 2:1-4; 6:9-13; 9:6-7; 10:20-22; 11:4, 9-12; 17:6; 37:31-33; 41:9; 42; 43; 44; 61:6. *Jeremiah* 3:17; 33:15-16. *Ezekiel* 20:38; 25:14. *Daniel* 12:1, 10. *Amos* 9:8-10. *Obadiah* 1:18. *Micah* 4:2-3; 5:8. *Zechariah* 8:20-23; 14:9. *Romans* 9:27-28; 11:1-5.

[345]. M. Higger, *The Jewish Utopia*, Lord Baltimore Press, Baltimore, (1932). R. H. Williams, *The Ultimate World Order—As Pictured in "The Jewish Utopia"*, CPA Book Publisher, Boring, Oregon, (1957?).

Nero, by which one declares an enemy insane or otherwise contemptible, in order to justify one's pre-existing dislike of the person so smeared, or one's desire to suppress the message the defamed person expresses. Max Nordau stated in his address to the First Zionist Congress in 1897,

> "No one has ever tried to justify these terrible accusations by facts. At most, now and then, an individual Jew, the scum of his race and of mankind, is triumphantly cited as an example, and contrary to all laws of logic, the example is made general. This tendency is psychologically correct. It is the practice of human intellect to invent for the prejudices, which sentiment has called forth, a cause seemingly reasonable. Probably wisdom has long been acquainted with this psychological law, and puts it in fairly expressive words: 'If you have to drown a dog,' says the proverb, 'you must first declare him to be mad.' All kinds of vices are falsely attributed to the Jews, because one wishes to convince himself that he has a right to detest them. But the pre-existing sentiment is the detestation of the Jews."[346]

Albert T. Clay documented the methods of the racist political Zionists in Palestine in 1921, in an article, "Political Zionism", *The Atlantic Monthly*, Volume 127, Number 2, (February, 1921), pp. 268-279, at 276-277 (this is an indication of what one can expect from Jewish fanatics around the world, when they anoint their Messiah),

> "The old resident Jews of Palestine certainly have other than religious grounds for their indifference toward the efforts of the Political Zionists. Last winter the Council of Jerusalem Jews appointed a commission of representative men holding leading positions, to visit parents who were sending their children to proscribed schools, in order to secure their withdrawal. Among these schools, which included those conducted by the convents and churches, some of which have existed in Jerusalem for a long time, are the British High School for Girls, the English College for Boys, and the Jewish School for Girls. In the latter, conducted by Miss Landau, an educated English Jewess, all the teachers are Jewish; most of the teaching is in the English language. This school, which is financed by enlightened Jews of England, was denounced more severely than the others, because, not being in sympathy with the programme of the Political Zionists, Miss Landau refused to teach the Zionist curriculum. She was even informed that her school would be closed.
> In a series of articles that appeared in *Doar Hayom*, the Hebrew daily paper, last December, it was stated that the parents who refused to comply with the requests of the Commission [of the Council of Jerusalem Jews] were to be boycotted, cast out from all intercourse with Jews, denied share

---

[346]. M. Nordau, "Max Nordau on the General Situation of the Jews", *The Jewish Chronicle*, (3 September 1897), pp. 7-9, at 8.

in Zionist funds, and deprived of all custom for their shops and hotels. 'Anyone who refused, let him know that it is forbidden for him to be called by the name of Jew; and there is to be for him no portion or inheritance with his brethren.' They were given notice that they would 'be fought by all lawful means.' Their names were to be put 'upon a monument of shame, as a reproach forever, and their deeds writte unto the last generation.' 'If they are supported, their support will cease; if they are merchants, the finger of scorn will be pointed at them; if they are rabbis, they will be moved far from their office; they shall be put under the ban and persecuted, and all the people of the world shall know that there is no mercy in justice.'

A month later the results of this 'warfare' were reviewed. We were informed that some Jews had been influenced, 'but others—and the greater number, and those of the Orthodox,—those who fear God—having read the letters [signed by the head of its delegates and the Zionist Commission] became angry at the 'audacity' of the Council of Jerusalem Jews 'which mix themselves up in private affairs,' have torn the letter up, and that finished it.'

Then followed a long diatribe against these parents, boys, and girls, in which it was demanded that the blacklist of traitors to the people be sent to 'those who perform circumcision, who control the cemeteries and hospitals'; that an order go forth so that 'doctors will not visit their sick, that assistance when in need, if they are on the list of the American Relief Fund, will not be given to them.' 'Men will cry to them, 'Out of the way, unclean, unclean.' ... They are in no sense Israelites.'

It is to be regretted that only these few paraphrases and quotations from the series of articles published can be presented here.

The work of the Councils Committee met with not a little success; pupils left schools, and teachers gave up their positions. Two instructors in the English College, whose fathers were rabbis, and a third, whose brother was a teacher in a Zionist school, resigned. Another refused to do so, and declared himself ready, in the interests of the Orthodox Jews, who were suffering under this tyranny, which they deplored, to give the fullest testimony to the authorities concerning this persecution. The administration, under Governor Bols, finally intervened, and at least no further public efforts to carry out their programme were made.

If, in this early stage of the development of Political Zionism, even the Palestinian Religious Jews already find themselves under such a tyranny, what will happen if these men are allowed to have full control of the government? And what kind of treatment can the Christian and th Moslem expect in their efforts to educate their children, if the Political Zionists are allowed to develop their Jewish state to such a point that they can dispense with their mandatory and tell the British to clear out? When such things happen under British administration, what will take place if the Jewish State is ever realized, and such men are in full control?"

Some relativists worship Albert Einstein as their hero and detest anyone who tells the truth about Einstein's career of plagiarism and the irrationality of Einstein's theorizations. These people believe that they have the right to defame anyone who disagrees with them and often invent spurious reasons to justify their hatred—and to change the subject from Einstein's failings to a personal attack against Einstein's critics. "The pre-existing sentiment is the detestation of" anyone who does not see Einstein as an infallible saint. It is a convenient political weapon to employ an *ad hominem* attack. The reasons for the dissent are, in this manner, disregarded, and the critic is stigmatized and forced to defend herself or himself, rather than her or his scientific findings, which are ignored and quietly removed from the public eye.

Yury Brovko has alleged that those who spoke out against relativity theory and Einstein in the Soviet Union ran the risk of severe political persecution. Yury Brovko, a critic of Einstein's claims to have originated the theory of relativity and a critic of the theory itself, alleges that there were many secret orders which effectively forbade criticism of Einstein in the U. S. S. R., and which forbade scientific journals, science departments and scientific organizations from receiving, considering, discussing or publishing literature which was critical of Einstein's theories.[347] American physics societies have also refused to consider for publication works critical of "fundamental theories", which is to say works critical of Einstein and "his" theory of relativity, or of quantum mechanics. Brovko refers to secret Orders of the Presidium of the Academy of Sciences of the USSR in 1964 and before, but does not give any specific references to such orders which your author could attempt to verify. Brovko wrote, *inter alia*,

"В 1964 году Президиум АН СССР издает закрытое постановление, запрещающее всем научным советам и журналам, научным кафедрам принимать, рассматривать, обсуждать и публиковать работы, критикующие теорию Эйнштейна."[348]

V. A. Bronshten stated in 1968,

"There is a sufficiently large group of pseudoscientists, who specialize in 'refuting' the theory of relativity. As a rule, the efforts of these 'refuters' only reveals their poor scientific literacy, although among them there are people with a university education."

---

**347.** Y. Brovko, "Einshteinianstvo—agenturnaya set mirovovo kapitala", Molodaia Gvardiia, Number 8, (1995), pp. 66-74, at 70. Юрий Бровко, "Эйнштейнианство — агентурная сеть Мирового капитала", Молодая гвардия, № 8, (1995), сс. 66-74; **and** Y. Brovko, "Razgrom einshteinianstvo", Priroda i Chelovek. Svet, Number 7, (2002), pp. 8-10. Юрий Бровко, "Разгром эйнштейнианства", Природа и Человек. Свет, № 7, (2002), сс. 8-10. <http://medograd.narod.ru/einstein.html>
**348.** Y. Brovko, "Einshteinianstvo—agenturnaya set mirovovo kapitala", Molodaia Gvardiia, Number 8, (1995), pp. 66-74, at 70. Юрий Бровко, "Эйнштейнианство — агентурная сеть Мирового капитала", Молодая гвардия, № 8, (1995), сс. 66-74. Special thanks to Karim Khaidarov for this citation!

"Есть довольно большая група гипотезоманов, специализировавшихся на «опровержении» теории относительности. Как правило, усилия этих «опровергателей» лишь отражают их низкую научную грамотность, хотя среди них попадаются и люди с высшим образованием."[349]

and,

"The so-called delirium of inventions and discoveries is one of the forms of paranoia. The nature of the disorder lies in the fact that the patient believes he has made an important invention or salient discovery, and that scientific-conservatives tragically cannot understand him. In this case the person remains completely normal in every other aspect of life, in the family, at work. [***] Thus, just in the year 1966, the Department of General and Applied Physics of the Academy of Science of USSR helped physicians to reveal 24 paranoiacs."

"Одной из форм паранойи является так называемый бред изобретений и открытий. Сущность его состоит в том, что больному кажется, будто он сделал важное изобретение или выдающееся открытие, и что вся беда в том, что его не могут понять ученые-консерваторы. При этом во всем остальном—в жизни, в семье, в работе—человек остается совершенно нормальным. [...] Так, только за один 1966 г. Отделение общей и прикладной физики АН СССР помогло медикам выявить 24 параноика."[350]

Lifshitz stated in 1978,
"It appears to me that there are two types of pseudoscientists. One of them — people with paranoid mental lapses, who absolutely believe in what they are saying. These are not scientific afferists, but are simply not completely normal people, whom you unfortunately encounter. They, as a rule, are occupied by fundamental questions: they refute quantum mechanics, the theory of relativity and so forth. However, they are completely normal when discussing other issues."

"Лжеученые, как мне кажется, бывают двух типов. Один из них — люди с параноидальными психическими сдвигами, они абсолютно верят в то, что сами говорят. Это не научные аферисты, а просто не в

---

[349] В. А. Бронштен. Беседы о космосе и гипотезах. Наука, Москва, (1968), стр. 206. V. A. Bronshten, *Besedy o kosmose i gipotezakh*, Nauka, Moscow, (1968), p. 206. Special thanks to my wife Kristina for her help in the translation!

[350] В. А. Бронштен. Беседы о космосе и гипотезах. Наука, Москва, (1968), стр. 198. V. A. Bronshten, *Besedy o kosmose i gipotezakh*, Nauka, Moscow, (1968), p. 198. Special thanks to my wife Kristina for her help in the translation!

полне нормальные люди, с которыми, к сожалению, приходится встречаться. Они, как правило, занимаются фундаментальными вопросами: опровергают квантовую механику, теорию относительности и т. д. Причем об остальных вещах они рассуждают нормально."[351]

In the same period of time, anyone who questioned the legitimacy of the Soviet State, or wished to leave it, was also considered psychotic—often dubbed "paranoid" and imprisoned in psychiatric prisons, even if he or she behaved in a completely sane, very normal way.[352] The same fate apparently befell many who dared to question the theory of relativity, or who called attention to Einstein's plagiarism. This recalls Trofim Denisovich Lysenko's tyrannical reign over the field of genetics and the murder, imprisonment and banishment of dissenting scientists in the Soviet Union.

The trial of Einstein's friend Friedrich Adler set a bizarre precedent for the charge of *per se* insanity for disagreeing with Einstein. Adler assassinated the Austrian Prime Minister Karl Graf von Stürgkh in 1916. Alder had written a work which is critical of the theory of relativity and the defense at his murder trial used this work as "proof" that he must be insane—but even Einstein did not maintain that that was true.[353] However, Einstein and his advocates did succeed in wrongfully stigmatizing any criticism of Einstein or the theory of relativity as if it were anti-Semitism, *per se*.[354] Kevin MacDonald argues in his book *The Culture of Critique*,[355] that Sigmund Freud planned to use psychoanalysis to rid the world of "anti-Semitism" Today, there are prominent persons in prison for the criminal offense of offending racist Jews.

## 5.4.2 Jewish Terrorism

---

[351]. И. Лифшиц, "Возможно ли 'невозможное'?" Литературная газета, № 24, (14 Июня 1978), стр. 13. E. Lifschitz, "Vozmozhno li 'Nevozmozhnoe'?", *Literaturnaia gazeta organ Pravleniia Soiuza sovetskikh pisatelei SSSR*, Number 24, (14 June 1978), p. 13. Special thanks to my wife Kristina for her help in the translation!

[352]. H. Fireside and Z. A. Medvedev, *Soviet Psychoprisons*, W. W. Norton & Company, New York, London, (1979), p. 140.

[353]. P. Frank, *Einstein, His Life and Times*, Alfred A. Knopf, New York, (1947), pp. 174-175.

[354]. "Prof. Einstein Here, Explains Relativity", *The New York Times*, (3 April 1921), pp. 1, 13, at 1.

[355]. K. MacDonald, *The Culture of Critique*, Praeger, Westport, Connecticut, London, (1998), pp. 115-117; **citing:** E. Jones, *Free Associations: Memories of a Psycho-Analyst*, Basic Books, New York, (1959); **and** F. Sulloway, "Freud as Conquistador", *The New Republic*, (August, 1979), pp. 25-31; **and** D. B. Klein, *Jewish Origins of the Psychoanalytic Movement*, Praeger, New York, (1981); **and** P. Gay, *Freud: A Life for Our Time*, W. W. Norton, New York, (1988); **and** L. Chamberlain, "Freud and the Eros of the Impossible", *Times Literary Supplement*, (25 August 1995), pp. 9-10; **and** K. MacDonald, *Separation and Its Discontents: Toward an Evolutionary Theory of Anti-Semitism*, Praeger, Westport, Connecticut, (1998).

In its article "Israel", the *Great Soviet Encyclopedia: A Translation of the Third Edition*, Volume 10, Macmillan, New York, (1976), pp. 477-484, at 478, wrote,

> "Thus, despite the UN resolution of Nov. 29, 1947, Israel expanded its territory to include four-fifths of the area of mandated Palestine. Both before the formation of Israel and the outbreak of the war and during the course of the war itself, Zionist terror led to the mass destruction of Arabs and the expulsion of nearly a million Arabs from the territory of Israel and from the Arab portion of Palestine that it had seized. The problem of Palestinian refugees emerged—a problem that, because of Israel's unaltering refusal to implement the UN resolution of Dec. 11, 1948 (on the right of refugees to return to their homeland or, if they choose, to receive material compensation), became one of the most important issues complicating the Middle East crisis. [The *Great Soviet Encyclopedia*, published in the 1970's at the time when the United Nations General Assembly Resolution Number 3379 declared that Zionism is a form of racism, detailed many of the Zionists' abuses and violations of international law. Refer also to its articles: "Anti-Semitism", "Jews", "Judaism", "Middle East Crisis", "Palestine", "Poale Zion", and "Zionism". *See also:* N. S. Alent'eva, Editor, *Tseli i metody voinstvuiushchego sionizma*, Izd-vo polit. lit-ry, Moskva, (1971). H. С. Алентьева, Редактор, Цели и методы воинствующего сионизма, Издательство Политической Литературы, Москва, (1971).—CJB.]

The political Zionists of the early Twentieth Century had a well deserved international reputation as murderers, torturers and terrorists.[356] The Jews of the Nineteenth Century had a reputation as revolutionary terrorists and assassins. Jewish terrorism continued through the Zionist "Sternists"[357] of the 1940's (who offered Hitler a military alliance between Zionists and Nazis based on the principle that Jews must be removed from Europe[358]) and Menachem Begin's terrorist Zionist Jews in the Irgun, through to the Jewish Zionist Meir Kahane,[359]

---

[356]. "Says the Zionists Disturb Palestine", *The New York Times*, (18 January 1914), Section 3, p. 3. D. Reed, *Somewhere South of Suez*, Devin-Adair Company, U. S. A., (1951), pp. 324-327.
[357]. *See:* "Bomb for Farran Kills Brother; Sternists Had Threatened Death", *The New York Times*, (4 May 1948), pp. 1, 17. *See also:* "Britons Indignant Over Bomb Outrage", *The New York Times*, (5 May 1948), p. 17:2. *See also:* "British Party Slain", *The New York Times*, (6 May 1948). p. 8. *See also:* "Parcel Bomb Sent to British General", *The New York Times*, (12 May 1948), p. 13. *See also:* "Anti-Zionist Tells of Threats on Life", *The New York Times*, (14 May 1948), p. 6. *See also:* "British Public is Warned", *The New York Times*, (14 May 1948), p. 10. *See also:* "Torture Reports Studied by Israel", *The New York Times*, (9 June 1950), p. 9.
[358]. D. Yisraeli, *The Palestine Problem in German Politics, 1889-1945* (Hebrew), Bar-Ilan University, Ramat-Gan, Israel, (1974), pp. 315-317.
[359]. R. I. Friedman, *The False Prophet: Rabbi Meir Kahane: from FBI Informant to Knesset Member*, Lawrence Hill Books, Brooklyn, New York, (1990).

and beyond to the present time.[360]

While the Sternists (led by Yitzhak Shamir) and the Haganah (led by David Ben-Gurion) were busy terrorizing British vessels and encampments, the Irgun (led by Menachem Begin) murdered 91 people at the King David Hotel and planned to murder the British Foreign Secretary Ernest Bevin. The Jews dressed up as Arabs when they bombed the King David hotel, in order to generate hatred towards innocent Arabs—not only did they murder innocent people, they blamed other innocent people for their crimes. They also planned to make the Jewish assassination of the British Foreign Secretary Ernest appear as if it had been committed by the Irish Republican Army, in order to hide the fact that Zionists were the true murderers.[361]

On 9 April 1948, Sternist and Irgun terrorists committed the Deir Yassin Massacre against defenseless Palestinians.[362] They murdered hundreds of helpless men, women and children.[363] The Jewish terrorists then stole the land of the dead Palestinians and chased off those who survived their attack, stealing their land and property, as well. The Israelis have repeated the Jewish atrocities

---

**360**. J. B. Bell, *Terror out of Zion: Irgun Zvai Leumi, LEHI, and the Palestine underground, 1929-1949*, St. Martin's Press, New York, (1977). L. Rokach and M. Sharett, *Israel's Sacred Terrorism: A Study Based on Moshe Sharett's Personal Diary and Other Documents*, Association of Arab-American University Graduates, Belmont, Massachusetts, (1986). B. Hoffman, *The Failure of British Military Strategy within Palestine, 1939-1947*, Bar-Ilan University Press, Ramat-Gan, Israel, (1983). J. Heller, *Stern Gang: Ideology, Politics and Terror, 1940-49*, F. Cass, Portland, Oregon, (1994). T. Segev, *One Palestine, Complete: Jews and Arabs under the British Mandate*, Henry Holt and Co., New York, (2001). B. Morris, *Righteous Victims: A History of the Zionist-Arab Conflict, 1881-2001*, Vintage Books, New York, (2001). J. Stern, *Terror in the Name of God: Why Religious Militants Kill*, Ecco, New York, (2003), pp. 85-106.
**361**. P. Day, "Jewish Plot to Kill Bevin in London", *The Sunday Times* (London), (5 March 2006); *TIMES ONLINE:* <http://www.timesonline.co.uk/article/0,,2087-2069967,00.html>
**362**. "Letter to the Editor", signed by Isidore Abramowitz, Hannah Arendt, Abraham Brick, Rabbi Jessurun Cardozo, Albert Einstein, Herman Eisen, M. D., Hayim Fineman, M. Gallen, M. D., H. H. Harris, Zelig S. Harris, Sidney Hook, Fred Karush, Bruria Kaufman, Irma L. Lindheim, Nachman Majsel, Seymour Melman, Myer D. Mendelson, M. D., Harry M. Orlinsky, Samuel Pitlick, Fritz Rohrlich, Louis P. Rocker, Ruth Sager, Itzhak Sankowsky, I. J. Schoenberg, Samuel Shuman, M. Znger, Irma Wolpe, Stefan Wolpe; dated "New York. Dec. 2, 1948."; published as: "New Palestine Party; Visit of Menachen Begin and Aims of Political Movement Discussed", *The New York Times*, (4 December 1948), p. 12.
**363**. D. A. Schmidt, "200 ARABS KILLED, STRONGHOLD TAKEN; Irgun and Stern Groups Unite to Win Deir Yasin—Kastel Is Recaptured by Haganah", *The New York Times*, (10 April 1948), p. 6. D. A. Schmidt, "ARABS SAY KASTEL HAS BEEN RETAKEN; JEWS DENY CLAIM; British Confirm That Control of Key Point on Palestine Road Has Shifted Anew MASSACRE IS DENOUNCED Both Sides Decry Killing of 250 by Terrorist Groups at Deir Yassin Friday ARABS SAY KASTEL HAS BEEN RETAKEN PALESTINE FIGHTING CONTINUES IN MANY AREAS", *The New York Times*, (12 April 1948), p. 1.

across Palestine, following the course laid out for them in *Exodus* 34:11-17,

"11 Observe thou that which I command thee *this* day: behold, I drive out before thee the Amorite, and the Canaanite, and the Hittite, and the Perizzite, and the Hivite, and the Jebusite. 12 Take heed to thyself, lest thou make a covenant with the inhabitants of the land whither thou goest, lest it be for a snare in the midst of thee: 13 But ye shall destroy their altars, break their images, and cut down their groves: 14 For thou shalt worship no other god: for the LORD, whose name *is* Jealous, *is* a jealous God: 15 Lest thou make a covenant with the inhabitants of the land, and they go a whoring after their gods, and do sacrifice unto their gods, and *one* call thee, and thou eat of his sacrifice; 16 And thou take of their daughters unto thy sons, and their daughters go a whoring after their gods, and make thy sons go a whoring after their gods. 17 Thou shalt make thee no molten gods."

Jews in Lithuania and Poland had acted in the same fashion during the Second World War. Perhaps taking their cue from Old Testament orders from the Jewish God to utterly destroy other Peoples' villages, leaving nothing left alive and no property intact (as but one example of many, *see:* I *Samuel* 15); Jews mass murdered the men, women, children and infants of Koniuchy (Kaniukai).[364] Many Jews welcomed the Bolsheviks into Poland and Lithuania and helped them to mass murder helpless Poles and Lithuanians. Jews were notorious for "denouncing" their Gentile neighbors to Communist authorities, who were often themselves Jewish. I *Samuel* 15:3 states,

"Now go and smite Amalek, and utterly destroy all that they have, and spare them not; but slay both man and woman, infant and suckling, ox and sheep, camel and ass."

The ultimate goal of Judaism is to enslave and exterminate all non-Jews (*Isaiah* 65; 66).

In 1948, the Zionist Sternists, under the leadership of Yitzhak Shamir, murdered Count Folke Bernadotte, whom the United Nations Security Council had appointed to mediate Palestinian-Israeli negotiations.[365] Count Bernadotte had rescued tens of thousands of Jews from the Nazis. These Jewish terrorists also hanged innocent Brits and wired their dead bodies with explosive booby-traps. They also sent letter bombs to British authorities and the Sternists murdered Lord Moyne British Minister of State and his driver in cold blood in a terrorist act.

Jewish Zionist terrorists, posing as native Gentiles, terrorized Jewish

---

**364**. C. Lazar, *Destruction and Resistance*, Shengold Publishers, New York, (1985). R. Cohen, *The Avengers*, Alfred A. Knopf, New York, (2000).
**365**. T. Schwarz, *Walking with the Damned: The Shocking Murder of the Man Who Freed 30,000 Prisoners from the Nazis*, Paragon House, New York, (1992). K. Marton, *A Death in Jerusalem*, Pantheon Books, New York, (1994).

populations in Egypt, Iraq, Hungary and Romania, in order to disparage those peoples and in order to force Jews to Palestine. Mossad agents infiltrated the Iraqi Government and instituted laws against Jews, and Jewish agents committed murderous terrorist acts against Jews in Iraq, in order to force the remaining Jews to emigrate to Palestine, just as Zionist Jews had put the Nazi régime into place and terrorized and murdered Jews in order to force Jews into Palestine.[366]

The Israeli Government has committed acts of war against the United States by bombing American interests in Egypt in 1954 with Israel's "Operation Susannah" in the "Lavon Affair"[367] and by attempting to sink the *U. S. S. Liberty* in 1967.[368] In both instances, the Israeli Government tried to lay blame on Egypt for the Israeli attacks on the United States, in an attempt to incite the United States to fight Israel's enemies. In the 1970's, the Israelis attempted to assassinate United States Ambassador to Lebanon, John Gunther Dean, as well as his entire family.[369]

---

366. There was allegedly an Arabic language book published in the 1950's under the title, "Venom of the Zionist Viper", which disclosed these facts. *See also:* N. Giladi, *Ben Gurion's Scandals: How the Haganah & the Mossad Eliminated Jews*, Glilit Pub. Co., Flushing, New York, (1992).

367. A. Golan, *Operation Susannah*, Harper & Row, New York, (1978). *See also:* D. Raviv, *Every Spy a Prince: The Complete History of Israel's Intelligence Community*, Houghton Mifflin, Boston, (1990). *See also:* V. Ostrovsky and C. Hoy, *By Way of Deception: A Devastating Insider's Portrait of the Mossad*, Stoddart, Toronto, (1990). V. Ostrovsky, *The Other Side of Deception: A Rogue Agent Exposes the Mossad's Secret Agenda*, Harper Paperbacks, New York, (1994). *See also:* I. Black and B. Morris, *Israel's Secret Wars: A History of Israel's Intelligence Services*, Grove Weidenfeld, New York, (1991). *See also:* S. Teveth, *Ben-Gurion's Spy: The Story of the Political Scandal That Shaped Modern Israel*, Columbia University Press, New York, (1996). *See also:* J. Beinin, *The Dispersion of Egyptian Jewry: Culture, Politics, and the Formation of a Modern Diaspora*, University of California Press, Berkeley, (1998).

368. W. Beecher, "Israel, in Error, Attacks U. S. Ship; 10 Navy Men Die, 100 Hurt in Raids North of Sinai Israelis, in Error, Attack U.S. Navy Ship 10 Navy Men Die and 100 Are Hurt Communications Vessel Is Raided From Air and Sea North of Sinai Peninsula", *The New York Times*, (9 June 1967), p. 1. N. Sheehan, "Sailors Describe Attack on Vessel; Israelis Struck So Suddenly U. S. Guns Were Unloaded", *The New York Times*, (11 June 1967), p. 27. "Israel Offers Compensation", *The New York Times*, (11 June 1967), p. 27. McClure M. Howland, "Families of Sailors" Letter to the Editor, *The New York Times*, (15 June 1967), p. 46. "Israel Accused at Hearing on U. S. Ship", *The New York Times*, (18 June 1967), p. 20. "U. S. Again Accused", *The New York Times*, (20 June 1967), p. 13. N. Sheehan, "Order Didn't Get to U. S. S. Liberty; Pentagon Reports Message Directing Ship Off Sinai to Move Arrived Late Ship 15.5 Miles Offshore", *The New York Times*, (29 June 1967), p. 1. *See also:* J. M. Ennes, *Assault on the Liberty: The True Story of the Israeli Attack on an American Intelligence Ship*, Random House, New York, (1979). *See also:* BBC Documentary, *Dead in the Water*, ( 21 August 2004, 7:00-8:10pm; rpt 1:50-3:00am)
<http://www.bbc.co.uk/bbcfour/documentaries/features/dead_in_the_water.shtml>

369. A. I. Killgore, "American Ambassador Recalls Israeli Assassination Attempt—With U.S. Weapons", *Washington Report on Middle East Affairs*, (November, 2002), p. 15.
<http://www.washington-report.org/archives/november02/0211015.html>

In her book *Israel's Sacred Terrorism*, Livia Rokach reproduced an excerpt from a 26 May 1955 entry in Moshe Sheratt's personal diary, which recounts his impressions of Moshe Dayan's plans to provoke the Arabs to respond by first attacking them, then stealing their land when they sought to defend themselves,

> "The conclusions from Dayan's words are clear: This State has no international obligations, no economic problems, the question of peace is nonexistent... . It must calculate its steps narrow-mindedly and live on its sword. It must see the sword as the main, if not the only, instrument with which to keep its morale high and to retain its moral tension. Toward this end it may, no—it must—invent dangers, and to do this it must adopt the method of provocation-and-revenge... . And above all—let us hope for a new war with the Arab countries, so that we may finally get rid of our troubles and acquire our space. (Such a slip of the tongue: Ben Gurion himself said that it would be worth while to pay an Arab a million pounds to start a war.) (26 May 1955, 1021)"[370]

Some Jews have long sought to destroy the Dome of the Rock and the Al Aqsa Mosque, and have recently persuaded Dispensationalist Christians to join them in the quest to destroy both so that the Jews can build a Jewish temple on the site. Under Jewish occupation, on 21 August 1969, arsonists inflicted heavy damage to the Al Aqsa Mosque. The United Nations Security Council condemned Israel for the attack in Resolution 271. In 2000, Ariel Sharon intentionally provoked Moslems by invading the Al Aqsa Mosque and Israeli police attacked Palestinians in the Mosque. Many Jews and Christian Dispensationalists have encouraged terrorist attacks against the Al Aqsa Mosque and the Dome of the Rock.

In 1968, Israel attacked a civilian airport in Beirut and destroyed numerous civilian aircraft. On 31 December 1968, United Nations Security Council Resolution 262 officially condemned the unprovoked Israeli attack on Lebanon. Numerous other United Nations Resolutions condemned Israel's repeated unprovoked and unjustifiable attacks on Lebanon, including resolutions 270, 279, 280, 285, 313, 316, 317, 332, 337, 347, 425, 427, 450, 467, 498, 501, 508, 509, 512, 513, 515, 516, 517, 518, 520, 521, and 587. In 1982, under Ariel Sharon's leadership, thousands of civilians were mass murdered in Lebanon in the Sabra and Shatila Massacre. In 1996, under Shimon Peres' leadership, Israel bombed civilians in Lebanon in operation "Grapes of Wrath". Many have accused Israel of fomenting the civil war between Christians and Moslems in Lebanon, which largely destroyed the most beautiful nation and city, Lebanon and Beirut, in the region. Israel also attacked helpless civilians in Jordan, perhaps most aggressively in 1968, and faced the condemnation of United Nations Security Council Resolutions 228, 248, 256 and 265. David Ben-Gurion once

---

[370]. L. Rokach, *Israel's Sacred Terrorism*, Third Edition, AAUG Press, Belmont Massachusetts, p. 41.

stated,

> "I proposed that, as soon as we received the equipment on the ship, we should prepare to go over to the offensive with the aim of smashing Lebanon, Transjordan and Syria. [\*\*\*] The weak point in the Arab coalition is Lebanon [for] the Moslem regime is artificial and easy to undermine. A Christian state should be established, with its southern border on the Litani River. We will make an alliance with it. When we smash the [Arab] Legion's strength and bomb Amman, we will eliminate Transjordan, too, and then Syria will fall. If Egypt still dares to fight on, we shall bomb Port Said, Alexandria, and Cairo. [\*\*\*] And in this fashion, we will end the war and settle our forefathers' accounts with Egypt, Assyria, and Aram."[371]

Lieutenant General Rafael Eytan, outgoing Chief of Staff of the Israeli Army, stated on 12 April 1983,

> "When we have settled the land, all the Arabs will be able to do about it will be to scurry around like drugged roaches in a bottle."[372]

In an article entitled, "An Israeli Mayor Is Under Scrutiny", *The New York Times* reported on 6 June 1989, on page 5,

> "Rabbi Yitzhak Ginsburg had offered biblical justification for the view that the spilling of non-Jewish blood was a lesser offense than the spilling of Jewish blood. 'Any trial based on the assumption that Jews and goyim are equal is a total travesty of justice,' he said."

Rabbi Yaacov Perrin was quoted by Clyde Haberman, in an article entitled, "Arafat Dismisses Rabin's Moves as 'Hollow'", *The New York Times*, (28 February 1994), p. 1. Rabbi Perrin stated,

> "One million Arabs are not worth a Jewish fingernail[.]"

---

[371]. D. Ben-Gurion, quoted in: M. Bar-Zohar, *Ben-Gurion: A Biography*, Delacorte Press, New York, (1978), p. 166.

[372]. D. K. Shipler, "Most West Bank Arabs Blaming U. S. for Impasse", *The New York Times*, (14 April 1983), p. A3; **and** "Israel's Military Chief Retires and Is Replaced by His Deputy", *The New York Times*, (20 April 1983), p. A8; **and** "The Israeli Army Signs a Political Truce", *The New York Times*, Section 4, (15 May 1983), p. 3. *See also:* A. Lewis, "Hope Against Hope", *The New York Times*, Section 4, (17 April 1983), p. 19; and "The New Israel; Away from the Early Zionist Dream", *The New York Times*, (30 July 1984), p. A21. *See also:* J. Kuttab, "West Bank Arabs Foresee Expulsion", *The New York Times*, (1 August 1983), p. A15. *See also:* A. Cowell, "Israel Frees More Prisoners, But Arabs Are Not Mollified", *The New York Times*, (4 March 1994), p. A10. *See also:* Y. M. Ibrahim, "Palestinians See a People's Hatred in a Killer's Deed", *The New York Times*, (6 March 1994), p. E16.

In an article "Begin and the 'Beasts'", *New Statesman*, Volume 103, Number 2674, (25 June 1982), page 12, Amnon Kapeliuk wrote of Menachem Begin, the Prime Minister of Israel,

> "The war in Lebanon cannot be interpreted, even by its most devoted proponents in Israel, as a war of survival. For this reason, the government has gone to extraordinary lengths to dehumanise the Palestinians. Begin described them in a speech in the Knesset as 'beasts walking on two legs'. Palestinians have often been called 'bugs' while their refugee camps in Lebanon are referred to as 'tourist camps'. In order to rationalise the bombing of civilian populations, Begin emotively declared: 'If Hitler was sitting in a house with 20 other people, would it be correct to blow up the house?'"

In 1982, Israelis massacred Palestinians in Beirut. The United Nations Security Council condemned Israel for the "criminal massacre" in Resolution 592. In 1986, Israeli soldiers opened fire on Palestinian students at Bir Zeit University. The United Nations Security Council condemned the attack in Resolution 592. In 1987, the Israeli Government instituted a policy under Yitzhak Rabin of smashing the bones of Palestinian demonstrators with rocks.[373] Israeli soldiers held helpless children and pounded heavy, jagged stones against their bodies until their limbs were crippled with compound fractures. On 25 February 1994, Benjamin C. Goldstein, a. k. a. Baruch Kappel Goldstein, murdered several people and injured many more in his terrorist attack against innocent Moslems who were peacefully praying in the Al-Ibrahimi Mosque during the holy month of Ramadan. Goldstein was a follower of Meir Kahane and a medical doctor who refused to treat Gentiles, because Maimonides forbade a Jewish physician from treating a Gentile unless under duress, and even then declared that a fee must be charged to the Gentile (Maimonides, *Mishneh Torah*, "Idolatry" 10:1-2).[374] More than 50 Palestinians were murdered and hundreds more were injured in the attack and its aftermath. The United Nations condemned the attack in Security Council Resolution 904.

In 1995, Yigal Amir assassinated Israeli Prime Minister Yitzhak Rabin in an attempt to end the peace process. Israel has legalized governmental political murders and the Israeli Government has brutally murdered and tortured many innocents. The program "Frontline" has produced a documentary *Israel's Next War*, which exposes the failed attempt of Jewish terrorists to set off a massive bomb at a Palestinian girls' school in 2002.[375] The Israeli Air Force bombed the Bahr el Bakar elementary school on 8 April 1970, mass murdering dozens of

---

[373]. Y. M. Ibrahim, "Palestinians See a People's Hatred in a Killer's Deed", *The New York Times*, (6 March 1994), p. E16.
[374]. A. Kizel, *Yediot Ahronot*, (1 March 1994). Y. Harkabi, *Israel's Fateful Hour*, Harper & Row, New York, (1988), p. 157.
[375]. http://www.pbs.org/wgbh/pages/frontline/shows/israel/

children and a teacher.[376] These are only a few of the countless atrocities the "Jewish State" has committed against innocent people.

Perhaps inspired by the accusations against Jews of poisoning wells in the 1300's, some Jews unsuccessfully attempted revenge against the Germans for the Holocaust after the Second World War by poisoning the water supply of Germany. They sought to kill at least six million Germans. Tom Segev wrote in his book *The Seventh Million: The Israelis and the Holocaust*,

> "Kovner therefore set six million German citizens as his goal. He thought in apocalyptic terms: revenge was a holy obligation that would redeem and purify the Jewish people. The group divided into cells, each with a commander. Their primary goal, Plan A, was 'to poison as many Germans as possible.' Plan B was to poison several thousand former SS men in the American army's POW camps. Reichman succeeded in infiltrating some members of the group into the Hamburg and Nuremberg water companies. Kovner went to Palestine to bring the poison—and, he hoped, to receive the blessing of the Haganah."[377]

Such leading figures in Israeli history as Menachem Begin, Yitzhak Shamer and David Ben-Gurion have been accused of terrorism, and/or of sponsoring terrorism, and/or of condoning terrorism. Jacob Bernard Agus wrote,

> "As the horrors of the Nazi 'final solution' were revealed after the war, the pitch of Jewish desperation reached unprecedented heights. The terrorist movements in Palestine against the British mandatory power were totally inconceivable before the war. Even veteran Jewish leaders were unable either to understand or to restrain the fury of the young terrorists, for whom the whole of Jewish experience was summed up in the raising of a gun with the slogan, *rak Kach*, 'Only thus!' The struggle of the terrorists, the desperation of the concentration camp graduates, and the military know-how of the European partisans shattered Arab resistance so effectively that nearly their entire population fled in panic."[378]

Begin brought his terrorist's mentality with him into Israel's top office. The racist State of Israel is the manifestation of this simplistic, genocidal and hate

---

[376]. R. H. Anderson, "30 PUPILS IN U.A.R. SAID TO DIE IN RAID; Teacher Is Also Reported Killed as Israeli Planes Strike a Delta Village Egyptians Report 30 Schoolchildren Killed in Israeli Raid on Village", *The New York Times*, (9 April 1970), p. 1, 14; *see also:* p. 15. R. H. Anderson, "Egypt Terms U.S. Cynical", *The New York Times*, (10 April 1970), p. 3. "Britain Deplores Deaths", *The New York Times*, (10 April 1970), p. 3.

[377]. T. Segev, *The Seventh Million: The Israelis and the Holocaust*, Hill and Wang, New York, (1993), p. 142.

[378]. J. B. Agus, *The Meaning of Jewish History*, Volume 2, Abelard-Schuman, New York, (1963), pp. 445-446.

driven mentality, which has existed at least as long as Judaism has existed. Michael Berenbaum wrote in his book, *After Tragedy and Triumph*,

> "Menachim Begin built upon this realization and constructed a usable past upon the twin pillars of antisemitism and the need for power. *Goyim* (literally, 'the nations') hate Jews, Begin maintained. In traditional language, Esau hates Jacob. According to Begin's worldview, Jews are a people that dwells alone. Power is essential. Powerlessness invites victimization. Jews must determine their own morality. The world's pronouncements toward the Jews mask—sometimes more successfully and sometimes less so—their genocidal intent. The desire to make the world *Judenrein* continues, and only fools would allow themselves to be deceived."[379]

*The New York Times* reported on 5 May 1948 on page 17,

> "While Scotland Yard directed an international search for the sender of the explosive parcel that killed Rex Farran, brother of Roy Farran, former Palestine police officer who was blacklisted by Jewish terrorists, official spokesman in the House of Commons voiced the indignation of the British people today at 'this wicked outrage.'"

Max Born wrote to the racist nationalist Albert Einstein on 22 May 1948,

> "I was very sad when the Jews started to use terror themselves, and showed that they had learned a lesson from Hitler. [***] Moreover, I detest nationalism of every kind, including that of the Jews."[380]

Zionist Jewish bankers have financed America's worst enemies including Great Britain, the Confederacy, Imperial Japan, Bolshevik Russia, Nazi Germany, etc. Zionist Jewish bankers are responsible for more American war casualties than any other group. Zionist Jewish bankers have deliberately caused America's worst recessions and depressions. They have corrupted the American media and American politics. Michael Collins Piper argues that Mossad agents were involved in the assassination of United States President John Fitzgerald Kennedy and that they wanted him dead because Kennedy opposed the Israeli nuclear weapons program, a program which is not in the best interests of the United States.[381] The Zionists have been a curse to America.

---

[379]. M. Berenbaum, *After Tragedy and Triumph: Essays in Modern Jewish Throught and the American Experience*, Cambridge University Press, (1990), p.7.
[380]. M. Born, *The Born-Einstein Letters*, Walker and Company, New York, (1971), p. 177.
[381]. M. C. Piper, *Final Judgment: The Missing Link in the JFK Assassination Conspiracy*, Wolfe Press, Washington, D.C., (1993).

## 5.5 Attempts to Prove the *Protocols* Inauthentic

*The London Times* published a series of articles in 1921, which relied upon an anonymous source "Mr. X" in contact with the *Times*' "Constantinople correspondent" Philip P. Graves. These articles set out to debunk the *Protocols* as a forgery. Graves claimed that the *Protocols* are a forgery, because they allegedly plagiarized Maurice Joly's *Dialogue aux enfers entre Machiavel et Montesquieu: ou, La politique de Machiavel au XIXe siècle*, A. Mertens, Bruxelles, (1864). Lucien Wolf, Herman Bernstein and many others have also claimed a forgery on the basis of plagiarism.[382]

Advocates of the alleged authenticity of the *Protocols* countered that the fact that sections of the *Protocols* were evidently plagiarized from Joly and others does not prove that the document was a forgery, only that its authors were students of, or plagiarists of the works of others, who deemed it inappropriate— or who had not yet had the opportunity—to name the sources for some of their statements. Others argued that all of these works had older common sources and it was to be expected that they should bear a resemblance to one another. Graves' articles and Zangwill's letter to the *Times* were as fantastic a conspiracy theory as the *Protocols* themselves in their allegations of Czarist conspiracies to defame the Jews, and in their reliance upon unnamed and unreliable sources.

The founder of modern political Zionism, Theodor Herzl, author of *The Jewish State* (*Der Judenstaat; Versuch einer modernen Lösung der Judenfrage*[383]) in 1896 was in some minds the alleged author of the *Protocols*. Herzl emphasized the fact that his book *The Jewish State* was not original, but instead drew from older sources. Herzl expressed racial mythologies found in the *Protocols* in Herzl's radical statements in his diaries and in his book *The Jewish State*. However, much that Herzl wrote was earlier published in Moses Hess' *Rom und Jerusalem*, Eugen Karl Dühring's *Die Judenfrage*, Leon Pinsker's *Auto-Emancipation*, and in the newspaper *Selbst-Emancipation*, which was published in Vienna from 1885-1886, and again from 1890-1893, and which

---

382. L. Wolf, *The Jewish Bogey and the Forged Protocols of the Learned Elders of Zion*, London, The Press Committee of the Jewish Board of Deputies, (1920); **and** *The Myth of the Jewish Menace in World Affairs; or, The Truth about the Forged Protocols of the Elders of Zion*, The Macmillan company, New York, (1921). *See also: Confrontation der "Geheimnisse der Weisen von Zion". ("Die Zionistischen Protokolle") mit ihrer Quelle "Dialogue aux enfers entre Machiavel et Montesquieu" der Nachweis der Fälschung.* ; *Dialogue aux enfers entre Machiavel et Montesquien*, Rechtsschutzabteilung des Schweizerischen Israelitischen Gemeindebundes, Basel, (1933). *See also:* H. Bernstein, *The History of a Lie, "The Protocols of the Wise Men of Zion"; a Study*, Ogilvie Pub. Co. New York, (1921); **and** *The Truth About "The Protocols of Zion"; a Complete Exposure*, Covici, Friede, New York, (1935).
383. T. Herzl, *Der Judenstaat; Versuch einer modernen Lösung der Judenfrage*, M. Breitenstein, Leipzig, Wien, (1896). English translation: *A Jewish State: An Attempt at a Modern Solution of the Jewish Question*, The Maccabæan Publishing Co., New York, (1904).

featured the same racist anti-assimilationist Zionist rhetoric one hears to this day. The fact that it drew from older sources does not render Herzl's book a forgery, nor a complete fabrication.

The New York Times also published many articles featuring John Spargo in early 1921, with the purpose of curbing the rise in anti-Semitism caused by the *Protocols* and the anti-Jewish articles published in THE DEARBORN INDEPENDENT. However, the defense against the *Protocols* was poorly managed, self-contradictory and factually incorrect; and many essays and pamphlets were published promoting the *Protocols* and arguing that they are authentic, which arguments, while sometimes unfair, exaggerated and factually incorrect, won out in the court of public opinion with tragic consequences.[384] There was often a

---

**384**. *The Cause of World Unrest*, G. Richards, ltd., London, G.P. Putnam, New York, (1920); which reproduces articles which first appeared in *The Morning Post* of London. **See also:** *World Conquest through World Government*, Britons Pub. Society, London, (1958/1920). **See also:** K. Paumgartten and L. Müller, *Juda: Kritische Betrachtungen über das Wesen und Wirken des Judentums*, Heimatverlag Leopold Stocker, Graz, (1920). **See also:** *El Peligro Judio*, C.T.S., Santiago de Chile, (1920). **See also:** B. Brasol, *The Protocols and World Revolution, Including a Translation and Analysis of the "Protocols of the Meetings of the Zionist Men of Wisdom"*, Small, Maynard & Company, Boston, (1920); **and** *Socialism Vs. Civilization*, C. Scribner's Sons, New York, (1920); **and** *The World at the Cross Roads*, Small, Mayhard & Co., Boston, (1921); **and** *The Balance Sheet of Sovietism*, Duffield, New York, (1922). **See also:** H. L. Strack, *Die Weisen von Zion und ihre Gläubigen*, C.A. Schwetschke & Sohn, Berlin, (1921). **See also:** E. Jouin, *Le Péril Judéo-Maçonnique*, Émile-Paul frères, Paris, (1921). **See also:** George Sydenham Clarke, Baron Sydenham of Combe, *The Jewish World Problem*, (1921). **See also:** L. Fry, *L'Auteur des Protocols Achad Ha-Am et le Sionisme*, Editions de la Vieille-France, Paris, (1921); **and** *Akhad-Khem (Asher Geintsberg); Tainyi vozhd' iudeiskii*, Tip. Presse, Berlin, (1922); **and** German translation by Th von Winberg, *Achad Cham (Ascher Hinzberg)*, München, (1923). **See also:** E. Jouin, *Le Péril judéo-Maçonnique: Les "Protocols" des Sages de Sion. Coup d'Oeil d'Ensemble*, Revue Internationale des Societés Secrètes, Paris, (1921). **See also:** H. Ford and S. Crowther, *My Life and Work*, Doubleday, Page & Company, Garden City, New York,(1922), pp. 250-252. **See also:** C. J. Gohier, *Le Peril Juif Comment le Conjurer*, (1923). **See also:** U. Gohier, *Criminelle Doctrine du Talmud*, Paris, (1923). **See also:** Brother Therapier, T.O.S.F., *The World's Unrest and the Jew (III)* Franciscan Friary, Montreal, (1923). **See also:** T. Fritsch, *Die Zionistischen Protokolle; das Programm der internationalen Geheimregierung. Aus dem Englischen übers. nach dem in Britischen Museum befindlichen Original*, Hammer-Verlag, Leipzig, (1924). **See also:** D. Eckart and A. Hitler, *Der Bolschewismus von Moses bis Lenin: Zwiegespräch zwischen Adolf Hitler und mir*, Hoheneichen-Verlag, München, (1924); English translation by W. Pierce: *Bolshevism from Moses to Lenin: A Dialog Between Adolf Hitler and Me*, Vanguard Books. **See also:** P. Copin-Albancelli, *La Guerre Occulte, les Sociétés Secrètes Contre les Nations*, Perrin, Paris, (1925). **See also:** A. Rosenberg, *Der Weltverschwörerkongress zu Basel*, Verlag F. Eher Nachf., München, (1927). **See also:** E. Jouin, *Le Péril Judéo-Maçonnique*, Revue Internationale des Sociétés Secrètes, Paris, (1927). **See also:** Imperial Fascist League, *The Era of World Ruin! (The Era of Democracy): The Claim of the Jews that They Installed Democracy for the Express Purpose of Ruining the Gentile World*, Imperial Fascist League, London, (1930's). **See also:** E. Cahill, Edward, *Freemasonry and the Anti-Christian Movement*,

Second Revised and Enlarged Edition, M.H.Gill, Dublin, (1930). *See also: 4 Protocols of Zion: (Not the Protocols of Nilus)*, Britons Pub. Society, London, (1931). *See also:* L. Fry, *Waters Flowing Eastward*, Éditions R.I.S.S., Paris, (1931). *See also:* A. Reiterer, *Das Judentum und die Schatten des Antichrist*, Styria, Graz, (1933). *See also:* E. Knauss, *Communism and the Protocols*, E. Knauss, Davenport, Iowa, (1933). *See also:* L. Müller, *Cianas gudro noslepumi ...*, J. Turks, Riga, (1933). *See also:* W. B. Riley, *Protocols and Communism*, Minneapolis, (1934). *See also:* E. N. Sanctuary, World Alliance against Jewish Aggressiveness, *"Are these things so?" Being a Reply to this Question Propounded by a Jewish High Priest of the First Christian Martyr 1900 Years Ago; a Study in Modern Termites of the Homo Sapiens Type*, Community Press, Woodhaven, New York City, (1934). *See also:* W. Creutz, *Les Protocoles des Sages de Sion: Leur Authenticité*, Les Nouvelles Éditions Nationales, Brunoy, (1934); English translation, *New Light on the Protocols: Latest Evidence on the Veracity of this Remarkable Document*, The Right Cause, Chicago, (1935). *See also: The Jewish Victory at Berne: Side Lights on the Verdict*, Christian Aryan Protection League, London, (1935). *See also:* U. Fleischhauer, *Die echten Protokolle der Weisen von Zion; Sachverständigengutachten, erstattet im Auftrage des Richteramtes V in Bern*, U. Bodung, Erfurt, (1935). *See also: The Berne Trail—Concerning the "Protocols of the Elders of Zion." Fourteenth Session; Tuesday, May 14, 1935—P.M.*, (1935). *See also:* H. Jonak von Freyenwald, *Der Berner Prozess um die Protokolle der Weisen von Zion. Akten und Gutachten*, U. Bodung, Bund Nationalsozialistischer Eidgenossen, Erfurt, (1939). *See also:* G. Schwartz-Bostunitsch, *Jüdischer Imperialismus*, O. Ebersberger, Landsberg am Lech., (1935). *See also:* R. E. Edmondson, *The Damning Parallels of the Protocol "Forgeries" as Adopted and Fulfilled in the United States by Jewish-Radical Leadership: A Diabolical Capitalist-Communist Alliance Unmasked*, New York, (1935). *See also:* S. Vász, *Das Berner Fehlurteil über die Protokolle der Weisen von Zion: eine kritische Betrachtung über das Prozessverfahren*, U. Bodung, Erfurt, (1935). *See also:* G. B. Winrod, *The Hidden Hand: The Protocols and the Coming Superman*, Defender Publishers, Wichita, Kansas, (1933); **and** *Antichrist and the Tribe of Dan*, Defender Publishers, Wichita, Kansas. *See also: A Protocol of 1935: Based on a Careful Study of the Present Day Jewish Activities*, Pan-Aryan Alliance, New York, (1935). *See also:* E. Engelhardt, *Jüdische Weltmachtpläne die Entstehung der sogenannten Zionistischen Protokolle; neue Zusammenhänge zwischen Judentum und Freimaurerei*, Hammer-Verlag, Leipzig, (1936). *See also:* Squire of Krum Elbow H. Spencer, *Our Neighbor and World Unrest*, Newport Historical Association, Newport, Rhode Island, (1936); **and** *Toward Armageddon / y the Squire of Krum Elbow*, Militant Christian Association, Charleston, (1936). *See also: Berner Bilderbuch vom Zionisten-Prozess um die "Protokolle der Weisen von Zion."*, U. Bodung-Verlag, Erfurt, (1936). *See also: The Ultimate (If): Concerning Something People Do Not Wish to Believe and Which Is So Near*, Zealous Enlightenment League, U.S.A., (1931). *See also: A Plot for the World's Conquest*, Britons Pub. Society, London, (1936). *See also:* G. Barroso, *Os Protocolos dos Sábios de Sião; o Imperialismo de Israel, o Plano dos Judeus para o Conquista do Mundo, o Código do Anti-Cristo, Provas de Autenticidade, Documentos*, Agencia Minerva, São Paulo, (1936). *See also:* K. Bergmeister, *Der jüdische Weltverschwörungsplan; die Protokolle der Weisen von Zion vor dem Strafgerichte in Bern*, U. Bodung, Erfurt, (1937); English translation, *The Jewish World Conspiracy; The Protocols of the Elders of Zion before the Court in Berne*, Sons of Liberty, Liberty Bell Publications, Hollywood, California, (1938). *See also:* L. Donoso Zárate, *La Verdad más Grande de la História: "Los Protocolos de los Sabios de Sión"*, Imprenta San José, Santiago de Chile, (1937). *See also: Out of Their Own Mouths; the*

deliberate confusion between the actions of some particular Jews, and all Jews, which unfair generalization was again and again pointed out, unfortunately with little success.

The Zionists continued to pretend that they spoke for all Jews and that they constituted a government for world Jewry. Adolf Hitler was one of the many Zionist anti-Semite stooges in the early 1920's, who asserted that the *Protocols* are genuine and represented a vast conspiracy and a threat that must be addressed.[385] Hitler used the *Protocols* as a means to put himself into power, so that he could fulfill the Zionist plans laid out in the *Protocols*. This was a common tactic of Zionists and Communists, who promoted a controlled opposition to their plans, which enabled them to fulfill them. Hitler was both a Zionist and a Bolshevist, and at war's end Eastern Europe, and very nearly all of Europe, turned Communist. Hitler and Stalin worked in collusion to make Europe ripe for a Communist takeover. At war's end, the Zionists were finally able to persuade the world's Jews to join them in founding a racist apartheid "Jewish State". Hitler succeeded in his goal to found this State.

## 5.5.1 Why Did Henry Ford Criticize the Jews?

Henry Ford's newspaper THE DEARBORN INDEPENDENT brought the attention of the American public to *The Protocols of the Learned Elders of Zion*. Many have noted that Ford showed no signs of bigotry before the spring of 1920, and the first anti-Jewish articles appeared in his newspaper on 22 May 1920, and 29 May 1920. It was seemingly inexplicable that Ford began so overwhelming an attack on Jews and reorganized his newspaper and his life to carry out this attack, with no chance for personal gain and no apparent reason other than a genuine belief that the *Protocols* were authentic in their message, if not authorship, and revealed the Jewish plan for world domination through Bolshevism and Zionism.

Ford did not state whether or not he believed that the *Protocols* were genuine, but he did state that they were an accurate reflection of real events that had occurred many years after the *Protocols* first appeared. On 17 February

---

*Jewish Will for Power and the Authenticity of the "Protocols."*, American Nationalist Press, New York, (1937). ***See also:*** H. de Vreis de Heekelingen, *Les Protocols des Sages de Sion Constituent-ils un Faux*, Lausanne, (1938). ***See also:*** *Hidden Empire*, Pelley Publishers, Ashville, North Carolina, (1938). ***See also:*** J. F. Norris, *Did the Jews Write the Protocols*, Temple Baptist Church, Detroit, Michigan, (1938). ***See also:*** *The Reign of the Elders*, In Two Volumes, (1938-1939). ***See also:*** F. Veloso, *Perante o Racismo, Novos Subsídios Acêrca dos "Protocols dos Sábios de Sião;"*, Livraria Portugália, Lisboa, (1939). ***See also:*** N. D. Zhevakhov, *Il Retroscena dei Protocolli di Sion: La Vita e le Opere del Loro Editore, Sergio Nilus e del Loro Autore Ascer*, Unione editoriale d'Italia, Roma, (1939). ***See also:*** G. Schwartz-Bostunitsch, *Jüdischer Imperialismus 3000 Jahre hebräischer Schleichwege zur Erlangung der Weltherrschaft*, T. Fritsch, Berlin, (1939). ***See also:*** H. H. Klein, *A Jew Exposes the Jewish World Conspiracy*, Sons of Liberty, Hollywood, (1970). ***See also:*** A. Baron, *The Protocols of the Learned Elders of Zion: Organized Jewry's Deadliest Weapon*, Anglo-Hebrew Pub., London, (1995).
**385**. A. Hitler, *Mein Kampf*, Houghton Mifflin, Boston, New York, (1971), p. 307.

1921, Henry Ford was quoted in *The New York World*,

> "The only statement I care to make about the Protocols is that they fit in with what is going on. They are sixteen years old, and have fitted the world situation up to this time. They fit it now."[386]

## 5.5.2 Controlled Opposition and "The Trust"

Henry Ford, and the articles in THE DEARBORN INDEPENDENT, repeatedly stated that Ford's campaign to inform the public of the dangers of: Bolshevism, Jewish control of the press, Jewish "power behind the throne" of numerous governments, and the power of racist Jewish financiers; was motivated by a genuine desire to help the Jews to overcome their prejudice against non-Jews, and to benefit society at large, but not out of hatred.[387] Some contemporary Jews believed that Ford was an *agent provocateur* for the Zionists, who had been promoting anti-Semitism for centuries as a means to keep Jews segregated from non-Jews, so as to preserve the "purity of the Jewish race".

THE DEARBORN INDEPENDENT reported on 11 September 1920 in an article entitled, "Does Jewish Power Control the Press?":

> "A sidelight on the first sentence above may be had from this Jewish statement regarding the British Declaration relating to Palestine: 'This Declaration was sent *from the Foreign Office to Lord Walter Rothschild.* * * * It came perhaps as a surprise to large sections of the Jewish people * * * But to those who were active in Zionist circles, the declaration was no surprise. * * * *The wording of it came from the British Foreign Office, but the text had been revised in the Zionist offices in America as well as in England. The British Declaration was made in the form in which the Zionists desired it.* * * *' pp. 85-86, 'Guide to Zionism,' by Jessie E. Sampter, published by the Zionist Organization of America.
> 
> 3. 'Literature and journalism are two most important educational forces, and consequently our government will become the owner of most of the journals. * * * *If we permit ten private journals, we shall organize thirty of our own, and so on. This must not be suspected by the public, for which reason all the journals published by us will be EXTERNALLY of the most contrary opinions* and tendencies *thus evoking confidence in them and attracting our unsuspecting opponents*, who thus will be caught in our trap and *rendered harmless.*'
> 
> This is most interesting in view of the defense now being made by so many Jewish journals. 'Look at the newspapers owned and controlled by

---

[386]. *Cf.* L. Fry, *Waters Flowing Eastward: The War Against the Kingship of Christ*, TBR Books, Washington, D. C., (2000), p. 109.
[387]. H. Ford and S. Crowther, *My Life and Work*, Doubleday, Page & Company, (1922), pp. 250-252, at 252.

Jews,' they say; 'see how they differ in policy! See how they disagree with each other!' Certainly, 'externally,' as Protocol 12 says, but the underlying unity is never hard to find.

Besides, one way of discovering who are the people that have knowledge of the Jewish World problem, of who can be convinced of it, or who will write about it, is just to start a paper which 'externally' seems to be independent of the Jewish Question. So deeply is this thought shared by even uneducated Jews that a rumor is today widespread in the United States that the reason for the present series of articles in THE DEARBORN INDEPENDENT is the desire of its owner to forward the Jewish World Program! Unfortunately, this scheme of starting a fake opposition in order to discover where the real opposing force is, is not confined to the Jewish Internationalists, although there is every indication that it was learned from them."

There might have been an *agent provocateur* behind Ford—in the person of Boris Brasol.[388] An agent who appeared from the East, Brasol, like the Zionist Nazi Alfred Rosenberg, directed attention to the *Protocols* from the East to the governments of the West. Just as the most virulent Christian zealots were often crypto-Jews, who attempted to hide their identities and use hatred of the Jews as a means to subvert Gentiles; the most virulent anti-Semites were often crypto-Jews or Jewish agents who used hatred of the Jews as means to accomplish the ends of Jewish leadership—Communist revolution and the formation of a "Jewish State".

*The New York Times* published an article entitled "Spargo Denounces Anti-Semitic Move" on 6 December 1920 on page 10, and paraphrased John Spargo,

"He attacked Mr. Ford for intolerance and said he was the 'tool' in this matter of men more able than himself."

*The New York Times* reported on 18 May 1922, on page 11, in an article entitled, "Says C. C. Daniels Aided Ford Crusade":

"Ford's fight on the Jews is ascribed by Hapgood to the fact that Ford was 'tricked' by Czarist sympathizers in the United States. He says Mr. Daniels [\*\*\*] was head of the detective agency which employed Boris Brasol,

---

**388**. B. Brasol, *The Protocols and World Revolution, Including a Translation and Analysis of the "Protocols of the Meetings of the Zionist Men of Wisdom"*, Small, Maynard & Company, Boston, (1920); **and** *Socialism Vs. Civilization*, C. Scribner's Sons, New York, (1920); **and** *The World at the Cross Roads*, Small, Mayhard & Co., Boston, (1921); **and** *The Balance Sheet of Sovietism*, Duffield, New York, (1922). **See also:** S. G. Marks, "Destroying the Agents of Modernity: Russian Antisemitism", *How Russia Shaped the Modern World*, Chapter 5, Princeton University Press, (2003), pp. 140-175; notes 354-358. **See also:** R. Singerman, "American Career of the *Protocols of the Elders of Zion*", *American Jewish History*, Volume 71, (September, 1981), pp. 48-78.

former investigator for the Russian secret service Black Hundred."

But was Brasol's interest really in restoring the Russian Monarchy, or was he an agent of the Zionists and Bolsheviks? We know today that most of the opposition to the Bolsheviks was controlled by the Bolsheviks themselves.

Communist leadership, who were disproportionately Jewish, created a plan which came to be known as "The Trust",[389] whereby they sent out supposed exiles from Bolshevist Russia to found and infiltrate anti-Communist organizations. These organizations actually served the interests of the Communists. Given that the Jews played such a disproportionate rôle in fomenting Communist revolution and in the leadership of Communist governments; anti-Communist organizations were often highly critical of the rôle Jews played in Bolshevism. We also know that crypto-Jews like "Sidney Reilly" (born Salomon Rosenblum) were agents of "The Trust".[390] Alfred Rosenberg, Boris Brasol and Paquita de Shishmareff may have been predecessors of this Jewish-Communist controlled opposition dubbed "The Trust".

Whether or not Henry Ford was intentionally promoting anti-Semitism as means to promote the Zionist movement and ultimately a Boshevik takeover of the United States remains an open question. It is more certain that Adolf Hitler was a Bolshevist Zionist. Rosenberg, Brasol, and Shishmareff—who wrote in defense of the authenticity of the *Protocols* and who assisted Brasol, may have sought to place Jewish Zionist Communists in power on a popular platform of anti-Communism and anti-Semitism. Such was the case with Adolf Hitler.

## 5.5.3 The Sinking of the "Peace Ship"

Henry Ford was a hardworking pacifist, who used his fortune to try to end the senseless slaughter of the First World War. Many criticized Ford for his pacificism.

Ford sued *The Chicago Daily Tribune* for libel on 7 September 1916 for an article "Ford is an Anarchist" published in *The Chicago Daily Tribune* on 23 June 1916 on page 6. Ford eventually won his libel suit and was awarded the nominal sum of six cents in 1919. *The Chicago Daily Tribune* had published articles claiming that Ford was ignorant of, and indifferent to, History.[391] The lawyers for the defense in the libel action questioned Ford about his knowledge

---

**389**. C. S. Viar, Editor, R. G. Rocca, *The Trust*, by the Center for Intelligence Studies, (1990). G. Brook-Shepherd, *Iron Maze: The Western Secret Services and the Bolsheviks*, Macmillan, London, (1998). R. B. Spence, *Trust No One: The Secret World of Sidney Reilly*, Feral House, Los Angeles, (2002).
**390**. R. B. Spence, *Trust No One: The Secret World of Sidney Reilly*, Feral House, Los Angeles, (2002).
**391**. C. N. Wheeler, "Close-Up View of Henry Ford and His Ideas", *The Chicago Daily Tribune*, (23 May 1916), pp. 1, 12. "Ford a Voice from the Dark", *The Chicago Daily Tribune*, (24 May 1916), p. 8.

of History and he was unable to state what rôle Benedict Arnold had played in history.[392]

Ford was ridiculed for being a pacifist during the First World War. The counsel for the defense tried to confuse Ford with the many meanings inherent in the euphemism "preparedness", a term warmongers used as a euphemism for their build-up to war. Ford knew that the term was used to disguise aggressive preparations for war—in Ford's mind, unnecessary war for profit brought on by Jewish bankers and Jewish controlled newspapers. Ford was not misled and the counsel for defense was frustrated in its efforts to manufacture contradictions in Ford's statements, which contradictions were instead due to the euphemisms Ford's critics employed to confuse and manipulate the public. They failed in their efforts to attribute their own inconsistencies to Ford.

Some Republicans ran Henry Ford as a Republican candidate for the Presidency in the Republican primaries of 1916.[393] The Prohibition Party also wanted Ford to run as their candidate.[394] Harry Bennett stated, "Henry Ford, in 1916, was perhaps better known to most Americans than their President."[395]

In 1915, Henry Ford, a vocal pacifist, pledged his entire fortune to his effort to end the war on humanitarian grounds[396] and organized the voyage of the "Peace Ship" on 4 December 1915, a mission to persuade the Europeans to end the war by Christmas. This vessel, which Ford had chartered, sailed to Northern Europe with a contingent of leading pacifists, who intended to meet with European leaders in order to bring about peace. Ford did not want America to enter the war, which war needlessly slaughtered millions of Europeans. Ford sought a just and humane peace. Republican candidate Theodore Roosevelt and others ridiculed Ford for his pacifist campaign to end the suffering of the war.[397]

---

**392**. "Odd Definitions Given by Ford in Libel Suit", *The New York Times*, (27 July 1919), p. 1. Benedict Arnold later became the subject of one of the articles in Ford's newspaper, "Benedict Arnold and Jewish Aid in Shady Deal", *The Dearborn Independent*, (15 October 1921).
**393**. "Name Ford for President", *The New York Times*, (12 December 1915), p. E3. "Ford Presses Smith in Michigan Primary", *The New York Times*, (4 April 1916), p. 3. "For Hughes after Ford", *The New York Times*, (19 April 1916), p. 9. "Ford's Lead Grows; Bryan Badly Beaten", *The New York Times*, (20 April 1916), p. 7. "Ford Willing to Run If People Call Him", *The New York Times*, (23 April 1916), p. 1, 8. "T.R. Has 1 Jersey Delegate", *The New York Times*, (27 April 1916), p. 4. C. W. Thompson, "Primaries No Clue to Presidential Nominees", *The New York Times*, (30 April 1916), p. SM5. E. O. Jones, "Ellis O. Jones Finds the Movement for Him Very Significant", *The New York Times*, (3 May 1916), p. 12. "Hughes in Lead with 224 Votes", *The New York Times*, (7 June 1916), p. 1.
**394**. "Want Ford for President", *The New York Times*, (9 July 1916), p. 5.
**395**. H. Bennett and P. Marcus, *We Never Called Him Henry*, Chapter 7, Gold Medal Books, Fawcett Publications, (1951), p. 7.
**396**. "Ford Willing to Give Fortune to End War", *The New York Times*, (24 November 1915), p. 1.
**397**. "'Grotesque,' Says J. R. Day", *The New York Times*, (1 December 1915), p. 3. "Ridiculed by A. B. Parker", *The New York Times*, (2 December 1915), p. 2. "Amen

Journalist Herman Bernstein and other passengers on the Peace Ship withdrew their support from Ford's mission.[398] Ford concluded that the Jewish bankers and their lackeys had torpedoed his attempt to end the war. Ford later ran for the Senate in Michigan in 1918 and lost in a race which resulted in investigations of election fraud.

Incumbent Democratic Presidential candidate Woodrow Wilson ran on the pacifistic slogan, "He kept us out of the war!" Henry Ford, the pacifist, life-long Republican and formerly Republican candidate, threw his support behind the Democrat Wilson on 27 September 1916 and eventually congratulated Wilson on his victory, confident that Wilson would keep America out of the war.[399] Republican candidate Theodore Roosevelt alienated many German-Americans, and took a strongly pro-British stance and openly called for American "preparedness" for war with Mexico and the Central Powers. Roosevelt attacked the "hyphenates", German-Americans, many of them Jews, who wanted to keep America out of the war. German-Americans represented the swing vote in key states and when Wilson announced that he would keep America out of the war, the Republicans determined that Roosevelt could not win the election. Roosevelt dropped out of the race and was replaced by Republican candidate Charles Evans Hughes, who had the approval of German-Americans—the allegedly traitorous "hyphenated Americans" Roosevelt had alienated.[400] Wilson, who was a Zionist, won the election and then betrayed the American People and brought them into the war at the behest of his Zionist blackmailers Louis Brandeis and "Colonel" House.

Robert Rutherford McCormick was President of *The Chicago Daily Tribune*. A staunch Republican, he had Republican roots running back to Abraham Lincoln through his maternal grandfather, *Tribune* owner and one of the founders of the Republican Party, Joseph Medill. *The Chicago Daily Tribune* did not shy away from politics. Abraham Lincoln and Joseph Medill, like Theodore Roosevelt, confronted pacifist opposition in the Civil War; so there was nothing new about their antagonism towards pacifism. Robert R. McCormick became a Colonel in the First World War, and his home and estate are now a very fine museum grounds, Cantigny, which houses the First Division Museum.

---

Corner Pokes Fun at Public Men", *The New York Times*, (4 December 1915), p. 8. "Roosevelt Urges Unity in Defense", *The New York Times*, (6 December 1915), p. 3. "Roosevelt Bitter in Assailing Wilson", *The New York Times*, (20 December 1915), p. 4.
**398**. "Ford Finds Peace Elusive", *The New York Times*, (21 December 1915), p. 1.
**399**. "Henry Ford Is out in Praise of Wilson", *The New York Times*, (15 September 1916), p. 3. "Henry Ford Urges Election of Wilson", *The New York Times*, (28 September 1916), p. 4. "No Ford Cash for Wilson Campaign" and "Mr. Ford Sees Victory", *The New York Times*, (3 October 1916), p. 1. "Henry Ford Gratified", *The New York Times*, (12 November 1916) p. 4.
**400**. "The German Favorite Son", *The New York Times*, (23 May 1916), p. 10. "The Hyphen Vote, the Silent Vote, and the American Vote", *The New York Times*, (2 October 1916), p. 10.

The Republican Charles Evans Hughes lost to the democratic incumbent Woodrow Wilson, who won, in part, because of his ability to peal off the vote of Americans of German descent in the Midwest based on the lie that he would keep America out of the war. Wilson soon brought America into the war against Germany, despite his campaign promises of continued non-involvement. These experiences embittered Henry Ford and he must have felt personally betrayed by President Wilson. Many of the articles which later appeared in THE DEARBORN INDEPENDENT from 1920-1927 took the form of personal attacks.

But was this what prompted Ford? As early as July of 1919, in his libel trial against *The Chicago Tribune*, Henry Ford agreed with the allegation that bankers and newspapers, "got [America] into the war for purposes of gain."[401] Ford attributed his views to discussions he had had with the two Jews Herman Bernstein and Rosika Schwimmer on the Peace Ship expedition in December of 1915 and January of 1916. *The New York Times* reported on 5 December 1921 on page 33,

### "FORD EXPLAINS ATTACKS
Caused by Statements Made to Him
by Jews on Peace Trip.
*Special to The New York Times.*

FLORENCE, Ala., Dec. 4.—Henry Ford today told reporters the fundamental reason why for the last two years he has attacked the Jew in his weekly magazine, The Dearborn Independent. He said that the course of 'instruction on the Jew which he intends to give the United States will continue for five years.'

'It was the Jews themselves that convinced me of the direct relation between the international Jew and war, in fact, they went out of their way to convince me,' he said.

'You remember the effort we made to attract the attention of the world to the purpose of ending the war through the medium of the so-called peace ship in 1915. On that ship were two very prominent Jews. We had not been to sea 200 miles before these two Jews began telling me about the power of the Jewish race, how they controlled the world through their control of gold and that the Jew, and no one but the Jew, could stop the war.

'I was reluctant to believe this and said so—so they went into detail to tell me the means by which the Jew controlled the war, how they had the money, how they had cornered all the basic materials needed to fight the war and all that, and they talked so long and so well that they convinced me. They said, and they believed, that the Jews had started the war; that they would continue it as long as they wished and that until the Jew stopped the war it could not be stopped. We were in mid-ocean and I was so disgusted that I would have liked to have turned the ship back.

'When I got back to the United States I still had in mind what the Jews

---

**401**. H. Ford quoted in: "Ford Says Press and the Bankers Got Us into War", *The New York Times*, (18 July 1919), p. 1.

had told me. In Europe, I had looked about quite a bit and I could see that a lot of the things the Jews had told me were so. Once at home, I set about investigating a bit, and the more I investigated the more I found to substantiate what the Jews had told me. I determined that the situation should be made clear to the people of the United States through publicity. But do you think I could get a newspaper to print it? Not on your life. It seemed there was no newspaper in the United States that dared print the truth.

'Then a funny thing happened just at this juncture. An old chap in Dearborn came to my office and wanted to sell the local paper, The Dearborn Independent, a weekly newspaper. The thought came to me like a flash. Surely some place in the United States there should be a publisher strong and courageous enough to tell the people the truth about war. If no one else will, I'll turn publisher myself. And I did.'

'How long will your paper continue to deal with the Jewish question?' he was asked.

'We've got a five years' course in sight, and we are going to tell the people, among other things, some American history that they don't teach in the schools. We will show indisputably that one of the great factors behind the Civil War, that brought it on and made peaceable settlement of the issues impossible, was the Jew. And that isn't the whole story either. There will be more than that.'

Mr. Ford and Mr. Edison spent Sunday morning looking over the site of dam No. 3 at Muscle Shoals, which is still to be started, and which, when built, will create a great reservoir for control of the back waters above the power plant. The afternoon was spent at a Southern barbecue at the home of E. A. O'Neal, head of the Alabama Farm Bureau."

Ford was later sued by Herman Bernstein, who claimed that Ford had named Bernstein as the source for some of the views expressed in THE DEARBORN INDEPENDENT, which Bernstein alleged included:

"That leading members of the Jewish faith precipitated the World War. 2. That in the middle of the war they switched their support to the Allies, selling out to the highest bidder, and that their price was the aid of the allied nations in restoring Palestine to the Jewish people as a national home. 3. That they murdered or caused the murder of the Russian Czar and his family. 4. That most of the dangerous and destructive theories of government abroad in the world are of Jewish origin. 5. That they have debased the professions, prostituted the arts and degraded sports and corrupted commerce. 6. That they control and dominate the press, finance, resources, institutions and politics of the United States, and prostitute the same to unlawful and iniquitous purposes and to their own aggrandizement and to the great injury of the civilized world. 7. That their alleged wealth and power as a race constitutes a threat to mankind."[402]

Ford was quoted in an "International News Service" interview on 5 January 1922, as stating,

> "The real reason why I printed these articles was because of what a Jew (Herman Bernstein) told me while I was crossing the ocean on the peace ship. He told me that if I wanted to end the war I should talk with the Jewish financiers who created it. I played ignorance and led him on. He told me most of the things that I have printed."[403]

Rosika Schwimmer, who was a very hardworking pacifist and who prompted Henry Ford to undertake the Peace Ship mission and was a leader on the voyage, was thought to be an agent of the Germans by the Norwegians, who rejected her and the Peace Ship mission.[404] This accusation reemerged in 1927.[405] The Danish Government also believed that the Ford mission was pro-German.[406] A great scandal ensued during the voyage, which caused problems in Holland.

Schwimmer was later cleared of the charges made against her with respect to the monies involved[407] and she claimed that she was the victim of subterfuge by Fannie Fern Andrews and Jane Addams. By all accounts Schwimmer was a brilliant and charming woman and had been an active feminist for years—as had Henry Ford.[408] On Schwimmer's return, she attempted to contact Ford, who ignored her for many years.[409] Since Schwimmer was the Jew who was closest to Henry Ford on the Peace Ship mission, it was alleged that she inspired much of the anti-Jewish material that later appeared in *THE DEARBORN INDEPENDENT* in two ways. First, it was alleged that she inspired the Peace Ship expedition,

---

**402**. "$200,000 Libel Suit Filed Against Ford", *The New York Times*, (19 August 1923), p. 2.
**403**. "$200,000 Libel Suit Filed Against Ford", *The New York Times*, (19 August 1923), p. 2. *See also: Jewish World*, (5 January 1922). This interview apparently appeared in a Detroit area newspaper on or about 5 January 1922.
**404**. "Ford Ill in Norway, Delays Peace Work", *The New York Times*, (21 December 1915), p. 1. "Demand Ford Expel Mme. Schwimmer", *The New York Times*, (22 December 1915), p. 3. "Ford's Peace plan Vetoed by Norway", *The New York Times*, (23 December 1915), p. 1.
**405**. "Denies Peace Ship Led to Ford Attack", *The New York Times*, (5 September 1927), p. 17. "Mme. Schwimmer on Stand in Suit", *The New York Times*, (27 June 1928), p. 16.
**406**. "Ford Peace Mission Died Belligerently", *The New York Times*, (1 February 1916), p. 4.
**407**. "Ford Exonerated Mme. Schwimmer", *The New York Times*, (29 June 1928), p. 12. "Peace SHip Leader Wins Suit for Libel", *The New York Times*, (30 June 1928), p. 12.
**408**. "Pacifists Can't Find $200,000 Ford Gift", *The New York Times*, (20 January 1916), p. 20. "Warned the Dutch Against Ford Party", *The New York Times*, (21 January 1916), p. 3.
**409**. "Mme. Schwimmer at Home", *The New York Times*, (18 May 1917), p. 13. "Woman Asks Ford to Vindicate Her", *The New York Times*, (4 September 1927), p. E1.

which failed, and Ford came to hate all Jews because Schwimmer was Jewish and Ford believed that the Jews had torpedoed the Peace Ship mission—Ford might even have believed that he was led to make the trip as means to humiliate him and discredit the pacifist movement.[410] Secondly, it was assumed that Schwimmer told Ford that powerful Jews were behind the war and that the war would not end until the Jewish bankers who had caused it wanted it to end. Interestingly, Schwimmer became good friends with Albert Einstein, who called her his "saving angel".[411]

Henry Ford praised Schwimmer years after the Peace Ship mission, so the first accusation was probably false, but Ford had since come under the influence of Louis Marshall, a very powerful Jewish leader, and it is possible that Ford's later statements in support of Schwimmer may have been scripted.[412] Henry Ford, though asked by Schwimmer to repudiate the second tacit accusation, never did.[413] It appears that Herman Bernstein and Rosika Schwimmer did indeed inform Henry Ford of "the Jewish Peril" on the Peace Ship voyage. The later Zionist betrayal of America and Germany, and the Bolshevik Revolution, must have confirmed for Ford that all he had been told was true.

*The New York Times* reported on 18 May 1922 on page 11 in an article entitled, "Says C. C. Daniels Aided Ford Crusade":

> "In quest of an explanation for Ford's continued attacks against the Jews, Hapgood says he finally went to Ford's plant, where he was told by one of Ford's employees that the motor car manufacturer was aggrieved by the failure of his peace ship expedition and further because it was suggested by a Jewess, Rosicka Schwimmer."

On 24 July 1923, in the "Topics of the Times" section of *The New York Times*, on page 20, it stated, *inter alia*,

> "MR. FORD says, the incidental, and to him highly satisfactory, effect of [the peace ship voyage of] teaching him a lot about war, its causes, the men who

---

**410**. "Lays Ford's Attacks to Peace Ship Fiasco", *The New York Times*, (13 July 1927), p. 9.
**411**. "Einstein Advocates Resistence to War", *The New York Times*, (15 December 1930), pp. 1, 12 at 12. *See also:* "Pacifists Demand Citizenship rights", *The New York Times*, (18 January 1931), p. 34. *See also:* "Sees Lasting Peace after Another War", *The New York Times*, (4 November 1931), p. 23. *See also:* "Mme Schwimmer Calls on Holmes", *The New York Times*, (22 January 1933), p. 23. *See also:* "Dr. Einstein Urges Hitler Protests", *The New York Times*, (17 March 1933), p. 15.
**412**. "Mme. Schwimmer Gets Ford's Reply", *The New York Times*, (18 September 1927), p. 9.
**413**. "Woman Asks Ford to Vindicate Her", *The New York Times*, (4 September 1927), p. E1. "Denies Peace Ship Led to Ford Attack", *The New York Times*, (5 September 1927), p. 17. "Mme. Schwimmer Gets Ford's Reply", *The New York Times*, (18 September 1927), p. 9. "Pacifist Disavows Influencing Ford", *The New York Times*, (28 June 1928), p. 18. "Ford Exonerated Mme. Schwimmer", *The New York Times*, (29 June 1928), p. 12.

brought it about, and the conditions from which it emerged. [***] But who [***] gave all of these valuable lessons to [Ford]? As Mme. ROSIKA SCHWIMMER seemed to be at least second in command, the chances are that it was she, and the kind of instruction she would give might not have been entirely trustworthy to anybody except her dear friends the Germans."

Schwimmer fought all such accusations made against her. The issue arose again in 1927-1928, when Ford distanced himself from the articles of THE DEARBORN INDEPENDENT. When criminal Jewish leaders ganged up on Ford and attempted to assassinate him, Schwimmer filed a law suit for libel against Fred M. Marvin. Rosika Schwimmer denied that she was the source of the information and allegations published in THE DEARBORN INDEPENDENT and held that Ford was anti-Jewish before taking the voyage with her. She requested Ford's help, knowing that he had been intimidated by Jewish leaders and was vulnerable.

*The New York Times* reported in an article, "Woman Asks Ford to Vindicate Her", on 4 September 1927, page E1,

"Mme. Schwimmer went on to say that she had been the object of abuse that wrecked her health and that the damage to her reputation had been further added to when it was declared that she, a Hungarian Jewess, was responsible for the anti-Jewish campaign of Mr. Ford which he recently ended by apology."

*The New York Times* reported in an article, "Mme. Schwimmer Gets Ford's Reply", on 18 September 1927, on page 9:

"Mme. Rosika Schwimmer [***] call[ed] on [Henry Ford] to exonerate her of charges [***] that she had been the original cause of his anti-Semitic campaign[.]"

Though Ford's secretary E. G. Liebold had responded to Schwimmer's letter, Ford did not deny that Schwimmer had been the source of his information. The 18 September 1927 article continued,

"Mme. Schwimmer said [***] that she regarded the letter as a partial vindication, but that the point of the anti-Jewish campaign had not been touched[.]"

Schwimmer stated that she would write Ford again asking for, "a 'point blank denial' of the insinuations relating to Jews." Ford did not repudiate the accusation.

On 28 June 1928 *The New York Times* reported in an article, "Pacifist Disavows Influencing Ford", on page 18, quoting Joseph T. Cashman, an attorney for the defense in a libel action Schwimmer had filed against Fred M. Marvin,

"'Will you admit that it was a matter of common gossip that Mr. Ford's association with you on the peace ship was the cause of his anti-Semitic propaganda?'

'Yes,' replied Mme. Schwimmer, 'but I have published three open letters to show that Mr. Ford preached anti-Semitism before I met him.'"

Schwimmer was quoted in *The New York Times*, in an article, "Denies Peace Ship Led to Ford Attack", on 5 September 1927, on page 17:

"'At my first meeting with Mr. Ford, at his plant at Detroit, no November, 1915,' said Mme. Schwimmer, in her apartment at 2 West Eighty-third Street, 'he amazed me by suddenly declaring, 'I know who caused the war— the German-Jewish bankers.' He slapped his pocket and went on, 'I have the evidence here. Facts. I can't give them out yet because I haven't got them all. But I'll have them soon.'"

Ford did not deny this claim, but it is difficult to draw any inferences from his failure to deny it, because at this time he had recently been intimidated by an attack on his life, and a public apology bearing his name had been published repudiating THE DEARBORN INDEPENDENT articles, which "apology" was manufactured by Jewish leaders and a sycophantic Jewish agent—as will be shown later on in this text.[414] Ford had faced a long-standing libel suit from Herman Bernstein, who was represented by Louis Marshall.[415] The suit claimed damages for the allegation that Bernstein had told Ford on the peace ship that Jews ruled the world and started the First World War. Louis Marshall boasted that Ford would sign anything Marshall told him to sign. Therefore, Ford was intimidated at the time he was asked to deny that any such statements were made to him on the peace ship, but even then failed to deny it. Perhaps Ford was constrained by the settlement of the law suits he had faced.

On the other hand, though Schwimmer's claim could have been fabricated from Ford's famous interview with Henry A. Wise Wood, which played a prominent rôle in his libel trial with *The Tribune*, Schwimmer's denials become

---

**414**. *The New York Times*, (9 July 1927), pp. 9, 12; *The New York Times*, (10 July 1927), p. 12. ***See also:*** H. Ford and L. Marshall, *Statement by Henry Ford Regarding Charges Against Jews, Made in His Publications: The Dearborn independent, and a Series of Pamphlets Entitled "The International Jew," Together with an Explanatory Statement by Louis Marshall, President of the American Jewish Committee, and His Reply to Mr. Ford*, American Jewish Committee, New York, (1927). ***See also:*** *Louis Marshall: Champion of Liberty; Selected Papers and Addresses*, Volume 1, The Jewish Publication Society of America, Philadelphia, (1957), pp. 321-389. ***See also:*** M. Rosenstock, *Louis Marshall, Defender of Jewish Rights*, Chapters 5-7, Wayne State University Press, Detroit, (1965). ***See also:*** H. Bennett and P. Marcus, *We Never Called Him Henry*, Chapter 7, Gold Medal Books, Fawcett Publications, (1951), pp. 46-56. ***See also:*** H. Bennett, *True* ("Man's Magazine"), (October, 1951), p. 125.

**415**. "Editor to Sue Ford; Hires Untermyer", *The New York Times*, (9 July 1923), p. 17. "Bernstein-Ford Suit Ends", *The New York Times*, (28 July 1927), p. 21.

more plausible when one considers that Ford may have met with David Starr Jordan just before leaving New York on the Peace Ship (*Oscar II*, a Norwegian vessel).[416] Though asked to attend the voyage, Jordan's name did not appear on the ship's roster.[417]

David Starr Jordan published *Unseen Empire: A Study of the Plight of Nations that Do Not Pay Their Debts*, American Unitarian Association, Boston, (1912); which critically analyzed the power of bankers to instigate, or to prevent, wars. Jordan was concerned that war was destroying the best genetic stock of humankind. Louis Marshall speculated that Jordan may have been the cause of Ford's campaign.[418] Both Jordan and Ford were very active in the pacifist movement. If Jordan put thoughts into Ford's head, perhaps even evidentiary papers into his pocket, it would not preclude the possibility that others soon reinforced those beliefs.

Any claims that Herman Bernstein and/or Rosika Schwimmer told Henry Ford on the Peace Ship that there was a "Jewish" plan to create the war for profit, and to acquire Palestine, and that Jews effectively owned the major governments of the world and corrupted civilization with the wealth of Jewish financiers; would appear to have been contradicted not only by Rosika Schwimmer's assertion that Ford was anti-Jewish before he met her, but also by Ford's statements immediately upon his return from the Peace Ship—were it not for statements Ford made soon thereafter in an interview with Henry A. Wise Wood.

Ford was quoted in *The New York Times* on 3 January 1916 on pages 1 and 6 in an article entitled, "Henry Ford Back, Admits an Error, Denies Deserting":

### "Changes Viewpoint of War.

'A marked change has come over my whole viewpoint since I went away,' he said. 'Before going to Europe I held the view that the bankers, militarists, and munitions manufacturers were responsible. I come back with the firm belief that the people most to blame are the ones who are getting slaughtered. They have neglected to select the proper heads for their Governments—the men who would prevent such chaotic conditions. In the great majority of cases the people select their rulers and then are afraid of them. They don't write enough letters to them and let them know their views.'

Asked if he thought a republic was not a more advisable form of government than a monarchy, the pacifist replied:

---

[416]. "Mrs. Boissevain Accepts", *The New York Times*, (30 November 1915), p. 6.
[417]. "Ford's Pilgrims, Ready to Sail, in Passport Tangle", *The New York Times*, (3 December 1915), p. 1.
[418]. Letter from L. Marshall to R. Gottheil of 8 January 1913, *Louis Marshall: Champion of Liberty; Selected Papers and Addresses*, Volume 1, The Jewish Publication Society of America, Philadelphia, (1957), pp. 322-323. Letter from L. Marshall to J. Spargo of 31 December 1920, *Louis Marshall: Champion of Liberty; Selected Papers and Addresses*, Volume 1, The Jewish Publication Society of America, Philadelphia, (1957), pp. 353-355.

'Yes, I think that is so. But France is a republic, and it doesn't elect the men who would prevent the nation preparing for war. And you see where France is now. The trouble is that citizens don't take enough interest in the government. But so far as neglecting government is concerned, I am one of the worst offenders. I have been a voter for thirty-one years, and during that time I have voted but six times. Then it was because Mrs. Ford drove me to do it.

'Formerly my idea was that in this country also the men behind the campaign for preparedness were the militarists and munition manufacturers. But I find the people who don't elect the right men are the ones to blame; they should express their own minds.'

Mr. Ford was asked if he had obtained expressions of sympathy with his peace movement from officials in the countries visited, and whether he had successful relations with them. He replied that he had 'seen others just as good.'

'If necessary I will go back,' he continued, 'and, if it will help matters, I will charter another ship. I went to Europe to show that I was willing to give something more than money to the cause, and I will go again if it will do any good. My absence has not hurt this movement any more than my absence from Detroit hurt my motor company. And as fine a delegation as you could find went from Sweden to Norway.'

### 'Get the People Thinking.'

Asked what he thought was the concrete result of his expedition he said:

'It's got the people thinking, and when you get them thinking they will think right.'

As to his plans for the future, Mr. Ford said:

'I haven't started in to work yet, but I don't think it would be wise to tell you more.'

'Do the newspapers think I am doing this for self-gratification or advertisement? I feel that I am simply a custodian of the money I got together. The people who are being slaughtered helped me to get it, and what I am willing to spend for them. Anyway, I think I feel that way. I have thought of it in every way. My business doesn't need any advertising.'

Mr. Ford said that the reports of serious dissensions were not based on fact. There was much diversity of opinion, he admitted, adding: 'But you know, we took over an absolute community, and I don't think a more jolly crowd could be found in the whole world.'

Mrs. Ford and Dean Marquis were present for a part of the interview, and the Dean interrupted to explain what had been termed the squabbles.

'Being a parson, I was used to the squabbles,' he said, with a smile. 'And so I was surprised at what was published in the newspapers.'

Mr. Ford explained that he never had intended Louis Lochner to be anything except secretary, and that Gaston Plantiff was the manager. If any one did not behave, he said, Mr. Plantiff stopped the payment of bills. Mr. Ford denied that any newspaper messages had been censored. The question of preparedness arose when he was asked about the President's message, and whether, now that he was home, he intended to join with Mr. Bryan in an attack on the Wilson programme.

'I am against preparedness of any kind,' he said, 'for preparedness is surely war. No man ever armed himself even with a knife and fork unless he intended to attack something, if only an oyster or a piece of meat. The President ought to find out what the people want. If they want to arm, they know what they will get—what Europeans are getting now—a rampage some day.'"

It appeared that Ford had disavowed any belief that there were corrupt forces preparing for war for profit. However, in Ford's mind there may have been no contradiction between his belief that: newspapers, and the bankers he believed (or was led to believe on the Peace Ship) corrupted the newspapers, polluted the minds of the public; and his belief that the onus was upon the public to make better decisions when electing their government officials—and Ford was planning to provide them with what he considered to be the truth in order to aid the public in making its decisions. Note Ford's statement, "I haven't started in to work yet, but I don't think it would be wise to tell you more." Ford may already have been planning to stir things up as President, Senator, or newspaper owner. Note the *Times* text quoting Ford,

"Mr. Ford was asked if he had obtained expressions of sympathy with his peace movement from officials in the countries visited, and whether he had successful relations with them. He replied that he had 'seen others just as good.'"

Ford, who was known for making odd statements, may have been implying that there were powers behind the thrones of these governments, or that he had spoken to persons who had convinced him that the governments were corruptly controlled and that he ought to speak to the people in charge, the bankers. Ford found it fortuitous when the owner of *THE DEARBORN INDEPENDENT* came to him and asked for advice on how to sell it. When America entered the war, Ford considered it his patriotic duty to pause his pacifist activities and stand behind the President for the duration of the war. Ford waited until the end of the war to begin his campaign to expose "the Jewish Peril" to the American People.

## 5.5.4 Ford Comes Under Attack—The War Against Pacificism

Ford's long campaign for peace angered many. On 8 May 1916, Henry Ford allegedly made statements (some of which he later denied having made—in particular, Ford denied that he stated that he would remove American flags from his factories) in an interview with Henry A. Wise Wood, a vocal advocate of "preparedness", who was prejudiced against Ford's pacifism[419]—Wood was a

---

[419]. "H. A. Wise Wood Out for Col. Roosevelt", *The New York Times*, (18 may 1916),

person who Ford stated appeared to be under the control of financiers[420]:

## "A WILD MENTAL JOURNEY WITH FORD.

History Is Myth, Two Bankers Invented This War, Flags Are Fatal and Preparedness Talk Is Eastern Scare Gas.

By HENRY A. WISE WOOD.

New York, May 15, 1916.

*To the Editor of The New York Times:*

On May 8, while in Detroit for the purposes of speaking on preparedness, I spent several hours with Henry Ford. I found Mr. Ford eager to talk about national defense, but unwilling to discuss it. While volleying his assertions with great rapidity, he refused to pause long enough to permit any one of them to be examined and dealt with. To facts which I submitted he responded with a brief word of dismissal or with a sweeping denial that they were facts; sometimes with the remark that he could not consider them because he himself did not know them to be facts.

In dealing with naval and military subjects his positions seemed to be that they were to be tossed aside, because a civilian in presenting them was not to be credited, nor a professional to be trusted. Therefore they were not open to discussion. By this simple mental operation Mr. Ford shut out of the conversation all naval and military affairs. The suggestion that, because of the results of this war or the situation in Mexico, we might eventually find ourselves in international difficulties from which, owing to our weakness, we might be unable easily to extricate ourselves, Mr. Ford pooh-poohed, saying that I was 'full of Eastern scare gas.'

When in our 'discussion' of a nation's need for defensive strength history was appealed to, Mr. Ford replied that he did not believe in history, that history was of the past and had no bearing upon the present, and that, there being nothing to be learned from it, history need not be studied nor considered. The American Revolution he refused to have touched upon, saying that the Revolution was 'tradition,' that he did not believe in tradition.

Coming to Mr. Ford's beliefs, which were given in fragments, with always his refusal to support them with evidence or to permit their analytical examination, these seemed to gather about a single thought. Mr. Ford's theory of wars—he granting no exceptions—is, or was on May 8, that they are created artificially by bankers. At the moment there are two bankers, but two, he believes, who are responsible for modern wars. If these be plucked then wars in our day will cease. Mr. Ford asserts he knows who these

---

p. 5.
[420]. "Odd Definitions Given by Ford in Libel Suit", *The New York Times*, (17 July 1919), p. 2.

bankers are and that he, personally, is going to see that the 'tooth is pulled.' He would not reveal the names of these bankers, nor explain the method by which he is to pull the tooth.

Mr. Ford asserted that he has found a permanent remedy for warfare, which he refused to reveal, saying that in due time I should learn what it is. This he said he would put into effect, but seemed unable to say when. When I sought to follow up these and other assertions equally vague I was invariably met by his refusal to divulge what he had in mind; I was abjured to wait and see. One clue to his thought may be got from his reply to my likening the external need for a defensive military force to the internal need for an armed police, which was that the police needed neither their clubs nor their revolvers; that the law could be enforced without any arms. Then, in the same breath, he asked if I was a Deputy Sheriff, saying that he and all of his men were Deputy Sheriffs, and that it was my duty also to be one.

When the word patriotism was touched upon Mr. Ford burst out with the assertion that he did not believe in patriotism, that no man is patriotic, and that the word patriotism is always the last resort of a scoundrel. To my inquiry as to what he would do in the event of war he replied that even if we were to be invaded he would not make a dollar's worth of arms for the United States. As I wished that there should be no mistake as to his meaning I put the question three times, and three times got the same answer.

Finally, I said: 'Mr. Ford, on your roof are three American flags. On seeing them it hurt me to think that beneath them there was a man who is spending vast sums, amassed under their protection, to ruin the defenses of his country, and lay it open to a possibly hostile world.' To this he replied: 'When the war is over those flags shall come down, never to go up again. I don't believe in the flag; it is something to rally around.'

In commenting upon my visit The Detroit Saturday Night aptly remarks: 'Understanding Henry Ford is more than a puzzle; it is a pursuit.'
HENRY A. WISE WOOD."

Whether Ford's accusations regarding bankers were true, partially true, or not at all true, Ford had revealed himself in May of 1916 to be the active enemy of some of the most powerful persons in the world—on pacifist grounds. Powerful people often have powerful friends, especially in the press, or with access to the press, or who can intimidate the press with threat of withdrawing advertising dollars. Much earlier, Arthur Schopenhauer and then Richard Wagner expressed pacificist sentiments similar to Ford's. They accused the Jews of being warmongers and war profiteers. Schopenhauer and Wagner were not alone in this belief. Jews have always been accused of being warmongers.[421]

Ford's aggressive attacks in THE DEARBORN INDEPENDENT did not go unanswered. *The London Times* made a concerted effort to discredit the

---

**421**. "Jewish Financiers", *The Jewish Chronicle*, (29 September 1922), p. 10.

*Protocols* in 1921.[422] John Spargo; whose name appeared in numerous articles in *The New York Times* in late 1920 and early 1921 attacking H. G. Wells and redressing attacks on Jews as well as discussing Bolshevism, Russia and Poland;[423] also attacked Henry Ford. *The New York Times* reported on 6 December 1920 on page 10:

## "SPARGO DENOUNCES ANTI-SEMITIC MOVE

Calls It Menace to American Democracy and to Christian Civilization Itself.

ATTACKS FORD AS A 'TOOL'

Resents Propaganda Blaming Jews for International Socialism and Bolshevism.

The anti-Semitic movement in Great Britain and the United States was denounced by John Spargo in an address on 'Anti-Semitism; a Menace to America,' before the Brooklyn Civic Forum in Public School 84, Glenmore and Stone Avenues, last night. Mr. Spargo said this movement was not a menace to the Jew alone, but a menace to American democracy and

---

[422]. "'Jewish World Plot.'", *The London Times*, (16 August 1921), pp. 9-10. "'Jewish Peril' Exposed'"", *The London Times*, (17 August 1921), pp. 9-10. "The Protocol Forgery", *The London Times*, (18 August 1921), pp. 9-10. "The End of the 'Protocols.'"", *The London Times*, (18 August 1921), p. 9. "The 'Protocols.'", *The London Times*, (20 August 1921), p. 9. "Forged Protocols ", *The London Times*, (20 August 1921), p. 7. "The Protocols ", *The London Times*, (22 September 1921), p. 6.

[423]. H. A. Jones, "Gentle Advice to 'My Dear Wells'", *The New York Times*, (5 December 1920), p. 103. J. Spargo, "H. G. Wells Lost in the Russian Shadow; First Article", *The New York Times*, (5 December 1920), p. 102. "Spargo Denounces Anti-Semitic Move", The New York Times, (6 December 1920), p. 10. J. Spargo, "When H. G. Wells Smokes the Opium of Utopia", *The New York Times,* (12 December 1920) p. XX3. "Spargo Says Wells Blames Bolshevism for Ills of Muscovites and Much of Europe's Misery", *The New York Times*, (15 December 1920), p. 2. H. A. Jones, "Yet Once More, My Dear Wells", *The New York Times*, (2 January 1921), p. X2. H. G. Wells, "Mr. Wells on His Critics", *The New York Times*, (6 January 1921), p. 10. J. Spargo, "Spargo Answers Wells", *The New York Times*, (10 January 1921), p. 9. "Issue a Protest on Anti-Semitism", *The New York Times*, (17 January 1921), p. 10. "Spargo Condemns Racial Antagonism", *The New York Times*, (22 February 1921), p. 10. H. Bernstein, "The New Anti-Semitic Workings", *The New York Times*, (3 April 1921), p. 43. "Spargo Would Let Ford Go on Talking", *The New York Times*, (26 November 1921), p. 9. Spargo was also seen in *The New York Times* in many articles not here referenced, discussing Bolshevism, Russia and Poland.

American ideals and institutions and a menace to Christian civilization itself.

Mr. Spargo said that anti-Semitic propaganda had tried to make it appear that the Jews were responsible for the international Socialist movement and for Bolshevism, both of which he denied. 'With this sort of propaganda those interested in the anti-Semitic movement hope to turn the rest of the world against the Jews,' he said. 'As a Socialist I resent the charge that we have consciously or unconsciously been the dupes of any conspiracy for the creation of any Jewish dictatorship.'

The anti-Semitic movement has gained headway in England and is even entrenched in the lobby of the House of Commons, Mr. Spargo said. He said he did not believe it existed in this country until he returned several weeks ago and found a copy of Henry Ford's Dearborn Independent on his desk. He attacked Mr. Ford for intolerance and said he was the 'tool' in this matter of men more able than himself.

'I am not defending the Jew,' Mr. Spargo said. 'I would not insult the Jew by assuming that he needs a demended. Anti-Semitism must not succeed. We shall right it until we have beaten it to its knees. We shall fight it, not for the Jew, but for America and America's value to the civilization of mankind."

Spargo was quoted in *The New York Times* on 22 February 1921 on page 10,

## "SPARGO CONDEMNS RACIAL ANTAGONISM

Denounces Propaganda of Anti-
Semitism as Treason
to America.

### ONLY PITY FOR HENRY FORD

Calls Him Poverty-Stricken Intellectually,
Morally and Spiritually
—Addresses Chicago Audience.

*Special to The New York Times.*

CHICAGO, Feb. 21.—John Spargo, Socialist author and formerly of the Industrial Relations Commission, spoke before 5,000 Chicago Jews at Sinai Temple tonight on 'The Jews and the American Ideal.' In referring to recent attacks on the Jewish race, Mr. Spargo said:

'Henry Ford is poverty-stricken intellectually, morally and spiritually. I regard him with profound and unmeasured pity. No more pitiful figure can be found in our history. With all his material wealth, he is poorer than the poorest wretch to be found in the bread lines of this city. His poverty of soul is so great that he is incapable of partaking of the American spirit.'

Mr. Spargo began his address by explaining that he was not a Jew and had investigated the anti-Semitic campaign because he felt that it was a monstrous thing which should be exposed. He sketched the history of Jewish immigration into this country, and maintained that the Jews had at no time been outranked by any other element of the citizenship in loyalty to American ideals. He continued:

'Yet we are witnessing the shameful spectacle of an organized campaign of hatred and calumny against the Jews of America, a campaign having for its object the creation of a terrible and dangerous antagonism between Americans, and antagonism founded upon racial and religious differences. Such a campaign cannot be accurately described as other than foul treason to America and a dangerous desecration of American ideals. It is not necessary to stigmatize that campaign; it is quite sufficient to describe it accurately for what it is. In prosecuting that campaign its leaders have not hesitated to seize upon the occasion of the anniversary of Lincoln's birth to besmirch his resplendent fame and glorious memory. Instead of seeing in the war of secession the result of a conflict of economic and political systems, these men—alien to America in soul if not in speech—have spread broadcast through the land the infamous charge that the fateful struggle was deliberately brought about by Jewish agents intriguing for the accomplishment of Jewish purposes.

'I do not insult my Jewish fellow-citizens by pretending to believe that this fantastic charge needs refutation. I refer to it only that I may voice my indignant protest against the infamous insult thus heaped upon the name and memory of Abraham Lincoln. If the charge were true, he whom we have loved to honor as the noblest and fairest exemplar of American ideals would have to be regarded either as a deliberate traitor compared to whom Benedict Arnold was a very patron saint of patriotism and loyalty, or as a poor silly dupe of others, a mere moron in fact. And whichever of these verdicts was rendered against Lincoln would have to be rendered against Seward and Chase and Welle and the rest of his advisers. No foul slander of America that emanated from the gutter press of Berlin during the war matched the infamy of this.

### Pity for Henry Ford.

'I do not abuse or condemn Mr. Henry Ford here today. On the contrary, I regard him with profound and unmeasured pity. No more pitiable figure can be found in our history. With all his material wealth he is poorer than the poorest wretch to be found in the bread line of this city. His poverty of soul is so great that he is incapable of partaking of the American spirit. He is poverty-stricken intellectually, morally and spiritually. I would rather be starving so that I envied the dogs their crusts, and homeless so that I envied the very rats their holes, but with an understanding love of American ideals in my heart, than be the responsible owner of The Dearborn Independent.

'In its attempts to poison the well-springs of American faith and inspiration The Dearborn Independent has retrieved from the sewers of the reactionary politics of Europe the so-called Protocols of the Wise Men of Zion. It professes that in publishing and distributing widely this notorious forgery it has only a patriotic motive, and that it is no part of its purpose to promote that hideous evil which we unscientifically call anti-Semitism, that evil of prejudice and hatred against the Jew as Jew. So professed the Bessarabetz of Kishinev, but pogroms resulted from its propaganda nevertheless. The success of the indecent and traitorous campaign of The Dearborn Independent would mean pogroms against the Jews in America, let there be no mistake upon that point. Fortunately, there is no likelihood of that success occurring, for the good sense of the gentle population of America is a bulwark against which the prostituted hirelings of the ignorant man of millions will spend themselves in vain. We shall beat anti-Semitism to its knees and crush it, because it is a menace to the America we love and an affront to everything in which we take pride.

### History of the Protocols.

'As many of you are aware, I have taken great pains to trace the origin and history of the so-called 'Protocols.' There is not the slightest doubt in my mind that they were deliberately concocted in the headquarters of the old secret police of Russia under Czarism as one of the means of combating the great struggle for democracy and self-government. This is made evident by the testimony of no less a person than the mysterious Nilus, reputed author of the book in which the protocols were first given to the world. Nobody has been able to produce this mythical personage; no responsible person has been found to testify to the actuality of his existence. If he could only be found and placed upon the witness stand and cross-examined, what a sight it would be for gods and men!

'In 1903 the first edition of a little book bearing his name appeared, a diatribe of such fanatical mysticism as Rasputin, of malodorous memory, might have written. In that book, despite its anti-Semitism, there was no reference to the protocols. In 1905 a second edition appeared containing the protocols. In that edition he tells us that the protocols came into his possession in 1901. He offered no explanation of his failure to use them or even to mention them in the first edition of his book in 1903, though they served his purpose so wonderfully well and had been in his possession for two years prior to its publication. I know the reason and will presently explain it to you.

'In that edition of 1905 Nilus told how the protocols came into his possession. He said that the protocols had been stolen by a woman from 'a highly initiated Freemason.' He said that the protocols were signed by representatives of the Thirty-third Degree of the Masonic Order of Zion. The name of the Freemason from whom the documents had been stolen was not given: the name of the woman thief was not given: the names of the Freemasons who signed them were not given. Not so much as a facsimile of a single page was offered as evidence of the authenticity of the documents. Indeed, Nilus naively admitted that he never saw the originals; that what had been handed to him was a manuscript purporting to be an 'authentic translation' of the documents stolen by the woman from the careless Freemason conspirator — evidently in some Swiss cabaret where the wine flowed freely. On the basis of such a flimsy story as that no judge or jury in the United States would convict a pickpocket. Yet The Dearborn Independent would convict three millions of our citizens of treachery to this republic upon that testimony.

'In 1917 appeared a new edition of the protocols, with a new introduction by the mysterious Nilus. Keep the date well in mind, together with that of the first publication of the protocols in 1905, for the dates are of the utmost significance. In this edition Nilus says of the protocols: 'This manuscript was called, 'The Protocols of the Zionist Men of Wisdom,' and it was given to me by the now deceased leader of the Tshernigov nobility, who later became Vice-Governor of Stavropol, Alexis Nicholaievich Sukhotin. I had already begun to work with my pen for the glory of the Lord, and I was friendly with Sukhotin because he was a man of my opinion, i. e., extremely conservative, as they are now termed.

'Sukhotin told me that he in turn had obtained the manuscript from a lady who always lived abroad. This lady was a noblewoman from Tshernigov. He mentioned her by name, but I have forgotten it. He said that she obtained it in some mysterious way, by theft, I believe. Sukhotin also said the one copy of the manuscript was given by this lady to Sipiagin, then Minister of the Interior, upon her return from abroad, and that Sipiagin was subsequently killed.

## Evidence Against Nilus.

'This story comes pretty close to convicting Nilus of being an agent of the Czar's Secret Police. Sukhotin, from whom he claims to have obtained the manuscript, was a notorious anti-Semite and leader of the Black Hundreds. Sipiagin, who is mentioned as having also had a copy of the manuscript, was also a bitter anti-Semite and one of the most infamous of the late Czar's bureaucrats. He was assassinated by Stephen Balmashev in 1902. Thus, if this story is true, Nilus is linked up in a very definite way with the secret agencies of the old regime. At the same time, it is worth while noting that Nilus names Sukhotin and Sipiagin only when they are dead and beyond questioning. He presents no evidence to substantiate his tale. He has 'forgotten' the name of the 'noblewoman from Tshernigov.' Criminologists would deduce from these two stories that the author belongs to a well-known criminal type.

'Let me call your attention to two interesting facts in connection with this story of 1917. The first is that Nilus omits all reference to his previous statement that the protocols were 'signed by representatives of Zion of the thirty-third degree.' The second is that having told us in 1905 that the friend who gave him the protocols in 1901 assured him that they had been 'stolen by a woman,' and told us in the introduction of 1917 that the friend from whom he received the documents was Sukhotin, who told him the name of the woman thief, which, however, he managed to forget, he adds an epilogue to the story in which he tells us that the protocols were actually stolen, not by a woman at all, but by Sukhotin himself! And that instead of having been stolen by a woman from a careless Freemason, Sukhotin stole them from a safe in Paris. His words are that the protocols 'were stealthily removed from a large book of notes on lectures' and that 'my friend found them in the safe of the headquarters office of the Society of Zion, which is situated at present in France.'

'Was ever liar more confused? First we have an unknown woman stealing the documents from 'one of the most highly initiated leaders of Freemasonry; next we have the documents presented as having been obtained by Sukhotin from a 'noblewoman from Tschernigov' whose name Nilus has forgotten; finally, we have this friend—i. e., Sukhotin—named as the thief. The woman thief disappears and the 'highly initiated Freemason' disappears. It is Sukhotin who is the thief, and he steals the protocols from a safe in Paris. So much for Nilus. I may add that I am assured—though I cannot vouch for the statement—that Sukhotin was not outside of Russia between 1890 and 1905.

'And now let me explain the significance of the dates of publication to which I have already referred: When the first publication of the protocols took place, in 1905, Russia was seething with revolution. When the second publication took place, in January, 1917, Russia was again seething with revolution. No one who is familiar with the history and practices of the Russian secret police and the Black Hundreds can have the slightest doubt that the publication of the protocols was in each case designed to create anti-Jewish uprisings to divert the minds of the Russian people from revolutionary agitation. That was a familiar method of the Czarist police and Black Hundreds. It was a backfire.

### Suppression of Evidence Charged.

'This then is the history of the protocols, a history of indecent forgery by the unscrupulous, conscienceless agents of Russian Czarism. It is upon materials so rotten and reeking with dishonor that this elaborate campaign is erected. I regret to have to say that those who are responsible for the publication and distribution of the protocols in this country—which includes not only Mr. Ford's paper, but publishing firms hitherto regarded as reputable—have been guilty of conduct as dishonest and dishonorable as the original concoctors of the protocols themselves. They have suppressed, deliberately and without the slightest explanation to the reader, passages from the original Russian publication of the protocols which would have made them the laughing stock of the English-speaking world.

'In 1895 a book was published in France which attempted to prove the existence of a world-wide conspiracy against Christian civilization. In that book the theory was advanced that the English people are all of the Jewish race, and that the British Government is the central force of this worldwide Jewish conspiracy. In his book Nilus reproduced this fantastic theory but, recognizing that it would cause the protocols to be laughed out of court, The Dearborn Independent, The London Morning Post and all the other publishers of the protocols in England and America have carefully deleted this part of the book by Nilus. The reason for the deletion is as obvious as the dishonor of it.

'Upon the strength of statements made in the protocols, The Dearborn Independent, The London Morning Post, and other organs of anti-Semitism have charged that the international Socialist movement is part and parcel of this vast conspiracy of Jewish world imperialism. Neither in the protocols themselves nor in any of the numerous comments upon them has any shred of evidence been adduced in support of this charge. As one who has given practically all his life to the Socialist cause, I indignantly repudiate the charge that I have either consciously served such a conspiracy or been ignorantly duped by it.

'The ignorance of Henry Ford upon all that pertains to American history is a matter of court record, and needs no demonstration here and now. Were he less ignorant of history, he would know that the charge thus leveled against the Socialist movement has been leveled against almost every great modern movement of protest. It was made against the Protestant Reformation, against the French Revolution, against Mazzini and his followers in Italy, against the German revolutionists of 1848, against trade unionism in England. Whether socialism is right or wrong, desirable or undesirable, is a question upon which honest men and woman may differ. It is a question to be answered upon its own merits in the American way. Whoever injects into the discussion of that question the passion engendered by racial and religious prejudices and hatreds is unworthy of America. He who propagates in this country antagonism to any race or creed represented in our citizenship, whether it be against Jews, Poles, Germans, Irish, English or negroes; or against Judaism, Catholicism or Protestantism, assails the very foundations of our most cherished and characteristic American institutions.

### 'Majority of Bolsheviki Not Jews.'

'The Dearborn Independent, like all the rest of the anti-Semitic press of both hemispheres, charges that Bolshevism in Russia and elsewhere is a movement instigated and led by Jews as part of the conspiracy to bring about the Jewish domination of the world. In support of this charge, the protocols are offered in evidence. The reasons for making the charge are quite obvious—Bolshevism is repugnant to the moral sense of the great mass of civilized mankind. It is the negation of virtuous principals which the enlightened of all races and all religions hold in reverence. It denies the ideal of government based upon the sanction of the governed and accepts that of government by brute force wielded by a few. To persuade the people of America that Bolshevism is essentially a Jewish movement, part of a conspiracy to reduce civilization to chaos and to prepare the way for a Jewish super-government of the world, would mean the uniting of all the rest of our population against the Jews. That is the object.

'In support of this most serious charge not a scintilla of credible evidence has been offered. It is true, of course, that there are Jews among the Bolsheviki in Russia, but it is equally true that the overwhelming majority of the Bolsheviki are not Jews, either racially or by religious faith and affiliation. It is also equally true that the anti-Bolshevist movement in Russia, that heroic struggle of democracy against an unspeakably brutal despotism, is very largely carried on by Jews.'

Mr. Spargo contradicted the statement of The Dearborn Independent that 'every commissar in Russia today is a Jew.' Enumerating Lenin, Tchitcherin, Krassin, Dzerzhinsky, Umarcharsky, Rykov, Kolontal, Borch-Brouyevich as non-Jews, he went on to assert that of the seventeen members of the Council of Peoples' Commissars only one, Trotsky, was a Jew, and while there were many Jews holding minor places in the Bolshevist régime, there were also serving in it many ex-officers of the Czar's army who were of Christian faith and for the same reason—because 'what else could they do?'

He went on to point out that Bolshevism was the negation of the faith and morals which constitute the strongest bond of the Jewish people, and cited the fact that the use of the Hebrew language had been prohibited under the Soviet, adding:

'There is not a single Jew connected with the Bolshevist movement in Russia in any prominent capacity who is not an apostate, having renounced all the faith and ties of Israel. There is not one of them who ever took the slightest part in the affairs of the Russian Jewry. As against this mere handful of apostate Jews, for every one of whom there are a hundred non-Jews among the Bolsheviki, we have the many millions of the Jewish population of Russia who are the innocent victims of Bolshevism. Hundreds of thousands of Jewish merchants and small business men, comprising a large part of the hated and persecuted bourgeoisie, have been ruined by the Bolsheviki, thousands of Jewish families have been deported from Soviet Russia, and are now dragging out a miserable existence as refugees in Siberia and elsewhere. Billions of Jewish wealth have been confiscated by the Bolsheviki. The Soviet Government has shot and is still shooting Jewish public men, lawyers, engineers, physicians, teachers and workmen, for participation in the struggle against Bolshevism. In view of these facts is it less than ridiculous to charge that Bolshevism is part of a Jewish conspiracy? Surely any intelligent person must see that the only hope for the success of any such conspiracy must lie in maintaining a Jewish solidarity in Russia which could only be attained, if at all, by devising some means of exempting the Jews from the suffering and oppression imposed upon the non-Jewish population.

'For the problems which arise from the presence in the same land of Jews and non-Jews, in large masses, solution must be sought and found by the best and ablest minds, Jewish and non-Jewish, working together in earnest co-operation, united by love of America and loyalty to its ideals and institutions. Because anti-Semitism makes that impossible, and thereby prevents the peaceful, wise and speedy solution of these difficult problems, I denounce it as treason to America and all that America stands for in our affections.'"

*The New York Times* reported on 26 November 1921 on page 9,

## "SPARGO WOULD LET FORD GO ON TALKING

Invite Him Here to Tell Why He
Opposes the Jews, Lecturer
Tells Audience.

SEES ORGANIZED CAMPAIGN

Socialist Author Says It Is Part of
International System With
Headquarters in Berlin.

Speaking on 'The Anti-Semitic Spirit in America,' at a meeting of the League for Political Education in the Town Hall yesterday morning, John Spargo, Socialist author and lecturer, said there was a campaign of organized anti-Semitism in this country which was part of an international system, with headquarters in Berlin, in so far as he was able to learn. It was not the business of the Jew as such but the duty of Jew and Gentile to combat this prejudice, he said. The situation called for diligence by the Christian in exposing the fallacies of the propaganda because he owed to the Jew precisely that measure of justice he would want to be shown to others who come to America to make their homes, Mr. Spargo argued.

Mr. Spargo reviewed the race prejudices which had existed in America in other years, and in his analysis of them said: 'It is always difficult to avoid suspicion of the different groups we have drawn from other countries where there has been a barrier of language, creed or customs.'

At the close of his address Mr. Spargo answered questions from the audience. One person asked what should be done with Henry Ford.

'Leave him alone,' replied Mr. Spargo, 'let him talk. Invite him to the Town Hall and let him tell you why he is opposed to the Jews, if he will.'

On the main topic of his lecture, Mr. Spargo said:

### The Jews and Columbus.

'We have always had the Jew with us, because essentially he is a wanderer. In years gone by we had the Jew only in numbers capable of assimilation. There were Jews interested in the voyage of Columbus, if we are to believe history. Certain there were Jews interested in the American Revolution. Washington knew several on whom he could depend and whose fortunes were at his disposal.

'It is a good thing to remember that there never was any time in the history of the country when it was possible to distinguish a citizen of Jewish birth from a citizen of non-Jewish birth. I say that, bearing especially in mind the accusation made against the attitude of the Jew in the great World

War. I went with Premier Clemenceau to visit the wounded of our men and one could distinguish no distinction of service to our country among them.

'We forget that the Jew comes to us virtually helpless. He doesn't speak our language; he doesn't understand our laws and customs. How is he going to know? He takes up his home among his own people who have preceded him. If he becomes successful and learns the ways of America he is likely to move elsewhere. Your task and mine is to see that in the administration of cities we do not permit our politicians to take advantage of the temporary condition of the peoples evolving into American citizens.'

Mr. Spargo dwelt on some of the hopeful signs of amicable relations among the people of America, in telling of Thanksgiving service in which Jews and Christians took part.

Taking up the existence of anti-Semitism in American, as already told, Mr. Spargo also said:

'I dislike to hear of Jewish organizations going to court for injunctions against Henry Ford and his Dearborn Independent. We cannot save ourselves from anti-Semitism by suppressing free speech. The only safe thing for Jew and Gentile to do is to let them come out in the open and not compel them to operate in subterranean channels.

**Pamphlets from Germany.**

'A few days ago a man came to New York from Yokohoma by way of San Francisco. He was introduced to a friend of mine to whom he said, 'See what I have come to do.' He exhibited pamphlets printed in most of the modern languages accusing the Jews of most every untoward event that has ever happened. He admitted that he had brought the pamphlets here for distribution. The pamphlets were printed in Yokohoma through funds provided by monarchist groups in Germany.

'This group desires the restoration of the old régime in Germany and Russia. If they are to succeed in Russia by a coup d'etat they must turn the peasant Russian men and women against those in authority. Nobody has suffered under Bolshevist rule quite as hard as the Jews, for they belonged to the small trading class which those now in authority set out to destroy. It is a libel against the Jews and a treason against America when people try to foster hatred because of what the Bolshevists did in Russia.'

'You and I as Americans worthy of Washington, Lincoln and Roosevelt must set ourselves against this attempt to divide our citizenry along the lines of religious and racial hatred. Let it go out to the world that every manifestation of this evil spirit will be deemed treason.'"

Spargo's efforts to discredit Ford and the *Protocols* were not very successful, and there are many reasons why he failed to achieve his aim. It must be borne in mind that Spargo's emotional flag-waving appeals to patriotism and his desire to link Henry Ford's activities to Germany came soon after the end of World War I, and many Americans had come to hate Germany. As a Marxist, Spargo was well aware of the value of "false consciousness" in appealing to the emotions of the public in order to avoid legitimate accusations of corruption.

Americans knew that corruption was rampant and Spargo should have made a less shallow, more substantive appeal to the public. Spargo should have recognized that Ford expressed legitimate concerns about the corruption that was occurring, and Spargo should have distinguished the criminal actions of the few, from the innocence of the many, and joined Ford in condemning the corruption, while chastising him for his overly general attacks on Jews.

John Spargo was long a socialist revolutionary, which put him in close company with many Jews. Spargo protested a little too loudly that he was not an apologist for the Jews, which revealed that he was not only an apologist, but a hypocrite as well—a man who could not be trusted.

It is interesting to note that Spargo places great emphasis on the dates of 1905 and 1917, but does not address the Jewish bankers' deliberate destruction of the Russian economy, their financing of the Japanese war against Russia and concurrent collusion to bankrupt Russia, their distribution of revolutionary propaganda to the Russian Army, and their funding of Bolshevik revolutionaries—all of which gave the Czar just cause to fight back against these Jewish bankers' war against him and the Russian State.

We know that the Jewish bankers attacked the Czar and the Russian people, because Jacob Schiff, a German Jewish financier who had emigrated to America and who headed the banking house of Kuhn, Loeb & Co., bragged that he had destroyed Russia, in *The New York Times* in 1917. These facts were well-known at the time.

Instead of simply making an *ad hominem* attack on Henry Ford, Spargo could have taken the opportunity to point out the injustice of generalizing the behavior of a few to the many innocent; and at the same have criticized Jacob Schiff's attack on the Russian People, which ultimately led to mass murder and countless other Bolshevist atrocities. Spargo did not mention the fact that Nilus complained in his book of 1905 that his earlier attempts to make the *Protocols* widely known were unsuccessful. Nilus only succeeded in popularizing the *Protocols* after events had fulfilled the plans set forth in the *Protocols*. Spargo was mistaken to believe that the *Protocols* appeared for the first time in 1905.

A more honest inquiry into the facts might have more successfully combated the harm the exposure of the *Protocols* caused to many innocent Jews. Instead of addressing the issues which were known to anyone who had read THE DEARBORN INDEPENDENT, Spargo largely relied upon personal attack to discredit people and alleged that there was a vast conspiracy to deceive the public with lies, allegations he tried to magically wave away with the American flag. It did not work. It was a poor attempt and both Spargo and THE DEARBORN INDEPENDENT played into the hands of the political Zionists, who savored the rapid rise in political anti-Semitism.

Spargo pointed to contradictions in the allegations that Jews were behind Bolshevism, whilst Jews suffered along side Gentiles from Bolshevik atrocities. He was right to assert that the majority of Jews were not Bolsheviks and that Jews could not be classified so narrowly by a single political stance. However, political Zionists saw the emancipation of the Jews of Russia by the Russian Revolution, which was soon taken over by the Bolsheviks, as a threat to the supposed purity of the Jewish "race". The Zionists had an incentive to attack Jews and cause their concentration and deportation in Bolshevik dominated lands, because the Zionists believed that Bolshevism potentially provided Jews with a sanctuary, which would result in assimilation that would be fatal to the "Jewish race". The Zionists and their anti-Semitic allies issued an international threat, that if the governments of the world failed to sponsor Zionism, all nations would suffer the terror of Bolshevism. The political Zionists viewed the anti-Semitism the terrors of Bolshevism provoked as a positive force which helped the Zionists to keep the Jews segregated against their will. Jews were indeed behind Bolshevism and it provided them with a means to oppress Gentiles and Jews in a way that would force segregation.

It was irrational to assert that the *Protocols* were the product of vast anti-Semitic conspiracy, and to concomitantly argue that the *Protocols* were forgeries on their face because they alleged a vast conspiracy of Jewish forces. Why was it that Gentiles were allegedly capable of conspiracies, but Jews were not? Such an argument left the public with no choice but to choose between two conspiracy theories. Many people decided that if this is way of the world, they had better side with their own kind. Most people were Gentiles.

John Spargo failed to note the fact that the United States Government took an active interest in the *Protocols* long before Henry Ford learned of their existence and the U. S. Government took the *Protocols* very seriously, because it believed that many of the events foretold in the *Protocols* had since come to pass, and that the world was in danger. The fantastic nature of the *Protocols*, which makes them appear to be fabrications on their face, is what convinced so many of their authenticity when actual events mirrored those foretold in its pages—for how else could anyone have known that such unprecedented things would come to pass, unless someone had planned them? Many asked, "Even if forgeries, forgeries of what?" It was difficult for many to believe that the *Protocols* were simply fabrications with no basis in fact. Though Spargo focused on discrediting Nilus, later attempts to debunk the *Protocols* considered Nilus to be an honest man who was duped by the Czar's secret police.

Henry Ford stated that he would not be persuaded to change his mind about the facts by emotional attacks aimed at discrediting him and the sources of his information, but which avoided addressing the indisputable factual record of events and published statements. Ford claimed that he was not motivated by prejudice and that should anyone be able to disprove the underlying facts and circumstances alleged in THE DEARBORN INDEPENDENT, or to discredit the logic used to draw the conclusions which were there drawn, he would disavow those contentions.[424]

John Spargo was initially a vocal and dogmatic advocate of Marxism. Spargo wrote *Karl Marx: His Life and Work*, B.W. Huebsch, New York, (1910), National Labour Press, Manchester, (1910); as well as many other books

---

[424]. H. Ford, *My Life and Work*, Doubleday, Page & Company, New York, (1923), pp. 250-252.

advocating Socialism and Marxism. He described himself as being far redder than the pink H. G. Wells.[425] Like "Colonel" Edward Mandell House, he publicly advocated many much needed social reforms and strongly supported women's rights. However, he did this as a means to gain the public's trust and later revealed that his true objective had always been to tear down society and make life unbearable for people so as to force them into revolution.[426]

While other Socialists went to prison for protesting the war that they alleged was fought not for the people of America but for the wealthy elites, Spargo took a turn to the right in 1917—the year the Zionists turned on Germany and brought America into the war—and began to support American intervention in the First World War on the side of the British. Though a member of the Socialist Party of the United States, he abandoned the Party in 1917, because it opposed American intervention in the European war. Spargo wrote in 1929, "I resigned from the Socialist Party, in 1917, because of the adoption by it of a policy of active opposition to the war."[427] He gave no reason for this move other than to say that the anti-war policy was "shameful", "stupid" and "thoroughly bad"—why he deemed it so, Spargo did not say—again he waved his hands and hoped the show was enough to end an argument.

Spargo's public statements were often emotional, not logical. He may simply have been a supporter of Zionism, which movement led America into the war. He may have felt a loyalty to England—he originally hailed from Great Britain. In any event, his primary interests were not those of the American proletariat. Some believed he was a crypto-Jew and he had a very Jewish appearance, though he asserted that he was a Gentile.

While Spargo began to support the war, most Socialists in America vocally opposed the "Imperialists' War". In order to suppress any expression of anti-war sentiment, President Wilson passed the Espionage Act, the Sabotage Act and the Sedition Act, which restricted free speech. Socialists were prosecuted under these laws, which obviously violated the First Amendment, despite the fact that Wilson's Supreme Court upheld their alleged constitutionality. Socialist leaders like Eugene V. Debs were sentenced to long prison terms under these illegal Acts. Others, like Emma Goldman, were deported to Russia.

Emma Goldman was a Russian Jew who had emigrated to America, where she agitated for anarchy and assassination. Her lectures inspired Leon F.

---

**425**. J. Spargo, "H. G. Wells Lost in the Russian Shadow; First Article", *The New York Times*, (5 December 1920), p. 102.
**426**. J. Spargo, "Why I Am No Longer a Socialist", *Nation's Business*, Volume 17, (February, 1929), pp. 15-17, 96, 98, 100; (March, 1929), pp. 29-31, 168, 170. Reprinted: *Why I Am No Longer a Socialist*, Chamber of Commerce of the United States, Washington, D.C., (1929).
**427**. J. Spargo, "Why I Am No Longer a Socialist", *Nation's Business*, Volume 17, (February, 1929), pp. 15-17, 96, 98, 100; (March, 1929), pp. 29-31, 168, 170; at page 98 of the February issue. Reprinted: *Why I Am No Longer a Socialist*, Chamber of Commerce of the United States, Washington, D.C., (1929).

Czolgosz to assassinate President McKinley. She disseminated Frankist [428] Nihilism in the United States. Her lectures discussed the sterilization of criminals, the alleged need for woman to not have children, the alleged need to end patriotism, the alleged need to destroy all government and the alleged need to destroy Christianity. She later agitated for Bolshevism in the United States. Bolshevism fell out of favor with Western Jews after the war when it became apparent that it did indeed lead to assimilation. When Goldman was deported back to Russia, she claimed that she had become disenchanted with the Bolshevik movement and with the tyrant Lenin's oppression of free speech.[429] She ended her years in luxury sponsored by the patronage of the immensely wealthy heiress Peggy Guggenheim. The rejection of Bolshevik brutality and the disenchantment of many Russians who had lived through the Revolution in Russia is captured in Alexander Blok's poem *The Twelve*. For Communists, Liberalism was only a means to attract initiates. They had no real desire to liberate the working class. Their desire was to destroy. Emma Goldman admitted that she had always known that Marxism would lead to tyranny. John Clayton quoted Emma Goldman in The Chicago Tribune on 18 June 1920 on the front page, in an article entitled, "Russian Soviet 'Rotten,' Emma Goldman Says",

> "'You're right, it is rotten,' she said. 'But it is what we should have expected. We always knew the Marxian theory was impossible, a breeder of tyranny. We blinded ourselves to its faults in America because we believed it might accomplish something."

"Big" Bill Haywood was sentenced to twenty years in Federal prison for encouraging workers to strike during the war. Robert Goldstein was sentenced to ten years in Federal prison for a making a movie about the American Revolution, *The Spirit of '76*, which depicted British soldiers firing upon Americans. Since Britain was our ally in the First World War, the Government

---

**428**. *See:* "Frank, Jacob, and the Frankists", *Encyclopaedia Judaica*, Volume 7 Fr-Ha, Encyclopaedia Judaica, Jerusalem, The Macmillan Company, New York, (1971), cols. 55-71. *See also:* G. Scholem, "The Holiness of Sin", *Commentary* (American Jewish Committee), Volume 51, Number 1, (January, 1971), pp. 41-70; reprinted: G. Scholem, "Redemption Through Sin", *The Messianic Idea in Judaism and Other Essays on Jewish Spirituality*, Schocken Books, New York, (1971), pp. 78-141. *See also:* G. Scholem, "The Holiness of Sin", *Commentary* (American Jewish Committee), Volume 51, Number 1, (January, 1971), pp. 41-70; reprinted: G. Scholem, "Redemption Through Sin", *The Messianic Idea in Judaism and Other Essays on Jewish Spirituality*, Schocken Books, New York, (1971), pp. 78-141; **and** *Sabbatai Sevi: The Mystical Messiah, 1626-1676*, Princeton University Press, (1973); **and** *Kabbalah*, New American Library, New York. *See also:* Rabbi M. S. Antelman, *To Eliminate the Opiate*, Volume 1, Chapter 10, Zahavia, New York, (1974). *See also:* H. Graetz, *Popular History of the Jews*, Volume 5, Fifth Edition, Hebrew publishing Company, New York, (1937), pp. 245-259.
**429**. E. Goldman, *My Disillusionment in Russia*, Doubleday, Page & Co., Garden City, New York, (1923); **and** *My Further Disillusionment in Russia*, Doubleday, Page & Co., Garden City, New York, (1924).

held that Goldstein's historically correct film was against the law. Goldstein, a man who exhibited great strength of character and the finest of American values, spent three years in prison and his career was destroyed. Pacifists like Henry Ford faced Federal Criminal prosecution if they continued to speak out against the war. Crypto-Jewish Communists/Socialists; including "Miss Rose Pastor", a Russian Jew, and Morris Hillquit, born Moses Hillkowitz in Riga, Latvia; were also prosecuted.[430] Note that most Americans were pro-German and anti-British, given the England was America's most common enemy in war, until Zionist propagandists turned America against Germany with lies and unconstitutional laws which made it illegal to be pro-German, Zionist laws which made it illegal to be honest.

Woodrow Wilson's actions were seemingly inexplicable, given that Wilson was long a pacifist, as was his first wife Ellen Axzon, who died on 6 August 1914. Wilson's Secretary of State, William Jennings Bryan, was also a pacificist, and he advocated American neutrality. Bryan helped Wilson to win his Presidential election. Wilson betrayed Bryan pacifism and his long terms efforts to prevent the Rothschilds from gaining control over America's money. On 9 July 1896 William Jennings Bryan gave a speech before the Democratic National Convention while running for President. He opposed the Jewish bankers who wanted control over America's money and spoke in expressly Christian term's,

> "No, my friends, that will never be the verdict of our people. Therefore, we care not upon what lines the battle is fought. If they say bimetallism is good, but that we cannot have it until other nations help us, we reply that, instead of having a gold standard because England has, we will restore bimetallism, and then let England have bimetallism because the United States has it. If they dare to come out in the open field and defend the gold standard as a good thing, we will fight them to the uttermost. Having behind us the producing masses of this nation and the world, supported by the commercial interests, the laboring interests and the toilers everywhere, we will answer their demand for a gold standard by saying to them: You shall not press down upon the brow of labor this crown of thorns, you shall not crucify mankind upon a cross of gold."[431]

Wilson's second wife, Edith Bolling Galt, whom he married on 18 December 1915, was a strong interventionalist. Wilson's friends won the Balfour Declaration and made great fortunes from the wars Wilson conducted[432]—wars anticipated in Zionist "Colonel" House's book *Philip Dru: Administrator*, B. W. Huebsch, New York, (1912). "Colonel" House was the Zionist agent who ran

---

**430**. B. J. Hendrick, "Radicalism among the Polish Jews", *The World's Work*, Volume 44, Number 6, (April, 1923), pp. 591-601, at 597.
**431**. W. J. Bryan, "Cross of Gold" quoted in: R. F. Reid, *Three Centuries of American Rhetorical Discourse*, Waveland Press, Prospect Heights, Illinois, (1988), pp. 601-606. A. C. Baird, *American Public Address*, McGraw Hill, New York, (1956), pp. 194-200.
**432**. S. D. Butler, *War Is a Racket*, Round Table Press, New York, (1935).

the Wilson administration.

Silas Bent published a review of the books *The Life of Woodrow Wilson*[433] by Josephus Daniels and *The True Story of Woodrow Wilson*[434] by David Lawrence under the caption "Career of the Creator of 'International Conscience'" in *The New York Times Book Review* 22 June 1924 on page 3, in which Bent wrote, among other things,

> "Mr. Lawrence quotes [President Woodrow Wilson] as calling the Colonel 'a monumental faker.' That was in private conversation. Mr. Wilson did not reply to his predecessor's attacks on him as a candidate.
>
> To Colonel E. M. House Mr. Lawrence gives credit for influence in naming the greater part of the first Wilson Cabinet. Mr. Daniels mentions Colonel House only in reference to the appointment of Albert. S. Burleson as Postmaster General. It was Colonel House, so Mr. Lawrence says, who first interested Mr. Wilson in banking reform. It was Colonel House who made a trip to Wall Street before the inauguration and reassured the most powerful bankers in this country about Mr. Wilson's views, telling them his intentions toward business and finance, so as to avert a threatened panic.
>
> The second Mrs. Wilson, according to Mr. Lawrence, was chiefly responsible for the break between her husband and Colonel House. She exercised an extraordinary influence and thought the Colonel was too much in evidence at Versailles. It was she, according to the same writer, who caused the break with Secretary Tumulty; but some of those who read Mr. Tulmuty's about himself and the President regarded that as abundant provocation."

Wilson, himself, stated in a campaign speech before he was elected for his first term as President,

> "Since I entered politics, I have chiefly had men's views confided to me privately. Some of the biggest men in the United States, in the field of commerce and manufacture, are afraid of somebody, are afraid of something. They know that there is a power somewhere so organized, so subtle, so watchful, so interlocked, so complete, so pervasive, that they had better not speak above their breath when they speak in condemnation of it."[435]

After the war, Spargo wrote numerous books and articles condemning the

---

[433]. J. Daniels, *The Life of Woodrow Wilson, 1856-1924*, The John C. Winston Company, Philadelphia, (1924).

[434]. D. Lawrence, *The True Story of Woodrow Wilson*, George H. Doran Company, New York, (1924).

[435]. W. Wilson, *The New Freedom: A Call for the Emancipation of the Generous Energies of a People*, Doubleday, Page & Company, Garden City, New York, (1913), pp. 13-14.

Bolshevists in Russia. Former pacifist and Marxist John Spargo was not alone in his post-war attacks on Bolshevism. Many Zionists were concerned that Bolshevism was leading to assimilation—and many Zionists like Einstein and Weizmann resented Rathenau for his assimilationist views. In Germany, vitriolic anti-Semite Theodor Fritsch alleged in 1922 that,

> "The Soviet government boasts in its own newspapers that since 1917 no fewer than 1,764,875 people have been slaughtered by [Bolshevism], among them 192,350 workers, 260,000 soldiers, 815,000 peasants, 155,250 intellectuals. The whole of Russian economic life has been destroyed; part [of the country] is transformed into a desert; and further millions have been consigned to starvation. We have never heard that Rathenau raised the slightest objection to the criminal regime. Rather, he entertains friendly relations toward the Soviet tyranny, ... "[436]

In 1929, John Spargo, the man who had protested so loudly against any implication that he had been duped into Socialism, published an article entitled, "Why I Am No Longer a Socialist", in the magazine *Nation's Business*. Though not attributing Socialism to a Jewish conspiracy, and maintaining that his motives had always been noble and pure, Spargo nevertheless believed that international Socialism was a dangerous delusion:

> "More than 20 years of my life were given to the advocacy of international Socialism and the work of upbuilding the Socialist movement. Today I am thoroughly convinced that the Socialist philosophy is unsound, the Socialist program dangerous and reactionary, and the Socialist movement a mischievous illusion. As sincerely and earnestly as I formerly proclaimed Socialism to be the greatest hope of mankind, though with less energy and strength, I now proclaim my conviction that only disaster could result from a serious and comprehensive attempt to carry the Socialist program into effect. [***] Deluded and misdirected in their aim as I believe them to be, the men and women who make up the Socialist movement are, by and large, as intelligent and as decent as other people, possessing their full share of the virtues and no more than their share of human frailty."[437]

The emotional and polemic nature of Spargo's attacks were typical of the religious zealotry and arrogance he affirmed were a part of his Socialist

---

[436]. T. Fritsch, in P. Lehmann, Editor, *Neue Wege aus Theodor Fritsch's Lebensarbeit: eine Sammlung von Hammer-Aufsätzen zu seinem siebzigsten Geburtstage*, Hammer-Verlag, Leipzig, (1922); English translation in: R. S. Levy *Antisemitism in the Modern World: An Anthology of Texts*, D.C. Heath, Toronto, (1991), pp. 192- 199, at 197-198.

[437]. J. Spargo, "Why I Am No Longer a Socialist", *Nation's Business*, Volume 17, (February, 1929), pp. 15-17, 96, 98, 100; (March, 1929), pp. 29-31, 168, 170; at pages 15-16 of the February issue. Reprinted: *Why I Am No Longer a Socialist*, Chamber of Commerce of the United States, Washington, D.C., (1929).

upbringing and propagandizing,

> "The comprehensiveness of the Marxian philosophy and the completeness and finality of its explanation of the social structure endowed the movement as a whole, and individual Socialists, with the superb audacity and splendid arrogance universally characteristic of the propaganda of the movement. [***] Like countless thousands of others, my life was consecrated to the cause as to a priesthood."[438]

Jean Paul Marat offered a model for these Socialist propagandists. Marat published the journals *L'Ami du Peuple* (*The Friend of the People*) and *Journal de la République Française* (*Journal of the French Revolution*) during the French Revolution and used them to make vitriolic personal attacks, which were effectively death warrants. Marat called for mass murder in the name of the people. He called for brutality and tyranny in the name of liberty, equality and fraternity.

Though Spargo wrote passionately of the alleged high morality which drove him to embrace Socialism with a religious devotion to its cause, he admitted that Socialism actively worked to undermine all that was good in society. He openly admitted that he was a part of this effort to inflict misery on the masses. Just as some political Zionists sought to subvert all good will toward Jews and to make the lives of Jews miserable in order to force them to Zionism, some Socialists deliberately subverted everything good in society in order to bring about its ruin and make way for their allegedly benevolent and Utopian tyranny. Burton J. Hendrick, who had just recently completed a series of articles on Jews in *The World's Work*,[439] warned in the early 1920's of the fact that the Polish and Russian Jews, who had emigrated to America, posed a threat to the American system and attempted to take over trade and labor unions in order to use the unions' membership to destroy the United States and make it a part of a world-wide soviet system run by Jews,

> "There are three divisions of Jews in the United States. These are the Sephardic Jews, the German Jews, and the Eastern or Polish Jews. The first two make up perhaps 500,000 of the more than 3,000,000 Jews in the United States. The last comprise more than 2,500,000; they comprise the vast bulk of our Jewish population. In previous articles the present writer has

---

**438**. J. Spargo, "Why I Am No Longer a Socialist", *Nation's Business*, Volume 17, (February, 1929), pp. 15-17, 96, 98, 100; (March, 1929), pp. 29-31, 168, 170; at page 96 of the February issue. Reprinted: *Why I Am No Longer a Socialist*, Chamber of Commerce of the United States, Washington, D.C., (1929).

**439**. B. J. Hendrick, "The Jews in America: I How They Came to This Country", *The World's Work*, Volume 44, Number 2, (December, 1922), pp. 144-161; **and** "The Jews in America: II Do the Jews Dominate American Finance?", *The World's Work*, Volume 44, Number 3, (January, 1923), pp. 266-286; **and** "The Jews in America: III The Menace of the Polish Jew", *The World's Work*, Volume 44, Number 4, (February, 1923), pp. 366-377.

emphasized the fact that about the only quality the Sephardic and German Jews have in common with the Polish Jew is a common religion. In all other respects, in history, ethnology, in physical and mental characteristics, they are absolutely different. Practically all students of Jewish history maintain that the Jews of Western and Eastern Europe are distinct races—as different as is an Englishman from a Sicilian or a German from a Slav. That the Western Jews represent a vastly higher stage of achievement in business, in politics, in literature and the arts than the Eastern, is the plain historic record. Practically all the great Jewish names that have become familiar to cultivated people—Spinoza, Mendelssohn, Heine, Disraeli, Ehrlich—are those of Western Jews. Such success as has come to American Jews in business and finance is confined, almost exclusively, to Jews of Western origin; such are the Seligmans, the Schiffs, the Kahns, the Warburgs, the Guggenheims. Is it true that in this matter of 'Americanization' this same distinction must be made? Is it a fact that, as a mass, the Spanish and German Jews become good Americans and that, as a mass, the Polish Jews do not? [***] [Polish Jews] always resented—as they do to-day—the idea that they were Poles or a part of the Polish State; they insisted on being Jews and nothing else. Nor does it seem to be the case that the Jews in Poland were compelled to lead a distinct existence by the Government as a part of an anti-Jewish policy; the Ghetto was their own creation and their own choice; the fact that they were able to enjoy this privilege and many others, was what made their sojourn in Poland so agreeable and so free from the persecutions to which they were subject in other countries. This seems to indicate that the lack of national feeling which the Polish Jews evince to-day is not the product of Russian persecution, but that it is a deep lying racial trait. Poland was perhaps the greatest 'melting pot' of the Middle Ages; it found no difficulty in absorbing Germans, Frenchmen, Englishmen, and Irish; but it never absorbed its Jews. For it seems the fact that the Polish Jews care no more for Poland to-day than did their medieval ancestors. As a mass they have shown no interest in a regenerated Poland; in the World War their support was thrown to Germany; and the present bitter anti-Jewish feeling in Poland to-day is explained by this pro-Germanism. Why is it that, whereas German, French, Spanish, and French Jews have demonstrated this nationalistic impulse, the Polish Jews have seemed to be so devoid of it? That is a question for the historian and the student of racial psychology. The training of this mass Polish mind, therefore, is not favorable to a quick understanding of and enthusiasm for American principles. Are there any manifestations of indifference and even unfriendliness in the daily life of the Polish Jews in New York? The first fact that impresses the inquirer, as he attempts to glance into the composite mind of metropolitan Jewry, is its reading matter. The thing that startles is that the Yiddish press of New York City is extremely socialistic. The great newspapers edited by Jews, published by Jews, and read by Jews, are preaching political principles whose success means the destruction of the American system of government. The great Yiddish newspaper of New York's East Side is

*Vorwarts* (The Forward), edited by Mr. Abraham Cahan, a Russian Jew of romantic personal history and of literary attainments of a high order—he has won wide recognition as a short story writer in English. *The Forward* has a daily circulation of 160,000 copies. It is one of the most successful and one of the most profitable newspapers in New York or in the United States. It is found in practically every Yiddish reading home and wields with its clientele an influence such as few English papers can boast with theirs. Its political principles are not found in the platform of the Republican or Democratic parties, in the Declaration of Independence, or in the Constitution of the United States. It draws practically no inspiration from American history. The lives of Washington, Franklin, Jefferson, Lincoln, and the other American immortals furnish its writers no examples. Its principles are derived from *Das Kapital* of Karl Marx. The wisdom or the folly of Socialism are not the issue here. The only point insisted on is that Socialism is not Americanism; it may be better or worse; but it is not the same. The triumph of Marxism means the destruction of every principle upon which the American state rests, and it makes ridiculous a century and a half of American history. It substitutes 'internationalism' for a robust American nationalism, 'the solidarity of the working classes' for the American allegiance to the central government, 'the dictatorship of the proletariat' for representative institutions. That a newspaper should exist advocating these doctrines is not especially significant; every opinion, in politics or theology, necessarily has its spokesman in so large and diversified a country as the United States; what is significant is that the newspapers preaching such doctrines, especially *The Forward*, should be the most widely read of all publications on the East Side. That, in order to obtain a large circulation with the Yiddish reading public, a newspaper should be obliged to preach the same principles that produced the Bolshevist Revolution in Russia is the thing that gives one pause. Let us imagine, for example, that the New York *Times*, the *Tribune*, and the *Evening Post* were constantly advocating the overthrow of the American Government and its substitution by a Socialistic state; that they were constantly denouncing American 'nationalism' and praying for the day when it would be superseded by international 'solidarity.' This would not necessarily mean that these newspapers represented a perverted mentality, for any man is free to believe these doctrines and to advocate them and need not be regarded as an abandoned soul because he does so. Such a policy would merely show that these journals, hitherto the upholders of American constitutionalism, had given up American principles and that they hoped for the overthrow of the American Government. Moreover—and this is the point—it would show that the English reading masses in New York City regarded Socialism as a better political system than the American Democracy. This one fact therefore, that the most influential and most largely circulated Jewish press of New York is devoted to Socialism, gives us that insight into the mass mind of the Polish Jew which is essential to any adequate comprehension of his present attitude toward the American state. If any one of the big English papers of New York should advocate such political principles, they would immediately lose their readers and pass out of existence; evidently the Yiddish press can keep its readers only by taking this stand. To those who still believe in the Constitution this fact is really appalling. This enthusiasm for the doctrines of Karl Marx, in preference to the doctrines of Washington and Jefferson and Franklin and Lincoln and Roosevelt, appears in other directions than in the daily press. Any one who attends a Socialist meeting in New York is immediately impressed by the fact that the audience is almost exclusively composed of East Side Jews. The great public meeting place established by Peter Cooper is a favorite headquarters for East Side radicalism. Practically all the orators of discontent who occupy soap boxes in the New York streets are unmistakably Eastern Jews. The mass meetings that are occasionally called in the interest of American recognition of the Russian Soviet Government are overwhelmingly Jewish in their composition. The behavior of European and American Socialists, when face to face with the European War, strikingly brings out the alien quality of American radicalism. Ever since the days of Karl Marx it has been a Socialist tenet that all wars are the products of capitalism; from this it necessarily follows that it is the duty of all Socialists in all countries to refuse to support their governments in war. This had been a doctrine of the First Internationale, but it went to pieces when the Franco-Prussian War broke out in 1870. The Second Internationale, organized on the ruins of the First, similarly made this rule of non-participation in nationalist wars one of the fixed stones in its edifice. Again the existence of such a principle did not affect the Socialists of Europe when the war began in 1914. The followers of Marx proved that their devotion to this idea was merely lip service; and that it had never seized their minds and their consciences. [***] There was one country, that is, in which the Socialists refused to support their government, and in which they actually took up a position of hostility. That country was the United States. The test of conflict disclosed that American Socialists were the only kind who remained faithful to their Socialistic creed. The American Congress declared war on Germany on April 6, 1917; the very next day the Socialist party of America met in congress at St. Louis and adopted a manifesto calling upon its followers to oppose the war. 'The Socialist party of the United States in the present grave crisis,' so read its proclamation, 'solemnly reaffirms its allegiance to the principle of internationalism and working class solidarity the world over and proclaims its unalterable opposition to the war just declared by the Government of the United States... . As against the false doctrine of national patriotism we uphold the ideal of international working class solidarity.' That the war was the handiwork of the capitalists, that American capitalists had forced the United States in, that German submarine warfare was not an invasion of American rights and that, 'in modern history there has been no war more unjustifiable than the one in which we are about to engage' —such were only a few of the sentiments contained in this document. These assembled Socialists pledged themselves to 'continuous, active, and public opposition to the war through demonstrations, mass petitions, and all other means in our power.' They voted to oppose 'all legislation for military or industrial conscription,' 'any attempt to raise money for payment of war expenses by taxing the necessaries of life or issuing bonds,' to organize workers 'into strong, class conscious, and closely unified political and industrial organizations, to enable them by concerted and harmonious mass action to shorten this war and establish lasting peace. [***] Thus the arrival of these Polish and Russian Jews introduced a new fact into the American population. For the first time the Socialists became powerful enough to elect an occasional member of Congress or of a state legislature. Even with these accessions Socialist voters have not been very numerous; yet the fact remains that the only considerable Socialistic bloc in this country is composed of these same Eastern Jews. [***] [Allen Benson], who had been the Socialist candidate for President in 1916, publicly explained the cause of his departure. 'The present foreign born leaders of the Socialist party,' he said, 'if they had lived during the Civil War, would doubtless have censured Marx for congratulating Wilson ... I therefore resign as a protest against the foreign born leadership that blindly believes a non-American policy can be made to appeal to many Americans.' [***] these radical teachings are part and parcel of the massmind of the Polish Jew. [***] They prove that the only sections of New York City which contain a large socialistic population are those in which the Polish Jew is the predominant element. The local election returns for fifteen years demonstrates the same fact. Whenever a Socialist is sent as a Congressman to Washington, an assemblyman to Albany, or an alderman to the City Hall, he always represents a district in which the population is almost exclusively composed of Polish Jews. [***] [T]he fact remains, however, that the chief opposition [Gompers, himself a Jew,] has met in his attempt to keep American Labor free from radicalism has come from Jews—almost exclusively of the Polish and Russian type. Up to 1914 the working classes in the clothing trades had never been very closely organized. The unions had existed for years and had engaged in many fierce strikes, but that lack of cohesion which is one of the failings of Jewish mentality had caused the members to hold their allegiance lightly and to become backward in paying dues. The great labor group in the clothing trades was the United Garment Workers of America, a union whose form of organization followed the accepted American standard. It was a union, that is, on simple craft lines; it existed to improve the general economic conditions of the workers; it proclaimed no political purpose, and certainly cherished no Socialistic or subversive programme. As such the United Garment Workers of America was affiliated with the American Federation of Labor and participated in all its conventions. It had accomplished many beneficial reforms, especially in the abolition of the sweatshop and improved working conditions. Its membership, naturally, was overwhelmingly Jewish, though there was then, as there is to-day, a considerable representation of Italian workers. For years the forces of radicalism had been seeking to capture the garment workers; in the year

1914 these elements, under the leadership of Sidney Hillman, one of the most revolutionary labor captains in New York, succeeded so far as to elect a group of radical delegates to the convention of the American Federation of Labor. Mr. Gompers's convention refused to admit these gentlemen because their announced programme was revolutionary and un-American. The Hillman cohorts therefore withdrew from the Hall, started a rump convention in another building, and organized a new union, called the Amalgamated Clothing Workers of America. The purpose of the new group was not disguised. It was blatantly radical. Its aim was to organize the clothing workers for political action; and it proposed to use the men of the clothing trades as a voting unit to destroy the present system of government as well as the present economic order and to plant in their place a condition not unlike that which prevails in Russia. Its constitution is full of the now familiar talk about 'class consciousness,' 'capitalism,' the 'ruling class' and the 'ruled class,' 'the constant and unceasing struggle,' 'craft unionism,' and the like. Its whole purpose is summed up in this section: 'The industrial and inter-industrial organization, built upon the solid rock of clear knowledge and class consciousness, will put the organized working class in actual control of the system of production and the working class will then be ready to take possession of it.' That is, the plan is for the one big union—the organization of all the workers, not on craft lines, but on class lines—this as the preparation for the day when the workers will themselves take possession of industry. The programme is thus that of the Soviet. [***] The attitude of the Amalgamated towards the American Government was sufficiently indicated by a banner borne in the streets of Boston during one of their strikes, with the following legend: 'To hell with the United States.'"[440]

John Spargo wrote that the Communists took a different tack in England where they simply sought to make life unbearable in order to make way for revolution,

"[T]he sooner the process of degradation is effected the better, for the sooner will the agony be over and the glorious consummation of Socialism be realized. [***] Haters of All Social Reforms. That logic controlled the policy of British Socialism in the days of my youth. That is why we busied ourselves distributing leaflets bearing the significant title, 'To Hell With Trade Unionism!' and appropriately printed in red. That also is why we inveighed against life insurance in our propaganda with all the bitterness of which we were capable. Life insurance was a protective device against poverty, an ameliorative measure designed to avert the poverty and degradation without which our Utopia could not be reached. In the same

---

[440]. B. J. Hendrick, "Radicalism among the Polish Jews", *The World's Work*, Volume 44, Number 6, (April, 1923), pp. 591-601.

spirit and under the compulsion of the same Marxian dogma we opposed every form of thrift, all philanthropy and social reforms calculated to lessen social misery and improve the conditions of life and labor. We regarded all these things with the hate and horror which religious fanatics might feel towards deliberate human thwarting of the clearly manifested design of God."[441]

The Communists used underhanded means to destroy Capitalistic society, and then criticized the Capitalists for the alleged failure of Capitalism to provide for the needs of the people, which the Communists had deliberately caused. The Communists did not care how many people they murdered, nor how much suffering they caused. They had no morals. Their only goal was to destroy society and in order to put their inhuman leaders into power.

## 5.5.5 Zionists Proscribe Free Speech

Most Americans initially opposed American involvement in the First World War and bore no ill will toward Germany. There were millions of German-Americans, many of them Jews. In addition, Americans did not like the British, against whom Americans had fought more wars than any other nation.

The Zionist Wilson administration opened a propaganda department aimed at vilifying Germany and any American who spoke out against America's intervention in the war on behalf of the Allies—truly on behalf of the Zionists. Many pacifists, Socialists and Germans in America suffered terribly as a result. H. C. Peterson and G. C. Fite detailed much of the tyrannical abuse in their book, *Opponents of War, 1917-1918*, University of Wisconsin Press, Madison, (1957). *The New York Times Current History: The European War*, In 20 Volumes, The New York Times Co., New York, (1914-1920), republishes many examples of the propaganda disseminated during the war to govern public opinion in America, and reproduces many contemporary cartoons from both sides of the conflict. Especially noteworthy are the anti-German, anti-Pacifist and anti-German-sympathisizer cartoons of the era. The Zionists converted America from a pro-German, anti-British nation; to a rabidly anti-German, pro-British nation.

George Creel, a muckraking journalist, headed the propaganda ministry in the United States, the so-called "Committee on Public Information". Libraries removed German books from their shelves. Orchestras refused to play Beethoven or Bach. Schools could no longer teach the German language to their students. Robert Paul Prager, a German, was lynched in Collinsville, Illinois, on 5 April 1918. By official decree, sauerkraut was to be called "liberty-cabbage". Iowa Governor William Harding issued a proclamation ordering that the

---

**441**. J. Spargo, "Why I Am No Longer a Socialist", *Nation's Business*, Volume 17, (February, 1929), pp. 15-17, 96, 98, 100; (March, 1929), pp. 29-31, 168, 170; at pages 96 and 98 of the February issue. Reprinted: *Why I Am No Longer a Socialist*, Chamber of Commerce of the United States, Washington, D.C., (1929).

speaking of any language other than English was forbidden on trains, in telephone conversations, or in public.

The propagandists published anti-German booklets and movies. From the beginning of the war, American and British newspapers and books published falsehoods accusing Germany of atrocities, which Germany had not committed.[442] The propaganda employed was extreme. For example, American pro-war propaganda posters, which urged Americans to buy war bonds, depicted a German soldier crucifying an Allied soldier. The scare tactics began early in the conflict. For example, on 3 September 1914, *The London Times* published a letter to the Editor from A. J. Dawe, which the *Times* captioned, "The Crime Of Louvain. Vivid Account By An Eye-Witness. A Ruthless Holocaust. The Real Horrors Of War." Note that the term "holocaust" was employed to vilify and dehumanize the Germans. The British sent over a lying propagandist Lord James Bryce to smear the Germans in America with his book J. Bryce, *Report of the Committee on alleged German outrages appointed by His Britannic Majesty's Government and presided over by the Right Hon. Viscount Bryce. Evidence and Documents laid before the Committee on alleged German outrages: (appendix to the Report).*, Printed Under the Authority of His Majesty's Stationery Office, London, (1915); which was reprinted in several languages and which was published in several English speaking nations including England, America, Canada and Australia.[443]

## 5.5.6 President Woodrow Wilson Becomes a Zionist Dictator

In America, Creel's propaganda office recruited 75,000 "four minute men" to give short propaganda speeches wherever crowds could gather. Seemingly unbiased Americans speaking their genuine beliefs, these propagandists promoted the war and vilified pacifists and Germans. The Zionist Wilson administration passed the Espionage Act, the Sabotage Act and the Sedition Act, which made it illegal to speak out against American involvement in the war. These acts were still enforceable when Spargo attacked Ford's patriotism, leaving Ford at a disadvantage when defending himself. In addition, the propaganda campaign against pacifists had had its effect on the American public. Both of these factors gave Spargo the courage to attack Ford in the underhanded

---

**442**. A. Ponsonby, *Falsehood in War-Time, Containing an Assortment of Lies Circulated Throughout the Nations During the Great War*, G. Allen & Unwin, ltd. London, E. P. Dutton, New York, (1928).
**443**. **French** *Rapport de la Commission d'Enquête sur les Atrocités Allemandes*, Darling & Son, London, (1915); **Italian** *Relazione della Commissione d'Inchiesta sulle Atrocità Tedesche*, Vincenzo Bartelli, Perugia, (1915), **Portugese** *Relatorio da Commissão sobre as Barbaridades Attribuidas aos Allemães, nomeada pelo Governo de Sua Magestade Britannica presidida pelo Visconde Bryce*, Thomas Nelson & Sons, Paris, Edimburgo, (1915); **Spanish** *Informe Acerca de los Atentados Atribuidos á los Alemanes, Emitido por la Comisión Nombrada por el Gobierno de su Majéstad Británica y Presidida por el muy Honorable Vizconde Bryce*, Thomas Nelson & Sons, Paris, Edimburgo, (1915).

way that he did.

The propaganda tactics Spargo used to attack Ford were reminiscent of Creel's "advertising" agency, though far less successful. Creel published propaganda all over the world and then he wrote a book about it in order to advertise himself, *How We Advertised America: The First Telling of the Amazing Story of the Committee on Public Information That Carried the Gospel of Americanism to Every Corner of the Globe*, New York, London, Harper & Brothers, (1920).

Creel's Committee on Public Information received the support of the head of British propaganda in America, Rt. Hon. Sir Gilbert Parker, Bart. Note that Parker admits that when the war started, Americans had little love for the British, who were America's most frequent enemy, and Americans felt no animosity towards the Germans. Parker boasts of the new unanimity of pro-Ally sentiment that he and Creel achieved in the United States. Parker does not mention the fact that the appearance of unanimity was achieved by undemocratic means—by making it illegal to speak out against the Allies, against the war, or on behalf of Germany. Note the statement that America stands nothing to gain by entering the war. Note also that the timing of these events appeared so fortuitous as to have been planned long in advance, and that Wilson had to trick the Democrats into going to war, and that Democrats would never have allowed the Republicans to have led them into the war. Zionists have an easy time controlling both sides in a two party system for the simple reason that politics is driven by money and media and the Zionists control both means to victory. In addition to being able to bring victory to one side, they often sponsor a controlled opposition and commit subterfuge of that opposition. Parker vilified Germany, but made no mention of the illegal Allied naval blockade of Germany, that resulted in the deaths of about 750,000 German men, women and children by starvation.[444]

Just as British propaganda made it appear uncivilized and unpatriotic to speak out in favor of peace (as Ford had done) and on behalf of the civil treatment of Germany, or to voice America's own interests; Relativists made it appear unethical and unscientific to speak out in favor of Einstein's predecessors and the open expression of the true history of the theory of relativity, or to express scientific arguments in opposition to Einstein's metaphysical mythologies. The same tactics and style of attack were often apparent among Communists, Zionists and "Relativists".

Parker published some of his propaganda in *Harper's Magazine*, Volume 136, Number 814, (March, 1918), pp. 521-531:

## "The United States and the War

---

[444]. "An die medizinischen Fakultäten der Universitäten der neutralen Welt und an den Präsident Wilson", *Psychiatrisch-Neurologische Wochenschrift*, Volume 21, Number 1/2, (12 April 1919), pp. 1-2. *See also:* J. R. Marcus, *The Rise and Destiny of the German Jew*, The Union of American Hebrew Congregations, Cincinnati, (1934), p. 93.

## BY RT. HON. SIR GILBERT PARKER, BART.

FOR the first time in its history the United States is engaged in a World War. It must be remembered that her only wars have been with Great Britain, with the Barbary pirates, with Mexico, with Spain, and with her own population. Idealistic always, her very first war had behind it the spirit of a great people; on the whole, it was a conflict between Britons and Britons. It was the principle of British freedom and independence in action; it was the soul of Hampton and William Penn and all the democratic nobility of the United Kingdom, which under distant skies was reasserting itself, reaffirming its faith in the ancient doctrine laid down by the barons when they wrested Magna Charta from King John. No one doubts now—and great numbers of British people in the time of the war, and most important statesmen of that day did not doubt, and said so in Parliament at Westminster, that the thirteen States were right in the action they took in the Revolutionary War; though great doubt is felt as to justification for the War of 1812.

Always firm and decisive, always alert and progressive, it was the United States that taught Europe how to subdue barbarism and sea-brigandage in the overseas expedition against the Barbary pirates. Of the rightness of heart and the strength of will of the American people, their whole history has been proof. They have lost nothing of their ancient qualities, even though they admit yearly to their shores a million aliens, of whom they absorb and train to American uses and principles the immense majority. Nothing is so remarkable as the power of the American commonwealth to absorb and inspire alien elements and heterogeneous peoples. Is it not wonderful to think that, with one-half at least of the whole population foreign in origin and descent, there is behind President Wilson and his Government a compact and loyal people?

And why? Because at bottom the intelligence and the spirit of the American people are idealistic, humane, and aspiring. I do not mean to say that the hundred millions of people of the United States are all moved by an immense humanitarian spirit; but I do, say that the majority are, or else the declaration of war against the Central Empires would never have been received with approbation. I believe profoundly that something far deeper than national, profit has moved the people of the United States to enter this war. Whatever may be thought of the motives of other nations fighting, only one thing can be thought of the motive of the United States. The Americans nave, nothing to gain by success in this war, except something spiritual, mental, manly, national, and human. They are in this war because they believe that the German policy is a betrayal of civilization. From August, 1914, there was a considerable percentage of the public who believed that the United States should, in the name of civilization, have officially resented the invasion of Belgium. Personally, I believe that it would have been extremely difficult for the United States to enter the war six months before she did. I was in the United States for some months on this trip. I have been from New York to San Francisco. I was at Washington when President

Wilson dismissed Count Bernstorff and heard him do so, and I am firmly convinced of this—that President Wilson committed his country to this war at the right moment—neither too soon nor too late. He had stopped up every avenue of attack by the pacifists and the jurists and the pedants and the pettifoggers.

Perhaps here I may be permitted to say a few words concerning my own work since the beginning of the war. It is in a way a story by itself, but I feel justified in writing one or two paragraphs about it. Practically since the day war broke out between England and the Central Powers I became responsible for American publicity. I need hardly say that the scope of my department was very extensive and its activities widely ranged. Among the activities was a weekly report to the British Cabinet on the state of American opinion, and constant touch with the permanent correspondents of American newspapers in England. I also frequently arranged for important public men in England to act for us by interviews in American newspapers; and among these distinguished people were Mr. Lloyd George (the present Prime Minister), Viscount Grey, Mr. Balfour, Mr. Bonar Law, the Archbishop of Canterbury, Sir Edward Carson, Lord Robert Cecil, Mr. Walter Runciman, (the Lord Chancellor), Mr. Austen Chamberlain, Lord Cromer, Will Crooks, Lord Curzon, Lord Gladstone, Lord Haldane, Mr. Henry James, Mr. John Redmond, Mr. Selfridge, Mr. Zangwill, Mrs. Humphry Ward, and fully a hundred others.

Among other things, we supplied three hundred and sixty newspapers in the smaller States of the United States with an English newspaper, which gives a weekly review and comment of the affairs of the war. We established connection with the man in the street through cinema pictures of the Army and Navy, as well as through interviews, articles, pamphlet etc.; and by letters in reply to individual American critics, which were printed in the chief newspaper of the State in which they lived, and were copied in newspapers of other and neighboring States. We advised and stimulated many people to write articles; we utilized the friendly services and assistance of confidential friends; we had reports from important Americans constantly, and established association, by personal correspondence, with influential and eminent people of every profession in the United States, beginning with university and college presidents, professors and scientific men, and running through all the ranges of the population. We asked our friends and correspondents to arrange for speeches, debates, and lectures by American citizens, but we did not encourage Britishers to go to America and preach the doctrine of entrance into the war. Besides an immense private correspondence with individuals, we had our documents and literature sent to great numbers of public libraries, Y. M. C. A. societies, universities, colleges, historical societies, clubs, and newspapers.

It is hardly necessary to say that the work was one of extreme difficulty and delicacy, but I was fortunate in having a wide acquaintance in the United States and in knowing that a great many people had read my books and were not prejudiced against me. I believed that the American people could not be driven, preached to, or chivied into the war, and that when they did enter it would be the result of their own judgment and not the result of exhortation, eloquence, or fanatical pressure of Britishers. I believed that the United States would enter the war in her own time, and I say this, with a convinced mind, that, on the whole, it was best that the American commonwealth did not enter the war until that month in 1917 when Germany played her last card of defiance and indirect attack. Perhaps the safest situation that could be imagined actually did arise. The Democratic party in America, which probably would not have supported a Republican President had he declared war, were practically forced by the logic of circumstances to support President Wilson when be declared war, because he had blocked up every avenue of attack.

There were some who said—and I heard them say it—that the breakage of diplomatic relations with Germany would not mean actual war. My reply was: 'It won't be the will of the United States to enter the war; it won't be a desire to fight. It will be the action of Germany—in stinging and lacerating the conscience of a great people.' The record was a terrible one. Every one knows that the Prussian military organization had thrown overboard all rules of war which centuries of civilization had produced and imposed; a solemn treaty, signed, was 'a scrap of paper,' hospitals and hospital-ships were proper food for the metal of guns and torpedoes. Gas and fire were used as war weapons—to the final injury of those who initiated their use. Prisoners, not by tens, but by thousands and scores of thousands, were treated shamefully, and the Belgian people, to the number of 300,000, were driven under the lash of slavery to the mines and factories of Germany and France, to set free men who could do duty in the German armies. The chambers of the German embassy in America were the breeding-places of crimes against the civil life of the United States, passenger-ships were sunk, factories were bombed or set on fire, all kinds of tricks were used to influence American opinion in England, and innocent lives by the scores of thousands were sacrificed. In France and Belgium towns and villages were wiped off the map for no military purpose, with no strategic intention, but with a vile and polluted barbarity, to break the spirit of a people or of peoples. America was shocked at the bombardment of helpless and undefended towns of England and Scotland by airships. Her spirit was abashed and shaken by the sinking of the *Lusitania*. She endured and yet endured. She waited and still waited, vainly believing that some spirit of remorse might stir Germany and change her course of action.

She awoke, however, to the fact that Germany's promises of reform, given to President Wilson after the sinking of the *Sussex*, in regard to the submarine were only given to gain time, to manufacture new types of submarines more powerful, and then with an insolence and a disdain worthy of Attila the Hun they announced indiscriminate attacks upon all shipping within the war zone. Also, Germany declared that she could allow only certain ships of the United States to sail, and on certain specified terms and conditions—and that only after a cry of indignation had gone up from the press of the United States. This was the final act which turned President Wilson from a pacifist into a warrior. And it is wholly in keeping with the spirit of Prussianism, that the Zimmerman note to Mexico, with its evil suggestions of treachery of Japan, and its declaration that New Mexico, Texas, and other American States and territory would be acquired again by Mexico, should have come at the critical moment when war was inevitable.

I had been in America through all these months of developing purpose and sentiment, and I had seen a whole people, who in January last had appeared to have grown indifferent to horror, suddenly amalgamate themselves, strip themselves of levity and indifference and the dangerous and insidious security of peace, into a great fighting force, which is not the less a fighting force because down underneath everything in the United States is a love of peace and devotion to the acquisition of wealth. None but a great fighting people could have, or would have, imposed conscription at the very beginning of the war. None but a skilled fighting people could have produced a Navy which silently and swiftly entered the war in the war zone within a week, and landed an army on the coast of France, with submarine-destroyers in those perilous seas, within two months of the declaration of war.

I speak of the Americans as a fighting people; I believe that this war will prove them to have everything that they have always had—courage, swiftness of conception, capacity to perform, and a lightning-like directness. The American nation has never been conquered. Like all democratic peoples, they are quick to anger, but slow to move; yet it must be remembered that out of the mass of conflicting views one great purpose can seize and hold the imagination and the capacity of the American people, just as the same elements seize and control the spirit of the people of England and France. I heard on many hands in the United States angry criticism of those in authority, but I heard it in England, and I saw it in France; and I know that England and France have renewed in this war the ancient great qualities of their peoples.

There has never been a war in the whole history of the world where so much courage was needed, and there has never been a war where so much dauntless courage has been shown. Think of what France was at the beginning of this war! Think of what England was! Officially, France was rotten when war broke out; officially, England was supine when war broke out, with this difference, however, that the small English Army was perfectly equipped and admirably appointed. The big English Navy was in perfect condition, while in France, as Germany knew, there was inadequacy of equipment for the army, and there were political difficulties which made the task of government and fighting Germany almost impossible. Where, I ask, is the official rottenness of France or England now? The truth is that nothing was rotten at the core.

England is not a republic, but she is the most democratic nation on earth, and that is saying much. What I mean is this: the British people can turn a Government out of office at a moment's notice, and king or monarchy cannot prevent it. The same thing exists in France; but here in America, with your written Constitution, your President and his Cabinet cannot be turned out in under four years. It may be that you are right in your system, but if the will of the people is the spirit of democracy, England, at any rate, is as much a democratic community as this country of the United States.

Now the United States is in the war, and I prophesy, with faith and confidence, that all that has made America great will make her do in this war what France and England and have done. Let me be a little explicit. I have heard many criticisms of the American Government from Americans themselves, but my comment has always been, Judge of a Government by what it does, and judge the American Government in time of war by what it does in time of war. It is well known that there had been no preparation on the part of the Army or Navy the United States for entrance into the war. Yet, when war was declared, there was instant and decisive action in both departments of the Army and the Navy.

The American Navy has done splendid work in relieving the British Navy from patrol work on the western side of the Atlantic, in the convoying of freight-ships and passenger-ships, and by sharing in the attacks upon the German U-boats in the war zone. The material assistance has been great—the moral assistance has been immense. No one could overestimate the moral effect of the entrance of the United States into the war. It must not be forgotten that she is the one nation about whose motives there could be no suspicion. She is in the war with no territorial or national ambitions—with nothing except the aspiration to fulfil the democratic principle: that all nations shall be allowed to work out their own salvation without fear or trembling—fear of punishment for right doing, and without trembling before the lash of tyranny.

The United States, true to its ancient faith, is out to defeat the loathsome purpose of Germany, which is the control of the world, the warping and suppression of small countries, and the application of the accursed Prussian doctrine of *Kultur* to all the rest of the world. The United States is in the war in the interests of civilization and humanity—for the right of every nation to live and have its being according to conscience and the laws of humanity. The United States is in the war because she believes she has the right to traverse the high seas, obeying the laws of warfare as laid down by the continued practice of many countries until the final codification by the Hague Conference. The United States is in the war in the protection of her own individual national rights; and those individual national rights are the properties of all countries; but the United States is also in the war because she believes that a republic which is the supreme democracy of the world should take her stand for the cause of civilization, which has been abused and despoiled by Germany. The United States is in the war for the cause of humanity. At the beginning she disbelieved that the German nation meant what Great Britain declared she did mean. But now, after every known law of warfare has been broken by Germany, she realizes the truth. And what is the truth? It is that the German people believe that Prussia and Prussian civilization should control the universe, and that it does not matter how that control is secured so long as it is got.

No more pernicious doctrine ever moved Pope or potentate in the Middle Ages. It is, in effect, Never mind how you do it so long as it is done! On that basis assassination would be a virtue. The United States has come to understand that when Germany passed a law preserving perpetual citizenship to her people, whatever other nationality they adopted, she was aiming at the heart of civilization. I have a brother who has become an American citizen. I think I should curse him to the uttermost death if he declined to take up sword or rifle to defend the United States in a war with Great Britain. I believe that is what all Americans feel. I did not know that my brother had become an American citizen until a year ago. It gave me a pang; but he did what was right. He was not entitled to make the United States his home, live by American energy, profit by American enterprise, and remain a Briton. Think, then, of what this foul principle of Prussia is. It would have me say to my brother, 'Be an American citizen, but remember that your real duty lies with the land of your birth, and when she calls, you must tear up your pledge and compact and sworn word and come back to the Union Jack.'

I wonder how many Americans know that all German-Americans are still Germans by law; and if they do know it, how they must resent the iniquity of the nation that makes of the law of naturalization a scrap of paper, to be torn up, like the sacred compact for the neutrality of Belgium!

The first act of Germany in this war was an act of perfidy, and I firmly predict that the last act will be an act of shame. She may succeed against Rumania, she may succeed against Russia, she may enter Petrograd with her armies, but so did the army of France in the time of Napoleon; and when I think of the millions of people in Russia, chaotic, undisciplined, uncontrolled, and yet aspiring, I still have a grim kind of satisfaction in knowing that if Russia has to be the momentary sacrifice, it is Germany that will be sacrificed in the end.

Lately I saw on a screen, at a theater in New York, pictures of hundreds of thousands of Russians accompanying victims of the Revolution to unconsecrated graves and without religious rites or ceremonies. However depressing such a scene may have been, the really startling effect produced upon my mind by this photography was that Russian life is without system, and that the poetic aspiration for a freer constitutional life is horribly handicapped by lack of knowledge and experience and the habit of control. The faces of the revolutionary leaders have few claims to consideration.

The Duma is as yet no more than a place of oratory. It has never had power or real authority, and, however great Kerensky or any other civilian leader may be, it must first be an army leader that will discipline that great nation into form. No civil dictator will be adequate for the task. I do not know what Mr. Root's views are, save from his public utterances, but I am quite certain that he realizes the truth of what I say—that Russia is in the melting pot, and from the crucible it must be the strong hand of a soldier that will pour out the liquid of order and civilization.

During the days I was in America I saw from my hotel window in New York two processions or parades of American regiments. The main effect upon my mind was a sense of lithe fitness and splendid discipline, which is much out of harmony with the general view of American organized life. I have known the United States for a great many years, and from the standpoint of acquaintance I should be able to judge of her with fairness and accuracy. The thing that has amazed and interested me most in my whole association with American life has been a sense of undiscipline in all the ordinary movements and activities in casual circumstances. But I believe there is no nation on earth that, in unusual circumstances, can pull itself together and get what it wants with precision and definiteness more than the United States. After all, the reason for this is simple. The American hates convention and is opposed to what he considers unnecessary discipline in ordinary life, but given the necessity for discipline in hazardous circumstances, he conforms to its rigidity with rare and manly skill.

I once stood between two Socialist labor members of the House of Commons at the Bar of the House of Lords, when King Edward VII. was opening Parliament with Queen Alexandra. One of these Socialist members had been very rebellious against the whole ritual of British legislative life, but on this occasion, at the moment when King Edward said in a quiet, conversational tone: 'Pray, my Lords, be seated,' and peers and peeresses in ermine and silks and coronets sank to their seats, this Socialist member turned to his friend and said, 'Jimmy, this'll take a lot of moving!'

To-day this Socialist member is a colonel in the British Army, and has bent to the logic of events all prejudice and spurious independence. His Socialistic principles are what they always were, but he has learned that traditions of a thousand years are powerful moral elements in the government of a people. So the average American. He is out against unnecessary form and discipline, but show him the necessity for it and his native independence makes his obedience to the necessity a very gallant and superbly confident thing. Democratic as the American citizen is, he bends to the pressure of events with a dignity and a vigor which make him a superb partner in international activity.

When people tell me that the United States can be of little use in this war I ask myself, 'What is *use?*' If the United States had not sent a man to France, her financial support of the Allies alone would be a throat-grappler for Germany. I believe the United States is spending twenty-four million dollars a day, but only eight millions of that is for her own military equipment—the other sixteen millions are for loans to the Allies. And if the test of the belligerents is power to endure, surely the wealth and resources of the United States settle that point.

If war is the test of endurance, only three things are necessary—men, money, and equipment. Unless Germany was able to defeat England and France before December of last year (1917), the *débâcle* of that country was sure. The United States can supply men, money, and equipment. She has over one hundred millions of people; she cannot be attacked by the armies of the enemy on her own soil; she has unlimited resources; her supply of men can be twelve millions, if necessary; her supply of money can be boundless, and there is no nation on earth that can excel her in organization for equipment.

Now, there is no chance, or there is the millionth chance, of Germany defeating France and England this year. She cannot do it in the winter-time, and when the summer has come the United States will have great numbers of men ready to take the field—probably 700,000. She has food, raw materials, and constructive skill. She has a capacity for applied science greater than any other nation fighting. I believe that with her aid the Entente Allies are as sure of winning this war as we are certain that the sun will rise and set to-morrow.

Great Britain has increased her acreage under wheat by one million acres, and all the products of her soil have been vastly increased. The United States has tremendously increased her production of foodstuffs, and when that genius for economic administration, Mr. Hoover, has been at work for another three months there will be an enormous curtailment of wastage in the Union. With one hundred millions of people, if there is a saving which represents five dollars per person for a year, there are five hundred million dollars contributed to the food-supply of the Allies.

The United States has not begun to appreciate her responsibilities and the dire necessity that faces her, but there is a quickness of apprehension in the American mind which is as good as brawn and muscle and the stolid and rigid insistence of the British people. It took us in Great Britain two and a half years to achieve conscription. It took the United States about two and a half months. There never was any real fight over the principle, and please to remember that this is a democratic country, and that when the Republic applied conscription in her Civil War there were bloody riots and an uprising of sections of New York. If it is true, and I know it is, that over seventy per cent. of the population of New York City is foreign-born, what a magnificent demonstration of democratic responsibility this application of conscription has been!

America is building ships in great quantities for the war service. She once had, proportionately to her population, the second greatest mercantile marine of the world. She lost that mercantile marine through no incapacity, but because she could make more money by investing her capital in industries and railway transportations. Now she is building 1,270 ships of 7,968,000 total tonnage, at a cost of $2,000,000,000, and by the middle of this year she will have a really great mercantile marine. This is in addition to almost 2,000,000 tons of shipping now building in American yards which has been commandeered by the Emergency Fleet Corporation.

Meanwhile, it must not be forgotten that all her shipping and all the German shipping that was in her ports have been seized for the use of the Entente Allies. Every day that passes strengthens and solidifies the Allies' engines of attack and defense. Every day that passes accelerates the intrepidity and the force of Allied aggression. Every day that passes lessens old antagonisms between Great Britain and the United States, and deepens in the American mind an appreciation of Britain's worth and valor.

The American is beginning to understand that in 1914 France—as France— might have been wiped from the international map had it not been for Britain and Britain's Navy and her 'contemptible little Army.' It is beginning to dawn upon the most prejudiced American mind that, in all the main departments of the war, Great Britain has borne, and is bearing, the overwhelming burden. France could not have fought so well without British money and British steel, British cloth, and the British Navy and Army; and Italy and Russia could not have carried on.

One does not need to say now that Great Britain was forced into the war by a spirit of honor, by the dictates of humanity and civilization, and not for commercial purposes. One does not need to say that if Great Britain had intended war she would not have rejected during so many years Lord Roberts's appeal for a national service army. All the records published prove that Great Britain was meant to be the victim of Prussian aggression.

Does the American public stop to remember who were the people in Great Britain who declared war? The Government in power at Westminster was a peace-loving Government, which had fought military and naval preparation with constant vigor and hatred. Who is Lloyd George, the present Prime Minister of Great Britain? He is a man whose life was in danger and who was assailed during the South African War because of his anti-war sentiments. I am certain that no intelligent human being will believe that the present Prime Minister of England is militaristic, just as I am certain that no sane American would call President Woodrow Wilson a man of war.

If the United States had not believed in Great Britain's *bona fides*, she would not have committed herself to this stupendous enterprise. Let all the world remember that Great Britain was the ancient enemy of the United States. Let the doubter recall that the United States has now linked hands with a nation whom at her Revolution she regarded as a tyrant and oppressor, as the ancient foe of liberty and democracy.

The War of the Revolution, that of 1812, and the American Civil War deepened the gulfs between the two great peoples, but, blessed be Providence, there are now no outstanding questions vexing England and the United States. We have settled the Maine boundaries dispute, the persistent Newfoundland fisheries question, the Oregon trouble, the Venezuela difficulty, the Civil War claims, the Panama anxiety, and now no vexed subject keeps us apart. What was accomplished at Manila toward making America a world power was exceeded infinitely there by the splendid action of Admiral Chichester and Britain's Navy in threatening the German naval forces, which drew the two nations together in a spirit of comradeship. If the United States disbelieved in Great Britain she would not be fighting in France and on the high seas. Never, in all the history of the two countries, was. there such a demonstration of understanding and friendship as when

Mr. Balfour was received in Washington, New York, and elsewhere. And let it here be said that Great Britain could have sent no one who would so have won the confidence of the American Government and people in the same way or to the same extent as Mr. Balfour. Whatever else this war may do, the greatest thing done for humanity and civilization has been to make these two nations one in the brotherhood of battle. Of this let every American be sure, that the closer comradeship of the two great peoples has not a single foe in Great Britain. Jealousy, envy, and a little malice there would always be between two great friendly rivals speaking the same language, but envy, jealousy, and a little harmless malice exist between States and cities of this Union and between countries of the British Empire. Never since the War of the Revolution had a British flag been hoisted on an American official building till last spring, and never had the same friendly compliment been paid to the American flag in England. But now they have waved together over Washington's tomb and over the House of Commons. Also, it should be remembered that the Society of Pilgrims, whose work of international unity cannot be overestimated, has played a part in promoting understanding between the two peoples, and the establishment of the American Officers' Club in Lord Leconfield's house in London with H.R.H. the Duke of Connaught as president, has done, and is doing, immense good. It should also be remembered that it was the Pilgrims' Society, under the fine chairmanship of Mr. Harry Brittain, which took charge of the Hon. James M. Beck when he visited England in 1916, and gave him so good a chance to do great work for the cause of unity between the two nations. I am glad and proud to think that I had something to do with these arrangements which resulted in the Pilgrims taking Mr. Beck into their charge.

I have sometimes been amazed at the hostility to Great Britain in certain portions of the United States and among certain sections of the people. Perhaps the real cause of this misunderstanding —for it is nothing else—is ignorance or forgetfulness of the facts of history. It is true that George III. endeavored to impose upon the American people the Stamp Act, just as the kings of France and Spain and Holland had imposed upon their colonies impositions for revenue, but it should not be forgotten by any American that King George III. failed, not only in America, but in Great Britain, his own country. Among his greatest enemies in this wretched business were Pitt, Fox, Rockingham, and Shelburne, and the operations of war in the United States on behalf of England were conducted by German mercenaries and a handful of the British professional Army, of whom a great many officers of standing and eminence refused to serve. It was impossible to raise an army of volunteers in England, and King George dared not attempt to raise a conscript army. Pitt declared in the House of Commons, when America refused to submit to the Stamp Act, that he rejoiced she had resisted. There was as great a fight in the British Parliament over the American war as there was in America itself on the field of battle. There is no British man to-day who is not opposed to George III. in what was perhaps the most insane and unwise national task ever undertaken by a British king.

It must not be forgotten that Benjamin Franklin, the representative of the United States in Paris, was in constant correspondence with British statesmen during the Revolutionary War, and the leaders of the opposition to King George in the British House of Commons were eager to give to the United States, as she was given in 1783, a status as a nation and not a province on the seacoast. The United States was given the Northwest Territory and the basin of the Ohio River to the Mississippi, so making possible the wonderful extension of power which has given to the American national life forty-eight States instead of the thirteen which fought King George. It should also be remembered that the Revolutionary War of the United States was a struggle of British men for rights which were being fought for in the British Parliament and against the last stand of British monarchical autocracy.

The United States is a warm friend of France, and properly so; but it must not be forgotten that the greatest enemy of American development was Napoleon Bonaparte, who considered all parliaments as chattering concerns, and, having grabbed from Spain the coast of the Gulf of Mexico, with New Orleans, the Middle West from the Mississippi to the Rockies, and established a base at Santo Domingo, ordered his Minister of Marine to furnish him with a full plan of conquest, and commanded the combined fleets of France and Spain to carry a French army to the shores of Louisiana. It must be remembered that the man who planned this maneuver was one of the greatest soldiers in history, and had an army which at that time was greater than any army in the world.

What saved the United States from this attack? Great Britain, and Great Britain only. The report of Mr. Rush, the American minister in London, contained the statement of Henry Addington, the British Prime Minister, that in case of war Great Britain would take and hold New Orleans for the United States. This is history. Who was the American President at the time? It was Thomas Jefferson, the great pacifist, whose firm despatch to Robert Livingston, in Paris, contained these words: 'The day that France takes possession of New Orleans we must marry ourselves to the British fleet and nation.' What was the result of this? Napoleon decided it was better to sell to the United States what would be certain to be lost, because he believed that the British fleet, supporting the United States, would take Louisiana from France—Louisiana, which he had forced from Spain.

The main cause of the War of 1812 was not the impressment of seamen from American boats by the Royal Navy, as is generally supposed, but the fact that both France and England had forbidden any neutral nation to trade with the other, and because of England's preponderating fleet she could make her blockade effective and Napoleon could not. The United States, therefore, joined what she considered the lesser of her enemies, France, in attacking the greater, England.

I have no doubt that many Americans regret the War of 1812 as most Britishers regret the acts of George III. which precipitated the Revolutionary War; but for nearly a hundred years the British Navy, and behind it the

British Government, has been the best friend that the United States ever had in its history. What Lafayette did for the United States was great and good, and what Great Britain did in 1824 was, in one sense, greater and better. It was George Canning, the British Foreign Minister, who informed the American minister of the intention of the Holy Alliance to attack representative government in both hemispheres, and offered the assistance of the British fleet in defending institutions won by valor, devotion, and power. It is remarkable that, when the purpose of the Holy Alliance was made clear, that the high contracting powers should 'use all their efforts to put an end to the system of representative government,' the Duke of Wellington immediately left the Congress at Verona. Soon after it was announced, Great Britain and the United States proclaimed that they could not see with indifference any South American territory transferred to any Power.

Then it was that the Monroe Doctrine became an accepted fact, but the United States could not have made it a fact unsupported and unprotected by the British Navy. It is no exaggeration to say that the policy and prosperity of the United States have had a free and fair run for over the last ninety years, because Great Britain, which had learned her great lesson in the American Revolutionary War, made her Navy the defender of the Monroe Doctrine. Perhaps the aged Jefferson's counsel to President Monroe on this matter is the best evidence of what I say. These were Jefferson's words:

The question presented by the letters you have sent me is the most momentous which has ever been offered to my contemplation since that of independence.... America, North and South, has a set of interests distinct from those of Europe. She should, therefore, have a system of her own, separate and apart from that of Europe.

One nation, most of all, could disturb us in this pursuit; she now offers to lead, aid, and accompany us in it. By acceding to her proposition, we detach her from the bands, bring her mighty weight into the scale of free government, and emancipate a continent at one stroke which might otherwise linger long in doubt and difficulty. Great Britain is the one nation which can do us the most harm of any one on all the earth; and with her on our side we need not fear the whole world. With her, then, we should most sedulously cherish a cordial friendship, and nothing would tend more to unite our affections than to be fighting once more, side by side, in the same cause.

It is wonderful to think that after these ninety-odd years the hope of Jefferson has been fulfilled. We are at last fighting once more 'side by side' in the same cause on the battle-fields of Europe, and against an enemy whose whole ambition has been to establish German control in the Western Hemisphere, as in Europe and in the East. No one knows better than President Wilson, who is a historian of high capacity, that what I say here is true. Monroe's letter to Jefferson, again quoted by Mr. Page, clearly indicates the initiative of Great Britain in the matter of the Monroe Doctrine. These are President Monroe's words:

They [two despatches from Mr. Rush, American minister in London] contain two letters from Mr. Canning suggesting designs of the Holy Alliance against the independence of South America, and proposing a cooperation between Great Britain and the United States in support of it against the members of that alliance.... . My own impression is that we ought to meet the proposal of the British Government.

Well, the Monroe Doctrine has been a success, and, at the tomb of Washington, Mr. Arthur Balfour, in effect, reaffirmed the friendly doctrine of George Canning, in which the British nation has as much interest, and for which it has as much honest affection, as the hundred millions of population of the United States.

I repeat that Great Britain is a friend of the United States in all that matters, and I believe that the present war, if it failed in everything else, will succeed in this it will bring shoulder to shoulder with a handclasp of understanding and a spirit of co-operation two great peoples without whom there is no real future for democracy in the world. The monarch of Great Britain has infinitely less power than the President of the United States, so far as the policy of his country is concerned. He is the head of the clan, as it were, the patriarch of the tribe, but his power is limited to a point where even Socialism says, 'This man cannot hurt his people politically; he can only hurt them socially and morally by his example.' It is impossible to discuss here the merits of our two systems of government; but one thing is clear, that the British Constitutional Monarchy is as democratic as the republican Constitution of the United States.

Of this thing I am sure: that the days of wilful misunderstanding between Great Britain and America are gone forever! And I like to think that when these banners of war are rolled up, and the terms of peace are signed, that the two most democratic nations on earth, the two most advanced in civilization and enterprise, will be working hand in hand for the political good of all the world.

For some months I saw the United States from many corners of the compass, and I state with unvexed confidence that a new spirit has entered the mind of the American people where Great Britain is concerned. They realize that England's severest critics are within her own borders; that her sternest monitors arc patriotic Britons; and that the burdens she has borne in this struggle to preserve civilization from disruption are beyond all comparison with those of the other belligerents. The thousand years' traditions of Great Britain belong also to the United States, because the foundations of American liberty and freedom had their origin in the principles embedded in the British Constitution. That is why members of the British Empire to-day can be proud of Washington, glad of Alexander Hamilton and Jefferson and Adams and Franklin, and be the faithful friend of President Monroe, whose doctrine could never have become valid and continuous without the British Navy. I feel bold enough to say that there is not a home in Great Britain that is not happier because the United States, the chief republic of the earth, is linked with us in the struggle for freedom and the small nations.

I was in the United States when all the great missions of the Allies arrived— Great Britain, France, Italy, Russia, Belgium, and now Japan. *And now Japan!* I emphasize these words because east and west in the United States, in San Francisco, in New York and Washington, I had found until very lately the most consuming distrust of the Government at Tokio and the people of Japan. It is, however, comforting to think that this mission of friendship from Japan is the direct result of the Zimmerman note. Whatever Japan's far purposes may be—laying aside all other considerations—it pays her better to be the friend of the Allies than the friend of Germany. I say it pays her better only because there are those who think that Japan in the politics of the world is out for gain. What could she gain by becoming the enemy of the United States, and, therefore, the enemy of England? Because, let this be understood, Japan knows her treaty of alliance with Great Britain does not include the possibility of war with the United States on the part of this Oriental Power. If Japan occupied the Pacific coast, her first immediate foe would be Great Britain, because British Columbia is on the Pacific coast, and Great Britain could not permit Japan or any other nation except the United States to seize or hold any portion of that littoral.

I believe that the anxieties of America have not been well based. I believe that the Japanese nation is as friendly to the United States as she is to Great Britain; and I also believe that, even on the lowest grounds of material benefit, Japan is true to her friendship with Great Britain and the Allies in this war. Far more dangerous is the German menace against the United States than the Japanese menace. And it must not be forgotten that the American Navy, whatever it is, exists to-day because Mr. William C. Whitney, the Secretary of the Navy in Mr. Cleveland's Cabinet, saw in German commercial invasion of South America a peril to the United States.

What the United States will do in this war is being shown from day to day—and this thing is sure, that even the German-American no longer believes that Germany is fighting a war of defense; but rather that she precipitated the war, and is only 'defending' herself because she failed in her first enterprise. I do not know to what extent the activity of the United States will expand, but I do know that if the war continues for another year the pinch of administration and losses in the field will stiffen the backs of the American people to the greatest effort that has ever been made in the history of the world."

Note that Parker, like "Colonel" House, advocated the instillation of a military dictator following a revolution (in Parker's case, in Russia) on the grounds that only a dictator could restore order. This was common practice in American and British foreign policy throughout the Twentieth Century. America installed many military dictators favorable to America and England. It justified the coup d'états by the notion that only a dictator could bring about a proposed democracy—a democracy that was often covertly suppressed by the intelligence agencies of both countries. The real goal was often to free up the natural resources and industry of the subject nation for exploitation by American and British corporations. "Lord Protector" Oliver Cromwell provided a model for the "logic" of installing a dictator in order to establish order.

Adolf Hitler expressed himself in an interview with Anne O'Hare McCormick published in *The New York Times* on 10 July 1933 on pages 1 and 6 in the same terms House used in his book on dictatorship. Hitler banned all political parties other than National Socialism, destroyed the parliament and passed the *Gleichschaltung* and the *Ermächtigungsgesetz* laws, all in the name of restoring and maintaining order. This was a common tactic of Zionist dictators including Cromwell, Napoleon, Wilson and Hitler—and George Bush. When asked which historical figure he most admired, Caesar, Napoleon or Frederick the Great, Hitler responded,

> "No, I admire Oliver Cromwell. I do not think the Commoner the greatest man that ever lived, but he saved England in a crisis similar to ours and saved it by obliterating Parliament and uniting the nation."

Cromwell, under petition from the Marrano Jews Menasseh ben Israel, David Abrabanel, Abraham Israel Carvajal, Abraham Coen Gonzales, and Jahacob de Caceres, permitted Jews to re-enter England over the objections of the Parliament. Hitler used his dictatorial power, enhanced by Jewish financiers and in cooperation with political Zionists, to force Jews to leave Germany. England would not then take Europe's Jews and it was the Zionists' hope that England would give them Palestine, which it eventually did do.

In reality most dictators after the French Revolution followed the example of Maximilien Marie Isidore Robespierre. Revolutionary dictators committed mass murder in the fascist governments the C. I. A. created and sponsored around the world, and in the Bolshevik nations of Europe and Asia. It should not be forgotten that Hitler was a socialist revolutionary, who began his political career as a Bolshevik. Hitler and Goebbels called for a worker's world revolution throughout the duration of the Nazi regime, and their speeches were often derivative of those of Trotsky (Bronstein). Apparently, the dictatorship of the proletariat could not be trusted to the proletariat and required an iron fisted tyrant in a totalitarian state. It is tragic that dictators promoted the ideals of liberty, equality and fraternity in order to gain power, and then subjugated the masses, promoted ignorance and suppressed dissent through violent means. However, it was perfectly in keeping with the Messianic prophecies of Judaism.

## 5.6 Why Did the Zionists Trouble the Jews?

In 1903, racist Zionist Israel Zangwill stated that the Jews' enemies were the Jews' friends. Zangwill implied that anti-Semitism would rescue the Jewish race from fatal assimilation and that the Zionist conferences signaled the Messianic Era,

## "ZIONISM AND THE FUTURE OF THE JEWS
THE SIXTH ZIONIST CONFERENCE GRAPPLING WITH POLITICAL QUESTIONS — A PASSION FOR PALESTINE THE JUDAIC ROMANCE — THE TENDENCY TOWARD DENATIONALIZATION AND THE HOPE OF RENATIONALIZATION.
BY
ISRAEL ZANGWILL

IN August the Sixth Zionist Congress met at Basle, and gathering strength with the years, and quickened by the horrors of Kishineff, this international Jewish parliament, numbering envoys from 'the four corners of the earth,' for the first time grappled with practical political proposals for the solution of the Jewish question. Delegates of South African millionaires took counsel with representatives of the rich American Jewry, and with these modern spirits conferred caftaned rabbis from Russia and sages from India and Persia. In the mere coming together of such an assembly the promised regathering of Israel is already literally accomplished. Eighteen centuries of dispersion have not succeeded in breaking the cohesion of the race; eighteen centuries of exile have not eliminated the passion for Palestine.

Here, surely, is a phenomenon unique in history. It may be profitable to examine briefly into the causes and conditions of this apparent miracle.

I

There is a many-sided symbolism in the dramatic picture of Jochanan ben Zakkai escaping from Jerusalem in a coffin, what time Titus and his legions hovered at the gates of the Holy City. For Jochanan bore in his own breast the seeds of the future, and saved Judaism from the fall of the Jewish State. The zealots of nationality preferred to meet the conquering Roman with grim suicide; Jochanan founded a school at Jamnia, under the protection of Titus. That disentanglement of religion from a *locale* which Jesus had effected for the world at large was in a minor degree effected, a generation after Him, for the Jews themselves by the mailed hand of Titus and the insight of the prudent sage. Possibly Jochanan had already outgrown 'the burnt offerings' which tied Judaism to the Temple; he may have felt already that Israel's greatness was spiritual, belonged to a category of force that could not, and should not, be measured against Rome's material might. However this be, his reconstruction of the Synhedrion, even in the absence of the hewn-stone hall of the Temple for it to meet in, and the subsequent conversion of the substantial sacrifices into offerings of prayer, made the salvage of Judaism more spiritual than the original totality. The unifying centre was no longer geographical, and the Jews became 'the People of the Book' in a far profounder sense than when they were the people of a soil, too. The law was never so obeyed in Bible times as it was when the record of these times became the all-in-all.

But this transformation was not achieved in one generation, nor without violent reactions. Scarce half a century after Jochanan ben Zakkai, the great rebel, Bar-Kochba (Son of a Star), beat back for a time the whole might of Rome, even the great general, Severus (hastily summoned from his task of quelling the less important revolt in Britain). And in the monstrous régime of religious persecution by which Hadrian avenged the difficult suppression of the uprising, the transformation of Judaism might well have been into paganism.

Nor was the transformation into mere spiritual Judaism ever effected radically. Two reactionary influences remained. Palestine still retained a certain authority over the Diaspora. Babylon soon asserted itself as the peer of Jerusalem, and later, with the movement of history and the great teachers, the spiritual hegemony shifted to Spain, to Cairo, to Poland. But underneath all this flux Jerusalem was still the Holy City. Secondly, the literary ritual substituted for the literal sacrifices did not profess to be more than a temporary necessity. The stubborn national spirit clung to the hope of glorious restoration. Rachel wept for her children, and comforted herself by the belief that they were not dead, but sleeping. As little as possible was changed of a liturgy enrooted in the Holy Soil, and thus it came to pass that in the narrow, sunless, stony streets of European ghettos shambling students and peddlers offered metaphorical first-fruits in ingenious lyrics, and celebrated the ancient harvest festival of Palestine in pious acrostics. Never was there such an example of the dominance of the word. Life was replaced by Literature. What wonder if the love of Zion grew mainly literary, so that even the passion of a Jehuda Halvei for Palestine has been dubbed more of the passion of a troubadour for a visionary mistress than a patriotism with its roots in reality.

Fantastic and factitious though this love of Zion was, yet, supplemented by eschatological superstitions, it made Jerusalem still the mystic City of God, still the capital of the Millennium, still the symbol of Israel's misery and Israel's ultimate regeneration. And, to this day, in the ghettos of New York and Philadelphia, the 'messenger of Zion' may be met on the trolley car, going his rounds, collecting the humble cents which enable graybeards to pore over moth-eaten Talmuds in the Holy City.

Thus, although Jerusalem has remained throughout the entire Christian era in the hand of foreign conquerors, the Jews have always retained some sense of being colonists whose mother city was in Asia. Some day it would be their own city again—but in God's good time, in a whirl of miracles! Hence, except under the ephemeral inspiration of pseudo-Messiahs, Zionism was never a matter of practical politics: it was a shadowy, poetic ideal, outside life; a romantic reminiscence. Old men went to Jerusalem to die—not to live. Its earth was imported—but to be placed in coffins. In practice, Jews have always been ardently attached to the country of their birth, and if they have seemed to remain apart, Ezra and Nehemiah are largely responsible, those zealots (more Mosaic than Moses) who stamped out marriages with other peoples, even when the strangers accepted Judaism. The very rabbis of the Talmud could not endorse this principle of compulsory mutual intermarriage, yet in practice it became the rule, and an institution designed in the fifth century before Christ to preserve the religion served in the Dark Ages of Christendom to preserve the race. Religion and race have, indeed, come to seem one and the same thing. And against this people, already doubly cut off from mankind, the Christian raised his material wall of separation, and created the ghetto.

But the ghetto fell at last, and separatist legislation tottered, and emancipation brought another development. With the liberal movements of the eighteenth century, Jews began to form part of the general life. The aspiration for Palestine was felt to be incongruous, even as a far-off religious ideal. Again it was proclaimed—by Moses Mendelssohn this time—that Judaism is larger than a land: that its future realm must be that of spiritual conquest. But in America, whither this doctrine spread in its broadest form, it was not followed by its logical outcome—by marriage outside the faith and the welcome of converts. Jewish life in the United States, instead of becoming expansive and spiritual, has drawn itself together in secular clubs. In Australia, on the other hand, where orthodoxy is still the professed creed, outside marriage has become frequent. In Germany, the notion that modern Judaism and Christianity are not very far apart has led many to baptism. A large minority everywhere—cultured, or rich, or callous—has succumbed to the general indifferentism of the modern world.

Thus, today Israel is face to face with a menace of disintegration more formidable than the legions of Titus.

To read the history of Israel is like reading a romance of perilous adventure written in the first person. Again and again the hero may be divided from death by a hair's breadth, yet we know that he will always come through safely, since is he not here, narrating? During the thirty centuries or so of his national existence, Israel has been perpetually stumbling on the verge of the abyss of annihilation, yet always he has recovered his footing. But Israel's serial is 'to be continued,' and who can say it will not 'end happily' after all?

II

As the century of Israel's disintegration closes, however, a new phenomenon meets our astonished eyes. It is 'Zionism.'

Zionism, in its latest official exposition, aims at securing a public legally assured home in Palestine for those Jews who are unable or unwilling to assimilate. It is not the movement that George Eliot's Mordecai dreamed, nor that which Rabbi Mohilewer of Russia initiated. The advent of Doctor Herzl has stamped Zionism with 'modernity.' In the Austrian journalist's first published scheme of a Jewish State, indeed, Palestine played no necessary part. Herzl, whose instrument of national regeneration is the bank, for dealing with the Sultan and subsidizing the selected immigrants, was never, despite the date of his advent, *fin de siecle* (which seems to imply a certain flippancy), but prophetically twentieth century. He would, if it were possible, lead back his people to Palestine by the moving sidewalk of the Paris Exposition. Withal a charming, magnetic, even poetic personality, a more diplomatic and domesticated Lassalle.

But the deeper issues and sequels of the movement will develop themselves with the material success, and the present leaders might quite conceivably be swept away by spiritual floods they have themselves let loose. The Orthodox Jewish Congregational Union of America, at the convention of June 8, 1898, while maintaining that 'the restoration to Zion is the legitimate aspiration of scattered Israel,' likewise declared, 'we reaffirm our belief in the coming of a personal Messiah.' The agents of political Zionism—men like Max Nordau, or Mandelstamm, the great Russian oculist, or Marmorek, of the Pasteur Institute—can no more control the religious future of Judaism than they can control the mystic interpretation which Christendom would put upon their success. Men are only instruments. And each must do the work he sees to hand.

At present, though orthodox rabbis are working amicably with ultra-modern thinkers, the movement is political, and more indebted to the pressure of the external forces of persecution than to internal energy and enkindlement. Yet in truth could any but a political cause unite the Jew of the East with the Jew of the West? And, viewed merely on its prosaic side, Zionism is by no means a visionary scheme. The aggregation of Jews in Palestine is only a matter of time—already they form a third of its population—and it is better that they should be aggregated there under their own laws and religion and the mild suzerainty of the Sultan than under the semi-barbarous restrictions of Russia or Rumania, and exposed to recurrent popular outbreaks. True, Palestine is a ruined country, and the Jews are a broken people. But neither is beyond recuperation. Palestine needs a people; Israel needs a country. If, in regenerating the Holy Land, Israel could regenerate itself, how should the world be other than the gainer? In the solution of the problem of Asia which has succeeded the problem of Africa, Israel might play no significant part. Already the colony of Rishon le Zion has obtained a gold medal for its wines from the Paris Exposition—which is not prejudiced in the Jew's favor. We may be sure the spiritual wine of Judea would again pour forth likewise—that precious vintage which the world has drunk for so many centuries. And, as the scientific activities of the colonization societies would have paved the way for the pastoral and commercial future of Israel in its own country, so would the rabbinical sing-song in musty rooms prove to have been but the unconscious preparation of the ages for the Jerusalem University.

But Palestine belongs to the Sultan, and the Sultan refuses to grant the coveted Judean Charter, even for dangled millions. Is not this fatal? No; it matters as little as that the Zionists could not pay the millions, if suddenly called upon. They have collected not two and a half million dollars. But there are millionaires enough to come to the rescue once the charter was dangled before the Zionists. It is not likely that the Rothschilds would see themselves ousted from their familiar headship in authority and well-doing. Nor would the millions left by Baron Hirsch be altogether withheld. And the Sultan's present refusal is equally unimportant because a national policy is independent of transient moods and transient rulers. The only aspect that really matters is whether Israel's face be or be not set steadily Zionward—for decades, and even for centuries. Much less turns on the Sultan's mind than on Doctor Herzl's. Will he lose patience? For leaders like Herzl are not born in every century.

III

Apart from its political working, Zionism forces upon the Jew a question the Jew hates to face.

Without a rallying centre, geographical or spiritual; without a Synhedrion; without any principle of unity or of political action; without any common standpoint about the old Book; without the old cement of dictory laws and traditional ceremonies; without even ghetto walls built by his friend the enemy, it is impossible for Israel to persist further, except by a miracle—of stupidity.

It is a wretched thing for a people to be saved only by its persecutors or its fools. As a religion, Judaism has still magnificent possibilities, but the time has come when it must be denationalized or renationalized."[445]

Racist Zionists were troubled by the fact that the Jews of Western Europe and America were assimilating into Gentile society. The Zionists feared that within a few generations the "Jewish race" would become impure and then extinct. Kerensky immediately emancipated the Jews after the Russian Revolution of 1917, and Lenin made anti-Semitism an offense punishable by death.[446] This opened the door to Jewish assimilation in the East and the further dilution of holy Jewish blood.

The Zionists believed that if they could form a racist apartheid "Jewish State" they could preserve the integrity of the "Jewish race". However, most Jews were not Zionists and few Jews were foolish enough to abandon their

---

**445**. I. Zangwill, "Zionism and the Future of the Jews", *The World's Work*, Volume 6, Number 5, (September, 1903), pp. 3895-3898.

**446**. V. I. Lenin, "Anti-Jewish Pogroms", *Collected Works*, Volume 29, English translation of the Fourth Russian Edition, Progress Publishers, Moscow, (1972), pp. 252-253. *See also:* D. Fahey, *The Mystical Body of Christ in the Modern World*, Browne and Nolan Limited, London, (1935), p. 251. *See also:* G. B. Shaw, *The Jewish Guardian*, (1931). *See also: Congress Bulletin*, American Jewish Congress, New York, (5 January 1940). *See also: The Jewish Voice*, (January, 1942). *See also:* G. Aronson, *Soviet Russia and the Jews*, American Jewish League against Communism, New York, (1949). *See also:* J. Stalin, "Anti-Semitism: Reply to an Inquiry of the Jewish News Agency in the United States" (12 January 1931), *Works*, Volume 13, Foreign Languages Publishing House, Moscow, (1955), p. 30. *See also:* S. S. Montefiore, *Stalin: The Court of the Red Star*, Vintage, New York, (2003), pp. 305-306. *See also:* "Anti-Semitism", *Great Soviet Encyclopedia: A Translation of the Third Edition*, Volume 2, (1973), pp. 175-177, at 176. *See also:* "Jews", *Great Soviet Encyclopedia: A Translation of the Third Edition*, Volume 9, Macmillan, New York, (1975), pp. 292-293, at 293. *See also:* N. S. Alent'eva, Editor, *Tseli i metody voinstvuiushchego sionizma*, Izd-vo polit. lit-ry, Moskva, (1971). Н. С. Алентьева, Редактор, Цели и методы воинствующего сионизма, Издательство Политической Литературы, Москва, (1971).

homes around the world and move to the desert in order to gratify the Rothschilds' desires to become King of the Jews. Most Jews did not oblige the racist Jews' desire to segregate them from the rest of humanity.

The Zionists believed that the only hope they had to keep the Jews segregated and to preserve the "Jewish race" was to put a virulently anti-Semitic dictator in charge of Europe, who would remind the Jews that they were Jews and force them into segregation so that they could then be forcibly expelled to Palestine.

## 5.6.1 The Zionist Myth of the Extinction of the "Jewish Race" Through Philo-Semitism and Assimilation

Hitler's propaganda asserted that both Capitalism and Communism were Jewish conspiracies to rule the world—Capitalism through alleged Jewish monopolies, high finance and decadence, and Communism through alleged Jewish revolution which destroyed the fabric of Western Civilization. Most Communists saw Socialism as an intermediary stage between Capitalism and the alleged true democracy of Communism. As an ideology, National Socialism, itself a socialist revolutionary movement, had much more in common with Communism than it did with Capitalism. Hitler was not bent on destroying Socialism, but rather promoting it in the undemocratic form of pure and final nationalistic racist Fascism—much like the Zionist David Ben-Gurion; and Hitler was determined that Germans should lead the world revolution as its alleged natural masters—much like Ben-Gurion's call for Jews to lead the world revolution, as God allegedly intended. Rac*ism* was the primary *ism* in Hitler's propaganda. For him, the state's primary function was the preservation of the "race". Much like racist Zionist Moses Hess, Hitler believed that the democratic and artificially international aspirations of Communism made it weak and diminished individual greatness for the sake of a sentimental and self-defeating idealism that largely only resulted in the "degeneration" of "pure" races. Hitler, like Stalin, wanted the masses to be uneducated and subjugated. He believed the masses are destined to be led, not to lead.

Max Planck was one of many leading scientists who dreaded Hitler's attacks on the German educational system. It seemed Hitler was out to destroy Germany by undermining the future of its youth and by leading Germany into perpetual war with nation after nation under the worst of conditions with almost no hope of ultimate victory. The Zionists had long hoped to destroy Germany, in which Jewish assimilation found its most comfortable home. Hitler provided the horrific stimulus which led a significant number of Jews into Zionism, a goal the Zionists, Christian and Jew, had not until then achieved, and which had remained as the only stumbling block to the fulfilment of their Apocalyptic dreams of a "restored" Israel—they did not care about what the majority of Jews wanted for themselves—as David Ben-Gurion stated in 1944 in the darkest days of the Holocaust in full knowledge that European Jewry (the Eastern "Red Assimilationist" and Western "rich assimilationist" Jews Ben-Gurion hated) had been decimated by the Nazis,

"One Degania [resident of the first communal settlement of Zionists in Palestine] is worth more than all the 'Yevsektzias' [Jewish Bolsheviks who sought to secularize Jews] and assimilationists in the world."[447]

In 1937—one year before *Kristallnacht*, Zionist Chaim Weizmann had fatalistically welcomed the idea that "only a remnant shall survive" and a had called "The old ones[... ] dust, economic and moral dust in a cruel world."[448] *Amos* 9:8-10 states,

"8 Behold, the eyes of the Lord GOD *are* upon the sinful kingdom, and I will destroy it from off the face of the earth; saving that I will not utterly destroy the house of Jacob, saith the LORD. 9 For, lo, I will command, and I will sift the house of Israel among all nations, like as *corn* is sifted in a sieve, yet shall not the least grain fall upon the earth. 10 All the sinners of my people shall die by the sword, which say, The evil shall not overtake nor prevent us."

*See also: Isaiah* 1:9; 6:9-13; 10:20-22; 11:11-12; 17:6; 37:31-33; 41:9; 42; 43; 44. *Ezekiel* 20:38; 25:14. *Daniel* 12:1, 10. *Obadiah* 1:18. *Micah* 5:8. *Romans* 9:27-28; 11:1-5. Zionist Nazis provided the Palestinian Zionists with a screen with which to sift out the assimilationist and Orthodox Jews of Continental Europe, and a sword with which to kill them.

Zionists feared that Capitalism was leading wealthy Jews to assimilate and that Communism would provide Jews with a sanctuary in which they would assimilate. Some had already argued in 1917 that the Russian Revolution made Zionism obsolete—a thought that terrified Zionist leader Chaim Weizmann, who otherwise had Socialist leanings. *The New York Times* reported on 23 December 1917 on page 7,

## "JERUSALEM FOR IDEALISTS.

Rev. Dr. Harris Discusses Effect of
Its Capture on Zionism.

The cause of Zionism as promoted by the capture of Jerusalem by the British was discussed by the Rev. Dr. Maurice H. Harris at the Temple Israel in Harlem yesterday.

'There will be less need now of a Jewish homeland,' said Dr. Harris, 'because the days of Jewish persecution are over. Whatever may happen in Russia and Rumania, we are satisfied that the era of the pale of settlement, anti-Jewish laws and pogroms has come to an end. Palestine will not appeal

---

[447]. D. Ben-Gurion, *Ba-Maarachah*, Volume 3, Tel-Aviv, (1948), pp. 200-211, English translation in A. Hertzberg, *The Zionist Idea*, Harper Torchbooks, New York, (1959), pp. 606-619, at 616.
[448]. C. Weizmann, *Chaim Weizmann*, V. Gollanez, London, (1945); quoted in A. Hertzberg, *The Zionist Idea*, Harper Torchbooks, New York, (1959), pp. 583-588, at 588.

to the enterprising on economic grounds, although it is offering opportunities to the farmers in the cultivation of oranges, barley, and olive oil. New harbors have been planned at Jaffa and Haifa, and a new railway is being carried to Port Said. With intensive cultivation, Palestine could maintain a population of 2,000,000 where there reside now but 600,000. But opportunities such as these can be found elsewhere and in greater abundance in this great Western Continent of North and South America.

'The Jew who bends his steps to Judea today will be the idealist who feels that 'not on bread alone doth man live.' He will not go there to make money, but because it is the Holy City. Jerusalem is still a name to conjure with. This great offer, whatever be its ultimate form, whether a dependent colony or an independent State, will enable our brethren to create for themselves a wholly Jewish environment. No longer a small minority living more or less on sufferance among an overwhelming majority of alien faiths, they will be able to impress their particular genius on the institutions of the country that will become theirs.'"

Even before World War I, racist Zionist Israel Zangwill voiced his concern that the emancipation of Russian Jews would lead to the "degeneration" of the Jewish race through interbreeding with allegedly inferior Slavs. Zangwill reiterated the common political Zionist theme, which alleged that anti-Semitism benefits Jews by maintaining their racial purity, and that philo-Semitism among Gentiles is destructive to the "Jewish race". Zangwill wrote in his booklet *The Problem of the Jewish Race*, Judean Publishing Company, New York, (1914), pages 7-8, 10-11, and 17-20; which was first published as an article, "The Jewish Race", *The Independent*, Volume 71, Number 3271, (10 August 1911), pp. 288-295, at 289-291, 293-295,

"But if from the Gentile point of view the Jewish problem is an artificial creation, there is a very real Jewish problem from the Jewish point of view—a problem which grows in exact proportion to the diminution of the artificial problem. Orthodox Judaism in the diaspora cannot exist except in a Ghetto, whether imposed from without or evolved from within. Rigidly professing Jews cannot enter the general social life and the professions. Jews *qua* Jews were better off in the Dark Ages, living as chattels of the king under his personal protection and to his private profit, or in the ages when they were confined in Ghettos. Even in the Russian Pale a certain measure of autonomy still exists. It is emancipation that brings the 'Jewish Problem.' It is precisely in Italy with its Jewish Prime Minister and its Jewish Syndic of Rome that this problem is most acute. The Saturday Sabbath imposes economic limitations even when the State has abolished them. As Shylock pointed out, his race cannot eat or drink with the Gentile. Indeed, social intercourse would lead to intermarriage. Unless Judaism is reformed it is, in the language of Heine, a misfortune, and if it is reformed, it cannot logically confine its teachings to the Hebrew race, which, lacking the normal protection of a territory, must be swallowed up by its proselytes. [***] Nor is there anywhere in the Jewish world of to-day any centripetal force to counteract these universal tendencies to dissipation. The religion is shattered into as many fragments as the race. After the fall of Jerusalem the Academy of Jabneh carried on the authoritative tradition of the *Sanhedrin*. In the Middle Ages there was the *Asefah* or Synod to unify Jews under Judaism. From the middle of the sixteenth to the middle of the eighteenth century, the *Waad* or Council of Four Lands legislated almost autonomously in those Central European regions where the mass of the Jews of the world was then congregated. To-day there is no center of authority, whether religious or political. Reform itself is infinitely individual, and nothing remains outside a few centers of congestion but a chaos of dissolving views and dissolving communities, saved from utter disappearance by persecution and racial sympathy. The notion that Jewish interests are Jesuitically federated or that Jewish financiers use their power for Jewish ends is one of the most ironic of myths. No Jewish people or nation now exists, no Jews even as sectarians of a specific faith with a specific center of authority such as Catholics or Wesleyans possess; nothing but a multitude of individuals, a mob hopelessly amorphous, divided alike in religion and political destiny. There is no common platform from which the Jews can be addressed, no common council to which any appeal can be made. Their only unity is negative—that unity imposed by the hostile hereditary vision of the ubiquitous Haman. [\*\*\*] The labors of Hercules sink into child's play beside the task the late Dr. Herzl set himself in offering to this flotsam and jetsam of history the project of political reorganization on a single soil. But even had this dauntless idealist secured co-operation instead of bitter hostility from the denaturalized leaders of all these Jewries, the attempt to acquire Palestine would have had the opposition of Turkey and of the 600,000 Arabs in possession. It is little wonder that since the great leader's lamentable death, Zionism—again with that idealization of impotence—has sunk back into a cultural movement which instead of ending the Exile is to unify it through the Hebrew tongue and nationalist sentiment. But for such unification, a religious revival would have been infinitely more efficacious: race alone cannot survive the pressure of so many hostile milieux—or still more parlous—so many friendly. [\*\*\*] In the diaspora anti-Semitism will always be the shadow of Semitism. The law of dislike for the unlike will always prevail. And whereas the unlike is normally situated at a safe distance, the Jews bring the unlike into the heart of every milieu and must thus defend a frontier-line as large as the world. The fortunes of war vary in every country, but there is a perpetual tension and friction even at the most peaceful points, which tend to throw back the race on itself. The drastic method of love— the only human dissolvent—has never been tried upon the Jew as a whole, and Russia carefully conserves—even by a ring fence—the breed she designs to destroy. But whether persecution extirpates or brotherhood melts, hate or love can never be simultaneous throughout the diaspora, and so there will probably always be a nucleus from which to restock this eternal type. But what a melancholy immortality! 'To be *and* not to be'—that is a question beside which Hamlet's alternative is crude. [***] But abolition of the Pale and the introduction of Jewish equality will be the deadliest blow ever aimed at Jewish nationality. Very soon a fervid Russian patriotism will reign in every Ghetto and the melting-up of the race will begin. But this absorption of the five million Jews into the other hundred and fifty millions of Russia constitutes the Jewish half of the problem. It is the affair of the Jews. [***] Moreover, while as already pointed out the Jewish upper classes are, if anything, inferior to the classes into which they are absorbed, the marked superiority of the Jewish masses to their environment, especially in Russia, would render *their* absorption a tragic degeneration."

As early as 1903, Zangwill wrote,

"At present, though orthodox rabbis are working amicably with ultra-modern thinkers, the movement is political, and more indebted to the pressure of the external forces of persecution than to internal energy and enkindlement. [***] Apart from its political working, Zionism forces upon the Jew a question the Jew hates to face. Without a rallying centre, geographical or spiritual; without a Synhedrion; without any principle of unity or of political action; without any common standpoint about the old Book; without the old cement of dictory laws and traditional ceremonies; without even ghetto walls built by his friend the enemy, it is impossible for Israel to persist further, except by a miracle—of stupidity. It is a wretched thing for a people to be saved only by its persecutors or its fools. As a religion, Judaism has still magnificent possibilities, but the time has come when it must be denationalized or renationalized."[449]

Zangwill was not alone in his beliefs. Racist Zionist Ignatz Zollschan worried that intermarriage and the emancipation of Russian Jews would tragically put an end to the "Jewish race". Zollschan stated at least as early as 1914,

"These four classes, however, which I have attempted to portray with a few bold strokes, are not fixed groups, but cross-cuts at at different positions, of a constantly flowing stream whose source to-day is in orthodox Judaism of eastern Europe, and which wends its way into the sea of Christianity. The process of infiltration of modern culture into Judaism goes on incessantly, and in the same manner, orthodox Judaism constantly yields to the members of the second tolerant class. The latter gradually yields to the class of reformers and freethinkers, and finally baptism, and especially intermarriage, leads the Jews to Christianity. These four classes can also be represented as four consecutive generations. Four or five generations

---

[449]. I. Zangwill, "Zionism and the Future of the Jews", *The World's Work*, Volume 6, Number 5, (September, 1903), pp. 3895-3898, at 3897-3898.

intervene between our own age and the time of Mendelssohn. It is a melancholy reflection, that hardly one of the Jews who lived at that time in Berlin has any Jewish descendants.

This process would also assume equally large dimensions in Russia, if the Jews were granted equal rights and if the Pale of Settlement were removed. The amelioration of the material conditions would remove the Ghetto environment which is one of the factors in preserving orthodox Judaism. But still more important would be the elimination of the second factor, namely, the keeping together of the Jews in one compact mass. If it were possible for the Russian Jews to spread themselves over the immense Russian Empire, the Jewish population in that country would not be denser than in western Europe. Thereby the progressive changes which exercise their destructive influences upon the western Jews would also apply to their Russian brethren. For the country that is more developed, serves as a picture of the future of the one that is less developed. Accordingly, eastern Jews will after some time apparently find themselves in the same position as the western Jews are to-day.

We may epitomise our conclusions from the processes described above, as follows: When the Jews in the diaspora became prosperous, assimilation which appears on the scene takes them away more or less from Judaism. It is mainly when they are oppressed, when they are in economically unfavorable conditions, that the Ghetto environment, in its old sense, is still retained. And although conditions to-day are not favorable in all countries, the beginning of this development can he recognized everywhere. Under favorable material conditions, and through the prevalence of secular education, Judaism, on account of its being scattered among nations of an alien race, is in danger of being disintegrated and destroyed, since the influence of ceremonial religion is waning."[450]

Jabotinsky advocated a racist *Blut und Boden* policy, before Hitler. In 1904, racist Zionist Vladimir Jabotinsky wrote, arguing that emancipation in Russia without the formation of a Jewish state would be a mistake and that he would rather see the Jews in a Ghetto, than see the Jews emancipated without a Jewish state,

> "[I]t is clear that the source of national feeling to be sought not in a man's education. And what is that? I contemplated this question and arrived at the conclusion that it lies in a man's blood. And I abide by this outlook even at present. That feeling of national ego is deeply ingrained in a man's 'blood'; in his racio-physical type, and in that alone. We do not believe that the independent spirit lies in the body; we believe that a man's spiritual outlooks are primarily determined by his physical structure. No education—neither

---

**450**. I. Zollschan, "The Significance of Mixed Marriage", *Jewish Questions: Three Lectures*, New York, Bloch Pub. Co., (1914), pp. 20-42, at 31-33.

the family or the surroundings, can transform a man on whom nature has bestowed a calm temperament into a stormy and tempestuous character and vice versa. The spiritual structure of a people reflects the physical type in a more pronounced and full-form than the spiritual outlook of the individual. The nation molds its national and spiritual character in that it adapts that character to its physical-racial type, and no other spiritual outlook on the basis of the physical type is possible. From the point of view of customs and manners, form of life changes of course as time goes on, but the national ego is to be traced not in customs and manner. And when we speak of the structure of a spiritual ego, we obviously have in mind something deeper. This something expresses itself at different times in various external manifestations, dependent on the period and on the social surroundings, but this 'something' in itself remains unchanged and immutable so long as the physical-racial type is preserved. For that reason we do not believe in spiritual assimilation. It is unconceivable, from the physical point of view, that a Jew born to a family of pure Jewish blood over several generations can become adapted to the spiritual outlooks of a German or a Frenchman. A Jew brought up among Germans may assume German customs, German words. He may be wholly imbued with that German fluid but the nucleus of his spiritual structure will always remain Jewish, because his blood, his body, his physical-racial type are Jewish. The basic features of his spirit are a reflection of the basic traits of his body. And a man whose body is Jewish cannot possibly mold within himself the soul of a Frenchman. The spiritual assimilation of peoples whose blood is different is impossible of effectuation. It is impossible for a man to become assimilated with people whose blood is different from his own. In order to become truly assimilated he must change his body. He must become one of them in blood. In other words, he must bring into the world through a whole string of mixed marriages, over a period of many scores of years, a great-great-grandson in whose veins only a minute trace of Jewish blood has remained, for only that great-great-grandson will be a true Frenchman or a true German by his spiritual structure. There is no other way. So long as we are Jews in blood, the sons of a Jewish father and mother, we may lie open to oppression, degradation and degeneration but not to the dangers of assimilation in the true sense of the word—assimilation in the sense of a complete disappearance of our spiritual ego. Such danger does not threaten us. There can be no assimilation so long as there is no mixed marriage. But the moment that the number of mixed marriages is on the increase, and account for the majority of marriages, only then will the children be half Jews in blood and so the first breach will be created for the inception of true and complete assimilation which can never be remedied. An increase in the number of mixed marriages is the only sure and infallible means for the destruction of nationality as such. All the nations that have disappeared in the world (apart from those, of course, who were completely massacred or who disappeared as a result of abnormal conditions of existence) were swallowed up in the chasm of mixed marriages. [***] In the First place, they said the Jews, at any rate in Russia, densely populate certain towns so that there is no ground to believe that they will all arise and scatter over the length and breadth of Russia when they will be allowed to do so. Large Jewish masses will remain living within the present 'pale of residence' and there they will by no means be such a negligible minority which will necessarily lead to an overwhelming increase of mixed marriages. I should like to reply to this argument as follows: Even at present, the Jews constitute only about 14% of the general population in the 'pale of residence.' If the gates of exit should be opened, this percentage would obviously be considerably reduced through emigration to other regions. True, the Jews constitute a much larger percentage of the urban population, nonetheless they are a minority also there. However, with the industrial development of the country, the stream of large numbers from the villages to the towns will increase, so as to double, or perhaps treble the number of non-Jewish residents in the towns, with the result that the Jews are likely to become a minority even in Berditchev. [***] [Y]our call will lead to the ancient grave of assimilation[.]"[451]

Before Zollschan, Zangwill and Jobotinsky, Communist Zionist Nachman Syrkin worried that Liberalism and Socialism were murdering the Jewish nation through assimilation. He feared that liberty, equality and fraternity led to a patriotic spirit in Jews for nations other than Israel. Syrkin dreaded the process of assimilation, which he saw stemming from the emancipation of Jews in the French Revolution and Napoleon's conquests, and accelerated by the loss of religiosity of the modern Jews of his day, as well as by Jewish involvement in Socialism. Indeed, Napoleon at one point appeared to mandate assimilation.[452] Syrkin advocated, "a true Jewish socialism, free of every servile trace of assimilation."[453] Syrkin stated in 1898, long before "Red Assimilation" in the Soviet Union became a reality,

> "To the Jewish socialists, socialism meant, first of all, the abandonment of Jewishness, just as the liberalism of the Jewish bourgeoisie led to assimilation. And yet, this tendency to deny their Jewishness was unnecessary, being prompted by neither socialism nor liberalism. It was a product of the general degeneration and demoralization of the Jews; Judaism was dropped because it conferred no benefits in the new world of free

---

**451**. V. Jabotinsky, "A Letter on Autonomy", T. Zohar, Editor, *Israel among the Nations: Selection of Zionist Texts*, World Zionist Organization Dept., Research Section, Jerusalem, (1966); reprinted in L. Brenner, *51 Documents: Zionist Collaboration with the Nazis*, pp. 7-20, at 10-14.
**452**. S. Schwarzfuchs, *Napoleon, the Jews, and the Sanhedrin*, Routledge & Kegan Paul, London, Boston, (1979), pp. 99-100.
**453**. N. Syrkin, under the nom de plume "Ben Elieser", *Die Judenfrage und der socialistische Judenstaat*, Steiger, Bern, (1898); English translation in A. Hertzberg, *The Zionist Idea*, Harper Torchbooks, New York, (1959), pp. 333-350, at 344.

competition."[454]

The Zionists crafted an alleged tautology of Jewish options in the age of enlightenment in order to justify their pre-existent racial prejudice. Non-Zionist Jews argued that the enlightenment would eventually end anti-Semitism. Zionists promoted anti-Semitic agitation to prevent the assimilation they believed followed from the enlightenment and emancipation. Moshe Leib Lilienblum succinctly iterated the three option theme of the Zionists at least as early as 1883:

"1. *To remain in our present state, to be oppressed forever, to be gypsies, to face the prospect of various pogroms and not be safe even against a major holocaust* [Note the term—CJB].

2. *To assimilate, not merely externally but completely within the nations among whom we dwell: to forsake Judaism for the religions of the gentiles, but nonetheless to be despised for many, many years, until some far-off day when descendants of ours who no longer retain any trace of their Jewish origin will be entirely assimilated among the Aryans.*

3. *To initiate our efforts for the renaissance of Israel in the land of its forefathers, where the next few generations may attain, to the fullest extent, a normal national life.*

Make your choice!"[455]

The Zionists saw the Nazis as their salvation. Since most Jews were choosing assimilationist option number two after the First World War, option three could only be achieved through option number one. Lenni Brenner wrote,

"Only the defeat of Nazism could have helped the Jews, and that could only have happened if they had united with the anti-Nazi working class on a programme of militant resistance. But this was anathema to the ZVfD [Zionist Federation of Germany] leadership who, in 1932, when Hitler was gaining strength by the day, chose to organise anti-Communist meetings to warn Jewish youth against 'red assimilation'."[456]

Karl Kautsky wrote in the second edition of *Rasse und Judentum*, published in English as *Are the Jews a Race?*, Chapter 11, "Pure Races and Mixed Races", International Publishers, New York, (1926):

---

[454]. N. Syrkin, under the nom de plume "Ben Elieser", *Die Judenfrage und der socialistische Judenstaat*, Steiger, Bern, (1898); English translation in A. Hertzberg, *The Zionist Idea*, Harper Torchbooks, New York, (1959), pp. 333-350, at 341.

[455]. M. L. Lilienblum, "The Future of Our People", in A. Hertzberg, *The Zionist Idea*, Harper Torchbooks, New York, (1959), pp. 173-177, at 177.

[456]. L. Brenner, *Zionism in the Age of the Dictators*, Chapter 3, Croom Helm, London, L. Hill, Westport, Connecticut, (1983), p. 32; Brenner refers to: D. Niewyk, *The Jews in Weimar Germany*, Louisiana State University Press, Baton Rouge, (1980), p. 30.

"WE cannot take leave of Zionism before discussing another one of its arguments, its last argument, which will lead us back to the question of race.

It may appear to be a paradox, but it is a fact, that not a few Jews look with some misgiving on the emancipation of the Jews in Eastern Europe. They understand, and rightly so, that this emancipation will extend into the east of Europe the assimilation of the Jews that has been going on in the west for some time. For when the artificial exclusiveness of the Jews is terminated, when the ghetto ceases to exist, their assimilation will become everywhere inevitable."

In 1922, Max Grunwald addressed Kautsky's work and reviewed several racial theories of Zionism in a series of articles, "Rasse, Volk, Nation" in the Jewish newspaper *Die Wahrheit* (Wien/Vienna).[457]

Kautsky noted in 1914 that the Zionists depended on the anti-Semite Houston Stewart Chamberlain for their racist Zionist ideology; referring to racist political Zionist Ignaz Zollschan's book *Das Rassenproblem unter besonderer Berücksichtigung der theoretischen Grundlagen der jüdischen Rassenfrage*, W. Braumüller, Wien, (1910). Constantine Brunner later emphasized the same point. Zollschan called for a "World Ghetto" (Theodor Herzl's phrase[458]) in Palestine in order to preserve the alleged racial purity of Jews. Though criticized by Kautsky, Zollschan's stance was lauded by the anti-Semitic segregationist Heinrich Class in 1912, further evincing the long-standing alliance between anti-Semites and Zionists,

"The Jews are members of an alien race who, despite partaking in the blessings of our culture, have not become Germans; they cannot do so in consequence of a fundamentally different outlook. Whoever sees Jews in this way will welcome the fact that among the Jews themselves a nationalistic movement, so-called Zionism, is gaining more and more adherents. We can only respect the Zionists. They admit openly and honestly that their nation *is* a nation whose basic traits are unalterable, surviving almost two thousand years of statelessness among other nations. They declare unconditionally that a real assimilation of the Jewish foreigners to the host peoples is impossible because of the natural law of race. This law is stronger than the outward will to adapt to the conditions of a foreign environment.

---

**457**. M. Grunwald, "Rasse, Volk, Nation", *Die Wahrheit* (Wien), Volume 38, Number 10, (12 May 1922), pp. 8-9; Number 11, (30 May 1922), pp. 9-10; Number 12, (9 June 1922), p. 6; Number 13, (23 June 1922), pp. 6-7; Number 14, (7 July 1922), p. 8; Number 15, (20 July 1922), pp. 7-8; Number 16, (11 August 1922), pp. 5-7; Number 17, (25 August 1922), pp. 5-7; Number 18/19, (21 September 1922), pp. 10-11; Number 20, (12 October 1922), p. 7; Number 22, (10 November 1922), p. 6.

**458**. T. Herzl, English translation by H. Zohn, R. Patai, Editor, *The Complete Diaries of Theodor Herzl*, Volume 1, Herzl Press, New York, (1960), p. 172.

The Zionists fully confirm what those who oppose the Jews on the standpoint of race have long maintained. Even though they are but a small troop in relation to the totality of their racial comrades, the truth that they proclaim can no longer be condemned to silence. *German and Jewish nationalists are of one opinion when it comes to the ineradicability of the Jewish race.* Who will then contest the right of the Germans to draw the necessary political consequences?"[459]

Lenni Brenner noted,

"What was needed was a popular Zionist version of the social-Darwinism which had swept the bourgeois intellectual world in the wake of Europe's imperial conquests in Africa and the East. The Zionist version of this notion was developed by the Austrian anthropologist Ignatz Zollschan. To him the secret value of Judaism was that it had, albeit inadvertently, worked to produce a wonder of wonders:

a nation of pure blood, not tainted by diseases of excess or immorality, of a highly developed sense of family purity, and of deeply rooted virtuous habits would develop an exceptional intellectual activity. Furthermore, the prohibition against mixed marriage provided that these highest ethnical treasures should not be lost, through the admixture of less carefully bred races... there resulted that natural selection which has no parallel in the history of the human race... If a race that is so highly gifted were to have the opportunity of again developing its original power, nothing could equal it as far as cultural value is concerned."[460]

Kautsky predicted that the Jews would disappear due to their assimilation following World War I, which emancipated the Jews of Russia. The First World War, which the Zionists planned would fulfill their dream of a Jewish state, instead rendered it obsolete, and they were the only group that had a vested interest in promoting discord in Europe, anti-Semitism and the segregation and expulsion of Jews. Others had learned that the emigration of large numbers of Jews from their country resulted in economic hardship, so the Zionists unwisely promised profits for all from racism. In 1881, the Nihilist Jews murdered Czar Alexander II, the great emancipator. Konstantine Petrovitch Pobiedonostsev (*also:* Constantin Pobedonostzeff), a man of Jewish appearance, won the favor

---

**459**. H. Class under the *nom de plume* D. Frymann, *Wenn ich der Kaiser wär': politische Wahrheiten und Notwendigkeiten*, Dieterich, Leipzig, (1912); English translation, R. S. Levy, "Daniel Freymann If I were the Kaiser (1912)", *Antisemitism in the Modern World: An Anthology of Texts*, D.C. Heath, Toronto, (1991), pp. 130-133, at 133.
**460**. L. Brenner, *Zionism in the Age of the Dictators*, Chapter 2, Croom Helm, London, L. Hill, Westport, Connecticut, (1983), p. 21; *citing:* I. Zollschan, *Jewish Questions: Three Lectures*, New York, Bloch Pub. Co., (1914), pp. 17-18

of Alexander III and "retaliated" with pogroms against the Jews, which, while certainly bad, were exaggerated in the international press. The alleged Czarist persecution of the Jews was used as a reason to sponsor the emigration of Jews to the West, which had a negative impact on the Russian economy. The Jewish population in the United States steadily rose from 200,000 in 1880, to several million by 1920. These were "Polish Jews" from the old Polish Empire, which had since been taken over by Russia—after the Shabbataian and Frankist Jews had largely destroyed Poland. The Sephardic and German Jews, who had settled in America, did not like these Eastern Jews, and sponsored legislation to prevent them from entering the country. They considered them to be of an inferior race and disposition, and would not intermarry with them.[461]

Albert Einstein's racist anti-assimilationist beliefs hailed from an ancient Jewish tradition of racism. Simon Dubnow wrote in 1905,

> "Assimilation is common treason against the banner and ideals of the Jewish people. [***] But one can never 'become' a member of a natural group, such as a family, a tribe, or a nation. One may attain the rights or privileges of citizenship with a foreign nation, but one cannot appropriate for himself its nationality too. To be sure, the emancipated Jew in France calls himself a Frenchman of Jewish faith. Would that mean, however, that he became a part of the French nation, confessing to the Jewish faith? Not at all. Because, in order to be a member of the French nation one must be a Frenchman by birth, one must be able to trace his genealogy back to the Gauls, or to another race in close kinship with them, and finally one must also possess those characteristics which are the result of the historic evolution of the French nation. A Jew, on the other hand, even if he happened to be born in France and still lives there, in spite of all this, he remains a member of the Jewish nation, and whether he likes it or not, whether he is aware or unaware of it, he bears the seal of the historic evolution of the Jewish nation."[462]

Dubnow argued from his Social Darwinist perspective that assimilated Jews were weeded out of the wonderful racist and tribal Jewish community in a process of natural selection, which strengthened the allegedly natural tendency of the Jewish community to be racist and tribal. Since assimilated Jews did not breed with racist Jews, but rather wandered off into other communities, only

---

[461]. B. J. Hendrick, "The Jews in America: I How They Came to This Country", *The World's Work*, Volume 44, Number 2, (December, 1922), pp. 144-161; **and** "The Jews in America: II Do the Jews Dominate American Finance?", *The World's Work*, Volume 44, Number 3, (January, 1923), pp. 266-286; **and** "The Jews in America: III The Menace of the Polish Jew", *The World's Work*, Volume 44, Number 4, (February, 1923), pp. 366-377; **and** "Radicalism among the Polish Jews", *The World's Work*, Volume 44, Number 6, (April, 1923), pp. 591-601.

[462]. S. Dubnow, *Die Grundlagen des Nationaljudentums*, Jüdischer Verlag, Berlin, (1905). English translation from K. A. Strom, Editor, *The Best of Attack! and National Vanguard Tabloid*, National Alliance, Arlington, Virginia, (1984), p. 60.

racist Jews would perpetuate the Jewish community, thereby creating a natural proclivity in the Jewish community to produce genetically racist Jews—which was a very good thing in Dubnow's mind. It is, therefore, easy to believe that these racist Jews organized to exterminate the assimilated Jews of Europe, thereby pruning off what they believed was a rotten limb of the Jewish family tree. Dubnow wrote in 1897,

> "While the mass of old-type orthodox Jews sees itself in practice as a religious nation and resists assimilation in the surrounding nations by the force of its faith, the assimilationist intelligentsia, on the other hand (mostly freethinkers or the neo-orthodox of the West), sees in Judaism only a religious community, a union of synagogues which imposes no national duties or discipline whatsoever on its members. According to this view, the Jew can become a member of another nation and remain a member of the Mosaic faith. He is a German Jew, for example, in the same way that there are German Protestants or German Catholics. It follows logically from this premise that a freethinking or non- religious Jew must be excluded from the community of Jews of the Mosaic faith. This corollary is usually glossed over so that whatever remains of Jewish 'unity' may not be disturbed. I shall discuss this doctrine, which was in vogue only a short time ago but has recently lost ground among its adherents, in greater detail in the following Letters. Here I only wish to point out that it contradicts both the traditional view of many past generations that the 'religious nation' must be kept pure, and the scientific view of the non-assimilability of the spiritual or cultural nation. This kind of doctrine comes neither from religion nor from science. It is the invention of naive ideologues, or calculating opportunists, who seek to justify by means of this artificial doctrine their desire to assimilate into the foreign environment in order to benefit themselves and their children. This is but a repetition of the process of natural selection and of the weeding out of those weak elements of the nation which are unable to bear the pressure of the alien environment."[463]

Long before the First World War, Voltaire stated in the end of Chapter 104 of his *Essai sur les Moeurs et l'Esprit des Nations, et sur les Principaux faits de l'Histoire Depuis Charlemagne Jusqu'à Louis XIII*, (1769); that should Gentiles—in Voltaire's view—become wise to the ways of Jews and prevent Jews from exploiting them, then rich Jews would abandon their religious superstitions and assimilate and the poor Jews would become thieves like Gypsies. According to Voltaire, whose work was well known, Jews would disappear through assimilation.[464] Again, the emancipation of Jews in Bolshevik

---

[463]. S. Dubnow, M. Selzer, Editor, "The Doctrine of Jewish Nationalism (1897)", *Zionism Reconsidered: The Rejection of Jewish Normalcy*, Macmillan, New York, (1970), pp. 131-156, at 146-147.
[464]. F. M. Arouet de Voltaire, *Histoire de Charles XII, Roi de Suède*, (1731); **and** *Dictionnaire Philosophique*, Multiple Editions; multiple English translations, including:

lands, and the assimilation of affluent Jews in capitalistic societies, greatly concerned the Zionists, who feared it would be the end of all Jews.

Before Voltaire, Spinoza noted that assimilation was causing the Jewish ethnicity to disappear. After Voltaire, Wellhausen, relying on Spinoza's observations, noted that emancipation was leading the Jews to assimilate and therefore to disappear—a fact that terrified the racist Zionists. Julius Wellhausen wrote in 1881,

> "The Jews, through their having on the one hand separated themselves, and on the other hand been excluded on religious grounds from the Gentiles, gained an internal solidarity and solidity which has hitherto enabled them to survive all the attacks of time. The hostility of the Middle Ages involved them in no danger; the greatest peril has been brought upon them by modern times, along with permission and increasing inducements to abandon their separate position. It is worth while to recall on this point the opinion of Spinoza, [*Footnote: Tract. Theol. Polit.* 0. 4, *ad fin.*] who was well able to form a competent judgment :—'That the Jews have maintained themselves so long in spite of their dispersed and disorganised condition is not at all to be wondered at, when it is considered how they separated themselves from all other nationalities in such a way as to bring upon themselves the hatred of all, and that not only by external rites contrary to those of other nations, but also by the sign of circumcision, which they maintain most religiously. Experience shows that their conservation is due in a great degree to the very hatred which they have incurred. When the king of Spain compelled the Jews either to accept the national religion or to go into banishment, very many of them accepted the Roman Catholic faith, and in virtue of this received all the privileges of Spanish subjects, and were declared eligible for every honour; the consequence was that a process of absorption began immediately, and in a short time neither trace nor memory of them survived. Quite different was the history of those whom the king of Portugal compelled to accept the creed of his nation; although converted, they continued to live apart from the rest of their fellow-subjects, having been declared unfit for any dignity. So great importance do I attach to the sign of circumcision also in this connection, that I am persuaded that it is sufficient by itself to maintain the separate existence of the nation for ever.' The persistency of the race may, of course, prove a harder thing to overcome than Spinoza has supposed; but nevertheless he will be found to have spoken truly in declaring that the so-called emancipation of the Jews must inevitably lead to the extinction of Judaism wherever the process is extended beyond the political to the social sphere. For the accomplishment of this

---

W. F. Flemming, *A Philosophical Dictionary*, Volume 6, Dingwall-Rock, New York, (1901), pp. 266-313; **and** *Essai sur les Moeurs et l'Esprit des Nations, et sur les Principaux faits de l'Histoire Depuis Charlemagne Jusqu'à Louis XIII*, Chapter 104, (1769); **and** *Philosophie Génerale: Métaphysique, Morale et Théologie*, Chez Sanson et Compagnie, Aux Deux-Ponts, (1792).

centuries may be required."⁴⁶⁵

Spinoza's observations are antedated by Biblical writings, which tell that God will punish assimilated Jews and pious Jews to remind all of Israel who God is. God punishes them with the sword and with fire and renders them ash. The punishment of the Jews through murderous anti-Semitism in order to drive them back to God is perhaps most strongly advocated in the books of *Deuteronomy* and *Ezekiel*, and in *Malachi* 4:1-6 it states,

> "1 For, behold, the day cometh, that *shall* burn as an oven; and all the proud, yea, and all that do wickedly, shall be stubble: and the day that cometh shall burn them up, saith the LORD of hosts, that it shall leave them neither root nor branch. 2 But unto you that fear my name shall the Sun of righteousness arise with healing in his wings; and ye shall go forth, and grow up as calves of the stall. 3 And ye shall tread down the wicked; for they shall be ashes under the soles of your feet in the day that I *shall* do *this*, saith the LORD of hosts. 4 Remember ye the law of Moses my servant, which I commanded unto him in Horeb for all Israel, *with* the statutes and judgments. 5 Behold, I *will* send you Elijah the prophet before the coming of the great and dreadful day of the LORD: 6 And he shall turn the heart of the fathers to the children, and the heart of the children to their fathers, lest I come and smite the earth *with* a curse."

American Zionist Richard Gottheil stated in 1898,

> "I KNOW that there are a great many of our people who look for a final solution of the Jewish question in what they call «assimilation.» The more the Jews assimilate themselves to their surroundings, they think, the more completely will the causes for anti-Jewish feeling cease to exist. But have you ever for a moment stopped to consider what assimilation means? It has very pertinently been pointed out that the use of the word is borrowed from the dictionary of physiology. But in physiology it is not the food which assimilates itself into the body. It is the body which assimilates the food. The Jew may wish to be assimilated; he may do all he will towards this end. But if the great mass in which he lives does not wish to assimilate him — what then? If demands are made upon the Jew which practically mean extermination, which practically mean his total effacement from among the nations of the globe and from among the religious forces of the world, — what answer will you give? And the demands made are practically of that nature."

Communist Zionist Nachman Syrkin wrote in 1898, referring to civil

---

**465**. J. Wellhausen, *Sketch of the History of Israel and Judah*, Third Edition, Adam and Charles Black, London, (1891), pp. 201-203.

assimilation as "national suicide",

> "The national suicide of the Jews would be a terrible tragedy for the Jews themselves, and that epoch would certainly be the most tragic in human history."[466]

The Zionists often repeated their alarmist rhetoric that Jews were in danger of extinction, not from anti-Semitism, but from philo-Semitism. At the turn of the century, Micah Joseph Berdichevski stated,

> "To be or not to be! To be the last Jews or the first Hebrews. Our people has come to its crisis, its inner and outer slavery has passed all bounds, and it now stands one step from spiritual and material annihilation. Is it any wonder that all who know in their hearts the burden, the implications, and the 'dread' of such an hour should pit their whole souls on the side of life against annihilation?"[467]

Ahad Ha'Am captured the spirit of panic some Zionists felt, in 1909,

> "To adopt a negative attitude toward the Diaspora means, for our present purpose, to believe that the Jews cannot survive as a scattered people now that our spiritual isolation is ended, because we have no longer any defence against the ocean of foreign culture, which threatens to obliterate our national characteristics and traditions, and thus gradually to put an end to our existence as a people. [***] We must secure our future by gathering the scattered members of our race together in our historical land (or, some would add, in some other country of their own), where alone we shall be able to continue to live as a people."[468]

Joseph Chaim Brenner stated in 1914,

> "And when we cry nowadays: 'If we do not become different—if now, the circumstances of our environment having changed, we do not really become a Chosen People—become, that is, like all other nations, each of whom is Chosen by itself—then we shall soon perish'; then what we mean is that we *shall* perish as a people—we *shall* die as a social entity."[469]

---

[466]. N. Syrkin, under the nom de plume "Ben Elieser", *Die Judenfrage und der socialistische Judenstaat*, Steiger, Bern, (1898); English translation in A. Hertzberg, *The Zionist Idea*, Harper Torchbooks, New York, (1959), pp. 333-350, at 343.
[467]. M. J. Berdichevski, "Wrecking and Building", in A. Hertzberg, *The Zionist Idea*, Harper Torchbooks, New York, (1959), pp. 293-295, at 293.
[468]. A. Ha-Am, "The Negation of the Diaspora", in A. Hertzberg, *The Zionist Idea*, Harper Torchbooks, New York, (1959), pp. 270-277, at 270-271.
[469]. J. H. Brenner, "Self-Criticism", in A. Hertzberg, *The Zionist Idea*, Harper Torchbooks, New York, (1959), pp. 307-312, at 307-308.

In 1917, Elisha Michael Friedman published the following article, which evinces the panic that had overtaken the Zionists, the belief that the Jewish 'race' would become 'extinct' through a process of assimilation, which had begun with the emancipation of the Jews in the French Revolution, and was continuing following the Russian Revolution. Friedman's article further evinces that the Zionists planned to use the First World War as an opportunity to argue that Jews were a nation deserving of official national status, not unlike many other small nations—and that it was the war which made Zionism appealing (note the common Zionist phrase "solution of the Jewish Question" to mean Zionism, which phrase the Nazis allegedly adopted in 1942—note further that it was the majority of Jews themselves who most strongly opposed Zionism and that the Zionists simply disregarded their wishes and sought to impose Zionism upon them through any and all means including war—note still further the Messianic belief that the Jews were inhibited from dominating humanity until restored to Palestine, at which time they would issue forth the Lord's proclamations onto humanity[470] in the same dictatorial fashion with which they demanded that Jews submit to Zionism, though they masked this desire with the more appealing assertion that they would offer benefits to humanity if only they were restored to Palestine, the benefit of their dictatorship over humanity—note even further still, the longing for segregation and the view that the Ghetto and enmity towards the Jew is the salvation for which the Zionists sought, that is to say that the Zionists created Nazism as a means to preserve the Jewish "race"):

## "ZIONISM AND THE AMERICAN SPIRIT"
### (*A New Perspective*)
### ELISHA M. FRIEDMAN

ZIONISM, for twenty centuries a religious yearning, and since twenty years a social program, did not appeal to the world at large until the advent of the great war. However, the attention that the minor peoples attracted during the course of the conflict set up a new standard in terms of which the Jewish problem might be reasoned out. Some, at least, of the blunders made in the treatment of the Jewish problem since the breaking up of the Ghetto, came from viewing it entirely as a theological problem instead of more broadly as a sociological one. But the tragedy of Belgium, the fate of Poland and the plea of the small nations, has furnished a new measure to apply to the whole Jewish problem.

Recent events have served to accentuate Zionism as an attempt at the solution of the Jewish question. The campaign in Palestine has dramatically brought the land of ancient Israel to the fore. Our own entry into the war, and the voice that we are to have at a coming peace conference, has given a peculiar turn to America's interest in the Zionist question.

Specifically, what is Zionism? Dating back as a hope, to the destruction of the Temple, and resuscitated as a project by its gifted leader, Theodore

---

[470]. M. Higger, *The Jewish Utopia*, Lord Baltimore Press, Baltimore, (1932).

Herzl, Zionism was formulated at the first International Zionist Congress in 1897 as a movement, aiming to secure for the Jewish people a publicly and legally assured home in Palestine. Much water has flowed to the sea since then. Ink has been spilled at and for the movement. However, the opposition was never on the part of non-Jews, strange to say, but only on the part of Jewish anti-Zionists, who either mistook the aims of the movement or had selfish fears as to their own status. However, twenty years of discussion have clarified thought on the subject, so that to-day it might be said that, regardless of political form, Zionism aims to preserve the Jewish people in their ancestral home that they may contribute, along with the other peoples of the world, to the enrichment of the world's culture. The Zionist community will affect not only the Jews who will return to Palestine after the war, but far more vitally, will it concern their scattered brothers in the various political states.

Not only because America numbers over a million Jews among her sons does the question interest us as Americans. In a more than selfish sense, America has a stake in the Zionist ideal. The righteous nation that fought for Cuba and then set her free, that alone of all the powers refused an unjust indemnity from China, that newly set for the world another example in high-minded rather than high-handed diplomacy in Colombia, that refused under powerful provocation to interfere with the liberty which the Mexican people were working out for themselves, and that entered the great war that 'the world might be made safe for democracy,' this friend of the small peoples has translated the square deal in terms of international affairs. It would be counter to every noble impulse to which America has given birth if she did not at an opportune moment, generously offer her aid toward the restoration of the Jewish people to a home and a center in Palestine. Because the ancient Hebrews were the first people that wrote democracy into its charter of government—the Bible, and because our republic was influenced at its birth by the Hebraic traditions that dominated New England, therefore when this ancient people is struggling to regain its position in the brotherhood of the world, America's interest in the freedom of small nations finds an added sanction.

## ZIONISM IN A NUT-SHELL

The emancipation of the Jew in Russia, while it may ameliorate the condition of the individual Jew, will not solve the problem of the Jewish people. Kicked and buffeted about for twenty centuries, it is now in danger of dissolution. The Jewish problem is not alone one of persecution. It involves as well the loss by an historic social group of its distinctive personality. The people that on its own soil produced the Bible has contributed nothing objective during two thousand years of dispersion, although it may have been the subject of an inspiring picture of persistence and martyrdom. It merely preserved itself. And when history brought to it political emancipation, it entered into spiritual sterility. Creature of persecution, the Jew, adaptable and imitative, assumed the hue of his surroundings with its decidedly materialistic tinge.

To-day, the Jewish people is slowly dying, culturally and socially. Lacking a home and a center of life, its religious reserves are being exhausted. The Jewish people may be contributing as individuals to the advance of civilization, but as a living, active, social group, they count for naught. In France, Italy and Spain, they have almost ceased to be. The Jews of England and Germany are following a similar course. Only the immigration from eastern Europe, hitherto the arena of persecution, is temporarily postponing—for but a few generations—the processes of decay of Jewish life in our own country.

The absorption of a scattered minority people is the inexorable law of history. Can the Jews hope to escape it? And if they will not, as they cannot, then emancipation will mean the complete dissolution, in Russia as well as in France, in the United States as well as in Italy, of this dispersed minor group.

Well, what of it? asks the anti-Zionist. The answer is—the harmony of world cultures. The world is the richer for the existence of a Belgian or a Polish people. Scatter them, and they will cease to produce Maeterlincks or Chopins, as the Jews have ceased to produce Isaiahs. Give the Jewish people Palestine, and a portion of them will produce distinctive and essential values to beautify and enrich human life. History proved it, when only forty-two thousand Jews returning to Palestine with Ezra, edited the Bible, and preserved the God idea, without which there would be to-day neither Christianity nor Mohammedanism. Indeed, the rest of the Jews, scattered over the world of that time were assimilated, but the nucleus in Palestine survived.

Without Zionism, without a center in Palestine, the Jews will, until they cease to exist, constitute an international irritation, as in the past—a problem in Germany as well as in Russia, or in any country where they as a scattered minority refuse to merge themselves completely and without qualification of blood or culture with the majority in every political state. And when they cease to be, as, without a center, they must, when the student will view them only as history, then the world will be the poorer, as it is for the passing of Greece and its art, or of Rome and its law, yes, poorer even as the world for the passing of the red man from this continent. At this perilous stage of his existence, the Jew has no other avenue of escape from dissolution but the reëstablishment by a portion of the people of a home and a center in Palestine. The disappearance of the non-Palestinian Jew will then be no loss to the world's cultures nor will his continued survival outside of Palestine be attended by any friction, as little as is the life of the Belgians in Russia or the Poles in England.

If only as a large social experiment Zionism should be tested out for its potentialities. For less than a century, the Jewish people have been freed from civil and political disabilities. Yet, in the train of emancipation, there followed various dangers. Released from pressure, the Jewish people have lost their distinctive spiritual bent, so that they no longer produce peculiar and essential social values, of any kind.

Worse, still, they are dying out. They are losing forever the power to create in a future new cultural values such as every people is capable of producing. The process of disintegration began in France after the French Revolution and in Germany after the razing of the Ghetto walls. The result is not sporadic or accidental in France or Germany, but continuous and inevitable everywhere—in England, in the United States, and, from now on, in Russia. The ferment of liberty will not spare the people that was hitherto encased within the walls of the Pale. During the process of disintegration, even, the Jews incur the prejudice of their fellowmen. Their death as a group is accompanied by all the pains of mortal dissolution—economic boycott in Poland, academic and military discrimination in Germany and social ostracism everywhere.

As a people, it is dying hard—a long, drawn-out and lingering death, for the basic law of existence is self-preservation. When a group becomes aware of approaching dissolution, it makes desperate efforts to live. Except for isolated cases, the scattered Jews will not readily merge their identity with the other peoples of the world: for, to do so would mean extinction, unless they previously established a center. This condition is unique with the Jews and does not hold for the members of any other people, for, when a Frenchman, Belgian, Pole or Irishman gives up his old connections, he leaves behind a great source of national life which can survive without his allegiance. Not so with the Jew or with any other dispersed group that has no territory.

Because his group is in continuous danger of dissolution, the Jew exhibits at all times a social psychology exhibited by other peoples only in times of war or other great dangers to the group. The lack of a center, *coupled* with the desire to continue to live, is the cause of the singular characteristics of the Jew. Loyalty in times of distress is a beautiful trait which is apotheosized in human relationships. Because the Jewish people, as a people, has always been in distress, down to this very day, its members have been keenly loyal to the group. Even though this loyalty is generalized and exhibits itself in many directions in relation to an employer, to an institution or to his native land, yet this trait in him alone is stigmatized as clannishness. Because, as a people, it dare not give up its identity, there has arisen against the entire group, regardless of the nobility of the character of any individual in it, a prejudice which varies in the degree of severity only with the breadth of vision of his neighbors. This anti-social feeling, in turn, develops a keen sensitiveness to criticism, a consciousness of self, and a lack of poise that is embarrassing. The Jew is also unique for his pride in his past. This is directly due to the fact that, as a creative social group, the Jewish people has a barren present, in striking contrast with its past. As individuals, baptized at times, the Jews may have enriched civilization out of all proportion to their number, in every field of human activity and in every country—in England, the Hersehels in astronomy, and Disraeli in statesmanship; in Germany, Marx in social reform, Herz in electricity, Ehrlich and Wasserman in medicine, and Mendelssohn in music, Ballin in commerce, and Harden in journalism; in Russia, Mendeleef in chemistry and Anotokolsky in art; in Holland, Spinoza in philosophy and Israels in painting—and so on, in France, Bergson; in Denmark, Brandes, and in Italy, Luzatti. [Jewish tribalism and racism caused more harm to progress than the individual contributions of Jews could compensate. Jewish self-aggrandizement and dogmatic insistence that their beliefs and heroes be worshiped set science, art and politics back throughout European and American History. Jews also have slackened the progress of humanity by promoting decadence and laziness in America and Europe—one must wonder if they fear competition, for their clannish in universities and the press clearly indicates that they, in general terms, do.] But, because as a people, as a social entity, it has produced little in the past two thousand years of dispersion, it harks back continually to a rich past as a source of pride. And, as Lyman Abbott put it, 'It is a poor present which shines only by the reflected glory of the past.' The Jew is singular in all these psychological traits, as he is peculiar also in the fact that his is the only living social group that has no center. If the Jewish people is permitted to reëstablish a normal group life in Palestine to save it from the ever-present threat of dissolution, its members will become normal like the rest of men.

The nations of the world have a selfish stake in the Zionist movement. If they would solve their Jewish problem, they must recognize the law of self-preservation of the group and aid in the restoration of a Jewish community in Palestine. If they fail to restore a part of the Jewish people to their ancestral home, they fail to get to the root of the problem, and leave unremoved, the international irritation of a homeless people that does not want to die, and therefore refuses to merge with the rest of the population. Diplomatic dilettantism, dallying with the symptoms of social maladjustments by legislating equality, or giving the Jew merely individual liberty, political or economic, will not solve the collective problem—the freedom of a group to live and express itself in accordance with its historic bent or its inherent inclinations.

## THE PROBLEM OF AMERICAN JUDAISM

The problem of American Judaism, as a writer in the magazines recently saw it, is not an isolated problem in itself. For it cannot be separated from the problem of the American Jew, just as one's opinion of a poem or a painting involves a judgment of its creator. One may decry this statement as a 'narrow racialism,' However, this would be absurd, for a world-noted scholar, Benjamin Kidd, in his 'Social Evolution,' calls attention to the generalization, that religion is the function of a social group. The 'people of the book' reflected its aspirations in the religion. Likewise the hopes for a restoration of his people are among the sublimest ideals which the prophets pictured.

If the Jewish religion in America is now colorless, it is because there is no unified Jewish community which can idealize its social aspirations. The contribution of the Jews, to the spiritual advance of humanity was made during the few hundred years when Israel was on its own soil and living a full, normal, social life. Twenty centuries of exile cannot boast of a single Moses, an Isaiah, or a Jesus, the products of a united people. For two thousand years the Jew has hibernated culturally. He has been living off his past. But now that all religion is being revalued and reinterpreted, the Jewish people, dismembered and scattered all over the globe, is powerless to adapt its spiritual heritage to modern life. The result is disintegration. The Jew cannot justify his further separate existence in a state of dispersion, except for the hope that he may be preserved until the day when his children again rebuild the Jewish group life. Reject Zionism as a future hope, not only to be prayed for, but to be realized at the earliest opportunity, and there cannot be found any justification for the persistence of a separate people.

Reform Judaism, was at one time anti-Zionistic. In rejecting the Palestine that either as a fact or as a hope united four thousand years of Jewish history, the theological reformers, in the flush of the cosmopolitanism of the early nineteenth century, had to find some justification for a further separate existence. So they constructed a 'mission theory,' by virtue of which the Jew was to act as a missionary to his fellow citizens and therefore the dispersion was interpreted to be a blessing and a state to be made permanent. This scheme is a perversion of Jewish history, for in thirty-five centuries there never arose a party that rejected Palestine as a fact or as a hope and yet survived. Time, the deadly foe of all error, has, in fifty years, shown the unreality of this excuse for a further separate existence of the Jewish people. So far from justifying a separate existence of the Jewish people, anti-Zionistic Reform Judaism has convincingly proven the logic and inevitableness of its disappearance, for, contrary to its intended aim, it has succeeded in cutting off from the Jewish people some of its finest families as the history of the Reform movement testifies.

History cannot furnish a single example of a people scattered among many others that has maintained its identity. The Jews were an apparent exception to this sociological law. The bonds of religion as an internal influence and the pressure of persecution as an external force, made possible for the Jew a sort of hot-house existence during twenty centuries of an immobile civilization. But formal religion is a weakening institution in a modern life, whose spirituality is universal and transcends geographical, racial or theological limitations. Correspondingly, persecution is lessening its rigors, and, since the beginning of the scientific era, life on this planet, far from remaining rigid, is become accelerated in its mobility. As a result, the Jewish people is rapidly undergoing the normal processes of assimilation, the merging of blood and the amalgamation of culture. It is following its erstwhile Greek and Roman contemporaries into oblivion.

Some anti-Zionists, and they never have been non-Jews—say that this dissolution is a desired consummation. Is it? Let us see. In the international harmony of cultures, each nation plays a distinct part. Eliminate from civilization the contributions of the English, French or German peoples and you impoverish it. Because Belgium gave birth to her characteristic literature, it is for the weal of civilization that she be regathered from exile.

Because Poland produced her peculiar poetry and music, the world will be enriched, if she is reëstablished. And so, because Israel, on its native soil and as a normal group, bore a Moses, an Isaiah and a Jesus, she should, if restored to her ancestral home, again produce leaders after her own kind to add her nuance to the harmony of the nations.

The intrinsic truth of Zionism may be seen in the fact that alone of all the movements in Jewry it was able, ultimately, to attract every section and party among the Jews, the Orthodox, the Conservative, the Reform Jew, the unchurched, nay, even the assimilationist, who believed that the destiny of the Jews lay in his disappearance. Many thoughtful non-Jews, among whom are Charles R. Crane, Norman Hapgood, and Alice Stone Blackwell, in this country, and H. G. Wells, Maxim Gorky and Bjornstjerne Bjornsen, abroad, viewed the matter as a social problem, which it largely is, and have come to the support of the Zionist movement. The Rev. Dr. Alexander Blackstone, an Episcopalian divine, antedated Herzl by several years in advocating the restoration of the Jews to Palestine.

Now, every new thought must fight its way to acceptance. The degree of opposition to it is a measure of its potency. But time is the ally of truth. 'The eternal years of God are hers.' So, while early Zionists preached against tremendous forces and under penalties which would ordinarily suppress all but those imbued with a great ideal, the last ten years have brought about a great change. When Louis D. Brandeis, who was fighting for justice in industrial relations, and who was all his life aloof from any Jewish interests, approached the problem, he viewed it not with the sentiment born in childhood associations and not with the bias of training, but as a problem of spiritual freedom, of the right of a fallen people again to stand erect with its fellow-peoples. Zionism appealed to him not from within, but from without; not as a personal affection, but as an abstract proposition. The winning in 1913 of Brandeis, the advocate of the 'square deal' in industry, was the turning-point in the struggle of Zionism for recognition. There had been won, in addition, Nathan Straus, among philanthropists, Julian W. Mack and Hugo Pam, of the bench, Eugene Meyer, Jr., in finance, and Stephen S. Wise and hosts of others in the Reform rabbinate. The tide had turned. Jacob H. Schiff, by reason of his prestige and leadership, was at one time the most damaging foe of Zionism. However, even he recently pinned his faith *in the hopes and aims* of Zionism. It is a tribute to the man that, in his advanced years, he retains the vigor of thought and the freshness of mind which enabled him to perceive the essential soundness of the movement he had been opposing and to re-adjust his views on it. And only yesterday, as it were, Adolph Lewinsohn, whose activities transcend creed, has likewise joined those that see in Zionism a solution to the Jewish question. The only opponents of Zionism left are a diminishing number of the radical rabbis, who, though not old, are of set mind, and with an unworthy consistency refuse to face the facts—the danger of disintegration of the scattered Jewish people in the present world ferment.

## IF THE BELGIANS OR POLES WERE DISPERSED

'Well,' says the man in the street, 'how does the matter affect me?' To this extent. If the Belgians or Poles were scattered from their ancestral hearths, they, too, would strive to maintain their group life. They, too, would become sensitive to criticism, self-conscious, proud of their past. They, too, would refuse to give up their identity among all the peoples in whose midst they were scattered, and they, too, would constitute a series of international irritations—problems to perplex statesmen and sociologists. And in this state of dispersion, there would form in their midst three parties—the assimilation party, the *status quo* party and the restoration party.

The assimilants, ever aware of the social maladjustments, would have the century-old struggle for survival end, by themselves disappearing as a people. This is a cult of cowardice and a program of flight from battle. Yet, even this policy has no significance unless it is carried out by all. But this is absurd, for you cannot expect millions of persons to abandon a tradition and deny a history which at one time was able to mould the life of mankind. Nor will a whole people reject the hope in its future—the prerequisite to social suicide. And here is the fundamental fallacy in the policy of Jewish assimilation. For, if only some advocate the dissolution of the group as the solution of the problem, they seem deserters of a losing cause, which needs their support. They are regarded as renegades by the world at large and by those that remain loyal, whose devotion is thereby intensified. Further, regardless of his own attitude in the matter, the outside world continues to identify the assimilationist Jew with his fellows. He is blamed for their faults, and pays the penalty in common with the rest of the group. Insofar as it affords no relief to the assimilationist and intensifies the loyalty of the great mass of a dispersed people, the policy of partial assimilation defeats its own ends. It is purposeless. It has been tested out, as a solution of the Jewish question, and has proven an eloquent failure.

Again, if the Belgians or Poles were scattered over the face of the earth, and, after centuries of persecution, were sharpened mentally to eke out a livelihood under difficult conditions, they, too, would, with the advent of a more humane era, become economically rooted to their native lands. Now, Prof. Seligman showed that the economic interpretation of history holds even in spiritual affairs. Accordingly, there should then develop a *status quo* group with a theory of living to fit in with the economic status of the established fugitives. Their leaders should, as did anti-Zionistic Rabbis, conveniently construct for them a philosophy to justify their dispersion. In view of the prejudice against them, they also might convince themselves about a destiny of spreading a mission of tolerance to the weak, which would possibly appeal to the original generation that escaped persecution, but not to their unscathed children. The subsequent generations would lose their attachment to the history and traditions of the group, and would desert it. In the scattered state, the hypothetical Belgians and Poles would no longer produce leaders and heroic figures, as the Jews have ceased to do so. Their cultural development would end. For a time they might move by the accumulated momentum of previous centuries. But, eventually, they would find themselves spiritual bankrupts and cultural anachronisms. And, reasoning theologically instead of sociologically, many people would overlook the fact that a scattered people is spiritually stagnant, that, at best, it can only preserve itself, and that only a normal group on its soil can generate its inherent and distinctive social values. And, possibly, some romantic and regretful young writer might also ask why some one of the scattered Belgians 'is not fired with that spirit which comes into the hearts of men' on their native Flemish soil, to thrill the world with a message of Belgian ideals.

And, finally, the hypothetical dispersed Belgians or Poles might develop a third party—the restorationists. In part, they might be idealists, who loved the history and traditions of a once-free Belgium. In part, they might be the persecuted Belgians or Poles in some benighted lands. Or they might even be righteous men and women, whether Belgian or not, who viewed the problem as one of social freedom or of the liberty of a repressed group. Then there might appear the scientist, to analyze the problem as one of an abnormal type in sociology, and to show that all the difficulties of the dispersed Belgians and Poles, the social maladjustments and the international irritations were due not to differences in belief, but to the attempt of a people to persist in a permanently scattered state, indeed, were due to the lack of a center and of a home.

This sociologist might show how all the parties, the assimilants, the *status quo* section, as well as the restorationists, would benefit by the reëstablishment of an unfettered community in their ancient home in Belgium or Poland. The restorationists among the Jews are the Zionists. They desire the rehabilitation of Palestine as a self-renewing and inexhaustible reservoir of Jewish life. This community could and would assume the responsibility of saving the people from dissolution. The non-Palestinian Jew could then merge, if he so chose, with any new social group, as completely as does the expatriated Dane or Swiss. Zionism would solve the assimilationist's problem, for it would relieve him of the ' back pressure' which now identifies him with his people and prevents his assimilation. The assimilationist Jew will under Zionism be an expatriate without the stigma of deserting a losing cause, for it will then no longer need his support.

For the status quo Jew, living in the present scattered state, who may want to maintain his historic connections,. the center in Palestine, with its newly-developing normal life, will invigorate the spent spiritual forces of Jewish life elsewhere. The status quo Jew may be the member of a free spiritual empire. Just as the Briton, 'overseas,' carried the English idea to the farthest corner of the globe, and in return brought back to his island home that broad tolerance for foreign cultures that has made England the world's colonizer, so also the Jew 'overseas' might be consuls of the spirit. He might justify his further scattered existence if he could exchange the products of a reinvigorated people in Palestine for all the cultural wealth of the nations to their mutual benefit. Further, a center in Palestine would serve as a potential alternative, the existence of which would create self-confidence and poise, the absence of which traits constitutes the common defect of the Jewish psychology to-day. Zionism will take the non-Palestinian Jew out of the class of social anomalies, and put him on a basis similar to that of the Swiss or the Dane, residing abroad, who lives unnoticed among all peoples and is never singled out either for blame or praise.

To the Palestinian Jew, nay, to the Jewish people, Zionism means the restoration to a free environment, with latitude for the development of any race endowments it may possess, To the progress of man it means the adding of another instrument to enrich, be it by ever so little, the cultural harmony of the nations. To the nations of the world it means the opportunity for atoning in one generous moment for the wrongs inflicted upon an unfortunate people for twenty centuries.

To us, as Americans, Zionism means the expression on the shores of the Mediterranean of the American spirit of fair play, of liberty for men and for nations. As the American chart of government inspired the leaders of the Latin-American republics, and guided the founders of the Commonwealth of Australia, so also the torch of civilization, burning so brightly on this hemisphere, may yet lend its light to the restored commonwealth at the junction of three continents. The Hebraic spirit of democracy was realized by the Puritans in our federation of states. Enriched by the genius of a great free people, the American idea may reinspire the cradle whence civilization sprung.

The great war, admittedly conceived in economic rivalries, has, however, taken on a higher aspect. It has stirred deep into the springs of human progress, A democracy, not only of individuals, but of groups and of nations, is the destiny toward which the struggle seems to be pointing, with statesmen as the pawns of a Higher Power. We may think Benjamin Franklin out of date, because he saw the finger of Providence in our Revolution. But that is the fault rather of our modern scientific spirit carried to an extreme. Our vision is narrowed to the field of the microscope. To many of us, however, there is something superhuman in the events that are shaping themselves under our near-sighted eyes. Time is fulfilling prophesy. In an off-corner of the stage, on which this mighty world drama is acting itself out, there is the Jewish people, just liberated in Russia, but about to be saved from the extinction that has been the counterpart of Jewish emancipation, by the 'remnant that will return' to the land of its fathers. The world may well join in the ancient prayer, 'May it come speedily in our days.'"[471]

Zionist Jacob Klatzkin stated,

"This belief in the impossibility of complete assimilation is one of the basic

---

[471]. E. M. Friedman, "Zionism and the American Spirit", *Forum*, Volume 58, (July, 1917), pp. 67-80; *reprinted as: Zionism and the American Spirit: A New Perspective*, University Zionist Society, New York, (1917).

tenets of Zionism. Lately this belief has sought support in the theory of race, which has been revived in certain scholarly circles. Even before the validity of this theory has been demonstrated, it has become the basis of many speeches on Zionism, which now use it as a quasi-scientific premise. [***] Our long survival in the Galut is certainly no proof of the impossibility of assimilation. The hold of the forms of our religion, which have served as barriers between us and the world for about two thousand years, has weakened and there are no longer any strong ghetto walls to protect a national entity in the Galut."[472]

## 5.6.2 The Zionists Set the Stage for the Second World War... and the Third

On 28 May 1921, THE DEARBORN INDEPENDENT published an article "Will Jewish Zionism Bring Armageddon?" which stated, *inter alia*,

"Zionism is challenging the attention of the world today because it is creating a situation out of which many believe the next war will come. To adopt a phraseology familiar to students of prophecy, it is believed by many students of world affairs that Armageddon will be the direct result of what is now beginning to be manifested in Palestine."

Jews dominated the Paris Peace Conference which imposed unjust terms on Germany. Leading and highly influential Jews in Germany stabbed Germany in the back and insisted that Germany accept the terms and pay the "reparations". The Jews who imposed severe and unjust sanctions on Germany at the end of the First World War knew that this would provoke a second world war and the rise of a Bolshevist régime in Germany, which would make a pact with the Soviets to destroy Eastern Europe. Racist political Zionist Israel Zangwill predicted in 1923 that Zionism would lead to an unprecedented world-wide conflagration.[473] He knew whereof he spoke. "Mentor" wrote in an article entitled "Peace, War—and Bolshevism" in *The Jewish Chronicle* on 4 April 1919 on page 7,

"It is a challenge to all the nations including the peoples who nourish liberty and freedom as precious principles, but who have passively allowed a state of affairs to grow and putrefy into the infamies of Russian Tsarism, the iniquity of Hungary, and the wickedness of German militarism; to the world

---

**472**. J. Klatzkin, *Tehumim: Ma'amarim*, Devir, Berlin, (1925); English translation by A. Hertzberg in his, *The Zionist Idea*, Harper Torchbooks, New York, (1959), pp. 316-327, at 320-321.
**473**. "Mr. Zangwill on Zionism", *The London Times*, (16 October 1923), p. 11. I. Zangwill, "Is Political Zionism Dead? Yes", *The Nation*, Volume 118, Number 3062, (12 March 1924), pp. 276-278.

that has suffered Society to fester into these and to break out into the prurient, gaping, sloughing, agonising tumour of such a war as that which is not ended, though it is suspended."

Lloyd George followed the Jewish method of calling on a war weary world to move towards world government as a means to secure peace, though world government was in truth, and in Jewish prophecy, a means for the Jews to secure the destruction of all Gentile Peoples. Note that Lloyd George's Zionist call for world government is speciously justified as a reaction to the Bolshevik quest for world government, such that the People of the world are left to choose between two paths to the same ultimate result, a Jewish dominated world government. The groundwork was also prepared for another world war, in that the battle lines were drawn and the alliances made to draw England and the United States into war with Germany on France's behalf—though ultimately when the Second World War came it was allegedly begun on Poland's behalf. Note that England, the United States and France were encouraged to be weak, such that when war came the Zionist Bolshevik Nazis would have the ability to overtake Continental Europe and herd together its Jews for forced deportation to Palestine. This also ensured a long and costly war the profits from which would pay for the rise of the "Jewish State". Note that Jews essentially bought up Germany after the First World War with the profits they had made during that war, and their economic advantage was especially strong because they had so viciously crippled the Gentile Germans. *The New York Times* wrote on 26 March 1922 on page 33 in the Editorial Section,

## "1918 PEACE VIEWS OF LLOYD GEORGE

Memorandum Written for Paris
Conference Published as
White Paper.

URGED JUSTICE TO ENEMY

Premier Also Insisted on Dealing
With Russian Situation—Bearing
on Genoa Conference.

Special Cable to THE NEW YORK TIMES.
LONDON, March 25.—An interesting document dating back to the time of the Paris peace negotiations was issued officially today in the form of a White Paper. It is a memorandum headed 'Some Considerations for the Peace Conference Before They Finally Draft Their Terms,' which was circulated by Premier Lloyd George at the Paris Peace Conference on March 25, 1919.

Extracts from this memorandum have been published, here and abroad, at various times in the form of quotations, and there is some speculation as to the reasons for its publication now, after the lapse of three years. The official explanation is that it is issued in response to repeated requests for its publication.

The memorandum opens by pointing out that it was comparatively easy to patch up a peace which would last for thirty years. What was difficult, however, was to draw up a peace which would not provoke fresh struggle when those who had had practical experience of what war meant had passed away.

## Plea for a Just Peace.

'You may strip Germany of her colonies, reduce her armaments to a mere police force and her navy to that of a fifth-rate power,' says Mr. Lloyd George. 'All the same, in the end, if she feels she has been unjustly treated in the peace of 1919, she will find means of exacting retribution from her conquerors. To achieve redress our terms may be severe; they may be stern and even ruthless; but at the same time they can be so just that the country on which they are imposed will feel in its heart it has no right to complain. But injustice and arrogance displayed in the hour of triumph will never be forgotten or forgiven.'

The memorandum goes on to urge the danger of transferring more Germans and Magyars to the rule of some other nation than can possibly be helped. Such action, it says, must sooner or later lead to a new war in the East of Europe.

'Secondly, I would say that the duration for the payments of reparation ought to disappear, if possible, with the generation which made war. The greatest danger that I see in the present situation,' Mr. Lloyd George proceeds, 'is that Germany may throw in her lot with the Bolsheviki and place her resources, her brains, her vast organizing power at the disposal of revolutionary fanatics whose dream it is to conquer the world for Bolshevism by force of arms. If Germany goes over to the Spartacists, it is inevitable that she should throw in her lot with the Russian Bolsheviki. Once that happens, all Eastern Europe will be swept into the orbit of the Bolshevist revolution, and within a year we may witness the spectacle of nearly 300,000,000 people organized into a vast Red army under German instructors and German Generals, equipped with German cannon and German machine guns and prepared for the renewal of the attack on Western Europe.

'I would, therefore, put it in the forefront of the peace that, once she accepts our terms, especially reparation, we will open to her the raw materials and markets of the world on equal terms with ourselves and will do everything possible to enable the German people to get upon their legs again. We cannot both cripple her and expect her to pay. It must be a settlement which will contain in itself no provocations for future wars, and which will constitute an alternative to Bolshevism because it will commend itself to all reasonable opinion as a fair settlement of European problems.

'The essential element in the peace settlement is the constitution of a League of Nations as an effective guardian of international right and international liberty throughout the world. The first thing to do is that the leading members of the League of Nations should arrive at an understanding between themselves in regard to armaments. It is idle to endeavor to impose permanent limitation of armaments upon Germany unless we are prepared similarly to impose limitation upon ourselves. The first condition of success for the League of Nations is a firm understanding between the British Empire and the United States and France and Italy that there will be no competitive building up of fleets or armies between them.

I believe that until the authority and effectiveness of the League of Nations has been demonstrated, the British Empire and the United States ought to give to France a guarantee against the possibility of a new German aggression.'

### Insists on Treating With Russia.

The concluding paragraph of the memorandum declares that the Peace Conference must deal with the Russian situation.

'Bolshevist imperialism does not merely menace the States on Russia's borders; it threatens the whole of Asia and is as near to America as it is to France. It is idle to think the Peace Conference can separate, however sound a peace it may have arranged with Germany, if it leaves Russia as it is today.'

### Timed for Genoa Conference?

As to the significance of the publication of the memorandum at the present time, one paper asks:

'Does the Prime Minister by publishing his memorandum after the lapse of three years and on the eve of the Genoa conference mean to indicate that there he is about to 'deal with the Russian situation' and to assist Germany 'to get upon her legs again'?' The Lloyd Georgian Daily Chronicle provides the answer. It says:

'The time has now come when the ideas of 1918 have a chance of being carried through. What seemed so original then is rapidly becoming common ground among those who are thinking seriously about politics, and Genoa points out the way.

'The document is remarkable in its anticipation of what has become the dominant sentiment among thoughtful people about the conditions of permanent peace in Europe. It is, in fact, an ideal introduction to the policy of Genoa. It proves that the Prime Minister's peace policy has been consistent, and that the principles of settlement for which he is working now are the same as those for which he was working three years ago.'

The Daily News, however, dissents from this view. It says:

'The contrast between the policy of December, 1918, and the policy for April, 1922, or between the policy proposed to the Allies and the policy ultimately adopted by them and vehemently defended by Mr. Lloyd George would be actually comical if its effects were not so appalling. Who shall say how great a share of the present ills of Europe and the world are due to this amazing instability of policy on the part of Britain's representative. If the policy of the memorandum, backed by America, had been adhered to by this country, what chance would the chauvinism of France have had against such a combination.'"

Racist political Zionist Israel Zangwill predicted in 1923 and in 1924, that Zionism would lead to an unprecedented world-wide conflagration.[474] He knew whereof he spoke. The Zionists Lloyd George and "Mentor" also realized at the end of the First World War that there would be second.[475]

In 1934, Zionist Marxist Berl Katzenelson warned against the nihilistic destruction sought by many Marxists,

> "History tells of more than one old world that was destroyed, but what appeared upon the ruins was not better worlds, but absolute barbarism."[476]

Henry Ford sought to curb the abuses of Bolsheviks, Socialists and financiers against the masses, which inevitably lead to depressions. Ford also sought to enlighten the public about the exploitation of the impoverished by financiers in periods of depression.

Years later, the Jewish financier Bernard Baruch, the descendant of slave traders and son of a member of the Klu Klux Klan, wrote passionately about the opportunities awaiting financiers during a depression and of the stupidity of the poor who failed to invest what they didn't have. Baruch wrote in his autobiography, in reference to the Depression of 1893,

> "I had never experienced a depression before. But even then I began to grasp dimly that the period of emergence from a depression provides rare opportunities for financial profit.
>
> During a depression people come to feel that better times never will come. They cannot see through their despair to the sunny future that lies behind the fog. At such times a basic confidence in the country's future pays off, if one purchases securities and holds them until prosperity returns.
>
> From what I saw, heard, and read, I knew that was exactly what the giants of finance and industry were doing. They were quietly acquiring interests in properties which bad defaulted but which would pay out under competent management once normal economic conditions were restored. I tried to do the same thing with my limited means."[477]

---

**474**. "Mr. Zangwill on Zionism", *The London Times*, (16 October 1923), p. 11. I. Zangwill, "Is Political Zionism Dead? Yes", *The Nation*, Volume 118, Number 3062, (12 March 1924), pp. 276-278.

**475**. "Peace, War—and Bolshevism", *The Jewish Chronicle*, (4 April 1919), p. 7. "1918 Peace Views of Lloyd George", *The New York Times*, (26 March 1922), Editorial Section, p. 33.

**476**. B. Katzenelson, *Ba-Mivhan*, Tel-Aviv, (1935), pp. 67-70; translated to English in A. Hertzberg, *The Zionist Idea*, Harper Torchbooks, New York, (1959), pp. 390-395, at 391.

**477**. B. M. Baruch, *Baruch: My Own Story*, Henry Holt and Company, New York, (1957), p. 92. *See also page 191.*

It was the depression of 1893 that made Jacob H. Schiff and Otto H. Kahn, of the banking house of Kuhn, Loeb & Co., immensely wealthy men. It led to Schiff's purchase, together with Edward H. Harriman, of the Sante Fe, Union Pacific, Northern Pacific, and Southern Pacific Railroads, among others.[478] Schiff used his ill-gotten gains to destroy the Russian Nation and bolster Imperial Japan, which soon became two of the most virulent enemies of the United States. The Harrimans used their fortune to finance the Nazi régime, a régime that killed many Americans.[479] Jewish financier Felix M. Warburg married Jacob H. Schiff's daughter—most of the Jewish bankers were related to each by blood and/or marriage.[480] The Warburgs, who descended from the Del Banco Jewish Venetian bankers, financed all combatants against one another in order to weaken Gentile governments and societies. As a Rothschild agent, Paul Warburg corrupted America by forming the Federal Reserve System. Max Warburg financed Trotsky and Lenin. The Warburgs also financed Adolf Hitler. Baruch owed much to Schiff, and to American depressions, from which they profited. The Zionists and Jewish bankers have been a curse to America.

Baruch was very powerful in the Wilson administration, and he, Wilson and "Colonel" Edward Mandell House were children of the Reconstruction South. Wilson betrayed and degraded the blacks who helped him to win the Presidency. The banking system Wilson created was one of the causes of the Great Depression. Bernard Baruch, Chairman of the War Industries Board, revealed in his autobiography that Nathan Rothschild's profiteering at Waterloo taught Baruch a method by which he could profiteer from war and that he was proud to

---

**478**. "Jacob H. Schiff and a Past American Era: A Joint Work of the Yankee and the Jew", *The World's Work*, Volume 41, Number 1, (November, 1920), pp. 19-20. B. J. Hendrick, "The Jews in America: II Do the Jews Dominate American Finance?", *The World's War*, Volume 44, Number 3, (January, 1923), pp. 266-286, at 279.

**479**. J. Buchanan, "Bush-Nazi Link Confirmed", *The New Hampshire Gazette*, Volume 248, Number 1, (10 October 2003); URL:

<http://www.nhgazette.com/articles/NN_Bush_Nazi_Link.html>

J. Buchanan and S. Michael, "'Bush-Nazi Dealings Continued Until 1951'—Federal Documents", The New Hampshire Gazette, Volume 248, Number 3, (7 November 2003); URL:

<http://www.nhgazette.com/articles/NN_Bush_Nazi_2.html>

J. Buchanan, *Fixing America: Breaking the Stranglehold of Corporate Rule, Big Media & the Religious Right*, Far West Publishing Company, Santa Fe, New Mexico, (2004). *See also:* E. Schweitzer, *Amerika und der Holocaust: Die verschwiegene Geschichte*, Knaur Taschenbuch Verlag, München, (2004).

**480**. B. J. Hendrick, "The Jews in America: II Do the Jews Dominate American Finance?", *The World's War*, Volume 44, Number 3, (January, 1923), pp. 266-286, at 278-280.

have done so in the Spanish-American War.[481] Baruch also claimed that his involvement in the foreign currency markets inspired him in his work with the League of Nations.[482] Baruch boasted of his manipulation of the stock market and told of the corrupt profits he made riding stocks up and down and of his ability to create monopolies by corrupt methods which are illegal today.[483] Smedley D. Butler demonstrated the enormous profits earned from war during the Wilson administration in his book *War Is a Racket*."[484]

Franklin Delano Roosevelt's son-in-law, Colonel Curtis B. Dall, wrote extensively on the subject of the Great Depression and Pearl Harbor and alleged that corruption by money interests was involved in both catastrophes.[485] Robert B. Stinnett has since been proven that FDR had foreknowledge of the Pearl Harbor attack.[486]

Ron Grossman capsulized newspaper publisher Colonel Robert R. McCormick's views on the subject,

> "Long after the defeat of Hitler and the Nazis, the Colonel told radio listeners that our GIs had fought and died in World War II 'not for the salvation of the United States' but because FDR had been hoodwinked by the British and Russians. Although he recognized the evil of Hitler, he opposed the U.S. getting involved overseas, right up to Pearl Harbor. He held that 'the United Nations was formed as a fake to fool people as to Roosevelt's real reason for going to war,' which was to make the world safe

---

**481**. B. M. Baruch, *Baruch: My Own Story*, Henry Holt and Company, New York, (1957), pp. 107-108. ***See also:*** A. Muhlstein, *Baron James: The Rise of the French Rothschilds*, Vendome Press, New York, (1982).
**482**. B. M. Baruch, *Baruch: My Own Story*, Henry Holt and Company, New York, (1957), pp. 87-88.
**483**. B. M. Baruch, *Baruch: My Own Story*, Henry Holt and Company, New York, (1957), pp. 114-116, 133-134.
**484**. S. D. Butler, *War Is a Racket*, Round Table Press, New York, (1935).
**485**. C. B. Dall, FDR, *My Exploited Father-in-Law*, Christian Crusade Publications, Tulsa, Oklahoma, (1967); **and** *A Tribute to Lincoln, Our Money-Martyred President: An Address in Springfield, Illinois*, Omni Publications, Hawthorne, California, (1970); **and** *Amerikas Kriegspolitik: Roosevelt und seine Hintermänner*, Grabert-Verlag, Tübingen, (1972); **and** A. J. Hilder and C. B. Dall, *The War Lords of Washington (Secrets of Pearl Harbor); an Interview with Col. Curtis Dall*, Educator Publications, Fullerton, California, (1972); **and** C. B. Dall, *Who Controls Our Nation's Federal Policies — and Why?*, Noontide Press, Los Angeles, (1973); **and** C. B. Dall and B. Freedman, *Israel's Five Trillion Dollar Secret*, Liberty Bell Publications, Reedy, West Virginia, (1977); **and** C. B. Dall, *Col. Dall Reports to the Board*, Liberty Lobby, Washington, D.C., Serial Publication, (1900's); **and** C. B. Dall and R. M. Bartell, *Liberty Lobby Progress Report*, Serial Publication, Liberty Lobby, Washington, D.C., (1970's) ; **and** C. B. Dall, *Colonel Dall Reports*, Serial Publication Liberty Lobby Washington, D.C., (1900's); **and** C. B. Dall and C. M. Dunn, *Ephemeral Materials*, 1957-, Liberty Lobby, Washington, D.C.
**486**. R. B. Stinnett, *Day of Deceit: The Truth about FDR and Pearl Harbor*, Free Press, New York, (2000).

for British imperialism and Soviet communism."[487]

Former Communist Douglas Hyde wrote in his book *Dedication and Leadership* of 1966,

> "When, therefore, the Communists speak of launching the world on the way to Communism in the period in which we are living, it is this that they mean—not the whole world with the exception of the United States, or the United Kingdom or whichever country, being your own, you may feel is proof against assault.
> Their aim is quite clear. They have never concealed it and it is something that is immensely meaningful to every Communist. It is a Communist world. In the past half-century they have achieved one-third of that aim. On any reckoning, that is a remarkable achievement, probably an unprecedented one. Nonetheless the world in which we live is still predominantly non-Communist. Twice as many people live in the non-Communist world as live under Communism. There is no basis here for defeatism."[488]

Former Communist Whittaker Chambers wrote in his book *Witness* of 1952,

> "Few Communists have ever been made simply by reading the works of Marx or Lenin. The crisis of history makes Communists; Marx and Lenin merely offer them an explanation of the crisis and what to do about it. Thus a graph of Communist growth would show that its numbers and its power increased in waves roughly equivalent to each new crest of crisis. The same horror and havoc of the First World War, which made the Russian Revolution possible, recruited the ranks of the first Communist parties of the West. Secondary manifestations of crisis augmented them—the rise of fascism in Italy, Nazism in Germany and the Spanish Civil War. The economic crisis which reached the United States in 1929 swept thousands into the Communist Party or under its influence. The military crisis of World War II swept in millions more; for example, a third of the voting population of France and of Italy. The crisis of the Third World War is no doubt holding those millions in place and adding to them. For whatever else the rest of the world may choose to believe, it can be said without reservation that Communists believe World War III inevitable."[489]

## 5.7 Henry Ford for President

---

[487]. R. Grossman, "The Colonel's World", *Chicago Tribune Magazine*, (27 March 2005), pp. 12-16, 20, 25-26, 28-29, at 14.
[488]. D. A. Hyde, *Dedication and Leadership: Learning from the Communists*, University of Notre Dame Press, (1966).
[489]. W. Chambers, *Witness*, Random House, New York, (1952), pp. 192-193.

Though John Spargo and others loudly decried Henry Ford, and though some had sued Ford and sought court injunctions to prevent the publication and distribution of THE DEARBORN INDEPENDENT and the book *The International Jew: The World's Foremost Problem*, which republished many of the anti-Jewish articles which appeared THE DEARBORN INDEPENDENT in the years 1920-1922, Ford's popularity steadily increased. In 1923, Henry Ford was becoming a serious contender for the Presidency of the United States of America. Ford made it quite clear that he intended to end the undemocratic power of the financiers and monopolies. It was then that Herman Bernstein, Samuel Untermyer and Louis Marshall began an all out war on Ford. If they had not succeeded, it is possible that Ford would have been elected President in 1933 following the stock market crash of 1929, and that Adolf Hitler would have had an ally in the White House.

Huey Long was another opponent of American involvement in the First World War. As a lawyer, Long successfully defended a man prosecuted under Wilson's "Espionage Act". Huey Long emerged as a Presidential candidate, who promised to curtail the corrupt power of the financiers, and who promised to defeat Franklin Delano Roosevelt. Roosevelt was a darling of the Communists.[490]

Long pledged to distribute the wealth. He directly and personally attacked the selfish power of Bernard Baruch and other top financiers. Though many of his liberal views mirrored those of the Socialist and Communist Parties, both parties denounced Long as a Fascist and smeared him as if he were another Adolf Hitler. They objected to Long's plan to distribute the wealth through the income tax, while maintaining the productive capabilities of Capitalism. They also objected to Long's alleged dictatorial control of the Government of the State of Louisiana. The Communists wanted to abolish private property, which is to say that they wanted to place property under the control of the Jews, as was prophesied in the Old Testament. Huey Long sincerely represented the interests of the working class and the Communists sincerely represented the interests of Jewish financiers.

In 1946, Robert Penn Warren (author of the racist and segregationist essay *The Briar Patch*,[491] which sought to prevent blacks from entering into competition with whites in the labor markets) posthumously attacked Huey Long

---

**490**. H. P. Long, *Every Man a King: The Autobiography of Huey P. Long*, National Book Co., New Orleans, (1933); **and** *My First Days in the White House*, Telegraph Press, Harrisburg, Pennsylvania, (1935). ***See also:*** T. H. Williams, *Huey Long*, New York, Knopf, (1969). ***See also:*** H. M. Christman, *Kingfish to America, Share Our Wealth: Selected Senatorial Papers of Huey P. Long*, Schocken Books, New York, (1985). ***See also:*** W. I. Hair, *The Kingfish and His Realm: The Life and Times of Huey P. Long*, Louisiana State University Press, (1991).

**491**. R. P. Warren, "The Briar Patch" in: Twelve Southerners, *I'll Take My Stand: The South and the Agrarian Tradition*, New York, London, Harper, (1930).

in a novel entitled *All the King's Men*.[492] The highly-talented Communist film director of Jewish descent, Robert Rossen, made Warren's book into a movie in 1949. As a "former" member of the Communist Party, Rossen was called before the House of Un-American Activities Committee and eventually told them the names of 57 other Communist Party members.[493]

In 1935, Dr. Carl Austin Weiss allegedly shot Huey Long and Long died soon thereafter due to the failure of his doctors to properly treat the gunshot wounds Weiss, and Long's own bodyguards, allegedly had inflicted on him. Immediately after Weiss allegedly shot Long, Long's bodyguards shot Weiss with at least 20 large caliber handgun rounds—perhaps as many as 60 rounds.[494] Weiss was very dead and very quiet.

It was alleged that Weiss had shot Long because Long had threatened to reveal Weiss' interracial family secrets.[495] If true, it is odd that Weiss believed he could save his family from embarrassment and keep secret facts hidden by shooting Huey Long, which was certain to embarrass Weiss' family and call attention to his family's secrets. Some believe that Huey Long's own bodyguards shot Long[496] and used Dr. Weiss as a "patsy".

THE DEARBORN INDEPENDENT succeeded in bringing political criticism of Jews to America, which had been relatively free of it until that point. The *Protocols* were derivative of the work of Adam Weishaupt (Marx plagiarized much of "his" philosophy from Plato, Weishaupt and Feuerbach[497]), Robespierre, Jean Paul Marat, Prince Klemens Lothar Wenzel Von Metternich, Marx, Maurice Jolly, Gougenot Des Mousseaux, Hermann Goedsche, Eugen Karl Dühring, Chabauty, Nietzsche, etc.[498] This was essentially already noted by

---

**492**. R. P. Warren, *All the King's Men*, Harcourt, Brace, New York, (1946).
**493**. A. Radosh and R. Radosh, *Red Star over Hollywood: The Film Colony's Long Romance with the Left*, Encounter Books, San Francisco, (2005).
**494**. H. B. Deutsch, *The Huey Long Murder Case*, Doubleday, Garden City, NewYork, (1963). D. Zinman, *The Day Huey Long was Shot, September 8, 1935*, I. Obolensky, New York, (1963). E. Reed, Requiem for a Kingfish, Award Publications/E. Reed Organization Baton Rouge, Louisiana, (1986). D. H. Ubelaker, "The Remains of Dr. Carl Austin Weiss: Anthropological Analysis", *Journal of Forensic Sciences*, Volume 41, Number 1,(1996), pp. 60-79.
**495**. H. B. Deutsch, *The Huey Long Murder Case*, Doubleday, Garden City, NewYork, (1963).
**496**. D. Zinman, *The Day Huey Long was Shot, September 8, 1935*, I. Obolensky, New York, (1963).
**497**.D. Fahey, *The Mystical Body of Christ in the Modern World*, Browne and Nolan Limited, London, (1935), p. 85.
**498**. A. Weishaupt, *Ueber Materialismus und Idealismus*, E.C. Grattenauer, Nürnberg, (1786); **and** *Apologie der Illuminaten*, Grattenauer, Frankfurth, (1786); **and** *Einleitung zu meiner Apologie*, Grattenauer, Frankfurt, (1787); **and** *Nachtrag zur Rechtfertigung meiner Absichten*, Frankfurt, Leipzig, (1787); **and** *Apologie des Misvergnügens und Ubels*, Frankfurt, Leipzig, (1787); **and** *Das verbesserte System der Illuminaten mit allen seinen Einrichtungen und Graden*, Grattenauer, Frankfort, Leipzig, (1787); **and** Kurze Rechtfertigung meiner Absichten: zur Beleuchtung der neuesten Originalschriften, Frankfurt, Leipzig, (1787); **and** *Nachtrag von weitern Originalschriften, welche die*

*Illuminatensette überhaupt, sonderbar aber den stifter verselben Adam Weishaupt, gewesenen professor zu Ingolstadt betreffen, und den der aus dem Baron Bassusischen schloss zu Sandersdorf, einem bekannten Illuminaten-Neste, vorgenommen visitation entdeckt, sofort aus churfürstlich höchsten befehl gedruckt, und zum geheimen archiv genommen worden sind, um solche jedermann auf Verlagen zur einsicht vorlegen zu lassen. Zwo abt.*, J. Lentner, München, (1787); **and** with A. Court de Gébelin, Philo, Sanchuniathon, *et al.*, *Saturn, Mercur, und Hercules: drey morgenländische Allegorien*, In der Montagischen Buchhandlung, Regensburg, (1789); **and** *Ueber Wahrheit und sittliche Vollkommenheit*, Montag und Weiss, Regensburg, (1793); **and** with A. Knigge, L. A. C. Grolmann, *Die neuesten Arbeiten*, München, (1794); **and** *Die neuesten Arbeiten des Spartacus und Philo in dem Illuminaten-Orden*, Illuminatenorden, München, (1794); **and** *Uber die Selbstkenntniss, ihre Hindernisse und Vortheile*, Montag und Weiss, Regensburg, (1794); **and** *Ueber die geheime Welt- und Regierungskunst ...* , F. Esslinger, Frankfurt, (1795); **and** *Ueber Wahrheit und sittliche Vollkommenheit, : Montag- und Weissische Buchhandlung*, Regensburg, (1793-1797); **and** Uber die Staats-Ausgaben und Auflagen: Ein philosophisch-statistischer Versuch, (1817). **For historic criticism of Weishaupt, see:** A. A. Barruel, *Mémoires pour Servir a l'Histoire du Jacobinisme*, De l'Imprimerie Françoise, Chez P. Le Boussonier, Londres, (1797-1798); English translation by R. Clifford: *Memoirs Illustrating the History of Jacobism*, Printed for the translator by T. Burton and co., London, (1798). **See also:** J. Robison, *Proofs of a Conspiracy Against All the Religions and Governments of Europe: Carried on in the Secret Meetings of Free Masons, Illuminati, and Reading Societies*, Printed for William Creech, and T. Cadell, Junior, and W. Davies, Edinburgh, London, (1797). H. Gruber, "Illuminati," *The Catholic Encyclopedia*, Volume 7, Robert Appleton Company, New York, (1910). **The Other Authors and Personalities:** "Metternich, Klemens Lothar Wenzel, Prince von", *The Catholic Encyclopedia*, Volume 10, Robert Appleton Company, New York, (1911), pp. 245-247. **See also:** K. Marx, *Manifest der komunistischen Partei (The Communist Manifesto)*, (1848), numerous editions and translations; **and** *Das Kapital*, numerous editions and translations. **See also:** R. Gougenot Des Mousseaux, *Le juif: le judaïsme et la judaisation des peuples chrétiens*, H. Plon, Paris, (1869); German translation: *Der Jude, das Judentum und die Verjudung der christlichen Völker*, Hoheneichen-Verlag, München, (1921). **See also:** H. Goedsche under the *nom de plume* Sir John Retcliffe, *Biarritz. Auf dem Judenfriedhof von Prag*, C.S. Liebrecht, Berlin, (1868); **and** *Um die Weltherrschaft: historisch-politischer Roman aus der Gegenwart*, C. Sigism, Berlin, (1876); **and** *Um die Weltherrschaft! Historischpolitischen Roman*, Carl Sigism, Liebrecht, Berlin, (1876); **and** *Das Geheimnis der jüdischen Weltherrschaft. Aus einem Werke des vorigen Jahrhunderts, das von den Juden aufgekauft wurde und aus dem Buchhandel verschwand*, Deutsches Wochenblatt, Berlin, (1919); **and** *Garibaldi*, Retcliffe-Verlag G.m.b.h., Berlin, (1926); **and** *Sir John Retcliffe's historisch-politischen Romane*, Rich. Eckstein Nachf., I.M. Bernhardi, Berlin, (1881-1891); **and** *Die Weltherrschaft der Juden "der neue Kabbalistische Sanhedrin"*, Nationaler Wirtschaftsdienst, Mannheim, (1932). **See also:** E. A. Chabauty, *Les Juifs, nos maîtres!: documents et développements nouveaux sur la question juive*, Société générale de librairie catholique, Paris, (1882); **and** under the *nom de plume* C. C. de Saint-André, *Francs-maçons et juifs: sixiéme age de l'église d'après l'Apocalypse*, Société General de Librairie Catholique, Paris, Bruxelles, (1880). **See also:** E. K. Dühring, *Die Judenfrage als Racen-, Sitten- und Culturfrage: mit einer weltgeschichtlichen Antwort*, H. Reuther, Karlsruhe, (1881); English translation by A. Jacob, *Eugen Dühring on the Jews*, Nineteen Eighty Four Press, Brighton, England, (1997); **and** *Der Werth des Lebens: Eine*

Aylmer Maude in 1920 in his response to the *Times* article "The Jewish Peril".[499]

This commonality of thought, however, was to be expected even if the *Protocols* were genuine and it should be noted that several of these men were agents of the Rothschilds and other leading Jews; such as Metternich, who worked for Salomon Rothschild,[500] and Robespierre who came to power after the "French" Revolution which was instigated by Moses Dobrushka (a. k. a. Junius Frey, a. k. a. Franz Thomas von Schoenfeld), who was Jacob Frank's nephew and successor as the false messiah of the Jews. For further such connections, *see:* L. Fry, *Waters Flowing Eastward: The War Against the Kingship of Christ*.

The Jewish mafia attempted to murder Henry Ford in 1927. The assassination attempt ended Ford's political ambitions. The murder of Huey Long was equally successful in ending his political ambitions. Franklin Delano Roosevelt, a Communist and pathological liar, enjoyed four terms as President of the United States.

## 5.8 The "Jewish Mission"

The Eighteenth Century philosophy of Moses Mendelssohn was seen by Jews and Gentiles alike as proposing a "Jewish Mission" or "Mission of the Jews"[501] to proselytize monotheism and the Jewish moral code to all the world. Mendelssohn stressed that Judaism was a religion, not a nation. Both Protestants and Jews were increasingly taking a "rationalist" approach to their religions, and attempted to distill their beliefs down into fundamental spiritual elements, which could be applied to all peoples and all times and which did not conflict with scientific facts.

Many Gentiles saw the "Jewish Mission" at its best as distasteful self-glorification by Jews, and at its worst as a movement for Jewish world domination.[502] Racist Zionists saw the Mendelssohnian "Jewish Mission" as an

---

*philosophische Betrachtung*, Eduard Trewendt, Breslau, (1865); **and** *Kritische Geschichte der Philosophie von ihren Anfängen bis zur Gegenwart*, Heimann, Berlin, (1869); **and** *Kritische Geschichte der Nationalökonomie und des Socialismus*, T. Grieben, Berlin, (1871); **and** *Die Ueberschätzung Lessing's und Dessen Anwaltschaft für die Juden*, H. Reuther, Karlsruhe, (1881); **and** *Sache, Leben und Feinde: Als Hauptwerk und Schlüssel zu seinen sämmtlichen Schriften*, H. Reuther, Karlsruhe, Leipzig, (1882); **and** *Der Ersatz der Religion durch Vollkommeneres und die Ausscheidung alles Judenthums durch den modernen Völkergeist*, H. Reuther, Karlsruhe, (1883); **and** *Die Parteien in der Judenfrage*, München, (1907).

**499**. A. Maude, Letter to the Editor on "The Jewish Peril", *The London Times*, (12 May 1920), p. 12.

**500**. L. Krieger, "Rothschild", *Encyclopedia International*, Volume 15, Grolier, New York, (1966), p. 577.

**501**. A. Hertzberg, *The Zionist Idea*, Harper Torchbooks, New York, (1959), p. 498.

**502**. P. W. Massing, *Rehearsal for Destruction: A Study of Political Anti-Semitism in Imperial Germany*, Howard Fertig, New York, (1967), pp. 278-287.

act of assimilation and Jewish racial suicide, which had to be restated in racial terms with the Jews as the dominant race. Mendelssohn's "Jewish Mission" became even more worrisome to those who did not wish to be governed by a universal tyranny of Jewish mysticism, when Moses Hess revealed that the "Jewish Mission" was to Zionists a racist biological theory in which Jews would reign as the brain of humanity and subjugate all the other inferior "races", who would be obliged to obey the Jews as mere organs of an allegedly divinely inspired Jewish will.

There was really nothing new in this racist Messianic vision dubbed the "Jewish Mission". It appeared in the Old Testament and its most vocal advocates have often been Christians, who have already fallen under the influence of the "Jewish Mission", and who too often view non-Christians as damned and evil. The movement for utopian Communism revealed itself in the Zionists' hands to be the proposed fulfillment of Jewish prophecies of Jewish world domination in the joyous millennium to come—a theme taken up by David Ben-Gurion, who spoke of world revolution, but who also spoke of certain Communists—apparently those who genuinely believed in its liberal and humanitarian precepts—as a threat to Zionism, which is a blatantly racist belief system.[503]

Ben-Gurion believed that politics fulfilled the role of Messiah in the modern world, in other words, that the Jewish people fulfilled the role of Messiah—a thought which had occurred to Moses Hess long ago. Ben-Gurion stated,

> "My concept of the messianic ideal and vision is not a metaphysical one but a socio-cultural-moral one... I believe in our moral and intellectual superiority, in our capacity to serve as a model for the redemption of the human race. This belief of mine is based on my knowledge of the Jewish people, and not some mystical faith; the glory of the divine presence is within us, in our hearts, and not outside us."[504]

David Ben-Gurion shared another of Moses Hess' convictions, the belief that only the Greeks and the Jews were great peoples, that the Greeks were lost, and that the Jews were superior to all the living. David Ben-Gurion was interviewed in 1948, and was asked if he believed that the United Nations boundaries of Israel would suffice to house the ten million Jews Ben-Gurion estimated would occupy Israel. Ben-Gurion doubted that it would, and the interview continued,

> "'We would not have taken on this war merely for the purpose of enjoying this tiny state. There have been only two great peoples: the Greeks

---

**503**. T. Segev, *The Seventh Million: The Israelis and the Holocaust*, Hill and Wang, New York, (1993), p. 129.
**504**. D. Ben-Gurion, as quoted in D. Duke, *Jewish Supremacism: My Awakening on the Jewish Question*, Free Speech Press, Covington, Louisiana, (2002), p. 345; Duke cites: A. Hertzberg and A. Hirt-Manheimer, "Relax. It's Okay to be the Chosen People", *Reform Judaism*, (May,1998).

and the Jews. Perhaps the Greeks were even greater than the Jews, but now I can see no sign of that old greatness in the modern Greeks. Maybe, when the present process is finished we too will degenerate, but I see no sign of degeneration at present.'

His voice took on a deeper tone:

'Suffering makes a people greater, and we have suffered much. We had a message to give the world, but we were overwhelmed, and the message was cut off in the middle. In time there will be millions of us—becoming stronger and stronger—and we will complete the message.'

'What is the message?' the reporter asked.

'Our policy must be the unity of the human race. The world is divided into two blocs. We consider that the United Nations' ideal is a Jewish ideal.'"[505]

Moses Hess and the other Jewish revolutionaries of 1848, to whom Benjamin Disraeli referred in 1844, were attempting to fulfill Judaic Messianic prophecy through political means. The *Encyclopaedia Judaica* writes in its article "Messianic Movements":

"In his letters Leopold *Zunz referred many times to the European revolution of 1848 as 'the Messiah.' Even many Jews who left the faith tended to invest secular liberation movements with a messianic glow. Martin *Buber expressed the opinion that the widespread Jewish activity in modern revolutionary movements stemmed both from the involvement of the Jew with state and his criticism of it through his messianic legacy (see *disputations).

Zionism and the creation of the State of Israel are to a large extent secularized phenomena of the messianic movements. The ideology of the Zionist religious parties, *Mizrachi and *Ha-Po'el ha-Mizrachi, tends to regard them—in particular the achievements of the State of Israel—as an *athalta di-ge'ulla* ('anticipating and beginning of redemption'), thus retaining the traditional concepts held by messianic movements in conjunction with the new secularized aspects of the State and its achievements."[506]

One must bear in mind that in Judaic prophecy the Jewish Messiah is a king who destroys the nations and religions with an iron scepter, and mass murders assimilated Jews and non-compliant Christians. According to the prophets, the Messiah would rule from Jerusalem, and demand the obedience of the enslaved Gentiles of the world. All of this would occur after a war to end all wars, the Holocaust of assimilated Jewry, and ingathering of Jews to Israel. The

---

[505]. *Time Magazine*, Volume 52, Number 7, (16 August 1948), p. 25.

[506]. "Messianic Movements", *Encyclopaedia Judaica*, Volume 11 LEK-MIL, Encyclopaedia Judaica, Jerusalem, The Macmillan Company, New York, (1971), cols. 1417-1427, at 1426-1427.

Communist, terrorist and racist Zionist first Prime Minister and Messiah of Israel David Ben-Gurion predicted in 1962 what he believed the world would be like in 1987. Ben-Gurion stated, among other revealing comments,

> "With the exception of the USSR as a federated Eurasian state, all other continents will become united in a world alliance, at whose disposal will be an international police force. All armies will be abolished, and there will be no more wars. In Jerusalem, the United Nations (a truly *United* Nations) will build a Shrine of the Prophets to serve the federated union of all continents; this will be the seat of the Supreme Court of Mankind, to settle all controversies among the federated continents, as prophesied by Isaiah."[507]

Communist dogma has many Messianic elements and proffers the ancient Jewish promise of an end to human struggles by the destruction of all nations and peoples but Israel. Meyer Waxman wrote in the "Translator's Introduction" to his English translation of Moses Hess' *Rome and Jerusalem*,

> "Hess's emphasis of creation gives to his philosophy an entirely new aspect, far exceeding in importance that of Spinoza. Spinoza, though employing the word creation, never conceived God as a real Creator, but endorses the mechanical view of the world, which sees in the universe a huge machine, working according to fixed laws, without aim and purpose. Hess, on the contrary, protests bitterly against this mechanical conception, and sees in the world a constant tendency toward creation, namely, the forming of things anew. The life of the world is not a mere blind operation of forces, but a development with a purpose and aim which will finally be realized. This aim is the harmony of all antagonistic elements, the reconciliation of all opposing forces, and the final peaceful cooperation of all for perfection and development. In this conception of reconciliation Hess shows the influence of Hegel's philosophy or *Synthesis*, which sees in the world of thought and life a constant process of opposition and reconciliation; but he employed it to better advantage than the master.
>
> The creative force of the universe is a vital force, and the entire universe a live being which is divided into three life spheres: the cosmic, organic and social or the human. There are no hard and fast lines separating them, but they are all parts of a great whole, one creative force called them into being. The world is all movement; there is nothing stable in it; all things were formed anew. Hess does not believe in the eternity of matter, nor in the constancy of atoms. The atoms were created as all other things in this world and are subject to growth and decay. Atoms are only centers of gravity from which creation proceeds, and corresponding to them, in other spheres, are the germs in the organic, and revelations of creative ideas in the social.
>
> Hess believes that this genetic conception is the real Jewish conception

---

507. *Look Magazine*, (16 January 1962), p. 20.

and points to the Biblical theory of creation. He was certainly right in his assertions. To look upon the world as a process of becoming and upon the creative force as vital, is a primary quality of Jewish thought and is best illustrated in Bergson. Comparing the view of Hess with that of the brilliant French-Jewish philosopher, we are struck with the similarity. Bergson, like Hess, struggles against the mechanical view of the world, and teaches a creative evolution constantly forming new productions, which are incalculable beforehand. Like Hess, he teaches the unity of the vital force which, though dividing itself into different forms, remains essentially one. There are undoubtedly differences between the two, but the fundamentals are the same with both of them; and, from a practical point of view, Hess's conception is far deeper and more fertile. Hess applies his philosophic thought to the social world, while Bergson remains in the middle of the road.

On the basis of the principles laid down by him in his view of the world, Hess constructed his philosophy of history. History, which embraces the social sphere of life is, according to him, not subordinate to Nature but on a par with it; it is dominated by the same laws and permeated with the same unified creative force. God reveals himself in history no less than in Nature; in this, he reminds us of the first Jewish national philosopher, Halevi,[*Footnote:* See the writer's article on Halevi in *The American Hebrew*, November 10, 1916.] and there is a divine plan in human affairs which is gradually unfolding itself in time.

Hess, like all thinkers of his time, was influenced in his conception of history by Hegel, whose principles he applied. History, like Nature, is a constant development, and is, of course, dominated by law, yet human freedom is preserved by the consciousness of our action. The development of history goes on in *dialectic* form, namely, forces opposing each other in earlier historical epochs are ultimately reconciled by a new synthetic epoch. Hess, viewing history as a part of the universal scheme, sees in its development an analogy to the development of Nature. In the former, as in the latter, there are three periods: rise, growth, and maturity, and there is also a corresponding similarity between the periods of these two spheres, which he elaborates fancifully in the tenth letter. The difference lies in this: that while Nature has already entered upon the third phase of its development, history is still striving toward it. Hess employs, as the means of conveying his ideas, the Biblical conception of Sabbath, which signifies 'rest' as well as 'completion.' Nature has already attained its Sabbath, but History is yet to attain it. The Sabbath of history, the period of maturity of human development, is the Messianic era of the Prophets. It is a time when all opposing and struggling forces of the social sphere will be harmonized and men will become morally free. But in order to comprehend the full significance of Hess's historical conception and his grand vision of the future, we must understand his view of Society and its strivings.

In his youth, when, in response to the impulses of his warm heart, he threw himself in the Socialist movement in order to attempt to alleviate human misery, Hess had no definite conception of human Society. He was swayed too often by different motives. Social life to him was only a constant antagonism between the collective body of society as a whole and its individual constituent members. Human history, he says somewhere in his writings, is a struggle actuated by two motives, egoism and love. In other words, there are two forces in Society, the disintegrating one, egoism, and the cementing force which binds one human being to the other, love. Hess always retained his belief in love as a moral factor and opens his book *Rome and Jerusalem* with a eulogy of it. As an escape from this eternal struggle, he proposed Communism, a state of Society which is bound to curb egoism and foster love. For a time, he swayed to Individualism. Under the influence of Feuerbach and Bauer, he wrote his *Philosophy of Action*, which advocated the freedom of the individual. But, even then, he was not an egoist. Later, again, under the influence of Marx, he became more a class-struggle socialist. But in all these social changes of his, Hess conceived Society only as an aggregate of individuals.

It was only later, as a result of his anthropological studies, that Hess came to the conclusion that Society is not a mere abstract idea but is composed of definite subdivisions known as races, each of which has definite hereditary mental and physical traits which are unchangeable. He then formed his organic conception of Society, entirely independently of Spencer, which is the corner-stone of his social and Jewish philosophy. Society, according to this conception, is an organic body composed of organs, the races. Each of these organs or races has a different function to perform for the benefit of the whole. It is in the performance of this function that the purpose of existence of the organ is realized; and there exists in every organ a natural tendency to perform the function.

Hess developed an elaborate historical scheme, according to which every historical race had or has a certain mission or function to perform. The important places in this scheme are reserved by him for the two antithetical nations, the Greeks and the Jews. To the Greeks, the world presented multiplicity and variety; to the Jews, unity; the former conceived Nature and life as *being*, namely, as an accomplished thing; the latter, as *becoming*, as a thing constantly being created. The Greeks, like Nature, which they represented, had reached their aim in life and had, therefore, disappeared from the world. The Jews, on the other hand, representing History, the constantly striving force, are still in existence, endeavoring to carry out their aim, to bring about in this sphere of social life the historical Sabbath, namely, the harmony of all social forces.

Judaism is a historical religion, a religion which has for its field of operation the social sphere, and which has discovered God in history, namely, the creative and reconciling principle in the life of humanity. The most characteristic point of Judaism, says Hess, in one of his later articles, [*Footnote: Die Einheit des Judenthums innerhalb der heutigen Religiosen Anarchie*, in the *Monatsschrift*, 1869.] is that it placed before human history its highest goal, the realization of universal law in Society. Judaism, he says in another place, is a humanitarian religion. According to its teachings, the life of the human genus is an organic process; it began with the family of the individual and will finally end with a family of nations. This, then, is the Jewish mission or function in Society, to realize the teachings of its great religion in practical life. The Jewish nation belongs to the creative organs of humanity. The Jews have taught humanity true religion, a religion which is neither materialistic nor spiritualistic, which has for its aim, unlike Christianity, not the salvation of the individual in the other world, but the perfection of social life in this world. And it is this function which they have to discharge to create for humanity new social values.

This function of Israel which, as a member of a great organism of Society, he is to perform, cannot be discharged anywhere else but in Palestine, where he will again be a nation possessing his own soil, a fundamental condition for living a regular normal social life. The regeneration of Judaism and Jewry is impossible in exile where it lacks the soil, the basis of a political life, and where there exists constant fear of disintegration. In exile, the Jews are unfruitful in all spheres, spiritually and economically. Jewish economic life, no matter how prosperous it may be in some countries, is abnormal; it lacks a basis, the soil; the Jews, therefore, cannot be creators and are only middlemen. It is only in their own land, where they will be able to produce new economic and social values, that they will continue to develop their greatest creation—Religion, which as a moral force will exert great influence upon humanity and thus bring about the realization of social harmony. In his attempt to lay the foundations of a positive view of Jewish life, Hess devoted considerable space to negative criticism of existing conceptions of Jewish life. His bitterest attacks are directed against the reformers and assimilators who deny Jewish nationality and substitute in its place an abstract indefinite teaching which they term, 'Mission.' Hess believes in a Jewish mission, but his mission is a natural function based on history and social life, while theirs is only a product of imagination and narrow vision. He attacks their ignorance of Jewish history and the misconception of the nature of Judaism as well as of Society in general, and ridicules their self-assumed rôle as the teachers of the nations. Their Judaism is only an empty shell, after the most important principles have been abandoned by them. The Orthodox Jews have, in his opinion, a much higher and truer conception of Judaism. They have retained in their ceremonies and prayers the kernel of Nationalism and the desire for Jewish restoration. Yet even they do not satisfy him entirely. Their inactivity and fossilized state irritate him. But he is optimistic. He believes that the spirit of regeneration will revive them and that they will finally furnish the material for a great National Movement. Hess also laid great hopes on Jewish science and expected it to become a great factor in the Jewish revival.

Hess developed a practical plan for the realization of his dream of Jewish restoration. He advocated the colonization of Palestine and the foundation of a Jewish Colonization Association. He dreamed that Jews, having been settled on the road to India and China, will become the mediators between Asia and Europe. For political support, he looked to his beloved France, the embodiment of freedom and the champion of oppressed nations. But he also dreamed of a Jewish Congress, demanding the support of the Powers for the purchase of Palestine, a dream quite prophetic in view of recent developments. He also foresaw a political situation resembling in its features the present state of affairs created by the war; he called it the last struggle between reaction and freedom. In some of his articles there are strikingly modern features.

Some of the dreams of this great visionary have partly come true. Let us gather confidence from the words of this modern seer, and hope that the glorious vision he foresaw for Israel will be realized in the coming period of history."

If we assume that there are no prophets who are divinely inspired to see into the future, we are led to conclude that it was the corrupt actions of disloyal Zionists which led to the fulfillment of Hess' "visions" through war and through genocide.

Some saw the "Jewish Mission" and Protestant Christian Evangelism as one movement toward fanatical degradation into a slavish mentality, or the worship of evil as the Frankists worshiped evil. The anti-Semites and Zionists found common joined forces to criticize the "Jewish Mission". Both resented the melding of the Jewish reformation with the Christian reformation, and both anti-Semites and political Zionists asserted that Jews were a "racial type" and a distinct nation, not a religion. In the introduction to the English translation of Moses Hess' racist treatise on Zionism *Rome and Jerusalem*, Meyer Waxman wrote,

"Emancipation was obtained, though not by means of Reform. It was achieved through the political and social circumstances of the revolutionary year 1848. But assimilation was not stemmed. The extreme spiritualization of Judaism of the radical reformers and the elimination of the National element, brought the new type of Judaism within dangerous approach to reformed Christianity, the line of demarcation between them becoming almost imperceptible. Many did not hesitate, therefore, to cross this line and enjoy the social advantages which the crossing afforded."[508]

Mendelssohn's "Jewish Mission" became the reform movement in German Jewry, which community of Jews had been experiencing turbulent times. Napoleon emancipated the Jews of what was to become the German Nation. This emancipation resulted in assimilation. The liberation of Germany from Napoleon resulted in the re-institution of discriminatory laws against Jews, which favored Jewish nationalism. The revolutions of 1848 again largely emancipated the Jews. Jewish racists were frustrated because they resented the indignity of

---

**508**. M. Hess, English translation by M. Waxman, *Rome and Jerusalem: A Study in Jewish Nationalism*, Bloch, New York, (1918/1943), p. 17.

discriminatory laws, but would not allow Jewish emancipation without a Jewish State, because emjancipation resulted in assimilation.

During Napoleon's philo-Semitic reign, some Jews betrayed him and encouraged all Jews to side against Napoleon and with an anti-Semitic Czar, because they feared that Napoleon's emancipation of the Jews was leading to assimilation.[509] The question naturally arises if Russian anti-Semitism was the work of such Jews and if the alleged anti-Semitism of some of the Czars came at the request of Jewish leaders—immensely wealthy Jewish leaders who held Russia's fate in their hands. A Jewish leader of the time, Shneur Zalman, who hated Gentiles, reasoned that,

> "If Bonaparte wins, the wealth of the Jews will increase and their positions will be raised. But their hearts will be estranged from their Father in Heaven. However, if Czar Alexander wins, then although the poverty of the Jews will increase and their position will be lower, their hearts will cleave to and be bonded with their Father in Heaven."[510]

Revolutionary forces battled Aristocratic forces in what was to become Germany, resulting in the Revolution of 1848 and both sides employed anti-Semitism as a means to garner popular support. Karl Marx and Moses Hess used anti-Semitism as a means to promote themselves and subvert Gentile society. Both Marx and Hess were Hegelians in the spirit of Feuerbach—and Bruno Bauer. Feuerbach taught that religion should be supplanted by the humanitarian view that mankind can, by its own nature, achieve the status formerly attributed to the "divine". For the Jews, this divine status meant the Messianic Era, when they would destroy the Gentile world. It occurred to them that they could attain Judaic prophetic goals by political means. These Socialists and Communists feigned atheism and Bauer and Marx while discussing the emancipation of Jews attacked Jews in general as religious, segregationist wealth accumulators. Like so many before them, they used anti-Semitism as means to control Gentile behavior which enabled them to accomplish Jewish ends. The German Revolution improved the condition of Jews in what was to become Germany and tended toward the amalgamation of the German Nation.

Another Hegelian, David Friedrich Strauss, published an influential treatise, *Das Leben Jesu, kritisch bearbeitet*, Tübingen, C.F. Osiander, (1835-1836); which taught that the Gospels are a mythology derived from Judaism. Communist[511] Mary Brabant Hennell began work on an English translation of

---

**509**. A. M. Dershowitz, *The Vanishing American Jew: In Search of Jewish Identity for the Next Century*, Little, Brown and Company, Boston, New York, Toronto, London, (1997), pp. 2-3.
**510**. *From:* A. Nadler, "Last Exit to Brooklyn: The Lubavitcher's Powerful and Preposterous Messianism", *The New Republic*, (4 May 1992), pp. 27-35, at 34. Nadler appears to quote from: N. Loewenthal, *Communicating the Infinite: The Emergence of the Habad School*, University of Chicago Press, (1990).
**511**. C. Bray and M. Hennell, *The Philosophy of Necessity, Or, the Law of Consequences*

Strauss' *Das Leben Jesu*, but she died in 1843. Charles Christian Hennell published *An Inquiry Concerning the Origin of Christianity*[512] in 1838 which, like many other works before and since, disputed the existence of Jesus Christ. Charles Hennell's sister Caroline Bray was married to the anti-Christian Communist Charles Bray. This group of intellectuals, which also included Robert Brabant and Elzabeth Rebecca Brabant Hennell and Sara Sophia Hennell, became close and influential friends to Mary Ann Evans, who published under the pen name "George Eliot", and who completed the English translation of Strauss' *The Life of Jesus: Or a Critical Examination of His History* in 1844-1846. "George Eliot" may have had love interests in Robert Brabant and Charles Bray.

"George Eliot" later published the Zionist novel *Daniel Deronda* in 1876,[513] which argued that Christians are essentially Jews—though not as noble. "George Eliot" was persuaded to write the Zionist novel by the racist Zionist Moses Hess, who was a very good friend of "George Eliot's" long term lover George Henry Lewes. "George Eliot" was an anti-Christian who studied Hebrew and the Talmud with her close friend, the noted scholar of the Talmud and of the Middle East, Emanuel Oscar Menahem Deutsch. She greatly enjoyed Gotthold Ephraim Lessing's *Nathan der Weise*, and her novel had many Frankist-like undertones, as did Lessing's work, which was based on the life of Moses Mendelssohn. One wonders if "George Eliot", whose ancestry was allegedly uncertain, discovered one day that she was of Jewish descent, or was told that she was—or had always known it. She privately rebelled against Zionism and may have discovered that Zionism ultimately means the destruction of all peoples but Jews.

In this era, Deist and Protestant Gentiles moved increasingly toward Judaism. Jewish reformists and Socialists, coming from the tradition of the Frankists, took the opportunity to promote the unity of reformed Judaism and reformed Christianity—Protestantism[514] as a unified front against Catholicism in the *Kulturkampf*; and, like the Frankists, many Jews pretended to convert to Christianity in order to gain rights and in order to subvert the Christian religion, which was increasingly returning to Judaism. Racist Zionists dreaded all of these forces which resulted in assimilation.

---

*as Applicable to Mental, Moral, and Social Science*, Longman, Orme, Brown, Green, and Longmans, London, (1841); **and** *An Outline of the Various Social Systems and Communities Which Have Been Founded on the Principle of Cooperation: With an Introductory Essay by the Author of "The Philosophy of Necessity"*, Longman, Brown, Green and Longmans, London, (1844); **and** *An Essay upon the Union of Agriculture and Manufactures, and upon the Organization of Industry*, Longman, Brown, Green, and Longmans, London, (1844).

**512**. C. C. Hennell, *An Inquiry Concerning the Origin of Christianity*, Smallfield, London, (1838); German translation: *Untersuchung über den ursprung des Christenthums*, Hallberger, Suttgart, (1840).

**513**. G. Eliot, *Daniel Deronda*, William Blackwood and Sons, Edinburgh, London, (1876).

**514**. "Judaism", *Great Soviet Encyclopedia: A Translation of the Third Edition*, Volume 2, (1973), pp. 311-312, at 312.

Mendelssohn was not out to advance the interests of the Gentiles, but to accomplish Judaic Messianic prophecies through the use of modern politics and modern science. All these Frankist movements, the Illuminati, "reformed Judaism", Communism, Bolshevism, etc. backfired on the Jewish racists. The Frankists kept their agenda well hidden, so well hidden that in the course of time even many Jews lost track of their original intentions. The Zionists reacted against the assimilation the Frankist movements had unintentionally caused, though they either misrepresented or misunderstood the racist intentions of the founders of those movements. Zionist Max Nordau wrote of the "Jewish Mission" of reformed Judaism,

"This gradually changed about the middle of the eighteenth century, when enlightenment first began to find its way into Jewdom, in the person of its first herald, Moses Mendelssohn, the popular philosopher. The faith of the Jews became more lukewarm; the educated classes, where they did not simply convert themselves to Christianism, began to regard the doctrines of their religion in a rationalist manner; for them the dispersion of the Jewish people was a final and unalterable fact; they emptied the conception of the Messiah and of Zion of every concrete meaning, and arranged for themselves a singular doctrine, according to which the Zion promised to the Jews was to be understood only in a spiritual sense, as the setting up of the Jewish monotheism in the whole world, as the future triumph of Jewish ethics over the less sublime and less noble moral teaching of the other nations. An American rabbi reduced this conception to the striking formula, 'Our Zion is in Washington.' The Mendelssohn teaching logically developed in the first half of the nineteenth century into the 'Reform,' which deliberately broke with Zionism. For the Reform Jew, the word Zion had just as little meaning as the word dispersion. He does not feel himself in any diaspora. He denies that there is a Jewish people and that he is a member of it. He desires only to belong to the people in whose midst he lives. For him Judaism is a purely religious conception which has nothing whatever to do with nationality. The land of his birth is his fatherland, and he will know of no other. The idea of a return to Palestine excites him either to indignation or to laughter. He answers it with the well-known, silly, would-be witticism, 'If the Jewish state is again set up in Palestine, I will ask to be its ambassador in Paris.'

The thinking Jew did not fail, however, to perceive, in the course of time, that Reform Judaism is a half measure, a compromise, which like every compromise, contains the germ of destruction, as it cannot for one instant resist logical criticism. Whom shall the Reform Judaism satisfy? The believing Jew? He rejects it with the greatest abhorrence. The unbelieving Jew? He despises it as hypocrisy and phrase-mongering. The Jew who really desires to break with his national past and to be absorbed by his Christian surroundings? For that Jew, Reform Judaism does not suffice; he goes a step farther, the step that leads to the baptismal font. Still less does it satisfy the Jew who desires to guard Jewdom against destruction and to preserve it as an ethnical individuality. For to him an openly expressed abandonment of all national aspirations is synonymous with a self-condemnation of the Jewish people to a perhaps slow, but sure, death. Reform Judaism without Zionism, that is to say, without the wish and the hope for a reassembling of the Jewish people, has no future. At the best, it can only be regarded as a somewhat crooked path that leads to Christianity. He who desires to reach that goal can find straighter and shorter routes.

II.

And so it has come about that the generations which had been under the influence of the Mendelssohnian rhetoric and enlightenment, of reform and assimilation, have, in the last twenty years of the nineteenth century, been followed by a new generation which seeks to take up a standpoint other than the traditional towards the question of Zion. These new Jews shrug their shoulders at that twaddle which has been the fashion among rabbis and *literati* for the last hundred years, and which boasts of a 'Mission of Jewdom,' said to consist in this, that the Jews must live forever in dispersion among the peoples in order to act as their teachers and models of morality, and to educate them gradually to pure rationalism, to a general brotherhood of mankind, and to an ideal cosmopolitanism. They declare the mission swagger to be either presumption or foolishness. They, more modest and more practical, demand only the right for the Jewish people to live and to develop itself, according to its abilities, up to the natural limits of its type. They have become convinced that this is not possible in dispersion, as, under that condition, prejudice, hatred, and contempt continually follow and oppress them, and either stint their development, or force them to an ethnical mimicry which necessarily makes of them, instead of original types with a right, to existence, mediocre or bad copies of foreign models. They therefore work methodically with a view to rendering the Jewish people once more a normal one, which lives on its own soil, and accomplishes all economical, intellectual, moral, and political functions of a civilized nation."[515]

Ardent Zionist spokesman Israel Zangwill wrote down many commonplace Zionist beliefs in 1911,[516] before World War I had begun: that Jews have a mission to convert the entire world to their beliefs, that the Jews are a superior race of God's chosen, that the emancipation of Jews in Russia would destroy the race and constitute a degeneration of a superior race into an inferior one by blending Jewish blood with Slavic blood, and that the persecution and antagonism of anti-Semitism were essential and necessary elements to the survival of the Jewish race and the creation of a Jewish nation-state and the loss of anti-Semitism increases the "problem" of maintaining a pure Jewish race.

---

**515**. M. Nordau, *Zionism and Anti-Semitism*, Fox, Duffield & Company, (1905), pp. 12-16.
**516**. I. Zangwill, *The Problem of the Jewish Race*, Judaen Publishing Company, New York, (1914); which was first published as an article, "The Jewish Race", *The Independent*, Volume 71, Number 3271, (10 August 1911), pp. 288-295.

Zangwill holds that Jews were better off segregated in the Ghettoes of the Middle Ages, than in emancipated Europe where they could assimilate. The mythologies of a master race and of racial degeneration through intermixing had both Jewish and Gentile adherents long before Zangwill, among them the early intellectual political critics of the Jews incuding Kant, Fichte, Bauer, Herder, Frege, Ghillany, Hegel, etc.[517] Later came Jewish and Gentile racists who promoted the idea of distinct Aryan and Jewish "races" including Disraeli,[518] Hess,[519] Gobineau,[520] Lassen,[521] Renan,[522] Hellwald,[523] Chamberlain,[524] List,[525]

---

**517**. P. L. Rose, *Revolutionary Antisemitism in Germany from Kant to Wagner*, Princeton University Press, (1990). R. W. Stock, *Die Judenfrage durch fünf Jahrhunderte*, Verlag Der Stürmer, Nürnberg, (1939).
**518**. B. Disraeli, *Coningsby; or, The New Generation*, H. Colburn, London, (1844), pp. 249-254. B. Disraeli, *Lord George Bentinck: A Political Biography*, Chapter 24, Third Revised Edition, Colburn, (1852), pp. 482-507. B. Disraeli, *Endymion*, D. Appleton and Company, New York, (1880), pp. 251-252.
**519**. M. Hess, *Rom und Jerusalem: die letzte Nationalitätsfrage*, Eduard Wengler, Leipzig, (1862); English: *Rome and Jerusalem: A Study in Jewish Nationalism*, Bloch, New York, (1918).
**520**. A. comte de Gobineau, *Essai sur l'Inégalité des Races Humaines*, Paris, Librairie de Firmin Didot, (1853-1855), Didot Frères, Rumpler, Hanovre, (1853-1855); English translation by A. Collins, *The Inequality of Human Races*,: William Heinemann, London, (1915); German translation by L. Schemann and P. Kleinecke, *Versuch über die Ungleichheit der Menschenracen*, F. Frommann, Stuttgart, (1902); An abridged translation to Czech by F. X. Lánský, *O Nerovnosti Lidských Plemen*, Orbis, Praha.
**521**. C. Lassen, *Indische Alterthumskunde*, H. B. Koenig, Bonn, Volume 1, (1847-1862), p. 414.
**522**. E. Renan, *Études d'Histoire Religieuse*, Fifth Revised Edition, M. Lévy frères, Paris, (1862), pp. 85-88, 130; **and** *Vie de Jésus*, Michel Lévy frères, Paris, (1863); English translation by C. E. Wilbour, *The Life of Jesus*, Carleton, New York, Michel Lévy frères, Paris, (1864); **and** *Histoire Générale et Système Comparé des Langues Sémitiques*, Fifth Revised and Enlarged Edition, Michel Lévy Frères, Paris, (1878), pp. 3-4.
**523**. F. v. Hellwald, *Culturgeschichte in ihrer natürlichen Entwicklung bis zur Gegenwart*, Lampart & Comp., Augsburg, (1875); **and** "Der Kampf ums Dasein im Menschen- und Völkerleben", *Das Ausland*, Volume 45, (1872), pp. 105ff., *see also: Das Ausland*, (1872), 901ff., 957ff.
**524**. H. S. Chamberlain, *Die Grundlagen des neunzehnten Jahrhunderts*, F. A. Bruckmann, München, (1899); English translation by J. Lees, *Foundations of the Nineteenth Century*, John Lane, New York, (1910).
**525**. G. v. List, *Das Geheimnis der Runen, mit einer Runentafel*, P. Zillmann Gross-Lichterfelde, (1907); English translation by S. E. Flowers, *The Secret of the Runes*, Destiny Books, Rochester, Vermont, (1988); Spanish, *El Secreto de las Runas*, Eds. Armanen, Santiago, Chile, (2000); Polish translation by E. Kulejewska, R*uny i ty: rozklady interpretacja zycie*, Studio Astropsychologii, (2002). G. v. List, *Die Armanenschaft der Ario-Germanen*, Verlag der Guido-von-List-Gesellschaft; In Kommission bei E.F. Steinacker, Wien, Leipzig, (1908); **and** *Die Ursprache der Ario-Germanen und ihre Mysteriensprache*, Guido-von-List-Gesellschaft, Wien, (1914); and *Der Übergang vom Wuotanismus zum Christentum*, G. von List, Berlin-Lichterfelde, (1926). ***See also:*** H. Kardel, *Adolf Hitler, Begründer Israels*, Verlag Marva, Genf, (1974);

Liebenfels,[526] Zollschan[527] and Rathenau.[528] Hitler's *Lebensraum* plan carried out under the supervision of the Nazi Governor-General Dr. Hans Frank (who

---

English translation *Adolf Hitler: Founder of Israel*, Modjeskis' Society Dedicated to Preservation of Cultures, San Diego, (1997).
**526**. J. Lanz-Liebenfels, *Katholizismus wider Jesuitismus*, Frankfurt, (1903); "Zur Theologie der gotischen Bibel", *Vierteljahrsschrift für Bibelkunde* (Lumen, Leipzig), Volume 1, (1903); **and** *Anthropozoon Biblicum*, Berlin, (1904); **and** *Theozoologie, oder Die Kunde von den Sodomsäfflingen und dem Götter-Elektron eine Einführung in die älteste und neueste Weltanschauung und eine Rechtfertigung des Fürstentums und des Adels...* , Moderner Verlag, Wien, (1905); **and** *Das Breve, Dominus ac redemptor noster: Aufhebung des Jesuitenordens durch Clemens XIV mit einer Einleitung und einem dogmatischkanonistischen Nachweis der Verwerklichkeit des S.J. versehen*, Neuer Frankfurter Verlag, Frankfurt a.M., (1906); **and** *Der Taxil-Schwidel: ein welthistorischer Ulk*, Neuer Frankfurter Verlag, Frankfurt a.M., (1906); **and** *Revolution oder Evolution? Eine freikonservative Osterpredigt für das Herrentum europäischer Rasse*, Ostara, Wien, (1906); **and** *Die Theosophie und die assyrischen "Menschentiere" in ihren Verhältnis zu den neuestan Resultaten der anthropologischen Forschung*, P. Zillmann, Berlin, (1907); **and** *Die Archäologie und Anthropologie und die assyrischen Menschentiere: mit einer Tafel*, P. Zillmann, Berlin, (1907); **and** *Der Affenmensch der Bibel: mit vier Tafeln und zehn illustrationen im Text*, P. Zillmann, Berlin, (1907); **and** *Die griechischen Bibel-Versionen. (Septuaginta und Hexapla.) Herausgegeben, mit Ammerkungen und deutscher Uebersetzung versehen*, Lumen, Leipzig, (1908); **and** *Die Lateinischen Bibel-Versionen (Itala und Vulgata) Volume 1, Genesis*, Lumen, Wien, (1909); **and** *Moses als Darwinist: eine Einführung in die anthropologische Religion*, F. Schalk, Wien, (1911); **and** "Kraus und das Rassenproblem", *Der Brenner*, Volume 4, (1913-1914); **and** *Rasse und dichtkunst ...* , Ostara, Mödling-Wien, (1916); **and** *Weltende und Weltwende*, Lorch, (1923); **and** *Grundriss der ariosophischen Geheimlehre*, Oestrich, (1925); **and** J. Lanz-Liebenfels, H. Reichstein **and** A. Hitler, *Das Buch der Psalmen teutsch: das Gebetbuch der Ariosophen, Rassenmystiker und Antisimiten. I. Band*, Herbert Reichstein, Düsseldorf-Unterrath, (1926); **and** J. Lanz-Liebenfels, *Der Weltkrieg als Rassenkampf der Dunklen gegen die Blonden*, Vienna, (1927); **and** *Theozoologie, oder Naturgeschichte der Götter*, J. Lanz v. Liebenfels, Wien, (1928); **and** *Theozoologie, oder Naturgeschichte der Götter*, J. Lanz v. Liebenfels, Wien, (1928); **and** *Bibliomystikon, oder: Die Geheimbibel der Eingeweihten; ariosophische Bibeldokumente und Bibelkommentare zu allen Büchern der heiligen Schrift, auf Grund der anthropologischen und archäologischen Forschungen und der arischen, klassischen und orientalischen Bibelversionen zusammengestellt*, In 10 Volumes, Pforzheim i. Baden, (1929-1934); **and** *Ostara österreichisches Flugschriften-Magazin freikonservativer Richtung*; (1905-1917 / 1926-1931); **and** *Praktisch-empirisches Handbuch der ariosophischen Astrologie*, Düsseldorf, (1926-1934). *See also:* H. Kardel, *Adolf Hitler, Begründer Israels*, Verlag Marva, Genf, (1974); English translation *Adolf Hitler: Founder of Israel*, Modjeskis' Society Dedicated to Preservation of Cultures, San Diego, (1997). *See also:* E. Hieronimus, *Lanz von Liebenfels: eine Bibliographie*, U. Berg, Toppenstedt, (1991). *See also:* W. Daim, *Der Mann, der Hitler die Ideen gab*, Ueberreuter, Wien, (1994).
**527**. I. Zollschan, *Das Rassenproblem unter besonderer Berücksichtigung der theoretischen Grundlagen der jüdischen Rassenfrage*, W. Braumüller, Wien, (1910); **and** *Jewish Questions: Three Lectures*, New York, Bloch Pub. Co., (1914).
**528**. W. Hartenau (W. Rathenau), "Höre, Israel!", *Die Zukunft*, Volume 18, (6 March 1897), pp. 454-462. W. Rathenau, *Reflexionen*, S. Hirzel, Leipzig, (1908).

was Hitler's lawyer and was of Jewish descent) to depopulate Slavic lands was not far in its hatred from the Zionists' hatred of the Slavs—the Zionists used the Germans as Esau's sword to kill off tens of millions of Slavs, under the guise of "anti-Semitism" the Jews had the deluded Germans kill off the Jews' Slavic enemy—under th guise of "anti-Bolshevism" the Jews had the deluded Germans kill off the Jews' Slavic enemy. The Jews had put both the Bolsheviks and the Nazis into power and led the Germans to believe that they were fighting Jewish interests, when all the while they were serving them.

## 5.9 Jewish Bankers Destroy Russia and Finance Adolf Hitler

Jacob H. Schiff worked against American interests and destroyed America's faithful Civil War ally Czarist Russia. Schiff was a prominent Jewish banker whose family was tied to the Rothschilds. John Hays Hammond, a man with intimate ties to high finance and close contacts in Russia,[529] gave testimony to the fact that Jewish bankers had ruined Russia in the Russo-Japanese war, in *The New York Times* on 18 November 1911, on page 2,

> "I, however, convinced them that there was no lack of friendliness toward Russians on the part of Americans, who remembered Russia's friendship to us at the time of our civil war. [***] Mr. Jacob H. Schiff has done more to accentuate the troubles of his co-religionists in Russia than any other one man, because of his boastful statement that the money of Jewish bankers had made it possible for Japan to wage a successful war against Russia."

*The New York Times* reported on 18 March 1917, in Section 2, on page 2,

> "JACOB H. SCHIFF REJOICES.
>
> **A Great and Good People Have Come Into Their Own, He Says.**
> By Telegraph to the Editor of THE NEW YORK TIMES.
> WHITE SULPHUR SPRINGS, W. Va., March 17.—May I through your columns give expression to my joy that the Russian nation, a great and good people, have at last effected their deliverance from centuries of autocratic oppression and through an almost bloodless revolution have now come into their own? Praised be God on high.
>
> JACOB H. SCHIFF."

---

[**529**]. *See also:* J. H. Hammond, *The Autobiography of John Hays Hammond; Illustrated with Photographs,* In Two Volumes, Farrar & Rinehart, incorporated, New York, (c1935); *reprinted:* Arno Press, New York, (1974), *see especially:* Chapter 23, "Russia", pp. 454-478.

In *The New York Times* on 24 March 1917 on pages 1-2, George Kennan explained how Jacob Schiff assisted Russia's enemies and how Schiff financed and trained Russian revolutionaries —Japan and the Soviet State which Schiff created became virulent enemies of the United States—enemies who came to power under Jacob Schiff's tutelage and financial patronage—Jewish bankers created the enemies of the United States and financed their wars against Americans,

## "PACIFISTS PESTER TILL MAYOR CALLS THEM TRAITORS

Socialists at Carnegie Hall Fail
to Make Russian Celebration
a Peace Meeting.

### RABBI WISE READY FOR WAR

Sorry We Cannot Fight with the
German People to Overthrow
Hohenzollerism.

### KENNAN RETELLS HISTORY

Relates How Jacob H. Schiff
Financed Revolution Propaganda
in Czar's Army.

The most violent clash between patriots and pacifists that has occurred in New York City since relations were broken with Germany marked the celebration of the Russian revolution held last night in Carnegie Hall. It was precipitated by Mayor Mitchel, whose declaration that we were about to go to war in behalf of the same kind of democracy that had freed Russia was met with a determined demonstration by pacifists, evidently previously organized, which threatened for a time to break up the meeting.

After the uproar had lasted for fifteen minutes, the Mayor, white with anger, stepped to the edge of the stage and shouted:

'This country is on the verge of war—' A loud chorus of 'No' greeted him, but above the tumult he made his voice heard with: 'And I say to you in the galleries that tonight we are divided into only two classes—Americans and traitors!'

'I hope they put you in the first ranks,' shouted a leader of the pacifists.

'You do me the greatest honor,' replied the Mayor, and the applause which followed, coupled with the ejection of some of the trouble makers, gave the Mayor's supporters the majority.

The meeting started in orderly fashion. The century old fight of Russian revolutionists was pictured in glowing words, matched by the promise of the Russia to be.

On the front of the speaker's stand hung a pair of leg irons, from a Siberian prison. They were unlocked. An authority on Russian affairs, George Kennan, told of how a movement by the Society of the Friends of Russian Freedom, financed by Jacob H. Schiff, had at the time of the Russo-Japanese war spread among 50,000 Russian officers and men in Japanese prison camps the gospel of the Russian revolutionists. 'And,' said Mr. Kennan, 'we know how the army helped the Duma in the bloodless revolution that made the new Russia last week.'

The galleries were largely filled with Socialists, downstairs an admission fee had been charged and the crowd was more orderly until awakened by the protestations of the pacifists.

Mayor Mitchel was introduced by Herbert Parsons, President of the Society of Friends of Russian Freedom, as a 'man of a race that has also struggled for freedom.' There were rumblings of trouble when a few voices in the galleries started to hoot the Mayor.

'We are gathered here,' the Mayor began, 'to celebrate the greatest triumph of democracy since the fall of the Bastile.' There were some cheers. 'America rejoices,' he said. 'How could she do otherwise when she sees power in Russia transferred from the few to the many, and in the country where there seemed the least hope of the cause of democracy triumphing.

'America, the great democracy, is proud tonight because democracy in Russia has supplanted the greatest oligarchy that remained on the face of the earth.' Then the Mayor stepped back and said:

'But I submit we have another reason to be proud. It is now inevitable, so far as human foresight can make a prediction, that the United States is to be projected into this world war and—'

'No! No!' rolled the chorus from the galleries.

There was quiet for an instant. Then the audience downstairs and in the boxes began to rise and a shout of 'Yes! Yes!' answered the galleries.

'The United States is for peace!' a voice from the gallery cried, and the tumult started anew. The ushers escorted some of the leaders of the disturbance out of the arena, and when the Mayor got partial order he said:

'We are to be projected into the war through no fault of ours, but because of conditions which have been thrust upon us—'

'No! No! No!' the galleries started again. Some one shouted an epithet at the Mayor, which brought, even from the galleries, shouts of 'Put him out! Choke him!'

'And when America does enter the contest,' shouted the Mayor, 'it will be to vindicate certain ideas as fundamental as those on which the Republic was builded, and among them will be the cause of democracy throughout the world. Let us be glad that, instead of fighting side by side with autocratic Russia, we shall be fighting side by side with democratic Russia.'

It was at this point that the galleries became so demonstrative that Mr. Mitchel told them they must be Americans or traitors.

'You are for America or you are against her,' he said, and here the Mayor made an indirect reference to the accusations he made against Senator Wagner. 'You are for America or against her, whether in private life or in legislative halls,' he said.

The Mayor then left the hall, followed by shouts of condemnation and of praise.

When the tumult had died down Rabbi S. S. Wise, a worker for world peace but not an extreme pacifist, was introduced.

'I feel it is my duty to say one word in support [hisses] and in reply to the Mayor. I would have this great audience know that I believe the Mayor was right—[This brought shouts of 'No. You're as bad as he is.']

'I am here to talk, and I'm going to talk,' shouted the Rabbi. 'If you don't like what I say, go; I am going to stay. The Mayor is right when he says we are on the verge of war. I pray God it may not come, but if it does the blame will not rest upon us, but upon that German militarism, which may it be given to the German people to overthrow as the Romanoffs have been forever overthrown.

'God knows we want peace. No man has ever fought and stood for peace as has Woodrow Wilson. [Cheers.] I do not believe that war is absolutely inevitable, but I thank God I am a citizen of a republic that has been patient.

'I am for peace, I say, but I would to God it were possible for us to fight side by side with the German people for the overthrow of Hohenzollernism.'

Then the rabbi praised the Russian revolution, but he ran into opposition when he said:

'At the risk of incurring the displeasure of those of you who have such bitter memories I hope that amnesty will be extended to the Czar himself. May God forgive the Czar.' [Shouts of 'No, never!'] 'May God forgive the monarch who never knew what mercy was.'

This was followed by shouts by a man in the gallery.

'I cannot forget,' continued the Rabbi, 'that I am a member and a teacher of a race of which half has lived in the domain of the Czar and as a Jew, I believe that of all the achievements of my people, none has been nobler than that part the sons and daughters of Israel have taken in the great movement which has culminated in the free Russia.'

It was after a review of the struggle of the Russian revolutionists, of whom he has been the leading American writer, that Mr. Kennan told of the work of the Friends of Russian Freedom in the revolution.

He said that during the Japanese-Russian war he was in Tokio, and that he was permitted to make visits among the 12,000 Russian prisoners in Japanese hands at the end of the first year of the war. He told how they had asked him to give them something to read, and he had conceived the idea of putting revolutionary propaganda into the Russian Army.

The Japanese authorities favored it and gave him permission. Later he sent to America for all the Russian revolutionary literature to be had. He said that one day Dr. Nicholas Russell came to him in Tokio, unannounced, and said that he had been sent to help the work.

'The movement was financed by a New York banker you all know and love,' he said, referring to Mr. Schiff, 'and soon we received a ton and a half of Russian revolutionary propaganda. At the end of the war 50,000 Russian officers and men went back to their country ardent revolutionists. The Friends of Russian Freedom had sowed 50,000 seeds of liberty in 100 Russian regiments. I do not know how many of those officers and men were in the Petrograd fortress last week, but we do know what part the army took in the revolution.'

Mr. Parsons then arose and said:

'I will now read a message from White Sulphur Springs sent by the gentleman to whom Mr. Kennan referred.' This was the message:

'Will you say for me to those present at tonight's meeting how deeply I regret my inability to celebrate with the Friends of Russian Freedom the actual reward of what we had hoped and striven for those long years! I do not for a moment feel that if the Russian people have under their present leaders shown such commendable moderation in this moment of crisis they will fail to give Russia proper government and a constitution which shall permanently assure to the Russian people the happiness and prosperity of which a financial autocracy has so long deprived them.

'JACOB H. SCHIFF'

This message from President Wilson was read:

'The American Ambassador in Petrograd, acting under instructions from this Government, formally recognized the new Government of Russia. By this act the United States has expressed its confidence in the success of and its natural sympathy with popular government. WOODROW WILSON'

Vladimir Resnikoff, the blind Russian baritone, sang a number of folk songs and the Symphony Orchestra, directed by Nikolai Sokoloff played Tschaikowsky's Symphony No. 4 in F minor and other selections. Miss Lillian D. Wald delivered a eulogy of Mme. Catherine Breshkovskaya, the Russian revolutionist, who had visited this country and who is now in Siberia, to be brought back at the age of 70 years to see in Petrograd the triumph of the cause for which she worked and suffered.

The following resolution was unanimously adopted:

*Resolved,* That the Mayor of the City of New York be requested to transmit the following cable to Professor Paul N. Milyoukoff, Minister of Foreign Affairs in the new Russian Government:

'Citizens of New York having at the call of the Society of the Friends of Russian Freedom assembled in mass meeting at Carnegie Hall on this 23d day of March, 1917, extend their congratulations to the Russian people upon the success of the revolution in Russia, and express their admiration for those who in the years gone by and those who in recent days have fought so bravely for liberty. They convey their earnest wishes for Russia's complete realization of self-Government, and declare their conviction that it will mean enduring friendship and co-operation between the Governments and peoples of Russia and the United States of America.'

At the close of the meeting the pictures of the revolutionary leaders were shown upon a screen, together with a picture of George Grey Bernard's statue of Lincoln which is to be placed in Petrograd.

## BREAK UP PACIFIST MEETING

Police Disperse Crowd Around Auto
of Orators in Wall Street.

The police stopped a pacifist street meeting in the Wall Street district yesterday afternoon after a big crowd had surrounded the speakers and had begun to dispute with them. Benjamin C. Marsh and other pacifist orators had been telling the crowd that the firm of J. P. Morgan & Co. and other financial interests were engineering a 'go-to-war' movement. Mr. Marsh spoke from an automobile.

'I am engaged in a fight against surrendering the Government to Wall Street,' he said. 'If the privileged class and their wealth were to be conscripted in case of war there would be no possibility of this country becoming involved.'

'What are you going to do about the German submarines?' some one in the crowd asked.

'I consider it more important to fight against special privileges than to engage in a war against poor, beaten Germany,' was the reply.

The crowd became unruly, and a police Lieutenant in charge of reserves made them move on before Mr. Marsh had finished speaking.

Dr. David Starr Jordan spoke against war yesterday at a meeting in Horace Mann Auditorium, Broadway and 120th Street, under the auspices of the Collegiate Anti-Militarism League and the Institute of Arts and Sciences.

Dr. Jordan, the Rev. Judah L. Magnes, Morris Hillquit, Arthur Le Soeur, James P. Maurer, and others will speak at a mass meeting of the Emergency Peace Federation in Madison Square Garden tonight. John F. Moors, President of the Boston Associated Charities, yesterday joined the 'unofficial commission' which is trying to find 'a way out' without war."

Rabbi Stephen S. Wise had been a member of the "Anti-Militarism Committee" which was formed to combat the "cult of preparedness" that sought "to stampede the nation".[530] He had been opposed to any talk of war.

*The New York Times* reported on 30 December 1917 on page 4 in an article entitled "KAHN ASKS ARMY OF 6,000,000 MEN":

"Jacob H. Schiff said that it now appeared reasonably sure that, at the end of this war, nationalities formerly subject would be freed and that, among

---

**530**. "Starts Open Fight Against Preparedness", *The New York Times*, (22 December 1915), p. 12.

them, Palestine would be restored to the Jews. He said that, although there had been much disagreement among the Jews of the world as to what was desirable for their future, they were now nearing an agreement and were preparing for the restoration of the Jewish State. In this situation he said that it was the duty of Jews to inquire into the reason why the Jewish nation had formerly fallen and been shattered, in order that the new Jewish State would stand. He asserted that their loss of country was originally due to their abandonment of their religion, and that a religious revival was the means of insuring the national future."

*The Jewish Communal Register of New York City 1917-1918* wrote of Jacob H. Schiff,

"Schiff, Jacob Henry, was born in 1847, at Frankfort-on-the-Maine, Germany. He received his education in the schools of Frankfort. In 1865 he came to America, where he settled in New York City. Here, he joined the staff of a banking house. In 1873, he returned to Europe where he made connections with some of the chief German banking houses. Upon returning to the United States, he entered the banking firm of Kuhn, Loeb and Company, New York, of which he later became the head. His firm became the financial re-constructors of the Union Pacific Railroad, and since then is strongly interested in American railroads. Mr. Schiff's principle of 'community of interests' among the chief railway combinations led to the formation of the Northern Securities Company, thus suppressing ruinous competition. The firm of Kuhn, Loeb & Co., floated the large Japanese War loans of 1904-05, thus making possible the Japanese victory over Russia. Mr. Schiff is director of numerous financial companies, among them the Central Trust Company, Western Union Telegraph Company, the National State Bank of New York. He is also vice-president of the New York Chamber of Commerce.

Mr. Schiff is widely known for his many philanthropic activities and for his interest in education. Of his numerous philanthropies only a few can be mentioned here. He founded the Chair in Social Economics at Columbia University; he presented the fund and the building for Semitic studies at Harvard, he is chairman of the East Asiatic Section of the Museum of Natural History of New York, which has sent out many expeditions for the study of Eastern history and conditions; he made donations to the various museums of the city, and presented the New York Public Library with a large number of works, dealing with Jewish subjects.

Mr. Schiff is the Jewish philanthropist par excellence. His philanthropies embrace every phase of the Jewish life. He is intensely interested in hospital work and is the president of the Montefiore Home, and a contributor to Mount Sinai Hospital and all other important Jewish hospitals of the city. He is profoundly interested in Jewish education and took a leading part in the reorganization of the Jewish Theological Seminary of America; he is also the founder of the Bureau of Education. In addition

Mr. Schiff is trustee of the Baron de Hirsch Fund and the Woodbine Agricultural School. He has provided the building and funds for the Young Men's Hebrew Association of New York City.

Mr. Schiff has always used his wealth and his influence in the best interests of his people. He financed the enemies of autocratic Russia and used his financial influence to keep Russia from the money market of the United States.

When last year, Mr. Schiff celebrated his seventieth birthday, all the factions of Jewry in the United States and elsewhere united in paying tribute to him."[531]

Elinor Slater and Robert Slater wrote in their book *Great Jewish Men*:

"Schiff also served as a director or advisor for many banks, insurance firms, and other companies. He helped float loans to the American government as well as to foreign countries. The most important was the two-hundred-million-dollar bond issue for Japan at the time of the 1904-1905 Russo-Japanese War. Furious with the Russians over their anti-Semitic policies, Schiff called the czarist government 'the enemy of government.' He was pleased to support the Japanese in their war effort. He also encouraged an armed revolt against the Czar. When the Japanese won the war, Schiff was presented with the Second Order of the Treasure, becoming the first foreigner to receive an official medal at the imperial palace.

In 1910 Schiff was one of several Americans who campaigned to revoke a commercial treaty with the Russians over their mistreatment of Russian Jews. Although the Russians sought him out for loans as well, he was steadfast in his refusals to grant them. Schiff made sure that no one else at Kuhn, Loeb underwrote Russian loans either. He did provide financial support for Russian-Jewish self-defense groups. It was only with the fall of the Czar in 1917 that Schiff dropped his opposition to underwriting the Russian government; he provided some support for the Kerensky government. But, angry at the Russians for refusing to honor the passports of American Jews, he successfully campaigned to abrogate the Russian-American Treaty of 1932. [***] During World War I Schiff and some of his American Jewish peers were assailed by the newer generations of Zionist leaning leaders for their indifference to Zionism. Schiff had indeed been a strong foe of Zionism, believing it a secular, nationalistic perversion of the Jewish faith and incompatible with American citizenship. He gave some funds to agricultural projects in Palestine, however, and by 1916 he had shifted his beliefs to be in favor of Zionist efforts, openly supporting the notion of a cultural homeland for Jews in Palestine."[532]

---

**531**. *The Jewish Communal Register of New York City, 1917-1918*, Second Edition, Kehillah, New York, (1919), pp. 1009-1010 (In the First Edition at 1018-1019).

**532**. E. Slater and R. Slater, "Jacob Schiff", *Great Jewish Men*, Jonathan David Publishers, New York, (2003), pp. 274-276, at 275-276.

Israel Zangwill wrote in 1911,

"[... ]Mr. Jacob Schiff financing the Japanese war against Russia and building up the American Jewry[.]"[533]

Jacob Henry Schiff was a financier who appeared to become a Zionist only after being intimidated by a Zionist smear campaign against him. However, Schiff had sponsored the rabid Zionist Rabbi Judah Magnes. Schiff funded the Russian Revolution and funded the Japanese against the Russians in their war. Schiff obstructed the Russians' access to international financing with which to fight the war, feed the Russian people and maintain the Russian economy. Many were amazed by Japan's ability to defeat mighty Russia. Schiff later showed no loyalty to anything other than the Zionists' cause.

He initially favored Germany in the First World War, since Schiff, like many American Jewish financiers, was born in Germany; and since Germany agreed to work toward the emancipation of Russian Jews and secure Palestine for the Zionists—actions Zionist Israel Zangwill defended in spirit, while Zangwill concurrently tried to bring America into the war on the side of England.[534] *The New York Times*, 22 November 1914, Section 5, page SM4, published a long article on, and interview with, Jacob Schiff together with a large portrait of the man glorifying him as a visionary of the war to end all wars; which article was entitled, "JACOB H. SCHIFF POINTS A WAY TO EUROPEAN PEACE; He Sets Forth the Disastrous Results to America That Would Follow the Complete Humiliation of Either Germany or England and Believes We Can Do Much to End This War and with It All War."[535] *The London Times* portrayed the interview with Schiff as pro-German propaganda on 23 November 1914, on page 8, and note the statement, "their line of attack is to secure a lasting peace", further note Schiff's call for a peace conference, long the ambition of the Zionists:

## "GERMAN PRESS CAMPAIGN

ADVANCE ON THE OLD METHOD.

---

[533]. I. Zangwill, *The Problem of the Jewish Race*, Judean Publishing Company, New York, (1914), p. 14; which was first published as an article, "The Jewish Race", *The Independent*, Volume 71, Number 3271, (10 August 1911), pp. 288-295, at 292.

[534]. I. Zangwill, "Zangwill Urges Jews to Support Allies", *The London Times*, (10 September 1914), frontpage; **and** "Mr. Schiff on Peace", *The London Times*, (25 November 1914), p. 9; **and** "The Voice of Jerusalem", *The London Times*, (2 December 1914), p. 9.

[535]. *See also:* "Consequences of the War", *The New York Times*, (22 November 1914), Section 3, p. 2; **and** "See Peace Campaign in Mr. Schiff's Talk", *The New York Times*, (23 November 1914), p. 3.

MR. JACOB SCHIFF'S VIEWS.

(FROM OUR OWN CORRESPONDENT.)

WASHINGTON, Nov. 22.

There are signs that the Germans are again planning to make a bid for American sympathy by peace talk. The *New York Times* publishes a long interview with Mr. Jacob H. Schiff, one of the leading German-American bankers, and a close friend of the German official representatives in the United States, which shows clearly that their line of attack is to secure a lasting peace.

Mr. Schiff argues that neither the Allies nor Germany should be allowed to score a smashing victory. A complete triumph for the Allies would hand over the world to England and her navies, while 'in the *rôle* of world-conqueror Germany would be a world-dictator and would indulge in a domination which would be almost unbearable to almost every other nation.' For the United States a complete British triumph would be especially disastrous. Probably the permanence of the Anglo-Japanese Alliance would saddle upon Americans the burden of a defensive militarism. If Germany won, the Monroe doctrine might, among other things, become a scrap of paper. Both England and Germany are patriotically resolved to fight until exhaustion supervenes. That means for Europe a prolonged period of bloodshed and misery. Hence for humanitarian and selfish reasons alike the United States is interested in ending the conflict. The United States should see whether she could not devise some sort of conference at which the belligerents could talk things over. It might perhaps be managed without an armistice.

I believe it to be not beyond the bounds of possibility that if this course could be brought about a way out of this struggle and carnage might be found, and I know I am not alone in this belief. The situation is unprecedented.... The peace must not be temporary. It must mark the ending of all war.... Towards this end America may help tremendously, and herein lies, it seems to me, the greatest opportunity ever offered to the American Press. Let the newspapers stop futile philosophizing on the merits and demerits of each case.... Let them begin stimulating public opinion in favour of rational adjustment of the points at issue.... Have we not the right to insist that the interests of neutral nations should be given some consideration by the nations whose great quarrel is harming us incalculably?

The moderation of Mr. Schiff's brief for Germany, his lamentation over the misery of the war, annotated as it is by accounts of suffering Flanders, his appeal to the humanitarian instinct of the American people, to their sympathy with the under-dog, to say nothing of his other points, all show a considerable advance of the Teutonic grasp of the American point of view since the Bernstorff manœuvres at the end of the summer. Even the *New York Times*, whose grasp of the basis of the issue, I have often pointed out, is particularly clear-visioned, while it thinks the plea is rather premature, hopes that in a few months, should one side or other score decisively enough to snatch from its enemies the hope of ultimate victory, the proposal of a conference might be opportunely pressed. It also expresses what is undoubtedly the general opinion over here, when it says:—

Whatever aims the belligerents in moments of heat and passion may profess, we here in America do not want to see Germany crushed; none of us want to see England crushed, or France or Russia. We have no wish to see any great people crushed. Such a result of the war would be an almost irreparable disaster, and we should share the loss.

The lessons of the above are fairly obvious. The peace campaign already launched by enterprising journalists, amiable pacifists, financiers worried by heavy German commitments, and by German propagandists, will sooner or later gain inconvenient strength. No pains must be spared to continue to advertise above-board our conception of the fundamental issues. It must be continually made clear that we are fighting against German militarism and not against the German people; that no peace can be lasting until the present German régime is crushed. Nor, judging from comment current here, is it enough simply to proclaim the fact.

Privately, Germans are trying to capitalize what they call the vindictive tone of certain British utterances. They draw attention, for instance, to the indiscriminate abuse of Germans as 'Huns' and of the way in which not only the Prussian contingent but the Bavarians, Wurtemburgers, &c., are bespattered with sneers. If, argue the German propagandists, such things really represent British opinion, how much reliance can be placed on British protestations that Prussian militarism is the only enemy? Does it not rather seem that Great Britain is embarked on a jealous crusade to crush utterly its dangerous rivals in the race for world supremacy?

*⁎* Mr. Jacob Henry Schiff, whose views are given above, is a native of Frankfurt-on-Main, where he was educated. He went to the United States in 1865 at the age of 18 and settled in New York. He is a member of the banking firm of Kuhn, Loeb, and Co., of which his son, Mortimer Schiff, is a partner."

Zionist spokesman Israel Zangwill, who was British but felt no loyalty to Great Britain because his only loyalty was to his fellow Jewish Zionists and their money—Zangwill ran to Schiff's defense. (In an aside, anti-Semite Eugen Karl Dühring had argued that Lessing was a poor writer and a plagiarist and that his promotion in Jewish circles was overblown and contrived.) Schiff proposed that the First World War be the war to end all wars, which became an international mantra after the war. The absolute end of all war heralded the Jewish Messianic Era in which the Jews would be "restored" to Palestine, where they would rule the world from Jerusalem. *Isaiah* 2:1-4 states,

"1 The word that Isaiah the son of Amoz saw concerning Judah and Jerusalem. 2 And it shall come to pass in the last days, *that* the mountain of the LORD's house shall be established in the top of the mountains, and shall be exalted above the hills; and all nations shall flow unto it. 3 And many people shall go and say, Come ye, and let us go up to the mountain of the LORD, to the house of the God of Jacob; and he will teach us of his ways, and we will walk in his paths: for out of Zion shall go forth the law, and the word of the LORD from Jerusalem. 4 And he shall judge among the nations, and shall rebuke many people: and they shall beat their swords into plowshares, and their spears into pruninghooks: nation shall not lift up sword against nation, neither shall they learn war any more."

*The London Times* printed a letter from Zangwill on 25 November 1914 on page 9,

## "MR. SCHIFF ON PEACE.

TO THE EDITOR OF THE TIMES.

Sir,—The interview with Mr. Jacob Schiff reported by your Washington Correspondent—the proposal for a permanent peace that shall end not only this war, but war—comes as the one gleam of light in the world's darkness. But why almost extinguish it under the head of 'German Press Campaign'? And why does he speak of Mr. Schiff's 'brief for Germany'? As one associated for many years in philanthropic work with this noblest of millionaires, I should like to testify that, despite his early associations with Germany, he is one of the most patriotic Americans I have ever known. Descended from a long line of Jewish Rabbis and scholars—one of his ancestors was Chief Rabbi of the Great Synagogue, London, in the 18th century—Mr. Jacob Schiff might himself have sat to Lessing for the portrait of 'Nathan der Weise,' and in proposing a conference to end Prussian militarism—and every other—he speaks not as the mouthpiece of Berlin, but with the voice of Jerusalem.

Yours faithfully,
Israel Zangwill

Jewish Territorial Organization, King's-chambers, Portugal-street, Nov. 23."

Zangwill was indeed familiar with Schiff's "philanthropy". Zangwill mentioned Schiff's involvement in the war between Russia and Japan in Zangwill's book, *The Problem of the Jewish Race*, Judean Publishing Company, New York, (1914), on page 14; which was first published as an article, "The Jewish Race", *The Independent*, Volume 71, Number 3271, (10 August 1911), pp. 288-295, at 292, "[... ]Mr. Jacob Schiff financing the Japanese war against Russia and building up the American Jewry[.]" Schiff provided approximately $20,000,000.00USD (non-adjusted) for the Russian Revolution.[536] Jacob

---

[536]. C. Knickerbocker, *New York Journal-American*, (3 February 1949). A. de Goulévitch, *Czarism and Revolution*, Omni Publications, Hawthorne, California, (1962),

Schiff's "philanthropy" ultimately cost the lives of tens of millions of Russians and subjected hundreds of millions more to Jewish repression which has yet to subside. The *Encyclopaedia Judaica*, Volume 14 RED-SL, Encyclopaedia Judaica, Jerusalem, The Macmillan Company, New York, (1971), cols. 960-962, at 961, states,

> "Schiff was prominently involved in floating loans to the government at home and to foreign nations, the most spectacular being a bond issue of $200,000,000 for Japan at the time of the Russo-Japanese War in 1904-05. Deeply angered by the anti-Semitic policies of the czarist regime in Russia, he was delighted to support the Japanese war effort. He consistently refused to participate in loans on behalf of Russia, and used his influence to prevent other firms from underwriting Russian loans, while providing financial support for Russian Jewish *self-defense groups. Schiff carried this policy into World War I, relenting only after the fall of czarism in 1917. At that time, he undertook to support the Kerensky government with a substantial loan."

The "anti-Semitic policies of the czarist regime in Russia" were the prohibition of racist Zionism, which the Czar prohibited because the Czar asked the Jews to integrate not segregate. The racism was Jewish, not Russian. The Czar was also confronted with murderous Jewish revolutionaries and Jewish led strikes that crippled the Russian economy and caused the Russian people to suffer and starve. But then, as now, Jews largely controlled the media and so Jews were able to blame the Czar for the wrongs Jews had done and for the racist segregationism Jews had insisted upon. In the Jewish media, the Czar became a racist for opposing Jewish racism and an enemy of the Russian People for trying to rescue them from the Jews who were out to destroy the Russian People.

Kerensky immediately emancipated the Jews after the Russian Revolution of 1917, so that they could take over the government, educational institutions, the press and other institutions of influence throughout the Russian Empire. Lenin made "anti-Semitism" an offense punishable by death, and thereby shielded all Jews from any criticism or accusation.[537] Jews then mass murdered

---

pp. 223-232. W. C. Skousen, *The Naked Capitalist: A Review and Commentary on Dr. Carroll Quigley's Book: Tragedy and Hope, a History of the World in Our Time*, Reviewer, Salt Lake City, (1971). J. Perloff, *The Shadows of Power: The Council on Foreign Relations and the American Decline*, Western Islands, Boston, (1988), p. 39. G. E. Griffin, *The Creature from Jekyll Island: A Second Look at the Federal Reserve*, American Opinion, Appleton, Wisconsin, (1995). p. 265.

**537**. V. I. Lenin, "Anti-Jewish Pogroms", *Collected Works*, Volume 29, English translation of the Fourth Russian Edition, Progress Publishers, Moscow, (1972), pp. 252-253. *See also:* D. Fahey, *The Mystical Body of Christ in the Modern World*, Browne and Nolan Limited, London, (1935), p. 251. *See also:* G. B. Shaw, *The Jewish Guardian*, (1931). *See also: Congress Bulletin*, American Jewish Congress, New York, (5 January 1940). *See also: The Jewish Voice*, (January, 1942). *See also:* G. Aronson, *Soviet Russia and the Jews*, American Jewish League against Communism, New York, (1949). *See*

educated Gentiles, and elevated Jews into positions of power and influence. Crypto-Jews in the government changed their names to Russian-sounding names. It was a crime punishable by death to reveal their true Jewish identity.

Prominent Jews had long advocated the use of tyrants following revolutions. The Bolsheviks Schiff put into power, after Kerensky, who was Jewish, failed to rule with an iron scepter, the Jewish Bolsheviks mass murdered millions of Russian Christians, destroyed Russian Orthodox Churches while leaving synagogues intact and pillaged, plundered and destroyed Russia for most of the Twentieth Century. Those many Jews who hated Russians had their revenge. Russian culture was largely destroyed in the process. Irreparable harm was done to the Russian people as a result of the mass murder of their best people and the introduction of carcinogens into their living environment. The famines and unemployment that the Jews blamed on the Czar, so as to cause the unrest which broke out in 1905, were instead due to Schiff and his Jewish financier friends. After Schiff's puppets came to power, they plundered Russia's vast wealth and sent back to the Jewish financiers, a process which continues to this very day. Such was Jacob Schiff's "philanthropy".

Before the Balfour Declaration, Jacob Schiff, a German-Jew who had emigrated to America, stated that he was not a Zionist, though he contributed to Jewish causes in Palestine in 1910,[538] and sponsored the rabid Zionist Judah Magnes. When the Zionists made a deal with the British Government to bring America into the war on the side of the Allies, Schiff found himself caught in several conflicts of interest. He did not commit wholeheartedly to Zionism. As has happened to so many, Jacob Schiff then became the victim of a Zionist smear campaign in the press, which included deliberate lies and threats. After being smeared with lies and distortions, and after being threatened, Schiff then assisted the Zionists and later became an ardent Zionist.[539] Whether or not this was mere theater is an open question. Einstein was told that Schiff was unreliable, apparently because Schiff was not an open Zionist and may have had some sentimental attachment to Germany. Einstein was told that the Warburgs, German Jewish financiers who later financed Hitler, were more reliable than Schiff the seemingly reluctant Zionist.[540] But Jacob Schiff, as fantastically

---

*also:* J. Stalin, "Anti-Semitism: Reply to an Inquiry of the Jewish News Agency in the United States" (12 January 1931), *Works*, Volume 13, Foreign Languages Publishing House, Moscow, (1955), p. 30. *See also:* S. S. Montefiore, *Stalin: The Court of the Red Star*, Vintage, New York, (2003), pp. 305-306. *See also:* "Anti-Semitism", *Great Soviet Encyclopedia: A Translation of the Third Edition*, Volume 2, (1973), pp. 175-177, at 176. *See also:* "Jews", *Great Soviet Encyclopedia: A Translation of the Third Edition*, Volume 9, Macmillan, New York, (1975), pp. 292-293, at 293. *See also:* N. S. Alent'eva, Editor, *Tseli i metody voinstvuiushchego sionizma*, Izd-vo polit. lit-ry, Moskva, (1971). Н. С. Алентьева, Редактор, Цели и методы воинствующего сионизма, Издательство Политической Литературы, Москва, (1971).

**538**. "Zionism in Palestine", *The London Times*, (19 July 1910), p. 7.
**539**. *Cf.* L. Fry, *Waters Flowing Eastward: The War Against the Kingship of Christ*, TBR Books, Washington, D. C., (2000), pp. 52-56.
**540**. Letter from A. Einstein to H. Zangger of 28 February 1919, *The Collected Papers of*

wealthy as he was, had little wealth or influence compared to the Rothschilds who ruled over him. The Rothschilds were the true force behind all of these inhuman intrigues.

Samuel Untermyer called for a boycott of Germany in 1933, and chastised Jewish bankers for financing Adolf Hitler and Nazism,

> "Revolting as it is, it would be an interesting study in psychology to analyze the motives, other than fear and cowardice, that have prompted Jewish bankers to lend money to Germany as they are now doing. It is in part their money that is being used by the Hitler régime in its reckless, wicked campaign of propaganda to make the world anti-Semitic; with that money they have invaded Great Britain, the United States and other countries where they have established newspapers, subsidized agents and otherwise are spending untold millions in spreading their infamous creed.
>
> The suggestion that they use that money toward paying the honest debts they have repudiated is answered only by contemptuous sneers and silence. Meantime the infamous campaign goes on unabated with ever increasing intensity to the everlasting disgrace of the Jewish bankers who are helping to finance it and of the weaklings who are doing nothing effective to check it."[541]

Fritz Thyssen,[542] Averill Harriman, George Herbert Walker and Prescott Bush (President George Herbert Walker Bush's father), also financed Hitler and Nazism.[543]

---

*Albert Einstein*, Volume 9, Document 7, Princeton University Press, (2004). On the role of bankers in the Nazi regime, see: H. Kardel, *Adolf Hitler, Begründer Israels*, Verlag Marva, Genf, (1974); English translation *Adolf Hitler: Founder of Israel*, Modjeskis' Society Dedicated to Preservation of Cultures, San Diego, (1997), *especially* pp. 50-52.
**541**. "Text of Untermyer's Address", *The New York Times*, (7 August 1933), p. 4. *See also:* "Untermyer Back, Greeted in Harbor", *The New York Times*, (7 August 1933), p. 4.
**542**. F. Thyssen, *I Paid Hitler*, Farrar & Rinehart, Inc., New York, Toronto, (1941).
**543**. J. Buchanan, "Bush-Nazi Link Confirmed", *The New Hampshire Gazette*, Volume 248, Number 1, (10 October 2003); URL:

<http://www.nhgazette.com/articles/NN_Bush_Nazi_Link.html>

J. Buchanan and S. Michael, "'Bush-Nazi Dealings Continued Until 1951'—Federal Documents", The New Hampshire Gazette, Volume 248, Number 3, (7 November 2003); URL:

<http://www.nhgazette.com/articles/NN_Bush_Nazi_2.html>

J. Buchanan, *Fixing America: Breaking the Stranglehold of Corporate Rule, Big Media & the Religious Right*, Far West Publishing Company, Santa Fe, New Mexico, (2004). *See also:* E. Schweitzer, *Amerika und der Holocaust: Die verschwiegene Geschichte*, Knaur Taschenbuch Verlag, München, (2004).

The attacks on Schiff no doubt intimidated other powerful and influential American and German Jews who were initially not Zionists—such as Louis Marshall. *The New York Times* reported Schiff's initial defiance on 5 June 1916, on page 6,

## "JACOB SCHIFF QUITS JEWISH MOVEMENTS

Hurt by Unjust Criticism, He
Tells Kehillah He Will Work
Alone for Reforms.

SPEAKS HIS VALEDICTORY

Says Attacks Were Based on
Misquotations That Made Him
Condemn Those He Defended.

Jacob H. Schiff informed the Kehillah at its seventh annual convention at the Hebrew Technical School for Girls yesterday that he had been hurt by recent attacks made upon him in connection with his efforts to help to solve the problems of his co-religionists, and that hereafter 'Zionism, nationalism, the Congress movement and Jewish politics in whatever form they may come up' would be a 'sealed book' to him.

'I shall continue to work for the uplift of my people,' he said in what he termed his valedictory. 'I shall continue to co-operate in all constructive work that is needed, and I shall continue to co-operate as far as I can in procuring full civic rights for our brethren in the war zone, especially in Poland, Russia, Rumania, and Palestine, for they are all flesh of my flesh and bone of my bone. But beyond this, my friends, my duty ends.'

Some of the criticism complained of by Mr. Schiff grew out of a speech made by him at the Central Jewish Institute recently, in which he was reported as having said that Jews in Russia brought many of their troubles on themselves because they kept apart as a separate people. Mr. Schiff later announced that he had not been correctly quoted, but the criticism continued. A minority group within the Kehillah, and certain Jewish newspapers, were charged with having made especial use of the speech at the Jewish Institute, largely because of their disagreement with the policies of the American Jewish Committee, of which Mr. Schiff is a member and of which Louis Marshall is President.

**Favored Quieter Plan.**

This minority group favored the calling of a 'Democratic Congress' of Jews in the United States to give immediate attention to the problems of Jews in the warring countries. The American Jewish Committee, on the other hand, advocated a quieter method and the approach of the subject through a conference which would not complicate existing troubles with hasty utterances.

Mr. Schiff was visibly affected while addressing the convention, and his voice trembled as he recounted the years of service he had devoted to the Jews of the United States and of other countries. He received a remarkable ovation at the conclusion of his speech, and ex-Justice Leon Sanders sprang to his feet with a resolution voicing complete confidence in Mr. Schiff, whom he described as 'the greatest Jew alive today.' This resolution was adopted on a rising vote, with only Z. Cutler, a delegate and a representative of a Jewish newspaper, opposing it. Mr. Cutler insisted on having his vote recorded, and was hissed.

A resolution to sever relations between the Jewish Kehillah and the American Committee was not adopted. Another resolution, also introduced by the minority group, providing for a discussion by the Kehillah of the movement to consider Jewish problems at a congress, was voted down. This was a double victory for those who agreed with the policies of the American Jewish Committee.

Mr. Schiff told the delegates that it was with the greatest regret that he had found it necessary to speak of himself to Jews of New York, and to the Jews of the country before whom he had been 'so maliciously maligned.'

### Mr. Schiff's Address.

'I have come here to deliver up the sword of dissension,' he said. 'I have lived for fifty-one years in New York. I am now almost at threescore and ten, and I believe ever since I have grown into manhood there has not a day passed that I have not been seeking the good of my people. Unfortunately, perhaps, the people of the City of New York and elsewhere have been, contrary to my bidding and even contrary to my protest, making a Jacob's coat for me. I say unfortunately because Jacob's coat, ever since the days of Joseph, has borne ill results, and, in my case, it is bearing ill results now. I hope the Yiddish press has able reporters here today, and I would ask them, if I may ask them anything, that they print in extenso what I am saying, if their reporters, as was their duty, at that meeting two weeks ago at the Central Jewish Institute, had taken down exactly what I said then instead of taking it secondhand from the secular press, there would, I believe, have been no need for me to stand before you here today. I want to read to you from a stenographic report exactly what I then said. It is not long. I shall read you only one paragraph, and I ask your patience:

Mr. Schiff, in speaking of the Jews in Russia and Poland, said: I am second to none in my feeling over oppression in Russia and Poland, not only for what they are suffering now, but for what they have suffered for the last fifty years. But it has occurred to me and it is considerable thought that I have given to this—that if the Jews of Russia and the Jews of Poland would not have been kept as a separate people by themselves, by discriminatory laws, the prejudices of persecution to which they have been subjected would not have reached the stage to which we all regret it has unfortunately come.

### Fight of Long Years.

'Now, my friends, there is not a word in this that I am not prepared to stand by. But instead of this, because one single reporter who probably—and who has since said so, I understand—did not grasp what this meant, represented that I made the Jews of Russia and Poland responsible for their persecutions, the Yiddish press launched against me a campaign of attack, maligned me, even threatened me, and continue it even now, although two or three days after that meeting, the correct stenographic report appeared, as I understand, in Yiddish in the Day, and in English in the American Hebrew. It made no difference to them; they ignored it, and they continue to ignore it now.

'Now, just think, to accuse me of such a crime. Think of it! I, who have for twenty-five years singlehanded struggled against the invasion of the Russian Government into American money markets, and to this day stave them off. Think of it! Who, as I, have been foremost in the past for agitation and insisted to the President of the United States—as some of you must know—that our treaty with Russia must be abrogated. Why did I say this treaty must be abrogated? Not that any one of us wants to go to Russia, but because others knew—and I knew—that whenever Russia would be compelled to open its doors free to the Jew, to the American Jew, and to the Jew of all nations, it would not be able to continue the restrictions against its own Jews, and to continue the Pale of Settlement which is at the bottom of all misfortune; and even if it has not come to it yet, friends, that will be the consequence.

'And these my accusers, not of this Yiddish press, but men who are here on sufferance, men who are refugees here because, unfortunately for them—and I am sorry for it—they cannot return to their homes at present as intended, and they write to the Jewish papers that I have furnished by my address munition to the Russian Government, which will be of more value to it than the munition which is furnished to them now, and the Russian Government will rejoice. No, my friends! The Russian Government will rejoice because you are battering down the man who has stood between persecution,—between anti-Semitism as far as his power goes—and the Russian Government.

### Attack Long Planned.

'Why am I attacked? I know, because I have been warned of it, and I have been warned from the inside of the Jewish press. I have been told time and again, and I have every reason to believe correctly, that if I did not stop my opposition to the Congress movement I would be first attacked, as perhaps the most conspicuous member of the American Jewish Committee, that the confidence of the Jewish people in me would be undermined, and I would be broken down, and this whole attack is only part of a very well conceived plan, and whatever I would have said, and if God Almighty would have laid the words in my mouth, I would have been maligned and attacked because it was part of a plan which has been very carefully worked out.

'Whosoever can assert that for the time he knows me, or who knows of me, I have ever denied myself to my people, have denied myself to their wants, have denied myself to any cause, that I have waited until Jewish problems have been brought to me instead of going after them in my desire to co-operate, that I have not given not only of my means, but day in and day out—and I may say night in and night out—have not given of myself, let him rise and accuse me.

'I may say this by way of valedictory: I have been hurt to the core, and hereafter Zionism, nationalism, the Congress movement, and Jewish politics in whatever form they come up, will be a sealed book to me. I shall continue to work for the uplift of my people; I shall continue to co-operate in all constructive work that is needed, and I shall continue to co-operate as far as I can in procuring full civic rights for our brethren in the war zone, especially in Poland, Russia, Rumania, and Palestine, for they are all flesh of my flesh and bone of my bone. But beyond this, my friends, my duty ends. I thank you for so patiently having listened to me, and I thank you for having encouraged me by your applause given to me.'

### Convinced in Sincerity.

Mr. Sanders, in introducing the resolution commending Mr. Schiff, said no one present could help being touched by or could question the sincerity of the statements made by Mr. Schiff. He said he had known Mr. Schiff for many years, and was convinced Mr. Schiff had not made the statement with which he was originally credited in the speech at the institute.

The Kehillah, before adjourning, adopted the following resolution, introduced by Maurice Simmons, Chairman of the Committee for the Protection of the Good Name of Immigrant Peoples, condemning discriminations in the National Guard because of religion or race:

> Resolved, That the Kehillah of New York City strongly condemns discrimination on account of race or religion in the National Guard of the State of New York, in the recruiting of members, or in the designation or election of its officers. Such discrimination is un-American and utterly opposed to the principles of the State Militia; and, further
>
> Resolved, That the National Guard of the State of New York should be regulated by necessary legislation or executive orders so that its membership and government should absolutely exclude any idea of private proprietorship or social club and the right to discriminate against men on account of their race or religion.

Mr. Schiff received many personal expressions of confidence and goodwill after his address."

The Congress Movement favored by Zionist Louis Dembitz Brandeis—was an attempt to unify Jews behind the Zionists, who were then unpopular among Jews. The Zionists created this Congress Movement so that at the close of the First World War the Zionists would have an organization in the name of which they could petition for the establishment of a Jewish State in Palestine at the peace conferences they planned would follow the war. The American Jewish Committee, and with it Jacob Schiff and Louis Marshall, seemingly opposed the

Zionists' strategy in the war, but were intimidated into following their course and were later converted to the cause. In 1918, Max Senior and Rabbi David Philipson organized a public meeting to oppose Zionism and the Balfour Declaration. Jacob H. Schiff, Oscar S. Straus[544] and Louis Marshall[545] asked Rabbi David Philipson and Max Senior not to oppose the Zionists. Schiff's letter to Philipson was quoted in *The New York Times* on 12 September 1918, on page 8:

## "SEES REFUGE FOR JEWS.
### Schiff Declines to Join Conference to Oppose Zionism.

The Zionist Organization of America gave out yesterday a letter written by Jacob H. Schiff to Dr. David Philipson of Cincinnati, Ohio, in which Mr. Schiff declared his opposition to anti-Zionist movements. Mr. Schiff asserted that even more than when he first ceased his opposition to the Zionist movement, he now felt that the creation of a Jewish homeland in Palestine was desirable. Declining Dr. Philipson's invitation to join a conference to organize an opposition to Zionism, Mr. Schiff said:

'I am very much afraid that conditions in Russia, Poland, Rumania, Austria, perhaps even Germany and elsewhere, are such that the outlook for the Jews there—and these form a vast majority of the Jewish population of the world—is far from being a favorable one, and that for reasons which would lead too far to go into here, but which by all those who want to use their eyes can be seen, considerable unhappiness, if not suffering, is likely in store in the countries I have named for the Jewish populations.

'American Israel alone, in co-operation with its English and French co-religionists, is in a position to effectually help this proposed creation of a centre where the Jew forced out by impossible conditions under which he may have to live in the Diaspora, shall be able to go with the assurance that he shall find very sympathetic surroundings and conditions under which he and posterity shall be willing to live.

'There can be no doubt that the success of these endeavors will have the most healthy and refreshing effect upon entire Israel, wherever in the world its members may be located, and the proposition which you bring forward that American Israel combine to oppose these efforts is in my opinion nothing less than preposterous.'

Mr. Schiff in the concluding paragraphs of his letter paid his respects to Dr. Philipson, but said that in organizing an opposition to Zionism Dr. Philipson was about to place himself at the head of a movement that is certain to fail.

---

**544**. Letter from O. S. Strauss to D. Philipson of 2 September 1918, quoted in: L. Fry, *Waters Flowing Eastward: The War Against the Kingship of Christ*, TBR Books, Washington, D. C., (2000), p. 54.
**545**. L. Marshall to J M. Senior, *Louis Marshall: Champion of Liberty; Selected Papers and Addresses*, Volume 2, The Jewish Publication Society of America, Philadelphia, (1957), pp. 721-723.

> The Zionist Organization of America announced yesterday a contribution by Bernard M. Baruch of the War Industries Board of $10,000 to the Palestine Restoration Fund."

Another source quotes more of the letter,

> "I believe I have heretofore explained to you the reasons which, soon after the outbreak of the Russian revolution, have induced me to change my former attitude towards the Zionist movement, and I have since become more and more convinced that it was in the best interests of our people that I did this."[546]

With the most powerful men in the American Government against him, "Colonel" House, President Wilson, Louis Brandeis and Bernard Baruch; and with the most powerful family in the world against him, the Rothschilds; one wonders what threats were used against Schiff and what offers were made to him to persuade him to change his mind.

The immense sums of money the financiers had at their disposal is mind boggling, and one wonders what could have been achieved had those funds been put to constructive purposes instead of ill purposes, or, had they been equitably distributed in a real democracy. Schiff, who headed the banking house of Kuhn, Loeb & Co., had given some $20,000,000.00USD (non-adjusted) of his own money to fund the destruction of the Russian government.[547] He was also able to corrupt the money markets of the world to prevent Russia's access to monies, which destroyed the Russian economy. Schiff achieved what Napoleon and Hitler could not—régime change in Russia followed by a replacement government of his choice. He did it with Jewish banks, not German tanks. Schiff accomplished his aim through an inhumane deception. Schiff destroyed the Russian economy, then, through propaganda, blamed the Czar for the terrible economic conditions Schiff, himself, had imposed upon the people of Russia. The financiers who corrupted governments and human affairs on an international scale produced the political climate which deliberately resulted in mass murder on the scale of tens—even hundreds—of millions of innocent lives lost. Schiff's Russian Revolution led to Stalin, and the financiers and Zionists behind

---

[546]. Letter from J. H. Schiff to D. Philipson of 5 September 1918, quoted in: L. Fry, *Waters Flowing Eastward: The War Against the Kingship of Christ*, TBR Books, Washington, D. C., (2000), pp.53-54.
[547]. C. Knickerbocker, *New York Journal-American*, (3 February 1949). A. de Goulévitch, *Czarism and Revolution*, Omni Publications, Hawthorne, California, (1962), pp. 223-232. W. C. Skousen, *The Naked Capitalist: A Review and Commentary on Dr. Carroll Quigley's Book: Tragedy and Hope, a History of the World in Our Time*, Reviewer, Salt Lake City, (1971). J. Perloff, *The Shadows of Power: The Council on Foreign Relations and the American Decline*, Western Islands, Boston, (1988), p. 39. G. E. Griffin, *The Creature from Jekyll Island: A Second Look at the Federal Reserve*, American Opinion, Appleton, Wisconsin, (1995). p. 265.

"Colonel" Edward Mandell House and his Zionist League of Nations led to Hitler. These men, who were in complete control of the American Government, were all enemies of the United States—just as Hitler, Goebbels and the other crypto-Jews who took over Germany were enemies of Germany. The "philanthropy" of Jewish financiers achieves their Messianic objectives. Jewish Messianic prophecies call on Jews to destroy all Gentile life, and to destroy the Earth. But what could humanity achieve if these Jewish financiers weren't so good to us?

These German-Jewish bankers installed a crypto-Jewish government in Germany, which not only ruined the lives of countless European Jews, but which infected the minds of innocent German children with hatred and a thirst for war which would ultimately result in their deaths, the death of their nation, their national heritage and their national honor. These Jewish bankers were a curse to all the nations and blessed none. While they stole the wealth of America, England, France, Germany, Russia, China, Japan, etc., they lived side by side with non-Jews in these countries and continually plotted to destroy them and placed their agents in power to subvert their economies, governments and religions. Germany could well have been the most productive and beneficial nation humankind has yet enjoyed—with the benefit of many well-meaning German Jews—had not ill-intentioned Internationalist and Zionist Jews deliberately destroyed it and corrupted it in their quest for Isaiah's "new earth", the Zionists' so-called "New World Order" (*Isaiah* 65:17; 66:22). Dare I say it, Germany was a victim of the Jewish religion and its mad adherents, and America will be next.

In addition to Jacob Henry Schiff and his son Mortimer; the family of Max, Paul, Felix and Fritz Warburg, were manipulative Jewish financiers in both World Wars, on both sides of both conflicts. Felix M. Warburg and Paul Warburg created and then headed the Federal Reserve[548] under President Woodrow Wilson.[549] Wilson's Svengali, "Colonel" House, wrote of how he would place a puppet dictator into power in 1912 in order to achieve this end in his book *Philip Dru: Administrator*.[550] That puppet dictator was Woodrow Wilson. The bankers made their plans for the Federal Reserve on Jekyll Island, Georgia, in 1910, and House helped to carry them out.[551] The man who drafted

---

**548**. E. C. Mullins, *Secrets of the Federal Reserve: The London Connection*, Bankers Research Institute, Staunton, Virginia, (1983); **and** *The Federal Reserve Conspiracy*, Common Sense, Union, New Jersey, (1954); **and** *A Study of the Federal Reserve*, Kasper and Horton, New York, (1952).
**549**. B. J. Hendrick, "The Jews in America: II Do the Jews Dominate American Finance?", *The World's Work*, Volume 44, Number 3, (January, 1923), pp. 266-286, at 281.
**550**. E. M. House, *Philip Dru: Administrator*, B. W. Huebsch, New York, (1912).
**551**. E. C. Mullins, *Secrets of the Federal Reserve: The London Connection*, Bankers Research Institute, Staunton, Virginia, (1983); **and** *The Federal Reserve Conspiracy*, Common Sense, Union, New Jersey, (1954); **and** *A Study of the Federal Reserve*, Kasper and Horton, New York, (1952).

the bankers' Jekyll Island plan, Paul Warburg supported the campaign of Wilson and Felix Warburg that of Taft, such that no matter who won the election the President would be friendly to the Warburgs. Max Warburg headed the German banking house of M. M. Warburg in Hamburg. Eugene Meyer was head of the War Finance Corporation.[552] Bernard Baruch was the Chairman of the War Industries Board. Many of the institutions and laws Wilson brought about under the influence of the financiers were quite similar to the institutions and laws Napoleon had begun under the influence of the Rothschilds.[553] These markets and laws again and again led to immense profits for financiers and to economic ruin for entire societies—even for humankind. Napoleon immediately faced opposition to his changes to the usury laws.[554]

The Warburgs and the Schiffs were related through marriage. The Warburgs and Jacob Schiff financed Trotsky and the Communist Revolution in Russia, as well as general revolution which led to Kerensky's rise and fall and the rise of Lenin's dictatorship and the Bolsheviks in 1917.[555] The Warburgs also financed Hitler in 1932,[556] and the Hungarian Jew Moses Pinkeles, a. k. a. Trebitsch-Lincoln,[557] financed Hitler, the NSDAP and its newspaper organ the *Völkischer Beobachter*, and many other Jewish financiers including Baron von Schroeder financed Hitler.[558] The NSDAP, after doing very poorly in an election, suddenly covered the nation with banners, posters and flags and advertised itself throughout the land in 1932. Their propaganda, uniforms, etc. must have cost a fortune. That fortune was provided by Jews who wanted to persecute other Jews

---

552. B. J. Hendrick, "The Jews in America: II Do the Jews Dominate American Finance?", *The World's Work*, Volume 44, Number 3, (January, 1923), pp. 266-286, at 281.
553. A. Muhlstein, *Baron James: The Rise of the French Rothschilds*, Vendome Press, New York, (1982).
554. P. B. Boucher, *Histoire de l'usure: chez les Egyptiens, les Juifs, les Grecs, les Romains, nos ancêtres et les Chinois, et considération sur les ravages qu'elle exerce actuellement en France ;dans laquelle l'auteur présente les moyens propres à la réprimer, prouve que l'édit de 1770, qui fixe l'intérêt de l'argent à 5 pour 100, est encore pleine vigeur, et donne une explications des articles 1153 et 1907 due Code Napoléon, relatifs au prêt à intérêt*, Chaignieau Jeune, Paris (1807).
555. A. C. Sutton, *Wall Street and the Bolshevik Revolution*, Buccaneer Books, Cutchogue, New York, (1974).
556. A. C. Sutton, *Wall Street and the Rise of Hitler*, GSG & Associates, San Pedro, California, (2002).
557. I. T. T. Lincoln, *The Autobiography of an Adventurer*, H. Holt and Co., New York, (1932). H. Kardel, *Adolf Hitler, Begründer Israels*, Verlag Marva, Genf, (1974); English translation *Adolf Hitler: Founder of Israel*, Modjeskis' Society Dedicated to Preservation of Cultures, San Diego, (1997), picture page between pages 35 and 36 and pp. 50-52, 62-63. B. Wasserstein, *The Secret Lives of Trebitsch Lincoln*, Yale University Press, (1988).
558. H. Kardel, *Adolf Hitler, Begründer Israels*, Verlag Marva, Genf, (1974); English translation *Adolf Hitler: Founder of Israel*, Modjeskis' Society Dedicated to Preservation of Cultures, San Diego, (1997), picture page between pages 35 and 36 and pp. 50-52, 62-63. See also: D. Irving, *Hitler's War and The War Path*, "Last updated Friday, April 13, 2001" <http://www.fpp.co.uk/books/Hitler/>, pp. *xxiv-xxv*.

and force them to Palestine against their will. Though the rise of the German economy in the early Nazi period is sometimes mistakenly attributed to the efficiency of Fascism, it was in fact due to a massive influx of investment capital provided by Jewish bankers. If anything, Hitler's régime was terribly corrupt and mismanaged the funds. *Papers Relating to the Foreign Relations of the United States, 1918, Russia*, Volume 1, File Number 862.20261/53, United States State Department Publication Number 222, 65th Congress, 3d Session, House Document Number 1868, United States Government Printing Office, Washington, D. C., (1931), pp. 373-376; bears witness to the Warburg transactions:

"DOCUMENT NO. 3

Circular November 2, 1914, from the Imperial Bank to the representatives of the Nya Banken and the agents of the Diskonto Gesellschaft and of the Deutsche Bank.

At the present time there have been concluded conversations between the authorized agents of the Imperial Bank and the Russian revolutionaries, Messrs. Zenzinov and Lunacharski. Both the mentioned persons addressed themselves to several financial men who, for their part, addressed themselves to our representatives. We are ready to support the agitation and propaganda projected by them in Russia on the absolute condition that the agitation and propaganda (carried on ?) by the above-mentioned Messrs. Z and L. will touch the active armies at the front. In case the agents of the Imperial Bank should address themselves to your banks we beg you to open them the necessary credit which will be covered completely as soon as you make demand on Berlin.

RISSER

Addition as part of document:

Z. and L. got in touch with Imperial Bank of Germany through the bankers (D?) Rubenstein, Max Warburg, and Parvus.

Note: L. is the present People's Commissioner of Education. Z. is not a Bolshevik, but a right Social Revolutionist and in the discard, whereabouts unknown. Parvus and Warburg both figure in the Lenin and Trotsky documents. P. is at Copenhagen. W. chiefly works from Stockholm.

[***]

DOCUMENT NO. 9

MR. RAPHAEL SCHOLNICKAN,
HAPARANDA.

Dear Comrade: The office of the banking house M. Warburg has opened, in accordance with telegram from the Rhenish Westphalian Syndicate, an account for the undertaking of Comrade Trotsky. The attorney [?] purchased arms and has organized their transportation and delivery track Luleå and Vardö to the office of Essen & Son in the name Luleå receivers and a person authorized to receive the money demanded by Comrade Trotsky.

J. FÜRSTENBERG

Note: This is the first reference to Trotsky. It connects him with banker Warburg and with Fürstenberg. Luleå is a Swedish town near Haparanda."

It was well known that financiers could affect the outcome of a war. The eleventh edition of *Encyclopædia Britannica* (1910) stated in its article "Anti-Semitism":

"Prince Bismarck himself confessed that the money for carrying on the 1866 campaign was obtained from the Jewish banker Bleichroeder, in face of the refusal of the money-market to support the war."

*The London Times* published a letter from "a member of the Vigilance Committee" on 26 November 1914 on page 9,

## "GERMAN-AMERICAN FINANCIERS AND THE WAR.

### TO THE EDITOR OF THE TIMES.

Sir,—Mr. Zangwill, in his praise of his co-religionist Mr. Jacob Schiff, of New York, in *The Times* of to-day, omits to point out that this is the second time that Jewish financiers have intervened at moments when Germany is in difficulties. It will be remembered that when the German attempt at Paris failed, Mr. James Speyer and his satellites began calling loudly for peace, and it is curious that just now, when the Germans have failed to take Warsaw and are still many miles from Calais, Mr. Jacob Schiff should be on the same tack.

The British public are getting alive to the operations of these financiers. It is fortunate that their machinations occasionally come to light, and one is grateful to Mr. Zangwill for the extra illumination he has cast upon their dark ways.

One knows now that every time the German cause is in difficulty we shall have fresh attempts to influence American neutrality. So far the pro-Germans in England and their organs in the Metropolitan Press have been wisely quiet. They are none the less being closely watched.

Yours faithfully.<br>A MEMBER OF THE VIGILANCE

COMMITTEE.
November 25."

Israel Zangwill published another letter in *The London Times* on 2 December 1914 on page 9,

## "THE VOICE OF JERUSALEM.

### TO THE EDITOR OF THE TIMES.

Sir,—If my friend Mr. Schiff speaks, as you say, with the voice of Berlin, then how splendid! For in that case what Berlin wants is 'the ending of all war.' Those are the words of Mr. Schiff which you report in your issue of the 23$^{rd}$ inst.—I have no other source of information. In your correspondent's own language:—'The line of attack is to secure a lasting peace.' In short, the admirable ultimatum of our statesmen is to be accepted:—

'No patched-up truce that would expose our children to a revival of the German menace.' Alas, I am only afraid that it is the voice of Jerusalem, and not the voice of Berlin.

Yours faithfully,
ISRAEL ZANGWILL.
Far End, East Preston, Sussex, Nov. 26."

Schiff was again in the foreground in 1917, when Jews lionized him as an instigator, and the financier, of the Russian Revolution, which succeeded just before President Wilson pushed for an American declaration of war against Germany. Benjamin Freedman asserted that there had been a meeting between the Zionists and the British government in October of 1916 and it was then that a deal was struck between them—Palestine for the Jews in exchange for America's involvement in the war on the side of the Allies. Louis Brandeis blackmailed President Wilson into accepting this deal, which cost countless American lives and prolonged the war, costing millions more lives, and which resulted in an unjust peace that led to the Hitler régime, which cost millions more lives. The Zionist Jews deliberately murdered some one hundred million people in the Twentieth Century through Communism alone,[559] deliberately disrupted and in many instances ruined the lives of billions of human beings, wasted vast resources which could have solved most of problems of the world had they been put to good use instead of applied to evil ends, all in order to force some few four or five million Jews into a land where they did not want to live. They are not done yet. Jewish prophecy demands that all other religions be prohibited, that all other cultures disappear, and eventually that all non-Jews and assimilated Jews be murdered. They will never lose sight of these goals.

---

[559]. S. Courtois, *et al., The Black Book of Communism : Crimes, Terror*, Repression, Harvard University Press, Cambridge, Massachusetts, (1999).

The deal made between Zionist Jews and Arthur Balfour was an illegal act, in that England had no right to determine the fate of Palestine and the Zionist Jews did not represent the will of the American People. Benjamin Freedman was a witness to the fact that Americans had been very pro-German up until that time, in part because German Jews did much to shape public opinion to make it pro-German. Saadia E. Weltmann wrote,

> "In general, there prevailed a feeling among Zionists that the World War and the peace conference might present a unique opportunity for the fulfillment of the Zionist aspirations in Palestine. But the diverse views of the Zionist leaders regarding the outcome of the War precluded any action by Zionist organs towards that end. At the beginning of the War, the majority of the European Zionist leaders expected a German victory. Some of them actually favored the Central Powers, largely as a result of their dislike of Russian antisemitic policies. A few, such as Nahum Sokolow, a Zionist representative in negotiating the Balfour Declaration, changed their minds in the course of the War. There were also some, especially in England and France, who believed in the victory of the Entente powers and were willing to stake on it the realization of the Zionist objectives. Among these was Professor Chaim Weizmann, who went as far as to refuse all contacts with the Berlin Executive and with the Copenhagen office.19"[560]

Freedman observed that after the Zionist Jews betrayed Germany and allied themselves with England, the German Jewish community and the Wilson Administration slandered and smeared the Germans with lies and distortions and criminalized pro-German sentiments in America.

Benjamin Freedman's charges are borne out by the historic record. As but one example among many, *The New York Times* reported on 18 January 1919 on page 4 (note that poet and Hitler apologist George Sylvester Viereck lived with, and had a homosexual relationship with the Jewish Zionist Ludwig Lewisohn.[561] Viereck was reputedly the grandson of Kaiser Wilhelm I and Edwina Viereck, and was the son of the Marxist Louis Viereck. George Sylvester Viereck was one of the chief pro-German propagandists in America during World War I, defended the Kaiser after World War I, was a devoted friend to Sigmund Freud and promoted Albert Einstein—as well as Adolf Hitler. Eustace Mullins stated that Viereck was flattered and pleased when Mullins told Viereck that Viereck had cost Germany victory in both world wars.[562] Just as the poet Ezra Pound

---

**560**. S. E. Weltmann, "Germany, Turkey, and the Zionist Movement, 1914-1918", *The Review of Politics*, Volume 23, Number 2, (April, 1961), pp. 246-269, at 253. Weltman cites: "Chaim Weizmann, *Trial and Error* (New york, 1949), pp. 148; 164-170."
**561**. Refer to the love letters between Lewisohn and Viereck in the "Ludwig Lewisohn papers, 1903-1980's" at the College of Charleston libraries, Special Collections, Third Floor, Mss 28, Box 1, Folders 1, 3 and 5.
**562**. Daryl Bradford Smith interview of Eustace Mullins of 25 January 2006, "The French Connection", *GCN LIVE*, http://www.iamthewitness.com.

propagandized for the Fascists in Italy, Viereck propagandized for the Nazis from the 1920's through the 1940's and served time in prison in America for his pro-Nazi activities. Viereck and Lewisohn remained friends after the Second World War—and the Holocaust.[563] William Jennings Bryan was Secretary of State under President Wilson. Both Bryan and Wilson, as well as Bryan's wife, and Wilson's first wife, were avowed pacificists, and advocated American neutrality. Wilson betrayed Bryan and America and brought the United States into the war as a result of Zionist blackmail.),

## "QUESTION DICKINSON, AGENT OF VIERECK
Senators Hear Letters Assailing
Wilson, Tumulty, Lansing,
and Others.

### TOLD NAVY 'SECRET ORDERS'

Writer Asserted They Were Against
Teutons—Explains 'Leak'
of Peace Note.
*Special to The New York Times.*

WASHINGTON, Jan. 17.—There was read today into the records of the Senate committee which is investigating German propaganda a large number of letters written by J. J. Dickinson, until Nov. 15 last a Captain in the army, to George Sylvester Viereck of New York, who during the period of American neutrality was one of the most active German propagandists in this country.

Most of these letters, according to the Army Intelligence Service, were really intended for Dr. Karl A. Fuehr, one of the propaganda chiefs sent to this country by the German Foreign Office. The Military Intelligence Service further alleges that the letters as a result were promptly transmitted by wireless to Berlin. The letters were all signed 'Josiah Wingate,' which Dickinson admitted was a nom de plume.

In his testimony before the letters were produced by Major E. Lowry Humes, Dickinson swore that at no time did he have reason to believe that he was employed by agents of the German Government. Until Bernstorff was ordered out of the country he had no inkling that Fuehr was one of the important cogs in the German propaganda machine. He said he worked

---

**563**. N. M. Johnson, "George Sylvester Viereck: Poet and Propagandist", *Books at Iowa*, Number 9, (November, 1968), URL:
http://www.lib.uiowa.edu/spec-coll/Bai/johnson2.htm
**and** *George Sylvester Viereck: Pro-German Publicist in America, 1910-1945*, Dissertation Thesis (Ph. D.), University of Iowa University of Iowa, Iowa City, Iowa, (1971); **and** *George Sylvester Viereck, German-American Propagandist*, University of Illinois Press, Urbana, Illinois, (1972).

simply as a Washington correspondent of Viereck's weekly, The Fatherland, and subsequently for the Transocean News Service, the German semi-official news organization, of which Dr. Fuehr was directing head in this country.

Dickinson was on the stand several hours. It never dawned upon him, he swore, until just before this country entered the war, that he had been 'duped.' After we entered the war, he said, he did all he could to help the Government build up a case against Viereck. Referring to the so-called 'peace note leak' [*The New York Times* reported on Bernard Baruch's involvement in this scandal at the time.[564]—CJB] of January, 1917, he said he was led to believe that he was in a way responsible.

He said he 'doped out' the situation correctly, and gave his deductions to John F. Harris of Harris, Winthrop & Co., 15 Wall Street, New York. He added that Bernard Baruch, who, he said, made $300,000 on steel common a result of his (Mr. Baruch's) foresight, had figured the situation out as he himself had done.

Dickinson said that in the controversy that followed the 'leak' he went to Chairman Henry of the House Committee on Rules and told him what he knew of the matter. He also communicated, he said, with Secretary Tumulty.

Various letters read into evidence were written in 1916. Dickinson admitted the authorship of all except one, which purported to report an interview with President Wilson at Shadow Lawn in October, 1916. Major Humes said that only a part of the letters were put into records.

### First Letter to Viereck.

The first letter from Dickinson to Viereck which was read into the record, dated June 4, 1916, bore a reference to Captain Guy Gaunt, then Naval Attaché of the British Embassy at Washington. In part it read:

'National Press Club,

'My Dear Mr. Viereck:

'Please note by the above that I am now receiving my mail at the National Press Club instead of the Army and Navy Club, as heretofore, the reason being that I find it more convenient to use the first-named club in doing my work than the latter.

'I learned yesterday from an authoritative source that the President has been informed that Secretary Lansing's attitude toward every newspaper man in Washington, who exhibits by his questions when calling at the State Department even a sense of fairness toward German interests, is growing more insulting every day. It is particularly marked in the case of the representatives, whether foreign or domestic, of the German-language press.

'Wilson, I know, is in a near-panic over the coming campaign. His desperation is perceptibly growing daily. This frame of mind may lead him to almost any outburst against Lansing or other Cabinet officers who may

---

**564**. "Hears Baruch Sold on Peace Note Tip", *The New York Times*, (4 January 1917), p.1. "American Board to Buy for Allies", *The New York Times*, (25 August 1917), p. 1. *See also:* "May Sieze Oil Rich Lands", *The New York Times*, (28 April 1918), p. 11.

fall under just criticism because of their unneutral attitude toward Germany. I had a long talk, somewhat startlingly frank, this morning with a Cabinet officer on this whole subject.

'In spite of denials from the White House recently of friction between Lansing and Wilson, I would not be at all surprised if Lansing would leave the Cabinet, possibly because of 'failing health,' within a few weeks. The Republican campaign managers are raking his Mexican relations and activities, past and present, with a fine-tooth comb. This the President knows, too. I confidently expect to have photographic copies of certain of his financial transactions with the Huerta Government at the City of Mexico within a couple of weeks. At any rate, I have been faithfully promised this by responsible Mexican representatives.

### 'Exposure of the Britisher.'

'I have been expecting to receive from you the promised resolutions on the Captain Gaunt affair. I have spoken to several members of Congress about the matter, men who have read with interest your exposure of the Britisher and who hope that the subject matter may be so presented in resolutions that they can handle them in some form in Congress.

'Schrader was with me several hours yesterday and doubtless will discuss with you several very interesting pointers I gave him for his next letter.

'I was not here when Bryan was last in Washington, but I have learned from two or three of his intimates who talked with him that he will give the Wilson cause only the most perfunctory support in the campaign. This will also mark the course of Speaker Clark. I do not know whether I told you in one of my last letters the story related by Mrs. Bryan to T. H. Pickford, a local Democratic magnate, of the immediate cause of her husband's precipitate retirement from the Cabinet. It was that Tumulty told a prominent German-American that Bryan was the sole cause of the Administration's anti-German policy. Pickford went to Tumulty with the story, and the atmosphere of the White House was blue with curses of the Bryans all the time Pickford was there. Pickford has since written to Mrs. Bryan a full account of his interview with Tumulty.

'This matter could be so worked up as to force Wilson to rid himself of Tumulty. What suggestions have you to make as to its handling? I believe it is too big an opportunity to be neglected. Mrs. Bryan possibly would be willing to come out in an open statement. She is a very able and a very determined woman. She loathes the whole Wilson outfit and especially Tumulty, the tumultuous. Faithfully yours,

'JOSIAH WINGATE.'

### On Eve of Convention.

Three days later, on June 7, Dickinson wrote that the Administration would 'remain excessively quiet on everything of domestic or international concern,' until after the result of the Republican National Convention in Chicago was known.

The next letter, dater June 8, 1916, contained an invitation to Viereck to come to Washington and meet Burleson, Tumulty, and Daniels. The letter indicated that the President would not receive the visitor, but 'Wingate' could introduce him to Tumulty, who would report everything he said to the President. He also touched on the punitive expedition into Mexico under Pershing in this letter.

In a letter of June 9, 1916, which also referred in the main to the impending Presidential campaign, Dickinson reported that he had talked with Secretary Tumulty, who 'manifested an unusually keen concern, asking me if I thought you would support the President or the Republican nominee at Chicago if he were other than Roosevelt.' Dickinson said he had been unable to answer so pointed a question, and added that he had also been unable to answer when Tumulty asked him 'whether or not you would direct the Fatherland (the pro-German Weekly of which Viereck was editor) along a neutral course in the campaign.' Continuing, Dickinson wrote:

'This only demonstrates how anxious the Administration people are growing over the question of the attitude of the German-American element in the forthcoming campaign. When I told Tumulty that you probably might make a visit to Washington shortly and that I should want to have him meet you and two or three others at luncheon, he was silent for a moment and said that it might be embarrassing all around, should he be seen with you. I ridiculed this strange declaration, and he finally said without explanation that you certainly ought to meet and talk with Burleson when you come here. However, I dare say that all he meant was that he would take the subject up with the President and be governed wholly by his chief's instructions.'

### In Doubt Over President.

In a letter of June 11, Dickinson wrote that he was still without information as to what the President would write into the Democratic platform 'on this subject,' his reference apparently being to the 'Americanism' question.

'He, (the President),' the letter continued, 'is naming the officers of his convention, is writing its platform, will man the National Committee through Tumulty and his son-in-law, McAdoo, and will run his own committee. What Bryan thinks of all this or intends to do about it I do not know now. I wrote Mrs. Bryan a letter today in the hope of obtaining some expression from her that might reflect her husband's mind.'

In this same letter, Dickinson prophesied that the Morgans would finance the Wilson campaign through Cleveland H. Dodge. He said that the politicians believed that the Standard Oil and Cowdray Oil interests would back Hughes.

On June 14, 1916, Viereck was informed by letter that 'by order of the President the War Department is preparing advertisements for 9,000 army trucks, in addition to 2,000 already to be bid for at the Depot Quartermaster's headquarters in New York on June 30.

'This is,' he observed, 'one of the most positive signs observable of Wilson's purpose to do something sensational before the Presidential campaign closes.'

On June 18, 1916, in a letter to Viereck, Dickinson wrote:

'* * * if you want to meet any of the folks here in high and responsible place I will attend to this end of the negotiations with pleasure. I would suggest that Untermyer, whom I know very well, be approached on the subject at once. I have no doubt at all that he would promptly and gladly respond. Fred Lynch told me recently that he had met you at Untermyer's Yonkers place several weeks or months ago. Samuel is a shrewd citizen and knows how to do things.'

### Suspected a Wilson 'Scheme.'

In a letter of June 23 Dickinson made reference to what he termed was 'further evidence of my conviction of a shrewdly devised scheme to tie us to the body of a corpse—England,' adding that this was propaganda 'started by the Wilson forces to place the blame for the extremely embarrassing situation in Mexico upon Germany.'

'Let us do something to reveal this whole damnable business and do it quickly,' he added. 'I am willing and anxious to serve in this cause in any capacity to which I may be assigned.'

'Nothing of the same relative importance has occurred since the opening of the war in Europe as the U-boat inquiry at Baltimore promises. If the Deutschland shall be captured or destroyed by a vessel of the allied powers the fault will be ours.

'Our navy has been secretly instructed to work against the interests of the Central Powers. A considerable element of the navy, whom I happen to know personally, is opposed to discrimination between the nations; but most of the element is favorably inclined toward the Teutonic element.

'If we can arrange to get together the various elements which in detail may be opposed to the British program, but which may indorse our general program, without admitting that they do so, I am confident that we may accomplish something worth while.'

Dickinson wrote on Aug. 20 that he knew that 'the Administration is anxious to catch Germany in a trap on the submarine question, and that we shall probably hear a great deal on this question before the votes are cast in November.' In this letter he also made reference to a conference the President had the previous day with the railroad executives.

### Wilson's 'Cunning and Craft.'

'Before he called these men of affairs into the conference,' he wrote, 'the President had prepared his statement, and he gave it to the newspapers through Secretary Tumulty while the conference was in session. In other words the President 'put one over' on the railroad executives and caught them napping. * * * This incident savors so of Wilson's cunning and craft that I think it could be used as a good text for an article in The Fatherland.'

Under date of Aug. 23, 1916, 'Wingate' wrote to Viereck:

'Here is a narrative that would be almost unbelievable if it were not for the fact that so many strange things have attended the Wilson foreign policies—not to say have influenced them. I obtained it recently from two

Democratic members of the Senate Committee on Foreign Relations:

'When the President was recently hard pressed by them to let them know what he was up to in Mexico—whether or not he intended eventually to intervene should he be re-elected—he told them that that eventuality would depend almost wholly on the conditions in Europe. He pointed out to them that he had announced a policy of broad and far-reaching neighborliness with all Latin America in his able speech two years ago, when he declared that never again, or at least so long as he was the responsible head of the Government, would the United States take a single foot of territory by conquest. * * *

'Now, said the President to my two friends at different times—I mean they were not with him at the same time—our word on this pledge has gone forth to the whole world, and it is doing us good in Latin America. Therefore, should be forced to intervene in Mexico, which would mean war, we could not in plain honor take a foot of Mexican territory as indemnity after we had overrung and conquered the country. We could only demand and levy a money indemnity.

'More than 50 per cent. of the productive wealth of Mexico, Wilson pointed out from statistics which he held in hand, was owned by foreigners, largely Americans, the next in holdings being the English and French. The levying of a money indemnity, therefore, would wring from 'our friends' the bulk of the extra taxation imposed through which to pay the indemnity. That would place a burden upon corporations in this country which own mines, ranches, &c., which it would be bad domestic politics to impose. It would also cause irritation in England and France.

'The Morgan-Guggenheim group are the largest owners of productive wealth in Mexico. Next to them comes the Lord Cowdray outfit in England.

'Need I tell any more of this remarkable story to enlighten you on the Wilson Mexican policy?'

### Disavows Shadow Lawn Letter.

The so-called Shadow Lawn letter, the authorship of which Dickinson denied, advancing the theory that it had been written by Viereck, was the last of the documents read into the record. It read in part as follows:

'Oct. 24, 1916.

'My Dear Mr. Viereck:

'At Shadow Lawn last Saturday the President initiated a conversation with me about you, which at least I regard as curious if not significant and of importance to you.

'He started the conversation by asking me how long I had known you personally and how well I knew you. I told him that while our personal acquaintance intercourse had extended over only two months, still I thought I knew you pretty well, mainly because I had for several years been very intimately associated with a German of your general type—the late Count Seckendorff—who temperamentally was a great deal like yourself, in that he was a man of punctilous honor and hence with strong inclinations always to be fair.

'Then the President asked me if I thought you were judicial-minded. I facetiously replied that you were a poet and that I had never known a poet of judicial mind.

'He then inquired with very apparent interest about what he called your 'equipment.' I dwelt upon your culture in a broad literary sense.

'While he was discussing your 'apparent' sense of fairness I related to him briefly the genesis of your statement for the press. I told him that you had in the original statement this assertion, 'an once of performance is better than a pound of promise,' and that you had elided this without any request or hint from me. This obviously pleased him very much.

'I infer—and my inference may be wide of the mark—that he has determined to appoint some sort of neutrality board after the election to aid him in reaching some new judgment in regard to our international relations in order that he may act within the new lights which may be thrown upon the subject.

'I was strongly tempted, of course, to ask him what he had in mind, but you can understand the sense of delicacy I felt when that thought was evolved in my mind.

### Attitude of Hyphenates.

'On the general subject of the hyphenates he seemed wholly at ease. He said he believed a year ago that their blood had been so heated against him that they were violently against him en masse. He added, however, he was convinced that their blood had cooled and that only their exclamatory leaders were in the main the only element that persistently took an unfair view of his conduct.

'He had on his desk while talking to me about you, a full copy of the statement you had prepared for the press in re the Ridder statement concerning Stone and Burleson. He remarked upon the fairness of its tone as illustrated by your assertion that you did not regard his Americanism as inferior to that of Hughes. Before I left him he looked around and said that he was sorry no stenographer had been present while he was talking to me so that what I had said concerning you might have been taken down.

'I remarked again that I was sorry he had replied at all to 'that crazy man O'Leary,' and he said that he had not dictated that statement in haste or heat, but that it was the result of very cool and careful thought on his part.

'I had almost forgotten to tell you that during the conversation the President said in effect that he wanted to know about you and others, who like yourself have individualized themselves in these troublesome times, because you might be useful 'when settlement time comes.''

Dickinson, in a statement to the committee, said he had served as a Major in Cuba in 1898 and had been commissioned soon after this country entered the European war as a Captain in the National Army. He said that his resignation became effective on Nov. 15 last.

The report of an investigation of his record was placed in evidence by Major Humes. In this report, signed by Brig. Gen. Marlborough Churchill of the Military Intelligence Service, General Churchill recommended that

Dickinson be discharged from the service by the President. His resignation followed and was accepted by President Wilson. Dickinson read into the record a letter which vouched for his loyalty and which was signed by Major Gen. Frank McIntyre of the General Staff.

J. M. Kennedy of Montana followed Dickinson on the stand. His testimony had to do with brewery and German activities, he said, had been active."

Jacob Schiff destroyed the Russian economy and caused Russia to lose its war with Japan in order to foster a revolution in 1905 which would bring about Jewish emancipation and Jewish domination of the Russian People. Schiff financed the Russian Revolution of 1917 towards the same end. When the Jews obtained dominion over Russia, the Jews oppressed and committed genocide against Russian Gentiles.

The Jewish revolutionaries behind the Russian Revolution believed that only a Communist Revolution would achieve the desired goal of emancipating the Jews of Russia, because Jews would dominate the Communist régime they would impose on the Gentile majority. In reality the only impediment to Jewish emancipation was Jewish racist nationalism. The Czar did not want an enemy State within Russian territory and the Czar offered the Jews complete freedom if only the Jews would abandon their racism and segregationism. Jewish Communist Zionist Nachman Syrkin stated in 1898,

> "In Russia, where Jews are not emancipated, their condition will not be radically altered through an overthrow of the present political regime. No matter what new class gains control of the government, it will not be deeply interested in the emancipation of the Jews. That emancipation will come to the Jews of Russia as 'manna,' or as a result of idealism and humanitarian principles, is inconceivable. Russian Jewry will attain its emancipation only in the future socialist state."[565]

Syrkin got his totalitarian Jewish Socialist State in Russia—much to the detriment of the majority of Russians and to the world, but ironically it led to "Red Assimilation", the assimilation of the Jews the Czar had wanted and the racist Jews had dreaded. Syrkin knew that assimilation followed emancipation in Western Europe, but he apparently pinned his hopes on the presumption that anti-Semitism would become so strong in Russia after the Jews had ruined the nation and mass murdered its People, and Russian Jews were so racist and segregationist, that the assimilation he knew followed emancipation after the French Revolution and Socialism in France, would not occur in Russia. When "Red Assimilation" did take place, Zionists again believed that they had the right and the duty to further ruin Russia and "rescue" Jews from themselves by putting

---

**565**. N. Syrkin, under the nom de plume "Ben Elieser", *Die Judenfrage und der socialistische Judenstaat*, Steiger, Bern, (1898); English translation in A. Hertzberg, *The Zionist Idea*, Harper Torchbooks, New York, (1959), pp. 333-350, at 346.

Hitler in power and keeping Hitler in power.

In Russia itself, the man behind Stalin's genocide and anti-Semitism, which caused the deaths of tens of millions of Christians and attempted to keep the Jews segregated, was an alleged "self-hating Jew",[566] Lazar Moiseyevich Kaganovich. American Communists, many of whom were ethnic Jews, largely turned a blind eye to these atrocities. Kaganovich was a Zionist who wanted to both punish assimilatory Jews and develop in them a keen interest in Zionism due to artificial anti-Semitism. Kaganovich was the power behind the throne of the Stalinist Regime, and he directed the genocide of the Ukrainians, as well as "Stalin's purges" and anti-Semitic campaigns. He was one of the world's worst genocidal Jewish mass murderers, worse even than the Zionist Bolshevist Adolf Hitler. The artificial anti-Semitism of Kaganovich and Hitler was part of the Zionists' strategy to force Jews to return to their roots.

Jewish Zionist Joachim Prinz wrote in his book *The Secret Jews*,

> "In Hitler's Germany, as so often before in Jewish history, persecution stimulated Jewish resilience and inspired a return to Jewish values. Oppression has repeatedly awakened the Jews' dormant resources and created contempt for the persecutor; the result has often been a renascence of Judaism. This is not to deny that many Jews did convert under the pressure of the Inquisition and the terror of the Gestapo. There were certainly many thousands of sincere converts who became devout Christians and totally gave up their Judaism. But the phenomenon, which may contain at least a partial answer to the riddle of the survival of the Jewish people, is that through centuries of persecution in each generation there have always been Jews who maintain their Jewishness in some way, and that to the present time their descendants manifest the memory of their ancestors' faith in their rituals and their lives.
>
> A more complicated aspect of this phenomenon occurred recently in Russia. At the turn of the century young Russian Jews, whose forefathers had suffered for decades under the czar's savage pogroms, were among the early converts to Communism and followed the lead of Marx, Trotsky and the other early Communist theoreticians—who themselves were Jews, though, of course, not observant Jews. To rid themselves of every vestige of their Jewish heritage and to demonstrate their allegiance to the new system, which scorned religion of any kind, some staged wild parties on the Day of Atonement, while the remnant of the faithful Jews were saying their prayers. (For those who wanted to retain their Jewish identity, early Communism provided a measure of religious freedom; some schools still taught Yiddish, many synagogues remained open.) The young Jewish students, marching under the red banner with their fellow Russians, were ecstatic about their sudden and glorious emancipation from the Pale of Settlement, those areas

---

[566]. S. Kahan, "Preface", *The Wolf of the Kremlin*, William Morrow and Company, Inc., New York, (1987).

of the country to which Jews had been confined since the end of the nineteenth century. They became super-Communists, freed from the daily degradation, the insults and the recurrent pogroms which had become part of the history of the Russian Jews under the czars. The new political dogma seemed to promise that this sort of persecution would never occur again.

The anti-Semitic brutality of the Stalin regime showed this Jewish euphoria to have been a fool's paradise. The Jewish schools were closed; most of the synagogues were boarded up. Hundreds of Jewish intellectuals and professionals, all fervent Communists, were exterminated in the purges. Soviet Jewry's Marranic period had begun. But it remained a rather quiet, even dormant form of secret Judaism until the creation of the State of Israel."[567]

Prinz appeared to strongly resent assimilated Jews, even at the late date he published *The Secret Jews*,

"The assimilated Jew of whom we speak is one of 'Jewish descent,' who may deny it, hide it or be ashamed of it. Like the Marrano, his Jewishness is the skeleton in his closet. He would prefer to associate with 'others' rather than cultivate his Jewishness. In many respects he is very much a modern Marrano. For although he is trying to keep his Jewish origin secret, he remains latently Jewish. There was a time when this type of Jew was a rarity. Vie are approaching the time when he may represent a majority of the Jewish community. Religious and secular ties are becoming less binding. A very large number of young Jewish people throughout the world have only tenuous ties with their Jewishness. But—and this is the problem which reminds us so much of the Marranos—*can Jewishness be forgotten?*"[568]

Perhaps the most compelling evidence that Soviet anti-Semitism was a ploy meant to force reluctant, assimilating Jews into Zionism against their will, was the fact that the most virulent anti-Semitic purges began after the failed attempt to create a "Jewish State" in the far Eastern regions of the Soviet Union, the Jewish Autonomous Oblast in Khabarovsk Krai in the districts of Birobidzhansky, Leninsky, Obluchensky, Oktyabrsky and Smidovichsky.[569] This plan failed, in part, due to the interference of some Zionist Socialists, who insisted that Palestine was the Jews' national home. An even earlier attempt to found a Jewish State in Russia in the districts of Homel, Witebsk and Minsk,[570] also failed, largely due to a lack of Jewish interest. The Zionists insisted that anti-Semitism alone could force the Jews to segregate. When the Zionists put

---

[567]. J. Prinz, *The Secret Jews*, Random House, New York, (1973), pp. 13-14.
[568]. J. Prinz, *The Secret Jews*, Random House, New York, (1973), p. 195.
[569]. "Jews", *Great Soviet Encyclopedia: A Translation of the Third Edition*, Volume 2, Macmillan, New York, (1973), pp. 292-293, at 293.
[570]. I. Zangwill, "Is Political Zionism Dead? Yes", *The Nation*, Volume 118, Number 3062, (12 March 1924), pp. 276-278, at 276.

Hitler in power, they had the needed impetus to force Jews to flee Europe and the Zionists attempted to steal Chinese territory for a "Jewish homeland" with the help of the Imperial Japanese under the "Fugu Plan".[571] Zionist Jews sought to establish a "Jewish State" in China, which had been taken over by the Imperial Japanese whom the Jews had been financing since the days when Jacob Schiff loaned them $200,000,000.00 in the Russo-Japanese War. The Zionists used the Imperial Japanese to destroy the Chinese government in preparation for the formation of a Jewish nation in China under the "Fugu Plan" in Manchuria or Shanghai. The Jews even promoted the *Protocols of the Learned Elders of Zion* to the Japanese as evidence as to how powerful they were. The "Fugu Plan" failed to attract enough Jews, even under Nazi pressure, and die hard Zionists wanted Palestine. The Zionists then arranged for war between the United States and Japan. When America declared war on Japan, Hitler, seemingly inexplicably, declared war on the United States ensuring the ultimate defeat of Germany. Hitler also went to war with the Soviets, which gave him access to large numbers of Jews the Zionists could then segregate and ready for deportation to Palestine.

Schiff's and the Zionists' war on Russia has caused the Russian people, Jew and Gentile, great suffering and loss of life for over a century. Both the Nazis and the Communists caused the Russians, and Slavs and Jews in general, to suffer genocide and prolonged tyranny at a time when the enlightenment promised far better things for humanity. In the minds of Cabalistic Jews, evil is good, and they celebrate the fact that they formed a racist apartheid "Jewish State" in Palestine by spilling oceans of blood. This racist State continually troubles the world and consumes vast resources which could otherwise be put to productive uses. The Jews in Israel regularly steal from the Palestinians and degrade and murder them. For Cabalistic Jews, evil is goodness.

Israel Zangwill was a prominent racist Zionist in Britain, who devoted his life to segregating Jews. Zangwill's statements prompt many questions regarding the motives and involvement of the Zionists in the persecution and concentration of Jews shortly before, during, and after the First World War. One might dismiss Zangwill's statements as rhetorical exaggeration expressed for effect, were these same points not so often repeated by Jewish racist political Zionists, both publicly and privately. Zangwill also states that most Jews of the period (unlike him) considered the notion of a Jewish state to be a "political perversion"; and, in the knowledge that the race-concept does not apply to humans, Zangwill maintains it anyway, for political purposes. The bragging of the Zionists was perhaps in small part a reaction to the denigration Jews had endured from Richard Wagner (who was perhaps himself of Jewish descent), Eugen Karl Dühring, Houston Stewart Chamberlain, and a host of others, though before Zangwill, Disraeli had made similar boasts, and there is no shortage of self-

---

**571**. M. Tokayer and M. Swartz, *The Fugu Plan: The Untold Story of the Japanese and the Jews During World War II*, Paddington Press, New York, (1979). D. Goodman and M. Miyazawa, *Jews in the Japanese Mind: The History and Uses of a Cultural Stereotype*, Free Press, New York, (1995).

glorification in the Old Testament. Israel Zangwill wrote in 1911:

## "The Problem of the Jewish Race

To sum up in a few thousand words a race which has energized for 4,000 years is a task which can only be executed, if at all, by confining oneself to elementals. And of these elementals the first and most important is the soul of the people. The soul of the Jewish race is best seen in the Bible, saturated from the first page of the Old Testament to the last page of the New with the aspiration for a righteous social order and an ultimate unification of mankind of which, in all specifically Jewish literature, the Jewish race is to be the medium and missionary. Wild and rude as were the beginnings of this race, frequent as were its backslidings, and great as were—and are—its faults, this aspiration is continuous in its literature even up to the present day. There is every reason to believe that the historic texts of the Old Testament were redacted in the interests of this philosophy of history, but this pious falsification is very different from the self-glorification of all other epics. Israel appears throughout not as a hero but as a sinner who cannot rise to his rôle of redeemer, of 'servant of the Lord'— that rôle of service, not dominance, for which his people was 'chosen.' The Talmud, the innumerable volumes of saintly Hebrew thought, the Jewish liturgy, whether in its ancient or its mediaeval strata, the 'modernist' platforms of reformed American Synagogues, all echo and re-echo this conception of 'the Jewish mission.' Among the masses it naturally transformed itself into nationalism, but even this narrower concept of 'the chosen people' found poetic expression as a tender intimacy between God and Israel.

'With everlasting love hast Thou loved the house of Israel, Thy people; a Law and commandments, statutes and judgments, hast Thou taught us... . Blessed art Thou, O Lord, who lovest Thy people, Israel.'

Such is the evening benediction still uttered by millions of Hebrew lips.

And the performance of this Law and these commandments, statutes and judgments, covering as they did the whole of life, produced—despite the tendency of all law to over-formality—at domestic ritual of singular beauty and poetry, a strenuous dietary and religious régime, and tender and self-controlling traits of character, which have combined to make the Jewish masses as far above their non-Jewish environment as the Jewish wealthier classes are below theirs. No demos in the world is so saturated with idealism and domestic virtue, and when it is compared with the yet uncivilized and brutalized masses of Europe, when, for example, the lowness of its infantile mortality or the heathiness of its school children is contrasted with the appalling statistics of its neighbors, there is sound scientific warrant for endorsing even in its narrowest form its claim to be 'a chosen people.'

This extraordinary race arose as a pastoral clan in Mesopotamia, roved to Palestine, thence to Egypt, and after a period of slavery returned to Palestine as conquerors and agriculturists, there to practice the theocratic code imposed by Moses (perhaps the noblest figure in all history), and to evolve in the course of the ages a poetic and prophetic literature of unparalleled sublimity. That union of spirituality, intellectuality and fighting-power in the breed, which raised it above all ancient races except the Greek, was paid for by an excessive individualism which distracted and divided the State. Jerusalem fell before the legions of Titus. But—half a century before it fell—it had produced Christianity and thus entered on a new career of world-conquest. And five centuries after the destruction of Jerusalem, its wandering scions had impregnated Mohammed with the ideas of Islam. Half the world was thus won for Hebraism in some form or other and the notion of 'the Jewish mission' triumphantly vindicated. A nucleus of the race, however, still persisted, partly by nationalist, instinct, partly by the faith that its doctrines had been adulterated by illegitimate elements and its mission was still unaccomplished, and it is this persistence to-day of a Hebrew population of twelve millions—a Jewdom larger than any that its ancient conquerors had ever boasted of crushing—which constitutes the much-discussed Jewish problem.

But there was a Jewish diaspora even before Jerusalem fell; settlements of Jews all around the Mediterranean, looking, however, to Jerusalem as a national and religious center. The Book of Esther is historically dubious, but it contains one passage which is a summary of Jewish history: 'And Haman said unto King Ahasuerus, There is a certain people scattered abroad and dispersed among the people in all provinces of thy Kingdom, and their laws are diverse from all people; neither keep they the King's laws; therefore, it is not for the King's profit to suffer them. If it please the King, let it be written that they may be destroyed.' The Jewish problem in fact, from the Gentile point of view, is entirely artificial. It springs exclusively from Christian or heathen injustice and intolerance, from the oppression of minorities, from the universal law of dislike for the unlike. In Russia, which harbors nearly half of his race, the Jew is confined to a Pale and forbidden the villages even of that Pale, he is cramped and crippled at every phase of his existence, he must fight for Russia but cannot advance in the Army or the Navy or the Government service, except at the price of baptism. Occasionally bands of Black Hundreds are loosed upon him in bloody pogroms, but his everyday existence has not even this tragic dignity. It is a sordid story of economic oppression designed to keep this mere four per cent. of the population from dominating Holy Russia. Ten years ago Count Pahlen's Commission reported that 'ninety per cent. of the Jews in the Pale have no stable occupation,' and if the Government enforces the Sunday Law recently passed by the Duma, it means that they will in many cases be forced to choose between their own Sabbath and semi-starvation. Already the ancient hope and virtue of the most cheerful of races are slowly asphyxiating in the never-lifting fog of poverty and persecution. A similar situation in Roumania, if on a smaller scale as affecting only a quarter of a million of Jews, is accentuated in bitterness by Roumania's refusal to fulfil the obligation of equal treatment she undertook at the Berlin Congress, and the passivity of the Powers in presence of violated treaties adds to the Jewish tragedy the tragedy of a world grown callous of its own spiritual interests. The Jews, whose connection with Roumania is at least fifteen centuries old, are not even classed as citizens. They are 'Vagabonds.' In Morocco the situation of the Jews is one of unspeakable humiliation. They are confined to a Mellah, and as the Moroccan proverb puts it, 'One may kill as many as seven Jews without being punished. The Jews have even to pickle the heads of decapitated rebels. Tested by the Judaeometer, Germany herself is still uncivilized, for if she has had no Dreyfus case, it is because no Jew is permitted military rank. Even in America with its lip-formula of brotherhood, a gateless Ghetto has been created by the isolation of the Jews from the general social life.

But if from the Gentile point of view the Jewish problem is an artificial creation, there is a very real Jewish problem from the Jewish point of view—a problem which grows in exact proportion to the diminution of the artificial problem. Orthodox Judaism in the diaspora cannot exist except in a Ghetto, whether imposed from without or evolved from within. Rigidly professing Jews cannot enter the general social life and the professions. Jews *qua* Jews were better off in the Dark Ages, living as chattels of the king under his personal protection and to his private profit, or in the ages when they were confined in Ghettos. Even in the Russian Pale a certain measure of autonomy still exists. It is emancipation that brings the 'Jewish Problem.' It is precisely in Italy with its Jewish Prime Minister and its Jewish Syndic of Rome that this problem is most acute. The Saturday Sabbath imposes economic limitations even when the State has abolished them. As Shylock pointed out, his race cannot eat or drink with the Gentile. Indeed, social intercourse would lead to intermarriage. Unless Judaism is reformed it is, in the language of Heine, a misfortune, and if it is reformed, it cannot logically confine its teachings to the Hebrew race, which, lacking the normal protection of a territory, must be swallowed up by its proselytes.

The comedy and tragedy of Jewish existence to-day derive primarily from this absence of a territory in which the race could live its own life. For the religion which has preserved it through the long dark centuries of dispersion has also preserved its territorial traditions in an almost indissoluble amalgam of religion and history. Palestine soil clings all about the roots of the religion, which has, however, only been transplanted at the cost of fossilization. The old agricultural festivals are observed at seasons, with which, in many lands of the Exile, they have no natural connection. The last national victory celebrated—that of Judas Maccabaeus—is two thousand years old, the last popular fast dates from the first century of the Christian era. The Jew agonizing in the Russian Pale rejoices automatically in his Passover of Freedom, in his Exodus from Egypt. Even while the tribal traits had still the potential fluidity of life, neither Greeks nor Romans could change this tenacious race. Its dispersion from Palestine merely indurated its traditions by freeing them from the possibility of common development. The religious customs defended by Josephus against Apion are still the rule of the majority. Even new traits superimposed by their history upon fractions of the race are conserved with equal tenacity. The Jews expelled from Spain in 1492 still retain a sub-loyalty to the King of Spain and speak a Spanish idiom, printed in Hebrew characters, which preserves in the Orient words vanished from the lips of actual Spaniards and to be found only in Cervantes.

This impotency to create afresh—which is the negative aspect of conservatism—translated itself after the final revolt of Bar-Cochba against the Romans early in the second century, into a pious resignation. The Jewish Exile was declared to be the will of God, which it was even blasphemous to struggle against, and the Jews, in a strange and unique congruity with the teachings of the prophet they rejected, turned the other cheek to the smiter and left to Caesar the things that were Caesar's, concentrating themselves in every land of the Exile upon industry, domesticity and a transmuted religion, in which realities were desiccated into metaphors, and the Temple sacrifices sublimated into prayers. Rabbinic opportunism, while on the one hand keeping alive the hope that these realities, however gross, would come back in God's good time, went so far in the other direction as to lay it down that the law of the land was the law of the Jews. Everything in short—in this transitional period between the ancient glory and the Messianic era to come—was sacrificed to the ideal of mere survival. The mediaeval teacher Maimonides laid it down that to preserve life even Judaism might be abandoned in all but its holiest minimum. Thus—under the standing menace of massacre and spoliation—arose Crypto-Jews or Marranos, who, frequently at the risk of the stake or sword, carried on their Judaism in secret. Catholics in Spain and Portugal, Protestants in England, they were in Egypt or Turkey Mohammedans. Indeed the *Dönmeh* still flourish in Salonika and provide the Young Turks with statesmen, the Balearic Islands still shelter the *Chuetas*, and only half a century ago persecution produced the *Yedil-al-Islam* in Central Asia. Russia must be full of Greek Christians who have remained Jewish at heart. Last year a number of Russian Jews, shut out from a university career, and seeking the lesser apostasy, became Mohammendans, only to find that for them the Trinity was the sole avenue to educational and social salvation.

Where existence could be achieved legally, yet not without social inferiority, a minor form of Crypto-Judaism was begotten, which prevails to-day in most lands of Jewish emancipation, among its symptoms being change of names, accentuated local patriotism, accentuated abstention from Jewish affairs, and even anti-Semitism mimetically absorbed from the environment. Indeed, Marranoism, both in its major and minor forms, may be regarded as an exemplification of the Darwinian theory of protective coloring. The pervasive assimilating force acts even upon the most faithful, undermining more subtly than persecution the life-conceptions so tenaciously perpetuated.

Nor is there anywhere in the Jewish world of to-day any centripetal force to counteract these universal tendencies to dissipation. The religion is shattered into as many fragments as the race. After the fall of Jerusalem the Academy of Jabneh carried on the authoritative tradition of the *Sanhedrin*. In the Middle Ages there was the *Asefah* or Synod to unify Jews under Judaism. From the middle of the sixteenth to the middle of the eighteenth century, the *Waad* or Council of Four Lands legislated almost autonomously in those Central European regions where the mass of the Jews of the world was then congregated. To-day there is no center of authority, whether religious or political. Reform itself is infinitely individual, and nothing remains outside a few centers of congestion but a chaos of dissolving views and dissolving communities, saved from utter disappearance by persecution and racial sympathy. The notion that Jewish interests are Jesuitically federated or that Jewish financiers use their power for Jewish ends is one of the most ironic of myths. No Jewish people or nation now exists, no Jews even as sectarians of a specific faith with a specific center of authority such as Catholics or Wesleyans possess; nothing but a multitude of individuals, a mob hopelessly amorphous, divided alike in religion and political destiny. There is no common platform from which the Jews can be addressed, no common council to which any appeal can be made. Their only unity is negative—that unity imposed by the hostile hereditary vision of the ubiquitous Haman. They live in what scientists call symbiosis with every other people, each group surrendered to its own local fortunes. This habit of dispersed and dependent existence has become second nature, and the Jews are the first to doubt whether they could now form a polity of their own. Like Aunt Judy in 'John Bull's Other Island,' who declined to breakfast out of doors because the open air was 'not natural,' the bulk of the Jews consider a Jewish State as a political perversion. There are no subjects more zealous for their adopted fatherlands: indeed they are only too patriotic. There are no Otto mans so Young-Turkish as the Turkish Jews, no American so spread-eagle as the American Jews, no section of Britain so Jingo as Anglo-Jewry, which even converts the Chanukah celebration of Maccabaean valor into a British military festival. Of the two British spies now confined in German fortresses one is a Jew. The French Jewry and the German reproduce in miniature the Franco-German rivalries, and the latter even apes the aggressive *Welt-Politic*. All this ultra-patriotism is probably due to Jews feeling consciously what the other citizens take subconsciously as a matter of course; doubtless, too, a certain measure of Marranoism or protective mimicry enters into the ostentation. At any rate each section of Jewry, wherever it is permitted entrance into the general life, invariably evolves a somewhat over-colored version of the life in which it finds itself embedded, and fortunate must be accounted the peoples which have at hand so gifted and serviceable a race, proud to wear their livery.

 What wonder that Jews are the chief ornaments of the stage, that this chameleon quality finds its profit in artistic mimicry as well as in biological. Rachel, the child of a foreign pedlar in a Paris slum, teaches purity of diction to the Faubourg St. Germain; Sarah Bernhardt, the daughter of Dutch Jews, carries the triumph of French acting across the Atlantic. A Hungarian Jew,

Ludwig Barnay, played a leading rôle in the theatrical history of Germany, and another, von Sonnenthal, in that of Austria. For if, like all other peoples, the Jews can only show a few individuals of creative genius—a Heine, a Spinoza, a Josef Israels, a Mendelssohn, etc.—they flourish in all the interpretative arts out of all proportion to their numbers. They flood the concert-platforms—whether as conductors, singers or performers. As composers they are more melodious than epoch-making. Till recently unpracticed in painting and sculpture they are now copiously represented in every gallery and movement, though only rarely as initiators. Indeed, the Jew is a born intermediary and every form of artistic and commercial agency falls naturally into his hands. He is the connoisseur *par excellence*, the universal art-dealer. His gift of tongues, his relationship with all the lands of the Exile, mark him out for success in commerce and finance, in journalism and criticism, in scholarship and travel. It was by their linguistic talents that the adventurous journeys of Arminius Vambery and Emin Pasha were made possible. If a Russian Jew, Berenson, is the chief authority on Italian art, and George Brandes, the Dane, is Europe's greatest critic, if Reuter initiated telegraphic news and Blowitz was the prince of foreign correspondents, if the Jewish Bank of Amsterdam founded modern finance and Charles Frohman is the world's greatest entrepreneur, all these phenomena find their explanation in the cosmopolitanism of the wandering Jew. Lifted to the plane of idealism, this cosmopolitan habit of mind creates Socialism through Karl Marx and Lassalle, an international language through Dr. Zamenhof, the inventor of Esperanto, a prophecy of the end of war through Jean de Bloch, an International Institute of Agriculture through David Lubin, and a Race Congress through Dr. Felix Adler. For when the Jew grows out of his own Ghetto without narrowing into his neighbor's, he must necessarily possess a superior sense of perspective.

As a physician the Jew's fame dates from the Middle Ages, when he was the bearer of Arabian science, and the tradition that kings shall always have Jewish physicians is still unbroken. Dr. Ehrlich's recent discovery of '606,' the cure for syphilis, and Dr. Haffkine's inoculation against the Plague in India, are but links in a long chain of Jewish contributions to medicine. Nor would it be possible to mention any other science, whether natural or philological, to which Jewish professors have not contributed revolutionizing ideas. The names of Lombroso for criminology, Benfey for Sanscrit, Jules Oppert for Assyriology, Sylvester for Mathematics, and Mendeleiff for Chemistry ('The Periodic Law') must suffice as examples.

In law, mathematics and philosophy, the Jew is peculiarly at home, especially as an expounder. In chess he literally sweeps the board. There is never a contest for the championship of the world in which both rivals are not Jews. Even the first man to fly (and die) was the Jew, Lilienthal.

But to gauge the contribution of the Jew to the world's activity is impossible here. To mention only living Jews, one thinks at random of Rothschilds with their ubiquitous financial and philanthropic activity, Sir Ernest Cassel financing the irrigation of Egypt, Mr. Jacob Schiff financing the Japanese war against Russia and building up the American Jewry, Herr Ballin creating the Hamburg-American Line, Maximilian Harden's bold political journalism, the Dutch jurist Asser at The Hague conference, or the American statesman and peace-lover Oscar Straus, the French plays of Bernstein, or the German plays of Ludwig Fulda, or the Dutch plays of Hyermanns, or the Austrian plays of Schnitzler, the trenchant writings of Max Nordau, the paintings of Solomon and Rothenstein, of Jules Adler and Max Liebermann, the archeologic excavations of Waldstein, Hammerstein building the English Opera House, Imre Kiralfy organizing our Exhibitions, Sidney Lee editing the Dictionary of English Biography, Sir Matthew Nathan managing the Post Office, Meldola investigating coal-tar dyes, the operas of Goldmark, the music-plays of Herr Oscar Straus and Humperdinck (Herr Max Bernstein), the learned synopses of Salomon Reinach, the sculpture of Antokolsky, Mischa Elman and his violin, Sir Rufus Isaacs pleading on behalf of the Crown, Signor Nathan polemizing with the Pope, Dr. Frederick Cowen conducting one of his own symphonies, Michelson measuring the velocity of light, Lippmann developing color photography, Henri Bergson giving pause to Materialism with his new philosophy of Creative Evolution, Bréal expounding the science of Semantics, or Herrmann Cohen his neo-Kantism, and one wonders what the tale would be both for yesterday and to-day if every Jew wore a yellow badge and every Crypto-Jew came out into the open, and every half-Jew were as discoverable as Montaigne or the composer of 'The Mikado.' The Church could not even write its own history; that was left for the Jew, Neander. To the Gentile the true Jewish problem should rather be how to keep the Jew in his midst—this rare one per cent. of mankind. The elimination of all this genius and geniality would surely not enhance the gaiety of nations. Without Disraeli would not England lose her only Saint's Day?

But the miracle remains that the Gentile world has never yet seen a Jew, for behind all these cosmopolitan types which obsess its vision, stand inexhaustible reserves of Jewish Jews—and the Talmudic mystic, the Hebrew-speaking sage, remains as unknown to the Western world as though he were hidden in the fastnesses of Tibet. A series of great scholars—Geiger, Zunz, Steinschneider, Schechter—has studied the immense Hebrew literature produced from age to age in these obscure Jewries. But there is a modern Hebrew literature, too, a new galaxy of poets and novelists, philosophers and humanists, who express in the ancient tongue the subtlest shades of the thought of to-day. And there is a still more copious literature in Yiddish, no less rich in men of talent and even genius, whose names have rarely reached the outside world.

And if the Jew, with that strange polarity which his historian Graetz remarked in him, displays simultaneously with the most tenacious preservation of his past the swiftest surrender of it that the planet has ever witnessed, if we find him entering with such passionate patriotism into almost every life on earth but his own, may not even the Jewish patriot draw the compensating conclusion that the Jew therein demonstrates the comparative superficiality of all these human differences? Like the Colonel's lady and Judy O'Grady all these peoples are the same under their skins—as even Bismarck was once constrained to remark when he saw Prussians and Frenchmen lying side by side in the community of death. Could Jews so readily assimilate to all these types, were these types fundamentally different? The primitive notion of the abysmal separateness of races can scarcely survive under Darwinism. Every race is really akin to every other. Imagine a Canine Congress debating if all these glaring differences of form, size and color could possibly consist with an underlying and essential dogginess. It is curious that Houston Chamberlain, the most eloquent champion of the race-theory and the Teutonic spirit, is himself an Englishman married to the daughter of Wagner (*alias* Geier) and that with quasi-Semitic assimilativeness he has written his book in German after a career as a writer in French.

Not only is every race akin to every other but every people is a hotchpotch of races. The Jews, though mainly a white people, are not even devoid of a colored fringe, black, brown or yellow. There are the Beni-Israel of India, the Falashas of Abyssinia, the disappearing Chinese Colony of Kai-Fung-Foo, the Judeos of Loango, the black Jews of Cochin, the negro Jews of Fernando Po, Jamaica, Surinam, etc., the Daggatuns and other warlike nomads of the North African deserts who remind us what the conquerors of the Philistines were like. If the Jews are in no metaphorical sense brothers of all these peoples, then all these peoples are brothers of one another. If the Jew has been able to enter itno all incarnations of humanity and to be at home in every environment, it is because he is a common measure of humanity. He is the pioneer by which the true race-theory has been experimentally demonstrated. Given a white child, it is the geographical and spiritual heritage—the national autocosm, as I have called it—into which the child is born that makes out of the common human element the specific Frenchman, Australian or Dutchman. And even the color is not an unbridgeable and elemental distinction.

Nor is it only with living races that the Jew has manifested his and their mutual affinity, he brings home to us his brotherhood and ours with the peoples that are dead, the Medes, the Babylonians, the Assyrians. If the Jew Paul proved that the Hebrew Word was universal, the Jews who rejected his teaching have proved the universality of the Hebrew race. One touch of Jewry makes the whole world kin.

The labors of Hercules sink into child's play beside the task the late Dr. Herzl set himself in offering to this flotsam and jetsam of history the project of political reorganization on a single soil. But even had this dauntless idealist secured co-operation instead of bitter hostility from the denaturalized leaders of all these Jewries, the attempt to acquire Palestine would have had the opposition of Turkey and of the 600,000 Arabs in possession. It is little wonder that since the great leader's lamentable death, Zionism—again with that idealization of impotence—has sunk back into a cultural movement which instead of ending the Exile is to unify it through the Hebrew tongue and nationalist sentiment. But for such unification, a religious revival would have been infinitely more efficacious: race alone cannot survive the pressure of so many hostile milieux—or still more parlous—so many friendly. The Territorial movement, representing the original nucleus of the Herzlian idea, is still searching for a real and not a metaphorical soil, its latest negotiation being with the West Australian Government.

But if the prospect of a territorial solution of the Jewish Question, whether in Palestine or in the New World appears remote, it must be admitted that the Jewish race, in abandoning before the legions of Rome the struggle for independent political existence, in favor of spiritual isolation and economic symbiosis, discovered the secret of immortality, if also of perpetual motion. In the diaspora anti-Semitism will always be the shadow of Semitism. The law of dislike for the unlike will always prevail. And whereas the unlike is normally situated at a safe distance, the Jews bring the unlike into the heart of every milieu and must thus defend a frontier-line as large as the world. The fortunes of war vary in every country, but there is a perpetual tension and friction even at the most peaceful points, which tend to throw back the race on itself. The drastic method of love—the only human dissolvent—has never been tried upon the Jew as a whole, and Russia carefully conserves—even by a ring fence —the breed she designs to destroy. But whether persecution extirpates or brotherhood melts, hate or love can never be simultaneous throughout the diaspora, and so there will probably always be a nucleus from which to restock this eternal type. But what a melancholy immortality! 'To be *and* not to be'—that is a question beside which Hamlet's alternative is crude.

It only remains to consider what part the world should be called upon to play in the solution of this tragic problem. To preserve the Jews, whether as a race or as a religious community, is no part of the world's duty, nor would artificial preservation preserve anything of value. Their salvation must come from themselves, though they may well expect at least such sympathy and help as Italy or Greece found in their struggles for regeneration. The world's duty is only to preserve the ethical ideals it has so slowly and laboriously evolved, largely under Jewish inspiration. Civilization is not called upon to save the Jews, but it *is* called upon to save itself. And by its treatment of the Jews it is destroying itself. If there is no justice in Venice for Shylock, then alas for Venice.

'If you deny me, fie upon your law!
There is no force in the decrees of Venice.'

Even from the economic standpoint Russia with her vast population of half-starved peasants is wasting one of her most valuable assets by crippling Jewish activity, both industrially and geographically. In insisting that Russia abolish the Jewish Pale I am pleading for the regeneration of Russia, not of the Russian Jew. A first-class ballet is not sufficient to constitute a first-class people. Very truly said Roditchev, one of the Cadet leaders, 'Russia cannot enter the temple of freedom as long as there exists a Pale of Settlement for the Jews.' But abolition of the Pale and the introduction of Jewish equality will be the deadliest blow ever aimed at Jewish nationality. Very soon a fervid Russian patriotism will reign in every Ghetto and the melting-up of the race will begin. But this absorption of the five million Jews into the other hundred and fifty millions of Russia constitutes the Jewish half of the problem. It is the affair of the Jews.

That the preservation of the Jewish race or religion is no concern of the world's is a conclusion which saves the honest Jew from the indignity of appealing to it. For with what face can the Jew appeal *ad misericordiam* before he has made the effort to solve his own problem? There is no reason why a race any more than a man should be safeguarded against its own unwisdom, and its own selfishness. No race can persist as an entity that is not ready to pay the price of persistence. Other peoples are led by their best and strongest. But the best and strongest in Israel are absorbed by the superior careers and pleasures of environment—even in Russia there is a career for the renegade, even in Roumania for the rich—and the few who remain to lead lead for the most part to destroy. If, however, we are tempted to say, 'then let this, people agonize as it deserves,' we must remember that the first to suffer are not the powerful but the poor. It is the masses who bear almost the entire brunt of Alien Bills and massacres and economic oppression. While to the philosopher the absorption of the Jews may be as desirable as their regeneration, in practice the solution by dissolution presses most heavily upon the weakest. The dissolution invariably begins from above, leaving the lower classes denuded of a people's natural defences, the upper classes. Moreover, while as already pointed out the Jewish upper classes are, if anything, inferior to the classes into which they are absorbed, the marked superiority of the Jewish masses to their environment, especially in Russia, would render *their* absorption a tragic degeneration.

But if dissolution would bring degeneracy and emancipation dissolution, the only issue from this delimma is the creation of a Jewish State or at least a Jewish land of refuge upon a basis of local autonomy to which in the course of the centuries all that was truly Jewish would drift. And if the world has no ethical duty to take the lead in this creation, it may yet find its profit in getting rid of the Jewish problem. Many regions of the New World, whether in America or Australia, would moreover be enriched and consolidated by the accession of a great Jewish colony, while to the Old World its political blessing might be many-sided. A host of political rivalries, perilous to the world's peace, center around Palestine, while in the still more dangerous quarter of Mesopotamia, a co-operation of England and Germany in making a home under the Turkish flag for the Jew in his original birthplace would reduce Anglo-German friction, foster world-peace and establish in the heart of the Old World a bridge of civilization between the

East and the West and a symbol of hope for the future of mankind."[572]

Israel Zangwill's racist tract corroborates much that appears in the *Protocols*. The Zionists exercised a grossly undue influence over the course of world events throughout the Twentieth Century, selfishly interfering in world events for the sake of a few million nationalists, but doing little to rescue millions of Europe's Jews during the Holocaust and the Stalinist purges. Unlike many other political Zionists, Einstein did make some effort to successfully rescue individuals from the Nazis, and by war's end had abandoned much of the political Zionist mythology he had initially espoused and disseminated, though Einstein also callously rejected some pleas for help, which prompts the question if Einstein, like so many racist political Zionists, placed more value on racist Zionist life than on assimilated Jewish life.

Israel Zangwill was a member of a long tradition of Jewish racism in Great Britain, which held that anti-Semitism benefitted the allegedly superior Jewish race. Zionist Joseph Chaim Brenner believed that the hostility towards Gentiles and the feeling of Jewish superiority commonly expressed in Jewish literature resulted from Jewish envy of Gentiles.[573]

Jewish racists also believed that racial integration would be the downfall of Gentiles of all races. The question arises as to what rôle Jewish racism played in the evolution of the modern liberal spirit of "racial integration" which is often promoted by Jewish liberals today, many of whom have the best of intentions and are philanthropic and loving persons.

Were there some darker souls who held the misguided view that they could degrade their enemies with a false Liberalism of racial integration? The question prompts itself as to whether or not the "Friendship of the Nations" of the Soviet Union with its long standing propaganda campaign for "race mixing" was intended to weaken the Russians' blood as revenge for their persecutions of the Jews and to render them easier to dominate. Stalin promoted "racial integration" in the sentimental film *Circus*, a motion picture released in 1936 directed by Grigori Alexandrov and starring Lyubov Orlova Benjamin, which like most Communist propaganda employed sentimentality as bait for a trap to lead people into intended harm. In the minds of racist Zionists, "race mixing" weakened the general population and the loss of a "race-based" national spirit left a people without a biological reason for existence. In addition, Houston Stewart Chamberlain wrote that miscegenation resulted in "chaos", weak strains of human beings who were in general incapable of competing with "pure races". His book was popular among Zionists and the English translation of it received a long and favorable review in the *Times Literary Supplement* of 15 December 1910, pp. 500-501. Before Chamberlain, racist Zionist Benjamin Disraeli wrote

---

[572]. I. Zangwill, *The Problem of the Jewish Race*, Judaen Publishing Company, New York, (1914); which was first published as an article, "The Jewish Race", *The Independent*, Volume 71, Number 3271, (10 August 1911), pp. 288-295.

[573]. J. H. Brenner, "Self-Criticism", in A. Hertzberg, *The Zionist Idea*, Harper Torchbooks, New York, (1959), pp. 307-312, at 310-311.

that human "races" could be weakened through "race mixing". Many have alleged that prominent Jews have long promoted liberal immigration policies and miscegenation in the American media, in order to open the gates to the immigration of Eastern European Jews, and to make it impossible, in their view, for European anti-Semitism to take over America, and to weaken American culture and render it incapable of competing with corrupt tribal and segregated Jewish American society. As is often the case, the ultimate source is found in the Old Testament, which teaches the Jews that Esau is angry with them and that they can profit by diluting the blood of Esau and lessen his capacity to fight.

Joseph Stalin was clearly not a philanthropist, and so we can safely conclude that his drive for miscegenation was not motivated by humanitarianism. He deliberately murdered intellectuals and degraded the genes of the Soviet peoples through the mass murder and the exile of their best citizens. Napoleon's wars and Hitler's wars also degraded the bloodlines of Europeans by killing off their best males of breeding age—and these effects were not unknown to Jewish racists, since they were known generally.[574] In addition, the Talmud at *Sanhedrin 37a* teaches the Jews the importance of the fact that taking the life of an individual can also signify the genocide of countless unborn descendants of that individual. The racist Jews who instigated countless wars and revolutions sought to exterminate the better part of the non-Jewish Peoples and leave them inferior and easily managed "races" forever, or at least until they were completely wiped out. The following article appeared in *The World's Work*, Volume 24, Number 6, (October, 1912), pp. 612-613,

## "EUGENICS AND WAR

ONE subject warmly discussed at the Congress of Eugenists recently held in London was the effect of war on national physique. Prof. Vernon Kellogg, of Leland Stanford, Jr. University, urged the necessity of peace for the development and maintenance of the best manhood. He declared that nothing could be more disastrous to the physical strength of a people than the direct selection of the most robust for work which carried them away from home, prevented their giving their vigor to children, and returned them, if at all, maimed, diseased, and exhausted. The prevalence of war, draining the country of its able-bodied men, brings with it an era of greatly lowered birth-rate and of the birth of weak and undersized children. This happened during the Napoleonic campaigns. When they were over, even though the survivors were decimated and wounded France entered on a period in which an inch was added to the wartime stature of its inhabitants.

Professor Kellogg's argument provoked replies from German and English military officers, who defended military service on the ground that it strengthened and developed the recruits. The German, a general, alluded to the physical strength and high spirits of the young soldiers he had seen marching through the streets of London. There can be no doubt that military

---

[574]. D. S. Jordan, *Unseen Empire; a Study of the Plight of Nations that Do Not Pay Their Debts*, American Unitarian Association, Boston, (1912).

exercise and discipline are beneficial to those brought under them—so long as they do not go to war. But the same exercise and discipline directed in other channels—in preparation for duties not destructive but efficient for prosperity—these would give the same result, as a by-product, while their chief purpose would not be wasted. Every advantage claimed for military service could be gained by training for war, not against other nations, but against the common foes of all. On the sole ground of the maintenance of a people's physical vigor, war is greatly to be deplored. It inevitably kills many, injures more, and at the best withdraws a large proportion of the most vigorous from fatherhood during their best years, while it leaves the weakest to transmit their deficiencies to the following generation."

Jews had long had access to European leaders, and given their networks of contacts throughout the world, could impress these leaders by forecasting events known to them by intercommunication with their colleagues, giving the illusion of an almost supernatural gift of prophecy to the leaders of Europe, whom they could then pit against one another for profit. If a "court Jew" knew of an opportunity, or could manipulate the markets to profit a leader, or could predict a war and its outcome, not based on insight, but based on inside information; it would make quite an impression on a naïve and gullible European leader, especially if the "court Jew" was able accomplish this seemingly miraculous feat time after time, while flattering the ego, and promoting the ambitions of the foolish leader. This would instill confidence in the leader, which could then be exploited at a critical time to take advantage of the leader's faith and trust to lead a nation into self-destruction through unwise investment, treaty or war. A "court Jew" often managed national loans. The powers which control capital and debt know what investments persons and nations will make in the future, which gives them inside information and the ability to stimulate or destroy a national economy. Whoever controls the press knows of events before the public. Anyone with a story to tell must first report it to the press. Therefore, the press knows of a great deal of inside information and knows of many scandals. The press can expose, suppress or utilize this information in a corrupt fashion.

Jews have long dominated both international finance and the mass media. Through tribal collusion, they can also regulate those interests which they place in Gentile hands, so as to remain in control behind the scenes. Zionist Jews and Jewish bankers used their control of the American Press to incite Americans into accepting Woodrow Wilson's efforts to make war with Germany without just grounds. Congressmen Moore and Callaway tried to warn the United States Congress that Wilson, who was under the control of Zionist Jews, together with the Jewish controlled Press of America were attempting to bring America into the First World War on false grounds. Their statements are captured in the *Congressional Record* for 9 February 1917,

> "Mr. MOORE of Pennsylvania. Mr. Chairman, the remarks Of the gentleman from Tennessee [Mr. AUSTIN] move me to say that, along with him and my other colleagues, I hope to see the President sustained in all

proper efforts to maintain the honor and dignity of this country. We are considering now one of the great war bills, and the most of us will vote for it even to the limit of those things asked for to sustain the President. While doing that and considering other war bills, it seems to me that we might say to ourselves—whether it is carried over the telegraphic lines to the people of the country or not—that there are many disturbing and conflicting rumors concerning war conditions which are asserted to-day and denied to-morrow. Yesterday we were informed that an American had been killed on the wrecked steamer *Turino*. His name was George Washington, and, of course, it would occasion a patriotic thrill the whole length and breadth of the country if it was true that George Washington had gone down at the hands of an enemy in foreign waters. But the newspapers had their say yesterday, and they had it again this morning, that this sure-enough American was killed, and therefore we ought to go to war with Germany.

Mr. BRITTEN. Will the gentleman yield?

Mr. MOORE of Pennsylvania. Yes.

Mr. BRITTEN. Did this man have any number?

Mr. MOORE of Pennsylvania. I do not know. He was an individual of color, but his taking off was supposed to be reason to cause war. Efforts have been made, desperate efforts have been made, since the President was here on Saturday last, to prove that we must go to war. The coasts of the world seem to have been raked to find some overt act to force the President to come in here and ask us to declare war. We have had very little but rumors, but we have had headlines galore, all with a view of stampeding the House and stampeding the country into an act of war. [Applause.] I rose to make this very brief statement because I do not want the people of this country to be deceived. I am satisfied that most of the people of the country want peace; peace with honor, of course. [Applause.] But they do not want to go into a dishonorable war, and they ought not to be forced into a war by the munition makers or the munition users of this or any other land. [Applause.]

Most of the dispatch headlines declaring that American ships have gone down, that American lives have been lost, that international laws have been violated have come from London, and London has been crazy with delight since it heard the glad tidings on Saturday last that the President had severed diplomatic relations with Germany. Coming from the Liberty Bell and Independence Hall district of the United States, I can not forget that we had trouble with London in 1776, and that we had trouble with London in 1812. I am not quite ready to accept all of these rumors that come out of London now without a grain of salt. London is a little more in need of American help just now than we are in need of the advice of London. I am not quite ready, therefore, to believe every damnable, pernicious, and lying report that comes out of London, or to accept it as an inducement to declare my country in a state of war. [Applause.]

On the night of the day that the President appeared here and informed the Congress of the fact that he had severed diplomatic relations with

Germany, we had newspaper 'extras' announcing in startling headlines that the *Housatonic* had gone down in violation of international law; there were great scare heads, and boys on the streets shouting it aloud. It was declared that American rights had been violated by a country with which we were on friendly terms up to that time. Yet the next day's newspapers announced in smaller type that the *Housatonic* was loaded with contraband, and even our State Department declared that there was no occasion for any warlike declaration in consequence of her sinking.

The CHAIRMAN. The time of the gentleman from Pennsylvania has expired.

Mr. MOORE of Pennsylvania. Mr. Chairman, I ask unanimous consent to proceed for five minutes more.

Mr. PADGETT. Mr. Chairman, I ask unanimous consent that debate upon the paragraph and all amendments thereto close in five minutes.

The CHAIRMAN. Is there objection to the request of the gentleman from Tennessee?

There was no objection.

The CHAIRMAN. Is there objection to the request of the gentleman from Pennsylvania?

There was no objection.

Mr. GORDON. Mr. Chairman, will the gentleman yield?

Mr. MOORE of Pennsylvania. Yes.

Mr. GORDON. Is it the contention of the gentleman that because a ship is loaded with contraband, Germany has the right to destroy the lives of passengers and crew?

Mr. MOORE of Pennsylvania. I made the statement that after all these headlines the State Department declared that there was no breach of international law. The people were being inflamed—

Mr. GORDON. But they did not say it was because the ship was loaded with contraband.

Mr. MOORE of Pennsylvania. I stated what the gentleman's own Secretary of State announced to the public—he was not as anxious as some newspaper editors are to rush into war.

Mr. GORDON. I agree with much of what the gentleman has said; but—

Mr. MOORE of Pennsylvania. I am not arguing the point of contraband at all. The gentleman is merely taking my time. I am trying to make a plain statement to the House as to the truth and the facts. The gentleman may be stampeded because certain things appear in the newspapers, but—

Mr. GORDON. Oh, don't you worry about my being stampeded. [Laughter.]

Mr. MOORE of Pennsylvania. I am making the statement that we see alarming headlines to-day indicating that we are on the verge of war because some 'overt act' has been committed, and the next day the whole thing is denied.

Mr. GORDON. I agree with the gentleman about that.

Mr. RAGSDALE. Mr. Chairman, will the gentleman yield?

Mr. MOORE of Pennsylvania. Yes.

Mr. RAGSDALE. Will the gentleman tell me what he thinks the duty of this Government ought to be if the German Government has taken charge of and forcibly restrained by order our ambassador in that country?

Mr. MOORE of Pennsylvania. The gentleman is carried away with the headlines.

Mr. RAGSDALE. No; he is not.

Mr. MOORE of Pennsylvania. If the gentleman will listen, I will demonstrate what fools some men are—not like the gentleman from South Carolina, of course—who believe everything they read. I was coming to that very point. For three days we have heard that our American ambassador, who was on excellent terms with everyone in high life in Germany, has 'been in captivity' and held for exchange. The gentleman believes that statement.

Mr. RAGSDALE. No; the gentleman does not.

Mr. MOORE of Pennsylvania. It is absurd upon its face. Though we have had it for three days, this morning's newspapers announce that Berlin is in conference with the American ambassador, that conferences have been going on in Berlin, and that the ambassador will be safeguarded out of Germany just as we are going to safeguard the German ambassador out of the United States. Oh, how easy it is for you to rush into war upon the say so of somebody who is interested in having war.

Mr. DYER. His passports have been issued to him.

Mr. MOORE of Pennsylvania. The ambassador is going to get out safely. Somebody wanted to inflame the American people by declaring that the American ambassador had been held in captivity. Absurd! We have given safe conduct to the German ambassador and are sending him home, and the Germans have been decent with the American ambassador. But at least 2 college professors and about 150 editors, more or less, yesterday declared—not that they were willing to enlist, for the barracks down here are waiting for men like them to come forward and enlist—but they declared in effect that they were willing to involve their country in war because 'the American ambassador was held in bondage in Berlin.' This morning the newspapers show that those editors and those college professors did not know what they were talking about, and that is what I am trying to say to the gentleman from South Carolina. The plain people should not be fooled. Mr. Chairman, how much time have I left?

The CHAIRMAN. One minute.

Mr. MOORE of Pennsylvania. In that one minute let me say, and I hope not to be interrupted again, that the *Housatonic* alarm has gone glimmering. The State Department seems to concede that the Germans were within their rights and that the *Housatonic* presents no casus belli. The next day we had the *California* sensation. Because this ship bore a good old American name everybody was made to suspect that it was an American ship, and that the Germans had perpetrated such an outrage as would force us to go to war.

After the sensation had thrilled the country we were quietly informed that the *California* was a British ship, sailing under the British flag, and that she had been given the warning required by international law. But a great deal is made of the fact that one American was aboard that ship. He may have been planted there to protect the cargo and to involve this country in an international warfare; I do not know, but the next day after the newspapers had worked the story of the American passenger to the limit, it developed that he was taken off the ship to a place of safety. It matters not that he was a colored man.

Mr. BRITTEN. And the ship was armed.

Mr. MOORE of Pennsylvania. Then, again, Mr. Chairman, the report went broadcast over the United States on the day after the President addressed Congress, that this Government had seized all the interned German ships. These reports were tempered here and there with the suggestion that the German sailors were endeavoring to destroy the property of their own country, but nevertheless it was broadly announced that our naval officers had seized this German property. I will not stop to discuss the moral aspect of this seizure except to say that there had been no declaration of war and that it was not clear why we should deliberately take this German property and appropriate it to the United States. Within a day or two the answer came from both the State Department and the White House that these German ships had not been seized, and that while this Government was taking certain precautions with respect to possible impediments to navigation, every courtesy was being shown the officers and men in charge of these German vessels. It was evident that some tall lying was done in this instance for the purpose of irritating Germany under very aggravating circumstances. Somebody evidently wanted Germany to commit an 'overt act' that would bring on a war. We ought to be on our guard against this dangerous 'rumor' business, whether it originates in London or the United States.

The CHAIRMAN. The time of the gentleman from Pennsylvania has expired.

The Clerk read as follows:

Maintenance, Bureau of Supplies and Accounts: For fuel; the removal and transportation of ashes and garbage from ships of war; books, blanks, and stationery, including stationery for commanding and navigating officers of ships, chaplains on shore and afloat, and for the use of courts-martial on board ships; purchase, repair, and exchange of typewriters for ships; packing boxes and materials; interior fittings for general storehouses, pay offices, and accounting offices in navy yards; expenses of disbursing officers; coffee mills and repairs thereto; expenses of naval clothing factory and machinery for the same; laboratory equipment; purchase of articles of equipage at home and abroad under the cognizance of the Bureau of Supplies and Accounts, and for the payment of labor in equipping vessels therewith, and the manufacture of such articles in the several navy yards; musical instruments and music; mess outfits; soap on board naval vessels; athletic outfits; tolls, ferriages, yeomen's stores, safes, and other incidental expenses; labor in general storehouses, paymasters' offices, and accounting offices in navy yards and naval stations, including naval stations maintained in island possessions under the control of the United States, and expenses in handling stores purchased and manufactured under 'General account of advances'; and reimbursement to appropriations of the Department of Agriculture of cost of inspection of meats and meat food products for the Navy Department: *Provided,* That the sum to be paid out of this appropriation, under the direction of the Secretary of the Navy, for chemists and for clerical, inspection, storeman, store laborer, and messenger service in the supply and accounting departments of the navy yards and naval stations and disbursing offices for the fiscal year ending June 30, 1918, shall not exceed $1,400,000; in all, $2,750,000.

Mr. MOORE of Pennsylvania, Mr. RAGSDALE, and Mr. CALLAWAY rose.

The CHAIRMAN. The Chair will recognize the gentleman from Texas, a member of the committee.

Mr. CALLAWAY. Mr. Chairman, I ask unanimous consent to insert in the RECORD a statement that I have of how the newspapers of this country have been handled by the munition manufacturers.

The CHAIRMAN. The gentleman from Texas asks unanimous consent to extend his remarks in the RECORD by inserting a certain statement. Is there objection?

Mr. MANN. Mr. Chairman, reserving the right to object, may I ask whether it is the gentleman's purpose to insert a long list of extracts from newspapers?

Mr. CALLAWAY. No; It will be a little, short statement, not over $2\frac{1}{2}$ inches in length in the RECORD.

The CHAIRMAN. Is there objection?

There was no objection.

Mr. CALLAWAY. Mr. Chairman, under unanimous consent, I insert in the RECORD at this point a statement showing the newspaper combination, which explains their activity in this war matter, just discussed by the gentleman from Pennsylvania [Mr. Moore]:

'In March, 1915, the J.P. Morgan interests, the steel, shipbuilding, and powder interests, and their subsidiary organizations, got together 12 men high up in the newspaper world and employed them to select the most influential newspapers in the United States and a sufficient number of them to control generally the policy of the daily press of the United States.

'These 12 men worked the problem out by selecting 179 newspapers, and then began, by an elimination process, to retain only those necessary for the purpose of controlling the general policy of the daily press throughout the country. They found it was only necessary to purchase the control of 25 of the greatest papers. The 25 papers were agreed upon; emissaries were sent to purchase the policy, national and international, of these papers; an agreement was reached; the policy of the papers was bought, to be paid for by the month; an editor was furnished for each paper to properly supervise and edit information regarding the questions of preparedness, militarism, financial policies, and other things of national and international nature considered vital to the interests of the purchasers.

'This contract is in existence at the present time, and it accounts for the news columns of the daily press of the country being filled with all sorts of preparedness arguments and misrepresentations as to the present condition of the United States Army and Navy, and the possibility and probability of the United States being attacked by foreign foes.

'This policy also included the suppression of everything in opposition to the wishes of the interests served. The effectiveness of this scheme has been conclusively demonstrated by the character of stuff carried in the daily press throughout the country since March 1915. They have resorted to anything necessary to commercialize public sentiment and sandbag the National Congress into making extravagant and wasteful appropriations for the Army and Navy under the false pretense that it was necessary. Their stock argument is that it is 'patriotism.' They are playing on every prejudice and passion of the American people.'"[575]

J. P. Morgan was a Rothschild agent,[576] and Louis Brandeis and Samuel Untermyer used Morgan and the debilitating panic of 1907 the Jewish bankers deliberately caused to make the American public clamor for banking reform.[577]

---

**575**. *Congressional Record: Containing the Proceedings and Debates of the Second Session of the Sixty-Fourth Congress of the United States*, Volume 54, Part 3, United States Government Printing Office, Washington, D. C., (27 January 1917-12 February 1917), pp. 2946-2948.

**576**. E. C. Mullins, *Secrets of the Federal Reserve: The London Connection*, Bankers Research Institute, Staunton, Virginia, (1983); **and** *The Federal Reserve Conspiracy*, Common Sense, Union, New Jersey, (1954); **and** *A Study of the Federal Reserve*, Kasper and Horton, New York, (1952).

**577**. **On Untermyer, see:** Corp Author: United States., Congress., House., Committee on rules., *Investigation of the Money Trust. No. 1-[2] Hearings Before the Committee on Rules of the House of Representatives, on House Resolutions 314 and 356. Friday, January 26, 1912.*, Washington, D. C., U. S. Govt. Print. Off., (1912). ***See also:*** Corp Author: United States., Congress., House., Committee on Banking and Currency., *Money Trust Investigation... Statistical and Other Information Compiled under Direction of the Committee.*, Washington, D. C., U. S. Govt. Print. Off., (1912). ***See also:*** A. P. Pujo and Arsène Paulin and E. A. Hayes. Corp Author: United States., Congress., House., Committee on Banking and Currency, *Money Trust Investigation. Investigation of Financial and Monetary Conditions in the United States under House Resolutions Nos.429 and 504, Before a Subcommittee of the Committee on Banking and Currency.*, Washington, D. C., U. S. Govt. Print. Off., (1913). ***See also:*** J. G. Milburn, W. F. Taylor, *Money Trust Investigation: Brief on Behalf of the New York Stock Exchange*, New York, C.G. Burgoyne, (1913). ***See also:*** J. P. Morgan, *Testimony of Mr. J. Pierpont Morgan and Mr. Henry P. Davison Before the Money Trust Investigation*, J.P. Morgan & Co., New York, (1913). **On Brandeis, see:** L. D. Brandeis, *Other People's Money and How the Bankers Use It*, F.A. Stokes, New York, (1914). ***See also:*** L. D. Brandeis, M. I.

It was a trap and the "reform" ultimately put in place the Federal Reserve System which created a private central bank that regulated the money supply and operated a fractional reserve banking system. The Jewish bankers finally had the system in place in America they had always sought. Senator and financier Nelson W. Aldrich, who was one of the infamous conspirators who helped draft the Federal Reserve Act on Jekyll Island confirmed that it was means to consolidate their power and reduce their competition, which had been growing in recent years,

> "Before the passage of this Act, the New York bankers could only dominate the reserves of New York. Now, we are able to dominate the bank reserves of the entire country."[578]

Congressman Charles A. Lindbergh Sr. was very aware of the fact that the bankers had deliberately caused the panic in 1907 in order to make the public clamor for banking reforms, banking reforms the bankers would draft which would give them complete control over the money supply and wipe out the lower level, but numerous, competing banks,

> "When the Aldrich-Vreeland Emergency Currency Bill was sprung on the House in its finished draft and ready for action to be taken, the debate was limited to three hours and Banker Vreeland placed in charge. It took so long for copies of the bill to be gotten that many members were unable to secure a copy until within a few minutes of the time to vote. No member who wished to present the people's side of the case was given sufficient time to enable him to properly analyze the bill. I asked for time and was told that if I would vote for the bill it would be given me, but not otherwise. Others were treated in the same way.
> Accordingly, on June 30, 1908, the Money Trust won the first fight and the Aldrich-Vreeland Emergency Law was placed on the statute books. Thus the first precedent was established for the people's guarantee of the rich man's watered securities, by making them a basis on which to issue currency. It was the entering wedge. We had already guaranteed the rich men's money, and now, by this act, the way was opened, and it was intended that we should guarantee their watered stocks and bonds. Of course, they were too keen to attempt to complete, in a single act, such an enormous steal as it would have been if they had included all they hoped ultimately to secure. They knew that they would be caught at it if they did, and so it was planned that the whole thing should be done by a succession of acts. The first three have taken place.

---

Urofsky and D. W. Levy, Editors, *Letters of Louis D. Brandeis*, In Five Volumes, State University of New York Press, Albany, New York, (1975). This set has numerous letters as well as editorial comment related to Brandeis and Untermyer's campaigns against some bankers.
**578**. N. W. Aldrich, *The Independent*, (July, 1914).

Act No. 1 was the manufacture, between 1896 and 1907, through stock gambling, speculation and other devious methods and devices, of tens of billions of watered stocks, bonds, and securities.

Act No. 2 was the panic of 1907, by which those not favorable to the Money Trust could be squeezed out of business and the people frightened into demanding changes in the banking and currency laws which the Money Trust would frame.

The Act No. 3 was the passage of the Aldrich-Vreeland Emergency Currency Bill, by which the Money Trust interests should have the privilege of securing from the Government currency on their watered bonds and securities. But while the act contained no authority to change the form of the bank notes, the U. S. Treasurer (in some way that I have been unable to find a reason for) implied authority and changed the form of bank notes which were issued for the banks on government bonds. These notes had hitherto had printed on them, 'This note is secured by bonds of the United States.' He changed it to read as follows: 'This note is secured by bonds of the United States or other securities.' 'Or other securities' is the addition that was secured by special interests. The infinite care the Money Trust exercises in regard to important detail work is easily seen in this piece of management. By that change it was enabled to have the form of the money issued in its favor on watered bonds and securities, the same as bank notes secured on government bonds, and, as a result, the people do not know whether they get one or the other. None of the $500,000,000 printed and lying in the U. S. Treasury ready to float on watered bonds and securities has yet (April, 1913) been used. But it is there, maintained at a public charge, as a guarantee to the Money Trust that it may use it in case it crowds speculation beyond the point of its control. The banks may take it to prevent their own failures, but there is not even so much as a suggestion that it may be used to help keep the industries of the people in a state of prosperity.

The main thing, however, that the Money Trust accomplished as a result of the passing of this act was the appointment of the National Monetary Commission, the membership of which was chiefly made up of bankers, their agents and attorneys, who have generally been educated in favor of, and to have a community interest with, the Money Trust. The National Monetary Commission was placed in charge of the same Senator Nelson W. Aldrich and Congressman Edward B. Vreeland, who respectively had charge in the Senate and House during the passage of the act creating it.

The act authorized this commission to spend money without stint or account. It spent over $300,000 in order to learn how to form a plan by which to create a greater money trust, and it afterwards recommended Congress to give this proposed trust a fifty-year charter by means of which it could rob and plunder all humanity. A bill for that purpose was introduced by members of the Monetary Commission, and its passage planned to be the fourth and final act of the campaign to completely enslave the people.

The fourth act, however, is in process of incubation only, and it is hoped that by this time we realize the danger that all of us are in, for it is the final proposed legislation which, if it succeeds, will place us in the complete control of the moneyed interests. History records nothing so dramatic in design, nor so skillfully manipulated, as this attempt to create the National Reserve Association,—otherwise called the Aldrich plan,—and no fact nor occurrence contemplated for the gaining of selfish ends is recorded in the world's records which equals the beguiling methods of this colossal undertaking. Men, women, and children have been equally unconscious of how stealthily this greatest of all giant octopuses,—a greater Money Trust,—is reaching out its tentacles in its efforts to bind all humanity in perpetual servitude to the greedy will of this monster.

I was in Congress when the Panic of 1907 occurred, but I had previously familiarized myself with many of the ways of high financiers. As a result of what I discovered in that study, I set about to expose the Money Trust, the world's greatest financial giant. I knew that I could not succeed unless I could bring public sentiment to my aid. I had to secure that or fail. The Money Trust had laid its plans long before and was already executing them. It was then, and still is, training the people themselves to demand the enactment of the Aldrich Bill or a bill similar in effect. Hundreds of thousands of dollars had already been spent and millions were reserved to be used in the attempt to bring about a condition of public mind that would cause demand of the passage of the bill. If no other methods succeeded, it was planned to bring on a violent panic and to rush the bill through during the distress which would result from the panic. It was figured that the people would demand new banking and currency laws; that it would be impossible for them to get a definitely practical plan before Congress when they were in an excited state and that, as a result, the Aldrich plan would slip safely through. It was designed to pass that bill in the fall of 1911 or 1912." [579]

This was not the first time the bankers had deliberately caused a financial calamity in order to cause the People of America to clamor for banking reforms, "reforms" which the bankers would draft and which would make the citizens of the United States the slaves of the Jewish bankers. When President Andrew Jackson sought to maintain a debt-free government and truly Federal control over the money, Nicholas Biddle and the Rothschilds conspired to create the panic of 1837. Biddle had previously deliberately caused the panic of 1819. Biddle bragged about his actions.

In 1802, Thomas Jefferson anticipated the Great Depression of the Twentieth Century when he stated in a letter to Albert Gallatin, Secretary of the Treasury,

"I believe that banking institutions are more dangerous to our liberties than standing armies... . If the American people ever allow private banks to

---

[579]. C. A. Lindbergh, *Banking and Currency and The Money Trust*, National Capital Press, Washington, D.C., (1913), pp. 92-98.

control the issue of their currency, first by inflation, then by deflation, the banks and corporations that will grow up around [the banks]... will deprive the people of all property until their children wake-up homeless on the continent their fathers conquered... . The issuing power should be taken from the banks and restored to the people, to whom it properly belongs."

In 1913, the creation of the Federal Reserve together with the creation of the Federal Income Tax made war an immensely profitable venture. The Jewish bankers had at last a means to tax the American People and heat up the economy and then collapse it in the Great Depression by contracting the money supply, which created a wonderful buying opportunity for them in that it forced others to sell and yet maintained the value of the bankers' money enabling them to buy up whatever they wanted to buy.

It appears that another trap is today being set for the American Public. Americans will be asked to chose between the gold standard as one panacea, or an international currency issuing from a central world bank as another panacea. Either option could ruin the nation. Poseurs serving the interests of the Jewish bankers, bankers who are driven by greed and religious fervor to place all of the wealth of the world in Jewish hands, will step forward and ridicule the bankers and the Federal Reserve and might even scapegoat all Jews including assimilated Jews. These propagandists will be the agents of the bankers themselves and they will offer up the poisoned fruit of the gold standard. Jewish bankers control most of the gold in the world and if America were to adopt the gold standard it would transfer America's wealth into the hands of Jewish bankers. America would lose its sovereignty to the prophesied Jewish world government and ultimately the gold will be melted down and shipped to Jerusalem severely contracting the money supply and destroying all Gentile economies (*Genesis* 47).

America's gold should be recovered by legal and military means and reparations and damages, as well as the principal and accrued interest stolen from the American economy by Jewish bankers should be recovered. However, the method of securing the lasting value of American money most likely to succeed is for the American Government to issue its own notes and so pay down the debt without accruing more debt. This cannot be done by adopting a gold standard.

J. P. Morgan served the interests of the Zionists by funding England in the war, which tied America to it in the minds of the public, and by financing the American war machine. He made immense profits doing it, most of which ended up in the hands of the Jewish bankers, who ultimately served Rothschild, King of the Jews. The newspapers were edited and staffed by a disproportionate number of Jews. At the end of Morgan's life, it was discovered that most of the monies thought to be controlled by him found their way back to the Rothschilds.

Another means of corrupting the press, one other than ownership, editorship and reporters, is the power of advertising. Jewish enterprises have often withdrawn their advertising from news sources which do not favor their perceived self-interests. This is ruinous to a newspaper. In addition, Jews boycott businesses which advertise in news sources they want shut down. The Jews have been expelled from many societies at many different times for many different reasons. Jewish tribal strategy is so corrupt, unethical and immoral that most Gentile societies, which cannot compete with Jewish corruption and still maintain their human dignity, and which refuse to degrade themselves by lowering themselves to the abnormal and inhuman standards of Jewish tribal behavior, find themselves with no option but to expel the Jews; which is exactly what Zionists have often wanted and is one reason why they so openly flaunt their corruption.

New York City Mayor John Francis Hylan believed that the bankers, directly or indirectly, owned the major newspapers. In 1918, a letter from Hylan to the President of the National Association of City Editors was published in *The New York Times* on 25 August 1918 on page 16,

## "HYLAN ATTACKS ALL NEWSPAPERS
Mayor Declares Confidence of
the Public in Them Has
Been Shaken.

### VANDERLIP DISPUTES THIS

Banker Tells City Editors He Would
Emigrate to Russia If Condition
Were True.

Mayor John F. Hylan, in a letter which was read last night at the dinner held at the Hotel Majestic of the National Association of City Editors, bitterly attacked the newspapers, saying that the confidence of their readers bad been shaken 'by misrepresentation, biased and untruthful news and editorials which had been and are at intervals appearing in the press.'

Frank A. Vanderlip, President of the National City Bank, who was one of the speakers at the dinner, promptly seized upon the Mayor's letter and asserted if he thought the conditions described by the Mayor were true he would consider emigrating to Russia. Mr. Vanderlip disputed the Mayor's assertions.

Mayor Hylan's letter was as follows:

City of New York.
Office of the Mayor.
Aug. 23. 1918.

Clyde P. Steen, Esq., President National Association of City Editors. Hotel Majestic. New York City:

Dear Mr. Steen: Your Invitation to be present and welcome the members of the National Association of City Editors at their annual banquet is received.

I have delayed answering, hoping that I might be able to arrange to be present and to personally extend a welcome on behalf of the city. I regret this is impossible. I am taking this opportunity to say a word to you.

The people of New York are highly honored to have such distinguished men in their midst who will attend your annual convention. As Chief Executive of the city, I wish to extend to you a warm and sincere greeting. I hope the result of your deliberations at your annual convention will meet the expectations of your association and result in benefits to the people throughout the country.

I would like to offer a word or suggestion, which I hope will be received in the spirit in which it is intended by the great men who control the destinies of the papers throughout the country. The people for many years past have looked to your association to guide and advise them in all matters of public importance and benefit. The daily readers have assumed that the papers they read are independent, unbiased, truthful, and fair in their articles and editorials. However, their confidence has been shaken by misrepresentation, biased and untruthful news and editorials which have born and are at intervals appearing in the press. They believe that the policy of the paper is controlled and influenced by certain interests that are more interested in the special privilege seeker than in the people. In many instances this is true, brought about, no doubt, by the financial condition of a particular paper, whose owners are unable to secure sufficient revenue from their paper to make a profit, and who are compelled to rely upon the subsidy furnished, in one form or another, by certain interests who are profiteering upon the people. This makes the paper a pliant tool of the interests and is used to mislead the people.

The management of the paper, with this policy in mind, sends out the news gatherer on a mission, with instructions. The facts gathered are distorted and the articles colored in accordance with instructions and in accordance with the prejudices of the individual news gatherer, thereby getting away from the purpose of disseminating fair and unbiased news. The editorial writer likewise colors his editorial to suit the Interests of the paper and his employer. The people in a small community quickly discover the gossip monger and the talebearer, and such person is discredited and has no standing in the community.

The people have discovered, particularly in New York, that practically all of the large newspapers are controlled by the special privilege seeking interests, and have as little regard and little respect for the truthfulness and fairness of such papers as they have for the gossip monger and trouble maker in a small community. This shaken confidence and the belief that the press is controlled to a great extent by those who are profiteering in the necessities of life, is causing great and most serious unrest among the people.

The policy of every paper in the country should be to present the facts as they find them, and not to attempt to bias and prejudice the minds of the people with untruthful and unfair editorials and news articles.

In order for the press to regain the confidence in the people they must first of all adopt a policy which will make their paper honest, fearless, and independent in the presentation of news. I sincerely hope that the great men who are connected with the papers of the United States will appreciate the necessity of regaining the confidence of the people, and use their influence against the profiteering interests that are controlling the necessities of life and exploiting the people.

Permit me to make this suggestion at this time: Would it not be wise for a return to the days when our writers and molders of public thought on matters affecting public questions appearing in the daily papers signed the same with their names? Very truly yours.

JOHN F. HYLAN, Mayor.

'When I hear of the low state of the public press as described by the Mayor, of the low state of justice as regards newspapers, I would look to Russia as a place to emigrate to, for it would be an improvement to live there,' Mr. Vanderlip said after the Mayor's letter had been read.

The occasion was the first dinner of the New York City Editors' Association, an organization formed under the auspices of the National Association of City Editors. The latter organization came into being, according to Clyde P. Steen, the President, at the suggestion of George Creel, Chairman of the Committee on Public information, so that the committee might have an organization to reach the bulk of the smaller editors of the country. The dinner was attended by a group of editors from up State."

Frank A. Vanderlip was one of the notorious conspirators on Jekyll Island who created the plan for the Federal Reserve Act which "Colonel" Edward Mandell House forced President Wilson to enact, despite Wilson's campaign promise to oppose such legislation. Paul Warburg drafted the plan and Senator and financier Nelson W. Aldrich attached his name to it in the first attempt to pass it. Vanderlip confessed to his crimes against the American People in an article entitled "The 'First-Name Club'" in the *Saturday Evening Post* in the edition of 9 February 1935, on page 25. George Creel was a muckraking journalist who became the chief propagandist for the Wilson Administration. He lied to the American Public and viciously defamed the German People in order to promote the Jewish bankers in their Zionist efforts to bring America into the First World War on the side of the British in exchange for the Balfour Declaration—a declaration written out to Lord Rothschild which the Zionists took as a blank check.

On 2 March 1922 on page 3 in an article entitled "Hylan Denounces Rule from Albany", *The New York Times* quoted Mayor Hylan,

### "Assails Big Newspapers.

'The present system permits big lawbreakers to escape punishment, provides constant opportunity for increasing the fields for public plundering and flouts the will of the majority, while legislation for the benefit of intrenched monopoly is smeared all over the statute books. And these interests are careful to see to it that they and their official trools receive clean bills of health when seeking popular favor. It is here that the subsidized press—the ever-ready and powerful ally of privilege, comes to the rescue. This help is never denied, for the sinister forces of greed and corruption influence, own or control practically all the newspapers throughout the country. Hence you may be sure that the journalistic pap dished out to the people is at all times of a character to make the people feel kindly disposed toward the hand-picked candidates who are secretly committed to the cause of the interests.

'While it is imperative to do everything possible to mitigate the consequences of political evils, the real solution of the difficulty lies in the removal of the causes, and so I say it would be a great day for the people of this State if we could but clean out the whole kit and caboodle of grasping interests, mercenary politicians and lick-spittol newspapers. These are the three heads of the hydra which must be lopped off together."

*The New York Times* wrote on 27 March 1922 on page 3,

## "HYLAN TAKES STAND ON NATIONAL ISSUES

Suggestion of a Presidential
Boom Is Seen in a Speech
Delivered in Chicago.

### CONDEMNS PACIFIC TREATY

Says International Bankers and
Standard Oil Constitute an
'Invisible Government.'

*Special to the New York Times.*

CHICAGO, March 26.—John F. Hylan, Mayor of New York City, in an address to the Knights of Columbus at the Hotel La Salle here tonight, declared that 'a little coterie of international bankers' virtually ran the United States Government for their own selfish interests, assailed 'invisible government' and the Rockefeller-Standard-Oil interests and predicted a 'whirlwind of public condemnation' for those Senators who voted for the ratification of the Four-Power Treaty, which he described as an 'awful act' and a departure from the policy of George Washington.

It was Mayor Hylan's maiden speech in Chicago on the occasion of his first visit to this city. His address was at the dinner of the Knights of Columbus following the initiation of 600 candidates to the fourth degree of the order.

Mayor Hylan spoke largely on national issues and his speech was considered by many present to mark the launching of his own Presidential boom, the suggestion for which was first put forward tentatively last month at Palm Beach by Commissioner Grover A. Whalen of his Cabinet, while others thought it was rather an amplification and endorsement of the utterances and theories of William Randolph Hearst, as presented almost daily in the Hearst papers.

While Mayor Hylan's speech was punctuated with occasional applause, it was not greeted with any unroarious display of approval. His audience was attentive, courteous and polite, but that was all.

### His Choice for President.

Mayor Hylan naturally did not mention himself for the Presidency, but he expressed the hope that both parties would nominate in 1924, 'men who are genuinely independent, men who have a little of the milk of human kindness in their souls, men of the type of Hiram Johnson, William Randolph Hearst and Rodman Wanamaker.'

With possible reference to his own political fortunes, Mayor Hylan urged complete religious tolerance in political action should never be founded on racial or religious impulse or alignment.

'We are all God's children, no matter in which religion we may chance to have been born,' he said. 'There is no room for bigotry in the free breezes of America and those who seek to instil it are unworthy the name of American.'

Quoting the late Theodore Roosevelt, he attacked 'invisible government,' which, he said, 'like a giant octopus sprawls its slimy length over city, State and nation,' and 'squirms in the jaws of darkness and is thus the better able to clutch the reins of government.'

Other points in Mayor Hylan's speech included a recital of events in the last two New York City Mayoralty elections, a demand that Europe pay its war debts to this country, a boost for the soldier bonus, advocacy of the referendum and recall 'used with discretion,' an ambiguous reference construed to favor beer and light wines and a protest against the prevailing heavy taxes.

### Assails Treaty Ratification.

Mayor Hylan pictured 'the flag that snapped proudly over Valley Forge and Bunker Hill' as drooping on its staff. 'For it has been decreed by a handful of Senators at Washington,' he continued, 'that the Stars and Stripes must flutter beside the standards of Great Britain and Japan if at any time the insular possessions of these empires in the region of the Pacific are in anywise threatened.

'The Senators who by their action have made the free and independent United States of America the prop of crumbling European or warlike Asiatic dynasties may live to regret the day and the deed that was done on it. As surely as the sun shines and the seasons come and go in this Republic founded by Washington and saved by Lincoln, those Senators will reap the harvest of the whirlwind of public condemnation which they have sown by this awful act of ratification.'

Mayor Hylan also attacked the New York newspapers which opposed him for re-election last Fall, and declared the 'kept' press did not support any candidate who did not have the approval of Wall Street and the traction interests.

'The hooting, gibinf and sneering at my candidacy and the tacking upon me of a nickname, which was an echo of the days when I used the pick and shovel and drove a locomotive, were most flagrant and disgraceful,' he added.

Beginning his speech with complimentary reference to the wartime and reconstruction work of the Knights of Columbus, Mayor Hylan launched almost immediately into an attack upon 'invisible government.'

'Some years ago,' he said, 'a sterling American, Theodore Roosevelt, condemned what he called 'invisible government.' He denounced as malefactors of great wealth and as enemies of the Republic those men of excessive fortune who were forever trying to grasp greater gain.

Names 'Head of the Octopus.'

'The warning of Theodore Roosevelt has much timeliness today, for the real menace of our republic is this invisible government which like a giant octopus sprawls its slimy length over city, State and nation.

'Like the octopus of real life it operates under cover of a self-created screen. It seizes in its long and powerful tentacles our executive officers, our legislative bodies, our schools, our courts, our newspapers and every agency created for the public protection.

'It squirms in the jaws of darkness and thus is the better able to clutch the reins of government, secure enactment of the legislation favorable to corrupt business, violate the law with impunity, smother the press and reach into the courts.

'To depart from mere generalizations, let me say that at the head of this octopus are the Rockefeller-Standard Oil interests and a small group of powerful banking houses generally referred to as the international bankers.

'The little coterie of powerful international bankers virtually run the United States Government for their own selfish purposes. They practically control both parties, write political platforms, make catspaws of party leaders, use the leading men of private organizations and resort to every device to place in nomination for high public office only such candidates as will be amenable to the dictates of corrupt big business. They connive at centralization of government on the theory that a small group of hand-picked, privately controlled individuals in power can be more easily handled than a larger group among whom there will most likely be men sincerely interested in public welfare.

'These international bankers and Rockefeller-Standard Oil interests control the majority of newspapers and magazines in this country. They use the columns of these papers to club into submission or drive out of office public officials who refuse to do the bidding of the powerful corrupt cliques which compose the invisible government.'

Mayor Hylan quoted the paper attributed to Dr. Frederick T. Gates of the General Education Board, which advocated educating rural children to remain in that station of life rather than training them for the professions.

'This is the kind of education the coolies receive in China,' Mr. Hylan said, 'but we are not going to stand for it in these United States. One of my first acts as Mayor was to pitch our, bag and baggage, from the educational system of our city the Rockefeller agents and the Gary plan of education to fit the children for the mill and factory.'

**Criticizes Our Entering War.**

Entrance of the United States into the World War was viewed by Mayor Hylan as a departure of doubtful wisdom from its traditional policy.

'In the second Wilson presidential campaign the slogan was 'He kept us out of war.' Shortly after the Administration entered upon its second term the cry 'to arms' was roared, and the free and independent United States of America was plunged into the seething cauldron of the European war.

'The slogan of the Harding campaign was 'No League of Nations.' Scarcely a year after this new national administration entered into office, a peace parley was called to effect an association of nations—which is the same as a League of Nations—to bind the Republic of the United States of America, pulsating with life, to the moribund monarchies of Europe.

'We have in this country a few Tories who are more interested in the welfare of foreign countries than they are in the United States Government. Some way ought to be found for dealing effectively with them.

'Our departure from the patriotic and wise admonitions of our far-sighted early patriots which led to our participation in the World War has taught thinking America a lesson, sad, bitter and costly.'

Mayor Hylan declared the United States should collect the ten billion dollars owed by her allies during the war, even though they showed no sign of willingness to pay. 'I for one,' he said, 'insist that the Government demand the return of principal and interest as soon as possible, so that at least part of these sums may be distributed to the soldiers of the United States and their families who are in need. Seventy-five thousand ex-service men are tramping the streets of the City of New York hungry and jobless, and on behalf of them and every other unemployed veteran, I sincerely hope that Congress will take this matter up and insist on an early settlement of at least part of the debts owing to the United States by these European countries.'"

On 9 December 1922, *The New York Times* quoted Hylan, "As the cities of the State of New York were organized to oppose Governor Miller last November, so Mayor Hylan plans a nation-wide cities bloc to fight against 'corporation and international bankers' in the Presidential election two years hence. [\*\*\*] We have got to get the cities together for the fight in 1924. There is going to be a battle then and a hard one to prevent the corporate interests and the great international bankers from dictating to the two old parties when the time comes for nominating a President."[580]

As Presidential candidate for the Progressive Party, Theodore Roosevelt gave a speech in August of 1912, in Oyster Bay, New York,"The Progressive

---

[580]. "Hylan Heads West to Push Cities Bloc", *The New York Times*, (9 December 1922), pp. 1-2.

Covenant With The People" (note that Roosevelt's allusion to an "invisible government" is similar to Walter Rathenau's declaration on 24 December 1912 in the *Wiener Freie Presse*, that "Three hundred men, each of whom knows all the others, govern the fate of the European continent, and they elect their successors from their entourage."[581]),

> "Political parties exist to secure responsible government and to execute the will of the people. From these great tasks both of the old parties have turned aside. Instead of instruments to promote the general welfare they have become the tools of corrupt interests, which use them impartially to serve their selfish purposes. Behind the ostensible government sits enthroned an invisible government owing no allegiance and acknowledging no responsibility to the people. To destroy this invisible government, to dissolve the unholy alliance between corrupt business and corrupt politics, is the first task of the statesmanship of the day. Unhampered by tradition, uncorrupted by power, undismayed by the magnitude of the task, the new party offers itself as the instrument of the people, to sweep away old abuses, to build a new and nobler government. This declaration is our covenant with the people and we hereby bind the party and its candidates, in state and nation, to the pledges made herein. With all my heart and soul, with every particle of high purpose that is in me, I pledge you my word to do everything I can to put every particle of courage, of common sense, and of strength that I have at your disposal, and to endeavor so far as strength has given me to live up to the obligations you have put upon me, and to endeavor to carry out in the interest of our whole people the policies to which you have today solemnly dedicated yourselves in the name of the millions of men and women for whom you speak. Surely there never was a fight better worth making than the one in which we are engaged. It little matters what befalls any one of us, who for the time being stand in the forefront of the battle. I hope we shall win, and I believe that if we can wake the people to what the fight really means, we shall win. But win or lose, we shall not falter. Whatever fate may at the moment overtake any of us, the movement itself will not stop. Our cause is based on the eternal principles of righteousness. Even though we who now lead may for the time fail, in the end the cause itself shall triumph. Six weeks ago, here in Chicago, I spoke to the honest representatives of a convention which was not dominated by honest men. A convention wherein sat, alas, a majority of men who, with sneering indifference to every principle of right, so acted as to bring to a shameful end a party which had been founded over half a century ago by men in whose souls burned the fire of lofty endeavor. Now to you men, who, in your turn have come together to spend and be spent in the endless crusade against wrong, to you who face the future resolute and confident, to you who strive

---

**581**. L. Fry, *Waters Flowing Eastward: The War Against the Kingship of Christ*, TBR Books, Washington, D. C., (2000), p. 108.

in a spirit of brotherhood for the betterment of our nation, to you who gird yourselves for this great new fight in the never ending warfare for the good of humankind, I say in closing what in that speech I said in closing: We stand at Armageddon, and we battle for the Lord."

Two key elements of Roosevelt's Progressive Party were iterated in the "Platform of the Progressive Party" on 7 August 1912,

**"The Old Parties**

Political parties exist to secure responsible government and to execute the will of the people.

From these great tasks both of the old parties have turned aside. Instead of instruments to promote the general welfare, they have become the tools of corrupt interests which use them impartially to serve their selfish purposes. Behind the ostensible government sits enthroned an invisible government, owing no allegiance and acknowledging no responsibility to the people.

To destroy this invisible government, to dissolve the unholy alliance between corrupt business and corrupt politics is the first task of the statesmanship of the day.

The deliberate betrayal of its trust by the Republican Party, and the fatal incapacity of the Democratic Party to deal with the new issues of the new time, have compelled the people to forge a new instrument of government through which to give effect to their will in laws and institutions.

Unhampered by tradition, uncorrupted by power, undismayed by the magnitude of the task, the new party offers itself as the instrument of the people to sweep away old abuses, to build a new and nobler commonwealth."

and,

**"Currency**

We believe there exists imperative need for prompt legislation for the improvement of our National currency system. We believe the present method of issuing notes through private agencies is harmful and unscientific.

The issue of currency is fundamentally government function and the system should have as basic principles soundness and elasticity. The control should be lodged with the Government and should be protected from domination manipulation by Wall Street or any special interests.

We are opposed to the so-called Aldrich currency bill, because its provisions would place our currency and credit system in private hands, not subject to effective public control."

Silas Bent published a review of the books *The Life of Woodrow Wilson*[582] by Josephus Daniels and *The True Story of Woodrow Wilson*[583] by David Lawrence under the caption "Career of the Creator of 'International Conscience'" in *The New York Times Book Review* 22 June 1924 on page 3, in which Bent wrote, among other things,

> "Mr. Lawrence quotes [President Woodrow Wilson] as calling the Colonel 'a monumental faker.' That was in private conversation. Mr. Wilson did not reply to his predecessor's attacks on him as a candidate.
>
> To Colonel E. M. House Mr. Lawrence gives credit for influence in naming the greater part of the first Wilson Cabinet. Mr. Daniels mentions Colonel House only in reference to the appointment of Albert. S. Burleson as Postmaster General. It was Colonel House, so Mr. Lawrence says, who first interested Mr. Wilson in banking reform. It was Colonel House who made a trip to Wall Street before the inauguration and reassured the most powerful bankers in this country about Mr. Wilson's views, telling them his intentions toward business and finance, so as to avert a threatened panic.
>
> The second Mrs. Wilson, according to Mr. Lawrence, was chiefly responsible for the break between her husband and Colonel House. She exercised an extraordinary influence and thought the Colonel was too much in evidence at Versailles. It was she, according to the same writer, who caused the break with Secretary Tumulty; but some of those who read Mr. Tulmuty's about himself and the President regarded that as abundant provocation."

Woodrow Wilson, himself, stated in a campaign speech before he was elected for his first term as President,

> "Since I entered politics, I have chiefly had men's views confided to me privately. Some of the biggest men in the United States, in the field of commerce and manufacture, are afraid of somebody, are afraid of something. They know that there is a power somewhere so organized, so subtle, so watchful, so interlocked, so complete, so pervasive, that they had better not speak above their breath when they speak in condemnation of it."[584]

Jacob Schiff, whose family had a long and intimate relationship with the Rothschild family, destroyed Russia through the collusive actions of

---

[582]. J. Daniels, *The Life of Woodrow Wilson, 1856-1924*, The John C. Winston Company, Philadelphia, (1924).

[583]. D. Lawrence, *The True Story of Woodrow Wilson*, George H. Doran Company, New York, (1924).

[584]. W. Wilson, *The New Freedom: A Call for the Emancipation of the Generous Energies of a People*, Doubleday, Page & Company, Garden City, New York, (1913), pp. 13-14.

international finance, which was disproportionately in the hands of Jewish financiers. The Bolshevists he put into power forestalled Russian progress for a century. Zionist Meir Kahane launched a secret war against the Soviet Union, attempting to provoke conflict between the Soviets and the Americans, in order to force the Soviet Union into sending Jews to Israel.[585] Israel needed to increase its Jewish population so as to change the demographics of the country and overwhelm the large native Palestinian population.

Kahane's actions could have brought the United States, N.A.T.O., the Warsaw Pact and the Soviet Union to war—had the potential to provoke World War III, but racist Jews are so selfish and so fanatical that they welcome the notion of a third world war which they see as necessary to fulfill Old Testament prophecy. There is today a rise in anti-Semitism in Russia and the Ukraine; and, given this history of Zionist agitation, the question arises, are Zionists agitating to provoke this anti-Semitism and yet again causing the Jews and Gentiles of Russia needless misery in order to promote their perceived Zionist self-interests? Zionists want to force Russian Jews to move to Israel, because the demographic situation still favors the Palestinians in Israel, which is by no means a democracy; and if Israel were to become a democracy, the Palestinians would effectively rule by swing vote and eventually by majority vote. When the Soviet Union broke apart, a Jewish mafia took over many of the profitable businesses of Russia and funneled the fortunes into the hands of Jewish financiers.[586] International finance grossly restricted the influx of investment capital into the former Soviet Nations preventing their successful transition into Capitalism, and the Jewish mafia discouraged the influx of foreign capital by manifesting rampant corruption that frightened off foreigner investors. Jews again are attempting to destroy Russia and Jews have scripted the Christian Zionists to openly call for war with Russia and for the nuclear genocide of the Russian People, as if such would be the fulfillment of the prophecies of *Ezekiel* chapters 38 and 39. Both before and after the reign of the Jewish "Red Terror", Russia, a nation with the greatest potential of any nation on Earth, was destroyed again and again by Jewish finance. *Malachi* 1:1-5 states,

"1 The burden of the word of the LORD to Israel by Malachi. 2 I have loved

---

**585**. R. I. Friedman, *The False Prophet: Rabbi Meir Kahane: from FBI Informant to Knesset Member*, Lawrence Hill Books, Brooklyn, New York, (1990), pp. 105-108.
**586**. M. R. Johnson, "The Judeo-Russian Mafia: From the Gulag to Brooklyn to World Dominion", *The Barnes Review*, Volume 12, Number 3, (May/June 2006), pp. 43-47. *See also:* D. Satter, *Darkness at Dawn: The Rise of the Russian Criminal State*, Yale University Press, New Haven, Connecticut, (2003). *See also:* R. I. Friedman, *Red Mafiya: How the Russian Mob Has Invaded America*, Little, Brown, Boston, (2000). *See also:* P. Klebnikov, *Godfather of the Kremlin: The Decline of Russia in the Age of Gangster Capitalism*, Harcourt, New York, (2000). *See also:* J. O. Finckenauer, *Russian Mafia in America: Immigration, Culture, and Crime*, Northeastern University Press, Boston, (1998). *See also:* S. Handelman, *Comrade Criminal: Russia's New Mafiya*, Yale University Press, New Haven, Connecticut, (1995).

you, saith the LORD. Yet ye say, Wherein hast thou loved us? *Was* not Esau Jacob's brother? saith the LORD: yet I loved Jacob, 3 And I hated Esau, and laid his mountains and his heritage waste for the dragons of the wilderness. 4 Whereas Edom saith, We are impoverished, but we will return and build the desolate places; thus saith the LORD of hosts, They shall build, but I will throw down; and they shall call them, The border of wickedness, and, The people against whom the LORD hath indignation for ever. 5 And *your* eyes shall see, and ye shall say, The LORD will be magnified from the border of Israel."

In recent times, Jews have fomented genocidal hatred toward Russians among Americans in the hopes that America will destroy Russia with a nuclear attack. Jews in Russia have likewise been fomenting genocidal hatred of Americans for a very long time. The Cold War was a Jewish creation meant to divide and destroy humanity. Americans and Russians must at long last confront this deliberate campaign by leading Jews to kill us all off.

Congressman Louis T. McFadden gave the following famous speech before the United States House of Representatives on 10 June 1932, which tells the story of how the Jewish bankers ruined Russia and delivered America into slavery, war and depression through their agent "Colonel" Edward Mandell House:

"Mr. McFADDEN. Mr. Chairman, at the present session of Congress we have been dealing with emergency situations. We have been dealing with the effect of things rather than with the cause of things. In this particular discussion I shall deal with some of the causes that lead up to these proposals. There are underlying principles which are responsible for conditions such as we have at the present time and I shall deal with one of these in particular which is tremendously important in the consideration that you are now giving to this bill.

Mr. Chairman, we have in this country one of the most corrupt institutions the world has ever known. I refer to the Federal Reserve Board and the Federal Reserve Banks. The Federal Reserve Board, a Government board, has cheated the Government of the United States and the people of the United States out of enough money to pay the national debt. The depredations and iniquities of the Federal Reserve Board and the Federal reserve banks acting together have cost this country enough money to pay the national debt several times over. This evil institution has impoverished and ruined the people of the United States; has bankrupted itself, and has practically bankrupted our Government. It has done this through the defects of the law under which it operates, through the maladministration of that law by the Federal Reserve Board, and through the corrupt practices of the moneyed vultures who control it.

Some people think the Federal Reserve banks are United States Government institutions. They are not Government institutions. They are private credit monopolies which prey upon the people of the United States for the benefit of themselves and their foreign customers; foreign and domestic speculators and swindlers; and rich and predatory money lenders. In that dark crew of financial pirates there are those who would cut a man's throat to get a dollar out of his pocket; there are those who send money into States to buy votes to control our legislation; and there are those who maintain international propaganda for the purpose of deceiving us and of wheedling us into the granting of new concessions which will permit them to cover up their past misdeeds and set again in motion their gigantic train of crime.

These 12 private credit monopolies were deceitfully and disloyally foisted upon this country by the bankers who came here from Europe and repaid us for our hospitality by undermining our American institutions. Those bankers took money out of this country to finance Japan in a war against Russia. They created a reign of terror in Russia with our money in order to help that war along. They instigated the separate peace between Germany and Russia and thus drove a wedge between the Allies in the World War. They financed Trotsky's mass meetings of discontent and rebellion in New York. They paid Trotsky's passage from New York to Russia so that he might assist in the destruction of the Russian Empire. They fomented and instigated the Russian revolution and they placed a large fund of American dollars at Trotsky's disposal in one of their branch banks in Sweden so that through him Russian homes might be thoroughly broken up and Russian children flung far and wide from their natural protectors. They have since begun the breaking up of American homes and the dispersal of American children.

It has been said that President Wilson was deceived by the attentions of these bankers and by the philanthropic poses they assumed. It has been said that when he discovered the manner in which he had been misled by Colonel House, he turned against that busybody, that 'holy monk' of the financial empire, and showed him the door. He had the grace to do that, and in my opinion he deserves great credit for it.

President Wilson died a victim of deception. When he came to the Presidency, he had certain qualities of mind and heart which entitled him to a high place in the councils of this Nation; but there was one thing he was not and which he never aspired to be; he was not a banker. He said that he knew very little about banking. It was, therefore, on the advice of others that the iniquitous Federal reserve act, the death warrant of American liberty, became law in his administration.

Mr. Chairman, there should be no partisanship in matters concerning the banking and currency affairs of this country, and I do not speak with any.

In 1912 the National Monetary Association, under the chairmanship of the late Senator Nelson W. Aldrich, made a report and presented a vicious bill called the National Reserve Association bill. This bill is usually spoken of as the Aldrich bill. Senator Aldrich did not write the Aldrich bill. He was the tool, but not the accomplice, of the European-born bankers who for nearly twenty years had been scheming to set up a central bank in this country and who in 1912 had spent and were continuing to spend vast sums of money to accomplish their purpose.

The Aldrich bill was condemned in the platform upon which Theodore Roosevelt was nominated in the year 1912, and in that same year, when Woodrow Wilson was nominated, the Democratic platform, as adopted at the Baltimore convention, expressly stated: 'We are opposed to the Aldrich plan for a central bank.' This was plain language. The men who ruled the Democratic Party then promised the people that if they were returned to power there would be no central bank established here while they held the reigns of government. Thirteen months later that promise was broken, and the Wilson administration, under the tutelage of those sinister Wall Street figures who stood behind Colonel House, established here in our free country the worm-eaten monarchical institution of the 'king's bank' to control us from the top downward, and to shackle us from the cradle to the grave. The Federal Reserve act destroyed our old and characteristic way of doing business; it discriminated against our 1-name commercial paper, the finest in the world; it set up the antiquated 2-name paper, which is the present curse of this country, and which wrecked every country which has ever given it scope; it fastened down upon this country the very tyranny from which the framers of the Constitution sought to save us.

One of the greatest battles for the preservation of this Republic was fought out here in Jackson's day, when the Second Bank of the United States, which was founded upon the same false principles as those which are here exemplified in the Federal Reserve act, was hurled out of existence. After the downfall of the Second Bank of the United States in 1837, the country was warned against the dangers that might ensue if the predatory interests, after being cast out, should come back in disguise and unite themselves to the Executive, and through him acquire control of the Government. That is what the predatory interests did when they came back in the livery of hypocrisy and under false pretenses obtained the passage of the Federal reserve act.

The danger that the country was warned against came upon us and is shown in the long train of horrors attendant upon the affairs of the traitorous and dishonest Federal Reserve Board and the Federal reserve banks. Look around you when you leave this chamber and you will see evidences on all sides. This is an era of economic misery and for the conditions that caused that misery, the Federal Reserve Board and the Federal reserve banks are fully liable. This is an era of financed crime and in the financing of crime, the Federal Reserve Board does not play the part of a disinterested spectator.

It has been said that the draughtsman who was employed to write the text of the Federal reserve bill used a text of the Aldrich bill for his purpose. It has been said that the language of the Aldrich bill was used because the Aldrich bill had been drawn up by expert lawyers and seemed to be appropriate. It was indeed drawn up by lawyers. The Aldrich bill was created by acceptance bankers of European origin in New York City. It was a copy and in general a translation of the statutes of the Reichsbank and other European central banks.

Half a million dollars was spent one part of the propaganda organized by those same European bankers for the purpose of misleading public opinion in regard to it, and for the purpose of giving Congress the impression that there was an overwhelming popular demand for that kind of banking legislation and the kind of currency that goes with it, namely, an asset currency based on human debts and obligations instead of an honest currency based on gold and silver values. Dr. H. Parker Willis had been employed by the Wall Street bankers and propagandists and when the Aldrich measure came to naught and he obtained employment from CARTER GLASS to assist in drawing a banking bill for the Wilson administration, he appropriated the text of the Aldrich bill for his purpose. There is no secret about it. The text of the Federal reserve act was tainted from the beginning.

Not all of the Democratic Members of the Sixty-third Congress voted for this great deception. Some of them remembered the teachings of Jefferson; and, through the years, there had been no criticisms of the Federal Reserve Board and the Federal reserve banks so honest, so out-spoken, and so unsparingly as those which have been voiced here by Democrats. Again, although a number of Republicans voted for the Federal reserve act, the wisest and most conservative members of the Republican Party would have nothing to do with it and voted against it. A few days before the bill came to a vote, Senator Henry Cabot Lodge, of Massachusetts, wrote to Senator John W. Weeks as follows:

NEW YORK CITY, *December 17, 1913.*

MY DEAR SENATOR WEEKS: * * * Throughout my public life I have supported all measures designed to take the Government out of the banking business * * *. This bill puts the Government into the banking business as never before in our history and makes, as I understand it, all notes Government notes when they should be bank notes.

The powers vested in the Federal Reserve Board seem to me highly dangerous, especially where there is political control of the Board. I should be sorry to hold stock in a bank subject to such domination. The bill as it stands seems to me to open the way to a vast inflation of the currency. There is no necessity of dwelling upon this point after the remarkable and most powerful argument of the senior Senator from New York. I can be content here to follow the example of the English candidate for Parliament who thought it enough 'to say ditto to Mr. Burke.' I will merely add that I do not like to think that any law can be passed which will make it possible to submerge the gold standard in a flood of irredeemable paper currency.

I had hoped to support this bill, but I can not vote for it as it stands, because it seems to me to contain features and to rest upon principles in the highest degree menacing to our prosperity, to stability in business, and to the general welfare of the people of the United States.

Very sincerely yours,

HENRY CABOT LODGE.

In 18 years which have passed since Senator Lodge wrote that letter of warning all of his predictions have come true. The Government is in the banking business as never before. Against its will it has been made the backer of horsethieves and card sharps, bootleggers, smugglers, speculators, and swindlers in all parts of the world. Through the Federal Reserve Board and the Federal reserve banks the riffraff of every country is operating on the public credit of this United States Government. Meanwhile, and on account of it, we ourselves are in the midst of the greatest depression we have ever known. Thus the menace to our prosperity, so feared by Senator Lodge, has indeed struck home. From the Atlantic to the Pacific our country has been ravaged and laid waste by the evil practices of the Federal Reserve Board and the Federal reserve banks and the interests which control them. At no time in our history has the general welfare of the people of the United States been at a lower level or the mind of the people so filled with despair.

Recently in one of our States 60,000 dwelling houses and farms were brought under the hammer in a single day. According to the Rev. Father Charles E. Coughlin, who has lately testified before a committee of this House, 71,000 houses and farms in Oakland County, Mich., have been sold and their erstwhile owners dispossessed. Similar occurrences have probably taken place in every county in the United States. The people who have thus been driven out are the wastage of the Federal reserve act. They are the victims of the dishonest and unscrupulous Federal Reserve Board and Federal reserve banks. Their children are the new slaves of the auction blocks in the revival here of the institution of human slavery.

In 1913, before the Senate Banking and Currency Committee, Mr. Alexander Lassen made the following statement:

But the whole scheme of the Federal reserve bank with its commercial-paper basis is an impractical, cumbersome machinery, is simply a cover, to find a way to secure the privilege of issuing money and to evade payment of as much tax upon circulation as possible, and then control the issue and maintain, instead of reduce, interest rates. It is a system that, if inaugurated, will prove to the advantage of the few and the detriment of the people of the United States. It will mean continued shortage of actual money and further extension of credits; for when there is a lack of real money people have to borrow credit to their cost.

A few days before the Federal Reserve act was passed Senator Elihu Root denounced the Federal Reserve bill as an outrage on our liberties and made the following prediction:

Long before we wake up from our dreams of prosperity through an inflated currency, our gold, which alone could have kept us from catastrophe, will have vanished and no rate of interest will tempt it to return.

If ever a prophecy came true, that one did. It was impossible, however, for those luminous and instructed thinkers to control the course of events. On December 23, 1913, the Federal reserve bill became law, and that night Colonel House wrote to his hidden master in Wall Street as follows:

I want to say a word of appreciation to you for the silent but no doubt effective work you have done in the interest of currency legislation and to congratulate you that the measure has finally been enacted into law. We all know that an entirely perfect bill, satisfactory to everybody, would have been an impossibility, and I feel quite certain fair men will admit that unless the President had stood as firm as he did we should likely have had no legislation at all. The bill is a good one in many respects; anyhow good enough to start with and to let experience teach us in what direction it needs perfection, which in due time we shall then get. In any event you have personally good reason to feel gratified with what has been accomplished.

The words 'unless the President had stood as firm as he did we should likely have had no legislation at all,' were a gentle reminder that it was Colonel House himself, the 'holy monk,' who had kept the President firm.

The foregoing letter affords striking evidence of the manner in which the predatory interests then sought to control the Government of the United States by surrounding the Executive with the personality and the influence of a financial Judas. Left to itself and to the conduct of its own legislative functions without pressure from the Executive, the Congress would not have passed the Federal reserve act. According to Colonel House, and since this was his report to his master, we may believe it to be true, the Federal reserve act was passed because Wilson stood firm; in other words because Wilson was under the guidance and control of the most ferocious usurers in New York through their hireling, House. The Federal reserve act became law the day before Christmas Eve in the year 1913, and shortly afterwards the German international bankers, Kuhn, Loeb and Co., sent one of their partners here to run it.

In 1913, when the Federal reserve bill was submitted to the Democratic caucus, there was a discussion in regard to the form the proposed paper currency should take.

The proponents of the Federal reserve act, in their determination to create a new kind of paper money, had not needed to go outside of the Aldrich bill for a model. By the terms of the Aldrich bill, bank notes were to be issued by the National Reserve Association and were to be secured partly by gold or lawful money and partly by circulating evidences of debt. The first draft of the Federal reserve bill presented the same general plan, that is, for bank notes as opposed to Government notes, but with certain differences of regulation.

When the provision for the issuance of Federal reserve notes was placed before President Wilson he approved of it, but other Democrats were more mindful of Democratic principles and a great protest greeted the plan.

Foremost amongst those who denounced it was William Jennings Bryan, the Secretary of State. Bryan wished to have the Federal reserve notes issued as Government obligations. President Wilson had an interview with him and found him adamant. At the conclusion of the interview Bryan left with the understanding that he would resign if the notes were made bank notes. The President then sent for his Secretary and explained the matter to him. Mr. Tumulty went to see Bryan and Bryan took from his library shelves a book containing all the Democratic platforms and read extracts from them bearing on the matter of the public currency. Returning to the President, Mr. Tumulty told him what had happened and ventured the opinion that Mr. Bryan was right and that Mr. Wilson was wrong. The President then asked Mr. Tumulty to show him where the Democratic Party in its national platforms had ever taken the view indicated by Bryan. Mr. Tumulty gave him the book, which he had brought from Bryan's house, and the President read very carefully plank after plank on the currency. He then said, 'I am convinced there is a great deal in what Mr. Bryan says,' and thereupon it was arranged that Mr. Tumulty should see the proponents of the Federal reserve bill in an effort to bring about an adjustment of the matter.

The remainder of this story may be told in the words of Senator GLASS. Concerning Bryan's opposition to the plan of allowing the proposed Federal reserve notes to take the form of bank notes and the manner in which President Wilson and the proponents of the Federal reserve bill yielded to Bryan in return for his support of the measure, Senator GLASS makes the following statement:

The only other feature of the currency bill around which a conflict raged at this time was the note-issue provision. Long before I knew it, the President was desperately worried over it. His economic good sense told him the notes should be issued by the banks and not by the Government; but some of his advisers told him Mr. Bryan could not be induced to give his support to any bill that did not provide for a 'Government note.' There was in the Senate and House a large Bryan following which, united with a naturally adversary party vote, could prevent legislation. Certain overconfident gentlemen proffered their services in the task of 'managing Bryan.' They did not budge him. * * * When a decision could no longer be postponed the President summoned me to the White House to say he wanted Federal reserve notes to 'be obligations of the United States.' I was for an instant speechless. With all the earnestness of my being I remonstrated, pointing out the unscientific nature of such a thing, as well as the evident inconsistency of it.

'There is not, in truth, any Government obligation here, Mr. President,' I exclaimed. 'It would be a pretense on its face. Was there ever a Government note based primarily on the property of banking institutions? Was there ever a Government issue not one dollar of which could be put out except by demand of a bank? The suggested Government obligation is so remote it could never be discerned,' I concluded, out of breath.

'Exactly so, GLASS,' earnestly said the President. 'Every word you say is true; the Government liability is a mere thought. And so, if we can hold to the substance of the thing and give the other fellow the shadow, why not do it, if thereby we may save our bill?'

Shadow and substance! One can see from this how little President Wilson knew about banking. Unknowingly, he gave the substance to the international banker and the shadow to the common man. Thus was Bryan circumvented in his efforts to uphold the Democratic doctrine of the rights of the people. Thus the 'unscientific blur' upon the bill was perpetrated. The 'unscientific blur,' however, was not the fact that the United States Government, by the terms of Bryan's edict, was obliged to assume as an obligation whatever currency was issued. Mr. Bryan was right when he insisted that the United States should preserve its sovereignty over the public currency. The 'unscientific blur' was the nature of the currency itself, a nature which makes it unfit to be assumed as an obligation of the United States Government. It is the worst currency and the most dangerous this country has ever known. When the proponents of the act saw that the Democratic doctrine would not permit them to let the proposed banks issue the new currency as bank notes, they should have stopped at that. They should not have foisted that kind of currency, namely, an asset currency, on the United States Government. They should not have made the Government liable on the private debts of individuals and corporations and, least of all, on the private debts of foreigners.

The Federal reserve note is essentially unsound.

As Kemmerer says:

The Federal Reserve notes, therefore, in form have some of the qualities of Government paper money, but, in substance, are almost a pure asset currency possessing a Government guaranty against which contingency the Government has made no provision whatever.

Hon. E. J. Hill, a former Member of the House, said, and truly:

\* \* \* They are obligations of the Government for which the United States has received nothing and for the payment of which at any time it assumes the responsibility looking to the Federal reserve to recoup itself.

If the United States Government is to redeem the Federal reserve notes when the general public finds out what it costs to deliver this flood of paper money to the 12 Federal reserve banks, and if the Government has made no provision for redeeming them, the first element of unsoundness is not far to seek.

Before the Banking and Currency Committee, when the Federal reserve bill was under discussion, Mr. Crozier, of Cincinnati, said:

In other words, the imperial power of elasticity of the public currency is wielded exclusively by these central corporations owned by the banks. This is a life and death power over all local banks and all business. It can be used to create or destroy prosperity, to ward off or cause stringencies and panics. By making money artificially scarce, interest rates throughout the country can be arbitrarily raised and the bank tax on all business and cost of living increased for the profit of the banks owning these regional central banks, and without the slightest benefit to the people. These 12 corporations together cover the whole country and monopolize and use for private gain every dollar of the public currency and all public revenue of the United States. Not a dollar can be put into circulation among the people by their Government without the consent of and on terms fixed by these 12 private money trusts.

In defiance of this and all other warnings, the proponents of the Federal reserve act created the 12 private credit corporations and gave them an absolute monopoly of the currency of the United States, not of the Federal reserve notes alone, but of all the currency, the Federal reserve act providing ways by means of which the gold and general currency in the hands of the American people could be obtained by the Federal reserve banks in exchange for Federal reserve notes, which are not money, but merely promises to pay money. Since the evil day when this was done the initial monopoly has been extended by vicious amendments to the Federal reserve act and by the unlawful and treasonable practices of the Federal Reserve Board and the Federal reserve banks.

Mr. Chairman, when a Chinese merchant sells human hair to a Paris wigmaker and bills him in dollars, the Federal reserve banks can buy his bill against the wigmaker and then use that bill as collateral for the Federal reserve notes. The United States Government thus pays the Chinese merchant the debt of the wigmaker and gets nothing in return except a shady title to the Chinese hair.

Mr. Chairman, if a Scottish distiller wishes to send a cargo of Scotch whiskey to the United States, he can draw his bill against the purchasing bootlegger in dollars; and after the bootlegger has accepted it by writing his name across the face of it, the Scotch distiller can send that bill to the nefarious open discount market in New York City, where the Federal Reserve Board and the Federal reserve banks will buy it and use it as collateral for a new issue of Federal reserve notes. Thus the Government of the United States pays the Scotch distiller for the whiskey before it is shipped; and if it is lost on the way, or if the Coast Guard seizes it and destroys it, the Federal reserve banks simply write off the loss and the Government never recovers the money that was paid to the Scotch distiller. While we are attempting to enforce prohibition here, the Federal Reserve Board and the Federal reserve banks are financing the distillery business in Europe and paying bootleggers' bills with the public credit of the United States Government.

Mr. Chairman, if a German brewer ships beer to this country or anywhere else in the world and draws his bill for it in dollars, the Federal reserve banks will buy that bill and use it as collateral for Federal reserve notes. Thus, they compel our Government to pay the German brewer for his beer. Why should the Federal Reserve Board and the Federal reserve banks be permitted to finance the brewing industry in Germany, either in this way or as they do by compelling small and fearful United States banks to take stock in the Isenbeck brewery and in the German bank for brewing industries?

Mr. Chairman, if Dynamit Nobel of Germany wishes to sell dynamite to Japan to use in Manchuria or elsewhere, it can draw its bill against the Japanese customers in dollars and send that bill to the nefarious open discount market in New York City, where the Federal Reserve Board and Federal reserve banks will buy it and use it as collateral for a new issue of Federal reserve notes, while at the same time the Federal Reserve Board will be helping Dynamit Nobel by stuffing its stock into the United States banking system. Why should we send our representatives to the disarmament conference at Geneva while the Federal Reserve Board and the Federal reserve banks are making our Government pay japanese debts to German munition makers?

Mr. Chairman, if a bean grower of Chile wishes to raise a crop of beans and sell them to a Japanese customer, he can draw a bill against his prospective Japanese customer in dollars and have it purchased by the Federal Reserve Board and Federal reserve banks and get the money out of this country at the expense of the American public before he has even planted the beans in the ground.

Mr. Chairman, if a German in Germany wishes to export goods to South America or anywhere else, he can draw his bill against his customer and send it to the United States and get the money out of this country before he ships or even manufactures the goods.

Mr. Chairman, why should the currency of the United States be issued on the strength of Chinese human hair? Why should it be issued on the trade whims of a wigmaker? Why should it be issued on the strength of German beer? Why should it be issued on the crop of unplanted beans to be grown in Chile for Japanese consumption? Why should the Government of the United States be compelled to issue many billions of dollars every year to pay the debts of one foreigner to another foreigner? Was it for this that our national-bank depositors had their money taken out of our banks and shipped abroad? Was it for this that they had to lose it? Why should the public credit of the United States Government and likewise money belonging to our national-bank depositors be used to support foreign brewers, narcotic drug vendors, whiskey distillers, wigmakers, human-hair merchants, Chilean bean growers, and the like? Why should our national-bank depositors and our Government be forced to finance the munition factories of Germany and Soviet Russia?

Mr. Chairman, if a German in Germany, wishes to sell wheelbarrows to another German, he can draw a bill in dollars and get the money out of the Federal reserve banks before an American farmer could explain his request for a loan to move his crop to market. In Germany, when credit instruments are being given, the creditors say, 'See you, it must be of a kind that I can cash at the reserve.' Other foreigners feel the same way. The reserve to which these gentry refer is our reserve, which, as you know, is entirely made up of money belonging to American bank depositors. I think foreigners should cash their own trade paper and not send it over here to bankers who use it to fish cash out of the pockets of the American people.

Mr. Chairman, there is nothing like the Federal reserve pool of confiscated bank deposits in the world. It is a public trough of American wealth in which foreigners claim rights equal to or greater than those of Americans. The Federal reserve banks are agents of the foreign central banks. They use our bank depositors' money for the benefit of their foreign principals. They barter the public credit of the United States Government and hire it out to foreigners at a profit to themselves.

All this is done at the expense of the United States Government, and at a sickening loss to the American people. Only our great wealth enabled us to stand the drain of it as long as we did.

I believe that the nations of the world would have settled down after the World War more peacefully if we had not had this standing temptation here—this pool of our bank depositors' money given to private interests and used by them in connection with illimitable drafts upon the public credit of the United States Government. The Federal Reserve Board invited the world to come in and to carry away cash, credit, goods, and everything else of value that was movable. Values amounting to many billions of dollars have been taken out of this country by the Federal Reserve Board and the Federal reserve banks for the benefit of their foreign principals. The United States has been ransacked and pillaged. Our structures have been gutted and only the walls are left standing. While this crime was being perpetrated everything the world could rake up to sell us was brought in here at our own expense by the Federal Reserve Board and the Federal reserve banks until our markets were swamped with unneeded and unwanted imported goods priced far above their value and made to equal the dollar volume of our honest exports and to kill or reduce our favorable balance of trade. As agents of the foreign central banks, the Federal Reserve Board and the Federal reserve banks try by every means within their power to reduce our favorable balance of trade. They act for their foreign principals and they accept fees from foreigners for acting against the best interests of the United States. Naturally there has been great competition among foreigners for the favors of the Federal Reserve Board.

What we need to do is to send the reserves of our national banks home to the people who earned and produced them and who still own them and to the banks which were compelled to surrender them to predatory interests. We need to destroy the Federal reserve pool, wherein our national-bank reserves are impounded for the benefit of the foreigners. We need to make it very difficult for outlanders to draw money away from us. We need to save America for Americans.

Mr. Chairman, when you hold a $10 Federal Reserve note in your hand you are holding a piece of paper which sooner or later is going to cost the United States Government $10 in gold, unless the Government is obliged to give up the gold standard. It is protected by a reserve of 40 per cent, or $4 in gold. It is based on Limburger cheese, reputed to be in foreign warehouses; or on cans purported to contain peas but which may contain no peas but salt water instead; or on horse meat; illicit drugs; bootleggers' fancies; rags and bones from Soviet Russia of which the United States imported over a million dollars' worth last year; on wine, whiskey, natural gas, on goat or dog fur, garlic on the string, or Bombay ducks. If you like to have paper money which is secured by such commodities, you have it in the Federal reserve note. If you desire to obtain the thing of value upon which this paper currency is based—that is, the Limburger cheese, the whiskey, the illicit drugs, or any of the other staples—you will have a very hard time finding them. Many of these worshipful commodities are in foreign countries. Are you going to Germany to inspect her warehouses to see if the specified things of value are there? I think not. And what is more, I do not think you would find them there if you did go.

Immense sums belonging to our national-bank depositors have been given to Germany on no collateral security whatever. The Federal Reserve Board and the Federal reserve banks have issued United States currency on mere finance drafts drawn by Germans. Billions upon billions of our money has been pumped into Germany and money is still being pumped into Germany by the Federal Reserve Board and the Federal reserve banks. Her worthless paper is still being negotiated here and renewed here on the public credit of the United States Government and at the expense of the American people. On April 27, 1932, the Federal reserve outfit sent $750,000, belonging to American bank depositors, in gold to Germany. A week later, another $300,000 in gold was shipped to Germany in the same way. About the middle of May $12,000,000 in gold was shipped to Germany by the Federal Reserve Board and the Federal reserve banks. Almost every week there is a shipment of gold to Germany. These shipments are not made for profit on the exchange since the German marks are below parity against the dollar.

Mr. Chairman, I believe that the national-bank depositors of the United States are entitled to know what the Federal Reserve Board and the Federal reserve banks are doing with their money. There are millions of national-bank depositors in this country who do not know that a percentage of every dollar they deposit in a member bank of the Federal reserve system goes automatically to American agents of the foreign banks and that all their deposits can be paid away to foreigners without their knowledge or consent by the crooked machinery of the Federal reserve act and the questionable practices of the Federal Reserve Board and the Federal reserve banks. Mr. Chairman, the American people should be told the truth by their servants in office.

In 1930 we had over half a billion dollars outstanding daily to finance foreign goods stored in or shipped between countries. In its yearly total, this item amounts to several billion dollars. What goods are those on which the Federal reserve banks yearly pledge several billions of dollars of the public credit of the United States? What goods are those which are hidden in European and Asiatic storehouses and which have never been seen by any officer of this Government, but which are being financed on the public credit of the United States Government? What goods are those upon which the United States Government is being obligated by the Federal reserve banks to issue Federal reserve notes to the extent of several billions of dollars a year?

The Federal Reserve Board and the Federal reserve banks have been international bankers from the beginning, with the United States Government as their enforced banker and supplier of currency. But it is none the less extraordinary to see those 12 private credit monopolies buying the debts of foreigners against foreigners in all parts of the world and asking the Government of the United States for new issues of Federal reserve notes in exchange for them.

I see no reason why the American taxpayers should be hewers of wood and drawers of water for the European and Asiatic customers of the Federal reserve banks. I see no reason why a worthless acceptance drawn by a foreign swindler as a means of getting gold out of this country should receive the lowest and choicest rate from the Federal Reserve Board and be treated as better security than the note of an American farmer living on American land.

The magnitude of the acceptance racket, as it has been developed by the Federal reserve banks, their foreign correspondents, and the predatory European-born bankers who set up the Federal Reserve institution here and taught our own brand of pirates how to loot the people—I say the magnitude of this racket is estimated to be in the neighborhood of $9,000,000,000 a year. In the past ten years it is said to have amounted to $90,000,000,000. In my opinion, it has amounted to several times as much. Coupled with this you have, to the extent of billions of dollars, the gambling in the United States securities, which takes place in the same open discount market—a gambling upon which the Federal Reserve Board is now spending $100,000,000 per week.

Federal reserve notes are taken from the United States Government in unlimited quantities. Is it strange that the burden of supplying these immense sums of money to the gambling fraternity has at last proved too heavy for the American people to endure? Would it not be a national calamity if the Federal Reserve Board and the Federal reserve banks should again bind this burden down on the backs of the American people and, by means of the long rawhide whips of the credit masters, compel them to enter another 17 years of slavery? They are trying to do that now. They are taking $100,000,000 of the public credit of the United States Government every week in addition to all their other seizures, and they are spending that money in the nefarious open market in New York City in a desperate gamble to reestablish their graft as a going concern.

They are putting the United States Government in debt to the extent of $100,000,000 a week, and with the money they are buying up our Government securities for themselves and their foreign principals. Our people are disgusted with the experiments of the Federal Reserve Board. The Federal Reserve Board is not producing a loaf of bread, a yard of cloth, a bushel of corn, or a pile of cordwood by its check-kiting operations in the money market.

A fortnight or so ago great aid and comfort was given to Japan by the firm of A. Gerli & Sons, of New York, an importing firm, which bought $16,000,000 worth of raw silk from the Japanese Government. Federal reserve notes will be issued to pay that amount to the Japanese Government, and these notes will be secured by money belonging to our national-bank depositors.

Why should United States currency be issued on this debt? Why should United States currency be issued to pay the debt of Gerli & Sons to the Japanese Government? The Federal Reserve Board and the Federal reserve banks think more of the silkworms of Japan than they do of American citizens. We do not need $16,000,000 worth of silk in this country at the present time, not even to furnish work to dyers and finishers. We need to wear home-grown and American-made clothes and to use our own money for our own goods and staples. We could spend $16,000,000 in the United States of America on American children and that would be a better investment for us than Japanese silk purchased on the public credit of the United States Government.

Mr. Speaker, on the 13th of January of this year I addressed the House on the subject of the Reconstruction Finance Corporation. In the course of my remarks I made the following statement:

In 1928 the member banks of the Federal reserve system borrowed $60,598,690,000 from the Federal reserve banks on their 15-day promissory notes. Think of it! Sixty billion dollars payable upon demand in gold in the course of one single year. The actual payment of such obligations calls for six times as much monetary gold as there is in the entire world. Such transactions represent a grant in the course of one single year of about $7,000,000 to every member bank of the Federal reserve system. Is it any wonder that there is a depression in this country? Is it any wonder that American labor, which ultimately pays the cost of all banking operations of this country, has at last proved unequal to the task of supplying this huge total of cash and credit for the benefit of the stock-market manipulators and foreign swindlers?

Mr. Chairman, some of my colleagues have asked for more specific information concerning this stupendous graft, this frightful burden which has been placed on the wage earners and taxpayers of the United States for the benefit of the Federal Reserve Board and the Federal reserve banks. They were surprised to learn that member banks of the Federal reserve system had received the enormous sum of $60,598,690,000 from the Federal Reserve Board and the Federal reserve banks on their promissory notes in the course of one single year, namely, 1928. Another Member of this House, Mr. BEEDY, the honorable gentleman from Maine, has questioned the accuracy of my statement and has informed me that the Federal Reserve Board denies absolutely that these figures are correct. This Member has said to me that the thing is unthinkable, that it can not be, that it is beyond all reason to think that the Federal Reserve Board and the Federal reserve banks should have so subsidized and endowed their favorite banks of the Federal reserve system. This Member is horrified at the thought of a graft so great, a bounty so detrimental to the public welfare as sixty and a half billion dollars a year and more shoveled out to favored banks of the Federal reserve system.

I sympathize with Mr. BEEDY. I would spare him pain if I could, but the facts remain as I have stated them. In 1928, the Federal Reserve Board and the Federal reserve banks presented the staggering amount of $60,598,690,000 to their member banks at the expense of the wage earners and taxpayers of the United States. In 1929, the year of the stock-market crash, the Federal Reserve Board and the Federal reserve banks advanced fifty-eight billions to member banks.

In 1930, while the speculating banks were getting out of the stock market at the expense of the general public, the Federal Reserve Board and the Federal reserve banks advanced them $13,022,782,000. This shows that when the banks were gambling on the public credit of the United States Government as represented by the Federal reserve currency, they were subsidized to any amount they required by the Federal Reserve Board and the Federal reserve banks. When the swindle began to fall, the bankers knew it in advance and withdrew from the market. They got out with whole skins and left the people of the United States to pay the piper.

On November 2, 1931, I addressed a letter to the Federal Reserve Board asking for the aggregate total of member bank borrowing in the years 1928, 1929, 1930. In due course, I received a reply from the Federal Reserve Board, dated November 9, 1931, the pertinent part of which reads as follows:

> MY DEAR CONGRESSMAN: In reply to your letter of November 2, you are advised that the aggregate amount of 15-day promissory notes of member banks during each of the past three calender years has been as follows:
> 1928 _____ $60,598,690,000
> 1929 _____ 58,046,697,000
> 1930 _____ 13,022,782,000

\* \* \* \* \* \* \*

Very truly yours,

CHESTER MORRILL, *Secretary*.

This will show the gentleman from Maine the accuracy of my statement. As for the denial of these facts made to him by the Federal Reserve Board, I can only say that it must have been prompted by fright, since hanging is too good for a Government board which permitted such a misuse of Government funds and credit.

My friend from Kansas, Mr. MCGUGIN, has stated that he thought the Federal Reserve Board and the Federal reserve banks lent money by rediscounting. So they do, but they lend comparatively little that way. The real rediscounting that they do has been called a mere penny in the slot business. It is too slow for genuine high flyers. They discourage it. They prefer to subsidize their favorite banks by making these $60,000,000,000 advances, and they prefer to acquire acceptances in the notorious open discount market in New York, where they can use them to control the prices of stocks and bonds on the exchanges. For every dollar they advanced on rediscounts in 1928 they lent $33 to their favorite banks for gambling purposes. In other words, their rediscounts in 1928 amounted to $1,814,271,000, while their loans to member banks amounted to $60,598,690,000. As for their open-market operations, these are on a stupendous scale, and no tax is paid on the acceptances they handle; and their foreign principals, for whom they do a business of several billion dollars every year, pay no income tax on their profits to the United States Government.

This is the John Law swindle all over again. The theft of Teapot Dome was trifling compared to it. What king ever robbed his subjects to such an extent as the Federal Reserve Board and the Federal reserve banks have robbed us? Is it any wonder that there have lately been 90 cases of starvation in one of the New York hospitals? Is there any wonder that the children of this country are being dispersed and abandoned?

The Government and the people of the United States have been swindled by swindlers de luxe to whom the acquisition of American gold or a parcel of Federal reserve notes presented no more difficulty than the drawing up of a worthless acceptance in a country not subject to the laws of the United States, by sharpers not subject to the jurisdiction of the United States courts, sharpers with a strong banking 'fence' on this side of the water—a 'fence' acting as a receiver of the worthless paper coming from abroad, indorsing it and getting the currency out of the Federal reserve banks for it as quickly as possible, exchanging that currency for gold, and in turn transmitting the gold to its foreign confederates.

Such were the exploits of Ivar Kreuger, Mr. Hoover's friend, and his hidden Wall Street backers. Every dollar of the billions Kreuger and his gang drew out of this country on acceptances was drawn from the Government and the people of the United States through the Federal Reserve Board and the Federal reserve banks. The credit of the United States

Government was peddled to him by the Federal Reserve Board and the Federal reserve banks for their own private gain. That is what the Federal Reserve Board and the Federal reserve banks have been doing for many years. They have been peddling the credit of this Government and the signature of this Government to the swindlers and speculators of all nations. That is what happens when a country forsakes its Constitution and gives its sovereignty over the public currency to private interests. Give them the flag and they will sell it.

The nature of Kreuger's organized swindle and the bankrupt condition of Kreuger's combine was known here last June when Hoover sought to exempt Kreuger's loan to Germany of one hundred twenty-five millions from the operation of the Hoover moratorium. The bankrupt condition of Kreuger's swindle was known here last summer when $30,000,000 was taken from the American taxpayers by certain bankers in New York for the ostensible purpose of permitting Kreuger to make a loan to Colombia. Colombia never saw that money. The nature of Kreuger's swindle and the bankrupt condition of Kreuger was known here in January when he visited his friend, Mr. Hoover, at the White House. It was known here in March before he went to Paris and committed suicide there.

Mr. Chairman, I think the people of the United States are entitled to know how many billions of dollars were placed at the disposal of Kreuger and his gigantic combine by the Federal Reserve Board and the Federal reserve banks and to know how much of our Government currency was issued and lost in the financing of that great swindle in the years during which the Federal Reserve Board and the Federal reserve banks took care of Kreuger's requirements.

Mr. Chairman, I believe there should be a congressional investigation of the operations of Kreuger and Toll in the United States and that Swedish Match, International Match, the Swedish-American Investment Corporation, and all related enterprises, including the subsidiary companies of Kreuger and Toll, should be investigated and that the issuance of United States currency in connection with those enterprises and the use of our national-bank depositors' money for Kreuger's benefit should be made known to the general public. I am referring, not only to the securities which were floated and sold in this country, but also to the commercial loans to Kreuger's enterprises and the mass financing of Kreuger's companies by the Federal Reserve Board and the Federal reserve banks and the predatory institutions which the Federal Reserve Board and the Federal reserve banks shield and harbor.

A few days ago, the President of the United States, with a white face and shaking hands, went before the Senate on behalf of the moneyed interests and asked the Senate to levy a tax on the people so that foreigners might know that the United States would pay its debt to them. Most Americans thought it was the other way around. What does the United States owe to foreigners? When and by whom was the debt incurred? It was incurred by the Federal Reserve Board and the Federal reserve banks when they peddled the signature of this Government to foreigners for a price. It is what the United States Government has to pay to redeem the obligations of the Federal Reserve Board and the Federal reserve banks. Are you going to let those thieves get off scot free? Is there one law for the looter who drives up to the door of the United States Treasury in his limousine and another for the United States veterans who are sleeping on the floor of a dilapidated house on the outskirts of Washington?

The Baltimore & Ohio Railroad is here asking for a large loan from the people and the wage earners and the taxpayers of the United States. It is begging for a hand-out from the Government. It is standing, cap in hand, at the door of the Reconstruction Finance Corporation, where all the other jackals have gathered to the feast. It is asking for money that was raised from the people by taxation, and wants this money of the poor for the benefit of Kuhn, Loeb & Co., the German international bankers. Is there one law for the Baltimore & Ohio Railroad and another for the needy veterans it threw off its freight cars the other day? Is there one law for sleek and prosperous swindlers who call themselves bankers and another law for the soldiers who defended the United States flag?

Mr. Chairman, some people are horrified because the collateral behind Kreuger and Toll debentures was removed and worthless collateral substituted for it. What is this but what is being done daily by the Federal reserve banks? When the Federal reserve act was passed, the Federal reserve banks were allowed to substitute 'other like collateral' for collateral behind Federal reserve notes but by an amendment obtained at the request of the corrupt and dishonest Federal Reserve Board, the act was changed so that the word 'like' was stricken out. All that immense trouble was taken here in Congress so that the law would permit the Federal reserve banks to switch collateral. At the present time behind the scenes in the Federal reserve banks there is a night-and-day movement of collateral. A visiting Englishman, leaving the United States a few weeks ago, said that things would look better here after 'they cleaned up the mess at Washington.' Cleaning up the mess consists in fooling the people and making them pay a second time for the bad foreign investments of the Federal Reserve Board and the Federal reserve banks. It consists in moving that heavy load of dubious and worthless foreign paper—the bills of wigmakers, brewers, distillers, narcotic-drug vendors, munition makers, illegal finance drafts, and worthless foreign securities, out of the banks and putting it on the back of American labor. That is what the Reconstruction Finance Corporation is doing now. They talk about loans to banks and railroads but they say very little about that other business of theirs which consists in relieving the swindlers who promoted investment trusts in this country and dumped worthless foreign securities into them and then resold that mess of pottage to American investors under cover of their own corporate titles. The Reconstruction Finance Corporation is taking over those worthless securities from those investment trusts with United States Treasury money at the expense of the American taxpayer and the wage earner.

It will take us 20 years to redeem our Government, 20 years of penal servitude to pay off the gambling debts of the traitorous Federal Reserve Board and the Federal reserve banks and to earn again that vast flood of American wages and savings, bank deposits, and United States Government credit which the Federal Reserve Board and the Federal reserve banks exported out of this country to their foreign principals.

The Federal Reserve Board and the Federal reserve banks lately conducted an anti-hoarding campaign here. Then they took that extra money which they had persuaded the American people to put into the banks and they sent it to Europe along with the rest. In the last several months, they have sent $1,300,000,000 in gold to their foreign employers, their foreign masters, and every dollar of that gold belonged to the people of the United States and was unlawfully taken from them.

Is not it high time that we had an audit of the Federal Reserve Board and the Federal reserve banks and an examination of all our Government bonds and securities and public moneys instead of allowing the corrupt and dishonest Federal Reserve Board and the Federal reserve banks to speculate with those securities and this cash in the notorious open discount market of New York City?

Mr. Chairman, within the limits of the time allowed me, I can not enter into a particularized discussion of the Federal Reserve Board and the Federal reserve banks. I have singled out the Federal reserve currency for a few remarks because there has lately been some talk here of 'fiat money.' What kind of money is being pumped into the open discount market and through it into foreign channels and stock exchanges? Mr. Mills of the Treasury has spoken here of his horror of the printing presses and his horror of dishonest money. He has no horror of dishonest money. If he had, he would be no party to the present gambling of the Federal Reserve Board and the Federal reserve banks in the nefarious open discount market of New York, a market in which the sellers are represented by 10 great discount dealer corporations owned and organized by the very banks which own and control the Federal Reserve Board and the Federal reserve banks. Fiat money, indeed!

After the several raids on the Treasury Mr. Mills borrows the speech of those who protested against those raids and speaks now with pretended horror of a raid on the Treasury. Where was Mr. Mills last October when the United States Treasury needed $598,000,000 of the taxpayers' money which was supposed to be in the safe-keeping of Andrew W. Mellon in the designated depositories of Treasury funds, and which was not in those depositories when the Treasury needed it? Mr. Mills was the Assistant Secretary of the Treasury then, and he was at Washington throughout October, with the exception of a very significant week he spent at White Sulphur Springs closeted with international bankers, while the Italian minister, Signor Grandi, was being entertained—and bargained with—at Washington.

What Mr. Mills is fighting for is the preservation whole and entire of the banker's monopoly of all the currency of the United States Government. What Mr. PATMAN proposes is that the Government shall exercise its sovereignty to the extent of issuing some currency for itself. This conflict of opinion between Mr. Mills as the spokesman of the bankers and Mr. PATMAN as the spokesman of the people brings the currency situation here into the open. Mr. PATMAN and the veterans are confronted by a stone wall—the wall that fences in the bankers with their special privileges. Thus the issue is joined between the host of democracy, of which the veterans are a part, and the men of the king's bank, the would-be aristocrats, who deflated American agriculture and robbed this country for the benefit of their foreign principals.

Mr. Chairman, last December I introduced a resolution here asking for an examination and an audit of the Federal Reserve Board and the Federal reserve banks and all related matters. If the House sees fit to make such an investigation, the people of the United States will obtain information of great value. This is a Government of the people, by the people, for the people, consequently, nothing should be concealed from the people. The man who deceives the people is a traitor to the United States. The man who knows or suspects that a crime has been committed and who conceals or covers up that crime is an accessory to it. Mr. Speaker, it is a monstrous thing for this great Nation of people to have its destinies presided over by a traitorous Government board acting in secret concert with international usurers. Every effort has been made by the Federal Reserve Board to conceal its power but the truth is the Federal Reserve Board has usurped the Government of the United States. It controls everything here and it controls all our foreign relations. It makes and breaks governments at will. No man and no body of men is more entrenched in power than the arrogant credit monopoly which operates the Federal Reserve Board and the Federal reserve banks. These evil-doers have robbed this country of more than enough money to pay the national debt. What the National Government has permitted the Federal Reserve Board to steal from the people should now be restored to the people. The people have a valid claim against the Federal Reserve Board and the Federal reserve banks. If that claim is enforced, Americans will not need to stand in the breadlines or to suffer and die of starvation in the streets. Homes will be saved, families will be kept together, and American children will not be dispersed and abandoned. The Federal Reserve Board and the Federal reserve banks owe the United States Government an immense sum of money. We ought to find out the exact amount of the people's claim. We should know the amount of the indebtedness of the Federal Reserve Board and the Federal reserve banks to the people and we should collect that amount immediately. We certainly should investigate this treacherous and disloyal conduct of the Federal Reserve Board and the Federal reserve banks.

Here is a Federal reserve note. Immense numbers of these notes are now held abroad. I am told that they amount to upwards of a billion dollars. They constitute a claim against our Government and likewise a claim against the money our people have deposited in the member banks of the Federal reserve system. Our people's money to the extent of $1,300,000,000 has within the last few months been shipped abroad to redeem Federal reserve notes and to pay other gambling debts of the traitorous Federal Reserve Board and the Federal reserve banks. The greater part of our monetary stock has been shipped to foreigners. Why should we promise to pay the debts of foreigners to foreigners? Why should our Government be put into the position of supplying money to foreigners? Why should American farmers and wage earners add millions of foreigners to the number of their dependents? Why should the Federal Reserve Board and the Federal reserve banks be permitted to finance our competitors in all parts of the world? Do you know why the tariff was raised? It was raised to shut out the flood of Federal reserve goods pouring in here from every quarter of the globe—cheap goods, produced by cheaply paid foreign labor on unlimited supplies of money and credit sent out of this country by the dishonest and unscrupulous Federal Reserve Board and the Federal reserve banks. Go out in Washington to buy an electric light bulb and you will probably be offered one that was made in Japan on American money. Go out to buy a pair of fabric gloves and inconspicuously written on the inside of the gloves that will be offered to you will be found the words 'made in Germany' and that means 'made on the public credit of the United States Government paid to German firms in American gold taken from the confiscated bank deposits of the American people.'

The Federal Reserve Board and the Federal reserve banks are spending $100,000,000 a week buying Government securities in the open market and are thus making a great bid for foreign business. They are trying to make rates so attractive that the human-hair merchants and distillers and other business entities in foreign lands will come here and hire more of the public credit of the United States Government and pay the Federal reserve outfit for getting it for them.

Mr. Chairman, when the Federal Reserve act was passed, the people of the United States did not perceive that a world system was being set up here which would make the savings of an American school-teacher available to a narcotic-drug vendor in Macao. They did not perceive that the United States were to be lowered to the position of a coolie country which has nothing but raw materials and heavy goods for export; that Russia was destined to supply the man power and that this country was to supply financial power to an international superstate—a superstate controlled by international bankers and international industrialists acting together to enslave the world for their own pleasure.

The people of the United States are being greatly wronged. If they are not, then I do not know what 'wronging the people' means. They have been driven from their employments. They have been dispossessed of their homes. They have been evicted from their rented quarters. They have lost their children. They have been left to suffer and to die for lack of shelter, food, clothing, and medicine.

The wealth of the United States and the working capital of the United States has been taken away from them and has either been locked in the vaults of certain banks and the great corporations or exported to foreign countries for the benefit of the foreign customers of those banks and corporations. So far as the people of the United States are concerned, the cupboard is bare. It is true that the warehouses and coal yards and grain elevators are full, but the warehouses and coal yards and grain elevators are padlocked and the great banks and corporations hold the keys. The sack of the United States by the Federal Reserve Board and the Federal reserve banks is the greatest crime in history.

Mr. Chairman, a serious situation confronts the House of Representatives to-day. We are trustees of the people and the rights of the people are being taken away from them. Through the Federal Reserve Board and the Federal reserve banks, the people are losing the rights guaranteed to them by the Constitution. Their property has been taken from them without due process of law. Mr. Chairman, common decency requires us to examine the public accounts of the Government and see what crimes against the public welfare have been or are being committed.

What is needed here is a return to the Constitution of the United States. We need to have a complete divorce of Bank and State. The old struggle that was fought out here in Jackson's day must be fought over again. The independent United States Treasury should be reestablished and the Government should keep its own money under lock and key in the building the people provided for that purpose. Asset currency, the device of the swindler, should be done away with. The Government should buy gold and issue United States currency on it. The business of the independent bankers should be restored to them. The State banking systems should be freed from coercion. The Federal reserve districts should be abolished and State boundaries should be respected. Bank reserves should be kept within the borders of the States whose people own them, and this reserve money of the people should be protected so that the international bankers and acceptance bankers and discount dealers can not draw it away from them. The exchanges should be closed while we are putting our financial affairs in order. The Federal reserve act should be repealed and the Federal reserve banks, having violated their charters, should be liquidated immediately. Faithless Government officers who have violated their oaths of office should be impeached and brought to trial. Unless this is done by us, I predict that the American people, outraged, robbed, pillaged, insulted, and betrayed as they are in their own land, will rise in their wrath and send a President here who will sweep the money changers out of the temple. [Applause.]"[587]

---

[587]. *Congressional Record: Proceedings and Debates of the First Session of the Seventy-Second Congress of the United States of America*, Volume 75, Part 11, United States Government Printing Office, Washington, D. C., (1 June 1932-11 June 1932), pp. 12595-12603.

# OTHER TITLES

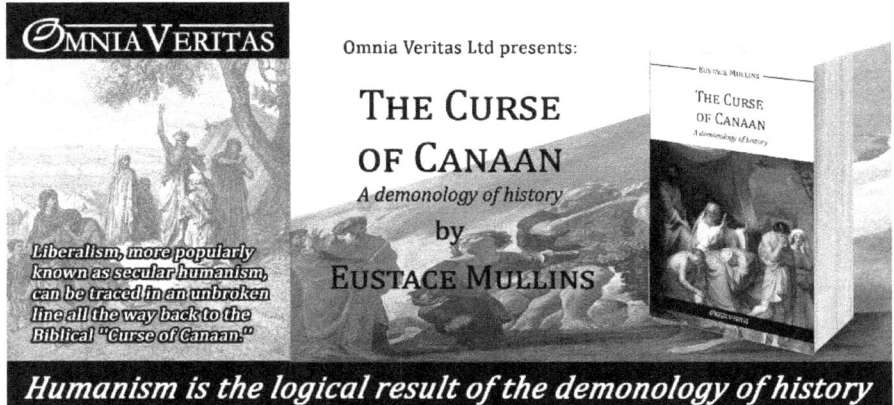

*The Protocols of the Learned Elders of Zion*

Omnia Veritas Ltd presents:

## MURDER BY INJECTION

by

EUSTACE MULLINS

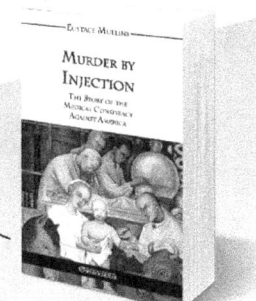

*The cynicism and malice of these conspirators is something beyond the imagination of most Americans.*

Ezra's interest in money as a phenomenon, in contrast to the usual attitude toward money as something to get, is a legitimate one.

Omnia Veritas Ltd presents:

## EZRA POUND
### THIS DIFFICULT INDIVIDUAL

by

EUSTACE MULLINS

*An illustration for his own monetary theories...*

Christ did not wish to be followed by robots and sleepwalkers, He desired man to awaken, and to attain the full use of his earthly powers.

OMNIA VERITAS LTD PRESENTS:

## MY LIFE IN CHRIST

BY

EUSTACE MULLINS

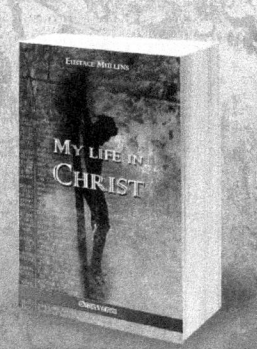

*THIS is the story of my life in Christ*

*The Manufacture and Sale* of Saint Einstein

**OMNIA VERITAS**

*I wish to tell of the things which have happened to me in my struggle against the forces of darkness.*

Omnia Veritas Ltd presents:

**COLLECTED ESSAYS**

EUSTACE MULLINS

*It is my hope that others will be forewarned of what to expect in this fight*

**OMNIA VERITAS**

OMNIA VERITAS LTD PRESENTS:

**CONVERSATIONS WITH JOHN F. KENNEDY**

BY EUSTACE MULLINS

*I engaged in lengthy disquisitions about the condition of man, the dangers apparent in his present estate, and what must be done to avert them...*

**The martyred President of the United States**

**OMNIA VERITAS**

OMNIA VERITAS LTD PRESENTS:

**SCARLET AND THE BEAST**

A HISTORY OF THE WAR BETWEEN ENGLISH AND FRENCH FREEMASONRY

*My research has revealed that there are two separate and opposing powers in Freemasonry.*

One is Scarlet. The other, the Beast.

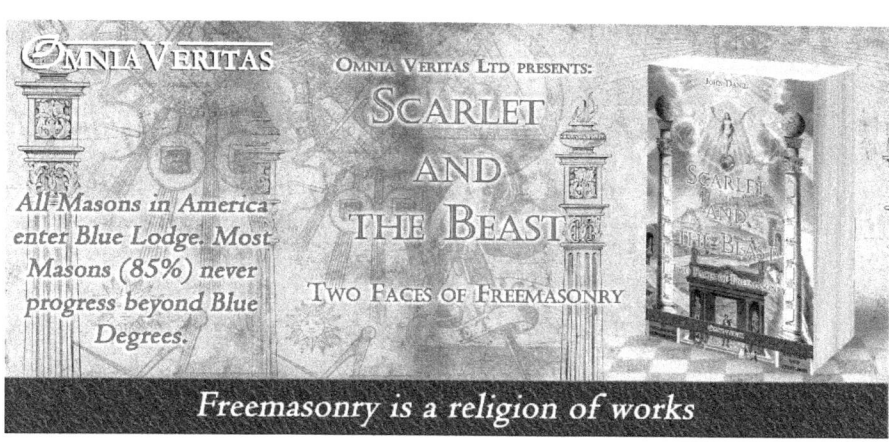
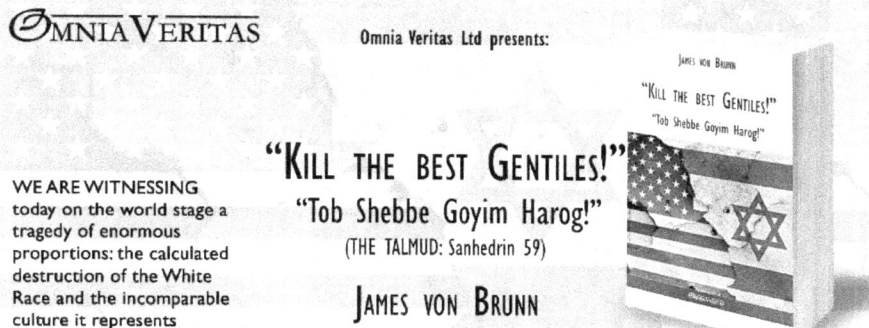

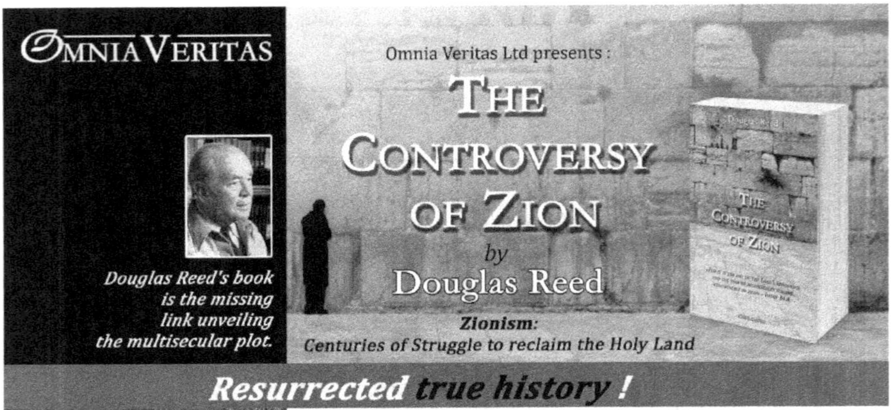
OMNIA VERITAS
Omnia Veritas Ltd presents :
THE CONTROVERSY OF ZION
by
Douglas Reed
Douglas Reed's book is the missing link unveiling the multisecular plot.
Zionism:
Centuries of Struggle to reclaim the Holy Land
Resurrected true history !

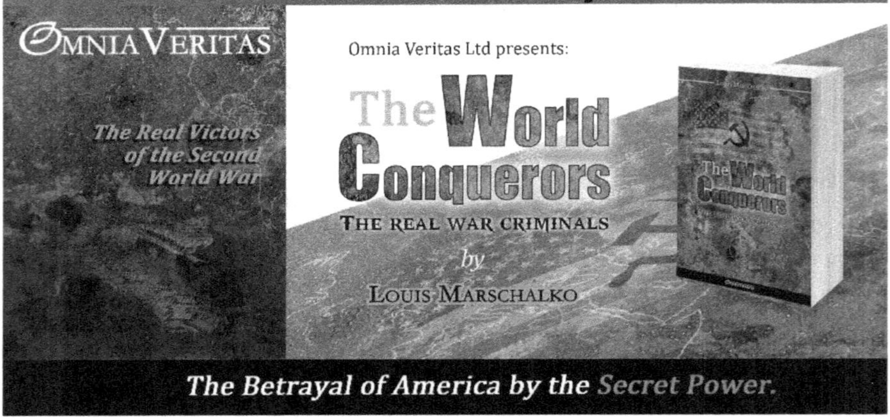
OMNIA VERITAS
Omnia Veritas Ltd presents:
The World Conquerors
THE REAL WAR CRIMINALS
by
LOUIS MARSCHALKO
The Real Victors of the Second World War
The Betrayal of America by the Secret Power.

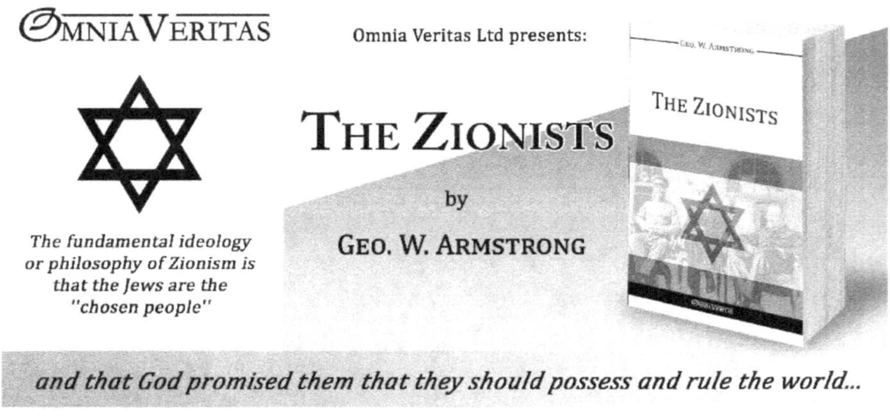
OMNIA VERITAS
Omnia Veritas Ltd presents:
THE ZIONISTS
by
GEO. W. ARMSTRONG
The fundamental ideology or philosophy of Zionism is that the Jews are the "chosen people"
and that God promised them that they should possess and rule the world...

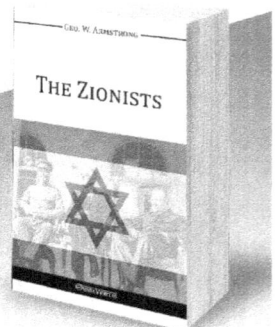

www.omnia-veritas.com

 https://www.instagram.com/omnia.veritas/

 https://twitter.com/OmniaVeritasLtd

www.ingramcontent.com/pod-product-compliance
Lightning Source LLC
Chambersburg PA
CBHW071358230426
43669CB00010B/1386